AAPG Treatise of Petroleum Geology Reprint Series

The American Association of Petroleum Geologists
gratefully acknowledges and appreciates the leadership and support
of the AAPG Foundation in the development of the
Treatise of Petroleum Geology.

Geophysics I

Seismic Methods

Compiled by
Edward A. Beaumont
and
Norman H. Foster

Treatise of Petroleum Geology
Reprint Series, No. 12

Published by
The American Association of Petroleum Geologists
Tulsa, Oklahoma 74101-0979, U.S.A.

Copyright © 1989
The American Association of Petroleum Geologists
All Rights Reserved

ISBN: 0-89181-411-6

American Association of Petroleum Geologists Foundation

Treatise of Petroleum Geology Fund*

Major Corporate Contributors
($25,000 or more)

Chevron Corporation
Mobil Oil Corporation
Oryx Energy Company
Pennzoil Exploration and Production Company
Shell Oil Company
Union Pacific Foundation

Other Corporate Contributors
($5,000 to $25,000)

Cabot Energy Corporation
Conoco Inc.
Marathon Oil Company
The McGee Foundation, Inc.
Phillips Petroleum Company
Texaco Philanthropic Foundation
Transco Energy Company

Major Individual Contributors
($1,000 or more)

C. Hayden Atchison
Richard R. Bloomer
A.S. Bonner, Jr.
David G. Campbell
Herbert G. Davis
Paul H. Dudley, Jr.
Lewis G. Fearing
James A. Gibbs
George R. Gibson
William E. Gipson
Robert D. Gunn
Cecil V. Hagen
Frank W. Harrison
William A. Heck
Harrison C. Jamison
Thomas N. Jordan, Jr.
Hugh M. Looney
John W. Mason
George B. McBride
Dean A. McGee
John R. McMillan
Rudolf B. Siegert
Robert M. Sneider
Jack C. Threet
Charles Weiner
Harry Westmoreland
James E. Wilson, Jr.

The Foundation also gratefully acknowledges the many who have supported this endeavor with additional contributions, which now total $12,000.

*Contributions received as of September 19, 1989.

Treatise of Petroleum Geology
Advisory Board

W.O. Abbott
Robert S. Agatston
John J. Amoruso
J.D. Armstrong
George B. Asquith
Colin Barker
Ted L. Bear
Edward A. Beaumont
Robert R. Berg
Richard R. Bloomer
Louis C. Bortz
Donald R. Boyd
Robert L. Brenner
Raymond Buchanan
Daniel A. Busch
David G. Campbell
J. Ben Carsey
Duncan M. Chisholm
H. Victor Church
Don Clutterbuck
Robert J. Cordell
Robert D. Cowdery
William H. Curry, III
Doris M. Curtis
Graham R. Curtis
Clint A. Darnall
Patrick Daugherty
Herbert G. Davis
James R. Davis
Gerard J. Demaison
Parke A. Dickey
F.A. Dix, Jr.
Charles F. Dodge
Edward D. Dolly
Ben Donegan
Robert H. Dott, Sr.*
John H. Doveton
Marlan W. Downey
John G. Drake
Richard J. Ebens
William L. Fisher
Norman H. Foster
Lawrence W. Funkhouser
William E. Galloway
Lee C. Gerhard
James A. Gibbs
Arthur R. Green
Robert D. Gunn
Merrill W. Haas
Robert N. Hacker
J. Bill Hailey
Michel T. Halbouty
Bernold M. Hanson
Tod P. Harding
Donald G. Harris
Frank W. Harrison, Jr.

Ronald L. Hart
Dan J. Hartmann
John D. Haun
Hollis D. Hedberg*
James A. Helwig
Thomas B. Henderson, Jr.
Francis E. Heritier
Paul D. Hess
Mason L. Hill
David K. Hobday
David S. Holland
Myron K. Horn
Michael E. Hriskevich
J.J.C. Ingels
Michael S. Johnson
Bradley B. Jones
R.W. Jones
John E. Kilkenny
H. Douglas Klemme
Allan J. Koch
Raden P. Koesoemadinate
Hans H. Krause
Naresh Kumar
Rolf Magne Larsen
Jay E. Leonard
Ray Leonard
Howard H. Lester
Detlev Leythaeuser
John P. Lockridge
Tony Lomando
John M. Long
Susan A. Longacre
James D. Lowell
Peter T. Lucas
Harold C. MacDonald
Andrew S. Mackenzie
Jack P. Martin
Michael E. Mathy
Vincent Matthews, III
James A. McCaleb
Dean A. McGee*
Philip J. McKenna
Robert E. Megill
Fred F. Meissner
Robert K. Merrill
David L. Mikesh
Marcus E. Milling
George Mirkin
Richard J. Moiola
D. Keith Murray
Norman S. Neidell
Ronald A. Nelson
Charles R. Noll
Clifton J. Nolte
Susan E. Palmer

Arthur J. Pansze
John M. Parker
Alain Perrodon
James A. Peterson
R. Michael Peterson
David E. Powley
A. Pulunggono
Donald L. Rasmussen
R. Randolf Ray
Dudley D. Rice
Edward C. Roy, Jr.
Eric A. Rudd
Floyd F. Sabins, Jr.
Nahum Schneidermann
Peter A. Scholle
George L. Scott, Jr.
Robert T. Sellars, Jr.
John W. Shelton
Robert M. Sneider
Stephen A. Sonnenberg
William E. Speer
Ernest J. Spradlin
Bill St. John
Philip H. Stark
Richard Steinmetz
Per R. Stokke
Donald S. Stone
Doug K. Strickland
James V. Taranik
Harry TerBest, Jr.
Bruce K. Thatcher, Jr.
M. Raymond Thomasson
Bernard Tissot
Donald Todd
M.O. Turner
Peter R. Vail
Arthur M. Van Tyne
Harry K. Veal
Richard R. Vincelette
Fred J. Wagner, Jr.
Anthony Walton
Douglas W. Waples
Harry W. Wassall, III
W. Lynn Watney
N.L. Watts
Koenradd J. Weber
Robert J. Weimer
Dietrich H. Welte
Alun H. Whittaker
James E. Wilson, Jr.
Martha O. Withjack
P.W.J. Wood
Homer O. Woodbury
Mehmet A. Yukler
Zhai Guangming

*Deceased

IN APPRECIATION...

The American Association of Petroleum Geologists and the AAPG Foundation gratefully acknowledge the contributions of the Society of Exploration Geophysicists to the Treatise of Petroleum Geology Reprint Series volumes on geophysics. The crucial role played by SEG in advancing exploration geophysics is universally recognized in the petroleum industry and is plainly documented by the many papers from *Geophysics* and *Geophysics: The Leading Edge of Exploration* reproduced in these volumes. The spirit of advancement and expansion in the science of geophysical exploration for hydrocarbons as well as the continuing synergism that melds the professions of geology and geophysics is exemplified by the permission granted by SEG to reproduce its papers in this series.

Although SEG and AAPG both have a long histo of independent activities and autonomous operatio there have been, and there continue to be, cooperati efforts to improve the professionalism of both geol gists and geophysicists. Previous activities, such as joi meetings, research conferences, and publication exemplify the ongoing cooperative efforts which crea results far more valuable than independent efforts l either group would have achieved. SEG's contributio to the earth sciences and to the Treatise of Petroleu Geology Reprint Series volumes on geophysics a gratefully acknowledged by AAPG, the AAPG Found tion, and the Advisory Board of the Treatise of Petr leum Geology.

INTRODUCTION

This reprint volume belongs to a series of that is part of the *Treatise of Petroleum Geology*. The *Treatise of Petroleum Geology* was conceived during a discussion we had at the 1984 AAPG Annual Meeting in San Antonio. When our discussion ended, we had decided to write a state-of-the-art textbook in petroleum geology, directed not at the student, but at the practicing petroleum geologist. The project to put together one textbook gradually evolved into a series of three different publications: the Reprint Series, the Atlas of Oil and Gas Fields, and the Handbook of Petroleum Geology; collectively these publications are known as the *Treatise of Petroleum Geology*. With the help of the Treatise of Petroleum Geology Advisory Board, we designed this set of publications to represent the cutting edge in petroleum exploration knowledge and application. The Reprint Series provides previously published landmark literature; the Atlas collects detailed field studies to illustrate the various ways oil and gas are trapped; and the Handbook is a professional explorationist's guide to the latest knowledge in the various areas of petroleum geology and related fields.

The papers in the various volumes of the Reprint Series complement the different chapters of the Handbook. Papers were selected on the basis of their usefulness today in petroleum exploration and development. Many "classic papers" that led to our present state of knowledge have not been included because of space limitations. In some cases, it was difficult to decide in which Reprint volume a particular paper should be published because that paper covers several topics. We suggest, therefore, that interested readers become familiar with all the Reprint volumes if they are looking for a particular paper.

Geophysics is an indispensable tool for geologists looking for and developing oil and gas fields. Because it lets us "see" into the subsurface, geophysics allows petroleum geologists to build better images of the subsurface than is possible using only surface geology and information from well bores. In the past, geophysics was the domain of the geophysicist, and the geophysicist alone acquired, processed, and interpreted geophysical data. During the past two decades, however, the technology of geophysics has exploded; at the same time, the petroleum industry has been forced to look for more and more subtle traps in more and more difficult terrain. This placed a tremendous burden on geophysicists, and they naturally looked to their colleagues, the geologists, for relief. At first, geologists only helped with interpretation. Today, however, geologists are also involved in helping geophysicists make decisions regarding acquisition and processing of data.

The choice of papers in these geophysics reprint volumes reflects this evolution. The papers were chosen to help geologists, not geophysicists, enhance their knowledge of geophysics. Math-intensive papers were excluded because those papers are relatively esoteric and have limited applicability for most geologists. Many of the papers included do contain mathematical equations, but they were selected because they are germane, and the math is presented at a level that, we trust, the majority of geologists are now comfortable with.

The number and distribution of the papers reprinted in these volumes reflect the current importance and uses of the different geophysical methods described in the papers. We have divided the topic of geophysics into four volumes. The first volume contains papers on Seismic Methods. Papers in this volume are concerned with seismic theory and are grouped into six sections: Seismic Methods, Seismic Rock Properties, Seismic Acquisition, Seismic Processing and Display, Seismic Velocities, and Migration. Volume II is subtitled Tools for Seismic Interpretation. Section titles in this volume are Synthetic Seismograms and Velocity Inversion; Seismic Modeling; Seismic Attributes: Amplitude, Frequency, Phase, Velocity; Shear Waves; Amplitude Variation with Offset; and Vertical Seismic Profiling. Volume III, Geologic Interpretation of Seismic Data, contains sections on Structural Interpretation and Stratigraphic Interpretation. The last volume is on Gravity, Magnetic, and Magnetotelluric Methods. It contains two sections: Gravity and Magnetic Methods, and Magnetotelluric Methods.

We would like to thank the various societies and publishers who gave us permission to reprint these papers, especially the Society of Exploration Geophysicists. We also wish to thank the members of Advisory Board of the Treatise of Petroleum Geology who suggested papers for these volumes, especially R. Randy Ray. Randy Ray is a geophysicist who was trained initially as a geologist. From a large list of proposed papers, he helped us select papers that would be both understandable and useful to a geologist exploring for and developing oil and gas fields.

Edward A. Beaumont
Tulsa, Oklahoma

Norman H. Foster
Denver, Colorado

Table of Contents

Geophysics I
Seismic Methods

Seismic Waveforms and Resolution

Aspects of seismic resolution.   R. E. Sheriff..1
How thin is a thin bed?   M. B. Widess..12
Complex seismic trace analysis of thin beds.   James D. Robertson and Henry H. Nogami..................18
The limits of resolution of zero-phase wavelets.   R. S. Kallweit and L. C. Wood..................27
Resolution comparison of minimum-phase and zero-phase signals.   M. Schoenberger..................40

Seismic Rock Properties

Formation velocity and density—the diagnostic basics for stratigraphic traps.   G. H. F. Gardner, L. W. Gardner, and A. R. Gregory..................48
Effect of water saturation on seismic reflectivity of sand reservoirs encased in shale. S. N. Domenico..................59

Seismic Acquisition

Whatever happened to ground roll?   Nigel A. Anstey..................70
Field techniques for high resolution.   Nigel A. Anstey..................76
Vibroseis' gentle massage obtains structural data safely, economically.   N. A. Anstey..................86
The Vibroseis system of seismic mapping.   Robert L. Geyer..................93
Vibroseis parameter optimization.   Robert L. Geyer..................112
Seismic data enhancement—a case history.   R. J. Graebner..................121

Seismic Processing and Display

Common reflection point horizontal data stacking techniques.   W. Harry Mayne..................151
Correlation techniques—a review.   N. A. Anstey..................163
The digital processing of seismic data.   Daniel Silverman..................192
Seismic data display and reflection perceptibility.   Frank J. Feagin..................208
Semblance and other coherency measures for multichannel data.   N. S. Neidell and M. Turhan Taner..................224
Predictive deconvolution: theory and practice.   K. L. Peacock and Sven Treitel..................241
Estimation and correction of near-surface time anomalies.   M. Turhan Taner, F. Koehler, and K. A. Alhilali..................257

Seismic signal processing.   Lawrence C. Wood and Sven Treitel..281

SEISMIC VELOCITIES

Seismic velocities from surface measurements.   C. Hewitt Dix...294
An analysis of stacking, rms, average, and interval velocities over a horizontally layered ground.
M. Al-Chalabi...314
Time-depth and velocity-depth relations in western Canada.   C. H. Acheson..........................332
Apparent velocity from dipping interface reflections.   F. K. Levin.....................................348
A velocity function including lithologic variation.   L. Y. Faust..355
Seismic data indicate depth, magnitude of abnormal pressures.   E. S. Pennebaker, Jr...................373
Synthetic sonic logs—a process for stratigraphic interpretation.   R. O. Lindseth......................379
Velocity spectra—digital computer derivation and applications of velocity functions.
M. Turhan Taner and Fulton Koehler...403
The effects of cracks on the compressibility of rock.   J. B. Walsh.....................................427

MIGRATION

Migration.   P. Hood...437
Two-dimensional and three-dimensional migration of model-experiment reflection profiles.
William S. French..517
A simple theory for seismic diffractions.   A. W. Trorey..530
Migration of seismic data from inhomogeneous media.   Les Hatton, Ken Larner, and
Bruce S. Gibson..553
Time migration—some ray theoretical aspects.   P. Hubral...570
The wave equation applied to migration.   D. Loewenthal, L. Lu, R. Roberson, and J. Sherwood............578
Wave-front charts and three dimensional migrations.   Albert W. Musgrave.............................598
Migration by Fourier transform.   R. H. Stolt..615

TABLE OF CONTENTS

GEOPHYSICS II
TOOLS FOR SEISMIC INTERPRETATION

SYNTHETIC SEISMOGRAMS AND VELOCITY INVERSION

The synthesis of seismograms from well log data.
R. A. Peterson, W. R. Fillippone, and F. B. Coker.

Inversion of seismograms and pseudo velocity logs.
M. Lavergne and C. Willm.

Well log editing in support of detailed seismic studies.
Brian E. Ausburn.

SEISMIC MODELING

Stratigraphic modeling: a step beyond bright spot.
E. V. Dedman, J. P. Lindsey, and M. W. Schramm, Jr.

Three-dimensional seismic modeling.
Fred J. Hilterman.

Interpretive lessons from three-dimensional modeling.
Fred J. Hilterman.

Stratigraphic modeling and interpretation—geophysical principals and techniques.
Norman S. Neidell and Elio Poggiagliolmi.

Synthetic seismic sections of typical petroleum traps.
Bruce T. May and Franta Hron.

SEISMIC ATTRIBUTES: AMPLITUDE, FREQUENCY, PHASE, VELOCITY

Application of amplitude, frequency, and other attributes to stratigraphic and hydrocarbon determination.
M. T. Taner and R. E. Sheriff.

Reflections on amplitudes.
R. F. O'Doherty and N. A. Anstey.

Velocity spectra and their use in stratigraphic and lithologic differentiation.
Ernest E. Cook and M. Turhan Taner.

Outlining of shale masses by geophysical methods.
A. W. Musgrave and W. G. Hicks.

Three-dimensional seismic monitoring of an enhanced oil recovery process.
Robert J. Greaves and Terrance J. Fulp.

SHEAR WAVES

Basis for interpretation of Vp/Vs ratios in complex lithologies.
Raymond L. Eastwood and John P. Castagna.

Evaluation of direct hydrocarbon indicators through comparison of compressional- and shear-wave seismic data: a case study of the Myrnam gas field, Alberta.
Ross Alan Ensley.

Relationships between compressional-wave and shear-wave velocities in clastic silicate rocks.
J. P. Castagna, M. L. Batzle, and R. L. Eastwood.

Direct hydrocarbon detection using comparative P-wave and S-wave seismic sections.
James D. Robertson and William C. Pritchett.

Vp/Vs and lithology.
R. S. Tatham.

Vp/Vs—a potential hydrocarbon indicator.
Robert S. Tatham and Paul L. Stoffa.

AMPLITUDE VARIATION WITH OFFSET

Plane-wave reflection coefficients for gas sands at nonnormal angles of incidence.
W. J. Ostrander.

VERTICAL SEISMIC PROFILING

Vertical seismic profiling—a measurement that transfers geology to geophysics.
B. A. Hardage.

The use of vertical seismic profiles in seismic investigations of the earth.
A. H. Balch, M. W. Lee, J. J. Miller, and Robert T. Ryder.

Prediction of overpressure in Nigeria using vertical seismic profile techniques.
S. Brun, P. Griveley, and A. Paul.

Offset source VSP surveys and their image reconstruction.
P. B. Dillon and R. C. Thomson.

Vertical seismic profiling technique emerges as a valuable drilling tool.
R. J. Roberts and J. D. Platt.

Table of Contents

Geophysics III
Geologic Interpretation of Seismic Data

Structural Interpretation

A process of seismic reflection interpretation.
J. G. Hagedoorn.

Interactive seismic mapping of net producible gas sand in the Gulf of Mexico.
Alistair R. Brown, Roger M. Wright, Keith D. Burkhart, and William L. Abriel.

Interactive interpretation of seismic data.
Anthony C. Gerhardstein and Alistair R. Brown.

Geologic interpretation of seismic profiles, Big Horn basin, part II: west flank.
D. B. Stone.

Stratigraphic Interpretation

Inferring stratigraphy from seismic data.
R. E. Sheriff.

Seismic stratigraphy, a fundamental exploration tool.
G. R. Ramsayer.

Seismic facies analysis concepts.
M. M. Roksandic.

Seismic signatures of sedimentation models.
J. C. Harms and P. Tackenberg.

Integration of biostratigraphy and seismic stratigraphy: Pliocene-Pleistocene, Gulf of Mexico.
John M. Armentrout.

The role of horizontal seismic sections in stratigraphic interpretation.
Alastair R. Brown.

Seismic stratigraphy and global changes of sea level, part 10: seismic recognition of carbonate buildups.
J. N. Bubb and W. G. Hatlelid.

Seismic interpretation of carbonate depositional environments
J. M. Fontaine, R. Cussey, J. Lacaze, R. Lanaud, and L. Yapaudjian.

Seismic expression of carbonate build-ups, northwest Java basin.
J. E. Burbury.

Field development with three-dimensional seismic methods in the Gulf of Thailand—a case history.
C. G. Dahm and R. J. Graebner.

Aspects of seismic reflection prospecting for oil and gas.
P. N. S. O'Brien.

Predictive isopach mapping of gas sands from seismic impedance: modeled and empirical cases from Ship Shoal Block 134 field.
Robert D. Woock and Alan R. Kin.

New seismic technology can guide field development.
J. P. Lindsey, M. W. Schramm, Jr., and L. K. Nemeth.

How hydrocarbon reserves are estimated from seismic data.
J. P. Lindsey and C. I. Craft.

Progress in stratigraphic seismic exploration and the definition of reservoirs.
Norman S. Neidell and John H. Beard.

Interpretation of depositional facies from seismic data.
J. B. Sangree and J. M. Widmier.

Seismic stratigraphic model of depositional platform margin, eastern Anadarko basin, Oklahoma.
William E. Galloway, Marshall S. Yancey, and Arthur P. Whipple.

Exploration for oil accumulations in Entrada Sandstone, San Juan basin, New Mexico.
Richard R. Vincelette and William E. Chittum.

Table of Contents

Geophysics IV
Gravity, Magnetic, and Magnetotelluric Methods

Gravity and Magnetic Methods

Gravity and magnetics for geologists and seismologists.
L. L. Nettleton.

Exploring for stratigraphic traps with gravity gradients.
Sigmund Hammer and Rodolfo Anzoleaga.

Measurement of gravity at sea and in the air.
Lucien J. B. LaCoste.

An approximate solution of the problem of maximum depth in gravity interpretation.
D. C. Skeels.

Use of gravity, magnetic, and electric methods in stratigraphic-trap exploration.
L. L. Nettleton.

The direct approach to magnetic interpretation and its practical application.
Leo J. Peters.

Magnetotelluric Methods

Basic theory of the magneto-telluric method of geophysical prospecting.
Louis Cagniard.

Processing and interpretation of magnetotelluric soundings.
G. Kunetz.

The magnetotelluric method in the exploration of sedimentary basins.
Keeva Vozoff.

Magnetotelluric responses of three-dimensional bodies in layered earths.
Philip E. Wannamaker, Gerald W. Hohmann, and Stanley H. Ward.

SEISMIC WAVE FORMS AND RESOLUTION

Aspects of Seismic Resolution

R. E. Sheriff
University of Houston
Houston, Texas

> Where two or more reflections are more closely spaced than a quarter wavelength, a common situation for stratigraphic features of interest in petroleum exploration, the same half-cycles of the embedded wavelet tend to add; we call this the "thin-bed" case. In contrast, where they are close but separated by more than a quarter wavelength, different half-cycles tend to add; this is the "thick-bed" case. Many characteristics of seismic reflections differ in thin- and thick-bed situations.
>
> On unmigrated seismic sections the limits of horizontal resolvability are imposed by Fresnel-zone considerations. However, on migrated sections other factors become important, such as noise on the unmigrated section, spatial aliasing, migration aperture, and uncertainties imposed by velocity, stacking, and two-dimensional assumptions.

INTRODUCTION

Resolution ultimately limits the amount of stratigraphic detail which can be extracted from seismic data. "Resolution" is the ability to tell that more than one feature is contributing to an observed effect.

Three aspects of resolution are examined in this paper. "Vertical resolution" concerns the ability to distinguish that more than one reflecting interface is contributing to an observed reflection. "Horizontal resolution" concerns the ability to distinguish between features which are displaced horizontally with respect to each other. Most of what has been written about horizontal resolution relates to the interpretation of unmigrated seismic sections, but the factors involved in the interpretation of migrated sections are different. Hence horizontal resolution is examined separately for unmigrated and migrated sections.

VERTICAL RESOLUTION

A definition of resolution as "the ability to distinguish that more than one" reflection is involved invites disagreement as to how resolvable limit is to be defined. One experienced in observing subtle effects of the interference of reflections will be able to distinguish situations where another without such experience cannot. Aspects evident on noise-free sections may be obscured in the presence of background noise. Subjectivity tends to enter when we endeavor to quantify whether a given situation is or is not resolvable, that is, when we define a "resolvable limit."

Kallweit and Wood (1982) discuss several definitions of resolvable limit as they apply to zero-phase (symmetrical) wavelets.

The most common definitions of resolvable limit are those attributable to Lord Rayleigh (who studied resolution as applied to visible light) and to Widess (1973). The values for resolvable limit given by these two definitions are, respectively, 1/4 and 1/8 of the dominant wavelength. Definitions given by others generally fall within this range. Widess convolved two spikes of equal magnitude but opposite polarity with a symmetrical wavelet and examined waveshape changes as the separation between the two spikes changed (Figure 1). He observed that separate distinguishable reflections resulted where the spikes were separated by more than a half wavelength and that the two reflections began to interfere constructively as the separation approached a quarter wavelength, producing an increase in reflection amplitude (called "tuning"). As the spikes approached an eighth wavelength separation the amplitude decreased and the waveshape changed slightly. Where the spikes were closer together than an eighth wavelength, the waveshape stabilized as the derivative of the waveshape from an individual reflector and the amplitude was nearly linear with the separation of the spikes.

The importance of the resolvable limit to stratigraphic interpretation becomes evident when we calculate the thickness in feet or meters which corresponds to a quarter wavelength. For a reflector with a velocity V and a wavelet with the dominant frequency f, the quarter wavelength is simply V/4f. Shallow reflections, perhaps at 5,000 ft (1,524 m) depth in a poorly consolidated sand-shale section, may involve a velocity of about 6,000 ft/sec (1,829 m/sec) and moderately high frequency, say 60 Hz, so the quarter-wavelength thickness is 25 ft (7.6 m). Deep reflections, on the other hand, are characterized by higher velocities, perhaps 15,000 ft/sec (4,572 m/sec), and lower frequency, say 15 Hz, increasing a quarter wavelength thickness to 250 ft (76 m). Wavelength ranges by a factor of roughly 10:1 over the range of petroleum objectives. Many stratigraphic situations of importance and many hydrocarbon reservoirs are thinner than these values, so the resolvable limit is important for stratigraphic interpretation.

The resolvable limit does not necessarily set an absolute threshold below which we cannot carry out seismic analysis, but rather it is a dividing point in the nature of evidences of features. Neidell and Poggiagliolmi (1977) and others analyze reservoir thickness from observations of amplitude variations where the reservoir is thinner than the Widess resolvable limit.

Where multiple reflectors are closer than a quarter wavelength, their contributions add within the first half-cycle whereas if they are farther apart than a quarter wavelength different half-cycles of the embedded wavelet interfere. Where the gross thickness is less than a quarter wavelength we have a "thin-bed" situation in contrast to the "thick-bed"

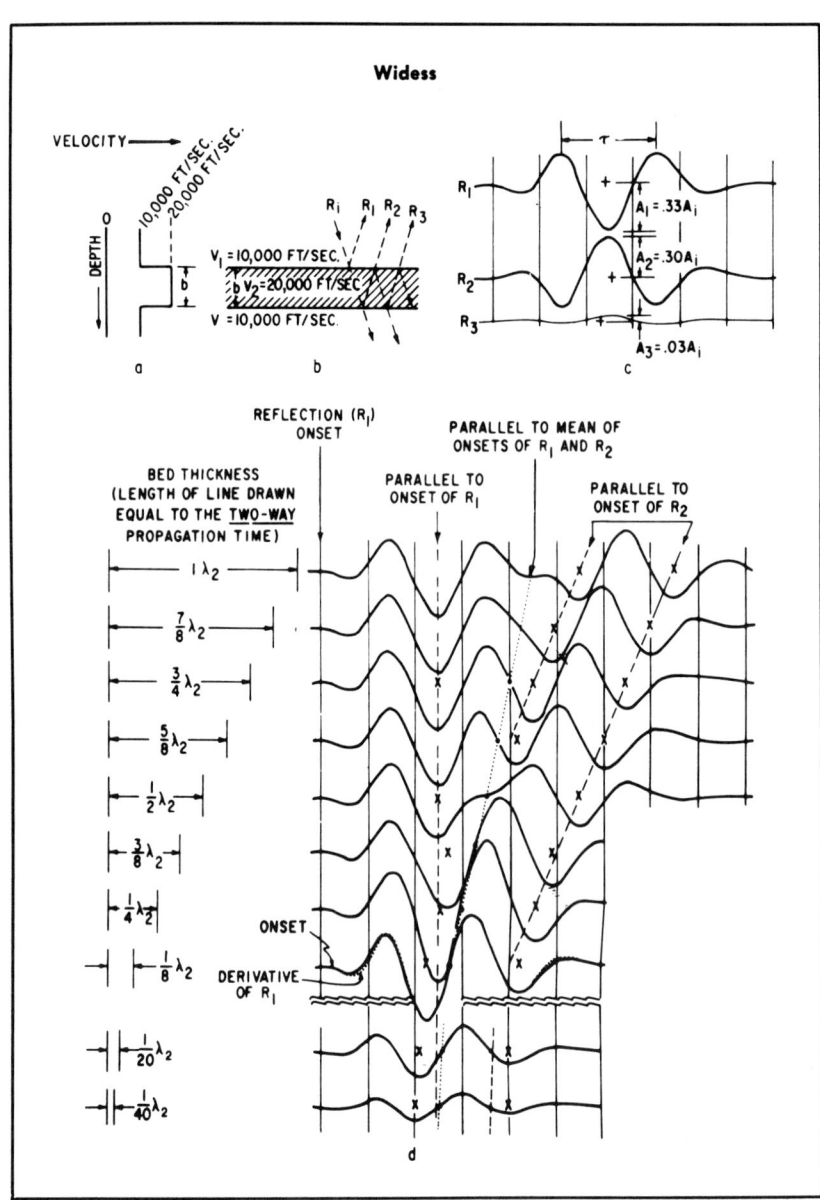

Figure 1. Reflection waveshape versus bed thickness for a symmetrical embedded wavelet (from Widess, 1973).

situation where the gross thickness exceeds a quarter wavelength. Seismic evidences differ depending on whether a situation involves thin or thick beds.

Figure 2 illustrates an interpretive problem in a thin-bed situation. Wavelet shapes are almost the same across the section but the amplitude is roughly proportional to the net sand thickness. The distribution of a reflecting member is not easily determinable. For example, for a reservoir sand, it is not clear whether it involves one member or two or more. Clearly this is an important fact for stratigraphic interpretation. Figure 2 is a one-dimensional synthetic seismogram in which each trace results solely from convolving model data with an assumed wavelet.

In the real earth the seismic process is more complicated than assumed in synthetic seismogram manufacture. Wave phenomena involve diffraction and other phenomena, and three-dimensional aspects clearly come into play. Mahradi (1983) constructed physical models over which he ran seismic profiles in the tank at the Allied Geophysical Laboratories of the University of Houston. Figure 3 shows one section he obtained across a model of variable reflector thickness in the thin-bed case; this can be compared with Figure 1 from Widess. Figure 4 shows a superposition of data from a series of such model seismic sections. The experimental data nicely support the predictions (Figure 5) from synthetic seismograms.

Figure 6 shows a model intended to verify a conclusion drawn from Figure 2 that the waveshape does not change as the number of reflectors changes in a

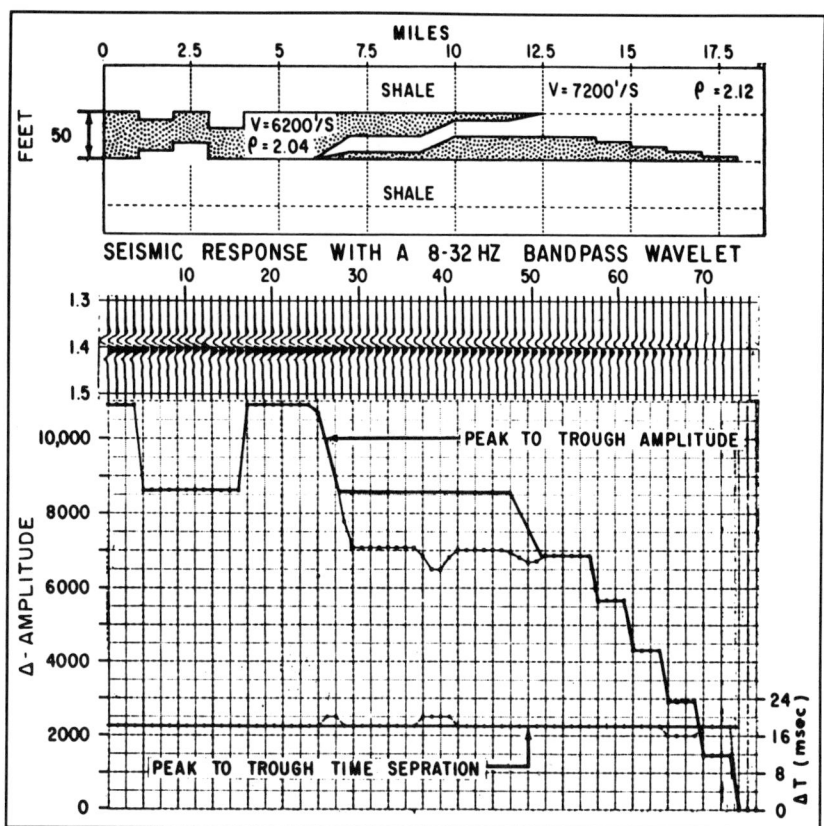

Figure 2. One-dimensional synthetic seismogram. The total model thickness is less than a quarter-wavelength (from Meckel and Nath, 1977).

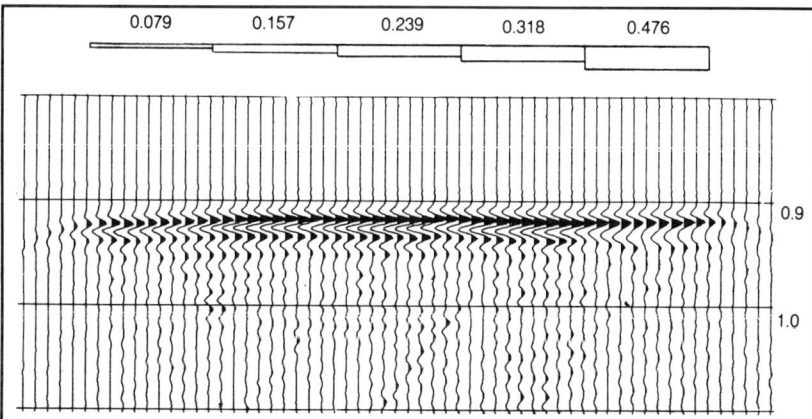

Figure 3. Seismic section across a model of variable thickness. Thicknesses are given as fractions of the dominant wavelength on the cross-section of the model. Below a quarter-wavelength amplitude changes dominate, above it, waveshape changes (from Mahradi, 1983).

thin-bed case, but that the amplitude is proportional to the net sand thickness. Mahradi's model materials had an unfortunately high reflection coefficient so transmissivity losses became important and had to be corrected for. However, the conclusion from the synthetics is supported. Figure 7 shows the same model in a thick-bed situation and the appearance is markedly different. The reflection from the second layer roughly cancels that from the first layer because of destructive interference. Figure 8 shows a model intended to verify another conclusion that can be drawn from Figure 2, that a reflection does not carry information about the distribution of reflectors in the thin-bed case. The reflection where we have a thin bed over a thick bed is nearly the same as where we have a thick bed over a thin one (allowing for transmissivity losses). This is in sharp contrast to the thick-bed situation shown in Figure 9.

Another important aspect of resolvability involves the wavelet bandwidth. Most model studies have used Ricker or similar zero-phase wavelets which do not realistically simulate an actual seismic wavelet. Obviously if a wavelet is very leggy, resolution deteriorates. A seismic wavelet

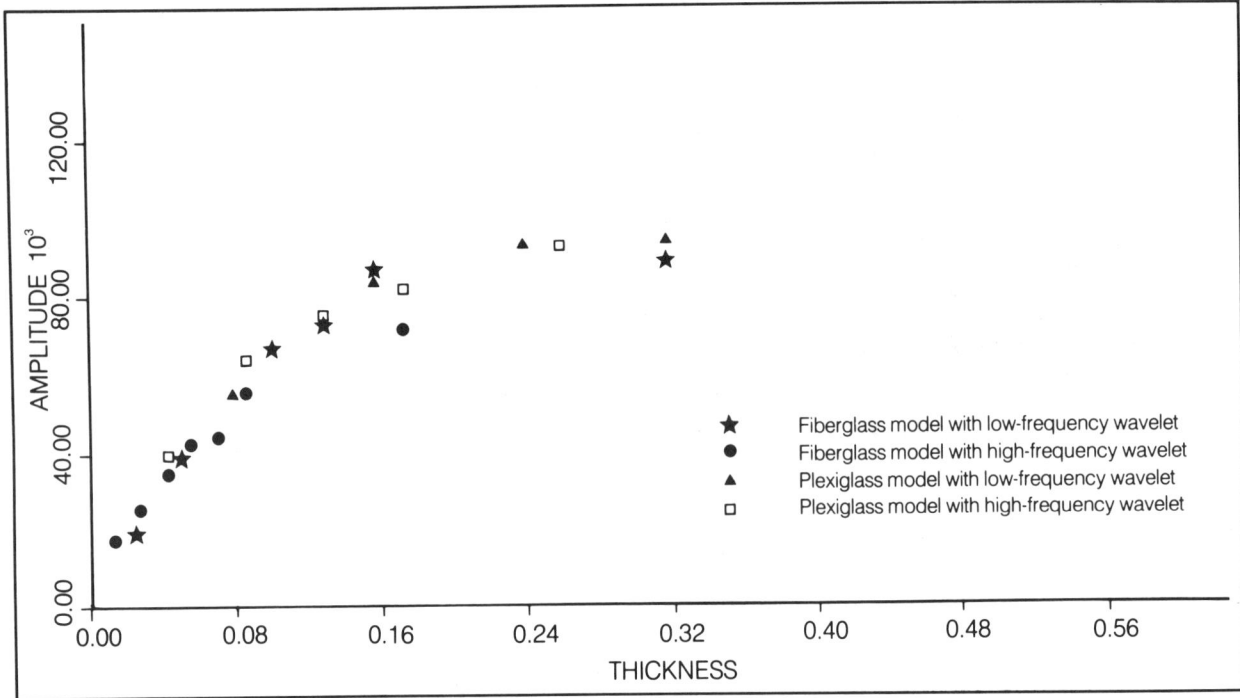

Figure 4. Amplitude as a function of bed thickness (as fractions of dominant wavelength) (from Mahradi, 1983).

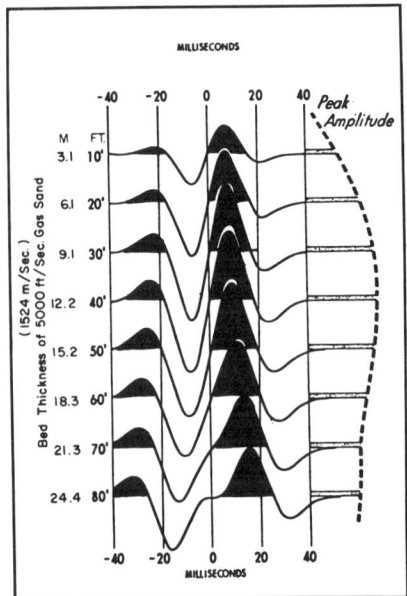

Figure 5. Synthetic seismogram for layer of variable thickness using a 25-Hz Ricker wavelet and plot of peak amplitude (from Neidell and Poggiagliolmi, 1977).

must have a bandwidth of at least one-and-a-half octaves to be reasonably compact and a broader bandwidth further improves a wavelet's shape, two-and-a-half octaves bandwidth being desirable. A bandwidth of one-and-a-half octaves means that the highest frequency components are 2.8 ($2^{1.5}$) times the lowest frequency component — for example a bandpass of 20-56 Hz. Kallweit and Wood (1982) showed that for zero-phase wavelets broader than about two octaves, the resolvability depends on the upper frequency limit. Expressed in terms of the upper frequency limit, the Rayleigh criterion (which Kallweit and Wood say is a practical limit) becomes one-third of the highest wavelength (as opposed to the dominant wavelength). Efforts to improve resolvability by improved acquisition techniques concentrate on increasing the high-frequency content.

The Rayleigh definition of resolvable limit as a quarter-wavelength thus provides a good separation point between seismic results in thin- versus thick-bed situations.

HORIZONTAL RESOLUTION ON UNMIGRATED SECTIONS

Horizontal resolution on unmigrated sections is usually described in terms of the Fresnel zone. The first Fresnel zone is the reflector area from which reflected energy reaches the detector within a half-cycle so that interference is constructive. If we are dealing with coincident source-receiver data (such as common-midpoint sections simulate), we may draw the wavefront which is tangent to a plane reflector at the "reflection point" and also another wavefront which leads this tangent wavefront by a quarter-wavelength (Figure 10a). The tail end of the reflection from the reflection point will be arriving at the detector as the front end of the first half-cycle from the zone's periphery, and the reflections from all intermediate points will thus arrive at the detector within the first half-cycle and thus constructively. Outside the first circular Fresnel zone is an annular ring (Figure 10b) of the second Fresnel zone from which the reflection arrives at the detector to interfere destructively with the reflection from the first zone, and then a third zone from which interference is constructive, and so on alternately. The contributions from zones outside the first alternatively cancel resulting in no net effect.

The magnitude of the Fresnel zone can be determined from the nomogram shown in Figure 11. The area effectively contributing to a reflection is fairly large. The outer parts of the Fresnel zone do not

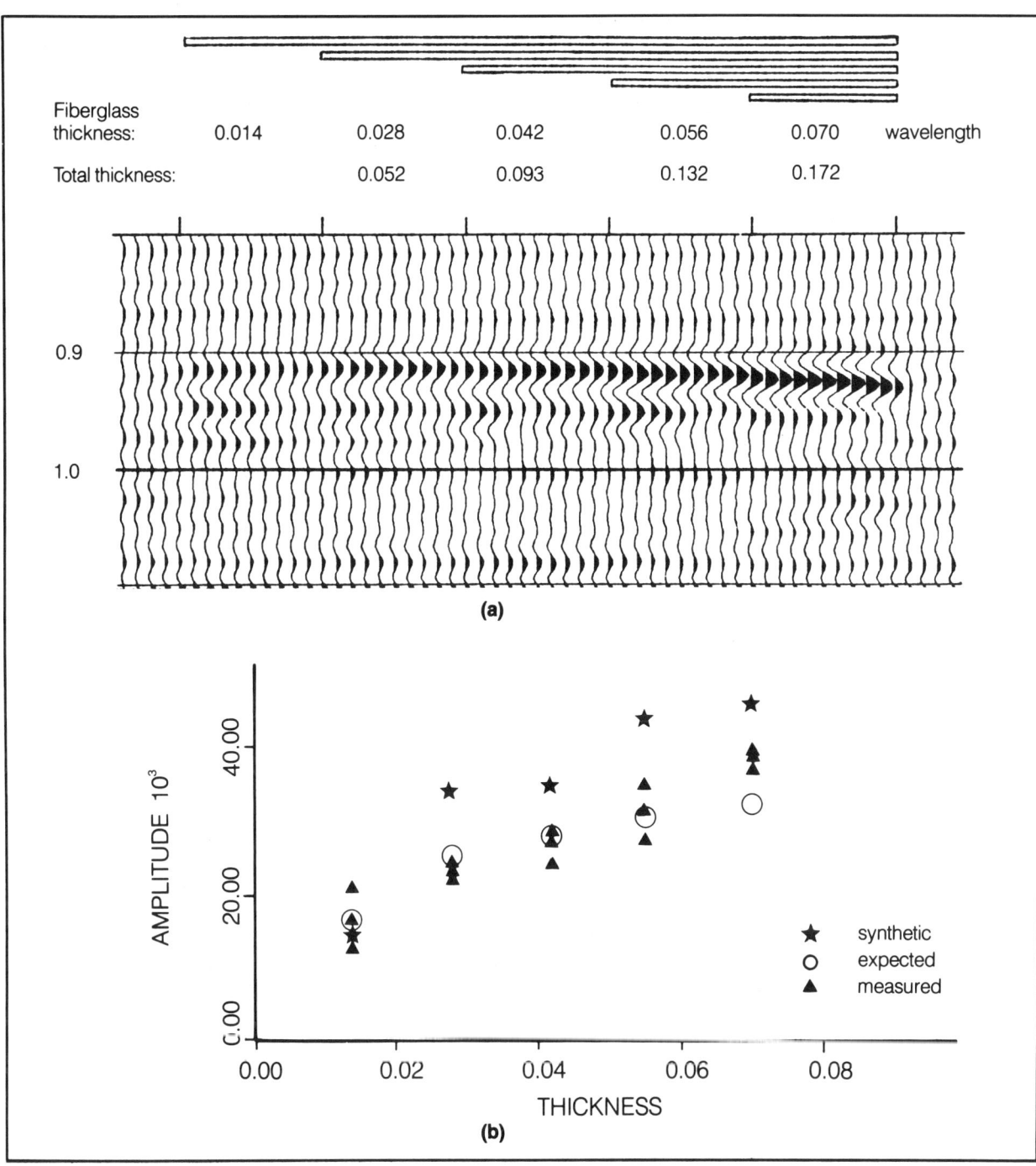

Figure 6. Seismic section (a) and plot of amplitudes (b) as a function of net fiberglass thickness (as fractions of the dominant wavelength). "Synthetic" refers to amplitudes measured on a synthetic seismogram that did not allow for transmissivity losses, "expected" are the synthetic results allowing for transmissivity losses (from Mahradi, 1983).

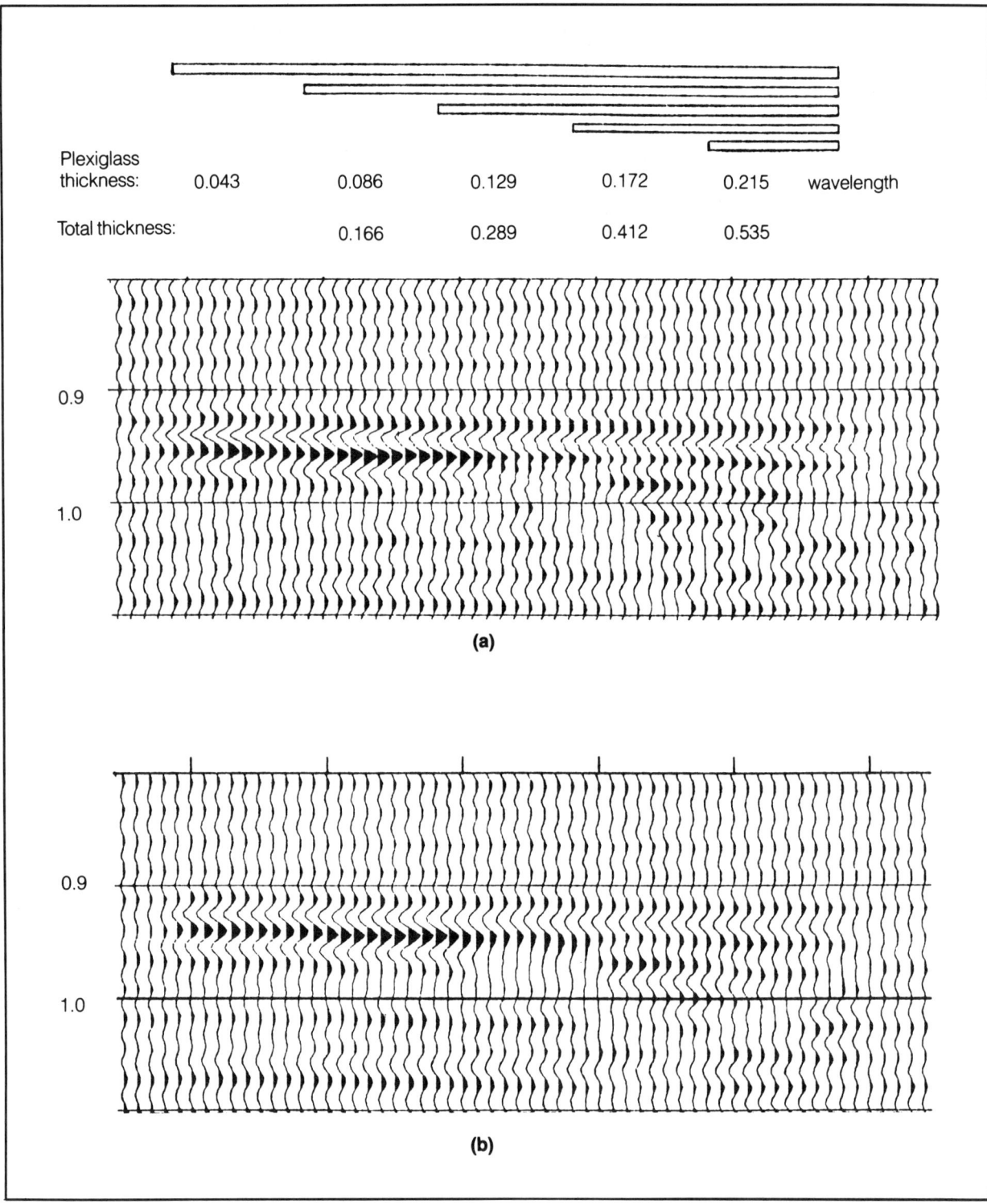

Figure 7. Seismic section along plexiglas model 2 with high-frequency wavelet. (a) Unmigrated; (b) migrated (from Mahradi, 1983).

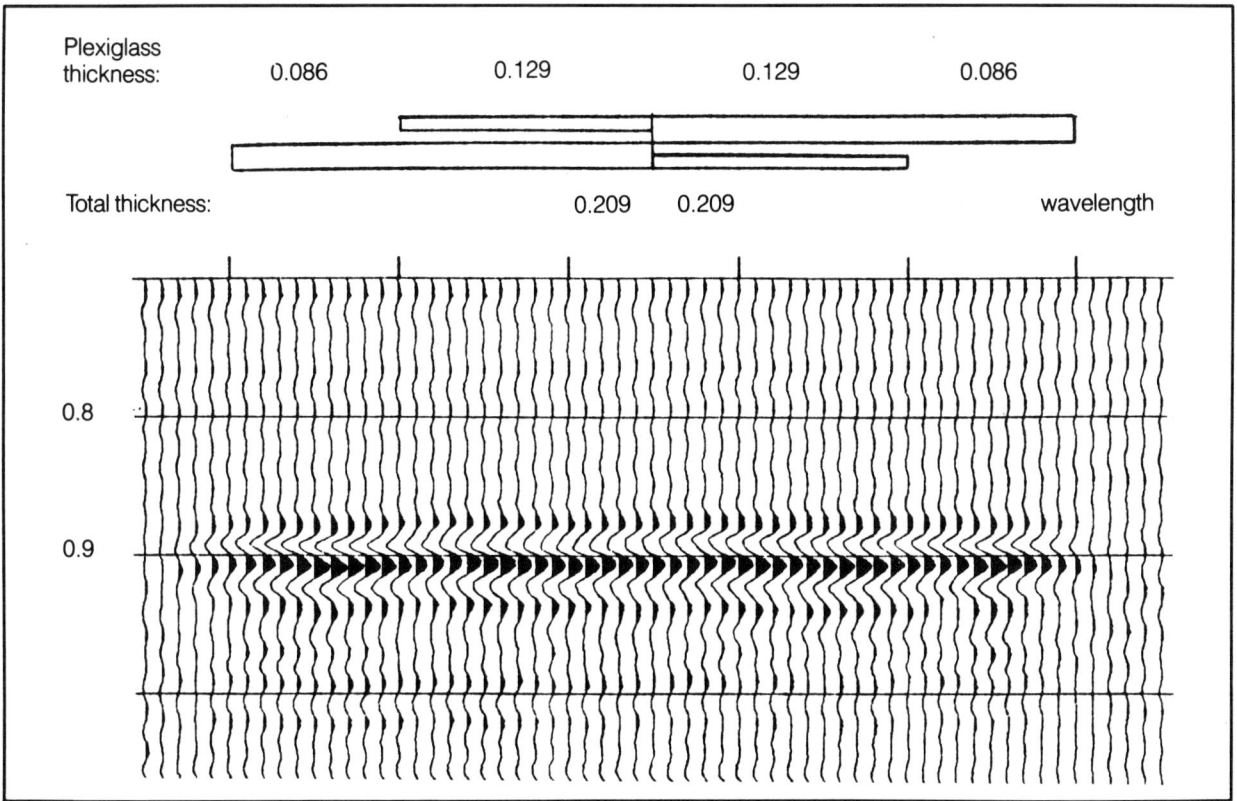

Figure 8. Seismic section along model with low-frequency wavelet (from Mahradi, 1983).

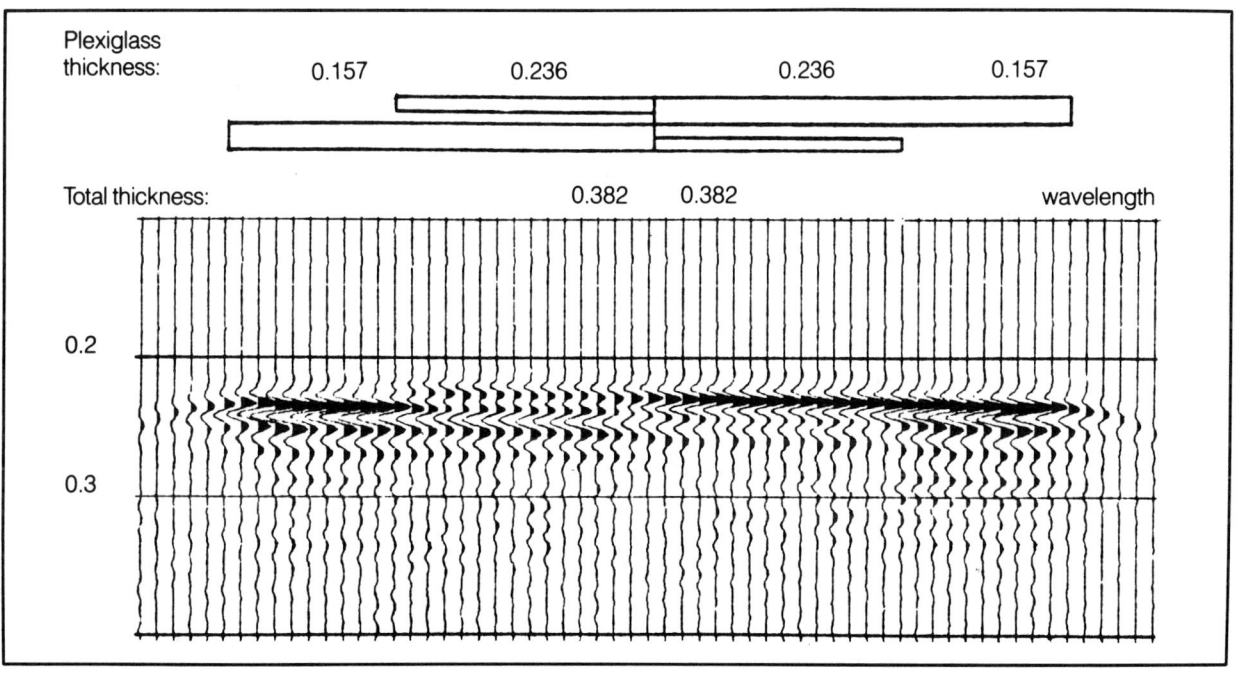

Figure 9. Seismic section along model with high-frequency wavelet (from Mahradi, 1983).

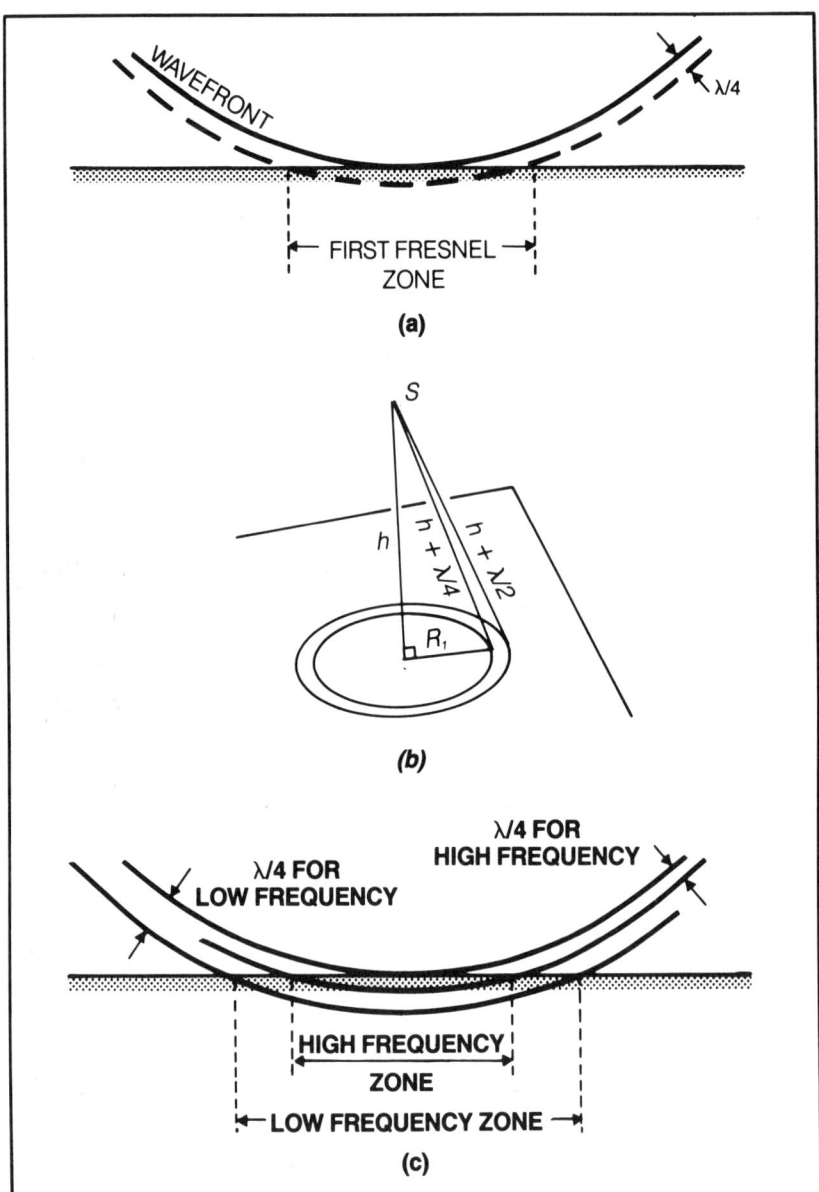

Figure 10. Fresnel zone. (a) How to determine the diameter of the first Fresnel zone for coincident source and detector (such as common midpoint sections simulate). (b) Showing the second Fresnel zone (shaded annular ring). (c) Fresnel zone size depends on frequency (or wavelength). From Sheriff, 1980.

contribute as much to a reflection as the inner parts of the zone, however, and sometimes an "effective" Fresnel zone is used which has half the radius of the Fresnel zone as here defined. This can be accommodated by treating the Fresnel-zone radius obtained from Figure 11 as the effective Fresnel-zone diameter.

Figure 12 shows the seismic response to reflectors whose dimensions are measured in Fresnel-zone units. The responses where reflectors are smaller than a Fresnel zone are similar in many regards and resemble simple diffractions. The patterns do contain information about the reflector size, for example amplitudes and diffraction tails are different, but these aspects are subtle. The response begins to show the reflector shape once the reflector reaches the size of the Fresnel zone, seen in Figure 12 as flattening of the top center of the pattern.

Figure 13 shows the response to a reflector which contains a hole three trace intervals wide, so there is no specular reflection for the three central traces. Nonetheless a reflection can be seen on these traces because the Fresnel zone laps onto the reflector nearby. This might simulate a reflection from a bed deposited everywhere on a shelf except where pinnacle reefs were growing, and thus indicates how difficult it would be to locate such reefs by interruptions in the continuity of surrounding reflections.

HORIZONTAL RESOLUTION ON MIGRATED SECTIONS

With migrated seismic data the Fresnel-zone dimensions effectively

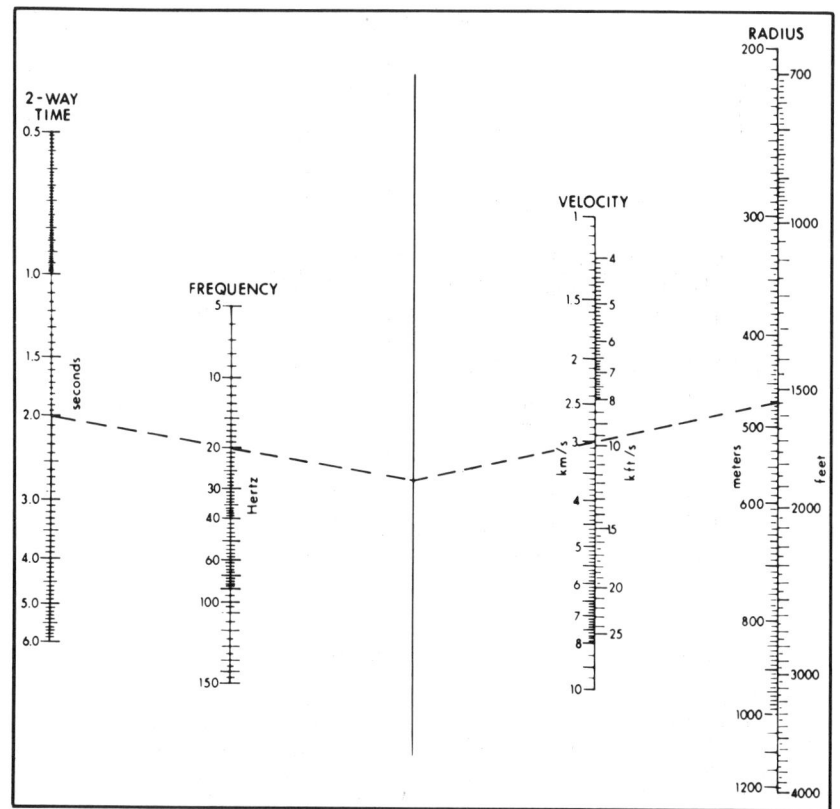

Figure 11. Nomogram for determining size of the Fresnel zone. A line connecting the arrival time and the frequency intersects the central line at the same point as a line connecting the average velocity and the radius of the zone. For example, a reflection at 2.000 sec with a dominant frequency of 20 Hz and an average velocity of 3 km/sec (1.9 mi/sec) represents a Fresnel zone of radius 470 m (1,542 ft). From Sheriff, 1980.

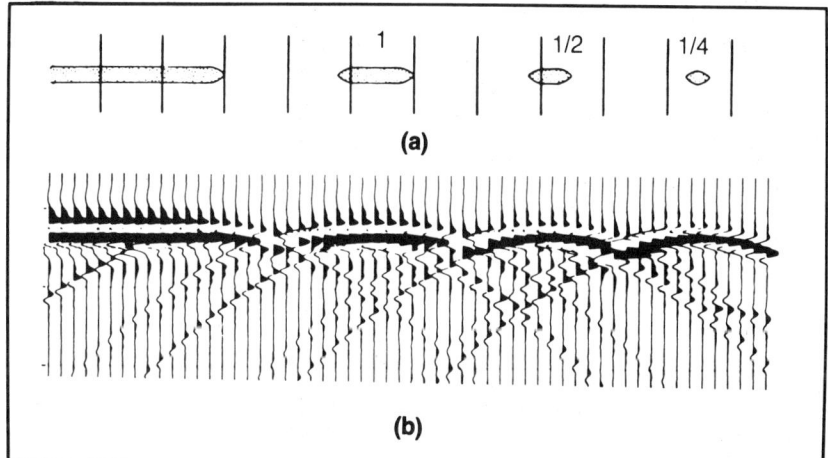

Figure 12. Reflections from strips of various widths measured in Fresnel zone units (from Meckel and Nath, 1977).

shrink so that the Fresnel-zone size no longer provides a useful criterion. Diffraction tails represent the part of the diffraction energy with a large horizontal component of travelpath, the aspect to be maximized to improve horizontal resolution. The patterns clearly differ for the two smaller reflectors in Figure 12.

Migration involves collapsing diffractions to a point and the complete diffraction must be included if the migration is to be complete.

Practical migration programs limit the aperture, the amount of data to be included in the migration process. The horizontal sampling interval sets a spatial aliasing limit to the significance of data traveling by paths with large horizontal components. Ground mixing (resulting from geophone and source array dimensions) limits how much of diffraction tails are recorded. Common-midpoint stacking further degrades the recording of diffraction tails. Thus recording and processing

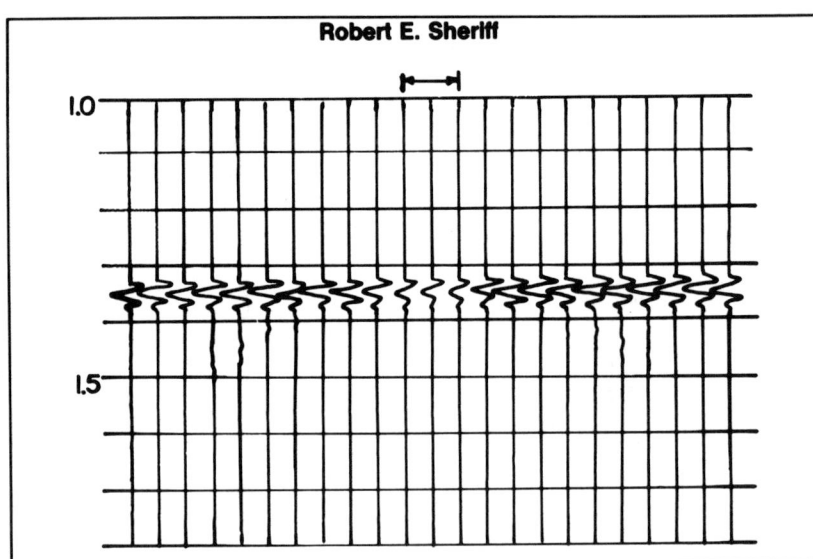

Figure 13. Reflection from a reflector containing a hole (as indicated by arrow at the top). From Sheriff, 1977.

parameters impose limits on horizontal resolution. For example, geophone intervals need to be much shorter than those used in current recording practice.

Migration works properly only on primary reflection and diffraction energy and migration smears-out noise, "smiles" being one of the more evident aspects. The migration of noise is one of the more important factors limiting horizontal resolution on migrated sections. Actual migration programs also involve approximations which further add to background noise.

Denham and Sheriff (1980) list the causes of loss of horizontal resolution (in order of importance) on migrated sections as:

1. Noise on unmigrated sections
2. Spatial sampling (aliasing problems)
3. Migration aperture (apparent dip)
4. Velocity errors
5. Stacking limitations
6. Two-dimensional assumption
7. Other approximations and errors.

None of these constitute inherent limitations, that is, proper recording and processing can reduce the objectionable aspects of any of these.

CONCLUSIONS

A quarter-wavelength thickness separates a thin-bed domain from a thick-bed domain and the appearance of features on seismic sections is different depending on whether they involve the one or the other. Many reservoirs and stratigraphic features fall within the thin-bed domain or straddle the quarter-wavelength region which makes understanding of how the characteristics differ very important in stratigraphic interpretation.

The factors affecting horizontal resolution on seismic sections differ depending on whether the sections have or have not been migrated. On unmigrated sections, regarding the Fresnel zone as the limit of horizontal resolution is probably too stringent by a factor of 2. On migrated sections, the most important factors controlling horizontal resolution are noise and spatial sampling.

REFERENCES CITED

Denham, L. R., and R. E. Sheriff, 1980, What is horizontal resolution: Paper presented at 50th Annual Meeting of the Society of Exploration Geophysicists.

Kallweit, R. S., and L. C. Wood, 1982, The limits of resolution of zero-phase wavelets: Geophysics, v. 47, p. 1035-1046.

Mahradi, 1983, Physical modeling studies of thin beds: University of Houston, Master's thesis, 100 p.

Meckel, L. D., and A. K. Nath, 1977, Geologic considerations for stratigraphic modeling and interpretation, in C. E. Payton, ed., Seismic stratigraphy—applications to hydrocarbon exploration: AAPG Memoir 26, p. 417-438.

Neidell, N. S., and E. Poggiagliolmi, 1977, Stratigraphic modeling and interpretation—geophysical principles and techniques: in C. E. Payton, ed., Seismic stratigraphy—applications to hydrocarbon exploration: AAPG Memoir 26, p. 389-416.

Sheriff, R. E., 1977, Limitations on resolution of seismic reflections and geologic detail derivable from them: in C. E. Payton, ed., Seismic stratigraphy—applications to hydrocarbon exploration: AAPG Memoir 26, p. 3-14.

——, 1980, Nomogram for Fresnel zone calculation: Geophysics, v. 45, p. 968-992.

Widess, M. B., 1973, How thin is a thin bed: Geophysics, v. 38, p. 1176-1180.

HOW THIN IS A THIN BED?†

M. B. WIDESS*

Based on reflective properties, a thin bed may be conveniently defined as one whose thickness is less than about $\lambda_b/8$, where λ_b is the (predominant) wavelength computed using the velocity of the bed. The amplitude of a reflection from a thin bed is to the first order of approximation equal to $4\pi Ab/\lambda_b$, where b is the thickness of the bed and A is the amplitude of the reflection if the bed were to be very thick. The equation shows that a bed as thin as 10 ft has, for typical frequency and velocity, considerably more reflective power than is usually attributed to it.

Editor's note: This paper is reprinted with the permission of the Geophysical Society of Tulsa. Although it is normally the policy of GEOPHYSICS *not to publish previously published material, an exception has been made for this paper because of the current renewed interest in seismic reflection amplitudes.*

INTRODUCTION

For an unknown reason the geophysical industry appears to have grown up with some misconceptions on the reflective properties of so-called thin beds. What is the reflective behavior of thin beds? How thin must the bed be before the reflection has negligible amplitude? Our purpose is to consider such questions in elementary terms.

GRAPHICAL ILLUSTRATION OF A TIME DERIVATIVE OF A WAVELET

Let us first examine the algebraic difference of two identical wavelets which are displaced slightly in time. Referring to Figure 1a, wavelets R_1 and $-R_2$ are identical except for the time difference ΔT between them. Vertical lines between R_1 and $-R_2$ mark the difference in amplitude at successive simultaneous times, and this difference R_d is plotted in Figure 1b. The following properties of R_d are evident. (1) Wavelet R_d has *zero* amplitude in each half cycle at a time close to midway between the times when R_1 and $-R_2$ are at their *maximum* amplitude. That is, there is a 90-degree phase shift between R_d and the mean of R_1 and $-R_2$, the phase being advanced in time. (2) Correspondingly, whereas R_1 exhibits an "M" form of character (when considering strongest peaks and trough), R_d has an "S" form of character (when considering strongest peak and trough). (3) The first maximum (a trough) of R_d arrives earlier than the first maximum (a trough) of R_1, and the last maximum (a peak) of R_d arrives later than the last maximum (a trough) of R_1. In total, R_d has a half cycle more than does R_1, and this incidentally is associated with a greater relative content of high frequency in R_d than in R_1.

The wavelet R_d is clearly the reflection from a thin bed, Figure 1c, when the acoustic impedance (product of velocity and density) in the medium above the bed is the same as that in the medium below the bed, R_1 being the reflection from the top interface and R_2 from the bottom interface (transmission loss and multiple reflections being neglected). The negative sign attached to R_2 in Figure 1a, of course, accounts for the phase inversion at the bottom interface in this example. The time displacement ΔT is equal to $2b/V_b$ for vertical incidence, where b is the thickness of the bed and V_b is the velocity of the bed.

Since R_d is the difference between identical wavelets that are displaced in time, R_d approximates the time derivative of R_1 when the time displacement is small. It is significant however that

---

† Reprinted from Proceedings of the Geophysical Society of Tulsa 1957–58. Manuscript received by the Editor May 14, 1973.
* Retired from Amoco Production Co., Houston, Tex. 77001
© 1973 Society of Exploration Geophysicists. All rights reserved

the *form* of the wavelet R_d still closely approximates the derivative of R_1 even when b is as great as one-eighth of the wavelength λ_b computed using the velocity in the bed ($\lambda_b = \tau V_b$, where τ is the predominant period of the wavelet). This is illustrated in the next figure to be discussed.

EFFECT OF BED THICKNESS ON REFLECTION CHARACTER AND TIMING

The traces in Figure 2d show reflections from a progessively thinner bed. As before, the velocity above the bed is the same as that below the bed, Figure 2a. The velocity of the bed itself is twice that of the superjacent and subjacent media. The wavelets R_1 and R_2 reflected from the upper and lower interfaces respectively, as well as the first-order multiple R_3, are shown in Figure 2c in terms of the amplitude A_i of the incident wavelet R_i. The relations are for vertical incidence, and density changes are neglected. The first-order multiple reflection is so weak that for our present purposes it could also have been neglected, as are the higher order multiple reflections. The traces in Figure 2d were derived arithmetically by compositing R_1, R_2, and R_3 in a time relation corresponding to the respective bed thicknesses. The traces exhibit interplay between reflections from the top and bottom interfaces of the bed, producing destructive interference for $b = \lambda_b/2$ and constructive interference especially for $b = \lambda_b/4$. Our attention however is to be directed toward the still thinner beds, where constructive interference is at first still active but with successively thinner

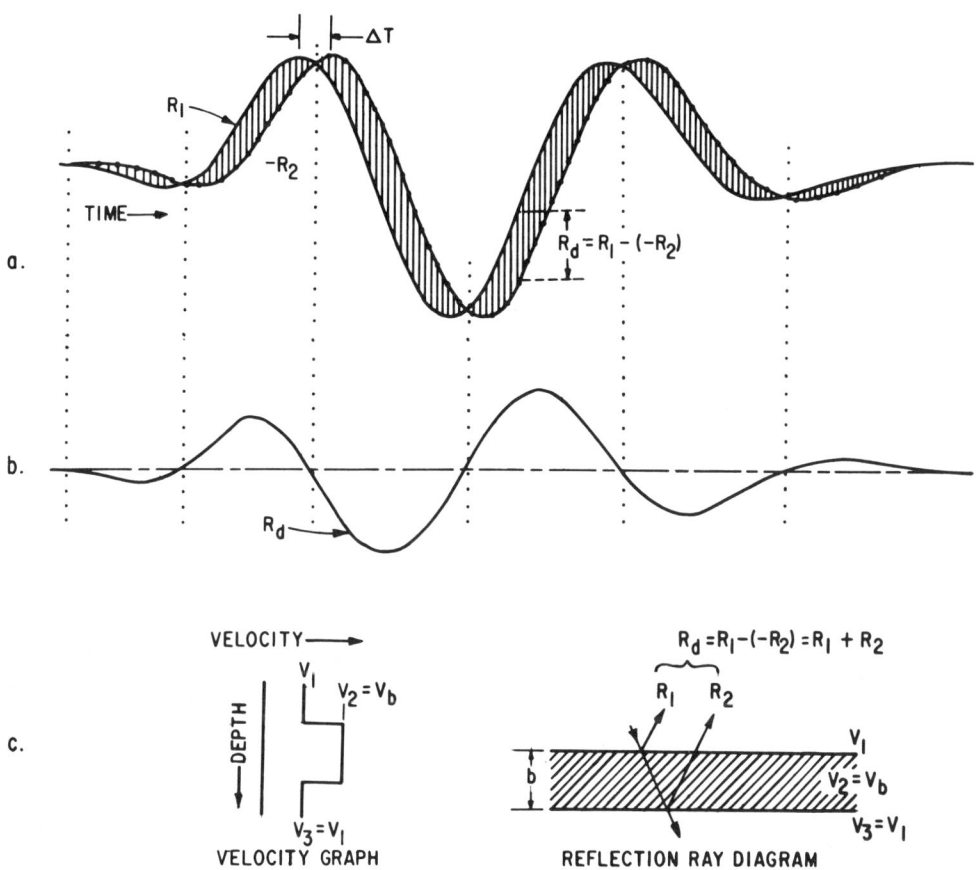

Fig. 1. The phase shift and change in character resulting from the difference of identical wavelets displaced slightly in time. a. Identical wavelets, R_1 and $-R_2$, displaced by time ΔT. b. Difference, R_d, between R_1 and $-R_2$. c. Reflection from a thin bed in which $V_3 = V_1$. R_1 and $-R_2$ are identical except for a time displacement. (For simplicity, the transmission loss and multiple reflections are neglected and density is considered uniform.)

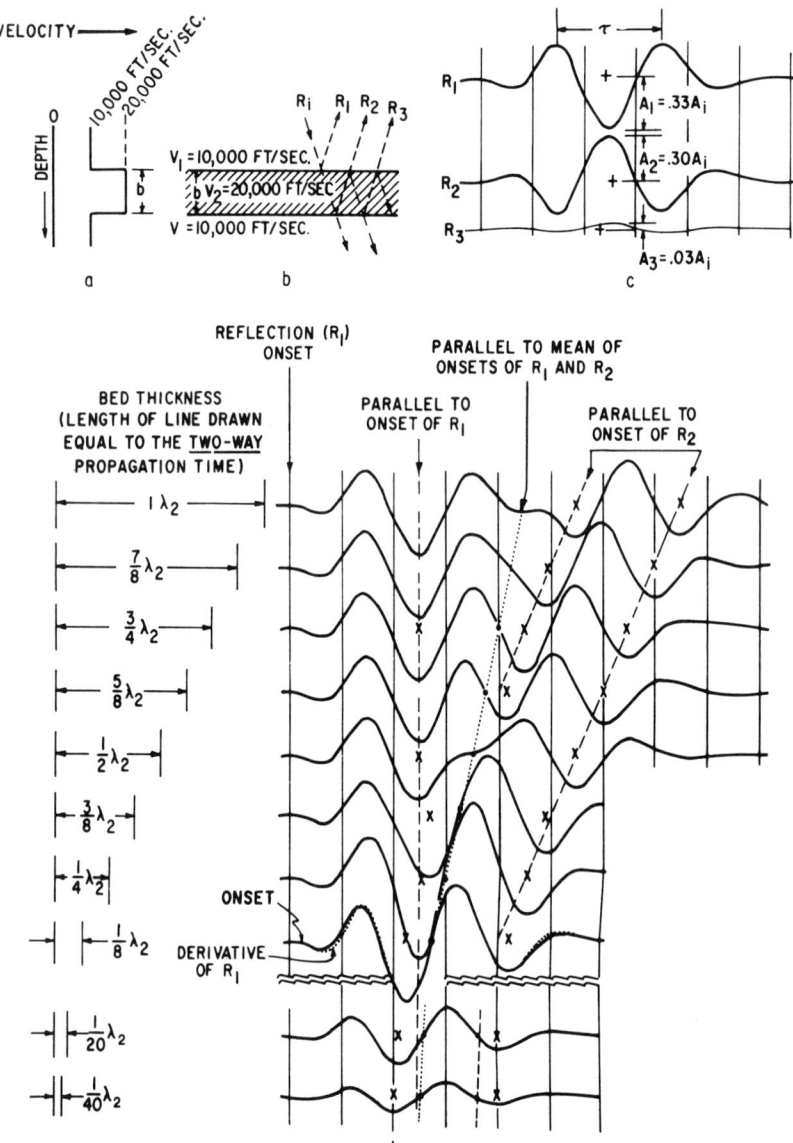

Fig. 2. Effect of bed thickness on the reflection. a. Velocity graph. b. Reflection ray diagram. c. Individual reflected waves that are composited using time delays computed from bed thickness. d. Form and relative timing of composite reflection as a function of bed thickness. × marks trough time. ○ marks zero-amplitude time ("center" of composite reflection). Timing line interval is $0.5\,\tau$. b = thickness of bed. τ = predominant period of incident wavelet. λ_2 = wavelength within bed ($\lambda_2 = V_2\tau$). Amplitudes for composite reflections are all relative to the same incident wavelet R_i.

beds, destructive interference proceeds to extinguish the reflection.

When the bed is very thin, the character of the reflection is that of the time derivative of the incident wavelet and the timing is dictated by the time to the center of the bed. That substantially the same character and timing exist for bed thickness as great as about $\lambda_b/8$ is demonstrated on the trace for that bed thickness. The time derivative of the incident wavelet is shown there by the dotted-line wavelet, and we see that this almost duplicates the reflection on that trace. (In drawing the time derivative wavelet, its amplitude was increased by a constant factor to match the amplitude of the reflection on the trace, and the onset of the time derivative wavelet was located at the mean of the onset times of R_1 and R_2.) Thus, insofar as bed thickness alone is concerned, the character of the reflections is indistinguishable for beds whose thickness is less than about $\lambda_b/8$. For that reason it is appropriate to define a thin bed as one whose thickness is less than about $\lambda_b/8$. Two-way time through a thin bed would then be less than about $\tau/4$. A bed that is thin for one frequency is, of course, not necessarily thin for a higher frequency.

EFFECT OF BED THICKNESS ON REFLECTION AMPLITUDE

To the first order of approximation the central portion of wavelet R_1 in Figure 1 may be treated as a sine wave whose maximum amplitude A is the mean between the amplitudes of the predominant peak and trough of R_1. This simplification permits an easy derivation of the approximate amplitude of reflection R_d from a thin bed. Referring zero time t_0 to the mean of the deep-trough times of R_1 and $-R_2$, the equations for the central portion of R_1 and $-R_2$ respectively are then

$$R_1 \cong - A \cos (t + b/V_b) 2\pi/\tau, \quad (1)$$

and

$$-R_2 \cong - A \cos (t - b/V_b) 2\pi/\tau, \quad (2)$$

where t is the time relative to t_0 and τ is the predominant period of the wavelet. By expanding the two equations and taking the difference, we obtain

$$R_d = R_1 + R_2 \\ \cong [2A \sin 2\pi b/\tau V_b] \sin 2\pi t/\tau. \quad (3)$$

The term in brackets is approximately the maximum amplitude A_d of wavelet R_d. To the first order of approximation in the case of a thin bed,

$$\sin 2\pi b/\tau V_b \cong 2\pi b/\tau V_b.$$

So that

$$A_d \cong 4\pi A b/\tau V_b.$$

Since

$$\lambda_b = \tau V_b, \text{ we have}[1] \quad A_d \cong 4\pi A b/\lambda_b. \quad (4)$$

Therefore for thin beds the amplitude of the reflection is approximately proportional to the thickness of the bed and inversely proportional to the wavelength.

We note that reflections from beds that are generally considered very thin are not necessarily restricted to small amplitudes. For example, if $b/\lambda_b = 1/20$, we have $A_d \cong 0.6A$. That is, in a typical case of a reflection whose predominant frequency is 50 hz and a bed whose velocity is 10,000 ft/sec, the wavelength λ_b is 200 ft, so that a bed whose thickness is only 10 ft would still have about 0.6 of the amplitude that would result if the bed were very thick. If the bed were to be only 5 ft thick, the factor would still be fairly large, i.e., 0.3 instead of 0.6. These magnitudes may be seen in the bottom two traces respectively of Figure 2d, comparing them with the top trace.

The above conditions apply only to thin beds for which the two media bounding the bed have the same acoustic impedance. The relations do not apply when the two bounding media have appreciably different acoustic impedance, since in that case not only is a thin bed involved but also an acoustic change in the absence of the thin bed. It is then generally sufficiently accurate to consider that the reflection from the bed is a composite of the following two reflections: (1) the reference reflection, i.e., the reflection which would result in the absence of the thin bed, and (2) the time-derivative type of reflection associated with the thin bed itself, for which the acoustic impedance above and below the bed is the same and is equal to the acoustic impedance of the medium which the bed replaces. The effect of the thin bed may then be reckoned in terms of the relative strength and phase relation between the reference reflection and the time-derivative reflection. The relation may be shown readily by vectorial representation.

---

[1] The exact equation for reflection from a single imbedded layer, considering simple harmonic waves and accounting for transmission loss (but not absorption loss), is given in Rayleigh (1945). The equation, adapted to our notation, is

$$A_d = A(1 + r)^2 [(2r \cot 2\pi b/\lambda_b)^2 + (1 + r^2)^2]^{-1/2},$$

where r is the ratio of acoustic impedances. The quantity A_d obtained from this equation differs from A_d in equation (4) by no more than only 12 percent when the bed is thin, $b/\lambda_d < \frac{1}{8}$, and when $\frac{1}{2} < r < 2$, the range of acoustic contrast usually encountered in practice.

RESOLVING POWER

A definition of the term "thin bed" involves the concept of resolving power. Resolving power is the ability to distinguish between the properties of two (or more) elements. The elements that we have been considering here are the reflecting interfaces of a bed. Resolving power is illustrated in Figure 2 as follows. When bed thickness b is large enough that the individual reflected wavelets from each of the two interfaces are completely separated in time, the trace on the record, of course, potentially yields maximum possible information for each of the interfaces. As the bed thickness diminishes, more and more of the energy becomes a composite for the two reflections. That is, there is successively less data for each of the reflections separately but more data in the form of combination of the two reflections. This trend continues until the thickness is equal to about $\lambda_b/8$. For this and still thinner beds, substantially the only information left is for the combination of the two reflections and, therefore, substantially none for the individual reflections.[2] At that point, therefore, resolving power may be said to be lost and the point may be loosely called the theoretical threshold of resolution. Practically, a number of other factors are involved that will determine the threshold of resolution. For example, with the presence of noise the broadening of the wavelet from $b=\lambda_b/8$ to $b=\lambda_b/4$ in Figure 2 may be obscured, thus forcing the threshold of resolution to the thicker bed. The threshold of resolution therefore depends not only on the predominant frequency of the incident wavelet but also on the signal-to-noise ratio. Still other factors include the form and duration of the incident wavelet, the degree to which this wavelet is known prior to the analysis, and the analytical tools used.

When the whole of a thin bed rather than the individual interfaces is to be considered, a different threshold of resolution is brought into play. For example, if either the velocity system or the bed thickness remains constant, the change of bed thickness or velocity, respectively, can be determined under favorable circumstances for a bed whose thickness is considerably less than $\lambda_b/8$. Measurements are then made on the change in reflection time and/or the change in reflection amplitude.

REFERENCE

Rayleigh, John William Strutt, 1945, The theory of sound: New York, Dover Publishing Co., v. 2, p. 88.

---

[2] The fact that information only on a combination of reflections is available means that substantially the same reflection can be obtained for a very wide assortment of thin beds, which in themselves may consist of different laminas. It may therefore be of interest to state that to the first order of approximation the only conditions which these beds must fulfill to qualify in this assortment are the following: (1) the overall bed must be thin, (2) the total area between the time versus velocity curve and the reference velocity line must be the same for the overall bed, treating the respective areas of the bed algebraically, relative to the reference velocity line; (3) the geometric center of this area must be at the same record time, again considering the respective areas algebraically relative to the reference velocity line.

Complex seismic trace analysis of thin beds

James D. Robertson* and Henry H. Nogami*

ABSTRACT

Displays of complex trace attributes can help to define thin beds in seismic sections. If the wavelet in a section is zero phase, low impedance strata whose thicknesses are of the order of half the peak-to-peak period of the dominant seismic energy show up as anomalously high-amplitude zones on instantaneous amplitude sections. These anomalies result from the well-known amplitude tuning effect which occurs when reflection coefficients of opposite polarity a half period apart are convolved with a seismic wavelet. As the layers thin to a quarter period of the dominant seismic energy, thinning is revealed by an anomalous increase in instantaneous frequency. This behavior results from the less well-known but equally important phenomenon of frequency tuning by beds which thin laterally. Instantaneous frequency reaches an anomalously high value when bed thickness is about a quarter period and remains high as the bed continues to thin.

In this paper, complex trace analysis is applied to a synthetic model of a wedge and to a set of broadband field data acquired to delineate thin lenses of porous sandstone. The two case studies illustrate that sets of attribute displays can be used to verify the presence and dimensions of thin beds when definition of the beds is not obvious on conventional seismic sections.

INTRODUCTION

The concept of treating a seismic trace as the real part of a complex function of time has been used in recent years to interpret both earthquake signals (Farnbach, 1975) and common-depth-point (CDP) reflection records (Taner and Sheriff, 1977; Sicking, 1978; Taner et al, 1979). Since a seismic trace is a causal time series, the imaginary part of the complex function can be computed directly from the seismic trace itself using a Hilbert transform. The real and imaginary (also called quadrature) parts are then the inputs which can be used to determine specific properties of the complex function such as the instantaneous attributes of amplitude, phase, and frequency. The advantage of this type of analysis is that the seismic signal is decomposed into functions which distinguish the amplitude information in the original trace from the angular (phase and frequency) information.

This paper examines the application of complex trace analysis to the exploration problem of detecting and quantitatively evaluating thin beds. A "thin bed" is defined in the manner of Widess (1973) as a bed whose thickness is substantially less than the dominant wavelength of the seismic pulse propagating through the bed. Complex trace analysis is applied both to a synthetic model of a wedge and to a set of broadband field data acquired to delineate thin lenses of porous sandstone.

PHYSICAL SIGNIFICANCE OF COMPLEX TRACE ATTRIBUTES

Previous publications have addressed both the methodology of computing complex trace attributes and apparent relationships between features of attribute displays and physical properties of subsurface geologic sequences (Farnbach, 1975; Taner and Sheriff, 1977; Sicking, 1978; Taner et al, 1979). Instantaneous amplitude (also called amplitude envelope or reflection strength) is equivalent to the envelope function of the seismic trace and is a robust, smoothed, polarity-independent measure of the energy in the trace at a given time. Instantaneous phase is the angle between the trace and its Hilbert transform at a given time and is an amplitude-independent estimate of the character of the trace. Instantaneous frequency is a sample-by-sample measure of the frequency in the trace and is equivalent to the time derivative of instantaneous phase.

A notable feature of the instantaneous frequency attribute is that absolute numerical values on an instantaneous frequency section sometimes can be used to extract significant information about the propagating seismic pulse. As an example, consider the 25 Hz zero-phase Ricker wavelet analyzed by Taner et al (1979) in Table 1 and Appendix B of their paper. In Table 1, the instantaneous frequency $w(t)$ of the wavelet is computed at $t = 2, 6, 10, \ldots$ msec. A supplemental computation to fill in the table at $t = 0$ msec produces

$$w(0) = 28.21 \text{ Hz}. \tag{1}$$

Now consider the analytical expression for the amplitude spectrum of the wavelet (Ricker, 1945),

$$A(\xi) = \xi^2 e^{-\xi^2}, \tag{2}$$

Presented at the 51st Annual International SEG Meeting October 15, 1981, in Los Angeles. Manuscript received by the Editor March 30, 1983; revised manuscript received October 6, 1983.
*ARCO Oil and Gas Company, P.O. Box 2819, Dallas, TX 75221.
© 1984 Society of Exploration Geophysicists. All rights reserved.

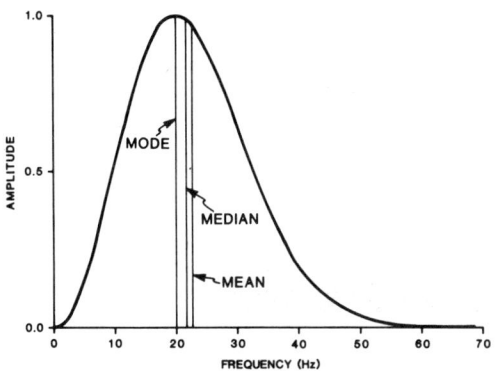

FIG. 1. Amplitude spectrum of 20 Hz Ricker wavelet.

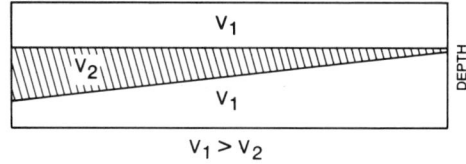

FIG. 2. Geometry of wedge model.

where $A(\xi)$ = amplitude spectrum, $\xi = f/f_1$, f = frequency in Hz, and f_1 = frequency in Hz corresponding to the maximum value of the spectrum, i.e., the mode. The center of gravity $\bar{\xi}$ of $A(\xi)$ is given by

$$\bar{\xi} = \frac{\int_0^\infty \xi A(\xi)\,d\xi}{\int_0^\infty A(\xi)\,d\xi}. \quad (3)$$

Upon substituting equation (2) into equation (3) and performing the integrations, the following result is obtained:

$$\bar{\xi} = 3/[2\Gamma(5/2)]$$
$$= 2/(\pi)^{1/2} = 1.1283, \quad (4)$$

where Γ is the gamma function. Therefore,

$$f_{\xi=\bar{\xi}} = 1.1283(25) = 28.21 \text{ Hz}. \quad (5)$$

Comparing equations (1) and (5), it is apparent that, for a zero-phase Ricker wavelet, the instantaneous frequency at the peak of the wavelet is equal to the frequency corresponding to the center of gravity of the wavelet's amplitude spectrum. This equivalence is a geophysical extension of the assertion made in the electrical engineering literature by Ackroyd (1970) that the instantaneous frequency of an analytic signal at a given instant of time is a measure of the center frequency corresponding to the normalized first moment of the power of the signal at that time. Thus, if the propagating pulse in a seismic section could be approximated by zero-phase Ricker wavelets, and if the geophysicist could identify occasional wavelet peaks in the section which were uncontaminated by noise or interference, the instantaneous frequencies at these samples would be direct estimates of the center of gravity of the pulse's amplitude spectrum. The example above uses a Ricker wavelet. The interpretational concept, however, applies to any zero-phase wavelet.

WEDGE MODEL

The response of the complex attributes to thinning beds will first be evaluated for a 20 Hz Ricker wavelet impinging on a model of a wedge. The amplitude spectrum of the wavelet is displayed in Figure 1, and the mode (20.0 Hz), median (21.8 Hz), and mean (22.5 Hz) are indicated on the figure. The wedge model itself is displayed in Figure 2. Acoustic velocity equals

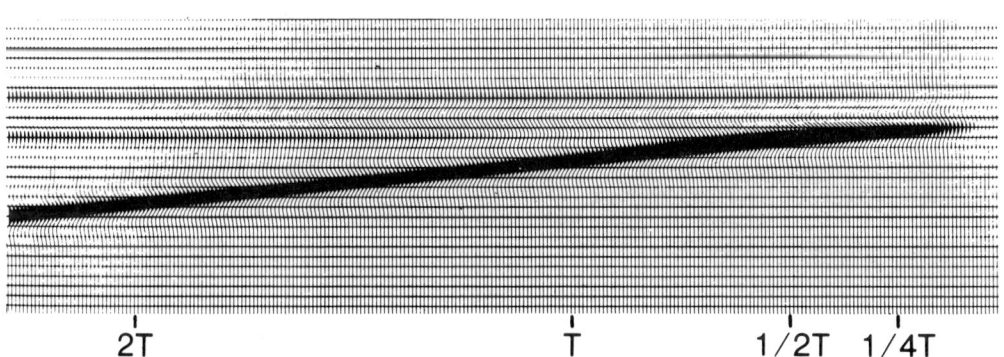

FIG. 3. Conventional seismic section of wedge model.

WEDGE MODEL
INSTANTANEOUS AMPLITUDE

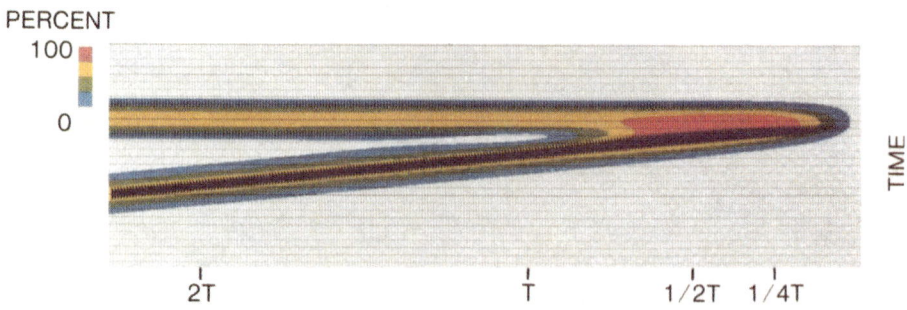

Fig. 4. Instantaneous amplitude section of wedge model.

WEDGE MODEL
INSTANTANEOUS PHASE

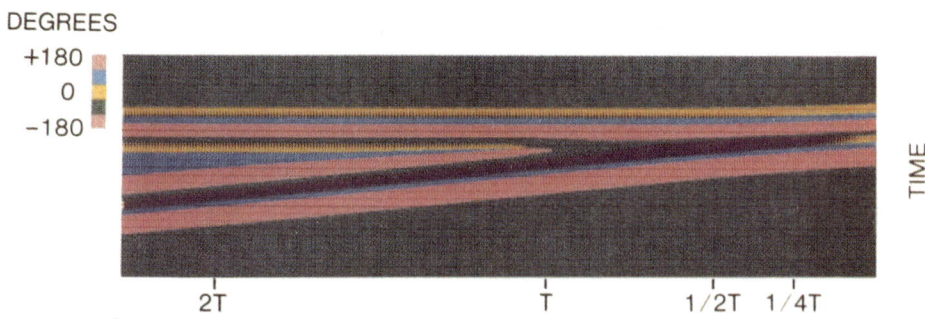

Fig. 5. Instantaneous phase section of wedge model.

WEDGE MODEL
INSTANTANEOUS FREQUENCY

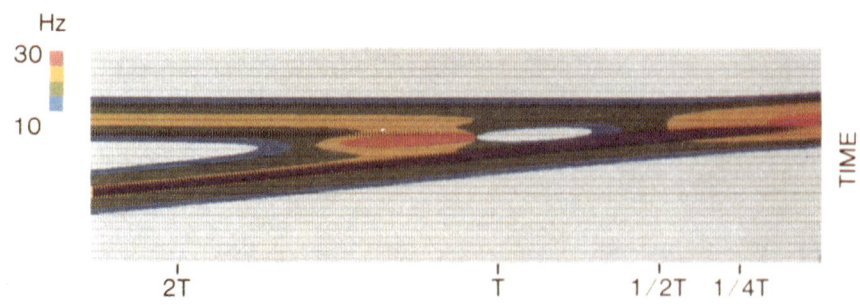

Fig. 6. Instantaneous frequency section of wedge model.

5000 ft/sec inside the wedge and 10,000 ft/sec outside the wedge, so the thinning structure is a low-impedance layer.

A synthetic seismogram of the wedge model is presented in Figure 3. The thickness of the wedge is marked along the bottom of the section in units of the period T, corresponding to the dominant frequency of the wavelet. The dominant frequency in this case is defined as the mean of the amplitude spectrum (22.5 Hz), so one period is approximately 44 msec. Where the wedge is thick, the top and bottom are obviously marked by the central peaks of the corresponding reflections. The reflections interfere with each other as the wedge thins, and the top and bottom of the wedge are no longer resolved by differential event time below about $T/4$, as pointed out by Widess (1973). As noted in Lindsey (1973), the thickness of the bed below about $T/2$ is resolved by the amplitude of the interference complex (Neidell and Poggiagliolmi, 1977). Amplitude reaches a maximum value at $T/2$ through constructive interference and falls off monotonically as the bed thins to zero thickness.

The synthetic seismic section has been converted to displays of the instantaneous attributes of amplitude, phase, and frequency. These displays are presented in Figures 4, 5, and 6. Half-period tuning clearly stands out on the instantaneous amplitude section, and the decrease in amplitude below a half period is easy to follow in color. The instantaneous frequency section complements the amplitude display by illustrating that a thinning bed tunes frequency as well as amplitude. Where the wedge is thick, values of instantaneous frequency at the central peaks of the top and bottom reflections are equal to the frequency of the center of gravity of the wavelet's amplitude spectrum as discussed above. When bed thickness approaches one period, there is an anomalous increase in instantaneous frequency along the center of the bed. The increase is followed, at about one period, by a clear and abrupt transition to a region of low instantaneous frequency. From a half period to a quarter period, the frequency attribute along the center of the wedge increases again. An anomalously high zone reappears when the bed becomes less than a quarter period thick. The

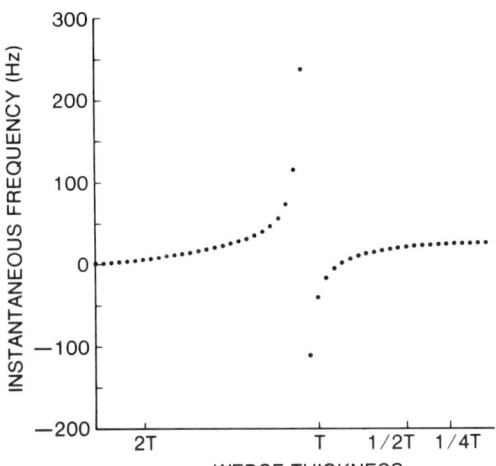

FIG. 7. Instantaneous frequency along a line bisecting the wedge model.

frequency pattern as a whole is visually striking. Additional modeling investigations have shown that the pattern is relatively insensitive to velocity variations as long as the polarities of the top and bottom reflection coefficients are not changed. It would appear that the combination of the instantaneous amplitude and frequency sections could be used with real data to obtain a more robust interpretation of bed thinning, particularly down to well below a half period, than could be obtained from conventional trace data alone.

The visual impact of the attribute displays is sensitive to color scaling. As an example, consider Figure 7 which is a plot

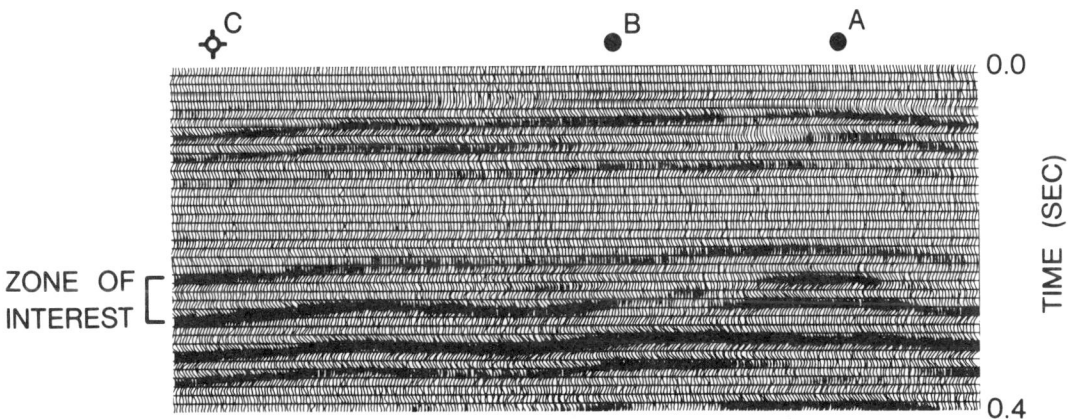

FIG. 8. Portion of a seismic section recorded over a stratigraphic sequence containing thin, low-impedance sandstones.

INSTANTANEOUS PHASE

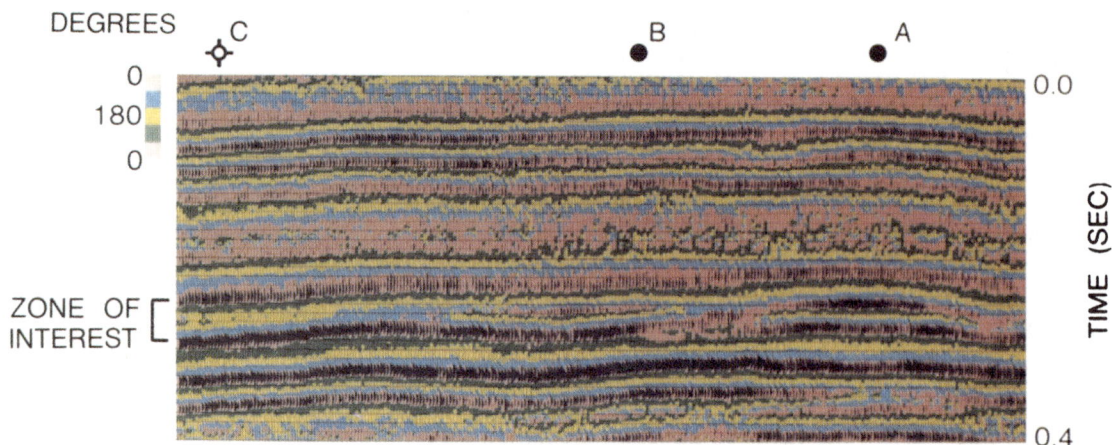

FIG. 9. Instantaneous phase display of seismic section in Figure 8.

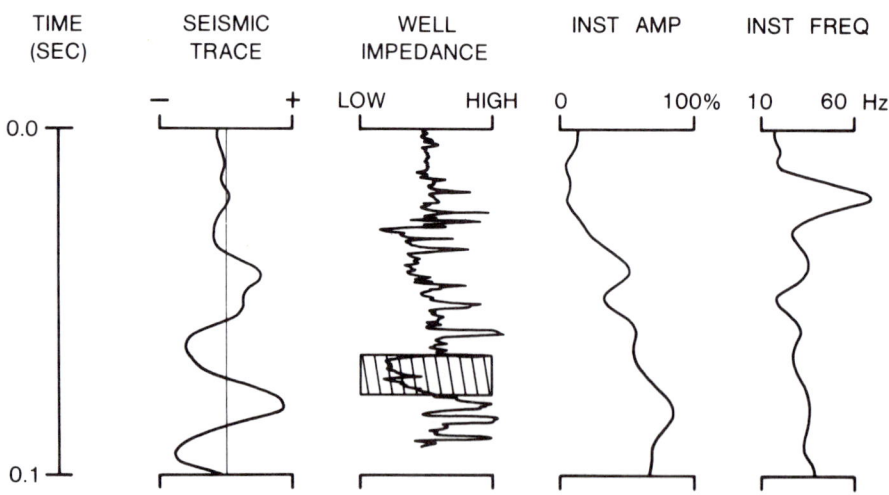

FIG. 10. Tie between surface and borehole data at well A.

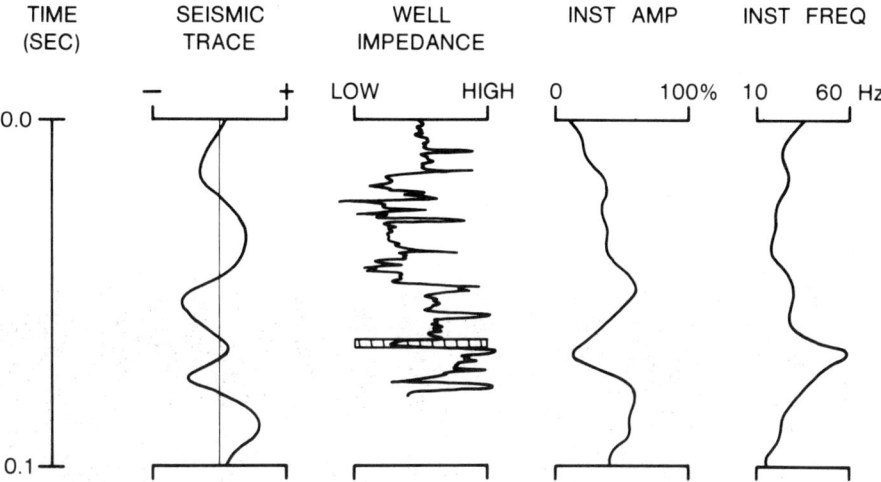

FIG. 11. Tie between surface and borehole data at well B.

of the values of instantaneous frequency along a line bisecting the wedge model. In Figure 6, both zones of tuned frequency are scaled to red for visual simplicity. The trade-off for the simplicity, however, is that the sharpness of the tuning at one period compared to the tuning below a quarter period, which is obvious in Figure 7, is not evident on the color display.

The instantaneous phase section (Figure 5), as expected, emphasizes the continuity of the wedge. The section also illustrates that the instantaneous frequency patterns are sensitive to and governed by lateral terminations of zero crossings in the real and imaginary traces.

THIN SANDSTONE EXAMPLE

A section of seismic data recorded to detect and define low-impedance lenses of porous sandstone is displayed in Figure 8. The vertical zone of interest is marked on the figure and extends over about 50 msec of two-way travel time. The events

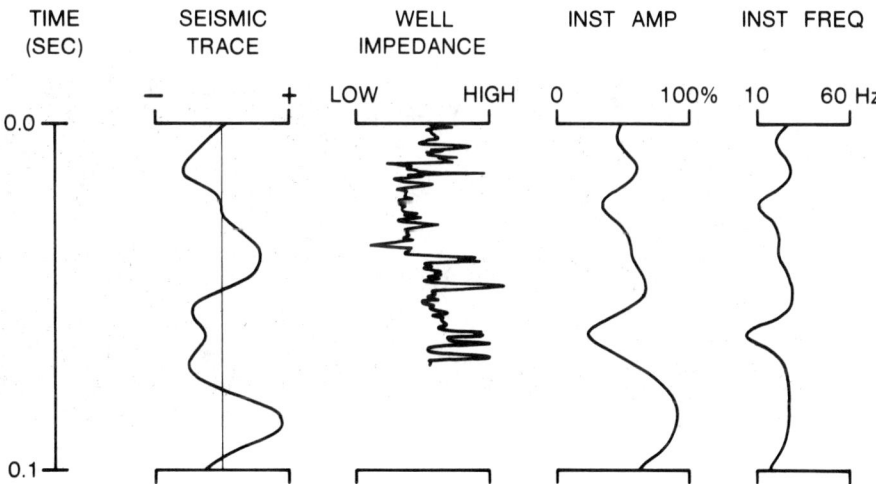

FIG. 12. Tie between surface and borehole data at well C.

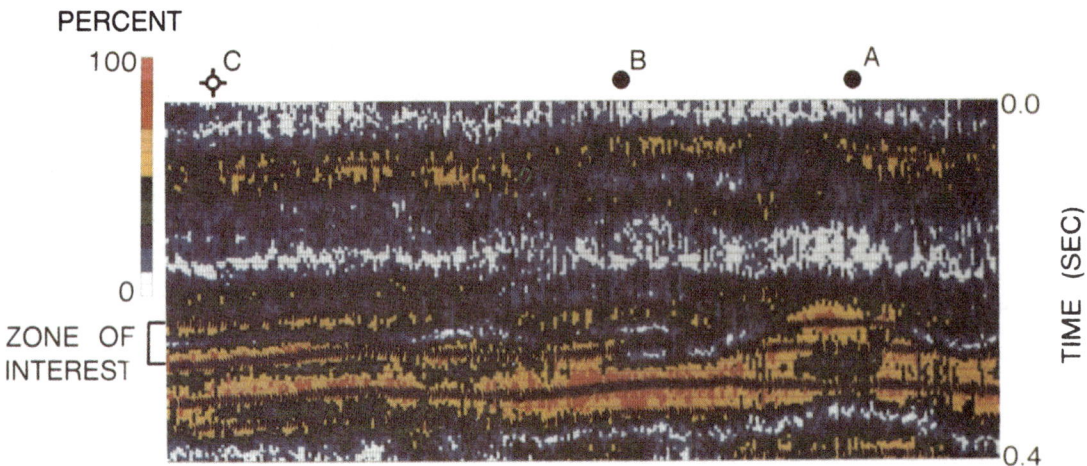

FIG. 13. Instantaneous amplitude display of seismic section in Figure 8.

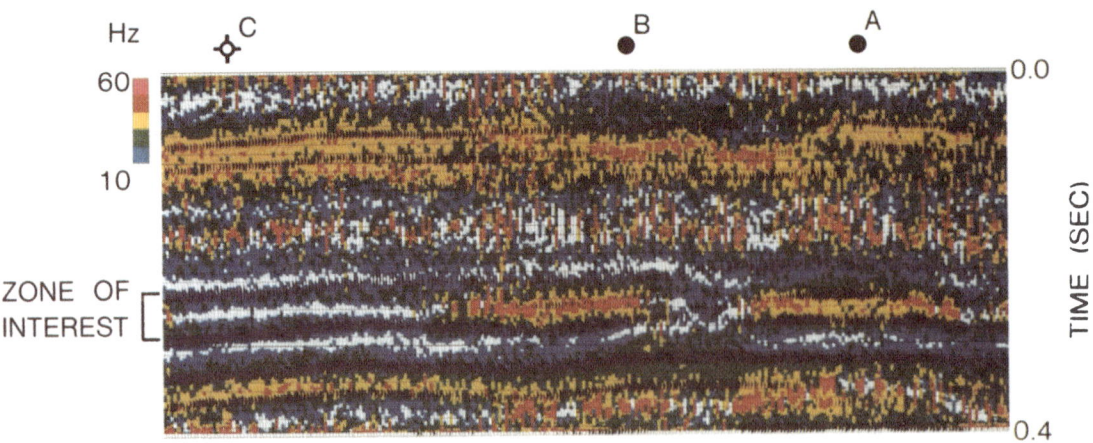

FIG. 14. Instantaneous frequency display of seismic section in Figure 8.

within the zone of interest exhibit lateral variations in character, amplitude, and arrival time. These variations are related to the presence and size of the low-impedance sandstone beds.

The section initially was converted to the instantaneous phase display shown in Figure 9. It is immediately apparent that this display emphasizes lateral continuity. Three distinct lateral anomalies are defined. The first lies at the left of the section where instantaneous phase varies by less than 90 degrees over about 20 msec of the zone of interest. The second lies at the center where the instantaneous phase vector rotates through an anomalous, additional 360 degrees inside the zone of interest. The third lies toward the right of the section and is similar to the second except it is vertically broader and laterally more abruptly terminated. It is apparent from Figures 8 and 9 that the instantaneous phase display assists in qualitatively delineating these anomalies.

A well has been drilled through each of the anomalies, and the next step in the interpretation is to analyze the well information. The well locations are plotted along the tops of the sections. Acoustic impedance in each well as determined from sonic and density logs is plotted in Figures 10 to 12, together with the seismic trace, instantaneous amplitude, and instantaneous frequency.

Well A contains a low-impedance sandstone about 90 ft thick. This thickness of sandstone is equivalent to about 15 msec of two-way travel time, which is one-half period of the dominant seismic energy. The well penetrates the rightmost phase anomaly (Figure 9) to the right of the center of that anomaly. As was illustrated by the wedge model, instantaneous amplitude in Figure 10 is tuned by the one-half period low-impedance bed. In this case, amplitude tuning occurs near the bottom of the bed, probably due to the influence of the fine layering in the overlying and underlying formations. The instantaneous frequency trace is relatively featureless, as was the wedge model at one-half period.

Instantaneous amplitude and frequency displays of the entire section (Figures 13 and 14) indicate that the sandstone bed maintains a thickness of approximately a half period for about 20 traces to the left of well A and for about 5 traces to the right. At the edges of this extent, the bed thins to a quarter period, as is shown by the decrease in instantaneous amplitude and increase in instantaneous frequency. The sandstone bed is still present but is less than a quarter period (45 to 50 ft) thick for another 20 traces on both sides as shown by the continuation of anomalously high instantaneous frequency. The anomalous instantaneous frequency finally terminates, indicating that the sandstone bed either has pinched out entirely or is so thin as to be lost in the geologic noise.

Well B (Figure 11) penetrates the extreme right edge of the anomaly located in the center of the instantaneous phase section (Figure 9). Amplitude is low along this anomaly (Figure 13), suggesting that a low-impedance sandstone a half period thick is not present. However, frequency is high (Figure 14), suggesting that a very thin low-impedance sandstone less than a quarter period thick might be present as a continuous lens. The presence of such a sandstone is confirmed by well B, which was drilled through an 18 ft section of the sandstone. The stratigraphy in well B is complex, as shown by the impedance log in Figure 11, and one would expect that a number of layers in the 50 msec zone of interest would contribute to tuning. It appears, fortunately, that the low-impedance sandstone produces the dominant tuning, even though it is thin. One would predict on the basis of the instantaneous phase and frequency sections that the sandstone is present, but continuously less than a quarter period thick, in a lens about 50 traces wide extending to the left from well B. The instantaneous frequency signature of the thin bed is an example of the anomalous frequency tuning observed by Sicking (1978) in a synthetic seismic model of interfingering lithologies.

The low-impedance sandstone is not present at all in well C (Figure 12). Instantaneous amplitude in the zone of interest is low, and one cannot distinguish between the presence and absence of the sandstone in wells B and C, respectively, on the basis of the amplitude attribute. In contrast, the stratigraphy is clearly defined on the frequency section. The anomalously high instantaneous frequency produced by the thin sandstone in well B is strikingly different from the low instantaneous frequency which characterizes the absence of sandstone in well C. Instantaneous frequency is low along the whole of the phase anomaly which well C penetrates, suggesting that there is no low-impedance sandstone present anywhere in this lens.

SUMMARY

Computer-modeled seismic data and broadband field data acquired to delineate complicated stratigraphy have been converted to displays of the instantaneous attributes of the complex seismic trace. The attribute sections enhance the interpretation of the conventional sections, not only by qualitatively highlighting specific properties of the conventional displays, but also by quantitatively defining wavelet characteristics like dominant frequency and stratigraphic variables like formation thickness.

An example of the quantitative use of complex attributes in wavelet definition is the phenomenon that the maximum instantaneous frequency of a zero-phase Ricker wavelet is synchronous with the central peak of the wavelet and equal to the frequency corresponding to the center of gravity of the wavelet's amplitude spectrum. Peak instantaneous frequency is thus a physically meaningful measure of the spectral content of a zero-phase Ricker wavelet. If the signal in a seismic section can be approximated by zero-phase Ricker wavelets and if the geophysicist can identify occasional wavelet peaks in the section which are uncontaminated by noise or interference, the instantaneous frequencies at these samples are direct estimates of the center of gravity of the spectrum of the propagating pulse.

An example of the quantitative use of attribute sections in seismic stratigraphy is their application to estimating the thickness of thin, porous sandstones. Low-impedance sandstones which are encased in high-velocity material and whose thicknesses are of the order of half the peak to-peak period of the dominant seismic energy show up as anomalously high-amplitude zones on instantaneous amplitude displays. These anomalies result from the well-known amplitude tuning effect which occurs when reflection coefficients of opposite polarity a half period apart are convolved with a seismic wavelet. As sandstone members thin to a quarter period of the dominant seismic energy, the thinning is revealed by an anomalous increase in instantaneous frequency. This behavior results from the less well-known but equally important phenomenon of frequency tuning by thinning beds. Instantaneous frequency reaches an anomalously high value when sandstone thickness is about a quarter period and remains high as the sandstone continues to thin. Thus, the instantaneous frequency section is a sensitive analytical tool for investigating stratigraphic sequences composed of very thin layers.

REFERENCES

Ackroyd, M. H., 1970, Instantaneous spectra and instantaneous frequency: Proc. Inst. of Electr. and Electron. Eng., v. 58, p. 141.

Farnbach, J. S., 1975, The complex envelope in seismic signal analysis: Bull., Seis. Soc. Am., v. 65, p. 951–962.

Lindsey, J. P., 1973, Modelling for lithology: Presented at Lithology and direct detection of hydrocarbons using geophysical methods, a symposium sponsored by the Geophysical Society of Houston, October 8 and 9.

Neidell, N. S., and Poggiagliolmi, E., 1977, Stratigraphic modeling and interpretation—Geophysical principles and techniques, in Seismic stratigraphy—Applications to hydrocarbon exploration: C. E. Payton, Ed., Am. Assoc. of Petr. Geol., memoir 26, Tulsa, p. 389–416.

Ricker, N., 1945, The computation of output disturbances from amplifiers for true wavelet inputs: Geophysics, v. 10, p. 207–220.

Sicking, C. J., 1978, Modeling with the complex trace: Presented at the 48th Annual International SEG Meeting, November 1, in San Francisco.

Taner, M. T., Koehler, F., and Sheriff, R. E., 1979, Complex seismic trace analysis: Geophysics, v. 44, p. 1041–1063.

Taner, M. T., and Sheriff, R. E., 1977, Application of amplitude, frequency and other attributes to stratigraphic and hydrocarbon determination, in Seismic stratigraphy—Applications to hydrocarbon exploration: C. E. Payton, Ed., Am. Assoc. of Petr. Geol., memoir 26, Tulsa, p. 301–327.

Widess, M. B., 1973, How thin is a thin bed?: Geophysics, v. 38, p. 1176–1180.

The limits of resolution of zero-phase wavelets

R. S. Kallweit* and L. C. Wood‡

ABSTRACT

This investigation deals with resolving reflections from thin beds rather than the detection of events that may or may not be resolved. Resolution is approached by considering a thinning bed and how accurately measured times on a seismic trace represent actual, vertical two-way traveltimes through the bed. Theoretical developments are in terms of frequency and time rather than wavelength and thickness because the latter two variables require knowledge of interval velocities. These results are compared with similar studies by Rayleigh, Ricker (1953), and Widess (1973, 1980). We show that the temporal resolution of a broadband wavelet with a white spectrum is controlled by its highest terminal frequency f_u, and the resolution limit approximates $1/(1.5\, f_u)$, provided the wavelet's band ratio exceeds two octaves. The practical limit of resolution, however, occurs at a one-quarter wavelength condition and approximates $1/(1.4\, f_u)$. The resolving power of zero-phase wavelets can be compared quantitatively once a wavelet is known in the time domain.

INTRODUCTION

Current efforts to increase the resolving power of the seismic method make it appropriate to examine closely our fundamental concepts of resolvability. Most reservoirs are small in the vertical dimension, and even those that are not thin out to zero thickness at the edges. A problem of major importance to an explorationist is defining thicknesses in a vertical sense and determining horizontal dimensions of reservoirs. This paper deals exclusively with vertical resolution of seismic reflection data and develops simple formulas for estimating the limits of resolution. Recent studies of wavelet resolution (Koefoed, 1981; Widess, 1980) pointed out the complexity of the problem. Koefoed proposed that there can be no single definitive measure of wavelet resolving power, only trade-off considerations between central lobe breadth, ratio of side lobe amplitude to maximum amplitude, and length of side-tail oscillations. Widess (1981), however, by defining resolving power as the maximum amplitude squared divided by the energy of the wavelet, ignored these trade-off considerations among wavelets having a common energy value. We approach the resolution problem by considering a thinning bed and how accurately measured times on synthetic traces represent actual, vertical two-way traveltimes through the bed.

Identifying individual reflections from the top and bottom of a thin bed differs from the problem of detecting the presence of the bed. It is necessary to identify two separate concepts: detection and resolution. Detection deals with recording a composite reflection with a sufficiently large signal-to-noise (S/N) ratio, regardless of whether the composite reflection can be resolved into the separate wavelets which compose it. Thus an event that is detectable may or may not be resolvable. Resolution is primarily a problem of frequency band, whereas detection (Farr, 1976) is principally one of acquisition. The discussion of resolution is restricted here to noise-free models of isolated thin beds, and the formulas derived here may be considered upper bounds on what can be achieved in practice.

HISTORICAL DEFINITIONS OF RESOLUTION

Let us begin by briefly examining Rayleigh's criterion of resolution, considering next a criterion developed by Ricker (1953), and finally reviewing the work of Widess (1973). In each case we relate theoretical limits of resolution to parameters that can be measured on the wavelet that is convolved with a reflectivity sequence.

Rayleigh's criterion

In the optical analogy, a point source is similar to a reflection spike, the optical instrument analogous to the earth, and the diffraction pattern plays the role of the band-limited wavelet. The resolving power of an optical instrument is its ability to produce separate images of objects lying close to each other. Diffraction patterns set an upper limit to resolution in a manner analogous to a propagating seismic wavelet (Figure 1). Thus, if the laws of geometrical optics were to apply perfectly, optical instruments could focus parallel light to a point image. This situation, however, is analogous to having a broadband Dirac-delta function (spike) for a seismic wavelet. Huygen's principle and the physical nature of light waves produce instead a diffraction pattern from the optical instrument, wherein the intensity of the central maximum has a finite width inversely proportional to aperture width. It becomes clear that images of two objects will coalesce into a single image and therefore not be resolved when their separation

---

Presented at the 47th Annual International SEG Meeting September 21, 1977, in Calgary. Manuscript received by the Editor April 2, 1979; revised manuscript received June 30, 1981.
* Amoco Production Company, P.O. Box 591, Tulsa, OK 74102.
‡ NORPAC Exploration Services, Inc., 6160 S. Syracuse Way, Englewood, CO 80111.
0016-8033/82/0701—1035$03.00. © 1982 Society of Exploration Geophysicists. All rights reserved.

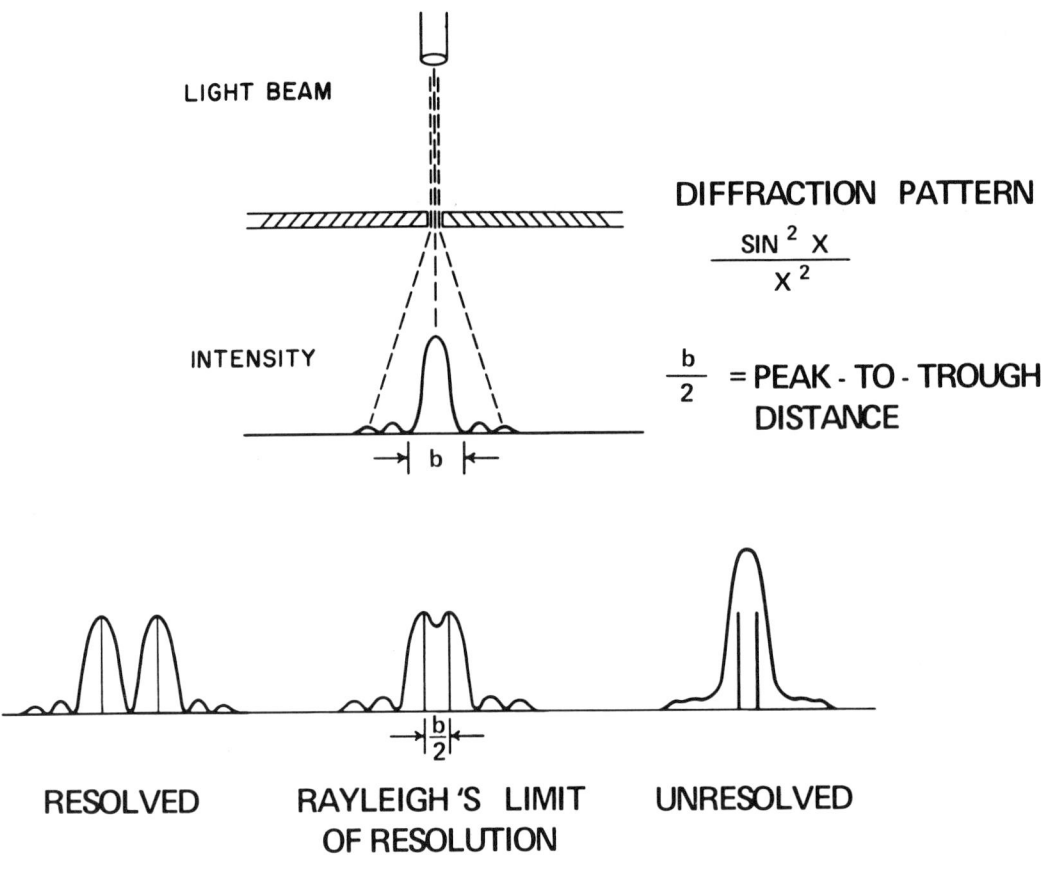

FIG. 1. Rayleigh's criterion states that the limit of an optical instrument to distinguish separate images of objects lying close together occurs when the two diffraction images are separated by a distance equal to the peak-to-trough distance of the instrument's diffraction pattern.

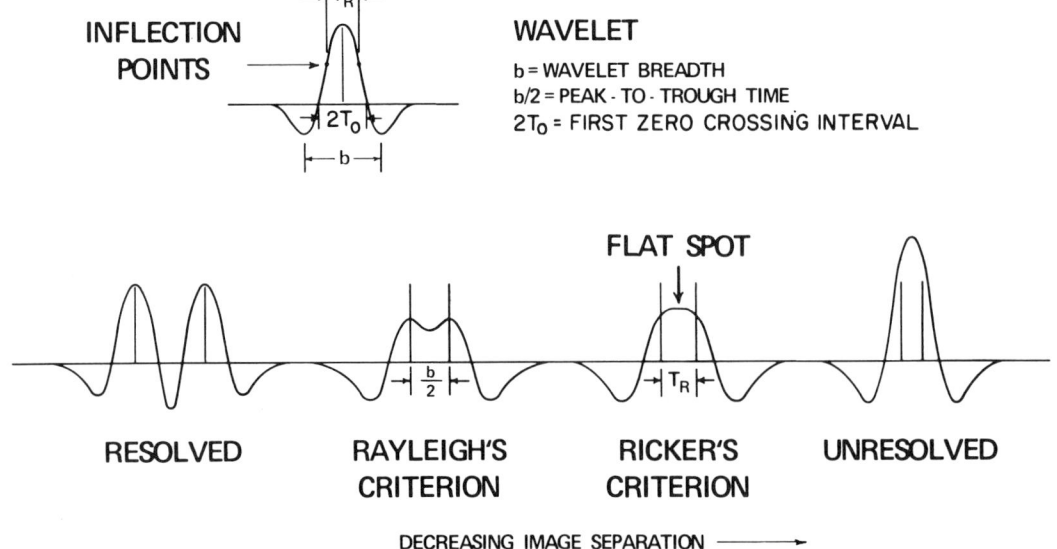

FIG. 2. Rayleigh's limit of resolution occurs when images are separated by the peak-to-trough time interval, whereas Ricker's limit occurs when they are separated by a time interval equal to the separation between inflection points.

becomes somewhat less than the width of the central diffraction maximum.

The criterion established by Rayleigh (Jenkins and White, 1957, p. 300) is to define the peak-to-trough separation ($b/2$), that is, the central-maximum to adjacent-minimum time interval of a diffraction pattern, as the limit of resolution. In other words, two point-source objects are said to be resolved when their separation is equal to or exceeds the peak-to-trough separation of the diffraction pattern of the optical instrument used for observation. Similarly, objects are said to be unresolved when their separation is less than the peak-to-trough separation as shown in Figure 1. Note also in Figure 1 that diffraction wavelet breadth or trough interval (b) and diffraction wavelet width or first zero-crossing interval ($2T_0$) are identical. This is not true of seismic wavelets containing negative side lobe energy, as seen in Figure 2. We will show that whereas wavelet breadth appears to be directly related to seismic resolution, wavelet width, for wavelets with white spectrums, does not appear to be. Thus Rayleigh's limit of resolvability relates to the first derivative (i.e., trough time) of the wavelet. Rayleigh recognized, however, that smaller separations could be detected with sensitive intensity measuring instruments such as microphotometers because the composite of the two diffraction patterns continues to exhibit two distinct central peaks at separations less than the peak-to-trough interval (Figure 2).

Ricker's criterion

Ricker (1953) studied the composite waveform as a function of separation and observed, as did Rayleigh, that central maxima exhibit two lesser peaks as separations decrease, merging finally into single major peaks with no subsidiary maxima. Ricker chose the limit of resolvability to be that separation where the composite waveform has a curvature of zero at its central maximum, i.e., a "flat spot" (Figure 2). Ricker (1953, p. 774) showed that the resolution limit can be determined by differentiating the wavelet twice. More simply, a flat spot or zero-curvature condition occurs when two spikes are separated by an interval equal to the separation between the inflection points on the central maximum (lobe) of the convolving wavelet.

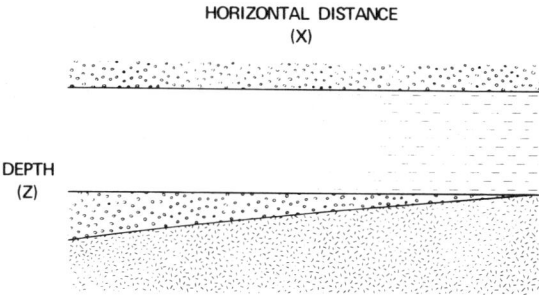

FIG. 3. The basis for two-term reflectivity spike models consists of a wedge of material bounded above and below by dissimilar material.

Ricker (1953) first discovered this elegant property and applied it to the case of equal-amplitude spikes with equal polarities. Widess (1973) devoted his study exclusively to the case of equal amplitudes and opposite polarities, which we discuss next.

Widess' criterion

Widess (1973) described a study of the composite waveform obtained by convolving a zero-phase wavelet with two spikes of equal amplitudes and opposite polarities. A fundamental observation is that as spike separations decrease, the effect of convolving a wavelet with two spikes of opposite polarities is one of differentiation.

Widess observed that as separations decrease, there is a point where the composite wavelet stabilizes into a good replica of the derivative waveform such that, for all practical purposes, there is no change in the peak-to-trough time but rather only a change in amplitude of the composite waveform. Widess concluded by

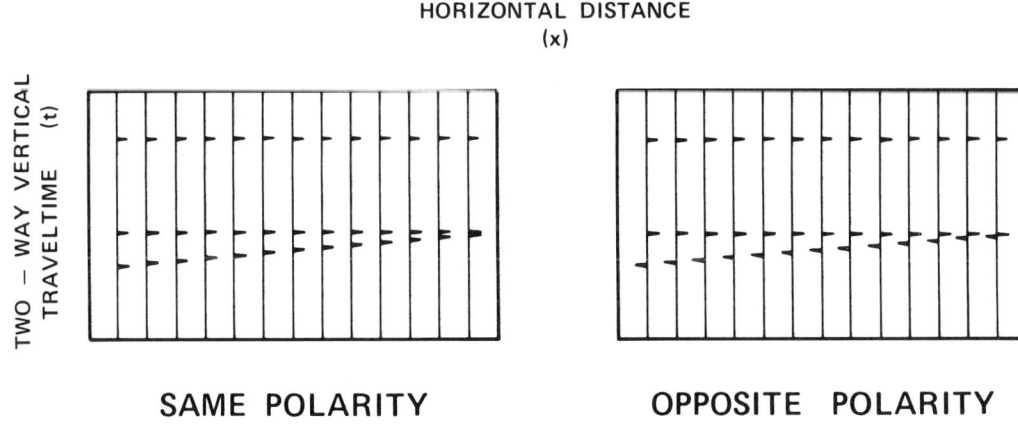

FIG. 4. The time-domain response of a wedge model (Figure 3) consists of two-term reflectivity sequences with equal or unequal spike amplitudes having either the same or opposite polarities.

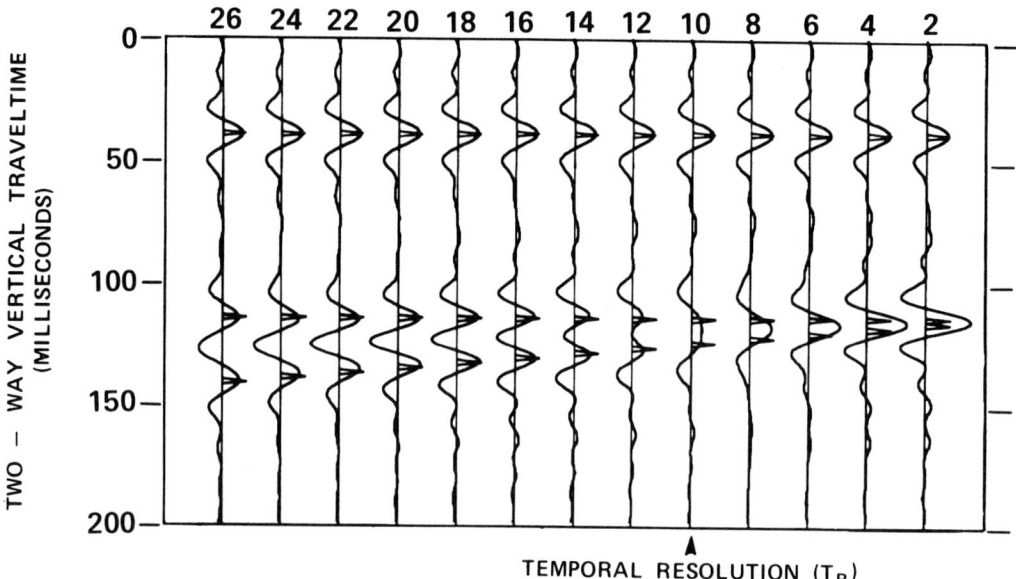

FIG. 5. The seismic response of a wavelet convolved with two spikes of equal amplitude and equal polarity is a composite wavelet that varies as a function of spike separation, i.e., layer thickness.

inspection that the limiting separation for wavelet stabilization occurs when the bed thickness (i.e., spike separation) is equal to 1/8 of a wavelength of the predominant frequency of the propagating wavelet. He remarked (1973, p. 1180) that beds thinner than 1/8 of a wavelength can be resolved in principle by measuring changes in amplitude of the composite reflection. Use of amplitudes for this purpose, however, is difficult in actual practice because one must normalize or calibrate amplitudes to a known bed thickness with a known zero-phase wavelet. (See discussions by Lindsey et al, 1976; Meckel and Nath, 1977; Neidell and Poggiagliolmi, 1977; Nath et al, 1977; Sheriff, 1977; Schramm et al, 1977; and Clement, 1976.) For Widess the 1/8-wavelength separation establishes the limit of resolvability of a thin bed because further decreases in spike separation do not appear to have corresponding changes in peak-to-trough times on a visual basis.

The main objective of this paper is to develop concepts of resolution which unify Rayleigh, Ricker, and Widess' viewpoints, thereby removing polarity considerations from resolvability.

THE BASIC MODEL

To assess the relative merits of the above resolution criteria in providing a measure of the ability to produce a seismic trace on which the reflections from the top and bottom of a thin bed could be picked visually for a meaningful estimate of bed thickness, we performed simple model studies of an isolated thinning bed.

The fundamental model consists of a wedge of material bounded above and below by dissimilar materials (Figure 3). The two-term reflectivity sequence associated with the wedge is shown in Figure 4 for the two cases of equal and opposite polarity reflections where it is assumed that the amplitudes of these reflections are equal. The corresponding zero-phase band-limited events are shown in Figures 5 and 7. The use of zero-phase wavelets over their nonzero-phase counterparts simplifies further developments because reflection arrival times correspond to peaks and troughs (Schoenberger, 1974; Berkhout, 1974).

Identical polarity

Observe in Figure 5 as Rayleigh did (Jenkins and White, 1957, p. 300; Ricker, 1953) that the central maxima exhibit two distinct peaks as separations decrease, merging finally into a single major peak. The limit of resolution, which we shall call "temporal resolution" (T_R), occurs when the true separation decreases to an interval where a flat spot or zero-curvature region appears on the central maximum of the composite waveform (Ricker, 1953). Ricker (1953) showed this limit can be determined by equating the second derivative of the convolving wavelet to zero as discussed in Appendix A. Temporal resolution is thus equivalent to the time separation between the inflection points on the central lobe of the convolving wavelet. Tuning thickness (Rayleigh's criterion), expressed in time $(b/2)$, is the point where apparent thickness (peak-to-peak time) is precisely the same as true thickness and can be determined by equating the first derivative of the convolving wavelet to zero as shown in Appendix B. It is equivalent to the wavelet's central peak-to-adjacent trough time.

Figure 6 shows apparent thickness (i.e., peak-to-peak time) and maximum absolute amplitude of the composite waveform as a function of true bed thickness for a 25-Hz Ricker wavelet. It can be seen that above tuning thickness, peak-to-peak time mea-

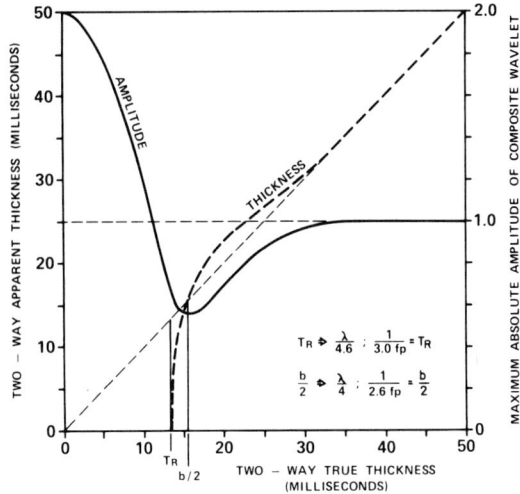

FIG. 6. Resolution and detection graphs for two spikes of equal amplitude and equal polarity convolved with a 25-Hz Ricker wavelet show apparent thickness (peak-to-peak time) and maximum absolute amplitude of the composite wavelet as a function of true thickness (spike separation).

surements are good approximations of bed thicknesses, whereas below tuning thickness, amplitude information must be used. This is shown by observing that the thickness curve crosses the 45-degree line at tuning thickness and then rapidly approaches a limiting value.

The waveforms in Figure 5 show that maximum composite waveform amplitudes decrease to a minimum and then increase for smaller separations. The minimum occurs at tuning thickness as seen in Figure 6; it is the difference between the peak absolute amplitude of the central lobe and the peak absolute amplitude of the adjacent side lobe of the wavelet. The amplitude curve then increases nonlinearly and finally doubles its value at small separations in the limit of zero thickness.

Opposite polarity

In Figure 7 there appears to be a minimum separation point where the composite wavelet stabilizes into a good replica of the derivative waveform beyond which there are no further significant changes in the peak-to-trough times. However, it can be seen from Figure 8, for a 25-Hz Ricker wavelet, that there is no point where the wavelet stabilizes into a good replica of the derivative waveform in the noise-free case except at the limiting value of zero thickness. Therefore, apparent wavelet peak-to-trough time stabilization is a poor criterion for determining true

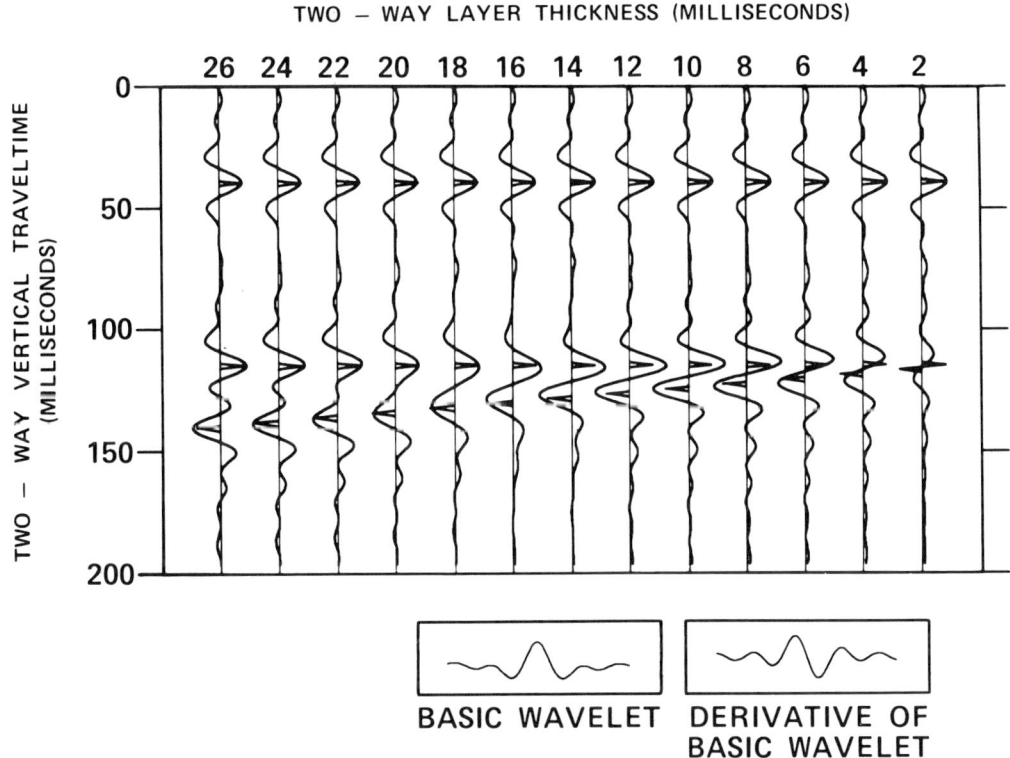

FIG. 7. Widess (1973) observed that a wavelet convolved with two spikes of equal amplitude and opposite polarity converges to the derivative of the convolving wavelet as spike separation decreases.

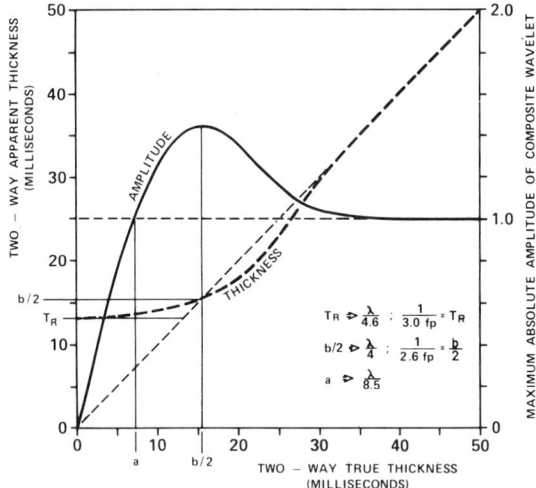

FIG. 8. Resolution and detection graphs for two spikes of equal amplitude and opposite polarity convolved with a 25-Hz Ricker wavelet show apparent thickness (peak-to-trough time) and maximum absolute amplitude of the composite wavelet as a function of true thickness (spike separation).

bed thickness. Observe the thickness curve as true thickness decreases from 50 msec to zero. At about 33 msec, the curve deviates downward from the 45-degree line, crosses back over the 45-degree line at precisely tuning thickness, and then asymptotically approaches a limiting value at zero true thickness equal to the peak-to-trough time of the derivative wavelet. Since this time can be obtained by differentiating the convolving wavelet twice with respect to time and equating to zero, this value is the same as the separation between the inflection points on the central lobe of the convolving wavelet or temporal resolution. The apparent stability of peak-to-trough composite waveform time below tuning thickness sets a minimum apparent bed thickness that varies in value between tuning thickness and wavelet temporal resolution which is a property inherent in the convolving waveform, and it is unrelated to true bed thickness.

The amplitude curve in Figure 8 shows that at tuning thickness, composite waveform amplitudes reach a maximum equal to the sum of the maximum absolute amplitudes of the central peak and adjacent side lobe of the convolving wavelet and then decrease nonlinearly to zero. We also note that at the true thickness separation of $\lambda_p/8.5$, the peak absolute amplitude of the composite waveform equals the peak absolute amplitude of the convolving Ricker wavelet.

Thus Ricker's (1953) zero-curvature criterion for temporal resolution can be applied to the two-term reflectivity series of equal strength and opposite polarity as well as to the case of equal amplitude and equal polarity. Ricker's criterion (temporal resolution) is thus generalized to apply without polarity constraints.

TEMPORAL RESOLUTION OF SEISMIC DATA

The above discussion shows that Rayleigh's limit is a wavelet's peak-to-trough time which corresponds with tuning thickness, whereas Ricker's limit, which is the time separation between the wavelet's inflection points, corresponds to temporal resolution.

A wavelet frequency band determines tuning thickness as well as temporal resolution. Appendices B and C derive simple formulas relating both tuning thickness and temporal resolution with spectral frequency band for Ricker and sinc wavelets, which are discussed in turn below.

Ricker wavelets

The Ricker (1953) wavelet (velocity type) is a useful wavelet to analyze for its resolving power because of its widespread use in seismic modeling and interpretation. Temporal resolution can be expressed either in terms of peak frequency or in terms of predominant frequency. Peak frequency f_p is defined as the frequency component with the largest amplitude, whereas predominant frequency f_p is the reciprocal of the trough-to-trough time or breadth about the central lobe. It is shown in Appendix B that temporal resolution for a Ricker wavelet is

$$T_R = \frac{1}{3.0\,f_p}, \quad (1)$$

and tuning thickness is given by

$$\frac{b}{2} = \frac{1}{2.6\,f_p}. \quad (2)$$

Sinc wavelets

A sinc wavelet has all frequencies present with equal amplitudes (i.e., a white spectrum) between a lower terminal frequency f_ℓ and an upper terminal frequency f_u. We refer to this frequency range as the wavelet's frequency band (f_ℓ, f_u). Other terms used are bandwidth $(f_u - f_\ell)$ and band ratio (f_u/f_ℓ). Examples of seismic data that might have a basic wavelet corresponding to a sinc wavelet are Vibroseis® recordings, wavelet processed records, and signature deconvolutions.

It is shown in Appendix C that temporal resolution (T_R) and tuning thickness $(b/2)$ for a sinc wavelet can be expressed as a function of the upper terminal frequency as

$$T_R = \frac{1}{(c_2/2)f_u}, \quad (3)$$

and

$$\frac{b}{2} = \frac{1}{c_1 f_u}. \quad (4)$$

Parameters c_1 and c_2 are functions of upper and lower terminal frequencies and therefore frequency band. The graph in Figure 9 shows parameters c_1 and $c_2/2$ versus band ratio, and we see that they converge to the asymptotic values 1.398 and 1.509, respectively, as band ratio increases. The rapid flattening of the curves beginning at around two octaves demonstrates that wavelets with flat amplitude spectra whose terminal frequencies embrace two or more octaves have inflection points and peak-to-trough times that depend mainly on f_u. Consequently, temporal resolution and tuning thickness may be expressed with excellent approximation as

$$T_R \cong \frac{1}{1.5\,f_u}, \quad (5)$$

and

$$\frac{b}{2} \cong \frac{1}{1.4\,f_u}, \quad (6)$$

®Trademark of Conoco, Inc.

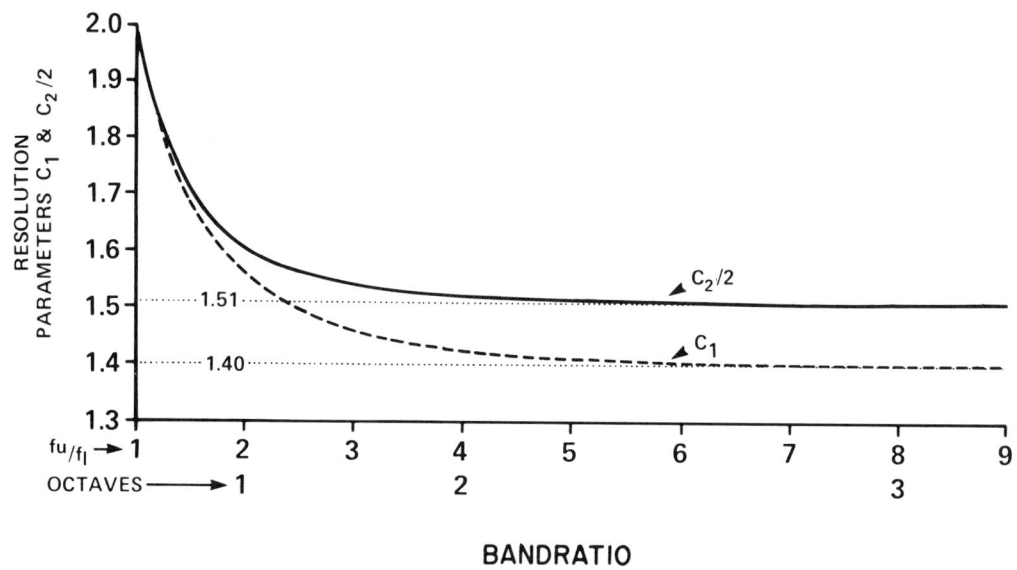

FIG. 9. Resolution parameters for tuning thickness (c_1) and temporal resolution ($c_2/2$) for sinc wavelets versus band ratio.

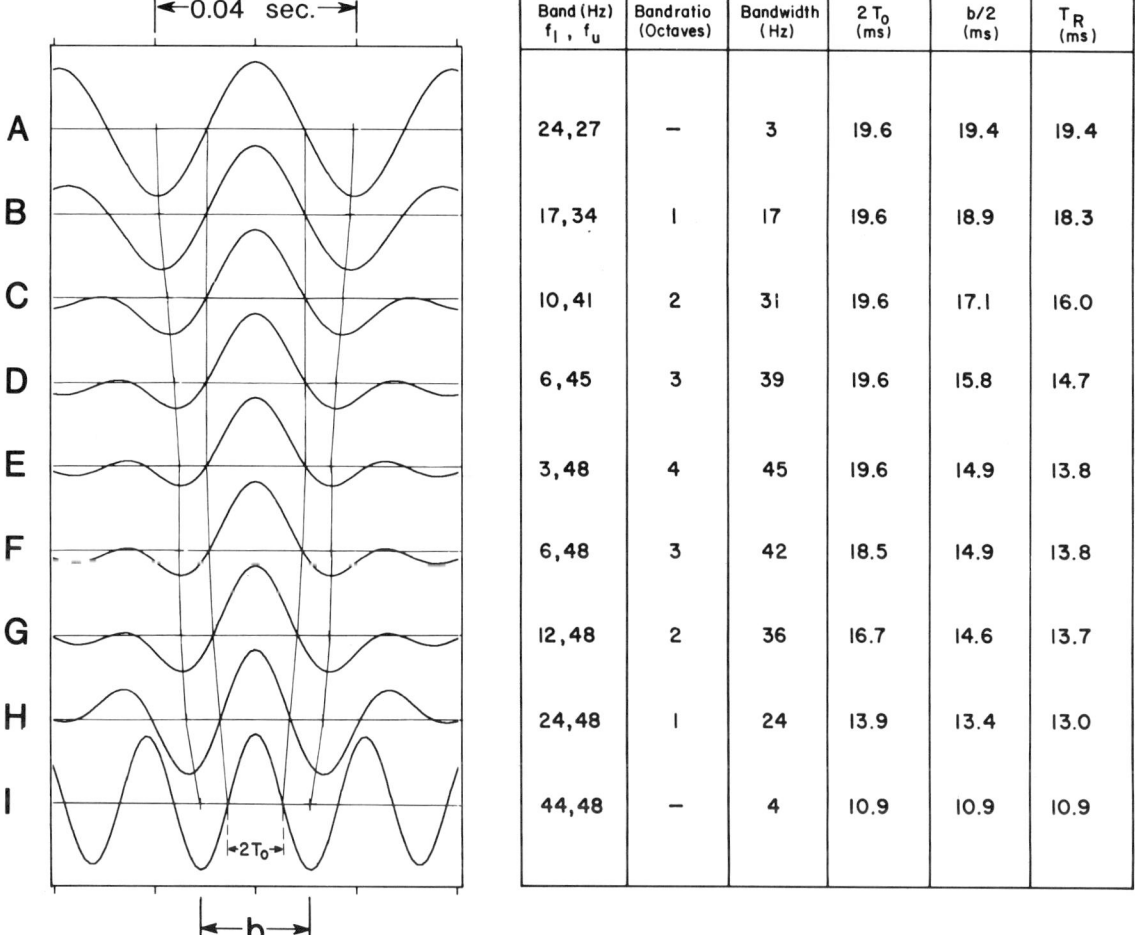

FIG. 10. A suite of wavelets shows the relationships between frequency band (f_ℓ, f_u), band ratio (f_u/f_ℓ), bandwidth ($f_u - f_\ell$), primary lobe zero-crossing interval [$2T_0 = 1/(f_u + f_\ell)$], tuning thickness ($b/2$), and temporal resolution (T_R).

for

$$\frac{f_u}{f_\ell} \cong 4.$$

Appendix C derives these relationships theoretically. For much of the data currently available with modern data processing and acquisition, these formulas are very useful interpretive tools.

There are other parameters in addition to temporal resolution and tuning thickness in the time domain, and terminal frequencies and band ratio in the frequency domain, that might be related to sinc wavelet resolution. These are central lobe zero-crossing interval ($2T_0$), where $2T_0 = 1/(f_\ell + f_u)$, midfrequency ($f_\ell + f_u)/2$, and bandwidth ($f_u - f_\ell$). These parameters are compared below for their relevance to resolution.

Figure 10 shows a series of sinc wavelets. Wavelets A through E have bandwidths that increase about a common midfrequency. Observe that $2T_0$ remains constant while tuning thickness ($b/2$) and temporal resolution (T_R) decrease with increasing bandwidth. This suggests that either resolution remains unchanged or increases with increasing bandwidth.

Next compare wavelets E through I where f_u is held constant and bandwidth is decreased. Now both $2T_0$, $b/2$, and T_R decrease with decreasing bandwidth, suggesting (contrary to the above) increasing resolution with decreasing bandwidth. Clearly, a

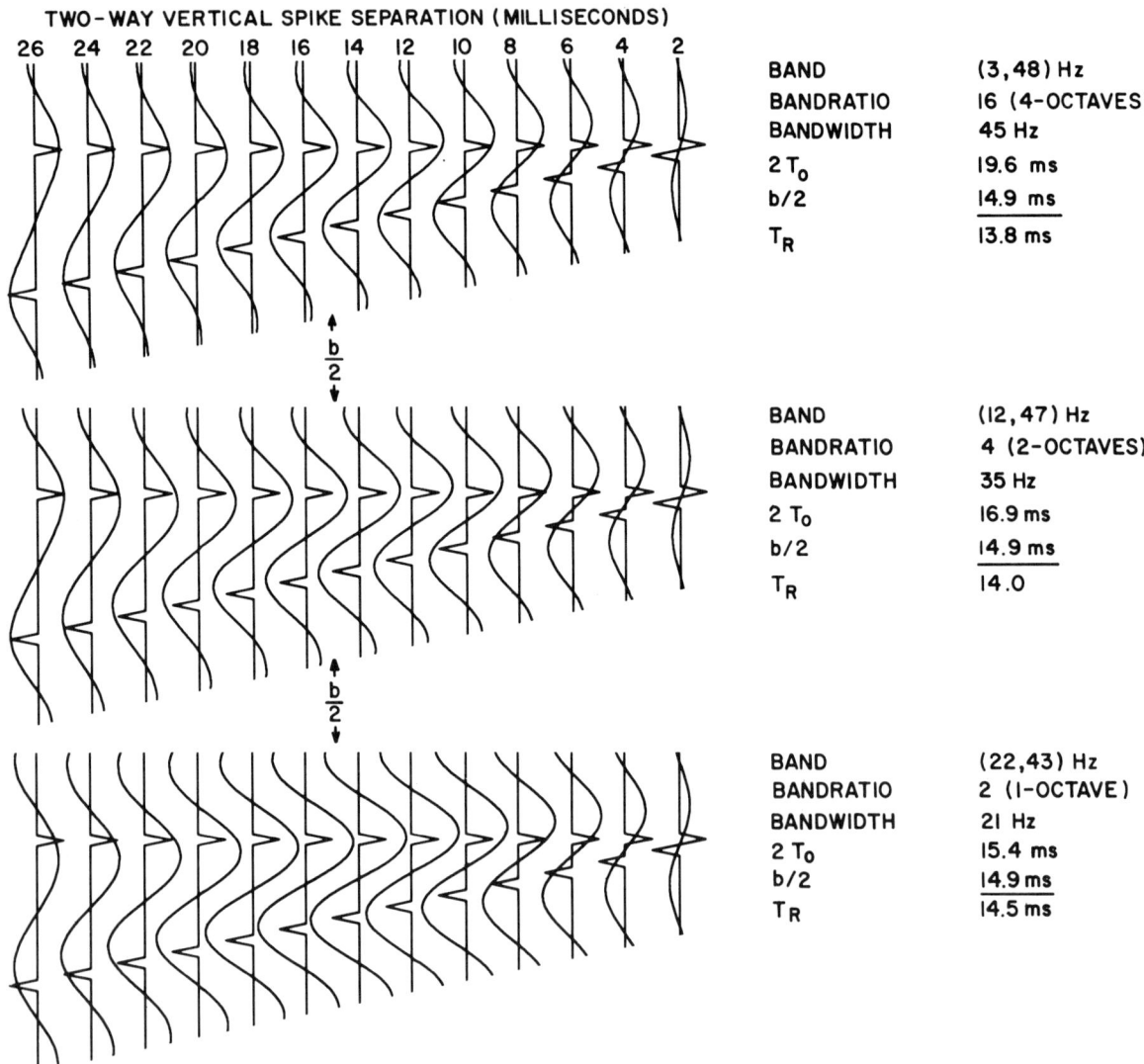

FIG. 11. The opposite polarity wedge model convolved with sinc wavelets having identical tuning thicknesses ($b/2$) show relationships of wavelet parameters.

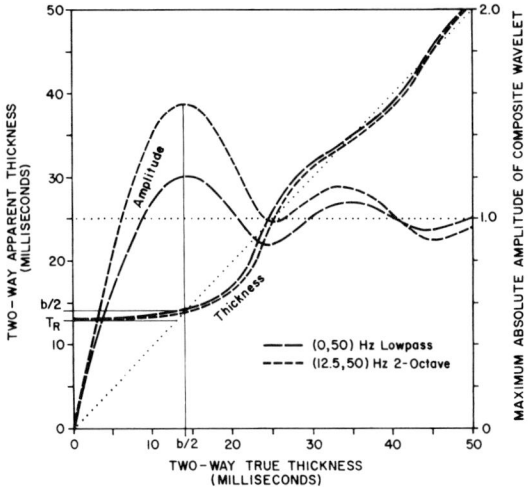

FIG. 12. Resolution and detection graphs for the opposite polarity case compare a low-pass sinc wavelet to a 2-octave sinc wavelet having identical upper terminal frequencies.

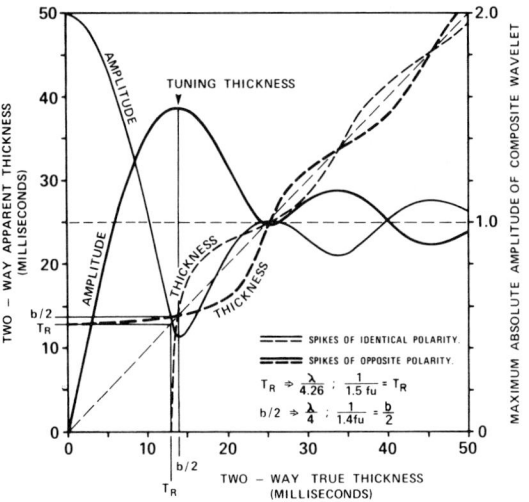

FIG. 13. Resolution and detection curves for a (12.5, 50) Hz sinc wavelet show tuning thickness ($b/2$) and temporal resolution (T_R) limits.

comparison of sinc wavelets per se yields little insight into resolution.

Returning to the basic wedge model, Figure 11 displays the sinc wavelets (3, 48) Hz, (12, 47) Hz, and (22, 43) Hz convolved with the opposite polarity case (Figure 7 wedge model). These wavelets have identical tuning thicknesses ($b/2$) and band ratios of 4 octaves, 2 octaves, and 1 octave, respectively. Observe that reflection peak-to-trough times between the middle and upper seismogram are virtually identical and vary slightly between the middle and lower models. Therefore wavelet resolving power, measured as a function of how accurately reflection peak-to-trough times on the traces represent actual two-way traveltimes through the thinning bed, is virtually the same for the 4-octave and 2-octave wavelets and close for the 2-octave and 1-octave responses.

In comparing the 2-octave wavelet to the 4-octave wavelet, we find that the only frequency-related parameter that does not change significantly is f_u. This suggests that sinc wavelet resolution is directly related to the upper terminal frequency and insensitive to band ratio provided band ratios are 2 octaves and greater. Figure 12 confirms this. A 2-octave wavelet and a low-pass wavelet with a common f_u are compared. The thickness graphs closely track for all relevant bed thicknesses and converge at temporal resolution (T_R).

Another consideration in wavelet resolution is interference effects of side lobe amplitudes. It is evident that side lobe amplitudes increase as band ratio decreases to become indistinguishable from the primary lobe at a monofrequency. The ability of the wedge model seismogram to reflect true bed thickness deteriorates rapidly at less than about a 1.5-octave band ratio, making resolution considerations in this range meaningless.

Comparison of sinc and Ricker wavelets

Consider the responses to the reflectivity series models with a (12.5, 50) Hz sinc wavelet. Figure 13 shows the time-amplitude plots for the two cases of equal top and bottom reflection coefficients with equal and unequal polarities. In the opposite polarity case, peak-to-trough time and normalized amplitudes are plotted against true thickness, whereas peak-to-peak measurements are made for the equal polarity case. It is apparent that the time-separation plot for both cases oscillates around the 45-degree line, and this deviation decreases as the true thickness increases. Figures 6 and 8 show similar plots for a 25-Hz Ricker wavelet. Note that the deviation of the time plot from the 45-degree line starts at a considerably lesser bed thickness. Since both of these sinc and Ricker wavelets have the same temporal resolution, this deviation is due to the side lobe effect (tuning) of the wavelet on the peak-to-trough or peak-to-peak measurements. By comparing synthetic traces generated with different convolving wavelets having identical temporal resolution, the effects of side lobe tuning on resolution and detection may be studied. This concept is of considerable importance since it establishes seismic events to be differentiated from artifacts created by wavelet tuning effects in a seismic section.

DISCUSSION

There is considerable vagueness in the current literature dealing with resolution concepts because of ambiguity in the definitions of the terms "frequency" and "wavelength." Predominant frequency corresponds to predominant wavelength, peak frequency relates to peak wavelength, while maximum frequency corresponds to minimum wavelength and minimum frequency to maximum wavelength. In applying the above resolution formulas to reservoir thickness calculations, it is imperative to distinguish clearly between these various types of frequencies. The term peak frequency used in conjunction with Ricker wavelets designates the frequency component having the largest value in the Fourier amplitude spectrum. Terminal frequencies f_u and f_ℓ define maximum and minimum frequencies for band-limited transients such as sinc wavelets. Widess (1973) defined predominant frequency to be that frequency obtained by computing the reciprocal of the time interval between the wavelet's two central side lobes; in other words, predominant frequency is obtained by reciprocating the wavelet's breadth time b. Referring to Figure 2, predominant frequency equals $1/b$. With these definitions in mind some of the confusion in establishing thin-bed thickness limits can be removed. Consider, for example, the model studied by Widess (1973) using a Ricker wavelet similar to that illustrated in Figure 7.

As shown in Appendix B, temporal resolution is given by

$$T_R = \frac{1}{3.0 f_p}, \qquad (7)$$

where f_p is the peak frequency, and tuning thickness is given by

$$\frac{b}{2} = \frac{1}{2.6 f_p}. \qquad (8)$$

Parameter $b/2$ may be used to relate peak frequency f_p to predominant frequency $\underline{f}_p$

$$\underline{f}_p = \frac{1}{b} = 1.3 f_p. \qquad (9)$$

Temporal resolution and tuning thickness for a Ricker wavelet can now be expressed in terms of predominant frequency with temporal resolution given by

$$T_R = \frac{1}{2.31 \underline{f}_p}, \qquad (10)$$

and tuning thickness by

$$\frac{b}{2} = \frac{1}{2 \underline{f}_p}. \qquad (11)$$

These correspond to limiting bed thicknesses of

$$\Delta Z_R = \frac{\lambda_p}{4.62} \qquad (12)$$

at temporal resolution and

$$\Delta Z_b = \frac{\lambda_p}{4} \qquad (13)$$

in the case of tuning thickness where $\lambda_p = V/\underline{f}_p$ is the predominant wavelength through a bed of interval velocity V. Similar relationships can be established for sinc wavelets.

Practical applications

We have shown that the parameters T_R and $b/2$ are directly related to resolution and may be equated to terminal frequencies and band ratio in the frequency domain. In particular, we have shown that provided a wavelet band ratio of at least 4 (2 octaves) is maintained, the resolving power of a sinc wavelet is directly related to f_u, but largely independent of band ratio beyond 2 octaves.

This observation has immediate applications in field acquisition and processing. Since the upper terminal frequency appears to be the controlling factor in resolution, every effort must be made to recover the highest frequencies commensurate with resolution objectives. Expanding band ratio beyond 2 octaves by lowering f_l is a worthy consideration if factors other than thin-bed resolu-

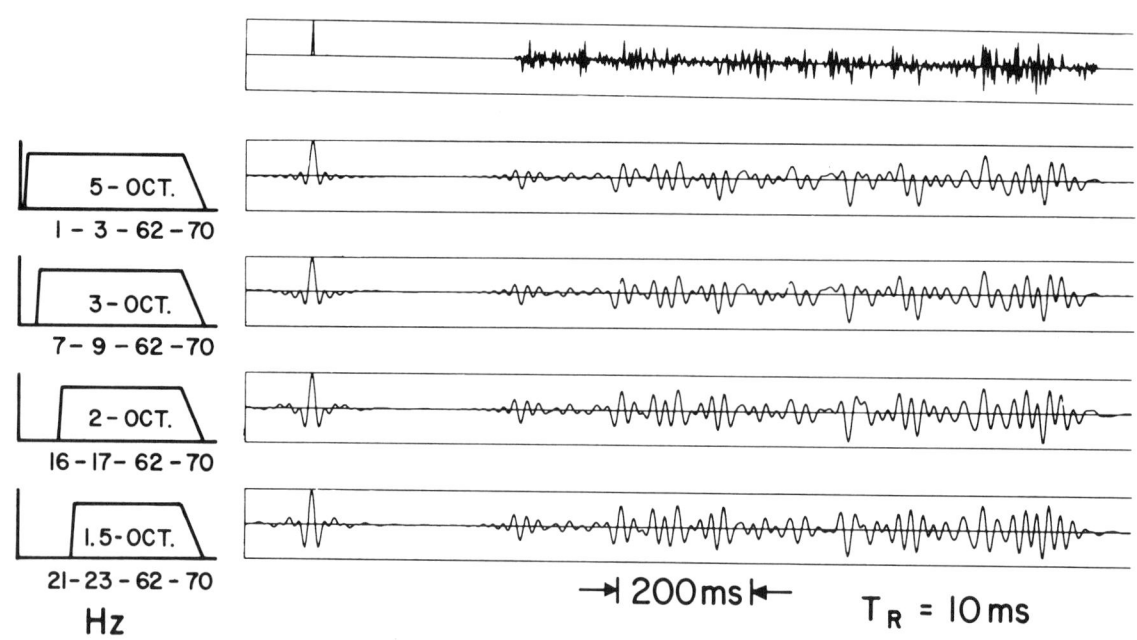

FIG. 14. A suite of synthetic seismograms produced by convolving a reflectivity sequence derived from a well log with Ormsby wavelets wherein the upper terminal frequency is kept constant while varying the lower terminal frequency.

tion are important. These other factors might include trace inversion studies and reduced side lobe amplitude tuning considerations. Figure 14 demonstrates these conclusions through a set of synthetic seismograms where the low-frequency sides of the convolving Ormsby wavelets (approximations to sinc wavelets) are varied while the high sides are kept constant. Observe that peak and trough times and number of reflections remain essentially invariant as band ratio is decreased. Reflection amplitudes increase slightly from 5 octaves to 2 octaves and more so from 2 octaves to 1.5 octaves. However, even at 1.5 octaves the increased amplitudes of events due to sidelobe tuning have not severely compromised one's ability to differentiate these events.

Reservoir thickness estimates involving clearly defined seismic events such as bright spots may benefit from these concepts. Consider a deconvolved section that has been zero-phased and filtered to a (10, 65) Hz band-pass so that tuning thickness becomes

$$\frac{b}{2} = \frac{1}{1.4 f_u} = \frac{1}{(1.4)(65)} = 11.0 \text{ msec},$$

and temporal resolution is

$$T_R = \frac{1}{1.5 f_u} = \frac{1}{(1.5)(65)} = 10.3 \text{ msec}.$$

Consider the thickness estimation of a gas-filled sand having an interval velocity of 8000 ft/sec. Tuning thickness is then $(11.0)(8)/2 = 44$ ft. Thickness estimates for noise-free models may be considered reliable for thicknesses greater than 44 ft and unreliable for thicknesses less than 44 ft. This is because beds thinner than 44 ft cannot be distinguished from each other on the basis of interval times alone. The maximum theoretical difference between the trough-to-peak times of a 44-ft bed and an infinitely thin one is less than 1 msec. Therefore, in this example no attempt should be made to estimate bed thicknesses on the basis of interval times alone for any beds having trough-to-peak times between 10 and 11 msec.

CONCLUSIONS

Resolution is an important aspect in the interpretation of seismic traces once the data have received proper handling from data acquisition through processing. Interpreting the convolutional model for stratigraphic purposes requires a zero-phase basic wavelet in order for peak-trough times to be valid measurements for estimating bed thicknesses to use in defining reservoir dimensions. The formulas and concepts presented here are immediately useful in defining limits of resolvability once signal frequency band has been established. These formulas represent upper bounds because they are based upon noise free models.

Below the tuning thickness limit, amplitude information encodes thickness variations provided the entire amplitude variation is caused by tuning effects, and amplitude calibration then permits net-pay thickness calculations for arbitrarily thin beds. The study, on the other hand, establishes in a quantitative manner the absolute limits of resolution when amplitude information cannot be used.

The literature contains diverse statements as to the limits of resolvability. The concepts developed above clarify and quantify the limits by showing the practical limit actually corresponds to Rayleigh's peak-to-trough time separation. This time interval in turn becomes the one-quarter wavelength value when the predominant frequency is used to convert to wavelengths. The most surprising result is that both Rayleigh's peak-to-trough limit and Ricker's inflection point criterion seem to depend mainly on the highest frequency present, within a high degree of accuracy, in the case of a band-limited, zero-phase wavelet with a flat amplitude spectrum extending 2 octaves or more in band ratio.

ACKNOWLEDGMENTS

The authors are indebted to J. R. Moffitt for development of computer programs and technical assistance, and to R. Hastings-James for invaluable aid in revising the manuscript. We also wish to thank Amoco Production Company for permission to publish this study.

REFERENCES

Berkhout, A. J., 1974, Related properties of minimum phase and zero-phase time functions: Geophys. Prosp., v. 22, p. 683–703.
Clement, W. A., 1976, A case history of geoseismic modeling of basal Morrow-Springer sands, Watonga-Chickasha trend, Geary, Oklahoma T13N, R10W: AAPG Special Memoir 26, p. 451–476.
Farr, J. B., 1976, How high is high resolution?: Presented at the 46th Annual International SEG Meeting October 26, in Houston.
Jenkins, F. A., and White, H. E., 1957, Fundamentals of optics: New York, McGraw Hill Publishing Co., 637 p.
Koefoed, O., 1981, Aspects of vertical seismic resolution: Geophys. Prosp., v. 29, p. 21–30.
Lindsey, J. P., Schramm, M. W., and Nemeth, L. K., 1976, New seismic technology can guide field development: World Oil, v. 182, p. 59–63.
Meckel, L. D., and Nath, A. K., 1977, Geologic considerations for stratigraphic modeling and interpretation: AAPG Special Memoir 26, p. 417–438.
Nath, A. K., Meckel, L. D., and Wood, L. C., 1977, Synergistic interpretation of the convolutional model: SEG Continuing Education Symposium.
Neidell, N. S., and Poggiagliolmi, E., 1977, Stratigraphic modeling and interpretation—Geophysical principles and techniques: AAPG Special Memoir 26, p. 389–416.
Ricker, N., 1953, Wavelet contraction, wavelet expansion and the control of seismic resolution: Geophysics, v. 18, p. 769–792.
Schoenberger, M., 1974, Resolution comparison of minimum-phase and zero-phase signals: Geophysics, v. 39, p. 826–833.
Schramm, M. W., Dedman, E. V., and Lindsey, J. P., 1977, Practical stratigraphic modeling and interpretation: AAPG Special Memoir 26, p. 477–502.
Sheriff, R. E., 1977 Limitations on resolution of seismic reflections and geologic detail derivable from them: AAPG Special Memoir 26, p. 3–14.
Widess, M. B., 1973, How thin is a thin bed?: Geophysics, v. 38, p. 1176–1180.
——— 1980, Generalized resolving power and system optimization: Presented at the SEG 50th Annual International SEG Meeting, November 19, in Houston.

APPENDIX A
A DISCUSSION OF RICKER'S RESOLUTION CRITERION

Consider a zero-phase wavelet convolved with two positive Dirac-delta functions separated by a time interval of $2T$. The wavelet complex $s(t, T)$ is

$$s(t, T) = w(t + T) + w(t - T). \quad \text{(A-1)}$$

A flat spot at the center of the wavelet complex corresponds to setting the second derivative with respect to time equal to zero and evaluating it at the origin of the coordinate system, i.e.,

$$w''(T) + w''(-T) = 0 \quad \text{(A-2)}$$

where double primes denote differentiating twice with respect to the wavelet's argument. Since zero-phase wavelets are symmetrical in time with symmetrical second derivatives, we obtain

$$w''(T) = 0.$$

A value of T corresponding to the inflection point on the main lobe of the wavelet satisfies the latter equation. In general there may be many values of T or roots satisfying this equation. However, zero curvature is satisfied by a value of T corresponding to the inflection point on the wavelet's main lobe, and it is the smallest value of T satisfying the zero-curvature criterion.

APPENDIX B
RESOLUTION OF RICKER WAVELETS

A Ricker wavelet $K(t)$ can be expressed analytically as

$$K(t) = [1 - 2(f_p \pi t)^2] \exp[-\pi f_p t)^2]. \quad \text{(B-1)}$$

Frequency f_p denotes the peak frequency in the wavelet's spectrum; it should not be confused with predominant frequency which is about 30 percent greater. Differentiating the expression for $K(t)$,

$$\frac{d}{dt} K(t) = 2(\pi f_p)^2 t [2(\pi f_p t)^2 - 3] \exp[-(\pi f_p t)^2], \quad \text{(B-2)}$$

and equating to zero yield the tuning thickness time as

$$\frac{b}{2} = \frac{1}{2.6 f_p}. \quad \text{(B-3)}$$

The above equation quantifies Rayleigh's resolution criterion for a Ricker wavelet.

Temporal resolution T_R is derived by solving for the separation between inflection points

$$K''(t) = (2\pi^2 f_p^2)^2 t^4 - 12(\pi f_p)^2 t^2 + 3 = 0, \quad \text{(B-4)}$$

yielding

$$T_R = \frac{1}{3.0 f_p}. \quad \text{(B-5)}$$

Both Rayleigh's and Ricker's criterion of resolution (inflection point separation and peak-to-trough time) depend solely on the peak frequency of the Ricker wavelet.

APPENDIX C
RESOLUTION OF SINC WAVELETS

A general expression for a sinc wavelet $w(t)$ can be derived by subtracting two low-pass sinc wavelets

$$w(t) = \frac{2f_u \sin 2\pi f_u t}{2\pi f_u t} - \frac{2f_\ell \sin 2\pi f_\ell t}{2\pi f_\ell t}, \quad \text{(C-1)}$$

where f_u and f_ℓ are upper and lower terminal frequencies, respectively. A change of variables simplifies the solution of first and second derivative normal equations. Without any loss of generality, define a real variable c such that the time variable t can be expressed as

$$t = \frac{1}{c f_u}, \quad \text{(C-2)}$$

and, as above, define r as the band ratio f_u/f_ℓ.

Setting the first derivative equal to zero yields

$$\frac{2\pi}{c} \cos \frac{2\pi}{c} - \sin \frac{2\pi}{c} - \frac{2\pi}{rc} \cos \frac{2\pi}{rc} + \sin \frac{2\pi}{rc} = 0, \quad \text{(C-3)}$$

and equating the second derivative to zero provides a similar but more complicated relationship. These two equations are the normal equations of our resolution problem. Peak-to-trough times and inflection points depend upon terminal frequencies f_u and f_ℓ, that is, they are a function of both parameters c and r. Parameter r describes the sinc wavelet's frequency band as a ratio, whereas parameter c expresses time as a function of the upper terminal frequency only. A normal equation can be solved by first choosing a band ratio r and then solving for values of c satisfying the equation; parameter c is then plotted as a function of r. Let c_1 denote the solution to the first-derivative normal equation and c_2, the solution to the second-derivative normal equation. Peak-to-trough time or tuning thickness is given by

$$\frac{b}{2} = \frac{1}{c_1 f_u}, \quad \text{(C-4)}$$

whereas inflection point times are equal to

$$\frac{T_R}{2} = \frac{1}{c_2 f_u}. \quad \text{(C-5)}$$

Since temporal resolution is the inflection-point time separation, it is equal to twice the inflection point time or

$$T_R = \frac{1}{(c_2/2) f_u}. \quad \text{(C-6)}$$

Each normal equation has an infinite number of roots. As in Appendix B, the smallest time or largest magnitude root in terms of parameter c corresponds to the central lobe complex, whereas the remaining roots correspond to side lobes. Consequently, the solution of interest is the largest root of c because it describes central lobe trough times and inflection points.

An iterative numerical method solves the normal equations for the largest value of c for any specified band ratio r. Resolution constants c_1 and $c_2/2$ for the first- and second-derivative normal equations as a function of r are plotted in Figure 9. Both parameters c_1 and $c_2/2$ converge to asymptotic values. Peak-to-trough and inflection-point times are constant, for all practical purposes, if r exceeds 2 octaves. The peak-to-trough asymptote is $c_1 = 1.398$, whereas the inflection point asymptote is $c_2/2 = 1.509$. Two simple equations describe Rayleigh's and Ricker's resolution criteria for low-pass sinc wavelets and approximate wavelets having band ratios greater than 2 octaves:

$$\frac{b}{2} = \frac{1}{1.398 f_u}, \quad \text{(Rayleigh)} \quad \text{(C-7)}$$

and

$$T_R = \frac{1}{1.509 f_u}. \quad \text{(Ricker)} \quad \text{(C-8)}$$

Note that both equations depend only on the upper terminal frequency.

RESOLUTION COMPARISON OF MINIMUM-PHASE AND ZERO-PHASE SIGNALS

M. SCHOENBERGER[*]

Despite their intuitive appeal, minimum-phase wavelets are not the shortest wavelets achievable on a seismic section. For several amplitude spectra typical of processed seismic sections, both minimum-phase and zero-phase wavelets are presented. In each case, several measures of length reveal that the zero-phase wavelet is shorter than the minimum-phase wavelet corresponding to the same amplitude spectrum. Furthermore, the zero-phase wavelet has smaller side lobes than the corresponding minimum-phase wavelet.

Synthetic seismograms were generated using both the zero-phase and minimum-phase signals as inputs. In each case, the seismogram generated with the zero-phase input signal had better resolution. This relation is demonstrated quantitatively and is also visually obvious on the seismograms.

In addition to comparing the wavelets' resolution capabilities, the accuracies permitted in estimating reflection times were compared. The zero-phase wavelets resulted in more accurate estimates of both reflection times and spacings.

INTRODUCTION

In a recent paper, Berkhout (1973) establishes a most interesting property of minimum-phase signals: "For a certain amplitude spectrum, the one-sided minimum-length signal and the signal with minimum-phase spectrum are identical signals." This has given rise to two thought-provoking questions: First, how does the length of a zero-phase signal, which is a two-sided signal, compare to the length of the minimum-phase signal with the same amplitude spectrum? This is a very practical question, since these are the most common alternatives available in seismic data processing. The second question is, what is the best definition of signal length? It is especially important that signals with the smallest "length" have the best resolution.

The processed seismogram that is seen by an interpreter consists of seismic wavelets at each reflection time. Over a suitably short time interval, the wavelet is relatively unchanged and the seismogram can be represented by

$$S(f) = R(f)W_f(f) + N'(f), \qquad (1)$$

where $S(f)$, $R(f)$, $W_f(f)$, and $N'(f)$ are the Fourier transforms of the seismogram, reflection coefficient train, wavelet, and additive noise, respectively (Robinson, 1957). In this context, the wavelet $W_f(f)$ is the source function filtered by the earth, the recording system, and the processing system.

If the wavelet $W_f(f)$ has insufficient resolution, the reflections will not be distinct but will be blurred and appear fewer in number. Although the reflection resolution is determined to a large extent by the bandwidth of $W_f(f)$, its phase characteristic is also significant.

Prior to deconvolution and band-pass filtering, the following variation on equation (1) can serve as a model:

$$S_R(f) = R(f)W(f) + N(f), \qquad (2)$$

where $S_R(f)$, $W(f)$, and $N(f)$ are now the Fourier transforms of the seismogram, wavelet, and additive noise, respectively, at this stage of processing. If the data are now deconvolved with the filter $1/\hat{W}(f)$ and band-pass filtered with $F(f)$, the resultant seismic wavelet will be

---

Paper presented at the 44th Annual International SEG Meeting, November 11, 1974, Dallas, Tex. Manuscript received by the Editor November 28, 1973; revised manuscript received May 16, 1974.
[*] Exxon Production Research Co., Houston, Tex. 77001
© 1974 Society of Exploration Geophysicists. All rights reserved.

$$W_f(f) = F(f)W(f)/\hat{W}(f). \quad (3)$$

Therefore, to the extent that $\hat{W}(f)$ is a good estimate of $W(f)$, the filtered wavelet will be simply the filter function corresponding to $F(f)$.

Of course, in practice, $W(f)/\hat{W}(f)$ will never be unity, and the seismic wavelet on the filtered section will be a distorted version of the filter operator. Inexact estimation of the $W(f)$ amplitude spectrum is partially responsible for this. However, phase differences between $W(f)$ and $\hat{W}(f)$ are even more likely. For example, if $\hat{W}(f)$ is a unit prediction distance filter, it is minimum-phase. However, $W(f)$ will not be minimum-phase if the source is nonminimum-phase or if distortion such as NMO correction occurs in data processing. Conversely, $\hat{W}(f)$ will be nonminimum-phase if it is a predictive filter with prediction distance greater than unity. Even if both are minimum-phase, they will have different phase functions if their amplitude spectra differ (even if this difference occurs outside the band-pass filter's passband).

Despite the reservations expressed above, the band-pass filter is a dominant factor in determining the seismic wavelet. To the extent that deconvolution flattens the spectrum, the amplitude spectrum of the wavelet $W_f(f)$ is that of the band-pass filter $F(f)$. Even if, in practice, the phase spectrum of the wavelet $W_f(f)$ is neither minimum- nor zero-phase, these may be regarded as goals achievable with ideal data processing. Perhaps it may be possible to approximate closely these extremes with realistic data processing.

The above discussion provides practical motivation for determining the resolution of zero-phase and minimum-phase signals. If minimum-phase signals have resolution superior to that of zero-phase signals, the processing should be designed to produce minimum-phase signals. Conversely, the superiority of zero-phase signals should motivate their production. At least under ideal circumstances, the production of these signals could be accomplished by suitable choice of the band-pass filter after deconvolution.

COMPARISON OF ZERO-PHASE AND MINIMUM-PHASE SIGNALS

Since the amplitude spectrum of the wavelet on a deconvolved and band-pass filtered seismogram may approximate that of the band-pass filter, some typical band-pass filter characteristics will

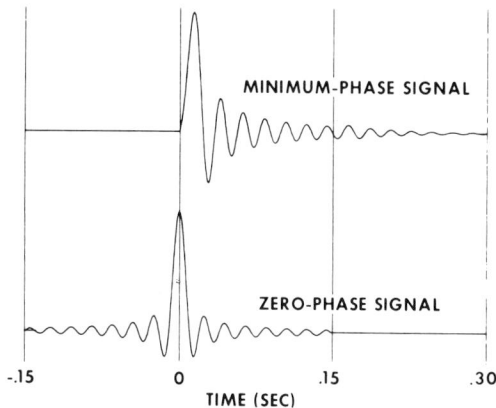

FIG. 1(a). The zero-phase signal and minimum-phase signal corresponding to a low-pass filter with a 50 hz cutoff.

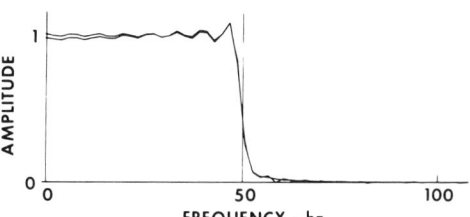

FIG. 1(b). Amplitude spectra of the signals of Figure 1(a).

be considered here. Each of Figures 1, 2, and 3 shows both the minimum-phase and zero-phase signals corresponding to an amplitude spectrum of a typical band-pass filter. The computed amplitude spectra are shown in part b of each figure. The minimum-phase signal is one-sided and is, thus, zero for time less than zero. In contrast, the zero-phase signal is not only two-sided; it is symmetric about zero time. The zero-phase signal corresponding to a given amplitude spectrum is easily calculated by taking the inverse Fourier transform of a function whose real part is the amplitude spectrum and whose imaginary part is zero. The inverse of the least-squares deconvolution operator calculated from the autocorrelation of the zero-phase signal is the corresponding minimum-phase signal (Robinson, 1957; White and O'Brien, 1973).

The "lengths" of these signals can be determined from the definition (slightly modified for two-sided signals) provided by Berkhout (1973):

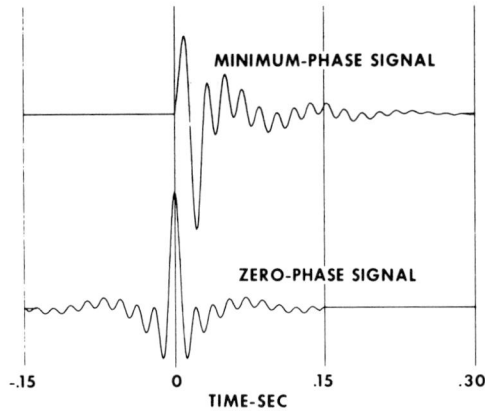

FIG. 2(a). The zero-phase signal and minimum-phase signal corresponding to a 10 to 60 hz band-pass filter.

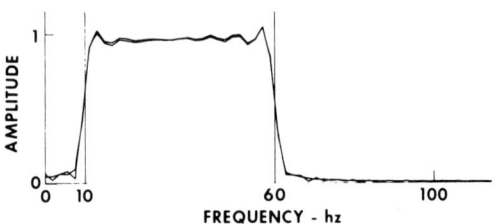

FIG. 2(b). Amplitude spectra of the signals of Figure 2(a).

$$L(S) = \frac{\Sigma W(|n|)S^2(n\Delta)}{\Sigma S^2(n\Delta)}, \quad (4)$$

where $L(S)$ represents the length of the signal S, Δ is the sample rate of the data, $W(n)$ is a nondecreasing function (which we will choose to be symmetric about $n=0$), $W(0)=0$, and the summation will be over all values of n (both positive and negative). Table 1 compares the lengths of the signals in Figures 1, 2, and 3 for two different weighting functions:

$$L_1(S) = \frac{\Sigma |n| S^2(n\Delta)}{\Sigma S^2(n\Delta)}, \quad (5)$$

$$L_2(S) = \frac{\Sigma n^2 S^2(n\Delta)}{\Sigma S^2(n\Delta)}. \quad (6)$$

The most obvious feature of Table 1 is that for both definitions of length in all the examples considered, the zero-phase signal is shorter—it has the smaller length.

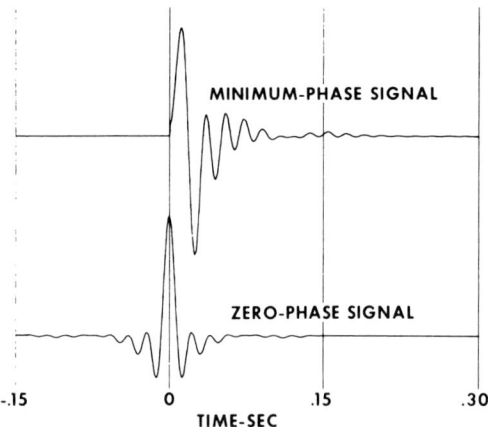

FIG. 3(a). The zero-phase signal and minimum-phase signal corresponding to a band-pass filter with unit response from 15 to 50 hz decreasing linearly to zero at 0 and 65 hz.

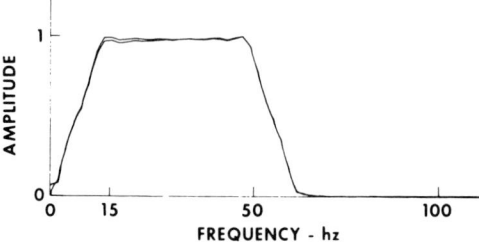

FIG. 3(b). Amplitude spectra of the signals of Figure 3(a).

However, it should be fairly obvious that the definitions of length given by equations (4) through (6) are not directly related to resolution. One of the problems with this class of definitions is simply that the signal lengths are measured with reference to time equal zero. The difficulty with this can be seen by comparing the signals of Figure 4. The time duration of signal (A) is much less than that of signal (B). However, since signal (A) is farther from the time origin, the weight function $W(n)$ will be larger at the time of its non-

Table 1. Length comparison of signals in Figures 1, 2, and 3 for two weighting functions.

Signals	$L_1(S)$ Zero-phase	$L_1(S)$ Minimum-phase	$L_2(S)$ Zero-phase	$L_2(S)$ Minimum-phase
Figure 1	5.2	20.2	151.	823.
Figure 2	8.3	25.9	305.	1258.
Figure 3	5.2	21.1	91.	612.

zero values, and, therefore, its length may be larger. Specifically, $L_1(A) = 0.11$, $L_1(B) = 0.025$, $L_2(A) = 0.012$, and $L_2(B) = 0.00083$.

The concept of resolution involves the relative magnitude of a signal's primary lobe and secondary lobes. Therefore, a more reasonable definition of length (for the purposes of resolution) would be referenced to the center of the signal's primary lobe. Specifically, let the "resolution length" be defined as

$$L^R(S) = \frac{\Sigma W(|n-x|)S^2(n\Delta)}{\Sigma S^2(n\Delta)}, \quad (7)$$

where x is the sample at which $S^2(n\Delta)$ is maximum [i.e., $S^2(x\Delta) \geq S^2(n\Delta)$ for all n], and $W(|n-x|)$ is a symmetric nondecreasing function about $n=x$; $W(0) = 0$. [A more general definition might be the minimum value with respect to x of the right side of equation (7). This is required in cases such as Figure 4 in which the maximum occurs at more than one sample value.] Using the following definitions,

$$L_1^R(S) = \frac{\Sigma |n-x| S^2(n\Delta)}{\Sigma S^2(n\Delta)} \quad (8)$$

and

$$L_2^R(S) = \frac{\Sigma (n-x)^2 S^2(n\Delta)}{\Sigma S^2(n\Delta)}, \quad (9)$$

we find that the resolution lengths of the signals of Figure 4 are $L_1^R(A) = 0.005$, $L_1^R(B) = 0.0125$, $L_2^R(A) = 0.33 \times 10^{-4}$, and $L_2^R(B) = 2.08 \times 10^{-4}$. Now

Table 2. Resolution length comparison of signals in Figures 1, 2, and 3.

Signals	$L_1^R(S)$		$L_2^R(S)$	
	Zero-phase	Minimum-phase	Zero-phase	Minimum-phase
Figure 1	5.2	9.1	151.	454.
Figure 2	8.3	11.9	305.	602.
Figure 3	5.2	9.5	91.	182.

A, the signal with the smaller time duration, has the smaller length. Table 2 uses definitions (8) and (9) to compare the resolution lengths of the signals of Figures 1, 2, and 3.

Despite the fact that the resolution lengths of the minimum-phase signals are considerably less than the lengths measured by the other definitions, Table 2 still verifies our earlier conclusions. Indeed, in all the examples considered, the zero-phase signal has smaller resolution length than the corresponding minimum-phase signal. However, it is worth noting that even these definitions lack an important property relative to resolution. Namely, zero-phase functions with the same bandwidth should have the same resolution length. Although all three signals in Figures 1, 2, and 3 have approximately the same bandwidth, their resolution lengths in Table 2 are different.

Definitions (7), (8), and (9) measure signal length relative to the signal's maximum value. Since this seems to be a good basis for comparing signals, Figure 5 displays the signals of Figures 1, 2, and 3 with each of the minimum-phase signals time shifted and, in the case of signals (B) and (C), inverted so that its maximum coincides with the maximum of the corresponding zero-phase signal. Technically, these time-shifted functions are no longer minimum-phase or even one-sided, but they can be referred to as such without confusion.

Interestingly enough, the time between zero-crossings of the primary lobe for the two signals is about the same in each of these cases. The differences between the two signals are fairly obvious. Although each minimum-phase signal has a sharp onset, its secondary lobes are much greater than those of the corresponding zero-phase signal. Table 3 contains the maximum side lobe level for each of the signals of Figure 5. The maximum side lobe level is the ratio of the maximum absolute value outside the primary lobe to the maximum value

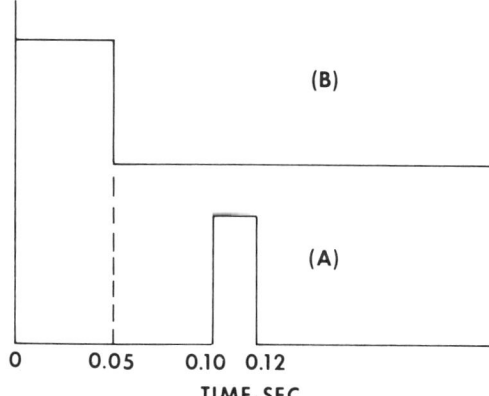

FIG. 4. Two rectangular pulses with different time durations. Pulse (A) has a smaller time duration but longer length.

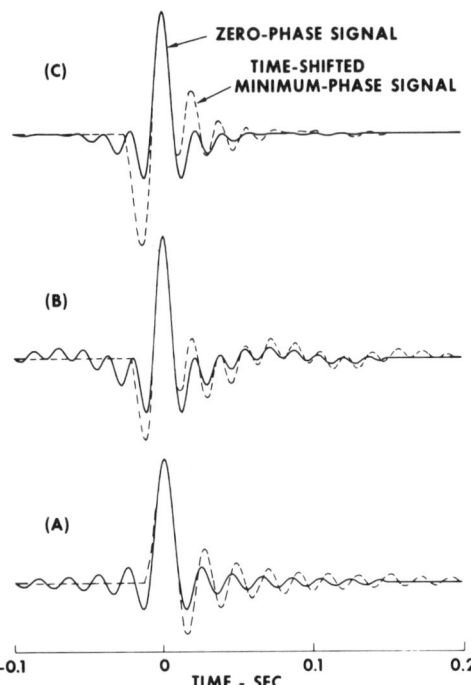

FIG. 5. Pairs of zero-phase and time-shifted minimum-phase signals corresponding to a given amplitude spectrum. (A), (B), and (C) correspond to the signals contained in Figures 1, 2, and 3, respectively.

within the primary lobe and is expressed as a percentage.

Table 3 expresses quantitatively what was obvious from Figure 5—the side lobe levels of the minimum-phase signals are considerably larger than those of the corresponding zero-phase signals. In the case of the signal (C) (or Figure 3), which incidentally may be the band-pass filter most typical of those in actual use, the side lobe level is 2.5 times larger for the minimum-phase signal. In this case, primary lobe and secondary lobe are almost equal in magnitude. For negative times, the zero-phase signal's amplitude is obviously infinitely greater than that of the minimum-phase signal, which is zero. Therefore, the trade-off seems to be between a two-sided signal with given amplitude side lobes or a one-sided signal with a sharp onset and side lobes often more than twice as large.

Since the ratio of the amplitude of the primary lobe to the largest side lobe of a signal is a measure of its resolution, it would seem that a zero-phase signal has better resolution than its minimum-phase counterpart. The same conclusion was drawn from Table 2 which compared the resolution lengths of the signals. However, any resolution measures that involve isolated signals are somewhat academic. The true measure of signal resolving power is our ability to distinguish with that signal two or more reflections that are close together.

Figures 6 and 7 are synthetic seismograms generated by convolving a reflection coefficient train with the zero-phase and minimum-phase signals of Figure 5. The reflection coefficients are all equal in magnitude and sign and occur in pairs. The shallowest pair is separated by 40 msec, the next pair by 20 msec, and the deepest pair by 16 msec. Since each of the signals has a bandwidth of approximately 50 hz, the shallowest pair should be easily resolvable, the middle pair marginally resolvable, and the deepest pair not quite resolvable.

Before analyzing these seismograms quantitatively, let us note some obvious differences. First, the same amplitude reflection coefficients (in each pair) result in the same amplitude zero-phase reflections but different amplitude minimum-phase reflections. Second, the peaks of the reflections occur at the proper time with the zero-phase reflections but are delayed with the minimum-phase reflections. Third, the side lobe levels are so high in the case of the minimum-phase seismogram that it is often difficult to pick the reflections from a long ringing signal; this is not true of the zero-phase seismogram. Fourth, the zero-phase signals contain nonrealizable precursors whereas the minimum-phase wavelets have sharp onsets.

To assist in comparing these sets of seismograms, a tabulation of the side lobe level associated with each of the reflection pairs is provided in Table 4. Reflection side lobe level (the ratio of the peak nonreflection amplitude to the smaller of the two peak reflection amplitudes) is another indicator of resolution.

Table 3. Maximum side lobe level for the signals in Figure 5.

	Side lobe level (percent of peak)	
Signals	Zero-phase	Minimum-phase
(A) Figure 1	21.0	43.0
(B) Figure 2	44.3	67.7
(C) Figure 3	35.5	90.5

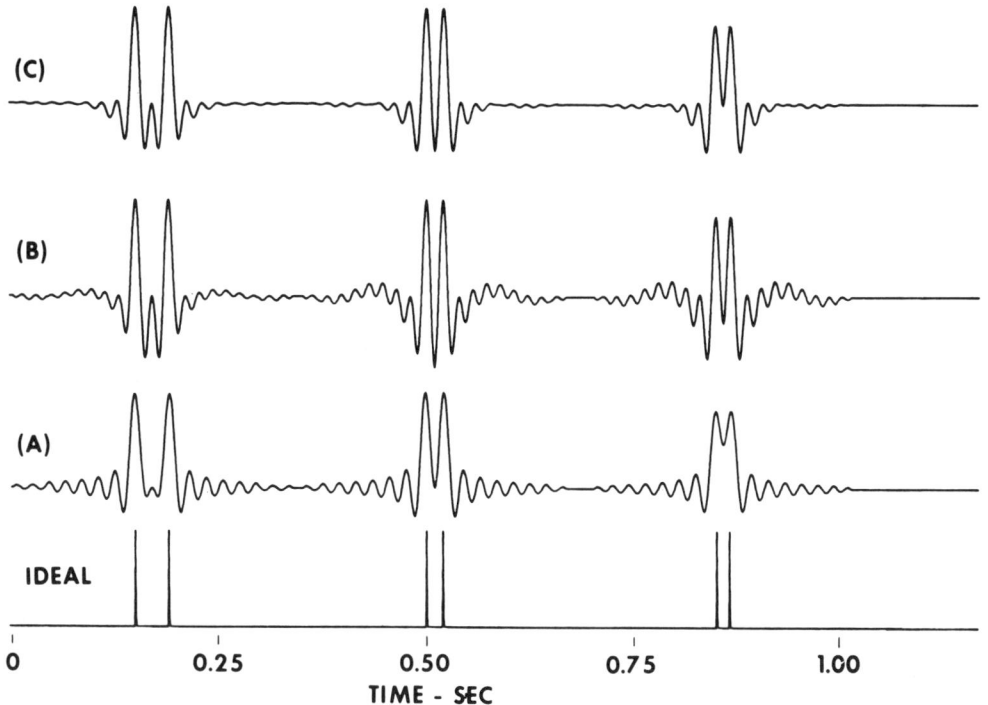

FIG. 6. Synthetic seismograms resulting from reflectors at .150, .190, .500, .520, .850, and .866 sec. The ideal trace results from using a spike as the input pulse. Traces (A), (B), and (C) result from using the zero-phase pulses of Figures 1, 2, and 3, respectively.

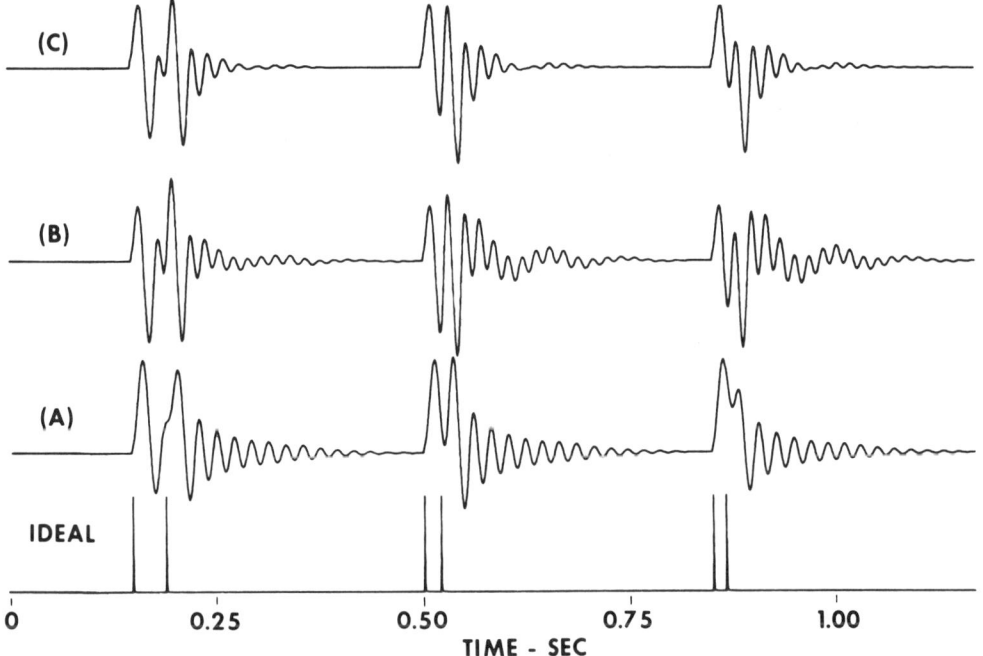

FIG. 7. Synthetic seismograms resulting from reflectors at .150, .190, .500, .520, .850, and .866 sec. The ideal trace results from using a spike as the input pulse. Traces (A), (B), and (C) result from using the minimum-phase pulses of Figures 1, 2, and 3, respectively.

Table 4. Reflection side lobe levels (percent)

	(A)		(B)		(C)	
	Zero-phase	Minimum-phase	Zero-phase	Minimum-phase	Zero-phase	Minimum-phase
Shallow	27.2	56.3	65.7	103.6	48.4	101.5
Middle	30.5	60.0	74.8	91.1	50.0	132.0
Deep	29.8	62.4	65.7	115.6	61.6	307.4

Now, with the aid of Table 4, let us examine Figures 6 and 7 more carefully. First, how well are the shallowest reflectors resolved? In each of the six seismograms, there are two lobes (excursions from zero) associated with the two shallow reflections but with different side lobe levels. The best seismogram in that sense is the zero-phase (A) with only 27 percent side lobes. The worst zero-phase seismogram is (B) with two 66 percent side lobes. However, each of the minimum-phase seismograms has larger corresponding side lobe levels, and the (B) and (C) minimum-phase seismograms have nonreflection energy even greater than the reflection energy.

The middle reflections also produce two distinct lobes in each of the six seismograms. However, if we did not know the form of the reflectors, we would be sorely pressed to detect the 0.52-sec reflector in the long ringing signal of minimum-phase (C). Indeed, in this case nonreflection energy is 30 percent greater than reflection energy.

The deepest pair of reflections can be discerned on all the zero-phase seismograms despite the inability of signal (A) to achieve adequate separation of the two reflections. However, in none of the minimum-phase cases is there anything but one long ringing signal. In (A), there is a slight indication of a separation between the deepest reflections—similar to but not as good as the zero-phase (A). Note that in the minimum-phase (C), nonreflection energy is three times greater than reflection energy.

For all three sets of signals and all three pairs of reflectors, Table 4 shows lower reflection side lobe levels associated with zero-phase signals. In addition to resolution or ability to distinguish between closely spaced reflectors, there are other desirable properties of signals. These include accuracy of depth estimation and accuracy of bed spacing estimation. Table 5 provides a tabulation of the ideal reflection times and the times at which the reflections on each of the six seismograms peaked.

Several interesting properties of these synthetic seismograms are illustrated by Table 5. First, the zero-phase seismogram reflection times (peak value of the reflection) are, for all of the examples, within one msec of the ideal. However, the minimum-phase reflections lag the ideal times by as much as 26 msec, and in no case is the lag less than 14 msec. Furthermore, in the case of the zero-phase seismograms, the time separation of reflections is exactly right for the two shallow pairs of reflections. For the deepest pair, which is not quite resolvable, the time separation is overestimated by 1 msec in (C) and by 2 msec in (A) and (B). However, the minimum-phase signals do not perform nearly as well. The shallowest reflections are separated by from 39 to 42 msec instead of 40 msec. The middle reflections are separated by 21 to 22 msec instead of 20 msec

Table 5. Reflection times

	(A)		(B)		(C)	
Ideal	Zero-phase	Minimum-phase	Zero-phase	Minimum-phase	Zero-phase	Minimum-phase
.150	.150	.164	.150	.172	.150	.175
.190	.190	.206	.190	.211	.190	.215
.500	.500	.514	.500	.521	.500	.524
.520	.520	.536	.520	.542	.520	.545
.850	.849	.864	.849	.870	.850	.873
.866	.867	.883	.867	.889	.867	.892

and the deepest reflections by 17 to 19 msec instead of 16 msec.

CONCLUSIONS

Because of the minimum length property of minimum-phase signals (Berkhout, 1973) as well as many intuitive notions, it is often supposed that the seismic section processed to have minimum-phase reflections has the best possible resolution. This is not the case. Berkhout's proof that the minimum-phase signal has minimum length is restricted to the class of one-sided functions. The two-sided signal with zero phase corresponding to the same amplitude spectrum as a minimum-phase signal actually has better resolution. This has been illustrated not only for a variety of quantitative measures of resolution but also by generating synthetic seismograms. The zero-phase seismograms have resolution that is visually superior to that of the corresponding minimum-phase cases.

In addition to their superior resolution, zero-phase signals also provide a more accurate indication of reflection time and, therefore, reflector depth. They also provide a more accurate indication of reflector spacing.

The phase property of the wavelet is largely determined by the band-pass filter applied subsequent to least-squares deconvolution. Therefore, a zero-phase band-pass filter is called for. Since such filters are in quite general use (mainly due to their ease of design and implementation), this investigation may be viewed as a verification of the correctness of present filtering practices.

REFERENCES

Berkhout, A. J., 1973, On the minimum length property of one-sided signals: Geophysics, v. 38, p. 657–672.

Robinson, E. A., 1957, Predictive decomposition of seismic traces: Geophysics, v. 22, p. 767–778.

White, R. E., and O'Brien, P. N. S., 1973, Estimation of the primary seismic pulse: Presented at the 35th E.A.E.G. Meeting, Brighton, England.

SEISMIC ROCK PROPERTIES

FORMATION VELOCITY AND DENSITY—THE DIAGNOSTIC BASICS FOR STRATIGRAPHIC TRAPS

G. H. F. GARDNER,* L. W. GARDNER,‡ AND A. R. GREGORY§

A multiplicity of factors influence seismic reflection coefficients and the observed gravity of typical sedimentary rocks. Rock velocity and density depend upon the mineral composition and the granular nature of the rock matrix, cementation, porosity, fluid content, and environmental pressure. Depth of burial and geologic age also have an effect.

Lithology and porosity can be related empirically to velocity by the time-average equation. This equation is most reliable when the rock is under substantial pressure, is saturated with brine, and contains well-cemented grains. For very low porosity rocks under large pressures, the mineral composition can be related to velocity by the theories of Voigt and Reuss.

One effect of pressure variation on velocity results from the opening or closing of microcracks. For porous sedimentary rocks, only the difference between overburden and fluid pressure affects the microcrack system. Existing theory does not take into account the effect of microcrack closure on the elastic behavior of rocks under pressure or the chemical interaction between water and clay particles.

The theory of Gassmann can be used to calculate the effect of different saturating fluids on the P-wave velocity of porous rocks. The effect may be large enough in shallow, recent sediments to permit gas sands to be distinguished from water sands on seismic records. At depths greater than about 6000 ft, however, the reflection coefficient becomes essentially independent of the nature of the fluid.

Data show the systematic relationship between velocity and density in sedimentary rocks. As a result, reflection coefficients can often be estimated satisfactorily from velocity information alone.

INTRODUCTION

The purpose of this paper is to set forth certain relationships between rock physical properties, rock composition, and environmental conditions which have been established through extensive laboratory and field experimentation together with theoretical considerations. The literature on the subject is vast. We are concerned primarily with seismic P-wave velocity and density of different types of sedimentary rocks in different environments. These properties govern occurrences of seismic reflections and variations of observed gravity. They thus have significant bearing upon the manner of use of these geophysical methods and their effectiveness in finding or delineating stratigraphic traps.

A stratigraphic trap connotes a porous and permeable reservoir rock which alters laterally on one or more sides into a nonpermeable rock by facies changes or a pinch-out. A particularly important example is a reef surrounded by rock with different properties. Reflection seismic represen-

---

Paper presented at the 38th Annual International SEG Meeting, October 3, 1968, Denver, Colo. and the 43rd Annual International SEG Meeting, October 24, 1973, Mexico City. Manuscript received by the Editor January 30, 1974.
* Gulf Research & Development Co., Pittsburgh, Penn. 15230.
‡ Retired, Austin, Tex. 78703; formerly Gulf Research & Development Co.
§ University of Texas, Austin, Tex. 78712; formerly Gulf Research & Development Co.
© 1974 Society of Exploration Geophysicists. All rights reserved.

tation of the convergence of seismic horizons bracketing pinch-outs may comprise very useful information. Also, relatively small lateral convergences in seismic transit times may reveal differential compaction, and lateral variation in transit time may indicate lateral change in the physical characteristics of the bracketed rock. Direct seismic reflection evidence, however, typically involves lateral variation of interference patterns of reflections from the top and bottom of a reservoir rock. Evidence furnished by gravity observations is even more subtle but can be applicable in some cases.

Many laboratory studies have been made to show how P-wave velocity in rocks is affected by pressure and fluid saturation; these studies principally use ultrasonic techniques. When the samples are cores recovered from wells, conditions that exist in the earth can be reproduced with a fair degree of realism. As a result, a number of significant relationships for P-wave velocities of rocks under different conditions of stress and fluid containment have been established.

A change in rock lithology or composition however cannot be simulated very satisfactorily in the laboratory. Consequently, no relevant relationships between different specific parameters are well established. Commonly then, we must resort to empirical correlations based on field data. Such correlations generally entail some unknowns, so they are satisfactorily applicable only for particular formations and environments.

An illustration of the wide range of P-wave velocities and lesser range of bulk densities for the more prevalent sedimentary rock types through a wide range of basins, geologic ages, and depths (to 25,000 ft) is given in Figure 1.

An additional consideration is a general tendency for velocity and density to increase with increase in depth of burial and with increase in age of formations as verified by Faust (1953). Successively deeper layers may differ materially in composition and porosity with accompanying marked local departures in velocity and density from progressive increase with depth.

An understanding of interrelationships between different rock properties and environmental conditions, however, requires recognition and consideration of the nature of rocks in general as being granular with interconnected fluid-filled

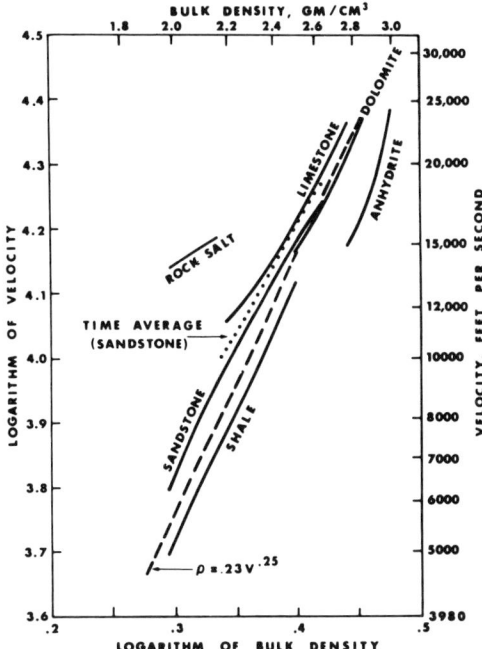

FIG. 1. Velocity-density relationships in rocks of different lithology.

interstices. As a consequence, porosity, mineral composition, intergranular elastic behavior, and fluid properties are primary factors. These factors are dependent upon overburden pressure, fluid pressure, microcracks, age, and depth of burial.

ROCK COMPOSITION AND GASSMANN'S THEORY

The three components which characterize the composition of rocks and the superscripts which indicate symbols of their properties are:

(1) the solid matter of which the skeleton or frame is built (index $\hat{\ }$);
(2) the frame or skeleton (index $^-$); and
(3) the fluid filling of the pores (index $\sim$).

Properties of the whole rock are indicated by symbols without superscripts.

In a classic paper, Gassmann (1951) showed that when a rock with its fluid is a closed system, grossly isotropic and homogeneous, the use of elementary elastic theory yields the following interrelationship between the rock parameters:

$$k = \hat{k}(\bar{k} + Q)/(\hat{k} + Q), \qquad (1)$$

where

$$Q = \tilde{k}(\hat{k} - \bar{k})/\phi(\hat{k} - \tilde{k}).$$

Definitions of the symbols are shown below.

NOMENCLATURE

Symbol	Quantity
M	Space modulus (or P-wave modulus)
μ	Rigidity modulus (or S-wave modulus)
k	Bulk modulus (or reciprocal of the compressibility)
ν	Poisson's ratio
ρ	Bulk density
ϕ	Fractional porosity
$\bar{F}$	Pressure on skeleton = Total external pressure less the internal fluid pore pressure = Net overburden pressure

Superscripts

- $^{-}$ Properties of the frame or skeleton (empty porous rock)
- $^{\wedge}$ Properties of the solid matter (grains or crystals) of which the skeleton is built
- $^{\sim}$ Properties of fluid occupying the pore space of the skeleton

Parameters of Time-Average Equation:

V	Velocity of fluid-saturated rock
V_F	Velocity of fluid occupying the pore space of the rock
V_M	Velocity of solid mineral of which the matrix of rock is built

He also noted that $\mu = \bar{\mu}$ and that $M = k + 4/3\, \mu$. White (1965) gives these relationships in the form:

$$M = \bar{M} + (1 - \bar{k}/\hat{k})^2 / (\phi/\tilde{k} + (1-\phi)/\hat{k} - \bar{k}/\hat{k}^2). \quad (1a)$$

It is also known that:

$$\rho = (1-\phi)\hat{\rho} + \phi\tilde{\rho}, \quad (1b)$$

and

$$P\text{-wave velocity} = \sqrt{M/\rho}. \quad (1c)$$

There is general agreement among experimenters that equations (1a) to (1c) satisfactorily predict the effect of different saturating fluids on P-wave velocity in most porous rocks, in spite of the obvious simplifications of the theory.

Gassmann's theory was extended by Biot (1956) to include the dynamic effects of relative motion between the fluid and the frame and also to take into account viscoelastic effects in part associated with the presence of microcracks. These refinements are unimportant for waves at seismic prospecting frequencies, and even at well-logging frequencies Gassmann's simple formula often is adequate.

Neither Gassmann's nor Biot's theory treats the effect of microcrack closure on the elastic behavior of rocks under pressure or the chemical effects such as the interaction between water and clay particles.

The parameters of the solid matter of the frame that enter equations (1), (1a), and (1b) are $\hat{k}$ and $\hat{\rho}$. Some typical values are listed in Table 1; additional data are available in Clark (1966).

The parameters of typical fluids are also known, and some are listed in Table 2.

Thus, characteristics of the solid matter and the pore fluid of many rocks can be assigned numerical values without much difficulty.

If values of $\bar{k}/\hat{k}$ are 0.5 or greater, with $\phi = 0.2$ and with $\hat{k}/\tilde{k} = 18$, the magnitude of the second term on the right of equation (1a) becomes about $.06\,\bar{k}$ or less. This means that if the frame or skeleton has relatively high elastic constants, its characteristics essentially govern the properties of the whole rock, regardless of the fluid filling. Those characteristics depend upon the elastic interactions between the grains, their bonding, and the presence of microcracks.

If, however, $\bar{k}$ and $\bar{\mu}$ are zero, μ and $\bar{M}$ also become zero, and equation (1a) becomes:

$$\frac{1}{M} = \frac{1}{k} = \frac{\phi}{\tilde{k}} + \frac{(1-\phi)}{\hat{k}}. \quad (1d)$$

Table 1. Bulk modulus and density of some minerals

Solid	Bulk modulus, $\hat{k}$ dynes/cm^2 × 10^{10}	Density, $\hat{\rho}$ gm/cm^3
α–Quartz	38	2.65
Calcite	67	2.71
Anhydrite	54	2.96
Dolomite	82	2.87
Corundum	294	3.99
Halite	23	2.16
Gypsum	40	2.32

Table 2. Bulk modulus and density of some fluids

Fluid	Bulk modulus, $\tilde{k}$ dynes/cm² × 10¹⁰	Density, $\tilde{\rho}$ gm/cm³
Water, 25°C (distilled)	2.239	0.998
Sea water, 25°C	2.402	1.025
Brine, 25°C (100,000 mg/L)	2.752	1.0686
Crude oil (1)	0.862	0.85
(2)	1.740	0.80
Air, 0°C (dry, 76 cm Hg)	0.000142	0.001293
Methane, 0°C (76 cm Hg)	0.001325	0.007168

This is the relationship applicable for mixtures of any two fluids. For clays and shales having very high water content, $\bar{k}$ and $\bar{\mu}$ approach zero, and the velocity of the formation approaches that of the fluid. It is noteworthy that for clays and shales the water is "bound" to the fine grained microstructure so that water content is a more pertinent term than porosity for use in describing the rock structure. Again, the properties of the rock frame are difficult to characterize. Permeabilities of clays and shales, of course, are very much lower than those of reservoir sands.

Table 3. Upper and lower theoretical bounds for velocity in aggregates of quartz crystals, with no porosity and no microcracks

	Voigt	Reuss
P-wave velocity, ft/sec	20,300	19,300
S-wave velocity, ft/sec	13,900	12,900

EFFECT OF MICROCRACKS

Some sedimentary rocks such as quartzites and most igneous rocks have almost no porosity. For these rocks the velocity is determined by the mineral composition, provided an extensive system of microcracks is not present. In the absence of microcracks, the elastic parameters can be accurately estimated by use of the theories of Voigt (1928) and Reuss (1929) and the known elastic constants of the crystals. It has been demonstrated by Hill (1952) that the Voigt and Reuss methods give upper and lower bounds to velocities for aggregates of crystals that are randomly oriented. A relevant example is provided by quartz crystals, and the results are given in Table 3.

The effect of microcracks can be illustrated by the behavior of gabbros upon heating, as demonstrated by Ide (1937). Also, our experimental data for a gabbro as depicted in Figure 2 shows that the untreated, dried rock has a P-wave

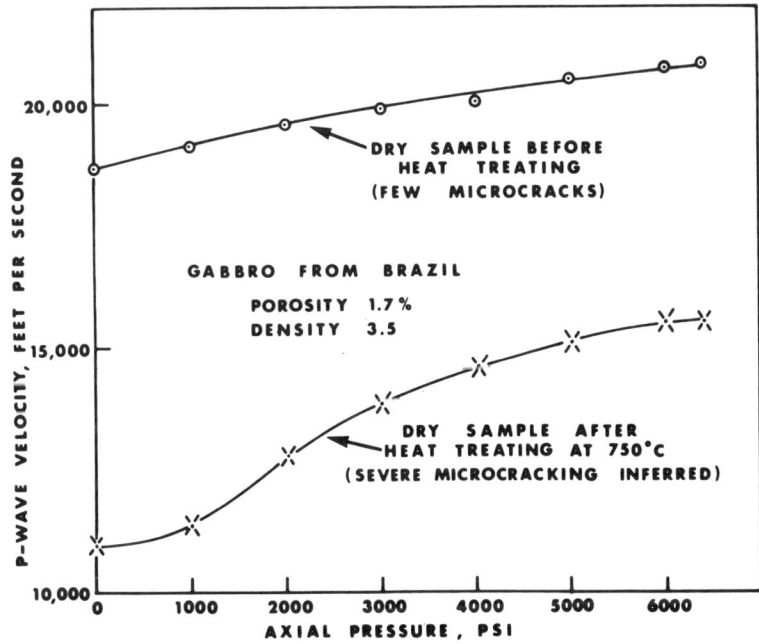

FIG. 2. Effect of microcracks on velocity of gabbro.

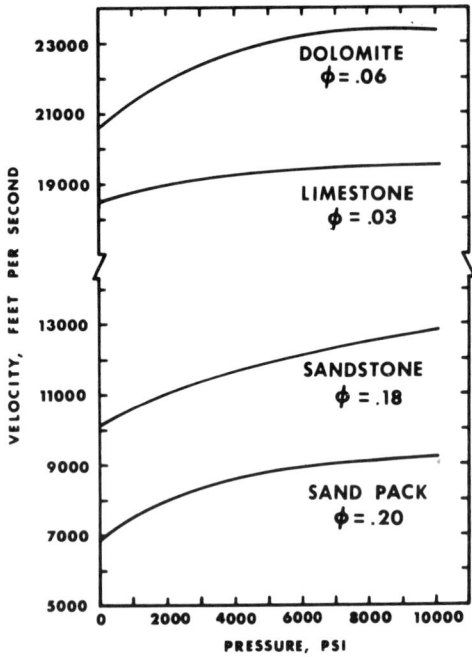

FIG. 3. *P*-wave velocity versus skeleton pressure for brine-saturated carbonates, sandstone, and sand pack.

velocity of 18,700 ft/sec which increases to 20,800 ft/sec with axial stress. In contrast, the treated rock (heated to 750°C and cooled) has a *P*-wave velocity of 11,000 ft/sec which increases with pressure but even at 6000 psi does not exceed 16,000 ft/sec. Heating is presumed to introduce a system of microcracks caused by the unequal expansion of the minerals. Pressure tends to close the cracks, but a very large pressure would be required to reestablish the original framework. The typical observation that nonporous rock velocity increases with pressure thus is attributable in substantial measure to the presence at low pressures of microcracks which are diminished at higher pressures. A similar relationship also holds true for the more porous rocks. Figure 3 shows examples of velocity versus pressure for rocks spanning a range of porosity and mineral composition. In the case of the packing of quartz grains, the "microcracks" are, presumably, the contacts between grains.

TIME-AVERAGE RELATIONSHIP

At pressures corresponding to those of deeper sediments in situ, the influence of variation in pressure on velocity becomes small, and then porosity and mineral composition alone determine velocity. Under this condition, a time-average relationship has been found empirically to interrelate velocity and rock parameters for a fairly wide range of porosities:

$$\frac{1}{V} = \frac{\phi}{V_F} + \frac{1-\phi}{V_M}. \quad (2)$$

The parameters V_F and V_M give the dependence on fluid velocity and mineral composition; ϕ is the fractional porosity. Wide experience with both in-situ determinations and laboratory experimentation supports the general applicability of this relationship for most sedimentary rocks, particularly when the fluid content is brine. This relationship is compatible with Gassmann's theory in recognizing that the elastic moduli of the frame increase as porosity decreases.

The parameter V_M is equal to the value of V as ϕ approaches zero. We have noted earlier that the theories of Voigt and Reuss can be used to estimate the velocity for this extreme case. For example, as given in Table 3, the velocity V_M for aggregates of quartz crystals should lie between 19,300 and 20,300 ft/sec. Such a value has been found to be very satisfactory for deep clean sandstones. A value of V_M between 22,000 and 23,000 ft/sec similarly is found to be applicable for deep limestones.

For formations at relatively shallow depths the influence of microcracks, and therefore pressure, cannot be ignored without introducing significant errors. The time-average equation can be retained if the parameter V_M is regarded as an empirical constant with a value less than the Voigt-Reuss values. In other words, we can assume that the traveltime (which is the quantity usually measured) is a linear function of porosity at any depth with the coefficients of the linear form to be chosen by consideration of suitable data.

As a special illustration of the use of the time-average equation, in Figure 4 we have plotted some laboratory data for cores from a depth of about 5000 ft. The cores were confined at a skeleton pressure of 3000 psi with brine in the pores to simulate the original environment. The two principal minerals in the rock (calcite and quartz in the form of tripolite) are intimately mixed in relative proportions varying from approximately 50 percent calcite–50 percent quartz to 80 percent calcite–20 percent quartz. A petrographic analysis of the cores indicated that the lower porosity samples (also from the upper part of the formation) appeared to have a continuous calcite matrix, whereas the higher porosity samples appeared to have a continuous quartz (tripolite) matrix. It can be seen that the lower porosity data points can be approximated by a time-average line with $V_M = 22,500$ ft/sec, which is a velocity suitable for a calcite matrix; the higher porosity data points can be approximated by a time-average line with $V_M = 19,200$ ft/sec, which is a velocity suitable for a quartz matrix. It is interesting that the data appear to separate on these two lines according to the mineral that is predominantly the continuous phase. In this investigation we found no correlation between velocity and the concentration by volume of the minerals.

It is also of interest to note that when the traveltimes at high pressure (10,000 psi) were plotted against porosity, no separation of the data along two lines could be detected. This may indicate that velocity measurements on cores at pressures appropriate to the depth of the formation contain more useful information than measurements at an arbitrarily high pressure.

OVERBURDEN AND FLUID PRESSURE

The overburden pressure is usually defined as the vertical stress caused by all the material, both solid and fluid, above the formation. An average value is 1.0 psi for each foot of depth, although small departures from this average have been noted. The fluid pressure is usually defined as the pressure exerted by a column of free solution that would be in equilibrium with the formation. The reference to a free solution is significant when dealing with clays or shales with which other pressures such as osmotic, swelling, etc., can be associated. The normal fluid pressure gradient is frequently assumed to be .465 psi for each foot of depth, although large departures from this value occur in high-pressure shales.

The skeleton or frame pressure of a rock is the total external pressure less the fluid pressure. The elastic parameters of the skeleton increase as the skeleton pressure increases, and a corresponding increase in velocity is observed. The increase in elastic parameters is attributable to the reactions at the intergranular contacts and the closure of microcracks as the skeleton pressure increases. Hence, when both overburden pressure and formation fluid pressure are varied, only the difference between the two has a significant influence on velocity.

A set of our velocity data that confirms this assertion is given in Figure 5. When the skeleton pressure $\bar{F}$ or the difference between overburden and fluid pressure is increased, the velocity increases; when the difference between overburden and fluid pressure remains constant, the velocity remains constant.

RECENT BASINS

In this section we consider a sedimentary basin which illustrates how P-wave velocity is affected by many of the factors discussed above. Wells in a number of young basins typically will penetrate successive layers of sand and shale that may range in age from recent to lower Eocene. This provides us an opportunity to study the effects of age, pressure, depth, porosity, and fluid content for a fairly constant matrix material.

The uppermost sedimentary layers are unconsolidated, and the porosity varies mainly with the grain size distribution and clay content. The velocity is only slightly greater than that of sea water. With increasing depth the velocity increases partly because the pressure increases and partly because cementation occurs at the grain-

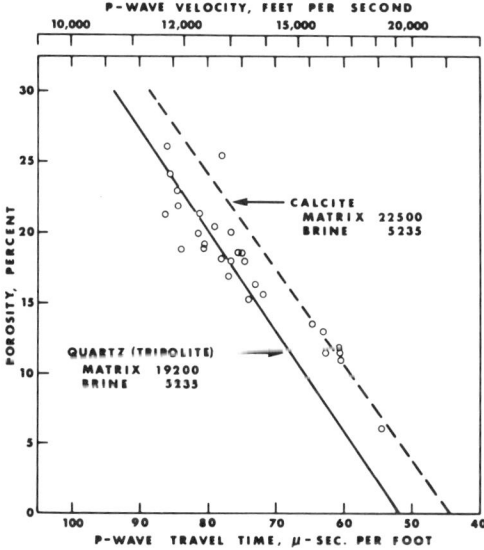

FIG. 4. Velocity versus porosity for samples of quartz-calcite rocks under 3000 psi confining pressure.

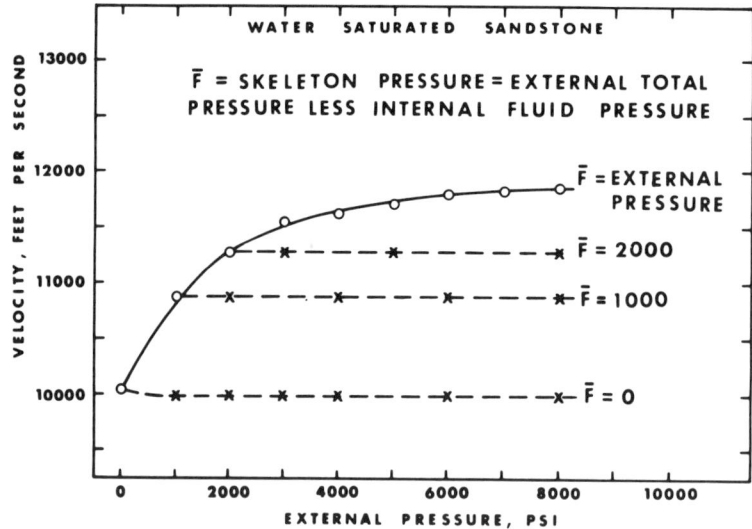

FIG. 5. Velocity through a water-saturated sandstone core as a function of skeleton pressure.

to-grain contacts. Cementation is the more important factor. It has been shown by Maxwell (1960) that the rate of cementation depends also on the rate of flow and the composition of fluids flowing through the pores as well as on temperature.

The rapid increase of velocity with depth normally continues until the time-average velocity is approached. Below this depth, the layers behave like other well-consolidated rock and the velocity depends mainly on porosity.

It is for the shallower layers that the fluid content, i.e., water, oil, or gas has an appreciable effect.

The solid curve in Figure 6 shows a representative curve of velocity versus depth for brine-saturated, in-situ sands based on some sonic log and electric log data. The dotted curve is from laboratory data for fresh, unconsolidated, water saturated packings of quartz sand grains at pressures corresponding to the depth. Thus, the dotted curve indicates what would happen to sands if they were buried without consolidation or cementation, and the divergence of these curves is attributable to these effects. The dashed curve shows the time-average velocity calculated using the average porosity read from well logs. At the shallower depths the actual velocity is less than the time-average, but below about 8000 ft the agreement is close.

Figure 7 illustrates the results for sands in more detail. For any depth the traveltime can be approximated by a linear function of porosity; below 8000 ft this linear function coincides with the time-average equation.

HIGH-PRESSURE SHALES

In the wells referred to above, the fluid pressure

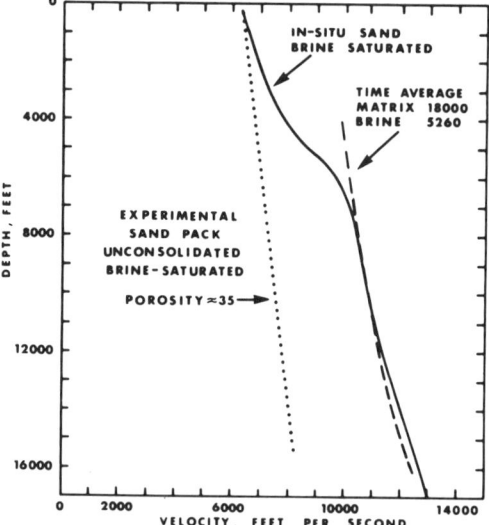

FIG. 6. Velocity as a function of depth showing consolidation effect for in-situ tertiary sands. For comparison, the velocities of experimental sand packs at pressures corresponding to these depths are also shown.

increased at a rate of about 0.5 psi per foot of depth. However, it sometimes happens that high fluid pressure zones are encountered, i.e., zones in which the fluid pressure is well above that given by the normal gradient with depth. Such wells provide an opportunity to study the relation between velocity, the difference between overburden pressure and fluid pressure, and consolidation with depth.

Hottmann and Johnson (1965) presented pertinent data for velocity measured in shales and the corresponding fluid pressure and depth. When there was no excess fluid pressure, they found that the interval traveltime ΔT, in microseconds per foot, decreased with depth Z, in feet, according to the formula

$$Z = A - B \log_e \Delta T, \quad (3)$$

where $A = 82,776$, and $B = 15,695$. They also give data from wells that penetrated zones with abnormally high fluid pressure. We have found that all these data can be correlated by the equation

$$\left(\frac{P_0 - P_F}{\alpha - \beta}\right)^{1/3} Z^{2/3} = A - B \log_e \Delta T, \quad (4)$$

where $P_0 =$ overburden pressure; $P_F =$ fluid pressure; $\alpha =$ normal overburden pressure gradient; and $\beta =$ normal fluid pressure gradient. Figure 8 illustrates this correlation.

For a normally pressured section equation (4) reduces to equation (3). One interesting feature of equation (4) is that both the pressure difference $P_0 - P_F$ and the depth Z are present. The factor $Z^{2/3}$ may be interpreted as the effect of increased consolidation with depth. For sands the effect of consolidation outweighs the effect of pressure.

UNCONSOLIDATED GAS SANDS

In a shale-sand sequence some sands may contain oil or gas, but the overlying shale may not contain either, except for the amount dissolved in water. The reflection coefficient at such an interface may be influenced appreciably by the fluids in the formation, if the depth is not too great. Gassmann's equation can be used to estimate the magnitude of the possible effect of the presence of oil or gas in a sand upon its velocity when the velocity of the brine-saturated sand is known.

By using appropriate values in equation (1a) for parameters other than the skeleton moduli of a given brine-saturated sand, we can determine the relative skeleton modulus $\bar{k}/\hat{k}$. The values of the parameters used are deduced in part from known wave velocities equations, (1b) and (1c), and a specification that $\overline{M} = S\bar{k}$ with S having a representative value of 2.0. Suitable values for $\bar{k}$ or $\bar{k}/\hat{k}$ can be used for different fluid filling of the pores, and the corresponding values of M and the P-wave velocity calculated using equations (1).

The results of some such calculations are given in Figure 9, where velocity is plotted versus depth for sands saturated with either brine, oil, or gas. The dashed curve illustrates typical values of velocity versus depth for shales saturated with brine. It is clear that in the first few thousand feet the reflection coefficient at the boundary between a shale and a sand will be significantly greater if the sand contains gas than if the sand contains brine. This observation might possibly be of practical significance when there is a lateral change from a brine-filled sand to a gas-filled sand. However, at considerable depths the reflection coefficient becomes almost independent of the nature of the fluid content.

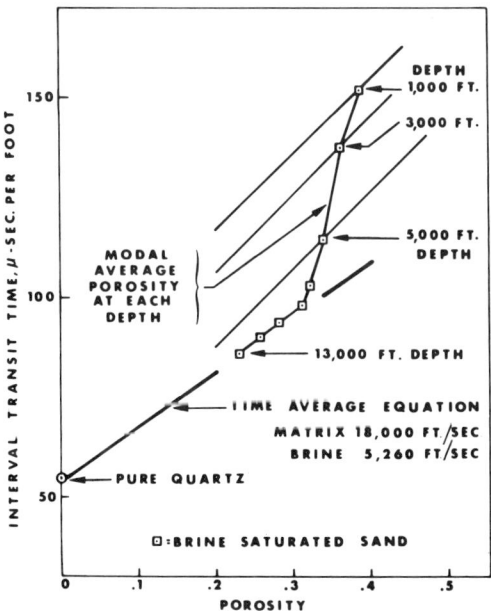

FIG. 7. P-wave transit times from well logs for sands in young sedimentary basins.

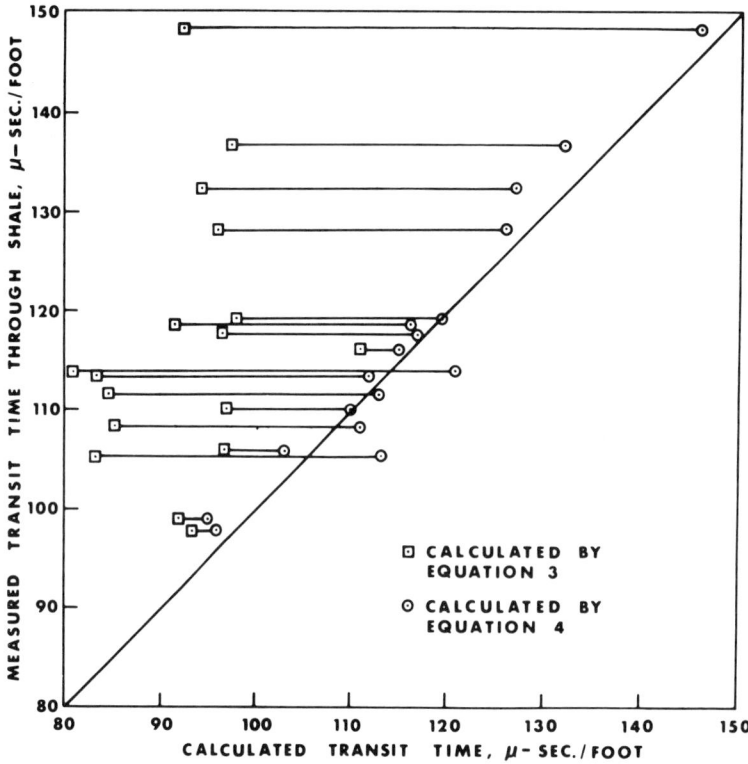

FIG. 8. Transit time relations for high-pressure shales [data from Hottmann and Johnson (1965)].

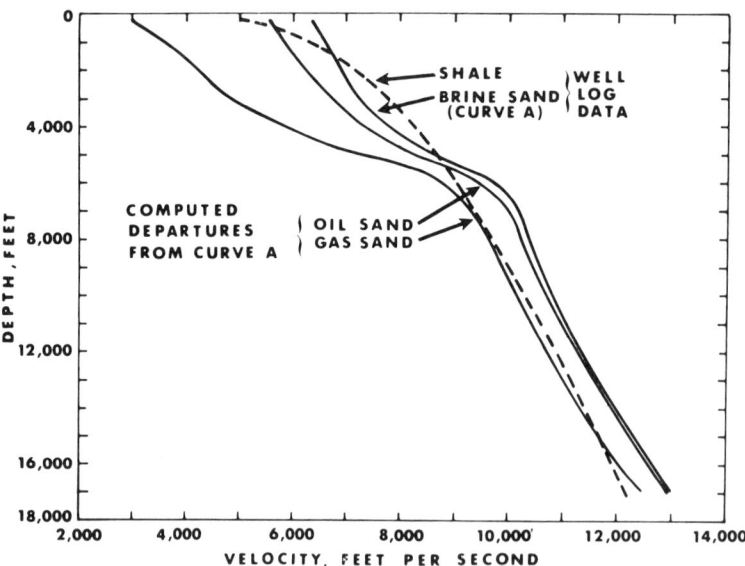

FIG. 9. Velocity versus depth for shale and for in-situ sands containing different fluids.

REFLECTION COEFFICIENTS

Acoustic impedance contrasts govern seismic reflection coefficients at a plane interface between two media. Thus,

$$R = \frac{\rho_1 V_1 - \rho_2 V_2}{\rho_1 V_1 + \rho_2 V_2} \quad (5)$$

gives the amplitude of the reflected wave when the incident wave has unit amplitude and is perpendicular to the interface.

Peterson et al (1955) showed that for practical purposes this formula can be approximated by

$$R = \tfrac{1}{2} \ln (\rho_1 V_1 / \rho_2 V_2). \quad (6)$$

The empirical relationship between density and velocity depicted in Figure 1,

$$\rho = .23 V^{.25}, \quad (7)$$

is a fair average for a large number of laboratory and field observations of different brine-saturated rock types (excluding evaporites). Combining this with Peterson's relation, we have

$$R = \tfrac{1}{2}(1.25 \ln V_1/V_2). \quad (8)$$

In a general way then, density commonly varies with velocity so that its effects upon reflection coefficients is fairly satisfactorily taken into account by multiplying the reflection coefficient due to velocity contrast by a factor 1.25. Departures from this rule exist as evidenced by scatter of observation points and may in some cases be significant. The multiplying factor 1.25 also increases the relative amplitudes of multiple reflections over that estimated using velocity contrasts alone.

CONCLUSIONS

We have attempted to show that the P-wave velocity in the upper layers of the earth (depths less than 25,000 ft) varies systematically with different factors, although an absolute prediction of velocity is seldom possible. The following summary of the effects appears to be generally valid.

(1) Gassmann's theory is typically valid for sedimentary rocks in interrelating elastic constants, densities, and P-wave velocities for different rock components and for the consolidated whole rock. An important component, however, is the frame or skeleton, which may have a wide range of elastic parameters. These parameters ordinarily can be characterized only through Gassmann's theory, and this fact limits the usefulness of the theory for practical applications.

(2) Microcracks can be present in a rock (within the rock skeleton) and materially reduce the P-wave velocity of the whole rock. Pressure can close them and cause the velocity to increase. The elastic parameters of rocks without microcracks can be estimated by using the theories of Voigt and Reuss and the known elastic constants of the crystals.

(3) The well known time-average relationship empirically relates velocity and porosity for a moderately wide range of rock types and formation fluids when the rock is under a substantial pressure.

(4) The effective pressure governing the elastic properties of the skeleton of porous sedimentary rocks is the difference between the total external pressure (or overburden pressure) and the internal fluid pressure. Increase in the skeleton pressure increases the elastic reactions at intergranular interfaces and the velocity of the whole rock.

(5) It is shown that in a recent basin, the increase in velocity of sands with depth is substantially greater than the increase in velocity of a sample of comparable sand subjected to pressure in the laboratory. The difference is mainly attributed to in-situ cementation of sand grains with geologic time.

(6) For high-pressure shales, the skeleton pressure is markedly subnormal and may approach zero. The associated velocities are also greatly reduced but have values consistent with the existing skeleton pressure.

(7) For unconsolidated shallow gas sands, some computations using Gassmann's theory indicate a substantial difference in P-wave velocity from that of the same sand filled with brine. Some possibility of recognizing this effect in seismic reflections exists.

(8) A simple systematic relationship exists between the velocity and density of many sedimentary rocks in situ. For these rocks the empirical relationship permits estimation of reflection coefficients from velocity information alone.

REFERENCES

Biot, M. A., 1956, Theory of propagation of elastic waves in fluid-saturated porous solid: Parts I and II: J. Acoust. Soc. Am. v. 28, p. 168–191.

Clark, Sydney P., Editor, 1966, Handbook of physical constants: GSA Memoir 97.

Faust, L. Y., 1953, A velocity function including lithologic variation: Geophysics, v. 18, p. 271–288.

Gassmann, F., 1951, Ueber die Elastizität poröser Medien: Vierteljahrsschrift der Naturforschenden Ges., Zürich, v. 96, p. 1–23.

Hill, R. W., 1952, The elastic behavior of a crystalline aggregate: Proc. Phys. Soc. (London), A65, p. 349–354.

Hottmann, C. E., and Johnson, R. K., 1965, Estimation of formation pressures from log-derived shale properties: J. Petr. Tech., v. 17, p. 717–722.

Ide, John M., 1937, The velocity of sound in rocks and glasses as a function of temperature: J. Geology, v. 45, p. 689–716.

Maxwell, John C., 1960, Experiments on compaction and cementation of sand: GSA Memoir 79, p. 105–132.

Peterson, R. A., Fillippone, W. R., and Coker, F. B., 1955, The synthesis of seismograms from well log data: Geophysics, v. 20, p. 516–538.

Reuss, A., 1929, Berechnung der Fleissgrenze von Mischkristallen auf Grund der Plästizitatsbedingung für Einkristalle: Z. Angew. Math. Mech., v. 9, p. 49, 58.

Voigt, W., 1928, Lehrbuch der Kristallphysik: Leipzig, B. G. Teubner.

White, J. E., 1965, Seismic waves: Radiation, transmission and attenuation: New York, McGraw-Hill Book Co., Inc., p. 132.

GEOPHYSICS

Volume 39 — December 1974 — Number 6

EFFECT OF WATER SATURATION ON SEISMIC REFLECTIVITY OF SAND RESERVOIRS ENCASED IN SHALE

S. N. DOMENICO*

Recent discoveries of the correspondence between seismic reflectivity and the presence of hydrocarbons in reservoirs have had a profound effect on petroleum exploration. This correspondence appears most pronounced in poorly consolidated sand reservoirs encased in shale. Reflectivity, defined as the absolute value of the reflection coefficient at the shale/sand-reservoir interface, is determined by the product of the longitudinal velocity and bulk density of the shale and, separately, of the sand reservoir. The formula for velocity taken from Geertsma shows that the square of the velocity in a sand reservoir varies inversely with bulk density and also inversely with fluid compressibility. As water saturation (fractional value of pore space occupied by water) increases in an oil sand, decrease in velocity caused by increasing bulk density is more than compensated by increase in velocity due to decreasing fluid compressibility. On the contrary, in a gas sand, the decrease in velocity due to increasing bulk density with increasing water saturation is not completely compensated by the increase in velocity due to decreasing fluid compressibility.

Appropriate values for shale density and velocity, for reservoir porosity, and for reservoir constituent densities and compressibilities are used to determine the reflectivity of a shale/oil-sand and, separately, a shale/gas-sand interface as a function of water saturation at depths of 2000, 6000, and 10,000 ft. At each depth, the reflectivity of a shale/oil-sand interface decreases moderately with increasing water saturation; whereas, the reflectivity of a shale/gas-sand interface decreases moderately from a completely gas-saturated reservoir to a water saturation of approximately 0.95, after which the reflectivity decreases abruptly and appreciably to the reflectivity of a completely water-saturated sand reservoir. Thus, a small quantity of gas (5 percent or less) increases reflection amplitude significantly, and reflection amplitude is not a simple linear measure of the amount of gas in the reservoir.

INTRODUCTION

Recent discoveries by explorationists of the correspondence between seismic reflection amplitudes and the presence of hydrocarbons in reservoirs have had a dramatic effect on petroleum exploration. Thus far, it appears that this correspondence is most pronounced in poorly consolidated sand reservoirs encased in shale common to Tertiary sedimentary basins consisting of rapidly deposited sequences of shale and sand layers. Examples are the U.S. Gulf Coast Basin, the basin extending offshore from eastern Trinidad, the South Caspian Sea Basin, the offshore western New Zealand Taranki Basin, the offshore north-

---

Paper presented at the 43rd Annual International SEG Meeting, October 22, Mexico City. Manuscript received by the Editor March 27, 1974.
* Amoco Production Co., Tulsa, Okla. 74102.
© 1974 Society of Exploration Geophysicists. All rights reserved.

west Australia Sahul Shelf, the extensive basin between Australia and New Guinea, and the myriad of eastern Asia basins from offshore Japan to Sumatra.

The purpose here is to examine the effect of water saturation in an oil sand and, separately, in a gas sand on the reflectivity for longitudinal seismic waves of the interface between the reservoir and the overlying shale layer. We define reflectivity R as the absolute value of the reflection coefficient, or ratio of the amplitude of the reflected wave A_r to the amplitude of the incident wave A_i, that is,

$$R = \left|\frac{A_r}{A_i}\right| = \left|\frac{\rho_b' V' - \rho_b V}{\rho_b' V' + \rho_b V}\right|, \quad (1)$$

where ρ_b is the bulk density of the sand reservoir, ρ_b' is the bulk density of the overlying shale, V is the longitudinal velocity in the sand reservoir, and V' is the longitudinal velocity in the overlying shale.

Equation (1) is valid only for plane wavefronts parallel to a plane interface. The reflection coefficient does vary with the angle of incidence of the wave at the reflecting interface, the variation depending upon the velocities and densities of the two layers (Knott, 1899) as well as their Poisson ratios (Koefoed, 1955). We consider here only normal incidence for which equation (1) is applicable.

In the sequel we shall use for the numerical values of bulk density ρ_b' and longitudinal velocity V' of the shale average values obtained from well logs. The bulk density of the sand reservoir ρ_b is simply the weighted-by-volume average of the constituent densities given by the equation

$$\rho_b = \phi S_w \rho_w + \phi(1 - S_w)\rho + (1 - \phi)\rho_s, \quad (2)$$

where ϕ is the fractional porosity of the sand, S_w is the fractional volume of the pores filled with water, ρ_w is water density, ρ_s is sand grain density, and ρ is gas density ρ_g or oil density ρ_0.

Longitudinal velocity of elastic waves in porous media is treated extensively in the literature. Unfortunately, laboratory measurements on natural rocks usually include only consolidated specimens; unconsolidated rock is often simulated by assemblages of spherical grains (e.g., alundum). Perhaps the most widely used expression for the velocity in porous media is that proposed by Wyllie et al (1956) called the "time average" equation, given by

$$V = \frac{\phi}{V_f} + \frac{1 - \phi}{V_s}, \quad (3)$$

where V_f is the velocity of the pore fluid, and V_s is the longitudinal velocity of the matrix solid material.

Equation (2) has proven useful under certain restricted conditions; specifically, it should be applied only for (a) sandstones subjected to high differential pressure[1] and (b) completely liquid fluids (i.e., no gas mixtures). Laboratory measurements have shown that velocity increases with increasing differential pressure, asymptotically approaching a terminal value (Wyllie et al, 1958). Near this terminal value of velocity, equation (2) provides a reasonable fit to experimental data and, thus, is used widely to estimate porosities of liquid-saturated sandstones at appreciable depths from observed acoustic well log velocities.

Equation (2) does not account for depth dependent properties such as fluid density and compressibility, compressibility of the pore space, and differential pressure. Gassman (1951) derived a theoretical expression for the longitudinal velocity in a hexagonal close packing of equal spheres with the pore space empty or filled with a fluid (liquid or gas). The packing at any depth is stressed by the weight of its overlying mass, and the elastic wave is assumed to produce variations from the initial state of stress and strain which are sufficiently small to allow application of Hooke's law. The medium is anisotropic with the vertical velocity greater than the horizontal velocity; both velocities increase with depth (i.e., differential pressure). Similarly, Brandt (1955) derived the longitudinal velocity in a model consisting of an aggregate of randomly stacked spheres of four different sizes. He compared theoretical values with experimental values from various sandstones, both dry and water-saturated. Again the velocity increases with differential pressure.

Perhaps the most comprehensive theoretical treatment of elastic wave propagation in a fluid-saturated porous solid is that by Biot (1956). He accounts for the viscosity of the fluid and assumes that the fluid is compressible and may flow rela-

[1] Differential pressure is the difference between the external (or overburden) pressure and internal fluid pressure (equal to the fluid hydrostatic pressure in normally pressured reservoirs).

tive to the solid, causing frictional forces. The theory is developed in two parts, namely, that which applies below and that applicable above a certain frequency which depends upon the kinematic viscosity of the fluid and the size of the pores. From Biot's theory, Geertsma (1961) developed basic equations for longitudinal velocity as a function of frequency, as well as for zero and infinite frequencies. For our purposes, we have applied Geertsma's equation for zero frequency, since the zero-frequency equation is applicable when wavelengths are much greater than pore dimensions, as is the case in the seismic reflection method of petroleum exploration. The expression appears also to agree with the variation in velocity due to fluid type and allows for variation with differential pressure. The longitudinal velocity V at zero frequency in a fluid-filled porous solid is given by Geertsma as

$$V^2 = \left[\left(\frac{\beta}{c_s} + \frac{4}{3} G_b\right) + \frac{(1-\beta)^2}{(1-\phi-\beta)c_s + \phi c_f}\right] \frac{1}{\rho_b}, \quad (4)$$

where c_s is the compressibility of the reservoir matrix material, c_b is the compressibility of the empty reservoir bulk material, β is the ratio c_s/c_b, G_b is the shear modulus of the reservoir bulk material, and c_f is the compressibility of the fluid.

We shall use equation (4) to compute the velocity V and equation (2) to compute the bulk density ρ_b of a gas- or oil-bearing unconsolidated sand reservoir as a function of water saturation S_w. These values together with average velocity and density values for shale will be used to derive the reflectivity R, given by equation (1), as a function of water saturation S_w.

ASSUMPTIONS

To illustrate the effect of water saturation in an unconsolidated sand reservoir on reflectivity of the shale/sand-reservoir interface, we obtained typical physical characteristics of the sand and shale at three depths, namely, 2000, 6000, and 10,000 ft. The bulk density and velocity for shale were obtained by averaging values observed on well logs. For this investigation the shale density and velocity are held constant at a given depth as the sand reservoir bulk density and velocity are varied by replacing oil or, separately, gas in the pore space with water.

We assume that the sand reservoir matrix consists only of quartz grains (i.e., the sand is "clean"), that pore sizes are randomly distributed, and that each pore contains the same proportions of oil and water or gas and water; that is, the fluid constituents are evenly distributed throughout the sand reservoir. Reservoir porosity which decreases with depth is typical of young Tertiary unconsolidated sand layers, and we assumed such a decrease in our models.

The gas in our calculations is assumed to be methane, and the oil is assumed to have an API gravity of 35 degrees and contain gas of gravity 0.60. The water contains 100,000 ppm of dissolved salt and produces a hydrostatic pressure gradient of 0.47 psi/ft. The pressure gradient due to the overburden (column of rock and contained fluid above the shale/sand-reservoir interface) is the commonly used 1.00 psi/ft. Thus, the differential pressure gradient affecting the sand reservoir is 0.53 psi/ft. Finally, the temperature is assumed to increase from a surface ambient temperature of 75°F at a rate of 0.01°F/ft.

SAND RESERVOIR DENSITY

For the assumed differential pressure, temperature, and fluid composition, the density of water ρ_w was obtained from a graph developed by Keenan and Keyes (1936) giving the variation in specific volume of water with temperature and pressure; the density of oil ρ_0 was derived from appropriate graphs published by Standing (1952); and the density of gas ρ_g was determined from a methane compressibility graph provided by the Natural Gas Processors Suppliers Association (1966). The density of the sand grains (quartz) ρ_s was assumed constant. Densities of the sand reservoir constituents for the assumed porosity, temperature, and differential pressure at each of the three depths are given in Table 1. Variation

Table 1. Sand reservoir characteristics

	Depth (ft)		
	2000	6000	10,000
Porosity (ϕ)	0.39	0.33	0.26
Temperature (°F)	95	135	175
Differential pressure (psi)	1060	3180	5300
Densities (gms/cm³)			
Sand grain (ρ_s)	2.650	2.650	2.650
Water (ρ_w)	1.097	1.089	1.083
Oil (ρ_0)	0.755	0.749	0.742
Gas (ρ_g)	0.023	0.103	0.156

of sand reservoir bulk density ρ_b with water saturation S_w is obtained by differentiation of equation (2);

$$d\rho_b/dS_w = \phi(\rho_w - \rho). \quad (5)$$

Thus, the bulk density increases linearly with the difference between water density ρ_w and oil density $\rho = \rho_0$ or the difference between water density and gas density $\rho = \rho_g$. (Porosity ϕ is constant at a given depth.) This difference is significantly greater for a gas sand than for an oil sand, since gas density is appreciably less than oil density (Table 1). The rate of change for a gas sand is 3.1, 2.9, and 2.7 times the rate of change for an oil sand at the depths of 2000, 6000, and 10,000 ft, respectively. Bulk densities of the oil sand and of the gas sand versus water saturation, at each of the three depths, are graphed in Figure 1.

SAND RESERVOIR VELOCITY

At a given depth, all of the parameters in the velocity function for the sand reservoir, equation (4), except the fluid compressibility c_f and the bulk density ρ_b remain constant as the water saturation S_w is varied. In addition to the bulk density ρ_b, evaluated in the preceding section, parameters which must be evaluated to determine the sand reservoir velocity at each of the three depths are the compressibilities of all constituents and the bulk shear modulus G_b, given by Geertsma (1961) as

$$G_b = \frac{3(1 - 2\sigma)}{2c_b(1 + \sigma)}, \quad (6)$$

where σ is Poisson's ratio and c_b is the previously defined bulk compressibility. The latter is given by

$$c_b = \phi c_p + c_s, \quad (7)$$

where c_p is the pore volume compressibility. Fluid compressibility c_f is the weighted-by-volume average of the constituent compressibilities given by

$$c_f = S_w c_w + (1 - S_w)c, \quad (8)$$

where c_w is the water compressibility and c is the oil compressibility c_0, in the case of the oil sand, and the gas compressibility c_g, in the case of the gas sand.

Compressibility of the reservoir matrix material c_s is assumed to be that of quartz and is constant. Water compressibility c_w was obtained from a graph published by Dodson and Standing (1944) giving water compressibility as a function of temperature and pressure. Oil compressibility c_0 was derived from graphs published by Standing (1952). Gas compressibility c_g was obtained from the methane compressibility graph provided by the Natural Gas Processors Suppliers' Association (1966) giving methane compressibility as a function of temperature and pressure. Of all the compressibilities required in the determination of velocity in an unconsolidated sand reservoir, pore-volume compressibility (i.e., the fractional change in volume of the pore space with increasing differential pressure) is the least known. For this study, pore-volume compressibility values were obtained from laboratory measurements of pore-volume compressibility versus differential pressure of sandstones reported by van der Knaap (1959). The values used here, listed in Table 2, are consistent with the upper limit of compressibilities given by van der Knaap, corresponding to semi-consolidated sandstones.

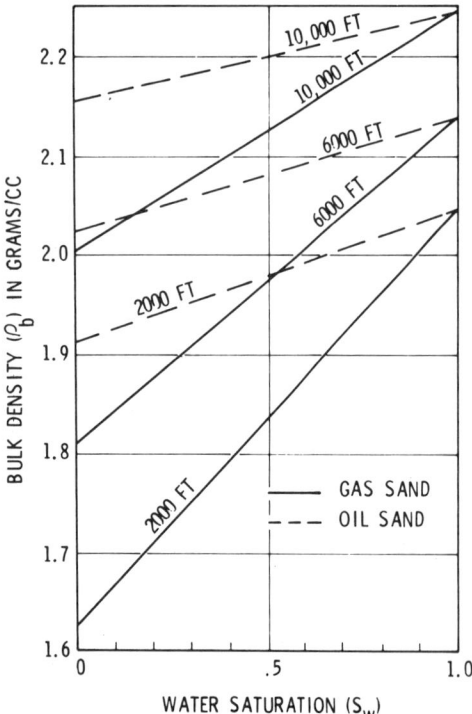

FIG. 1. Bulk density ρ_b as a function of water saturation S_w for gas and oil sands at depths of 2000, 6000, and 10,000 ft.

Table 2. Sand reservoir Poisson's ratio and constituent compressibilities

	Depth (ft)		
	2000	6000	10,000
Poisson's ratio (σ)	0.39	0.24	0.24
Compressibilities (ft^2/poundal) *			
Matrix material (c_s)	$3.72 \cdot 10^{-11}$	$3.72 \cdot 10^{-11}$	$3.72 \cdot 10^{-11}$
Water (c_w)	$6.82 \cdot 10^{-10}$	$6.25 \cdot 10^{-10}$	$6.18 \cdot 10^{-10}$
Oil (c_o)	$1.51 \cdot 10^{-9}$	$2.22 \cdot 10^{-9}$	$3.09 \cdot 10^{-9}$
Gas (c_g)	$2.37 \cdot 10^{-7}$	$7.14 \cdot 10^{-8}$	$3.89 \cdot 10^{-8}$
Pore-volume (c_p)	$1.64 \cdot 10^{-9}$	$0.82 \cdot 10^{-9}$	$0.58 \cdot 10^{-9}$
Bulk (c_b)	$6.82 \cdot 10^{-10}$	$3.07 \cdot 10^{-10}$	$1.90 \cdot 10^{-10}$

* Compressibility dimensions are the inverse of pressure dimensions. One ft^2/poundal = $6.72 \ 10^{-2}$ cm^2/dyne or to $0.463 \ 10^{-4}$ psi^{-1}.

Perhaps the most difficult parameter to establish for an unconsolidated sand reservoir is Poisson's ratio σ. This ratio can vary from a value slightly less than 0.5 (corresponding to a fluid medium of zero rigidity) for a highly porous sand reservoir at a shallow depth to a value near 0.1 at depths exceeding 10,000 ft. Values used here are those required to obtain velocities from equation (4) equal to the average of water-sand velocities observed on velocity well logs. Thus, Poisson's ratio was used as an adjustment factor after all other parameters of equation (4) (densities, compressibilities, and porosities) were established as described above. It is gratifying that Poisson's ratios given in Table 2 for the three depths appear reasonable.

Variation of fluid compressibility c_f with water saturation S_w is obtained by differentiation of equation (8);

$$dc_f/dS_w = -(c - c_w). \qquad (9)$$

Thus, the fluid compressibility decreases linearly at a rate given by the difference between water compressibility c_w and oil compressibility $c = c_0$ or, separately, gas compressibility $c = c_g$. From Table 2 it is noted that gas compressibility is from two orders of magnitude (2000-ft depth) to one order of magnitude (10,000-ft depth) greater than oil compressibility, which in turn is from approximately two (2000-ft depth) to five (10,000-ft depth) times water compressibility. Thus, replacement of gas with water will affect sand reservoir velocity to a much greater degree than will replacement of oil with water. The rate of change of fluid compressibility, given by equation (9), for a gas sand is 229, 43, and 16 times the rate of change for an oil sand at the depths of 2000, 6000, and 10,000 ft, respectively. Fluid compressibility versus water saturation for the oil sand and the gas sand at each of the three depths is graphed in Figure 2. Fluid compressibility is plotted on a logarithmic scale in order to display all curves on one graph. On a linear scale, the fluid compressibility curves would be straight lines.

Having obtained values for all parameters in equation (4), we now may determine the velocity of the oil sand and of the gas sand as a function of water saturation at each of the three depths. Results are given in Figure 3. At each of the three

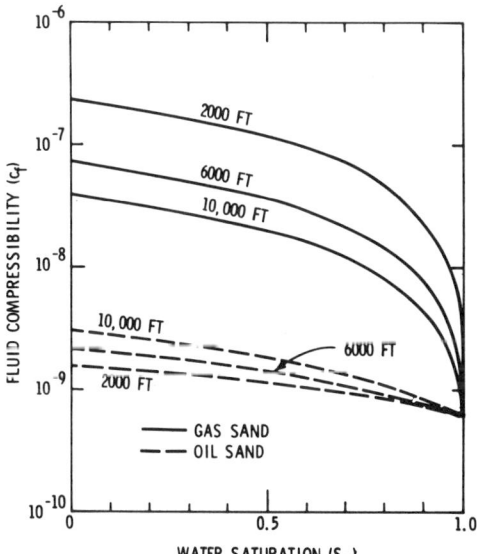

FIG. 2. Variation of compressibility c_f with water saturation S_w for gas and oil sands at depths of 2000, 6000, and 10,000 ft.

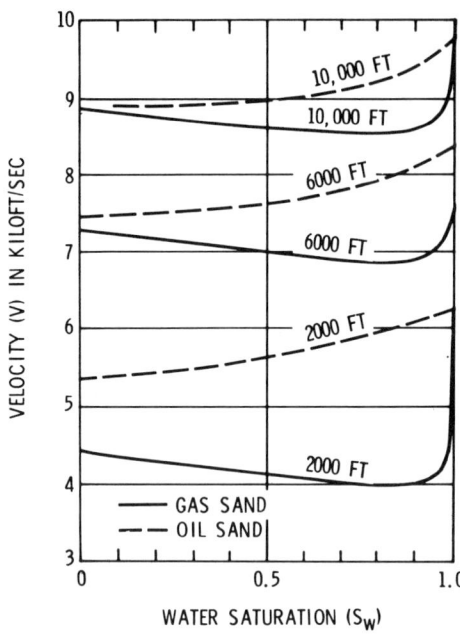

FIG. 3. Longitudinal velocity V as a function of water saturation S_w for gas and oil sands at depths of 2000, 6000, and 10,000 ft.

depths the oil-sand velocity increases monotonically from zero water saturation ($S_w=0$) to complete water saturation ($S_w=1.0$), the total change decreasing with depth. On the contrary, the gas-sand velocity *decreases* as water saturation increases from zero to a water saturation of approximately 0.8 and then increases to the velocity of a fully water-saturated sand reservoir, the largest change occurring as the last 2 or 3 percent of the pore space is occupied by water.

RELATIVE EFFECT OF BULK DENSITY AND FLUID COMPRESSIBILITY ON SAND RESERVOIR VELOCITY

The effect of c_f and ρ_b on the longitudinal velocity V is determined by inspection of equation (4).

$$V^2 = \left[\left(\frac{\beta}{c_s} + \frac{4}{3}G_b\right) + \frac{(1-\beta)^2}{(1-\phi-\beta)c_s + \phi c_f}\right]\frac{1}{\rho_b}, \quad (4)$$

where $\beta = c_s/c_b$ as defined above. The right-hand side of equation (4) may be considered a product of two functions squared; namely, a function of fluid compressibility $f(c_f)$ and a function of bulk density $f'(\rho_b)$, given respectively, by

$$f^2(c_f) = \left(\frac{\beta}{c_s} + \frac{4}{3}G_b\right) + \frac{(1-\beta)^2}{(1-\phi-\beta)c_s + \phi c_f} \quad (10)$$

and

$$f'^2(\rho_b) = \frac{1}{\rho_b}. \quad (11)$$

Then the longitudinal velocity V at any S_w is

$$V = f(c_f) \cdot f'(\rho_b), \quad (12)$$

and the longitudinal velocity V_w of a water-sand reservoir (i.e., $S_w = 1.0$) is

$$V_w = f_w(c_f) \cdot f'_w(\rho_b). \quad (13)$$

The ratio of the longitudinal velocity V for any S_w to the longitudinal velocity V_w for a water-sand reservoir is, from equations (12) and (13),

$$V/V_w = [f(c_f)/f_w(c_f)] \cdot [f'(\rho_b)/f'_w(\rho_b)] = F(c_f) \cdot F'(\rho_b). \quad (14)$$

This ratio for a gas sand is

$$V_g/V_w = F_g(c_f) \cdot F'_g(\rho_b) \quad (15)$$

and for an oil sand is

$$V_0/V_w = F_0(c_f) \cdot F'_0(\rho_b). \quad (16)$$

The functions $F_g(c_f)$ and $F'_g(\rho_b)$ for a gas sand and the functions $F_0(c_f)$ and $F'_0(\rho_b)$ for an oil sand, together with their products, are plotted against water saturation in the graph of Figure 4 for a depth of 2000 ft. Since V^2 is inversely proportional to ρ_b, the ratios F'_g and F'_0 decrease with increasing S_w to a final value of 1.0; F'_g decreases at a greater rate. The square of the longitudinal velocity V^2 also varies inversely with c_f. Since c_f decreases with increasing S_w, the ratios F_g and F_0 increase with increasing S_w. In the case of the oil sand, the *increase* in velocity due to decreasing c_f more than compensates for the *decrease* in velocity due to increasing ρ_b, as shown by the curve for the product $F_0 \cdot F'_0$. However, for a gas sand c_f does not decrease at a sufficient rate to overcome the effect of increasing ρ_b, as shown by the curve for the product $F_g \cdot F'_g$, and the net result is a decrease in velocity to a minimum value at approximately $S_w = 0.85$. Above $S_w \cong 0.93$, the velocity increases rapidly and appreciably to the velocity of a water-sand reservoir ($S_w = 1.0$). The curves for the products $F_0 \cdot F_0'$ and $F_g \cdot F_g'$, of course, are the normalized versions of the oil and gas sand velocity curves, respectively, for a depth of 2000 ft (from Figure 3).

SHALE BULK DENSITY AND VELOCITY

As mentioned above, shale bulk densities and velocities were obtained by averaging values from well logs. Values used here for the three depths chosen to illustrate the effect of water saturation on reflectivity of the shale/sand-reservoir interface are given in Table 3.

REFLECTIVITY OF THE SHALE/SAND-RESERVOIR INTERFACE

The reflectivity of the upper interface of a sand reservoir encased in shale is defined as the absolute value of the ratio of the reflected wave

Table 3. Shale bulk density (ρ_b') and velocity (V')

	Depth (ft)		
	2000	6000	10,000
Density (gm/cm³)	2.14	2.30	2.40
Velocity (ft/sec)	5850	8210	9660

amplitude A_r to the incident wave amplitude A_i [equation (1)]. Given the shale densities and velocities and having obtained the bulk density and velocity of the sand reservoir as a function of water saturation [equations (2) and (4), respectively], we may derive the reflectivity as a function of water saturation. Results for the three chosen depths are shown in Figure 5. Although the absolute value of the ratio A_r/A_i is graphed, it should be mentioned that this ratio is actually negative except for high values of S_w at a depth of 2000 ft.

In Figure 5 it can be seen that the reflectivity of the shale/oil-sand interface at each of the three depths decreases without abrupt changes as S_w

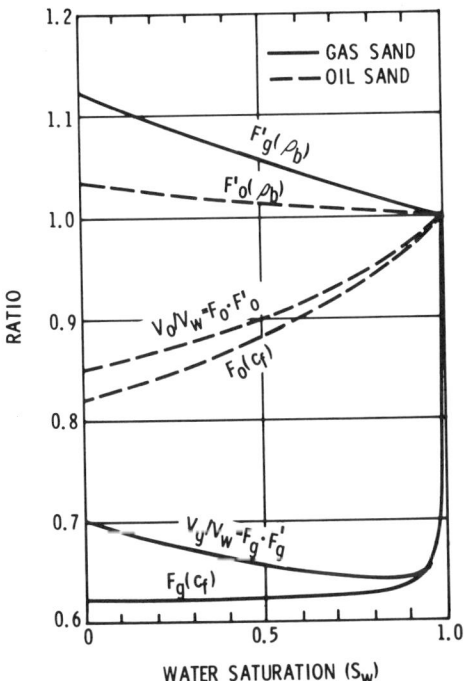

FIG. 4. Functions $F(c_f)$, $F'(\rho_b)$, and the ratio of velocity V to the velocity V_w of a fully water-saturated sand ($S_w = 1.0$) as a function of water saturation S_w for a depth of 2,000 ft. Functions for gas and oil sands are indicated by subscripts "g" and "o", respectively.

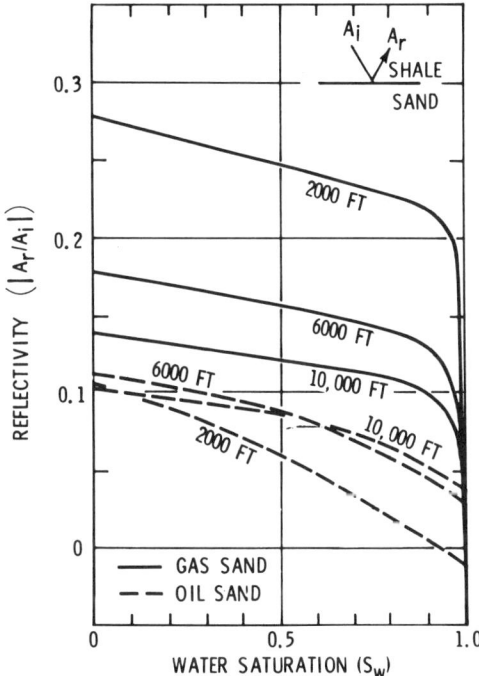

FIG. 5. Absolute value of the ratio of reflected wave amplitude A_r to incident wave amplitude A_i at shale/gas-sand and shale/oil-sand interfaces as a function of water saturation S_w for depths of 2000, 6000, and 10,000 ft.

Table 4. Variation in reflectivity of the shale/gas-sand interface with water saturation

S_w	Reflectivity					
	2000 ft	(change)	6000 ft	(change)	10,000 ft	(change)
0	.275		.177		0.136	
		(.059)		(.047)		(.038)
0.9	.216		.130		0.098	
		(.216)*		(.104)		(.067)
1.0	.012		.026		.031	

* This is the change from $S_W = 0.9$ to $S_W \simeq 0.99$ at which the reflectivity is zero.

increases. By contrast, the largest change in reflectivity of the shale/gas-sand interface occurs at water saturations above 0.9, where the reflectivity decreases abruptly to the reflectivity of a completely water-saturated sand reservoir ($S_w = 1.0$). This characteristic of the shale/gas-sand interface, of course, is attributable to the variation with water saturation of density and velocity in the gas sand as shown in Figures 1 and 3, respectively. The abruptness of the change in reflectivity of the shale/gas-sand interface is illustrated further in Table 4.

The change in reflectivity from $S_w = 0.9$ to $S_w = 1.0$ ranges from 1.8 times (at 10,000 ft) to 3.7 times (at 2000 ft) the change in reflectivity from $S_w = 0$ to $S_w = 0.9$. Thus, a small quantity of gas in the unconsolidated sand reservoir (say, 5 percent of the total pore space) increases markedly the reflectivity of the shale/gas-sand interface. It follows then that variation in reflection amplitude is not a simple linear indicator of gas quantity.

PREVIOUS INVESTIGATIONS RELATING TO THE EFFECT OF GAS ON VELOCITY

Others have reported the effect of gas in a medium on the velocity of the medium. Wood (1949) considers the variation of sound velocity in an air-water mixture with variation in fractional volume of the constituents. From the fundamental equation for velocity,

$$V = \sqrt{\frac{E}{\rho}}, \quad (17)$$

where $E =$ elasticity[2] and $\rho =$ density, he derives

[2] The elasticity E for a fluid is simply the incompressibility (ordinarily expressed as "k") or the reciprocal of the compressibility c used in this paper. For a solid $\phi = 0$, $c_s = c_b = c$, $\rho_b = \rho$, and $\beta = 1$ in our notation. The

the equation for velocity in an air-water mixture. The density ρ of such a mixture is given by

$$\rho = x\rho_1 + (1 - x)\rho_2, \quad (18)$$

where $x =$ proportion of air (equivalent to porosity ϕ), $\rho_1 =$ density of air, and $\rho_2 =$ density of water.

The elasticity E is given by

$$\frac{1}{E} = \frac{x}{E_1} + \frac{(1 - x)}{E_2}, \quad (19)$$

where $E_1 =$ elasticity of air and $E_2 =$ elasticity of water. Since E is the reciprocal of compressibility, equation (19) is simply the weighted-by-volume average of the constituent compressibilities. Substituting equations (18) and (19) into equation (17), Wood obtains

$$V = \left\{ \frac{E_1 E_2}{[xE_2 + (1-x)E_1][x\rho_1 + (1-x)\rho_2]} \right\}^{1/2}. \quad (20)$$

For air the constants in cgs units are $\rho_1 = 0.0012$ gms/cm^3, and $E_1 = 1.2 \cdot 10^6$ dynes/cm^2; for water they are $\rho_2 = 1.0$ gms/cm^3, and $E_2 = 2.25 \cdot 10^{10}$ dynes/cm^2.

Using these constants in equation (20) he obtains the graph (shown in Figure 6) of velocity versus fractional volume of air in water. We note

---

velocity V, given by equation (4), reduces to

$$V^2 = \left(\frac{1}{c} + \frac{4}{3}G_b\right)\frac{1}{\rho_b}.$$

In more common nomenclature $G_b = \mu$, the modulus of rigidity, and $1/c = k$, the incompressibility. Then for a solid,

$$V^2 = \left(k + \frac{4}{3}\mu\right)\frac{1}{\rho},$$

and for a liquid ($\mu = 0$),

$$V^2 = k/\rho.$$

that as the amount of water mixed with air increases, the velocity decreases from the velocity in air (316 m/sec) to a minimum (36 m/sec) and then increases gradually, approaching the velocity in water (1500 m/sec). The minimum velocity is in the vicinity of one-tenth air by volume in the mixture.

In a similar manner, Lester (1932) considers the effect of air in an air-earth mixture on the velocity of elastic waves in the mixture. He demonstrates that inordinately low velocities in the near-surface "weathered" layer are attributable to inclusion of air in a free state in this layer. Field measurements of near-surface seismic velocities in a Gulf Coast region near Houston, Tex., demonstrated that velocity in the upper 8 ft is appreciably less than the underlying velocity of approximately 5200 ft/sec (1580 m/sec) and approaches 550 ft/sec (168 m/sec) at the surface. Assuming that the Gulf Coast surface gumbos may be treated as fluids, he uses equation (20) to determine the variation of velocity with variation in fractional volume of the constituents in the

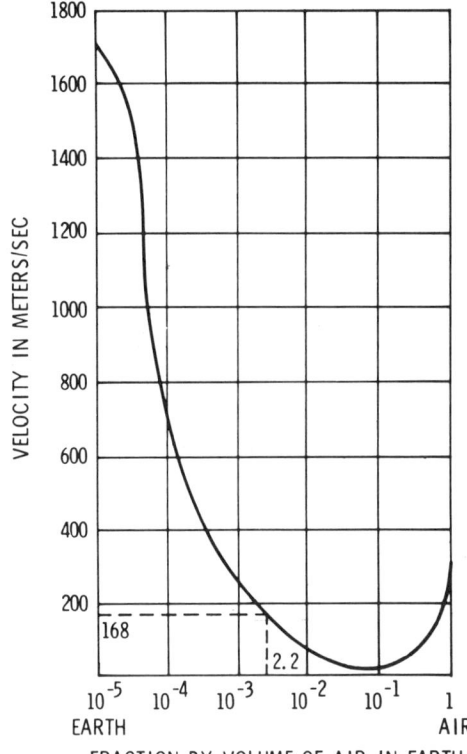

FIG. 7. Sound velocity in an air-earth mixture as a function of air volume. From Lester (1932).

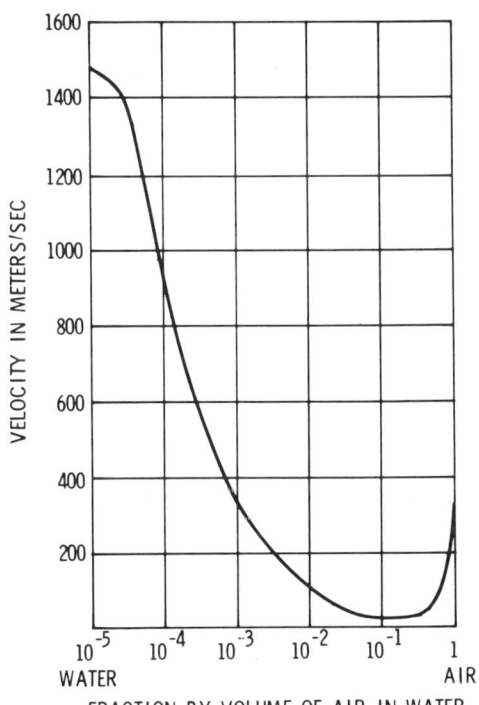

FIG. 6. Sound velocity in an air-water mixture as a function of air volume. From Wood (1949).

air-earth mixture. He uses the same constants for air (i.e., ρ_1 and E_1) as given above, and for earth he uses $\rho_2 = 1.9$ gms/cm³, $E_2 = 5.58 \cdot 10^{10}$ dynes/cm².

The corresponding graph of velocity versus fractional volume of air in the air-earth mixture is given in Figure 7. As for the air-water mixture, the velocity in the air-earth mixture decreases from the velocity of air to a minimum (26 m/sec) at a fractional volume of air equal to one-tenth, increasing gradually to the velocity in solid earth (1670 m/sec) as the fractional volume of earth increases. The observed sound velocity at the surface of the air-earth mixture (168 m/sec) corresponds to a fractional air volume of only $2.2 \cdot 10^{-3}$. Results reported by Wood (1949) and Lester (1932) show a similar variation in velocity to that derived above for gas sands with water saturation.

Although we have considered only unconsolidated sand reservoirs thus far, there is some evidence that gas in consolidated porous sandstones

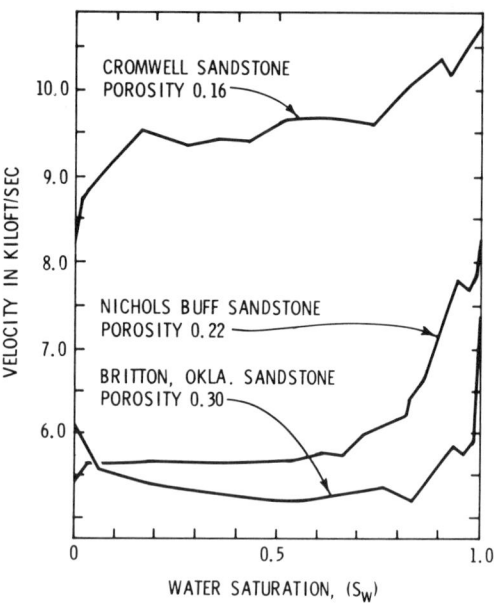

FIG. 8. Velocity in consolidated, porous sandstones with water and air saturation. From Wyllie et al (1956).

may cause longitudinal velocity to vary in a highly nonlinear manner with water saturation. Wyllie et al (1956) measured velocity in three sandstone cores with various saturations of air and water in the pore space. The graph of velocity versus water saturation S_w for each sandstone is shown in Figure 8. Velocities were measured under atmospheric pressure; there was no attempt to simulate in-situ conditions. Nonetheless, velocity in the Nicholls Buff and Britton, Okla. sandstones appears to decrease with increasing water saturation to S_w values in the vicinity of 0.7 to 0.8, thereafter abruptly increasing to the velocity in a fully water-saturated sandstone. The effect of gas (air) on the Cromwell sandstone velocity is less dramatic (possibly due in part to the lower porosity); however, the nearly horizontal portion of the curve between $S_w \cong 0.2$ and $S_w \cong 0.7$ suggests that gas is effective in suppressing an increase in velocity as water saturation increases.

CONCLUSIONS

It has been shown that the seismic reflectivity (absolute value of reflection coefficient) of unconsolidated sand reservoirs encased in shale varies with water saturation. Pertinent assumptions are that the sand reservoir matrix consists of quartz grains and that pore sizes and fluid constituents (oil and water, and gas and water) are evenly distributed. The reflectivity of a shale/oil-sand interface decreases moderately with increasing water saturation whereas the reflectivity of a shale/gas sand interface decreases moderately from a completely gas-saturated reservoir to a water saturation of approximately 0.95, after which the reflectivity decreases abruptly and appreciably to the reflectivity of a completely water-saturated sand reservoir. Thus, a small quantity of gas (say 5 percent) increases reflection amplitude significantly, and additional quantities of gas will not change the reflection amplitude significantly.

The peculiar manner in which reflectivity of a shale/gas-sand interface varies with water saturation is attributable to the variation of the sand reservoir velocity and bulk density with water saturation. Bulk density increases as water displaces oil or gas in the reservoir pore space, the rate of increase being appreciably greater as gas is displaced. The velocity in a sand reservoir decreases with increasing bulk density and increases with decreasing fluid compressibility as water saturation increases. For an oil sand the decrease in velocity attributable to increasing bulk density is more than compensated by the increase in velocity due to decreasing fluid compressibility. For a gas sand the decrease in velocity due to increasing bulk density with increasing water saturation is at a much greater rate due to the replacement of low-density gas by much higher density water. The fluid compressibility in a gas sand, although decreasing at a much greater rate with increasing water saturation than in an oil sand, does not change at a sufficient rate to overcome the decrease in velocity caused by increasing bulk density until the water saturation increases to approximately 0.8.

Others have noted the effect of gas in a medium on the velocity in the medium. Sound velocity in air-water and air-earth mixtures decreases from velocity in air to a minimum at a fractional air volume of one-tenth, increasing gradually thereafter to sound velocity in water or solid earth, respectively. Previously published experimental data indicate that as water displaces gas in the pore space, the velocity in consolidated porous sandstones may vary in a manner similar to the computed velocity in unconsolidated sand reservoirs.

ACKNOWLEDGMENTS

The author wishes to express his appreciation to Amoco Production Co. for permission to publish this material and also to staff members of Amoco's Research Department for valuable assistance. He wishes especially to acknowledge the efforts of M. E. Arnold, who compiled the appropriate sand reservoir constituent compressibilities and densities, and G. L. Rethford, who performed the necessary computations.

REFERENCES

Biot, M. A., 1956, Theory of propagation of elastic waves in a fluid-saturated porous solid: I. Low-frequency range, and II. High-frequency range: J. Acoust. Soc. Am., v. 28, p. 168–191.

Brandt, H., 1955 A, study of the speed of sound in porous granular media: J. Appl. Mech., v. 22, p. 479–486.

Dodson, C. R., and Standing, M. B., 1944, API drilling and production practice: Am. Petr. Inst., p. 173.

Gassman, F., 1951, Elastic waves through a packing of spheres: Geophysics, v. 15, p. 673–685.

Geertsma, J., 1961, Velocity-log interpretation: The effect of rock bulk compressibility: Soc. Petr. Eng. J., v. 1, p. 235–248.

Keenan, J. H., and Keyes, F. G., 1936, Thermodynamic properties of steam: New York, John Wiley and Sons, Inc.

Knott, C. G., 1899, Reflection and refraction of elastic waves with seismological applications: Phil. Mag., v. 48, p. 64–97.

Koefoed, O., 1955, On the effect of Poisson's ratios of rock strata on the reflection coefficients of plane waves: Geophys. Prosp., v. 3, p. 381–387.

Lester, O. C., 1932, Seismic weathered or aerated surface layer: Bull. AAPG, v. 16, p. 1230–1234.

Natural Gas Processors Suppliers' Association, 1966, Engineering data book: Tulsa, Okla., p. 162.

Standing, M. B., 1952, Volumetric and phase behavior of oil field hydrocarbon systems: New York, Reinhold Publishing Corp.

van der Knaap, W., 1959 Non-linear behavior of elastic porous media: Trans. AIME, v. 126, p. 179–186.

Wood, A. B., 1949, A textbook of sound: London, G. Bell and Sons, Ltd., p. 361–362.

Wyllie, M. R. J., Gregory, A. R., and Gardner, G. H. F., 1958, An experimental investigation of factors affecting elastic wave velocities in porous media Geophysics, v. 23, p. 459–493.

Wyllie, M. R. J., Gregory, A. R., and Gardner, L. W., 1956, Elastic wave velocities in heterogeneous and porous media, Geophysics, v. 21, p. 41–70.

REFERENCES FOR GENERAL READING

Belyayeva, A. I., Rayoza, O. E., and Semenova, S. A., 1968, Certain results of dynamic analysis of reflected waves in Kuleshovka oil field: Intl. Geol. Rev., v. 10, p. 1067–1072.

Brandt, H., 1960, Factors affecting compressional wave velocity in unconsolidated marine sand sediments: J. Acoust. Soc. Am., v. 32, p. 171–179.

Dickman, E., and Wierczeyko, E., 1970, A possible method for determining the extent of spread of the gas in an aquifer storage by seismic techniques: 11th Intl. Gas Conf., Moscow.

Eden, H. F., and Felsenthal, P., 1973, Elastic wave propagation in granular media: J. Acoust. Soc. Am., v. 53, p. 464–467.

Garanin, V. A., and Rogoza, O. I., 1967, Effect of certain formations with noncommercial gas shows on the nature of reflected wave recordings: Geol. Nefti Gaza, v. 11, p. 50–54.

Gardner, G. H. F., and Gregory, A. R., 1968, Formation velocity and density—The diagnostic basics of stratigraphic traps: Unpublished, presented at SEG Seminar, The Seismic Future in Stratigraphic Trap Exploration, March 22, Fort Worth, Tex.

Gerasimov, M. E., and Alekhin, S. V., 1968, Reflection of elastic waves from oil- and gas-saturated formations: Neftegaz. Geol. Geofiz., no. 5, p. 56–60.

Gibson, F. W., 1970, Measurement of the effect of air bubbles on the speed of sound in water: J. Acoust. Soc. Am., v. 48, p. 1195–1197.

Gierasimov, M. J., 1972, The influence of the existence of oil and natural gas reservoirs on the character of seismic records: Nafta, v. 28, p. 154–156.

Hicks, W. G., and Berry, J. E., 1956, Fluid saturation of rocks from velocity logs: Geophysics, v. 21, p. 739–754.

Pan, P. H., and de Bremaecker, J. C., 1970 Direct location of oil and gas by the seismic reflection method: Geophys. Prosp., v. 18, p. 712–727.

Sergeev, L. A., and Churlin, V. V., 1963, Development of the principles for direct oil and gas deposit search using seismic prospecting methods: YINITI (All Union Inst. Scient. and Tech. Information), Moscow.

Sieck, H. C., 1973, Gas-charged sediment cones post possible hazard to offshore drilling: Oil and Gas J., v. 71, p. 148–163.

Telezhenko, V. P., Konoreykin, B. A., Tsivinskoyo, Y. V., Frolova, L. A., Doroginilskoyo, L. M., and Kim, A. J., 1971, Certain features of dynamic characteristics of reflected waves in the Ust'-Balyk oil field: Intl. Geol. Rev., v. 13, p. 1218–1224.

Torvik, P. J., 1970, Note on the speed of sound in two-phase mixtures: J. Acoust. Soc. Am., v. 48, p. 432–433.

Wyllie, M. R. J., Gardner, G. H. F., and Gregory, A. R., 1962, Studies of elastic wave attenuation in porous media: Geophysics, v. 27, p. 569–589.

Zemtsov, E. E., 1962, Reflecting capability of the water-oil and water-gas contacts of some fields in the Krasnodar region: Razved. i Promysl. Geofiz., no. 46.

Seismic Acquisition

Part 1:

Whatever happened to ground roll?

Reprinted by permission of the Society of Exploration Geophysicists from *Geophysics: The Leading Edge of Exploration*, v. 5, no. 3 (March 1986), p. 40-45.

By NIGEL A. ANSTEY
Geophysical Consultant
Boston, Massachusetts

(Editor's Note: In August 1985, the SEG Workshop on Seismic Field Techniques was held in Monterey, California, under the chairmanship of Dr. Elmer Eisner. This talk was given as the keynote address; it is reproduced substantially verbatim. Part 2, Field techniques for high resolution, *follows in the next issue of* THE LEADING EDGE.*)*

Ladies and gentlemen, first let me acknowledge, quite cheerfully, that I have absolutely no business kicking off a workshop on field techniques. For although I have always been drawn to field techniques, almost all my present everyday work is as an interpreter. So, if when I talk about field techniques, I sound out of touch, out of date, out of the picture . . . well, I am. I even live in Boston.

Needing as I do an update on recent advances in field techniques, what am I hoping to hear?

I realize, of course, that the experts will be announcing solutions to problems that I didn't even know we had — and I promise to be suitably impressed by all that. But there are a few problems which I *do* know we have, and I would be overjoyed to hear that they are solved also.

You know, I'm a simple chap, and I love simple things. Things that are clever, but simple. Like the wheel, and the weed whacker, and hang gliders. For that reason, I like electric motors, but I abominate the internal combustion engine. Too damn complicated. When I stand and look at a Vibroseis vibrator, I get the same feeling — too damn complicated. So I am hoping to hear someone announce a new type of vibrator, simple enough that simple chaps can understand it, and simple enough that simple chaps can keep it working.

For that matter, I get the same feeling with marine airgun arrays. All those problems that come with using different-size guns — incomprehensible directional responses, heaven-knows-what spectrum for the source-generated noise . . . Would someone please announce the death of the air-gun array, so I don't have to make the effort any more?

What else am I hoping to hear? That someone has finally cleared up all the confusion and contradictions about what happens in the *near field of the Vibroseis vibrator* (or, rather, in the near field of an array of three or four vibrators). You remember the problem: what *is* the shape of a Vibroseis reflection? In the 1960s we used to tell interpreters that the reflection pulse on a Vibroseis record was basically a zero-phase pulse, modified only by the earth and the geophone. This was so because we locked the particle *velocity* of the baseplate to the control signal, and because our geophones also measured particle velocity, in the far field. Then we hesitated, remembering that (in a fluid, at least) the particle velocity changes phase by 90 degrees as the pulse propagates through the near field. So perhaps our reflections were actually skew-symmetric pulses, 90 degree pulses. What to do? First we got out all the old classical papers again. Then, looking at all that mathematics, we decided it might be better just to put a geophone down a hole, and measure it. But researcher A disagreed with researcher B, and B with C, and the Russians disagreed with everybody, and — unless I've missed some recent announcement — we *still* don't know what shape our Vibroseis reflections are. And that's awful; we really should know. So I am delighted to see that, according to the program, all this confusion will be completely cleared up. This very afternoon.

Anything else I'm hoping to hear? Well, you know what the hi-fi salesmen say: the strong link in the hi-fi chain is the amplifier, and the weak link is the speakers. So spend your money on the speakers. In seismic instrumentation, I suppose the strong link is again the amplifier, and the weak link is the geophone. So if I have a concern with our instruments, it's in the geophone. Oh, of course it's much better than it used to be, but does that mean that its harmonic distortion is so small that we can forget it? Under real conditions of plant, and real conditions of tilt? I hope that someone will set my mind at rest, on . . . let's see . . . tomorrow, or perhaps Friday? Good.

So those are some of the solutions which I hope to take away with me from this workshop — in addition, as I said, to the solutions of all the other problems that I didn't know I had.

But the program also talks of give and take. The operative word, Elmer tells us, is *sharing*. I can take away something, but I must also bring something.

So, I thought, what can I bring — what do I have to say that's new? Well, I do have something to say, but, oddly enough, I have no idea whether it's new or not. In part, of course, this is because I know that if I live in Boston I must be out of touch. In the other part, it is because what I have to say is too simple and too "obvious" to be new.

Let me tell you about it.

In addition to being a working interpreter, I do some teaching. Specifically, I freelance for IHRDC, in guiding the geophysics portion of their Video Library. And you know that there is nothing like teaching to make you *organize* what you know, and to make you realize (oh my goodness!) what you *don't* know. Time and again it happens. You understand the basic concepts, you apply the conclusions every day, but when you try to explain it to somebody else — something sticks, something won't come out right. So I have learned much from having to teach. Often, I find myself going back, filling in gaps in my knowledge which I should have filled (and meant to fill) many years ago.

One such thing I always meant to do was to work out what actually happens to the ground roll in the stack. Well, of course, it gets attenuated. But is this a purely random, *statistical*, $\sqrt{N}$-type of attenuation . . . or is there some systematic *order* to it?

The answer — which is my piece of sharing today — is that there is indeed order to it.

You know what happens whenever you make what you think is an announcement. Somebody jumps up and says, "I thought of that years ago." "No you didn't." "Yes I did." "No you didn't." "Yes I did." Well, there will be none of that here. What I have to say is so obvious that I cannot expect that I am the first person to notice it. Indeed, my immediate reaction, on noticing it, was one of shame — that it should have taken this old plodder so long to realize what everybody else must have known since the discovery of fire. Yet, I don't know of any publication of it. Of course, that could be my fault again — for missing it in the literature. But then, I think to myself, how little of what we actually *do* is in the literature! And conversely . . . oh researchers! how little of what is in the literature we actually *do*! Then again, this view of how ground roll gets attenuated in the stack leads to a set of criteria for multiple coverage in the field, and many modern sections which pass across my desk violate those criteria. So maybe the criteria are *not* widely known. Anyway, here goes. Whether it's ancient history, or hot off the press, here it is.

The received wisdom, as taught in textbooks and papers being published today, is that ground roll — or, in a wider sense, source-generated noise — should be attenuated by *arrays* . . . by geophone and source arrays. We are taught to make noise tests in the field, to establish the troublesome range of wavelengths in the ground roll, and to design arrays for which this range of wavelengths is down among the sidelobes. Thus traditional array design is based on, and directed at, the source-generated noise.

Wherever the source-generated noise includes long wavelengths (in particular, in ground-roll country) this traditional approach leads to *long* arrays. However, we do not like long arrays. They attenuate the first breaks, lose the high frequencies at early times and long offsets, lose the high frequencies in rough terrain, and — most serious — lose the high frequencies from dipping reflectors. Long arrays are weighty and tanglesome in the field. And they are conceptually out of tune with modern practice, in that they make a final decision — an irrevocable exclusive decision — in the field.

Where the longest noise wavelengths are associated with the lowest frequencies, we can shorten the necessary arrays by excluding these low frequencies with a low-cut filter. Then, with shorter arrays, we are more likely to preserve our *high* frequencies — on shallow events, on dipping events, and on the first breaks. In other words, we can maintain the high frequencies if we sacrifice the low frequencies. But in the modern world we do not wish to do either; the low frequencies are critically important to inversion, and to the recognition of transitional reflectors, and to polarity estimation. Therefore we seek to be rid of this need to compromise. We seek a method of obtaining extremely long arrays for the suppression of even the lowest ground-roll frequencies, while presenting short arrays to the first breaks, and the far groups, and the steep dips. Further, of course, we seek arrays which are easy to lay.

All of these objectives are satisfied by what I call the *stack-array approach* to array design. The stack-array approach *junks* the traditional approach to array design. It has no concern with the wavelengths of the ground roll, and therefore it needs no noise tests. It says: "To hell with long arrays. To hell with tangled strings. To hell with array responses, to hell with complexity, to hell with Chebyshev." (Most of all, to hell with Chebyshev.) It says: just select a sensible group *interval* on the usual considerations of having enough traces to see the geologic feature we're looking for, and of avoiding spatial aliasing — and then just set the effective array length equal to the group interval. The group *length* equal to the group *interval*. Always.

The effect of this is to provide a uniform succession of geophones along the spread — equally spaced, equally weighted, even, regular, continuous. This is our fundamental starting-point. All right, nothing unfamiliar in that.

Now we consider a field record, with first breaks, and reflections, and (in this context) ground roll (Figure 1). Normally we think of the ground roll as a wave train on a *trace* — a time function at one instant of space (Figure 2a). Equally well, we could think of it as a wave train in *space* — a snapshot of the surface at an instant of time (Figure 2b). Then from that instant of time to some later instant of time, the ground roll moves in the fashion of Figure 3.

Now let's do a terrible thing — let's just *add* all the traces of the field record. At each instant of time, we are *adding all the geophones along a line the full length of the spread*. In effect, we are forming *one long array*, perhaps a thousand or more equally-spaced geophones in one array whose length is the length of the spread. The

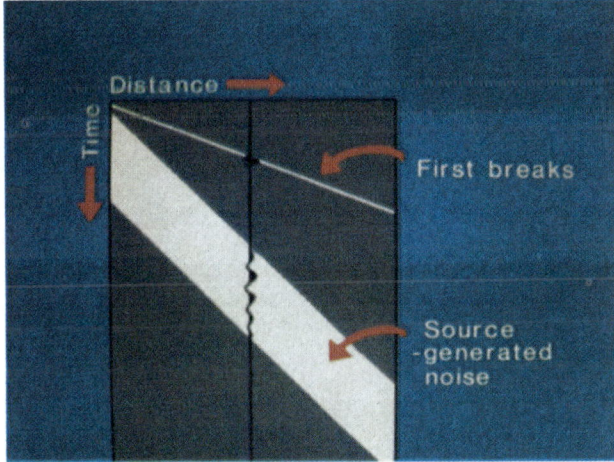

Figure 1. A rudimentary field record, with one trace shown in full. Normally we think of the waves as traces, that is, as functions of time.

array is many times longer than the longest wavelength in the ground roll, and so the ground roll is *zapped*.

Of course, the reflections are zapped also, so that won't do. We have the problem of dipping reflections, and we have the problem of normal moveout. To remove the problem of dip first, we turn (as usual) from the field record to the *gather*.

Let's start with the gather derived from an end-on spread — group length equal to the group interval — and let's set the source interval to *half* the group interval. Then, of course, from the stacking diagram — a fold of stack equal to the number of channels (Figure 4a). Six channels, six fold — six traces on the field record, six traces on the gather. Indeed if the field record looks like Figure 1, the gather looks the same. Now let's draw the arrays, the groups, on the stacking diagram (Figure 4b). We see that if the groups are even, continuous, end-to-end on the field record, then they are even, continuous, end-to-end on the gather (Figure 4c). So adding all the traces of the gather (that is, stacking) yields one continuous uniform array, having the same length as the spread (Figure 4d). Again, the ground roll is zapped.

Then the problem of normal moveout. Well, when you think about it, it really isn't much of a problem. First, because at the offsets and times where the ground roll arrives, the NMO is small anyway. And second, because

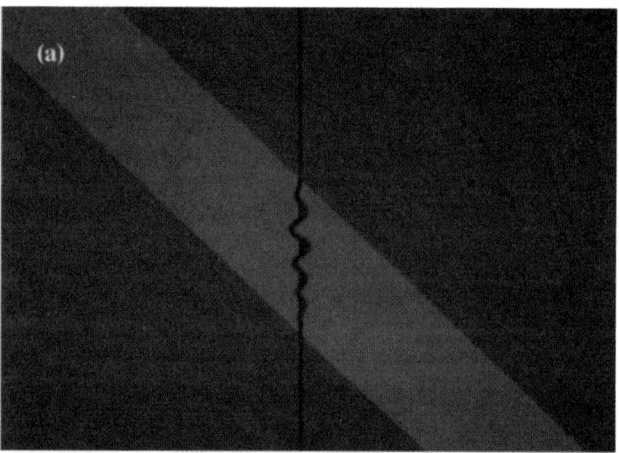

Figure 2. Ground-roll wave train as a function of time, at an instant of space (a), and as a function of space, at an instant of time (b).

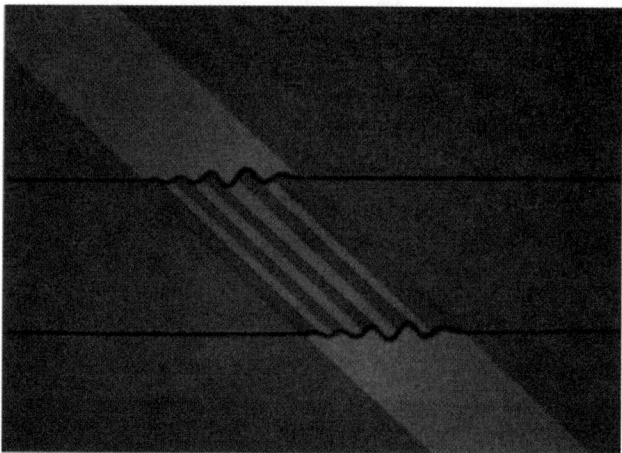

Figure 3. Ground-roll wave train as a function of space and time.

the physical array across the gather is still uniform and continuous — it's just that, when we sum along the moveout hyperbola, the sample of ground roll put into the sum jumps forward a few milliseconds, in time, between groups. At the frequencies of the ground roll, and for the smooth nature of the NMO curve, the ground roll remains substantially zapped.

Now, all that was for an end-on spread with source interval equal to half the group interval. Let's consider a more usual case for land work — a split spread, again with the group length equal to the group interval, but now with the source interval *also* equal to the group interval. Then, in this situation, it is desirable to shoot *between* the groups — not on the flags. Now the fold is only half the number of channels (Figure 5a). The groups — although still end-to-end and continuous in the field (Figure 5b) — are no longer end-to-end and continuous across the gather (Figure 5c). Indeed they are regularly *on-off-on-off* across the gather. But then if we look at the other side of the gather, we have the same effect, but now *off-on-off-on*. We can see the effect of stacking the gather if we flip the lower side to the upper side (Figure 5d); the offsets interleave to give, once again, a continuous even array stretching the whole length of *one* side of the split spread. Again, the ground roll is zapped.

So both of the spread geometries we've just considered — the end-on spread with source interval half the group interval, and the split spread shot between the groups with source interval equal to the group interval — both satisfy the basic stack-array criterion: the multiple coverage must be such that there is *an even, continuous, uniform succession of geophones across the gather*. Not across the spread — across the *gather*. When this is so, the operation of stacking forms a long uniform array the length of the spread, and the ground roll is zapped.

Let's see what happens if we violate this criterion.

Suppose, for example, that we halve the fold. In the end-on case, instead of using a source interval of half the group interval, we use a source interval *equal* to the group interval (Figure 6a). (You see lots of crews doing that.) Then the stack-array is *on-off-on-off* all across the gather, and there is no suppression — no suppression at all — of ground-roll wavelengths equal to two group intervals

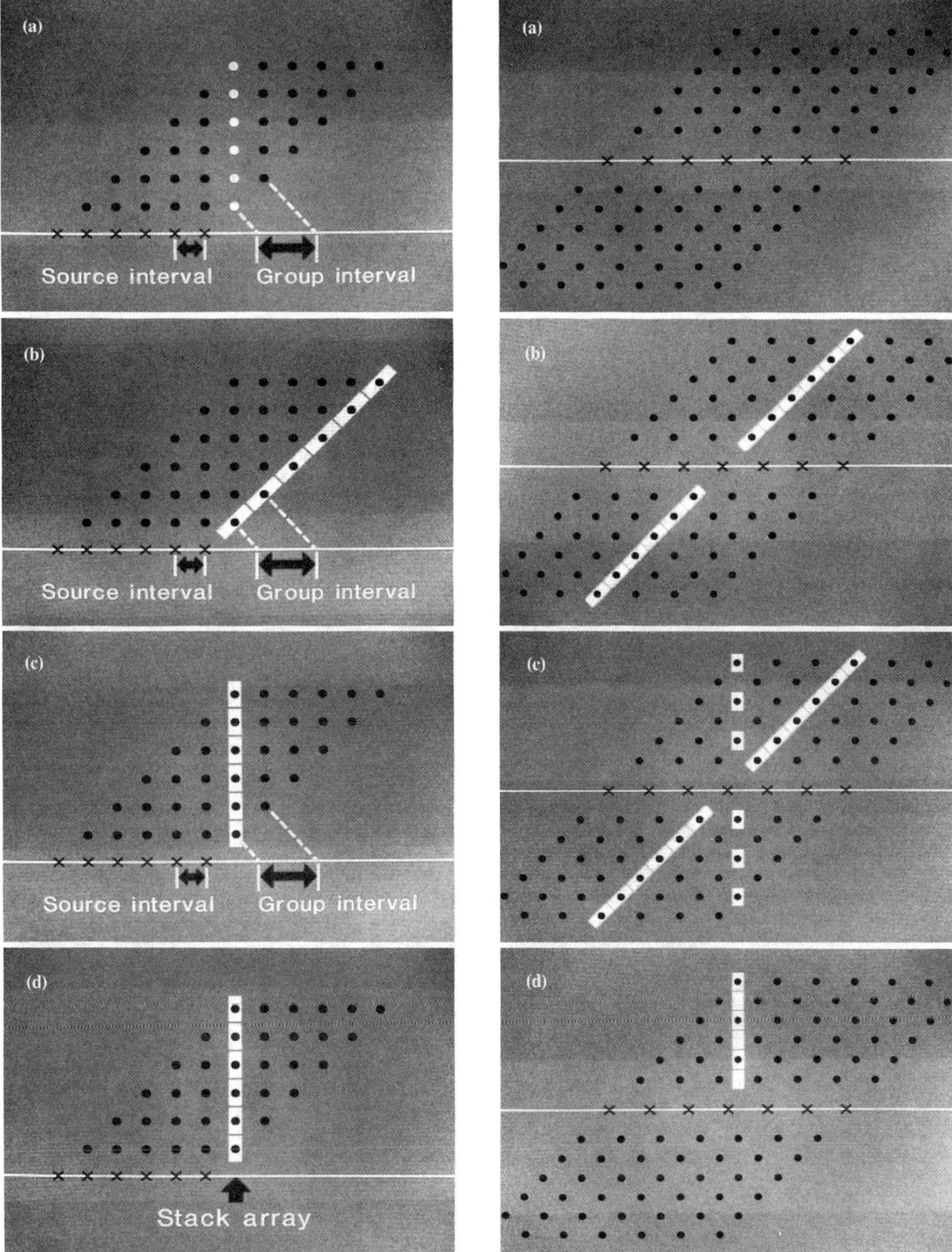

Figure 4. (a) Stacking diagram for an end-on spread. (b) End-to-end groups shown in the common-source direction. (c) End-to-end groups shown in the common-midpoint direction. (d) The addition of the gathered traces forms the stack array.

Figure 5. Counterparts of Figure 4 for a split spread shot between the groups. The groups interleave to form a continuous stack array the length of one side of the spread.

(Figure 6b). Clearly, this is a bad thing to do.

If we draw the stacking diagrams, we can quickly see that several other much-practiced techniques also violate the stack-array criterion, in particular, end-on spreads with a source interval equal to or greater than the group interval, split spreads with a source interval of double the group interval, split spreads of any type shot on the flags, and — check it for yourself — *any* dynamite shooting in which the arrays are *not* equal in length to a group interval (or an integral multiple of the group interval).

In fact, for the field-man's preference of a group length equal to one group interval, there are *only two* multiple-coverage techniques which satisfy the basic stack-array criterion. This allows us to restate the criterion as two criteria: for end-on spreads, the source interval must be *half* the group interval; for split spreads, the source interval must be *equal* to the group interval, and shot between the flags. Those are *the* two techniques.

(Incidentally, if we are prepared to *mix* traces, either in the processing or on the ground, then there are other arrangements which satisfy the basic stack-array criterion. However, in all cases these other arrangements involve some sacrifice — either of steep dips or of resolution. After this sacrifice, the results are probably no better than we would have obtained with one of these two preferred arrangements and fewer channels; we've paid good money for nothing.)

So, if lots of crews are out there right now violating these criteria, what happens? Well, perhaps there's not too much source-generated noise, and the results are fine anyway. Perhaps the noise wavelengths are short, and short groups are sufficient to attenuate it — fine again. But maybe some wavelengths of the noise come right through the stack unattenuated, and give the appearance of low-frequency multiples on the final section . . . we've all seen that. Then the processors do some tests and find the results are much better if they apply f-k filtering on the shot records, or if they apply a low-cut filter. *Nobody recognizes those spurious events as an error in the design of the multiple coverage.* Or the processors see some herringbone effects on the section (we've all seen that too), which they find can be suppressed by f-k filtering after stack. Again, nobody recognizes it as an error in the design of the multiple coverage.

OK — a few other points about the stack-array approach.

First, it really is quite distinct and different from the traditional approach to array design. In the traditional approach, the length of the arrays (the groups) is based on the *longest apparent wavelength of the source-generated noise*. In the stack-array approach, the length of the groups is just the group interval, which is *half the shortest apparent wave-*

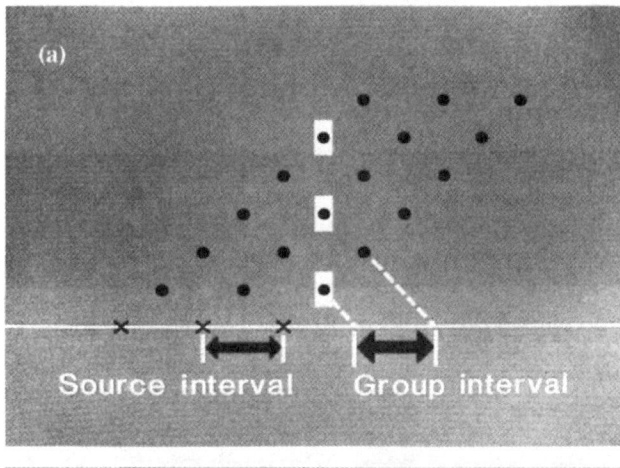

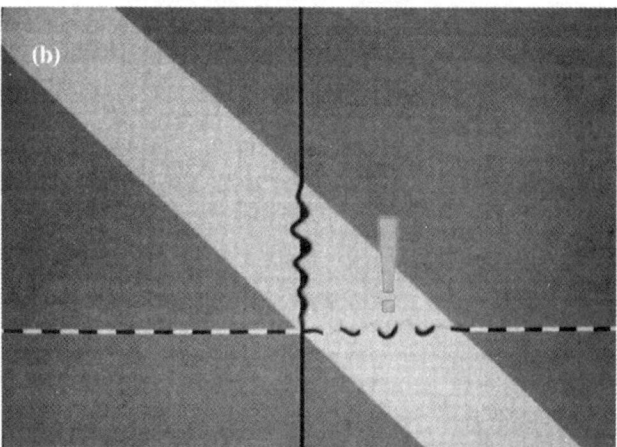

Figure 6. End-on spread with source interval equal to the group interval (a) yields no suppression of ground-roll wavelengths equal to two group intervals (b).

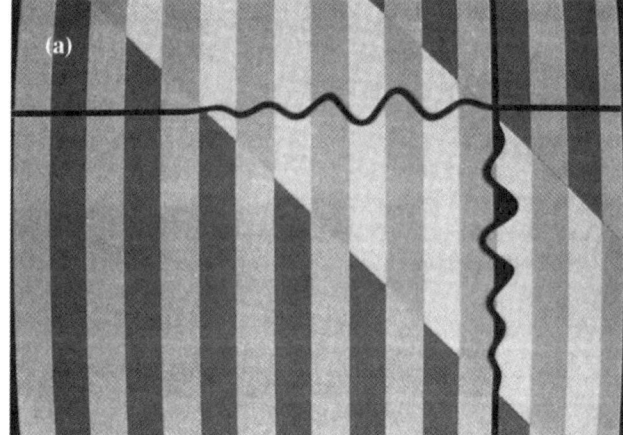

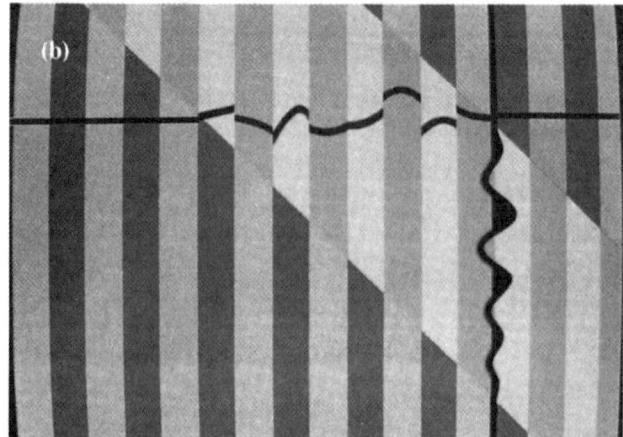

Figure 7. On a field record (a), the traces display the physically-propagating wave train across the spread. On a gather (b), differences in the ground roll from shot to shot disturb the picture of the wave train.

length of the signal. Thus the stack-array approach really does require no ground-roll tests, no noise spreads, no lost time, and no compromises between low frequencies and high frequencies.

Second, it is true that the stack array does necessitate a high fold of stack. So if this is unthinkable (for example, in portable dynamite work in the jungle), we abandon the stack-array approach, and go back to traditional arrays. But the stack-array approach is *ideal* for work with a *mobile* source, and in particular, of course, it is ideal for Vibroseis.

Third, with Vibroseis we are likely to be using both source and geophone arrays. Interestingly, the stack-array criterion (if otherwise satisfied) remains satisfied if *either the source array or the geophone array* is equal to the group interval. So, if your boss is a stickler for traditional arrays — no problem, just take the convenience of a geophone group length equal to the group interval, and then space out the vibrators to give him his traditional source array. And that source array can also be an areal array, if there is noise scattered back from the side.

Fourth, in the stack-array approach, the suppression of the ground roll does not occur until the stack. Does this mean that we take records just swamped with ground roll, and somehow muster the courage to say, "Don't worry, it'll all be systematically killed in the stack"? No, it does not. With the stack-array approach, we are totally free to raise the natural frequency of the geophones, and to use a ground-roll filter in the field, and thus to remove the ground roll from the field records. But doesn't that negate one of the advantages of the stack-array approach — that it does not require any sacrifice of the low frequencies? No again — because the response of the geophone, and of the low-cut filter, does not change the ratio — the *ratio* — of the low-frequency signal to the low-frequency noise. We can apply the low-cut filtering in the field, to get good recording and a usable monitor record . . . then zap the ground roll in the stack, and finally use a determined inverse filter to recover the low frequencies. Thus we can use rugged 14 Hz geophones, and 12 Hz or even 18 Hz ground-roll filters, and still maintain a final processed bandwidth down to 5 Hz.

Next, the stack array is as long as the spread, and many times longer than the longest wavelength in the ground roll. It contains many, many traces, but it does have an end. What happens when the ground roll gets to the end of the gather? Well, we get a few minor side-lobes. If we feel strongly about these, we can give a little less weight to the last trace or two — taper the stack at the ends. The same, too, if we have a gap in the gather on account of rivers and houses — just taper the edges.

Finally, in the context of the stack array, we have to recognize that there *is* a difference between a very long array formed across the gather and a very long array formed across the spread. Thus on a field record the traces display the actual passage of a physically propagating wave train across the spread (Figure 7a), and we can be totally confident that the addition of these traces would kill the ground roll dead. But on a gather, each trace comes from a *different source,* in a *different place;* perhaps the amplitude of the ground roll is different, or its velocity, or its frequency. In the worst case, the samples of ground roll recorded by the traces at any one time could be quite random (Figure 7b). Then what would the stack array give? It would no longer kill the ground roll dead, it would just give a statistical $\sqrt{N}$-type attenuation. Here, then, we see the range of suppression provided by the stack array. If the ground roll is substantially *constant* from shot to shot, the ground roll is zapped. If the ground roll is extremely *erratic* from shot to shot, the ground roll is reduced only by the square root of the fold of stack. With in-betweens in between.

You know, when I came into this game, the major enemy of the geophysicist — the devil himself — was called ground roll. Everywhere the question was the same: what can we do about ground roll? Then, with surface sources, the ground roll actually got *worse!* But nowadays — whoever talks about ground roll? About half the crews out there take no account of ground roll in their arrays; geophone frequencies go lower and lower, and ground-roll filters are very unfashionable. So whatever happened to ground roll?

In the heyday of fancy geophone arrays — Chebyshev and all that — a wise old man whispered to me that the best that arrays could do (however fancy) was to reduce the ground roll by 15 dB. Say to 20 percent. We agreed just now that the worst the stack array could do — if the ground roll was entirely random from shot to shot, or indeed if the stack-array criterion was violated in any way — was to reduce the ground roll by the square root of the fold of stack. At 25-fold, therefore, the worst-case stack array is about as good as traditional arrays. And that, I think, is whatever happened to ground roll. When we got to 24-fold, we found — probably quite empirically — that we didn't need long arrays any more. But what we did not notice (or at least what I did not notice) was that by walking the extra inch — designing the multiple coverage to satisfy the stack-array criterion — we could have done even better.

So there is the stack-array approach to array design. As I said earlier, I doubt very much whether I am the first to formalize the stack-array criterion. So if you got there first, 15 years ago, be my guest — feel free to claim it. But then, be prepared to be outdone by the guy who got there 25 years ago. And, of course, there's always Harry Mayne, whose thinking may have got this far 35 years ago. As always, Harry is hard to beat.

Just one other point about the stack-array approach — if your company is a sponsor of IHRDC's Video Library, you already have the details of all this in-house, in the manual for module GP305, on array design. **LE**

Nigel A. Anstey has participated in all facets of the seismic exploration method. Following his graduation from the University of Bristol in England in 1948, he worked for Seismograph Service Ltd. until 1968 when he joined Seiscom Delta, Inc. In 1975 he became an independent consultant. Anstey is recipient of many awards from the SEG, EAEG and AAPG (see TLE, January 1986, p. 26). In addition to his métier as a seismic interpreter, Anstey is a geophysical instructor and writer of international renown. He is also the guiding force behind the geophysics portion of the Video Library of the International Human Resources Development Corp. to which he refers in his article.

Part 2:
Field techniques for high resolution

By NIGEL A. ANSTEY
Geophysical Consultant
Boston, Massachusetts

(Editor's Note: In August 1985, the SEG Workshop on Seismic Field Techniques was held in Monterey, California. This talk, of which Part 1, Whatever happened to ground roll?, *appeared in the last issue of* THE LEADING EDGE, *was given as the keynote address; it is reproduced substantially verbatim.)*

Let me turn now to a topic which I, as an interpreter, am just delighted to see on the program — the *interrelation* between field work, processing and interpretation, and in particular the constraints imposed on each by the others. Specifically, I would like to address the constraints on the field work imposed by two strongly-felt needs of the interpreter today — the need for better resolution, and the need for improved delineation of the faults.

Let's formulate the resolution problem first. Figure 1a, we'll say, is the earth, and one point of acoustic contrast within it. We all know that the seismic response to such a point — which response (Figure 1b) should be a single spike on a single trace — is in fact a diffraction pattern of wiggly pulses (Figure 1c). We know that it is the function of the *decon* process to turn the wiggly pulses into spikes (Figure 1d), and the function of the *migration* process to collapse the diffraction pattern to a single trace (Figure 1b).

We also all know, to our chagrin, that these processes are imperfect — that the decon does an imperfect job, (Figure 1e), and then the migration does an imperfect job (Figure 1f) — and that therefore the seismic response to a point is *not* a single spike on a single trace. Our final resolution, finite both vertically and laterally, is defined by the *imperfections* of the deconvolution and migration processes. In this sense the problem of resolution is the problem of improving decon and migration... and this is of the first importance, in a workshop on field techniques, because that improvement must start (indeed, can only start) *in the field*. Decon can work only at those frequencies where the signal is above the noise — which is decided, once and for all, in the field. Migration imposes requirements on the layout of the lines, and on the group interval — which are decided, once and for all, in the field.

The interesting thing about the design of a field technique for a high-resolution prospect is that it is *easier* than the design of a general-purpose field technique. By the stage of a high-resolution survey, we have done the reconnaissance survey and the semidetail survey. We know what the target is, and we have fewer compromises to make. Let's see how it goes.

First let's stress that point about the line locations. The single most important factor determining the success of a 2-D high-resolution survey is the layout of the lines. From the interpretation of the last survey, we position the lines intelligently... lines positioned specifically for *valid velocities*... lines positioned specifically for *valid migration*... lines positioned specifically to detail the *faults*... lines positioned specifically to give the measure of *strike-slip* movement... lines positioned specifically for *fluid contacts*. And all of them positioned to tie at places where the reflections are good. I keep saying all this to my clients,

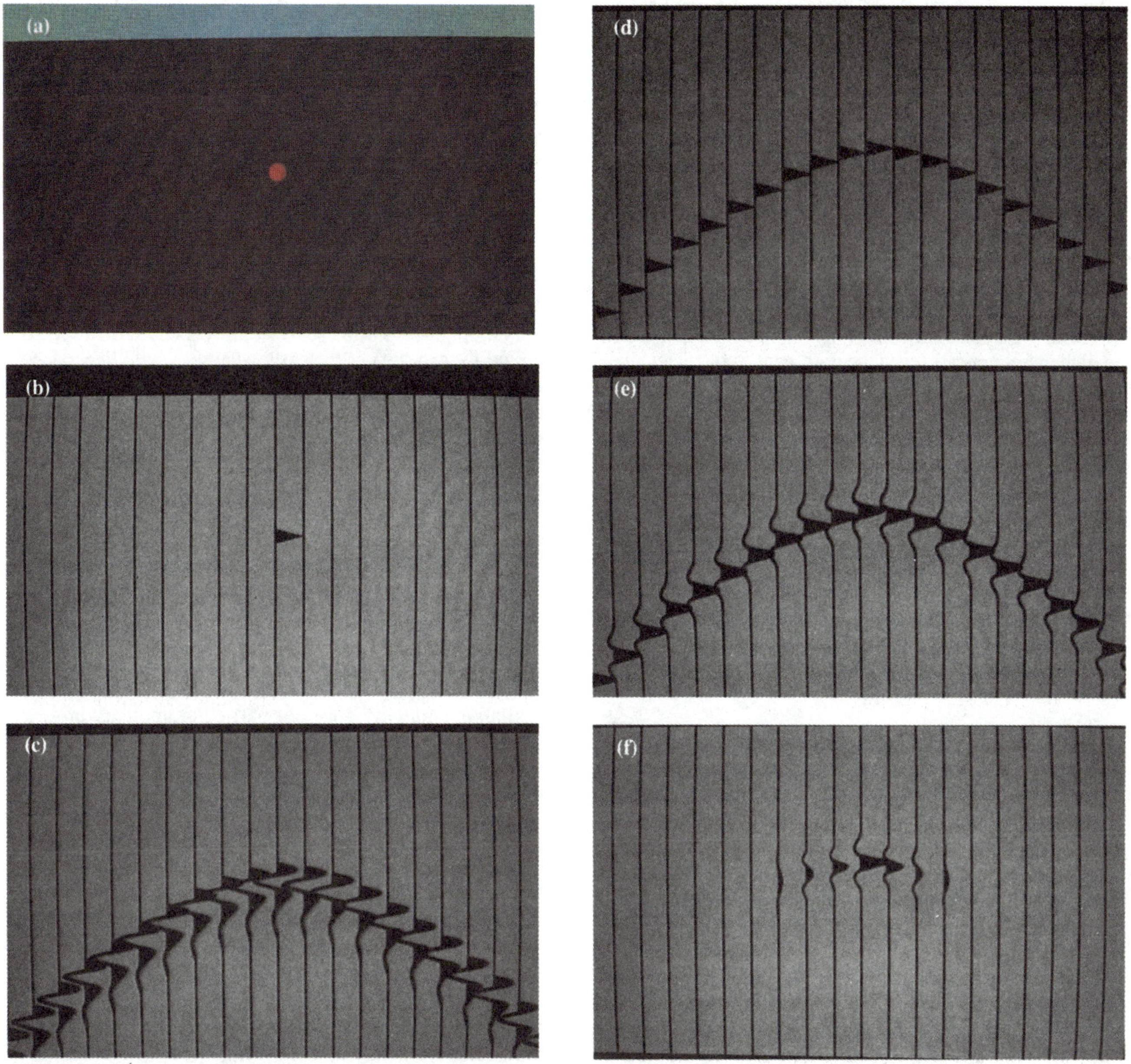

Figure 1. The seismic response to a point contrast in the earth (a) should be a single spike on a single trace (b). In fact, it is a diffraction pattern of wiggly pulses (c), which we try to spike by decon (d) and collapse by migration (to b). But the decon process is imperfect (e), and so is migration (f).

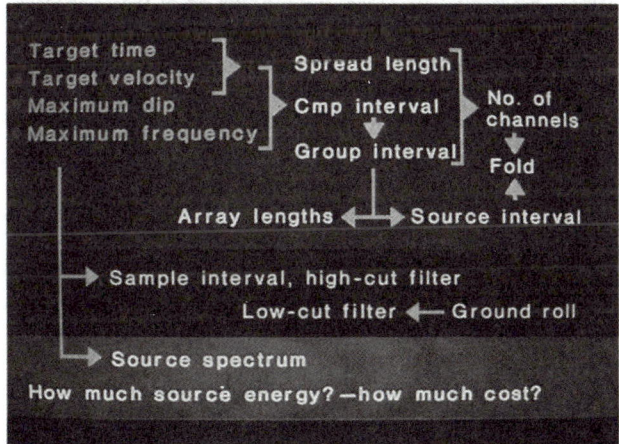

Figure 2. Design procedure for a high-resolution survey.

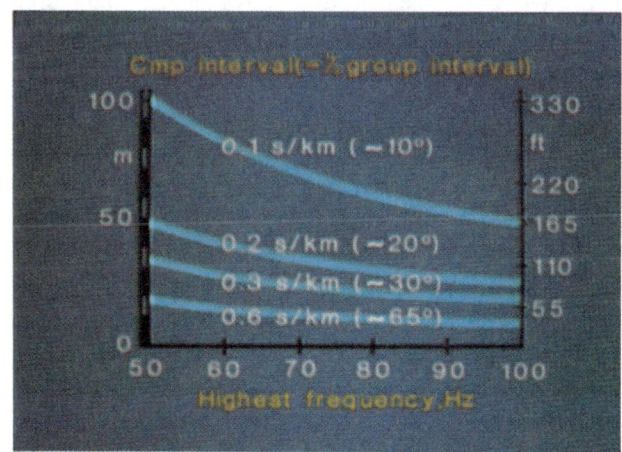

Figure 3. Typical relation between the maximum frequency and the necessary common-midpoint interval.

and they keep saying they really would prefer just to shoot north-south, east-west. Ah, well.

Anyway, with the line locations chosen, we turn to the choice of field technique. The design procedure requires *four decisions, one judgement* and *one measurement.*

First, the four decisions: target time, target velocity, maximum dip, maximum frequency.

Target time, target velocity — easy, from last year's survey.

Maximum dip. The first thing to do, obviously, is to go back to the structural maps from last year's survey, and to read off the maximum dip (in ms/km) from the contours. But we are also careful to check the *sections* — looking for any evidence of reflections at steeper dips, where the higher frequencies (even the dominant frequencies) have been lost by the long arrays or dip-restrictive processing of the previous work.

Then the maximum frequency. In a mature province, where we know the stratigraphic setting of the reservoir, this is dictated by the synthetics from neighboring wells — what do we need to resolve the top and base of the reservoir? Or to measure such-and-such an interval thickness? In other provinces, where the play may be less specific, we go through this design procedure for two or three *choices of maximum frequency,* and then look at the cost-effectiveness — balancing the cost of seismic against the cost of drilling, or the cost of this prospect against one which needs less resolution.

OK then — let's go through the design procedure (Figure 2). First, target time and target velocity give us the target depth, which gives us the spread length. Then from the target velocity, the maximum dip and the maximum frequency, we compute the common-midpoint interval to give adequate spatial sampling for migration. Typically we can *graph* the relation (Figure 3) — CMP interval on the vertical axis, chosen maximum frequency on the horizontal axis, and maximum dip as a parameter — first for a modest dip, about 10 degrees, then for 20, 30, 65 degrees.

Then, since for all preferred spread arrangements the CMP interval is half the group interval, we have the group interval. That and the spread length give the number of channels (Figure 2 again). Since any high-resolution survey is certain to be using the stack-array approach, we have the array length — equal to the group interval. (Incidentally, if we have goofed our estimate of the maximum dip, it is true that this exposes us to spatial aliasing — but that we can fix later, by mixing adjacent traces in the processing, if we have to. Let's assume we made a good estimate of the maximum dip, and carry on.)

Then, again following the stack-array criterion, we select a source interval which gives a uniform stack array across the gather — *half* the group interval for an end-on spread, *equal* to the group interval for a split spread shot between the groups.

Now, having defined the number of channels, the group interval, and the source interval, we have defined the *fold.* That's the whole spread geometry, completely defined, with no judgements to make and no tests to run, from the initial four decisions.

At this stage, with the intelligent layout of the lines and the intelligent definition of the spread geometry, we have done everything we can do in the field to get good migration, and therefore good lateral resolution.

Next, the instruments. The maximum frequency, of course, decides the sample interval and the high-cut filter. The low-cut filter is no longer a critical setting — we'll set it when we get to the field, feeling free to adopt whatever setting gives a good monitor record. Then, as we agreed earlier (see Part 1), we'll take out the effect of that filter later, in the processing. OK — that's the instruments.

Now we have to choose a source. You remember what we said: we have to make four decisions (which we've made), one judgement, and one measurement. Well, this is the judgement — the source.

For a high-resolution survey tightly focused on a specific target, we can see fairly clearly what we would like to have: other things being equal, we would like a source with lots of poop at the high frequencies (up to but not beyond our chosen maximum frequency), a fair amount of poop down to the very low frequencies, and — if getting these means some sort of sacrifice — less poop at the middle frequencies. But of course other things are not equal. There are considerations of cost and reliability, and the environment; there are considerations of portability, and work along roads, and work in the Arctic; and there are considerations of flexibility, because a source optimized to get *penetration in a reconnaissance survey* is usually quite different from one optimized to get *bandwidth in a high-resolution survey.* That is why we still have different sources available, and why — for our particular application — we have to make a judgement between them.

With the judgement made, we come to . . . the measurement.

How much source energy? At sea, how much compressor power, how many guns? In dynamite work using small charges (for we need small charges, usually, for high resolution), how many small charges? In Vibroseis, how many vibrators, how many sweeps?

It is a critical question, because the answer decides the final success of the decon, and hence decides the vertical resolution. We recall again that decon can extend the reflection bandwidth only at those frequencies where the *signal exceeds the noise.* And that is what we are deciding here. We recall also that, once we leave the field, the only methods left to us for improving the signal-to-noise ratio are *the mute and the stack;* none of the other fancy processing techniques can improve the signal-to-noise without sacrificing resolution in some way. So everything hinges on that final field question — how much source energy?

The most interesting case to consider in deciding how to answer this question is Vibroseis with nonlinear sweeps. But there are an infinity of nonlinear sweeps, so let us restrict immediately to *intelligent* nonlinear sweeps — not gung-ho, think-of-a-number nonlinear sweeps.

One touted practice in the intelligent design of nonlinear sweeps is to use the nonlinearity to *compensate* frequency-selective losses which are known to exist — such as the indifferent frequency response of the vibrator itself, or an average response for the vibrator-ground coupling, or indeed the absorption in the earth.

Consider the vibrator itself, for a moment. The mechanical response of a traditional general-purpose vibrator looks something like the lower (violet) curve in

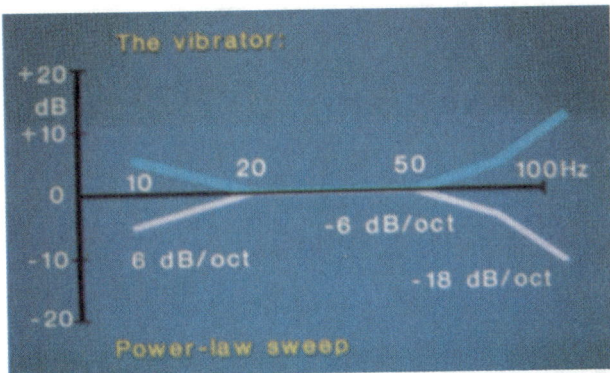

Figure 4. Illustrative response for a vibrator without force control, and the nonlinear-sweep response required to compensate it.

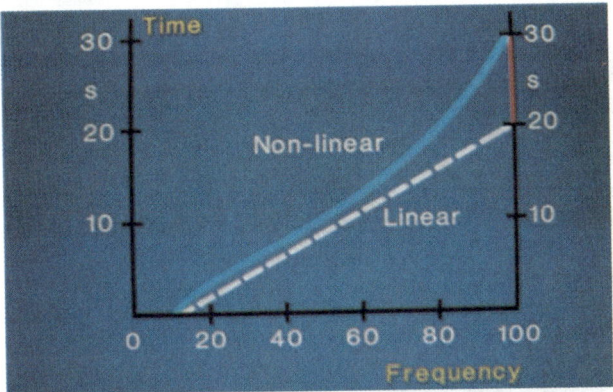

Figure 5. Nonlinear sweep calculated from Figure 4.

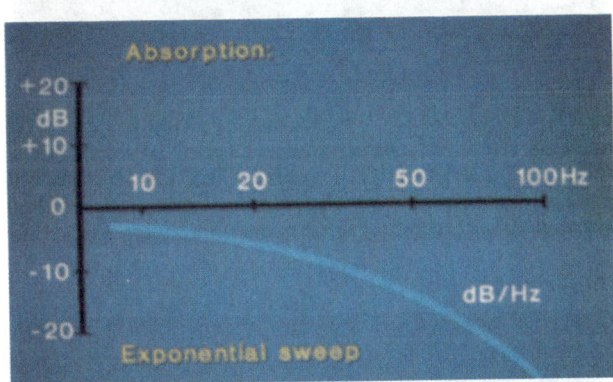

Figure 6. Illustrative response for the absorptive earth at target level.

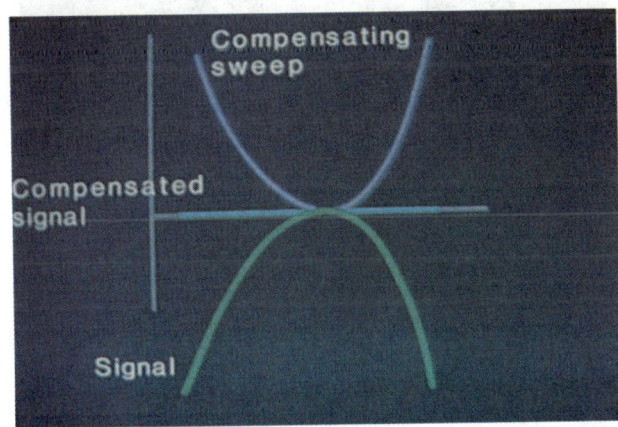

Figure 7. A postulated signal spectrum, the spectrum of the compensating nonlinear sweep, and the resulting flat output.

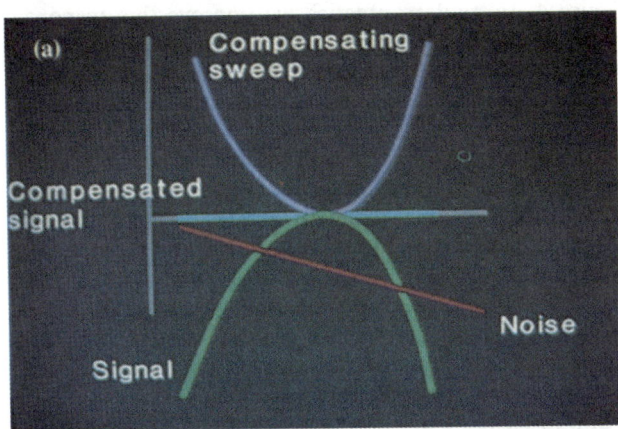

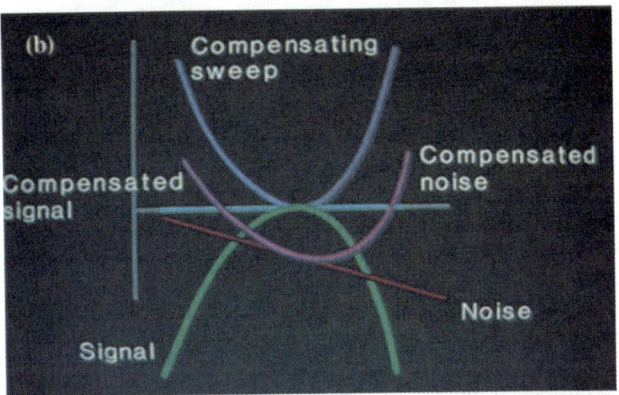

Figure 8. Representative noise spectrum (a) as it is modified (b) by the nonlinear sweep.

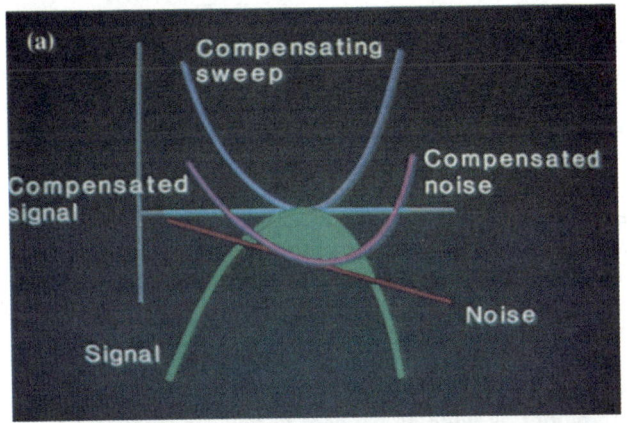

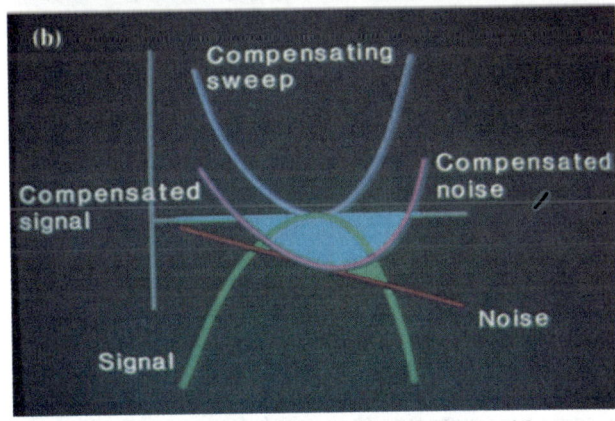

Figure 9. Signal-to-noise bandwidth (full color) without compensation (a) and with compensation (b).

Figure 4; in modern practice we improve this response by ground-force control, but on most surfaces the droop at low frequencies must be preserved to avoid baseplate chatter. In any case, the compensation of this response gives us a good illustration of the design of nonlinear sweeps. We design the nonlinear sweep to have the *compensating* response — the upper (turquoise) curve. Since the slopes of the response are so-many dB/octave, we know that the correct nonlinearity, for each segment of the response, follows a power law — we use *dB/octave* sweeps. The frequency-time relation to compensate this vibrator is then that of Figure 5, and we can quickly see (in red) how much extra sweeping time is necessary to give that compensation (relative to a linear sweep).

Similarly, if we are prepared to estimate the *earth absorption,* down to target level, then we can design a sweep to compensate it. In Figure 6, we are guessing an absorption of 0.3 dB/Hz; because the response is so-many dB/Hz, rather than so-many dB/octave, the sweep is now a *dB/Hz* sweep.

In this way, for any measured or postulated reflection spectrum, we can design a nonlinear sweep which pre-compensates the losses, and yields a flat reflection spectrum at target level (Figure 7). However — and here my voice continues to cry in the wilderness — this *compensation* is not the optimum thing to do. The crews that are out there compensating the reflection spectrum have forgotten to ask *what the nonlinear sweep is doing to the noise.* They have forgotten that a nonlinear sweep which raises the signal, at any frequency, by a factor K, also raises the noise, at that frequency, by a factor $\sqrt{K}$.

In Figure 8a we add a representative noise spectrum. Now, as the nonlinear sweep dwells longer at low and high frequencies, it also takes in a longer segment of noise — more noise — at those frequencies (Figure 8b). The square-root relation. The compensating sweep *does* improve the signal-to-noise bandwidth, of course — from Figure 9a to Figure 9b. But that is an expensive way to get that bandwidth. It is always cheaper — it always takes less sweeping time — to use a nonlinear sweep which goes *beyond* compensation — even in the limit, to invert the signal spectrum (Figure 10a). The noise spectrum now rises even more at the low and high frequencies — but it remains *below the signal* (Figure 10b). Then we correlate against a sweep whose *amplitude* follows the postulated signal spectrum . . . and we have signal-to-noise bandwidth as wide as we care to pay for (Figure 11), at guaranteed minimum cost.

With this approach, we can see how to answer that fundamental question, "How much source energy?" Indeed, if last year's survey also used Vibroseis (which may be linear Vibroseis) we can even answer it in advance, before we go to the field. We go back to last year's survey, and restack a critical piece of line with fewer sweeps (or fewer fold) — ½, ¼, ⅛ of what was used then — and a very tight filter right at the peak of the reflection spectrum at target level. Thus we can decide how many sweeps would have been required *to obtain just-adequate signal-to-noise at the peak of the spectrum.* Dead easy. Then there's the answer: for the new high-resolution survey *we must spend the same time sweeping those peak frequencies,* and then correspondingly more (as we have just computed) at the

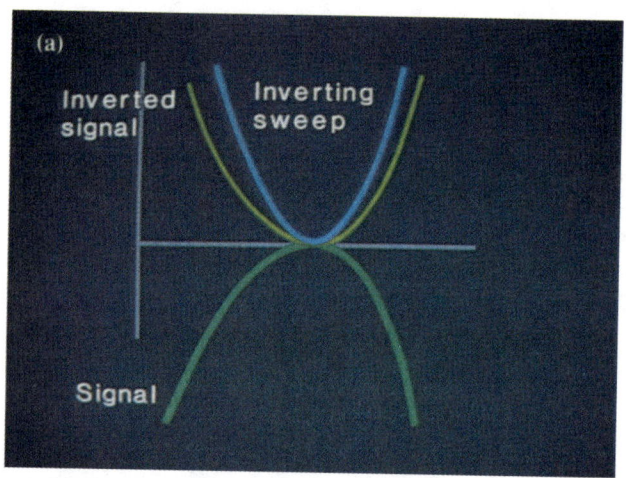

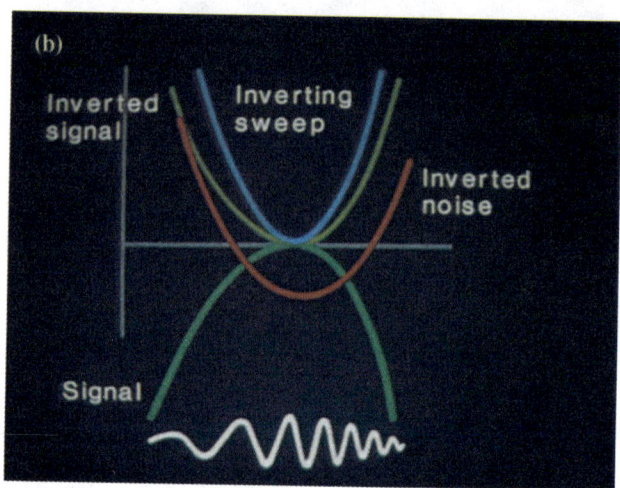

Figure 10. The use of a more nonlinear sweep to invert (rather than compensate) the signal spectrum (a), and its effect on the noise (b). Also shown is the amplitude modulation of the sweep used in correlation to restore a flat signal spectrum.

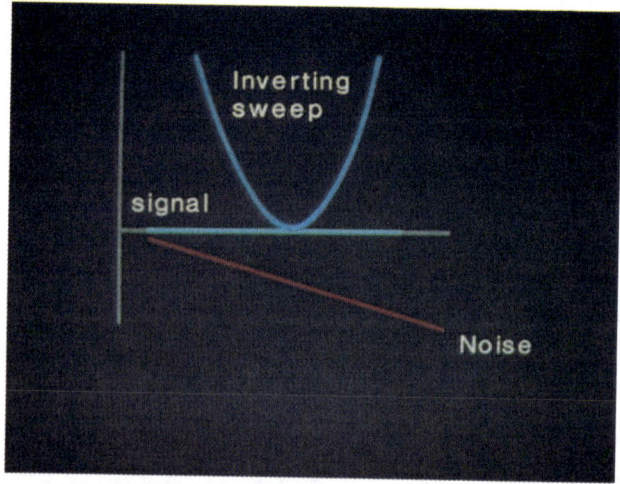

Figure 11. Final effect of the inverting sweep of Figure 10; the signal spectrum is flat, and is everywhere above the noise. Although the cost of this approach is high, it is the minimum possible to achieve a given signal-to-noise bandwidth.

lower and higher frequencies.

With that done, we can work out how much it will cost to achieve our desired maximum frequency. Then we review that initial maximum-frequency decision, as we said earlier, and ask whether the resulting survey will be cost-effective. If not, we lower the maximum frequency, and go through the complete design procedure for the new value.

So there we are — a rational step-by-step routine for the total determination of the field technique for a high-resolution survey. By making four intelligent decisions, one intelligent judgement, and one intelligent measurement on last year's survey, we can design the total field technique before we go to the field. Of course we refine it when we're there, but we can certainly get very close before we start.

For me — the simple chap who likes simple things — there is great pleasure in this rational design of the field technique. In a world which passes inevitably from the general to the specialized, and from the simple to the complex, what a pleasure it is to find that a very complex matter — the design of the field technique — can actually be made simpler by advancing knowledge!

Before I go on to my final topic, I should point out, again for the sponsors of the Video Library, that the details of all this material are spelled out in the manual of module GP 311, on improving the resolution.

Now to that other cry from the heart of the interpreter: give us better delineation of the faults!

If you are not an interpreter, I cannot overstress to you the importance of faults in today's exploration. Indeed, can anyone even remember four-way dips? More and more of our prospects depend on faults. We need to know the *timing* of those faults. We need to know whether they have been *conduits* in the past. We need to know whether they are still conduits, or whether (and when) they have become seals. We need to know about recrystallization, and precipitation, and cementation. We need to know how far out from the fault plane can such changes be expected. We need to know about rejuvenation of the faults, about breakage of the seals, and about reformation of those seals. And, if we make a discovery, then to ensure the maximum ultimate recovery from the field we need to know where are the conduits, and where are the seals, *before we start to produce the field* — not before we start enhanced recovery.

Traditionally, in seismics, we have sought to optimize our *reflectors* — our bedding planes, our unconformities. Then our faults we obtained by inference, indirectly, as breaks in those reflectors. In this sense, improved fault delineation follows from improved vertical and lateral resolution, as we have just been discussing. Fine. But let's look beyond that. Let's look, in particular, at the fault plane itself, and the fault-plane reflection from it. For years, our long arrays, our unsophisticated stacking and our limited migration have effectively destroyed all fault-plane reflections except those at low angle. Today we can certainly *record* vertical fault planes, if we are prepared to pay for it. In the processing, vertical fault planes remain more of a problem. But in the majority of petroleum prospects we don't ask to see *vertical* fault planes; either the faults are inclined by nature (like growth faults, and antithetic faults, and thrust faults), or — while they may be vertical

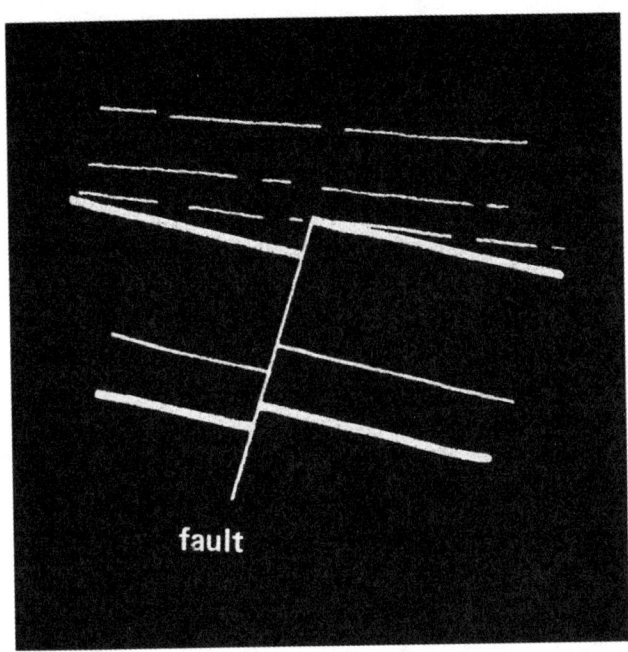

Figure 12. A simple fault model.

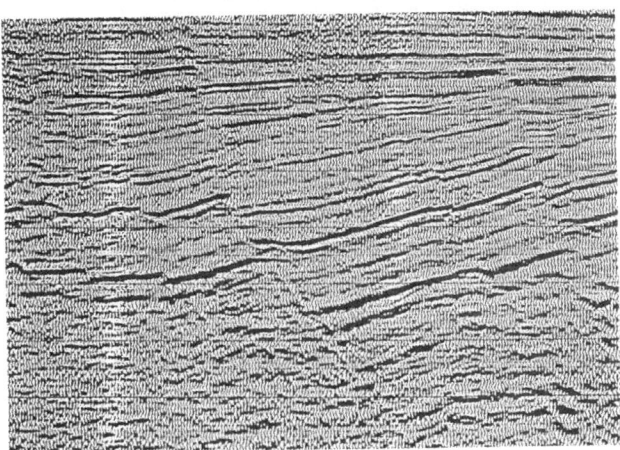

Figure 13. Fault-plane reflection from thick lithologic units with a large throw (from Bally, Seismic Expression of Structural Styles, courtesy AAPG).

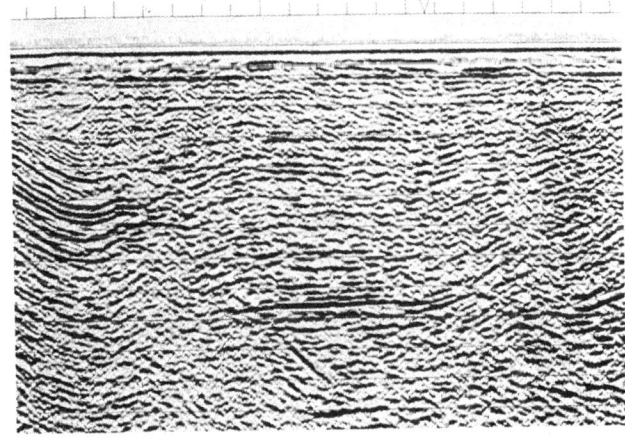

Figure 14. Another example of thick lithologic units and large throw (courtesy Merlin Profilers Ltd.).

Figure 15. Thin lithologic units and a throw large enough to juxtapose sequences of different age (Bally, op.cit., courtesy AAPG).

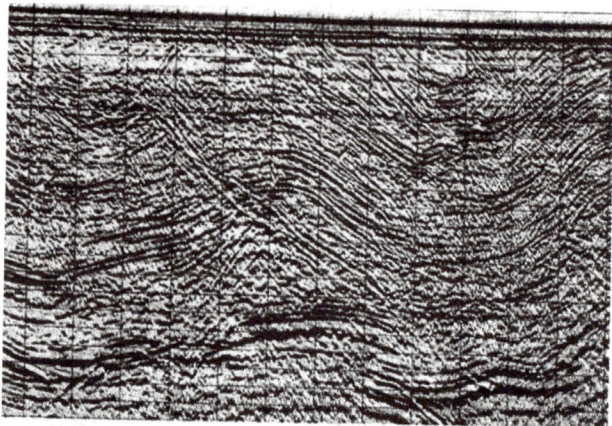

Figure 18. Good reflections from thrust planes (Bally, op.cit., courtesy AAPG).

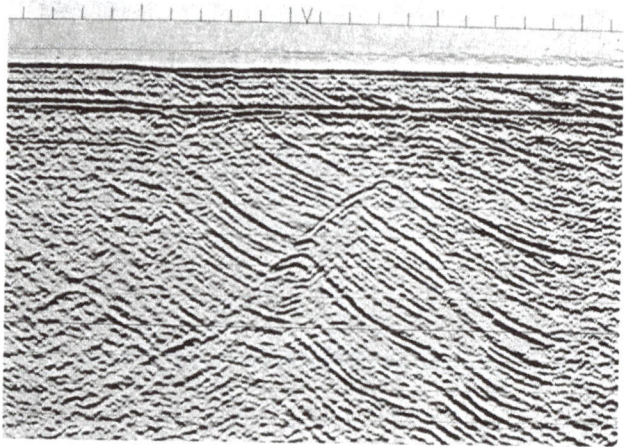

Figure 16. Here, differences due to lithology locally exceed those due to age (courtesy Merlin Profilers Ltd.).

Figure 19. A simple normal fault. What happens, seismically, at the fault plane?

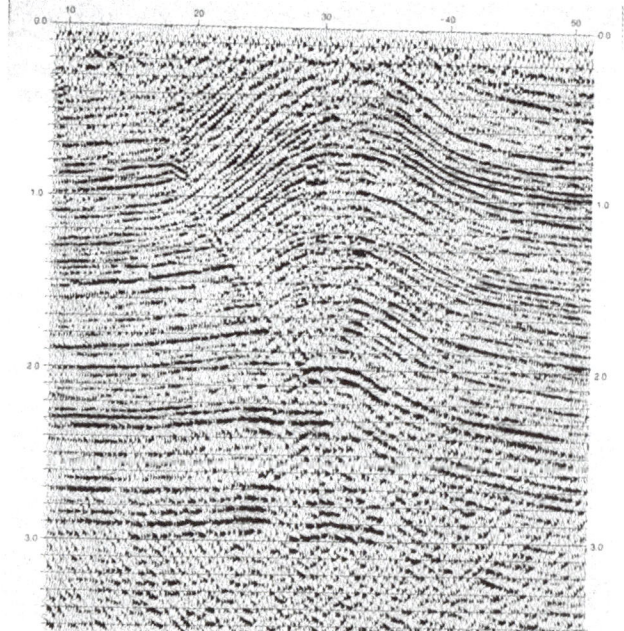

Figure 17. Fault-plane reflection which suggests the inadequacy of the simple model (courtesy GECO).

at shallow depth — they become inclined at target depth (like extensional normal faults). Therefore we'll probably get most of what we want if we can handle 60 or 70 degrees — and modern processing is virtually there.

So now is the time to ask: as we begin to do it right, what can we expect to see?

Let's start with a simple model (Figure 12), which assumes acoustic continuity across the fault. Then the first important variable is the throw of the fault relative to the thicknesses of the lithologic units. If the bold lines are the top and base of a major hard lithologic unit, and the thin line is a bedding reflector within that unit, then obviously we get a positive reflection coefficient from the upper part of the fault plane . . . then zero . . . then a small positive . . . then zero . . . then a negative. And the up-and-down extent of the fault-plane reflecting segments represents the throw of the fault.

If the lithologic units are thin, and the throw is small, then from this simple model we cannot realistically expect to see a fault-plane reflection. There's not enough area to generate a useful reflection, and in any case the positive segments of the reflection are offset by the negative segments.

The first way to get a significant fault-plane reflection

Figure 20. The norm is, of course, a fault *zone*.

Figure 21. An example with both thick and thin lithologic units and both large and small throws, but with fault-plane reflections of surprisingly constant character. Are the faults therefore active gas conduits? (courtesy Western Geophysical Company).

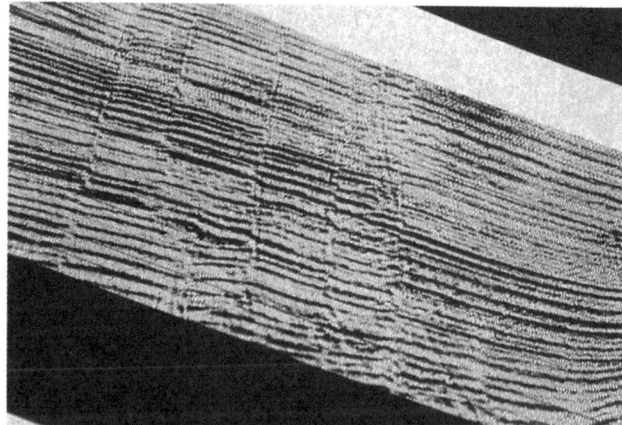

Figure 22. Foreshortened view of Figure 21.

— still with this simple model — is to have thick lithologic units and large throws. In this case, we get what we expect — a fault-plane reflection (Figure 13). Rather weak and low-frequency, because it's a steep event on an old section shot with long arrays. And not quite in the right place, because it used an old migration program. But — a fault-plane reflection. *Not* a reflected refraction.

Figure 14 is the same story, basically. But now it's easier to see that the extent of the fault plane reflection represents the throw of the fault.

The second way to get a significant fault-plane reflection — still with our simple model — is to stay with thin lithologic units, but to have a throw large enough to juxtapose sequences of distinctly different age (Figure 15). Then if the effect of age is greater than the effect of lithologic change from bed to bed, we obtain a fault-plane reflection of *constant polarity* — always positive for a normal fault or a growth fault, always negative for a reverse fault. So we see no fault-plane reflection high in the section, where the throws are small, but a fault-plane reflection of constant polarity, strong and well-defined, deep in the section where the throw is large.

Figure 16 is another example — thin layers, large throw. Again constant polarity to the fault-plane reflection, except in one place, where the impedance contrast between beds exceeds that between younger and older rocks.

But then . . . Figure 17. Now we have a fair fault-plane reflection, still low-frequency, of course, because of the arrays and the migration, but a fair fault-plane reflection — even though we don't have thick lithologic units — and the fault-plane reflection is clearest where the throw is least. Is our simple model inadequate? Is there some other agency at work, other than the simple juxtaposition of impedance contrasts across the fault?

And when we see another like it (Figure 18), we have to ask: is it something to do with the fact that these are thrust faults?

The fundamental question, clearly, is: what happens, physically, at the fault plane? So we put on our boots, and go to look at some faults.

First a simple normal fault (Figure 19) — an extensional situation, with the right-hand block dropped down. Now, is our simple model correct? Is there acoustic continuity across the fault plane? Well, not everywhere, for sure. Some grains must be in contact — highly stressed intimate contact — and may well be recrystallized. Others have no contact at all, and see only a free surface. On balance, with these fault voids containing air, wouldn't we expect a fair negative reflection from the fault plane? And a reflection

Figure 23. Thrust fault and its brecciated zone.

Figure 24. Closer view of Figure 23.

Figure 25. A change of processing techniques, across the splice, has destroyed a good fault-plane reflection (Bally, op.cit., courtesy AAPG).

from the *entire* fault plane, almost without regard to the local impedance contrasts across the fault?

But if the fault voids contain water, then of course we would expect much less of a fault-plane reflection, with the reflections from the two sides of the fault almost canceling. Much closer, now to the simple model. And if the fault voids are cemented solid, forming a seal, then we are right back to the simple model.

Larger faults, of course, generate multiple slivers — the slivers themselves faulted, fault planes diverging and recombining — a fault *zone* (Figure 20). Again we would be certain that if the voids contain air we have, not a free surface, of course, but perhaps a half-free surface. Emphatically not our simple model.

And the same if the voids contain gas — if the fault is an active gas conduit. But if they are full of water, or oil, or cement, then we are back to the simple model.

So, what is Figure 21? Take your choice. There are thin layers and thick layers, small throws and large throws. But all of that makes no difference — we have clear fault-plane reflections of *surprisingly constant character*. Here . . . hold the section up, and squint along the fault planes (Figure 22). Surprisingly constant character. The conclusion? Well, what do you think? I think they are *active gas conduits*.

Now let's look at a thrust fault (Figure 23). Dissimilar rocks on the two sides, but now the possibility of a zone of pulverized rock flour in the fault plane; even, as in this case, brecciated rocks from somewhere else in the section — a third rock type.

Acoustically . . . what? Let's come in tighter (Figure 24). Well, probably low impedance in the brecciated zone, and therefore a negative-positive doublet reflection, representing a conduit. But if the breccia were cemented hard and tight . . . perhaps a positive-negative doublet, representing a seal.

So is that what we are seeing in Figure 17 from the thrust plane? And which is it — negative-positive, a conduit; or positive-negative, a seal? Well, we don't know. But what we do know is that the fault-plane reflection maintains a remarkably constant character, and that there must be half a chance of answering that question — conduit or seal — if we really tried. With such a prize before us we surely have to try.

So what should we do in our search for fault-plane reflections? First, perhaps, we'll be careful not to raise management's hopes too much. (We've been there before, haven't we?) Second, we'll design our field work to record those fault-plane reflections to 60, or 70, or even 90 degrees. Third, we'll remember to tell the processors what we've done — lest their first move is to decimate our lateral resolution in the processing. And fourth, by the ingenuity of the researchers and the skill of the processors, we'll carry on trying to image everything to the right place.

Figure 25 is a mischievous one, to remind us how much we depend on the skill of the processors. Just look at that fault-plane reflection across the splice! Now you see it . . . now you don't.

So there we are. May we see many more fault-plane reflections in the future, and may we gradually work out what they mean. **LE**

THESE THREE SERVO HYDRAULIC VIBRATORS are operating in synchronism. The units, owned by Seismograph Service Corp., can operate along roads or in cities as the vibrators cause no damage.

Vibroseis' gentle massage obtains structural data safely, economically

System can use whatever frequencies the earth can transmit well

BY N. A. ANSTEY
Seismograph Service Ltd.*
Holwood, Keston, Kent, England

TO ESTABLISH common ground between the conventional and Vibroseis® methods, it is of interest to review briefly the operation of the conventional (explosive) seismic method.

Since a normal geophone measures the velocity of earth motion, let us imagine a perfect velocity detector placed at several points along the path of the seismic wave. In Fig. 1a, the shape of the signal very close to the charge is seen. The amplitude here is enormous; much of the energy of the charge is wasted in crushing the rock, up to a radius of many feet from the explosion. Thereafter, the propagation is more predictable, and the pulse shape changes gradually, over several thousand feet of travel, to the general form of Fig. 1b. It is then detected at the surface of the ground, and normally it is filtered electrically to obtain the general form of Fig. 1c. This then is the basic unit of which the seismic record is composed; each reflector returns one such pulse, and the final record is the superimposition of the pulses corresponding to all conceivable earth paths.

It is common knowledge that the recorded seismic pulse cannot be said to have "a frequency"; it contains a band of frequencies. These different frequencies existing in the pulse, and their relative proportions comprise what is termed the spectrum of the pulse. The spectra which correspond to the pulses of Fig. 1a-c are shown as Fig. 1d-f respectively. From Fig. 1d, we see that the sharp "spike" of Fig. 1a contains all frequencies equally, whereas Fig. 1b illustrates that the action of the earth on the spread-

®Registered trademark and service mark of Continental Oil Co. *A subsidiary of Seismograph Service Corp., Tulsa, Okla.

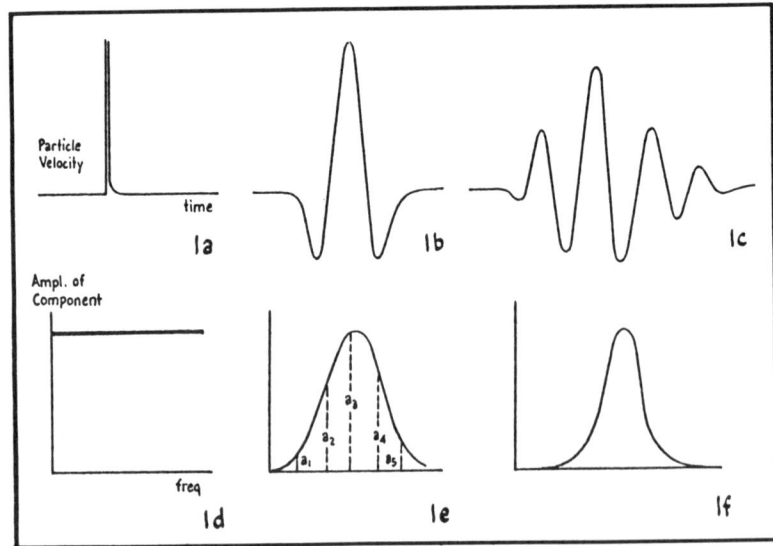

WHAT HAPPENS when a perfect detector is placed at different points along the path of a seismic wave? In the examples above, (a) is very close to the signal, (b) after several thousand feet of travel the pulse changes, and (c) after electrical filtering, it looks like this. The lower sequence (d) to (f), shows the corresponding frequency spectra of (a) to (c). Fig. 1.

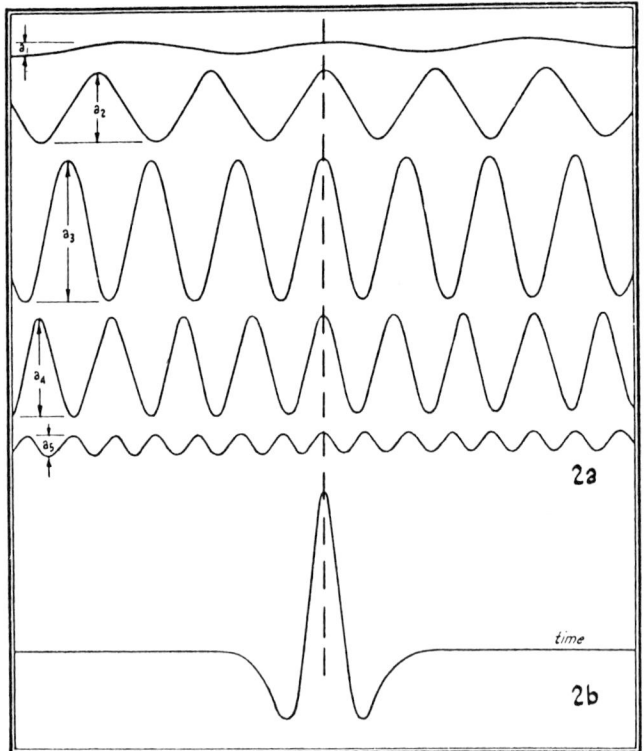

RELATIVE PHASE of the different frequency components within the spectrum must be known. This is shown in (a) above. Adding together all these sine waves, the resultant is the pulse shape of (b). Fig. 2.

ing wavefront has robbed the pulse of its very high and very-low frequency components. When filters are used in the instruments, the high and low frequencies are cut still further, and the pulse of Fig. 1c might then have the spectrum of Fig. 1f.

This treatment implies a measure of interchangeability or equivalence between a pulse and its spectrum. However, one must also know the relative phase of the different frequency components within the spectrum. This is illustrated in Fig. 2a, where several continuous sine waves of frequencies spaced across the spectrum of Fig. 1e are drawn to have the amplitudes indicated by that spectrum. Note that all the sinusoidal components are in phase at just one time. Adding together all these sine waves, it is seen that the resultant is large at the in-phase point, that away from this point the amplitude decreases and reverses to give us the pulse shape of Fig. 2b, and that at times remote from the in-phase point the sinusoidal components interfere to give us zero signal before and after the pulse.

Therefore, the shape of a pulse can be expressed completely in terms of the amplitude and phase of its sinusoidal frequency components. In particular, the action of the dynamite can be visualized as the injection into the earth of a wide band of frequencies, with a certain phase relation.

From this it follows that **any method of injecting a band of frequencies into the earth, with a certain phase relation, represents a possible exploration method.**

The basic Vibroseis system. The basis of the Vibroseis method is the application of these frequencies consecutively, followed by operation on the results to restore them to the same kind of pulses that would have been obtained if the frequencies had been applied simultaneously. Therefore, the output from the Vibroseis system is essentially a normal seismic record. A Vibroseis record has the same appearance as a conventional record because it is the same thing; it has the same primary reflections, the same multiple reflection, the same diffractions — in short, the same earth paths. What is more, all the effects of the geometry of the spread (offset profiles, long patterns, steep dips) are identical, since these are features of the earth paths.

The signal which is used to introduce a band of frequencies consecutively into the earth is a simple

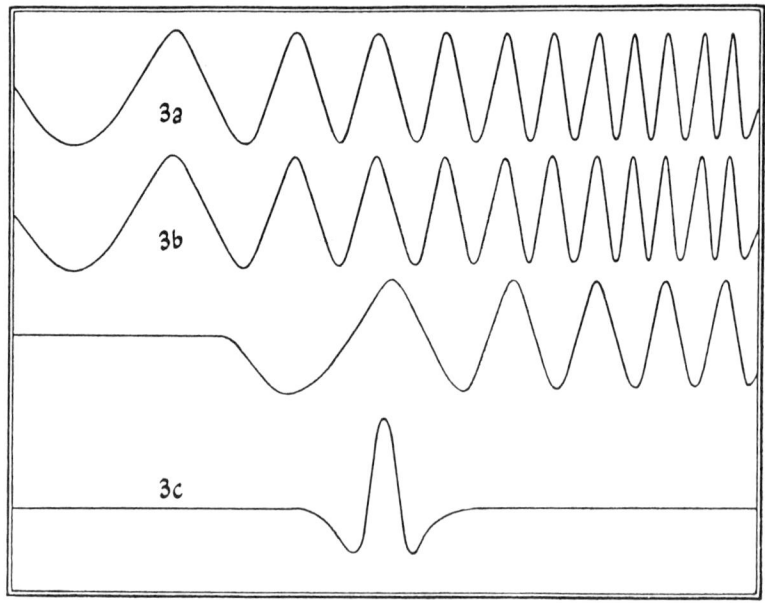

A SIMPLE-SWEEP SIGNAL is used to introduce a band of frequencies consecutively into the earth (a). Two identical swept-frequency signals are aligned in (b). A plot of over-all coherence, as a function of the shift of the second trace relative to the first results in the pulse of (c), is identical to Fig. 2b. Fig. 3.

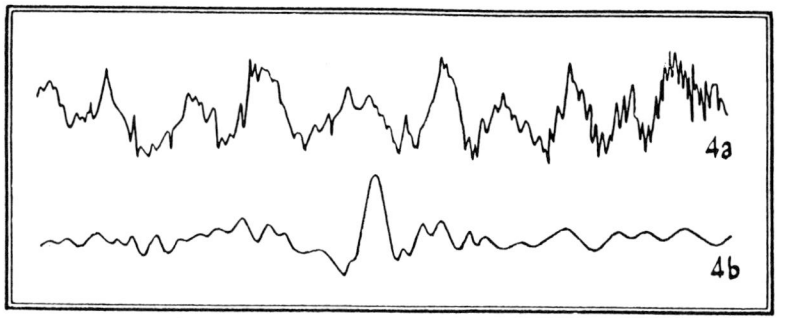

A SWEPT-FREQUENCY SIGNAL buried under four times its own amplitude of noise is shown in (a). Correlating this noisy trace with the noise-free swept-frequency trace results in (b). Fig. 4.

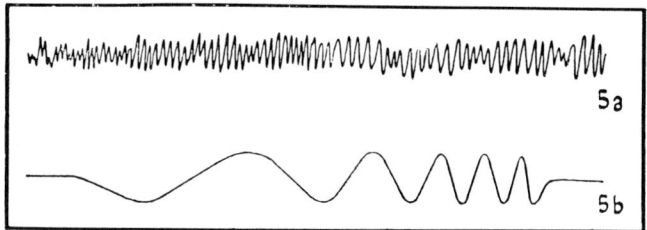

WIND NOISE is eliminated by strict filtering of the components of noise outside the chosen sweep. The high-frequency signal (a) has no coherence with the swept-frequency signal (b). Hence the correlation process has two important purposes: it provides a filter ideally suited to the signal, and discriminates against noise by using both duration and amplitude of the signal. Fig. 5.

sweep, typically from 15 to 90 cps and of about 7 seconds' duration. The important point to notice about this signal (and evident from the abbreviated illustration of Fig. 3a) is that it is unique; **nowhere does any part of the signal repeat itself.**

The operation of recovery of pulses from this continuous sweptfrequency signal is termed correlation. Those intimately concerned with the Vibroseis technique will need to study this subject carefully, but for the moment the principle can be established using little more than intuition. Observe the two identical swept-frequency signals aligned in Fig. 3b. Then it is obvious that any quantitative measure of over-all similarity or coherence of the two traces will have a high value. If now one trace is kept fixed and the other moved a few cycles to the right or left, it is seen that often where one trace is positive the other is negative, and that for every place where both are positive, another place can be found where both are negative. The resultant over-all coherence therefore tends to zero. Repeating this assessment of over-all coherence for different shifts of the second trace relative to the first is another way of performing the same type of operation used in Fig. 2 to synthesize a pulse from its sinusoidal components. In fact, a plot of over-all coherence as a function of the shift of the second trace relative to the first results in the pulse of Fig. 3c, which is identical to that of Fig. 2b. This fact drives home the strict equivalence of the impulse method (represented by Figs. 1b, 1e, and 2) and the continuous-sweep method (represented by Fig. 3).

In this treatment the over-all coherence between two identical traces as one of them is shifted in time has been assessed; such a process is termed autocorrelation. In the practical field case, obviously, the signal received at the geophones is not identical to the ones sent out; it must represent the superposition of signals from many earth paths, and it will also have wind and other noise masking it. Let us first consider the effect of the noise.

In conventional work using explosives, it is obviously preferable to have the reflected pulses appreciably larger than the wind noise.

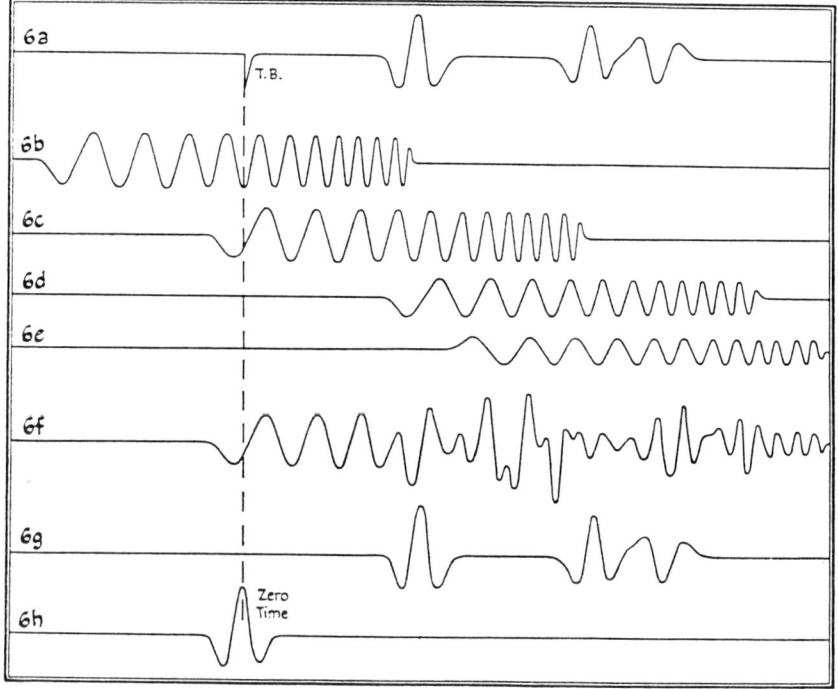

CONSIDER A CONVENTIONAL SEISMIC RECORD with a shot near the surface and three reflectors at depth (a). An uncorrelated Vibroseis record from reflectors is shown in (f). This record is composed of the three arrivals shown in (c), (d), and (e). This field record is impossible to interpret by eye. Yet when correlated with the transmitted signal in (b), (g) is obtained. The time break of (b) is similarly correlated to obtain (h). Fig. 6.

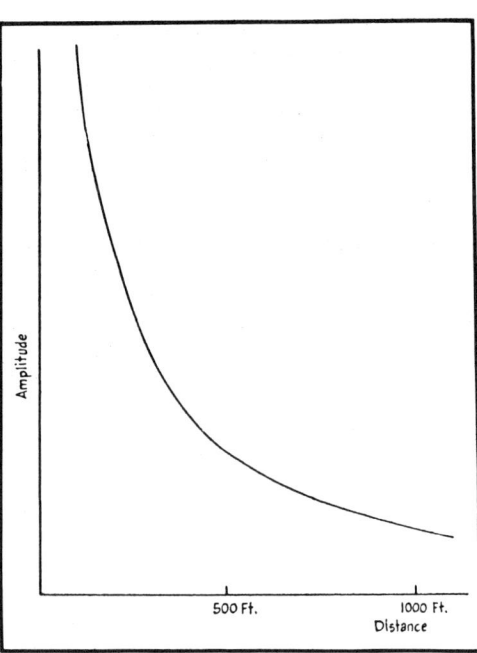

AMPLITUDE of horizontal arrivals is plotted as a function of distance from the source. Fig. 7.

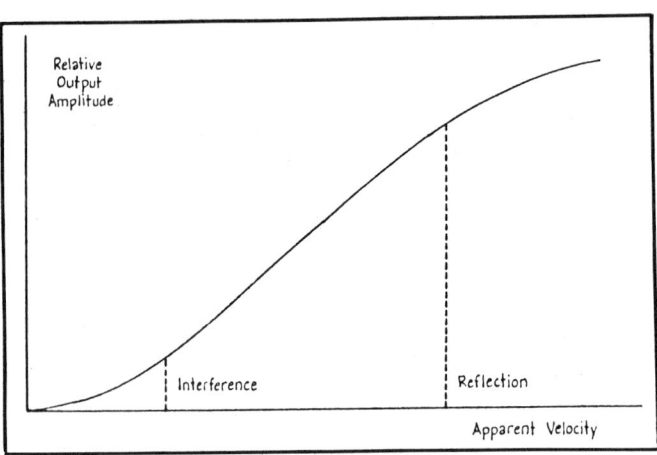

GENERAL FORM is a plot of effective output of a linear pattern as a function of apparent velocity. Fig. 8.

It is usually estimated that a reflection can be picked across a number of traces with fair reliability, if its amplitude is at least equal to the wind noise (i.e., if the signal-to-noise ratio is at least 1:1). Now consider a continuous sweep from 20 to 80 cps, lasting 6 seconds, and therefore containing 300 cycles. Loosely, each half cycle may be viewed as a little pulse; then, recalling that the correlation process looks not for one little pulse but for all together, it is evident that to get an appreciable output from the correlator all or most of these half cycles must be in the right place. It is easy to see that this discriminates against incoherent noise; in fact improvements of 15:1 in signal-to-noise ratio are commonplace.

In Fig. 4a a swept-frequency signal buried under four times its own amplitude of noise is shown. By correlating this noisy trace with a same but noise-free swept-frequency trace, an output can be expected to be at least as good as Fig. 4b.

The correlation process therefore offers a means of using the **duration** of a signal (as well as its amplitude) to discriminate against noise.

In the Vibroseis method, then, there are three major advantages over impulse methods (whether the latter use explosives or dropping weights). These are:

1. Power is not wasted generating frequencies which the earth will not transmit.
2. Power is not wasted crushing rock.
3. The correlation process permits distinguishing s i g n a l s much weaker than the wind noise, and which are invisible to the eye.

It is these features which allow the enormous chemical energy of a charge of dynamite to be replaced by a comparatively weak vibrator.

As a corollary to the fact that in the Vibroseis method, frequencies are not generated which are not used, it is legitimate to impose strict filtering on those components of wind noise which are outside the chosen sweep. In fact this is done automatically i n t h e correlation process, for it is easy to see that the high-frequency signal of Fig. 5a has no coherence with the swept-frequency signal of Fig. 5b, and so any correlation process designed to assess that coherence will give no output. The correlation p r o c e s s therefore has two important features: it discriminates against noise by providing a filter ideally suited to the signal being used, and it discriminates against noise by using the duration of the signal as well as its amplitude.

If one now considers the simple case of a shot near the surface and three reflectors at depth, a conventional seismic record will be as in

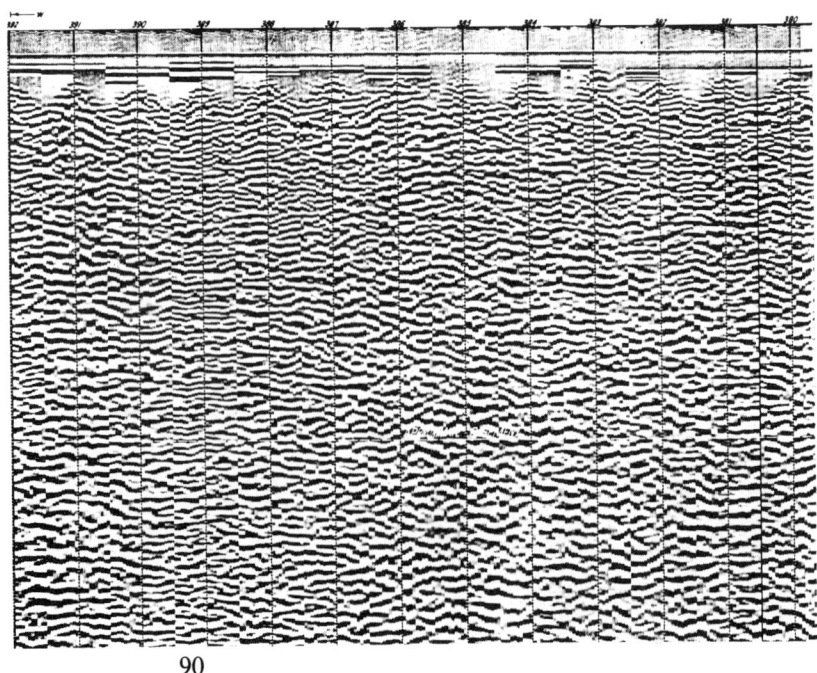

Fig. 6a, comprising a time-break and three primary pulses (which may overlap if the reflectors are closely spaced). The Vibroseis record made in the field under these circumstances would appear as in Fig. 6f, where 6b represents the signal transmitted into the earth, 6c represents that reflected from the first reflector, 6d and 6e those from the second and third reflectors, and 6f the superposition of 6c, d, and e. This field record would be impossible to interpret by eye.

However, when correlated with a trace identical to the transmitted signal (6b), moving this trace all the way down the field record and assessing the correlation as it varies with time, we obtain the record of Fig. 6g. To obtain the time break, the transmitted signal (6b) is similarly correlated with itself, yielding an autocorrelation pulse as in 6h. The center of this pulse (called "zero time") corresponds to the conventional time break.

This treatment for the case of three reflectors may now be extended to the practical field case of many hundreds of reflectors, without any basic change; the superposition of the swept-frequency signals from many reflection paths does not impede the correlation process.

The Vibroseis system in the field. The swept-frequency control signal is first recorded on the field tape; the initial and final frequencies, and the sweep rate, are selected appropriate to the field problem. This signal is then played back from the field tape, amplified, and transmitted by radio to the vibrators. The vibrators may be hydraulic or electromagnetic; in either case they are capable of exerting a total thrust of several tons on the earth, and of following faithfully the control signal transmitted to them. Once the signal is communicated to the earth, the effect of the earth paths is exactly as it would be for the impulse method.

Thus, in addition to the desired downward-traveling energy, the normally undesired horizontally traveling energy is present. This may represent surface waves or shallow refractions; such unwanted signals differ from wind noise in that they have the same form as the transmitted signal, and therefore they will be rigorously detected by the correlation process. Since this is an earth-path effect, the techniques used to minimize the horizontal arrivals are exactly those which are customarily used in impulse methods, namely:

1. Offset of the spread from the source.
2. Use of geophone patterns.
3. Use of source patterns.

If the amplitude of the horizontal arrivals as a function of distance from the source is plotted, a curve of the general form of Fig. 7 is obtained. The region in which the curve becomes markedly less steep is typically 500 or 600 ft from the source, and so it is usual in Vibroseis operation to place the spread at least this distance from the vibrators. In fact, for all normal oil-prospecting applications, the angles of incidence on the shallowest reflectors of interest are still reasonable with offsets greater than this, and so considerations of practical convenience often lead to an offset distance equal to one spread length (i.e., typically 1,320 ft).

To yield a further improvement in the ratio of the reflected signals to the horizontal signals, it is normal practice to utilize geophone patterns. The apparent velocity of a wave front across a geophone pattern is simply the length of the pattern divided by the time to cross it; for the horizontal arrivals this may range from 1,100 ft/second for an air wave, through to 3,000 ft/second for a surface wave, and to perhaps 12,000 ft/second for a shallow refraction. The apparent velocity of a horizontal reflection at 0.6 second, measured 2,000 ft from the source, is typically between 25,000 and 30,000 ft/second, and it is rare indeed to find a practical problem which does not give a ratio of at least 3:1 between the apparent velocities of the shallowest reflector of interest and horizontally traveling interference.

Fig. 8 shows the general form of a plot of the effective output of a linear pattern as a function of apparent velocity, for a typical bank of frequencies and for a certain length of pattern. Clearly, then, one should choose the length of pattern so that a 3:1 ratio of apparent velocities corresponds to the steepest part of the curve; typically this will give 5:1 improvement in the ratio of reflection to horizontal arrivals. If then a source pattern is made of the same dimensions and type as the geophone patterns, the cascade effect gives 25:1 improvement in the ratio of reflection to horizontal arrivals.

The combination of these three techniques (offset spreads, geophone patterns, and source patterns) yields adequate discrimination against the horizontally traveling signals. Of course there is nothing in these techniques which is not also common to impulse methods. As stated

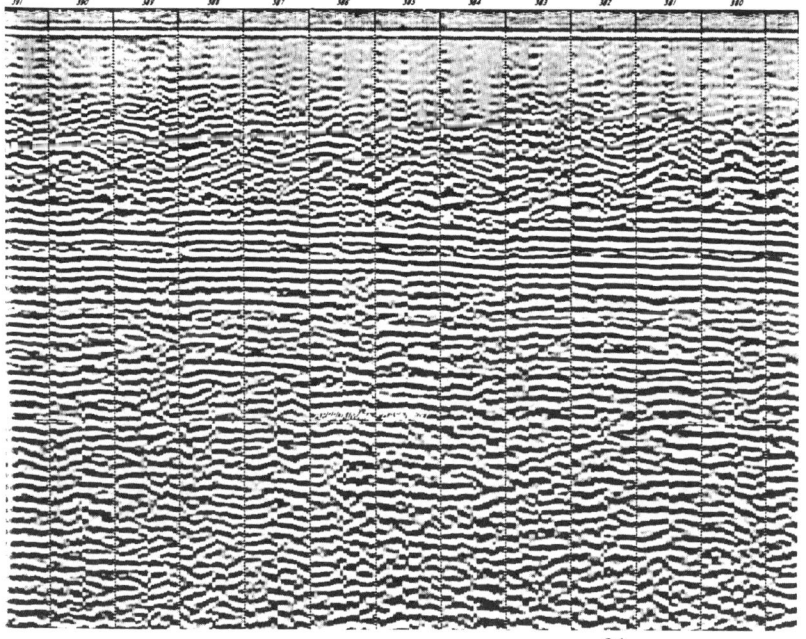

THERE ARE DIFFERENCES in compositing. Compare the Vibroseis larger-scale compositing on the right to the section on the left. Costs per mile are comparable. Fig. 9.

before, any matter concerned with earth paths will have comparable features on both impulse and Vibroseis systems. Thus Vibroseis records do not normally show strong "first breaks"; neither would conventional records shot with the same source and spread geometry.

The use of source and geophone patterns implies compositing, or the summation of signals which have substantially the same reflection path. The ease with which this can be done is one of the main features of the Vibroseis system. Even in a fairly good area it is usual to take 30 sweeps; and if two vibrators and 20 geophones per trace are used, each trace of the final record represents 2 × 30 × 20 or 1,200 reflection paths. In some areas where the near surface is heterogeneous, the conventional explosive method is defeated unless a very great amount of compositing is undertaken, and this becomes uneconomical; the Vibroseis system, however, can composite a large number of reflection paths per trace and still maintain a very good production rate. An example of the difference which compositing can make in such an area is given in Fig. 9; the improvement which the Vibroseis section shows is probably due almost entirely to the large-scale compositing. The two sections represent comparable cost per mile of profile.

The economic advantage of the Vibroseis system becomes more and more marked as the solution of the seismic problem requires more and more compositing. This is a particular feature of the fact that the vibrators are rigorously slaved to a master control signal; their number may be increased without limit, as the problem requires, to maintain the optimum balance between equipment cost and production rate.

In the past, large-scale compositing has introduced a difficulty in the field recording equipment; the process required many magnetic tapes or much transcribing. This information-storage problem has been solved by the narrow-head recorder, which was developed as a part of the Vibroseis system. The magnetic heads of this recorder are effectively one-fifteenth of the width of standard heads, and may be indexed across the tape to occupy 15 discrete positions within the width of one standard head. Thus the geophone signal for each of 15 sweeps may be recorded side by side with a narrow head, and composited immediately by playback from a standard head. The normal 40-channel recorder can therefore record 30 sweeps into 20 geophone patterns, without transcribing. If greater storage is required, the composite of the first 15 sweeps may be transcribed on a second narrow head, and then erased from the original recordings; this allows 225 sweeps to be recorded into 20 geophone patterns.

The magnetic drum normally makes one revolution in approximately 13 seconds; of this the sweep itself occupies approximately 7 seconds. The noise generated by the vibrator trucks in moving from one point of their pattern to the next is incoherent noise, and this movement is permissible during the remaining 6 seconds (i.e., while reflected signals are still being recorded). Thus the drum revolves continuously, and each time the control sweep signal is transmitted the vibrator trucks are in the new position. Therefore, 30 sweeps from each of two or three vibrators (typically the equivalent of one conventional shot point) can be recorded in less than 7 minutes.

The remainder of the equipment used in the field is well known; the geophones, cables, radios, and power supplies are normal commercial units. The recording amplifiers are considerably simpler than conventional amplifiers, in that they need little filtering and no automatic volume control. The compositing operation may be monitored visually as the signal builds up, and no photographic processing is necessary in routine field operation. The quality of the field records, of course, cannot be assessed by the eye; the amplitude of the received signal is adjusted in the field to insure optimum drive to the tape, but meaning of signal can be elicited only by correlation process.

The correlation system. The correlation process can be accomplished by a rigorous mathematical approach or by a practical analog solution. In the practical application, a correlation of the data recorded at 10 geophone stations can be accomplished in less than 2 minutes with a high degree of accuracy.

The output of the correlator is generally a magnetic tape which contains the seismic information in equivalent form as obtained in conventional methods.

Summary. The Vibroseis system is an immensely practical system for obtaining, with many practical operating advantages, the well-established type of record.

When making a comparison with conventional explosive work, the obvious operating advantages deriving from the abolition of drilling, explosives, and hole loading become apparent. Also, a marked speed advantage results in areas where the drilling is difficult. Those areas where the explosives method produces poor results are usually associated with unfavorable near-surface conditions; in these cases many samples of each reflection path are necessary, and Vibroseis scores by its ability to permit low-cost large-scale compositing.

When making a comparison with weight-dropping work, the Vibroseis method enjoys two major advantages. First, the vibrators operate synchronously, and their number may be increased without limit. If the problem requires considerable compositing and the production rate becomes poor, the rate may be restored merely by the addition of vibrators. Second, the Vibroseis system can use whatever frequencies the earth will transmit well. It is not restricted to the very low and narrow spectrum normally associated with weight-drop techniques.

In addition, the vibrations do no damage; the work may therefore proceed along roads and even through cities.

Therefore, the Vibroseis system has marked practical advantages over the established techniques, in operating facility and in the cost of useful geophysical information. These practical advantages are stressed because it is on these that any new method must commend itself. However, every geophysicist who is concerned with the problem of extracting the maximum stratigraphic information from the signal which comes out of the earth will realize immediately the value of knowing what signal went into the earth, and will be quick to see what promise this yields for the future. **End**

THE VIBROSEIS®* SYSTEM OF SEISMIC MAPPING

by

ROBERT L. GEYER**

INTRODUCTION

It is most important to recognize that VIBROSEIS exploration utilizes an engineered system that is based on a specific choice of input signal frequencies. Concurrent with the choice of input — *Pilot Signal* — frequencies is the obligation to support that choice of pilot signal by an equally careful choice of the other system parameters such as the source and receiver pattern geometry, group interval, and total spread length. The purpose of this essay is to review the basic principles and demonstrate some of the relations between the operating parameters.

There have been many changes in field equipment and processing systems since the first public disclosure of the VIBROSEIS system in Ponca City, April 10, 1958, and the paper by Crawford, Doty, and Lee in the February, 1960, issue of GEOPHYSICS. However, the principles are the same in spite of the changes from rotating weight to large truck-mounted hydraulic vibrators and from analog magnetic recording to digital field recording and data processing. The main results of such improvements are improved signal quality and much more reliable and effective field operations. To our knowledge, there has been no published tutorial description or analysis of the VIBROSEIS system as presently used after more than 10 years of field experience and development. We hope that this paper will fill that void.

BASIC PRINCIPLES

The VIBROSEIS input pilot signal is typically a swept-frequency sinusoid that lasts about six or seven seconds. The signal energy is partly reflected and partly transmitted at each elastic boundary within the earth. The transit time from the signal generator to the reflecting boundaries and back to the signal detector is generally much less than the length of the input signal. The long reflected signals overlap in time so that realistic examples tend to obscure rather than demonstrate clearly the basic principles. Therefore, we have chosen for our example not only a very simple earth model but also a very short pilot signal.

The earth model and input signal are shown in Figure 1. The earth model has three elastic boundaries or reflecting interfaces. They are labeled R_1, R_2, and R_3. The reflection amplitude, which depends on the elastic contrast across the interface, can be represented graphically by a line. If we place the lines that represent the amplitudes of reflections R_1, R_2, and R_3 above or below a horizontal line (to represent posi-

---

®*Registered trademark and service mark of Continental Oil Company.
**Seismograph Service Corp., 1969.

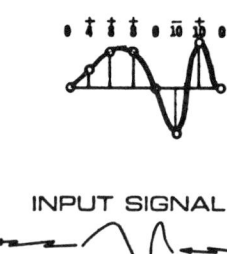
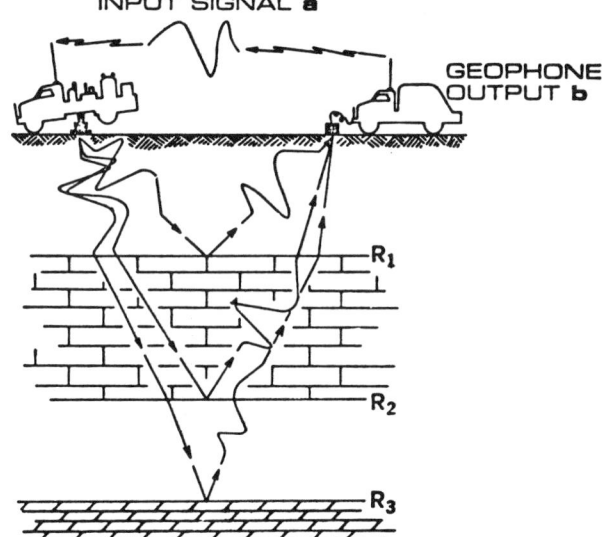

(a)

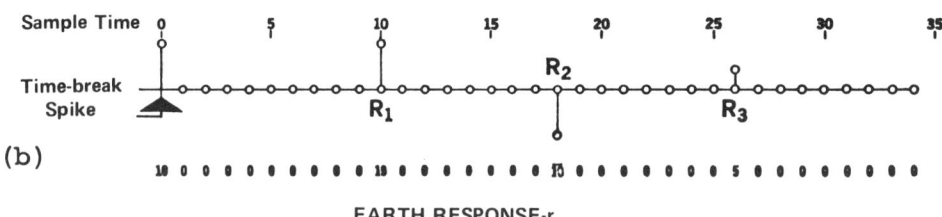

(b)

FIG. 1.—(a) Sketch of earth model with three primary reflections of input wavelet a; (b) Earth response function r, i.e. spike synthetic seismic trace.

tive and negative polarity, respectively) and space them in proportion to their transit time we will have a time-amplitude graph. To this graph we add a line to mark the time the signal was initiated and a series of little circles to represent equal intervals of transit time. The resulting graph is a digitized, spike, synthetic seismic trace that shows the earth response to primary reflections only. This earth response trace is shown in Figure 1b.

With the earth response established, we can construct the reflection signal that the geophone sees by replacing each of the spikes on the spike trace with a replica of the input signal that has the same amplitude and polarity as the spike. The mathematical name of the replacement or substitution process is CONVOLUTION. The result for our model is shown in Figure 2c. Each replica of the input signal can be clearly distinguished because the spacing of the spikes was made greater than the length of the signal. The line was broken between the zero-time reference and geophone output to emphasize the fact that in practice they are recorded as separate signals.

As mentioned above, the usual VIBROSEIS signal is many times longer than the interval between reflections. Thus, individual reflections cannot be distinguished in the geophone output and another process is required to compress the signal to a relatively narrow wavelet or pulse. There are a number of different ways to go about compressing the pulse. The procedure used to compress the VIBROSEIS signal is called crosscorrelation. Conceptually, the geophone output is searched for replicas of the input signal, which may or may not have the same polarity, and, in general, will not have the same amplitude. The searching is done mathematically by computing a number that represents the degree of similarity between the input signal and the geophone output as one is displaced with respect to the other. When the relative position of the input signal and geophone output are shifted successively by one sample interval and the similarity factor, i.e. the correlation coefficient, is computed for each position and the resulting series of numbers is plotted graphically the result is a correlogram. If a signal is crosscorrelated with itself the result is the autocorrelation of the signal.

The result of crosscorrelating the input signal a, with the geophone output b, is shown in Figure 2e. The autocorrelation of input signal a is shown in Figure 2f. Note that the effect of the crosscorrelation has been to replace each replica of input signal a with an autocorrelation of a that has the same amplitude and polarity as the original spike of the earth response r, Figure 2a.

The basic VIBROSEIS principles can be summarized as follows.

A specified short symmetrical wavelet is substituted for the earth response by means of a two-step process: (1) a unique signal, which is longer than the transit time of the deepest reflection, is transmitted into the earth where it is convolved with the earth response to form a geophone output and (2) the long reflected signals are compressed by

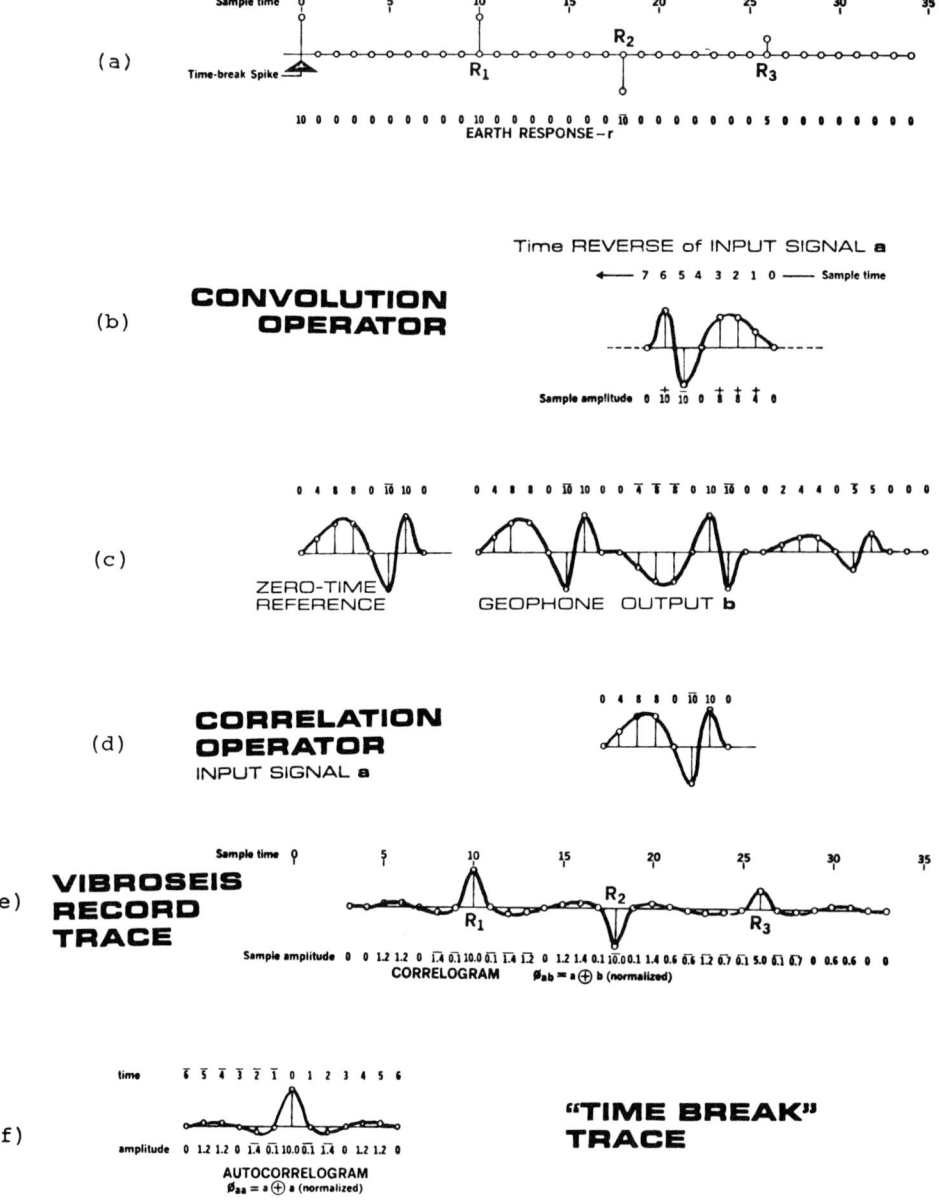

FIG. 2.—Convolution and Correlation Functions. (c) a convolved with r; (e) crosscorrelated with b.

crosscorrelation into short symmetrical wavelets that are comparable to the wavelets from an impulsive seismic source. The characteristics of the specified short wavelet are determined by the nature of the input signal.

The principles of the VIBROSEIS system and the digital operations of convolution and crosscorrelation are demonstrated by a "slide rule" from which Figures 1 and 2 are taken.

CONVOLUTION AND CORRELATION

The slide rule to demonstrate convolution and correlation is based on the theory of convolution as described by Robinson* and an equivalent theory for correlation.

The VIBROSEIS input signal and resulting geophone output are continuous (analog) functions within certain time limits. As soon as the analog signals are sampled and converted to a sequence of numbers they become digitized discrete time functions. When two such signals are convolved or crosscorrelated, the arithmetic is simply the sums of the diagonal terms of a binary cross-product table (matrix) taken in increasing time sequence with respect to one of the signals (ordinarily the longest).

For example, let the discrete time function

$$a = (a_0, a_1, a_2), \qquad (1)$$

where the subscripts denote the time sequence of the sample amplitude a, represent the input signal a. Let a second discrete time function

$$r = (r_0, r_1, r_2, r_3) \qquad (2)$$

represent the earth response.

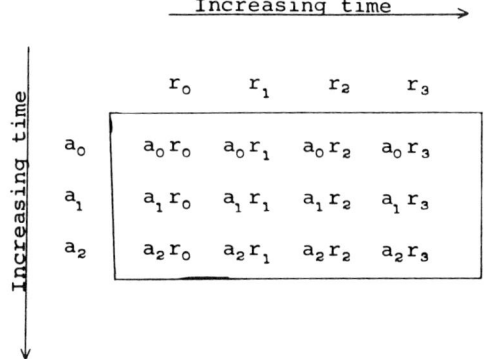

FIG. 3.—Binary cross product table (matrix) for time series a and r (after Robinson).

---

*Robinson, E. A., "Statistical Communication and Detection . . . " Charles Griffin and Company, Ltd., London, 1967.

The geophone output, which is a convolved with r, is

$$g = a * r = (g_0, g_1, g_2, \text{etc.})$$

where $*$ is the symbol for convolution of discrete time functions and the amplitudes g_0, g_1, g_2, etc. are read from the binary cross-product table shown in Figure 3. The entries in the table are the products of the time-series coefficients shown in the margins. Note that the time sequence of r increases from left to right and of a from top to bottom.

The convolution coefficients (g_0, g_1, g_2, etc.) are the sums of the cross-products in the "up-to-the-right" diagonals of the table taken in increasing time sequence with respect to time series r. Figure 4 shows how each of the convolution coefficients is related to the diagonals of Figure 3.

Note that as shown on Figure 4, when the cross-product terms are arranged in the sequence of increasing time of r the terms of a are in the sequence of decreasing time. That is, the a coefficients are in the reverse time sequence. This shows why, when we illustrate convolution graphically (as on the slide rule), the convolution operator is the time reverse of the time series that is to be substituted for each term in the series with which the operator is to be convolved.

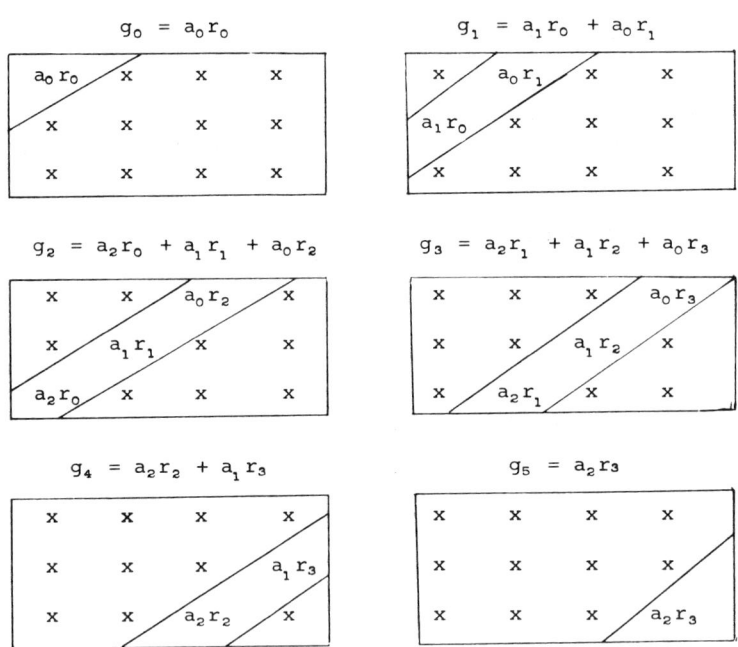

FIG. 4.—Convolution cross-product terms for successive "up-to-the-right" diagonals of the matrix shown in Figure 3. Note that the sense of the sequence of the subscript numbers of a is reversed with respect to the subscript r.

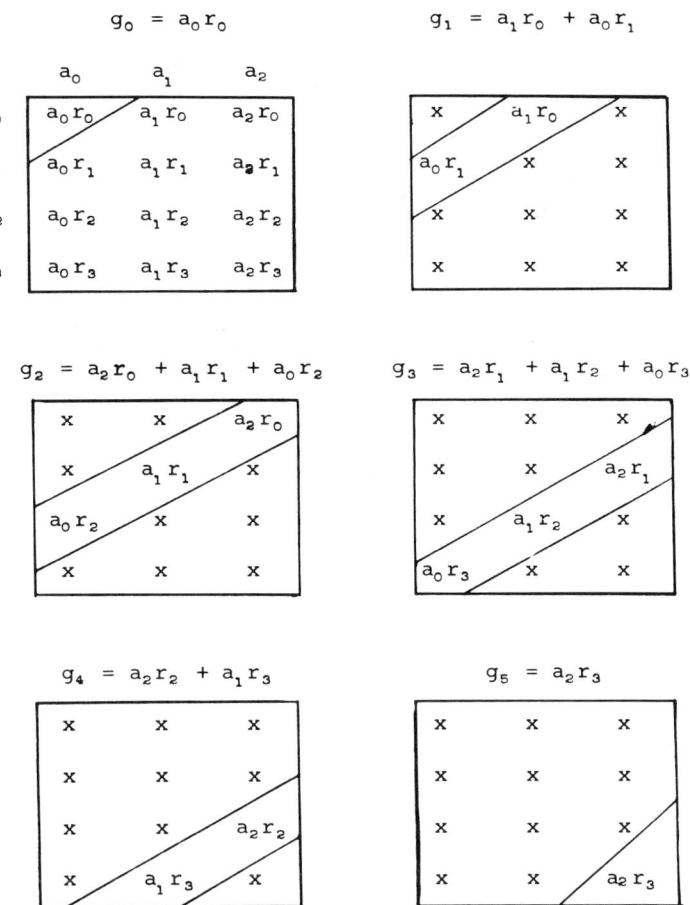

FIG. 5.—Binary cross-product matrix with locations of a and r interchanged. Note that the convolution terms are identical with those on Figure 4.

Convolution is a commutative process as shown in Figure 5 where a and r are interchanged in the binary cross-product table.

By reading the diagonals down-to-the-left in Figure 5, the sums for g_0, g_1, g_2, g_3, g_4, and g_5 are easily recognized as being identical to the values shown in Figure 4. Thus, it has been demonstrated that

$$a * r = r * a. \qquad (4)$$

Correlation, which is also demonstrated on the slide rule, uses the same binary cross-product table as in Figure 3 but the opposite set of diagonals as shown in Figure 6. The crosscorrelation of a and r is

$$\phi ar = a \oplus r = (\phi_0, \phi_1, \phi_2, \text{etc.}), \qquad (5)$$

where $\oplus$ is the symbol for correlation of discrete time functions. The coefficients ϕ_0, ϕ_1, ϕ_2, etc., are the sums of the binary cross-product terms in successive down-to-the-right diagonals of the cross-product matrix. Note that, on Figure 6, when the cross-product terms are written in the sequence of increasing time of r the terms of a that occur in each product pair are likewise in the sequence of increasing time. Let us interchange the locations of a and r in the binary cross-product matrix. The result is shown in Figure 7.

The coefficients of ϕ_{ra} and ϕ'_{ar} are tabulated in Table 1 for comparison as columns 1 and 2. Examination of the values of $a \oplus r$ and $r \oplus a$ shown in columns 1 and 2, respectively, reveals the relation shown in column 3. Namely, $r \oplus a$ is the time reverse of $a \oplus r$. If

$$\phi_{ar} = a \oplus r \tag{6}$$

then

$$\phi_{ra} = r \oplus a \tag{7}$$

and

$$\phi_{ra} = \phi_{ar(-t)} \tag{8}$$

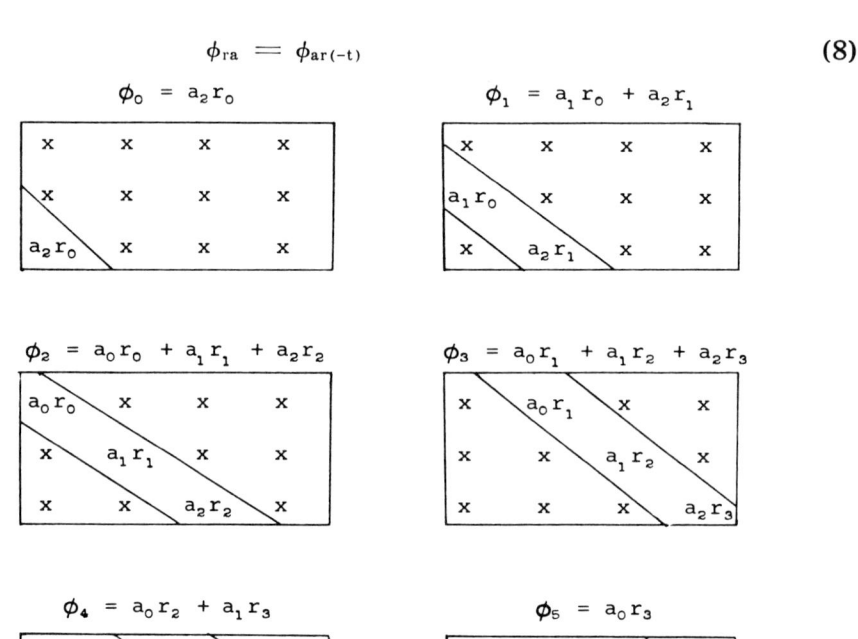

FIG. 6.—Crosscorrelation cross-product terms from successive down-to-the right diagonals of the binary cross-product matrix of Figure 3. Note that the sense of the sequence of the subscripts of a is the same as of r.

THE VIBROSEIS® SYSTEM OF SEISMIC MAPPING

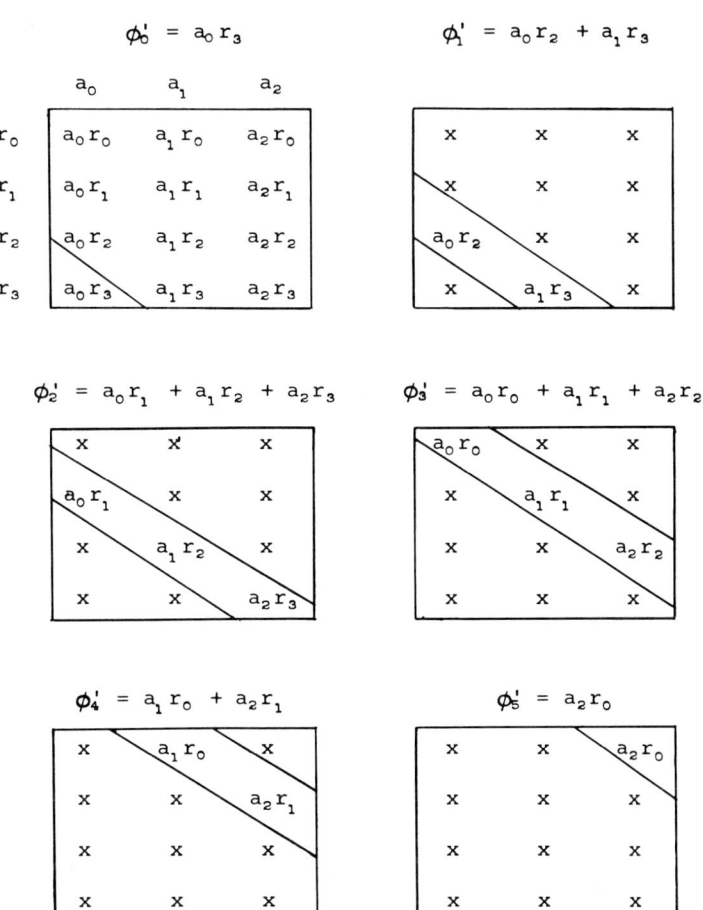

FIG. 7.—Crosscorrelation cross-product terms from matrix arrangement of Figure 5. Note that the combinations of terms in each down-to-the-right diagonal matches the terms in the diagonals of Figure 6, if the latter are taken in the reverse order.

ϕ_{ar}	$\phi'_{ar} = \phi_{ra}$	
$\phi_0 = a_2 r_0$	$\phi'_0 = a_0 r_3$	$\phi_0 = \phi'_5$
$\phi_1 = a_1 r_0 + a_2 r_1$	$\phi'_1 = a_0 r_2 + a_1 r_3$	$\phi_1 = \phi'_4$
$\phi_2 = a_0 r_0 + a_1 r_1 + a_2 r_2$	$\phi'_2 = a_0 r_1 + a_1 r_2 + a_2 r_3$	$\phi_2 = \phi'_3$
$\phi_3 = a_0 r_1 + a_1 r_2 + a_2 r_3$	$\phi'_3 = a_0 r_0 + a_1 r_1 + a_2 r_2$	$\phi_3 = \phi'_2$
$\phi_4 = a_0 r_2 + a_1 r_3$	$\phi'_4 = a_1 r_0 + a_2 r_1$	$\phi_4 = \phi'_1$
$\phi_5 = a_0 r_3$	$\phi'_5 = a_2 r_0$	$\phi_5 = \phi'_0$

TABLE 1.—Crosscorrelation cross-product terms form Figures 6 and 7 tabulated in columns 1 and 2, respectively. The equivalency is summarized in Column 3.

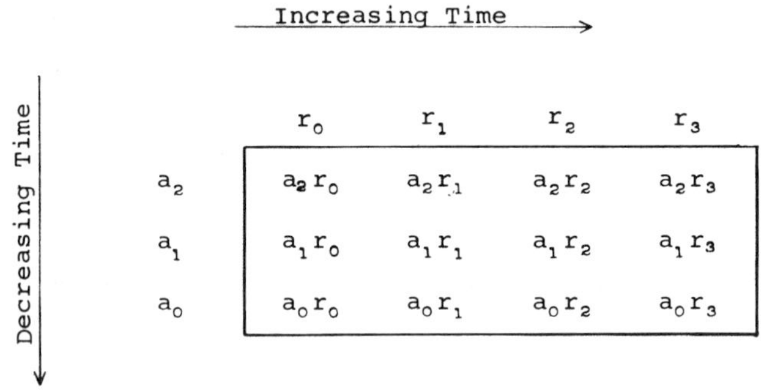

FIG. 8.—Binary cross-product matrix for $r \oplus a$. The terms in successive down-to-the-right diagonals are identical to the terms in the successive down-to-the-left convolution diagonals in Figure 4.

where the subscript (-t) denotes that the time series is reversed. Thus, the comparisons of columns 1 and 2, Table 1, show that crosscorrelation is not commutative. Note that when the terms in ϕ_{ar} and ϕ_{ra} are arranged in an increasing sequence of r the a terms are also in an increasing sequence.

Now, since convolution turns the indices of one factor around without changing the sequence of the output terms and crosscorrelation of ϕ_{ra} puts the output terms in the reverse time sequence of ϕ_{ar}, what would happen if the sequence of one set of terms was changed and then convolved with the other sequence? For example, let us reverse the order of a's in Figure 4 and place them as shown in Figure 8. The terms of $a_{(-t)} * r$ are identical to the terms of ϕ_{ar} as listed in Table 1, columns 1 and 2. Thus, by demonstration,

$$a \oplus r = a_{(-t)} * r \tag{9}$$

and, as convolution is commutative

$$a \oplus r = r * a_{(-t)}. \tag{10}$$

INPUT-SIGNAL SPECIFICATIONS

The simple slide rule example showed that the idealized result of the VIBROSEIS system of seismic mapping is to convolve the autocorrelation function of the input signal (the autocorrelation pulse) with the earth response. That is, for each "spike" (which represents the reflection coefficient at the boundary between contiguous earth layers that have different elastic properties) in the earth response we can substitute a replica of the autocorrelation pulse with the appropriate amplitude and polarity. If we know the earth response or, in other words, if we have a "spike" synthetic trace we can get an idealized seismic trace by convolving an autocorrelation pulse with the synthetic spike trace. The characteristics of the autocorrelation pulse depend on the basic VIBROSEIS pilot-signal specifications, namely amplitude spectrum and duration of the input signal. Thus, theoretically, we can choose an input signal that will give the best resolution of the desired reflections. Inasmuch as the reflections may be the response to several adjacent interfaces the sharpest (most spikelike) autocorrelation pulse will *not necessarily* give the best seismic resolution. An input signal that has a restricted bandwidth (amplitude spectrum) can often give better resolution, i.e. reflection quality, than a broadband input signal. Therefore, an important VIBROSEIS tool is the knowledge of the fundamental relations between the input signal characteristics and the autocorrelation pulse shape. This knowledge is basic to the selection of input signals for testing or production.

Input signal characteristics

One popular VIBROSEIS input signal, which we will call an LSF input signal, is a constant amplitude (except for leading- and trailing-end taper) sinusoid that changes frequency linearly with time from some beginning frequency f_1 to some terminal frequency f_2. The bandwidth Δ is the difference between f_1 and f_2. If we assume that $f_2 > f_1$ then

$$\Delta = (f_2 - f_1) \text{ Hz (or cps)}. \tag{11}$$

The length of the signal in seconds is represented by T. The product T·Δ is a very useful characteristic of the input signal and is called the dispersion (D). The rate of change of frequency Δ/T is represented by k. It is also convenient to define the midfrequency of the input signal as

$$f_0 = \tfrac{1}{2}(f_1 + f_2) \tag{12}$$

and the ratio of the terminal frequencies as

$$R_f = f_2 / f_1 \tag{13}$$

Note that the bandwidth ratio, R_f, is often expressed in "octaves" and we speak of a one-octave or two-octave, etc., input signal. In octave notation

$$R°_f = \frac{\log R_f}{\log 2} = \frac{\log R_f}{0.30103} \quad \text{(octaves)} \quad (14)$$

A typical linearly swept frequency signal is shown schematically by the solid wiggle line in Figure 9. The "boxcar" envelope that is assumed for discussion of signal characteristics is indicated by the rectangle. In general, the small deviation of the envelope of the actual input signal (output of the vibrator's radio receiver) from the ideal boxcar does not appreciably affect the shape of the autocorrelation wavelet as discussed below.

The analytic characteristics of the linearly swept frequency-modulated sinusoidal input signal are thoroughly examined by Klauder, et al* in a classic discussion of chirp radars.

Klauder represents the LSF or "chirp" signal by the real part of the complex waveform (Klauder Equ. (1))

$$\varepsilon_1(t) = \text{rect}\left(\frac{t}{T}\right) \exp\left[2\pi i \left(f_0 t + \frac{kt^2}{2}\right)\right], \quad (15)$$

where the rectangular ("boxcar") envelope of the signal is expressed by the function rect z and is defined by the relations

$$\text{rect } z = \begin{cases} 1, & \text{if } |z| < \tfrac{1}{2} \\ 0, & \text{if } |z| > \tfrac{1}{2} \end{cases} \quad (16)$$

(The pulse *rect* has unit time duration and unit height and is centered at $t = 0$. Its spectrum is given by its Fourier transform, $\dfrac{\sin \pi f}{\pi f}$. The characteristics of the pulse rect are discussed by Robinson.**)

If we let S be the real part of $\varepsilon_1(t)$ then

$$S = \text{rect}\left(\frac{t}{T}\right) \cos 2\pi t \left(f_0 + \frac{kt}{2}\right) \quad (17)$$

where, as previously noted, t is referred to $t = 0$ at the center of the pulse; T is the duration of the pulse; $f_0 = \tfrac{1}{2}(f_2 + f_1)$; and $k = \dfrac{\Delta}{T}$.

Although the VIBROSEIS pilot signal is generally defined by its terminal frequencies, f_1 and f_2, and the duration, T, the expression (17)

---
*Klauder, J. R., Prince, A. C., Darlington, S., and Albersheim, W. J., "The Theory and Design of Chirp Radars," The Bell System Technical Journal, Vol. 39, No. 4 (July, 1960), pp. 745-808.
**Robinson, E. A., Op. Cit., pp. 58-61.

is a little easier to calculate. Therefore, f_o and k are evaluated first and used to evaluate S.

The amplitude spectrum of the pulse (15) is obtained from its Fourier transform, which requires a tedious evaluation of complex Fresnel integrals well beyond the scope of this discussion. The interested reader will find an adequate discussion by Klauder.

Klauder wavelet characteristics

The autocorrelation of waveform (15) is given by Klauder's equ. (43) as

$$\varepsilon_{m1}(t) = \frac{1}{\pi k t} e^{2\pi i f_o t} \sin \pi (ktT - kt^2) \qquad (18)$$

Equ. (18) is very useful as it can be used to easily compute the autocorrelation functions of the LSF input signal without resorting to the expensive process of crosscorrelating the signal with itself. For simplicity in discussion, the waveform represented by (18) will be called the *Klauder wavelet* (cf. Ricker wavelet). A very important application of the *Klauder wavelet* is the arbitrary waveform used to produce

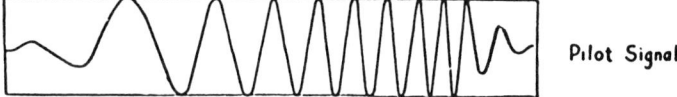
Pilot Signal

$$\mathcal{R}_e \; rect\left(\frac{t}{T}\right) e^{2\pi i \left(f_o t + \frac{kt^2}{2}\right)} = S$$

$$S = \frac{t}{T} \cos 2\pi \left(f_o t + \frac{kt^2}{2}\right)$$

$$= \cos(2\pi f_o t)\cos(2\pi k t^2/2) - \sin(2\pi f_o t)\sin(2\pi k t^2/2)$$

FIG. 9.—The typical VIBROSEIS LSF pilot-signal waveform and the theoretical rectangular ("boxcar") envelope.

synthetic VIBROSEIS seismic traces by convolution with the spike response obtained from an acoustic velocity log. Some of the characteristics of the Klauder wavelet are illustrated in Figure 10.

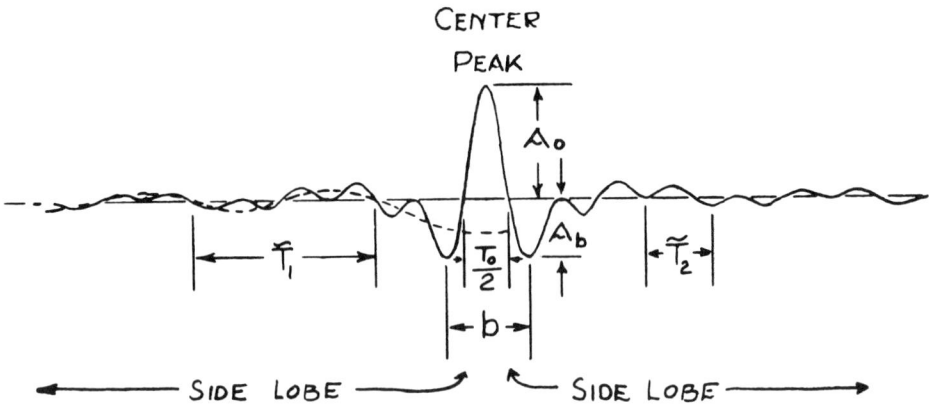

$$K = \mathcal{Re}\left[\frac{1}{\pi k t} e^{2\pi i f_0 t} \sin \pi (ktT - kt^2)\right]$$

FIG. 10.—The "Klauder Wavelet," theoretical autocorrelation of the LSF pilot-signal obtained by evaluating Klauder's eq. (43).

The example in Figure 10 has the typical shape of the autocorrelation of a VIBROSEIS input signal that has a two-octave bandwidth. This particular Klauder wavelet was chosen for several reasons. It is one of the best for illustrating the wavelet characteristics because, in many areas, the maximum useful bandwidth of the input signal is about two octaves. And the center peak is sufficiently prominent to provide an acceptable approximation of a spike on the VIBROSEIS record (correlogram).

As can be seen, the wavelet is symmetrical about a vertical line through the center peak. If we define b_0 (the breadth of the center peak) as the absolute value of the time interval between the first + and − zero crossing points on either side of the center peak we find that

$$b_0 = \frac{T_0}{2} \qquad (19)$$

where T_0 is the period of the midfrequency f_0. Thus, we can use this same pulse shape for any *two-octave* Klauder wavelet (i.e. autocorrelation of the LSF sinusoidal input signal) by adding vertical time lines with the proper spacing. For example, if f_0 is 50 cps b_0 will be 10 milliseconds. Therefore, the 10-ms time lines should be spaced at integral multiples of the center-peak breadth b_0. Or, if f_0 is 25 cps b_0 will be 20 ms and the 10-ms time lines should be spaced at integral multiples of $\frac{1}{2}b_0$.

We can use the Klauder wavelet in synthesizing VIBROSEIS data as we use the Ricker wavelet in synthesizing seismic data from impulsive sources. Therefore, we define the Klauder wavelet breadth as the absolute value of the time interval between the first troughs on either side of the center peak. The pulse breadth is an easy-to-measure index of the mid- or peak-frequency of either the Ricker or Klauder wavelet. However, in neither case is it the period of the midfrequency. Although b in the Klauder wavelet is approximately twice b_0 for signals of one octave bandwidth or less, b is always less than $2b_0$ and the difference increases as the pilot-signal bandwidth increases.

The portion of the wavelet outside of the center peak is called the side lobe. The side lobe extends to the end of the wavelet and, therefore, is almost as long as the original input signal. (The autocorrelation function of a digital signal n samples long will have 2n-1 samples. Another useful evaluation of the input signal characteristics is given by the relative amplitude of the center peak and the side lobes. If the amplitude of the center peak represents the signal on the VIBROSEIS record then the amplitude of the side lobes represents a signal-generated noise. Thus the centre peak-to-side lobe amplitude ratio represents a theoretical intrinsic signal-to-noise ratio (A_0/A_{SL}) inherent in the specifications of the input signal.

The envelope of the wavelet varies with time. The most rapid decrease in envelope amplitude occurs near the center peak because of the factor $\frac{T}{\Delta \pi t}$. At great side-lobe times the amplitude of the envelope decreases more slowly. The autocorrelation pulse side lobes from narrow-band (e.g. less than one-half octave) input signals have "beats" of decreasing maximum amplitude that may still be only 20 db below the center peak at a time of 20b after the center peak (e.g. about .700 sec after the center peak for a 1/3-octave signal and $f_0 = 30$ cps). If we define the amplitude of the first trough outside the center peak as A_b then A_b/A_0 represents one of the most useful measures of the "spikiness" of the autocorrelation pulse. Theoretically $A_b/A_0 = 1$ where $f_2/f_1 = 1$, A_b/A_0 decreases to about one-half when f_2/f_1 is about four or one-quarter. In the case of the two-octave example $f_2/f_1 = 4$ and A_b/A_0 is about one-half.

The side lobes of the two-octave autocorrelation pulses show rather closely the normal high-frequency "ripple" in phase with the center peak and the underlying low-frequency upon which the ripple appears to be superimposed. This underlying low frequency appears to be 180° out of phase with the high frequency at the center peak time. (This apparent contradiction of the zero phase relation at zero time is related to the fact that frequencies from zero to f_1 have been omitted from the input signal. The dashed low-frequency line does not represent the phase relation of f_1. It is simply the "bias" upon which the high frequency f_2 appears to oscillate.) Where these two periods, T_1 and T_2, can be recognized they give an excellent "rule-of-thumb" criteria for determining the approximate f_1 and f_2 of the input VIBROSEIS signal. The average of f_1 and f_2, (f_0), should be very close to the value computed from the relations

$$f_0 = \frac{1}{2b_0} \text{ or } \widetilde{f_0} = \frac{1}{b} \qquad (20)$$

The effect of the bandwidth of the LSF input VIBROSEIS signal on the autocorrelation pulse is shown in Figure 11. These autocorrelation

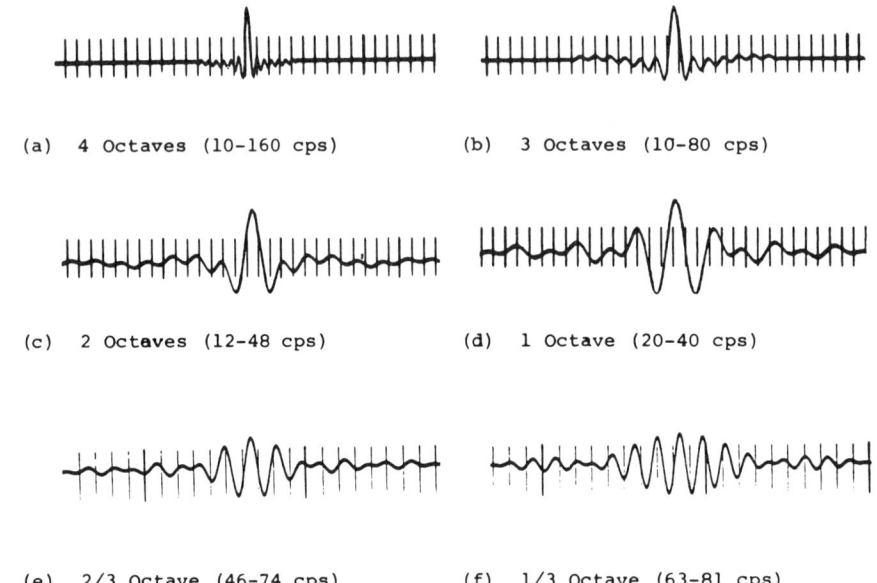

(a) 4 Octaves (10-160 cps)   (b) 3 Octaves (10-80 cps)

(c) 2 Octaves (12-48 cps)   (d) 1 Octave (20-40 cps)

(e) 2/3 Octave (46-74 cps)   (f) 1/3 Octave (63-81 cps)

FIG. 11.—Klauder wavelets that demonstrate the effect of pilot-signal bandwidth on the autocorrelation pulse shape.

pulses are for LSF input signals of (a) four, (b) three, (c) two, (d) one, (e) two-thirds, and (f) one-third octave bandwidths. The terminal frequencies are indicated by parentheses so that the pulse shape can be related to the 10-ms time lines.

Input signals in the one- to two-octave range are among the most practicable. Broader bandwidths, which have an autocorrelation pulse that is more spike like, are more difficult to achieve in routine field practice. Input signals with narrower bandwidths have "multi-legged" autocorrelation pulses that give the VIBROSEIS record a "ringing" appearance.

It is very important to avoid mistaking the ringing caused by input signals that are too narrow with seismic ringing such as encountered frequently in marine operations. One means of distinguishing between the two kinds of ringing is a comparison of the autocorrelation of the received seismic signal. If the ringing is caused by an LSF signal that has too narrow a bandwidth the autocorrelation will resemble the pulses in Figures 11e and 11f. If the ringing is caused by seismic reverberation, the autocorrelation will have a strong side lobe at some time t_r after the center peak where t_r is the period of the reverberation. The characteristics of the autocorrelation pulses from marine reverberation are illustrated and discussed by Anstey.*

Signal-to-noise (S/N) characteristics of the LSF input signal and crosscorrelation

Up to this point, the only noise that has been discussed is the "side lobe" noise inherent in the LSF input signal bandwidth. In any seismic system there are many sources of the interference that tend to mask the "desired earth-response" signal. In this context, we define *signal* as the primary reflections from the elastic interfaces that we wish to map. *Noise,* therefore, is all of the non-signal energy that is recorded. This non-signal energy includes electrical "pickup" and instrument noise as well as all of the non-signal energy in the geophone output voltage.

The VIBROSEIS system gives us two stages for improving our S/N. The first stage is the choice of the input signal bandwidth, terminal frequencies, and duration. The second stage is the crosscorrelation process, which can further improve the S/N.

There are many kinds of seismic noise, which we may roughly classify as ambient (non-source-generated noise) and source generated noise. One of the most common and troublesome ambient noises is the 60-cps interference from electrical transmission lines (i.e. hi-line noise). Another form of ambient noise is caused by wind, especially when the seismometers are planted in grassy, brushy, or wooded areas. Airborne sound from highways, aircraft, or stationary machinery are the other main types of ambient noise.

The S/N on VIBROSEIS records (correlograms) can be improved with respect to mono-frequency signals by omitting that frequency in the input signal as, for example, by using a signal that sweeps linearly from 15 to 56 cps to reduce 60 cps hi-line interference. The crosscorrelation process provides a basic S/N improvement when the monofrequency is within the input signal bandwidth. Landrum* developed the following relations, (20) and (21), which permit a practical prediction of the S/N

---

*Anstey, N. A., "Correlation Display Pinpoints Seismic Multiples," Seismograph Service Limited, reprinted for circulation by Seismograph Service Corporation, 1966.

*Landrum, Ralph A., Jr., 1967, "Extraction of Signals from Random Noise by Crosscorrelation," paper at 37th Annual International Meeting of the Society of Exploration Geophysicists in Oklahoma City, Oklahoma.

improvement to be expected, and are accurate within a range of a few dB.

$$\text{S/N improvement} = 20 \left[\log_{10} \sqrt{D}\right] \text{ dB} \qquad (21)$$
$$f_1 < f_n < f_2$$

where $D = T \cdot \Delta$, f_n = the noise frequency, and f_1 and f_2 are the terminal frequencies of the input signal. When f_n is the same as one of the terminal frequencies of the input signal the S/N improvement is twice as good, i.e.,

$$\text{S/N improvement} = 20 \left[\log_{10} 2\sqrt{D}\right] \text{ dB} \qquad (22)$$

The S/N improvement increases rapidly, generally speaking, from that given in equation (22) as the noise frequency departs farther from the signal pass band.

The crosscorrelation process is likewise a powerful tool for improving the S/N when the noise has a random pattern as produced by wind or traffic noise. The improvement resulting from crosscorrelation is

$$\text{S/N improvement} = 20 \left[\log_{10} \sqrt{T \cdot W}\right] \text{ dB} \qquad (23)$$

where T is the length of the input signal in seconds, W is the bandwidth in cps of the noise (i.e. $W = f_{n2} - f_{n1}$),

$$f_{n1} < f_1, \text{ and } f_{n2} > f_1 \qquad (24)$$

In general, f_{n1} will be the lower limit of the recording system, such as the seismometer cutoff frequency, for example. And, f_{n2} will be the upper limit of the recording system such as, for example, the cutoff frequency of the anti-alias filter in the amplifier. Note that, as long as W contains Δ entirely, the S/N improvement depends on the length of the input signal and is independent of the input signal frequencies or bandwidth.

Although it is not possible to express the S/N improvement quantitatively with respect to source-generated noise in general formulas, such as equations (21), (22), and (23), the same principles apply qualitatively. The most appropriate example is the improvement in S/N with respect to Rayleigh-type surface waves. In essence, such waves have very narrow frequency bandwidths. Thus, the most effective attenuation technique is to avoid the surface-wave frequencies in the input signal specification. For example, if the dominant surface-wave bandwidth is 14-18 cps choose 20 cps as the f_1 of the input signal.

In practice, of course, it is not quite that simple. In the first place the surface-wave frequencies must be determined by field tests at sites representing extremes in the velocity-layering characteristics. These tests determine the total bandwidths of the surface-wave interference to be expected over the prospect. When the surface-wave bandwidth overlaps the bandwidth of the desired signals such frequency discrimination becomes less effective as the overlap increases. However, if the field tests that give the frequencies of the surface waves are designed to give also the phase velocities, the range of wavelengths to be attenuated can be computed. Thus, source and receiver patterns can be designed to provide maximum attenuation for the noise wavelengths that the selected input signal will generate.

The foregoing discussion has highlighted the principles of the VIBROSEIS system of seismic exploration that lead to the selection of an input signal that will both enhance the desired reflection signals and give optimum attenuation of the undesired interference from ambient and source-generated noise. It must be emphasized that the VIBROSEIS system is not a panacea for all seismic ills. It is an engineered system that, more than any other seismic system, can be tailored to the specific signal and noise characteristics that represent the earth response of the prospect.

PART I

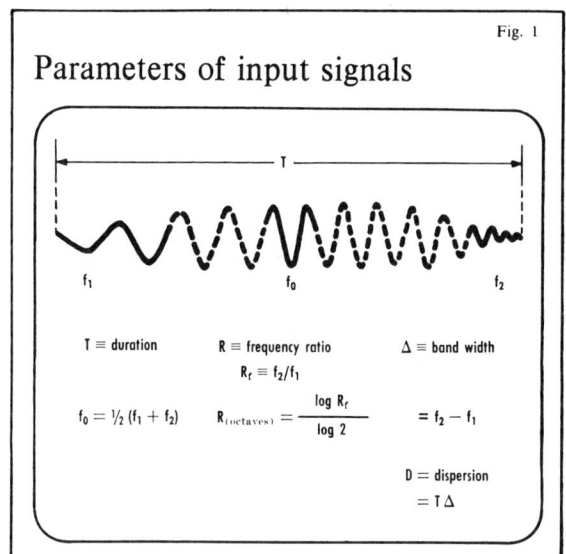

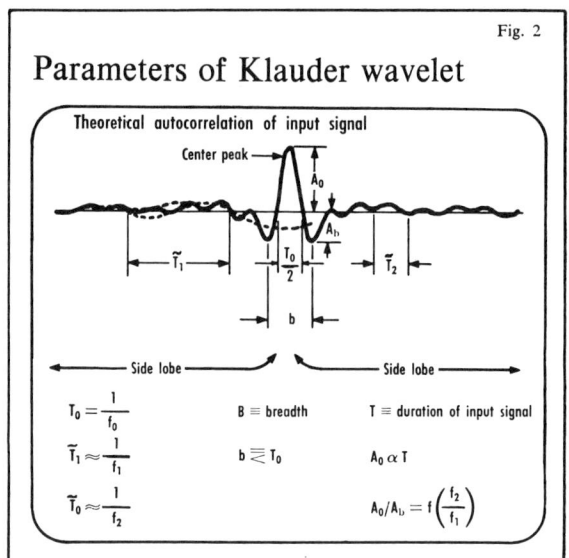

Vibroseis parameter optimization

Robert L. Geyer
Personnel Training Manager
Seismograph Service Corp.

"It is my personal opinion that the Vibroseis system has more potential than any of the new field systems available. And yet it is the most widely misused system that I know of today," the author declared at the 1969 SEG convention in Calgary.

One of the most common reasons for this misuse is because people do not really understand the parameters involved in the Vibroseis system of seismic mapping. We must define these parameters and show how they should be optimized with respect to the near-surface geology and deep geologic objective horizons.

Parameters of the input signal. The input signal is a swept frequency sinusoid of length or duration T as shown in Fig. 1. The beginning frequency we call f_1 and the ending frequency we fall f_2.

The average of the beginning and ending frequencies, which we call f_0, is the frequency at the center of the signal. In Fig. 1, the ends of the input have been tapered, as they are in normal practice, to avoid the transients that we would get from the ideal boxcar envelope described in the early chirp radar literature.[1]

The beginning and ending frequencies define the bandwidth of the input signal. The bandwidth can be expressed as a simple ratio (f_2 divided by f_1), a ratio expressed in octaves, or as the difference Δ between f_2 and f_1 in hz. Klauder calls the product $T \cdot \Delta$ the dispersion D. The dispersion is one of the measures of the power, or more properly, the correlation power in the collapsed signal.

We have named the theoretical autocorrelation of the linear-swept frequency signal (Klauder's Equation 43) the Klauder wavelet.

The Klauder wavelet has essentially all the characteristics of the true autocorrelation function but requires much less machine time to compute. The wavelet shape depends on the input signal frequencies.

For example, the bandwidth determines the sharpness of the peak. More specifically, however, the frequency ratio determines the basic shape of the Klauder wavelet, as well as of the true autocorrelation function.

The shape, or character of the wiggles, if we ignore the time scale, is identical for all wavelets that have the same frequency ratio. If we change the frequency ratio, we change this shape.

The wavelet breadth b, as defined in Fig. 2, is the time interval between the two first negative side lobes. This definition is analogous to the wavelet breadth as defined for the Ricker wavelet.

The Klauder wavelet breadth is always less than or equal to the period T_0 of the frequency f_0, the average frequency of the input signal. The zero crossing interval for the central peak of the autocorrelation function is exactly half of the period T_0.

The Klauder wavelet has a low frequency underlying the high-frequency jitter of the side lobes, indicated by the dashed line on Fig. 2, which appears to be 180° out of phase with the central peak.

This low frequency is approximately the same as the lowest frequency of the input signal. The superimposed high-frequency jitter is essentially the same as the high frequency end of the input signal.

These observations indicate one way in which the shape of the autocorrelation function is directly related to the frequencies in the input signal.

The ratio of the amplitude of the center peak, A_0, to the amplitude of the first side lobes, A_b, is a function of the ratio of the highest to the lowest frequency in the signal.

Thus, there is a double check on the input signal frequencies. If the

---

Paper presented at SEG convention, Calgary, 1969. Awarded best presentation award Research Session.

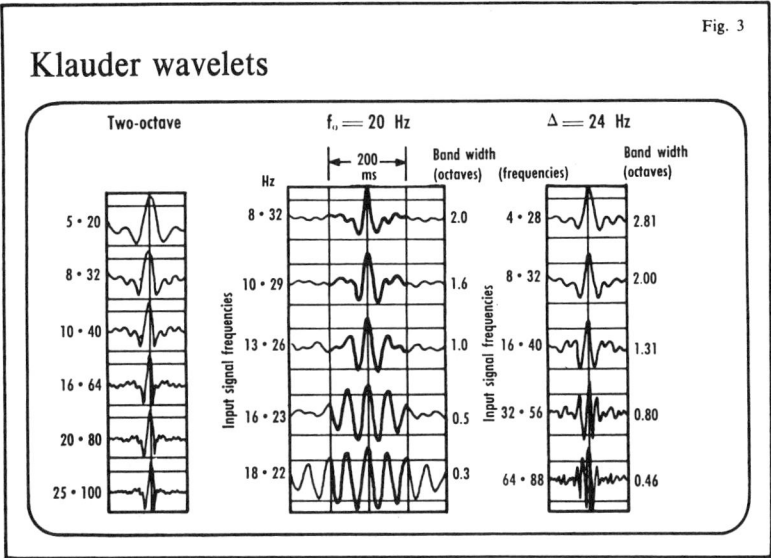

Fig. 3

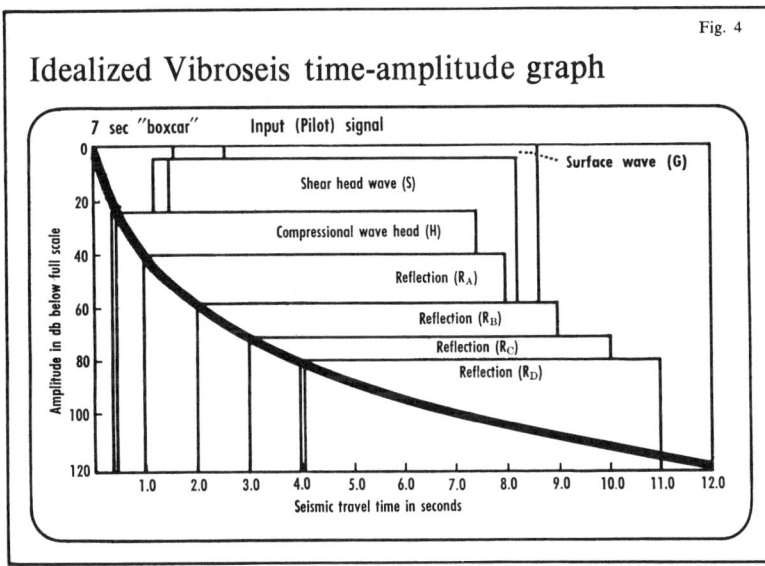

Fig. 4

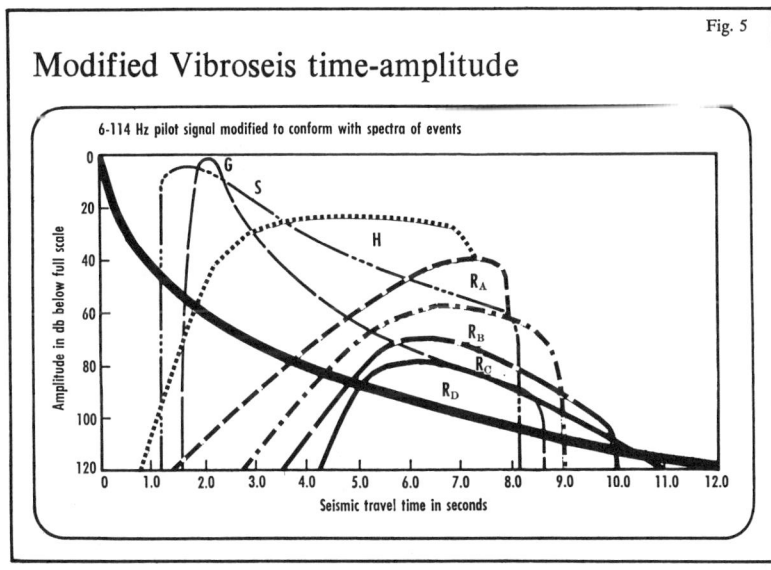

Fig. 5

high-frequency jitter and the low underlying frequency cannot be measured to determine the beginning and ending frequencies, the central wavelet breadth and the ratio of the amplitude of the central peak to the amplitude of the first side lobe can be measured to get an approximation of the frequency ratio.

It is much more important, however, to realize that the frequency relations are more useful for us in reverse. In other words, we can predetermine the shape of the Klauder wavelet by the proper choice of the input signal frequencies.

The effect of input signal frequencies on the shape is shown in Fig. 3 for different bandwidth ratios and differences. On the left are two-octave signals. The wavelets have essentially identical shapes except for the time scale itself.

The higher frequency signals at the bottom are compressed in time. But the shape, i.e. the relative amplitude of each successive up or down excursion, is the same.

The central group of Klauder wavelets in Fig. 3 have the same central frequency, namely, $f_n = 20$ hz. The input frequencies range from 8-32 hz at the top to 18-22 hz at the bottom.

Notice that the smaller the ratio to high-to-low frequency, the more legs there are in the wavelet and the more the Klauder wavelet appears to ring.

All of the Klauder wavelets on the right in Fig. 3 have the same bandwidth, namely 24 hz. At the top is the wavelet for a 4-28 hz signal, which has a bandwidth of about 2½-octaves, and a good wavelet shape. Below that is an 8-32 hz wavelet from a two-octave signal, which also has a good wavelet shape.

The 16-40 hz, 1⅓-octave wavelet, has about the same amplitude on the first side lobe as it has on the central peak. When the frequency ratio drops to 8/10 of an octave, as in the 32-56 hz wavelet, there are more legs. When the ratio drops to 4/10 of an octave there are many legs.

Here the breadth of the wavelet envelope is roughly the same for all of the autocorrelation functions that have the same delta, but the frequencies within that envelope change.

In other words, if we consider that the Klauder wavelet is basically the modulation of a carrier frequency f_0, the carrier frequencies are emphasized in each one of these signals.

Thus, if we know ahead of time what shape we would like to have our seimic wavelets take, theoretically, we can strive for that shape by the proper choice of the input signal frequencies. Ordinarily, we would like to choose an input signal that would produce a seismic wavelet (a Klauder wavelet, if you like) that would give the best resolution of the geologic problem. To find this best wavelet, we make a spike synthetic seismic trace that represents the geology of the objective horizons. Then we convolve the spike synthetic trace with Klauder wavelets having different input frequency ranges and different center frequencies to find the wavelet that gives the best resolution of the geologic objective.

The linear-swept frequency signal whose autocorrelation function is the same as the best Klauder wavelet is, theoretically, the best signal to use in the field to define the subsurface geologic objective.

If there are a number of different geologic objectives, each requiring a different ideal wavelet shape, then we must compromise in specifying the input signal frequencies. Ordinarily, we would concentrate on the geologic zone with the greatest economic value.

There is another aspect of the Vibroseis system which we tend to forget or overlook when we consider the change in reflection amplitude with record time and that is the relation of the very long input signal to the very short pulse of the usual dynamite system.

Fig. 4 shows the idealized Vibroseis time-amplitude graph of the reflections of a 7-sec boxcar input signal. The reflections are shown as a series of rectangles of which the upper left corners fall on the attenuation curve.

The attenuation curve represents the decay in amplitude of compressional reflections with record time. It is an idealized curve based on Gulf Coast data.

However, the shape of the curve not only depends upon the geologic province in which we are making our study, but it also depends upon the bandwidth of our recording.

In particular, it is sensitive to the low-frequency end imposed by the natural frequency and damping of the seismometers that are used.

If we use lower frequency seismometers, we'll get a curve with a slower rate of attenuation with time. If we use higher frequency seismometers, we reduce some of the low frequencies of the deeper reflections and the curve will drop off at a faster rate.

In Fig. 4, the usual reflections from a dynamite explosion would give events that would last little longer than the width of the vertical lines at the left end of each rectangle. That is, at the time scale used, the breadth of the reflection wavelet is barely greater than the breadth of the line until the time reflection R_D at 4 secs.

Compressional and shear headwaves and Rayleigh type surface waves are similarly indicated by 7-sec-long rectangles on Fig. 4. Notice that the relative-amplitude rectangles for the shear and compressional headwaves completely cover almost all of the reflections.

Therefore, if we record on digital tape with a broad frequency band at full scale, it is the amplitude of shear headwave (S) and the Rayleigh type surface wave (G) that determine the gain in the amplifier rather than the amplitude of the reflections that we would like to record.

If we record under the circumstances indicated in Fig. 4, with no additional attenuation from source or receiver patterns, we find that our reflection R_D is more than 80 db below the full-scale recording level and therefore would not be recorded under any circumstances on the usual digital field tape which has a resolution of about 78 db.

A similar graph, Fig. 5, in which the rectangular input wave shape has been modified to agree with the specific amplitude spectra of the individual seismic events, shows that the surface wave is very sharply peaked at a particular time related to the arrival time of a very narrow band low-frequency part of the input signal.

It is those early low frequencies that cause a resonant surface wave. The sheer headwaves also peak near the beginning of the record and drop off rapidly with later time.

The compressional headwave is a broad-band event and therefore has high amplitude throughout almost the entire recorded time. The reflections are weaker and as the last part of the reflection of the Vibroseis signal is higher frequency than the beginning part, we see that later and later portions of the uncorrelated Vibroseis signal are emphasized as we look at shallower and higher frequency reflections. As a result, in the neighborhood of 7 to 8 sec, there is a large buildup or hump in the overall record amplitude, which is related to the compressional reflections.

As a matter of fact, if we were

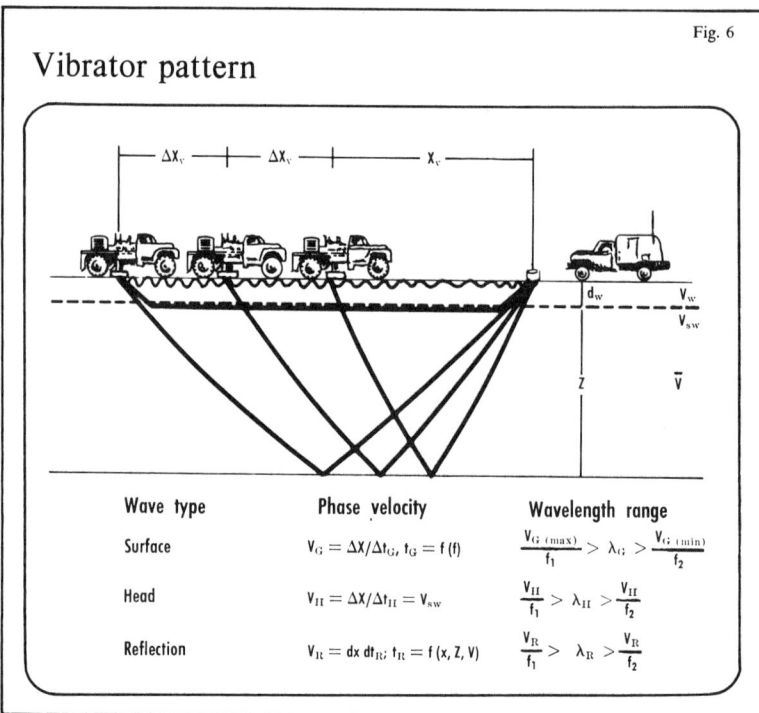

Fig. 6

Vibrator pattern

Wave type	Phase velocity	Wavelength range
Surface	$V_G = \Delta X/\Delta t_G, t_G = f(f)$	$\frac{V_{G(max)}}{f_1} > \lambda_G > \frac{V_{G(min)}}{f_2}$
Head	$V_H = \Delta X/\Delta t_H = V_{sw}$	$\frac{V_H}{f_1} > \lambda_H > \frac{V_H}{f_2}$
Reflection	$V_R = dx\,dt_R; t_R = f(x, Z, V)$	$\frac{V_R}{f_1} > \lambda_R > \frac{V_R}{f_2}$

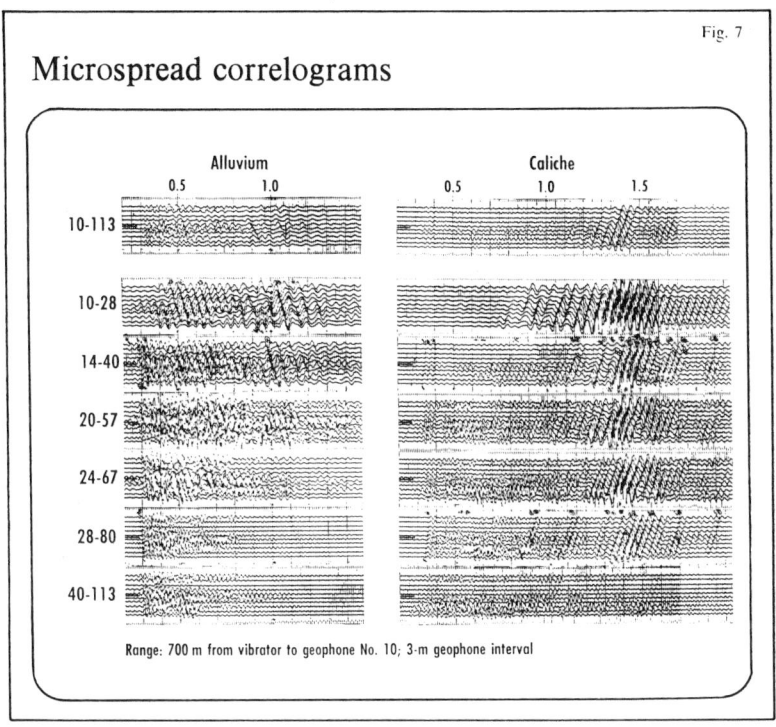

Fig. 7

Microspread correlograms

Range: 700 m from vibrator to geophone No. 10; 3-m geophone interval

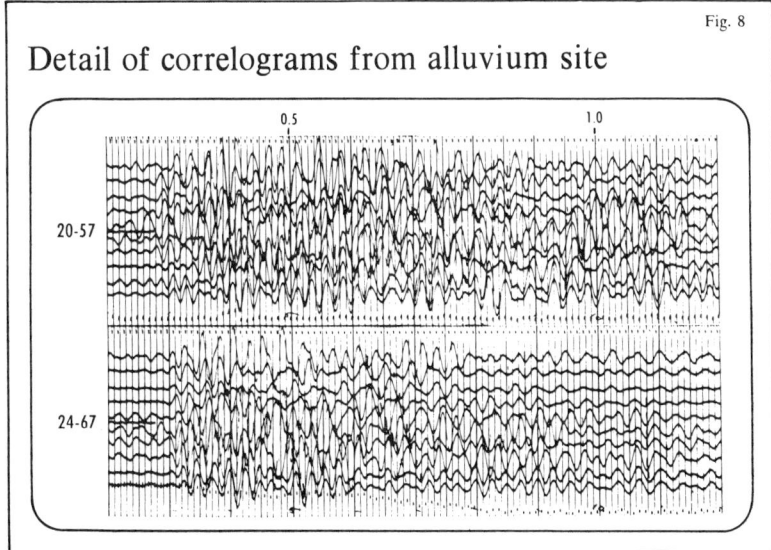

Fig. 8

Detail of correlograms from alluvium site

to discard the energy before about 4 sec in this broad-band example, we would still have basically all the reflection energy needed. In addition, we would gain about 25 db with which to bring the headwave up to full scale and concurrently raise the level of all reflections by the same amount. Therefore, by omitting the low end of the frequency band we improve the recoverability of the deep reflections.

When possible, graphs such as this should be made during the initial field-test to determine what frequencies should be used to emphasize the objective reflections and to minimize the effects of compressional headwaves, surface waves, and shear waves.

Practically speaking, we should pick a linear-swept frequency signal in which the lowest frequency starts after the peak frequencies of the ground roll and shear waves.

As indicated, this choice would give us perhaps 24 db increase in signal amplitude at the outset due to the improvement in signal/noise ratio. The amplitude of the compressional head- wave would still be much higher than we would like. Therefore, the vibrators should be spaced to provide a source pattern that would attenuate wavelengths in the headwave so that the seismometers would receive a lower-amplitude headwave and higher-amplitude reflections and thus improve the signal/noise ratio still further.

This reduction in the amplitude of the headwave is one of the most important techniques for achieving maximum signal/noise ratio in Vibroseis work.

Fig. 6 shows the range of wavelengths and simple velocity-frequency relations for the main types of seismic waves including the objective reflections.

For example, the phase velocity of the surface wave is determined by dividing the distance between the detectors or the vibrators by the wave travel time.

In the transposed method illustrated, the vibrators are moved and the time is recorded separately for each vibrator location. In the normal (opposed method) the receivers are spaced close together and recorded separately. Both transposed and opposed techniques permit us to identify the waves.

Surface waves are dispersive, thus the wave velocities are functions of the wavelength. Dominant wavelengths are a function of the thickness of the low-velocity layer. Therefore, for a given element spacing in the pattern, the velocity is a function of the input signal frequency. We divide the maximum velocity by the smallest input frequency to determine the largest wavelength and we divide the smallest velocity by the largest frequency to determine the smallest wavelength.

These two computations give the wavelength range that we would have to contend with in a particular area. The velocities are measured on test records. The beginning and ending frequencies are determined by synthetic record studies and other criteria.

The headwaves will ordinarily have one velocity, the subweathering velocity, unless there is enough distance between the source and the farthest receivers to allow several refraction branches of multiple weathering layers to be recorded.

In the single velocity case, we

simply divide the headwave velocity by the beginning low frequency to find the longest wavelength and by the ending high frequency to find the shortest wavelength. This computation gives a range of wavelengths for the headwave, which has been shown to be one of the most important causes of poor Vibroseis signal/noise ratio.

The reflection wavelength, of course, depends on the slope of the travel-time curve, which can be expressed as (dx/dt_r) where (t_r) is a function of the source to receiver distance x, the depth z and the average velocity v to the reflectors. Thus, at one particular receiving location, we can say that the reflection wavelengths will vary between the reflection velocity divided by the lowest frequency and the reflection velocity divided by the highest frequency.

Our problem in parameter optimization is that of choosing, if possible, a spacing for the vibrators that will do the best job of attenuating the headwaves and will affect least the reflection wavelengths.

The spacing between vibration points or seismometers should be small enough to attenuate the shorter wavelengths in the surface waves.

Fig. 7 shows standard microspread correlograms.

The vibrators were at one spot. A single vibration was adequate for each record to determine the kind of interference produced by the different input pilot signals.

These correlograms are from two recording sites about 600 m apart. The set on the right was made on alluvium of unknown thickness.

Only on the 10-28 hz record is there a severe noise problem. Shear waves are developed on the early part and Rayleigh type surface waves in the middle and later part. The 14-40 hz record has appreciable surface-wave amplitudes but the 20-57 hz record has relatively little. The higher frequency records have very little evidence of surface waves.

On the right set, made on a surface outcrop of caliche, there is a long band of interference waves starting with the shear waves and ending with a very resonant set of Rayleigh type surface waves.

These waves appear on all of the correlograms including the one from the 28-80 hz signal. There is even some slight evidence of the interference on the 40-113 hz record.

Tests, such as illustrated in Fig. 7, give the range of wavelengths (by computation) that would be encountered at one particular distance from the vibrators to the recorders. Of course, similar tests must be made at the near and far ends of the production reflection spread to determine the total expected range of interference wavelengths.

A common and important Vibroseis phenomenon is illustrated by the 20-57 and 24-67 hz records. The detail is enlarged by the 20-57 and 24-67 hz records. The detail is enlarged in Fig. 8. Notice at about 0.55 sec on the 20-57 hz record that although the amplitude is too large for comfort there is a very strong, nearly vertical event. This is a shallow reflection.

At the same record time, on the 24-67 hz record, there are only surface waves, or low-velocity interference waves. This is a typical example that shows how a relatively small change in the input signal frequency can make a very large change in the reflection continuity, or the signal/noise ratio of the received data. Usually it is the low-frequency end of the signal that has this influence.

END PART 1

PART II—Conclusion

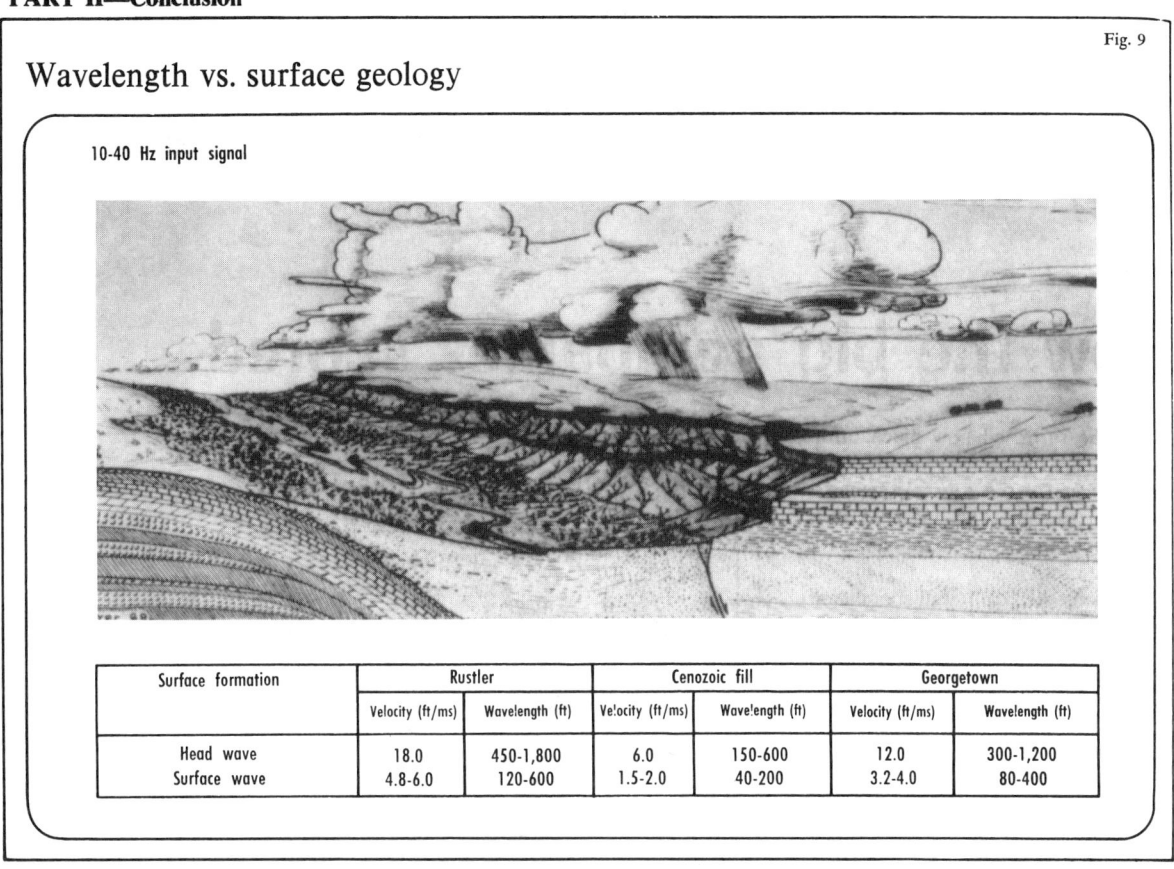

Fig. 9

Wavelength vs. surface geology

10-40 Hz input signal

Surface formation	Rustler		Cenozoic fill		Georgetown	
	Velocity (ft/ms)	Wavelength (ft)	Velocity (ft/ms)	Wavelength (ft)	Velocity (ft/ms)	Wavelength (ft)
Head wave	18.0	450-1,800	6.0	150-600	12.0	300-1,200
Surface wave	4.8-6.0	120-600	1.5-2.0	40-200	3.2-4.0	80-400

Vibroseis parameter optimization

ROBERT L. GEYER
Seismograph Service Corp.

Now that we have examined typical data from one particular location, let's take a brief look at the wavelengths characteristic of typical geology in West Texas and New Mexico.

Fig. 9 has a compressed schematic geologic section from the Rustler Hills on the left to the Edwards plateau on the right.

The Rustler Hills comprise a high-speed anhydrite-dolomite outcrop underlain by salt and anhydrite.

The velocity in the Rustler is 1,800 fps, which gives a wavelength range from 450 to 1,800 ft for the 10-40 hz input pilot signal.

In the Cenozoic fill, a thick section of detrital material, the average compressional wave velocity is about 6 fpms, and therefore the wavelengths are about a third as large as for the

Vibroseis is a trademark of Continental Oil Co.

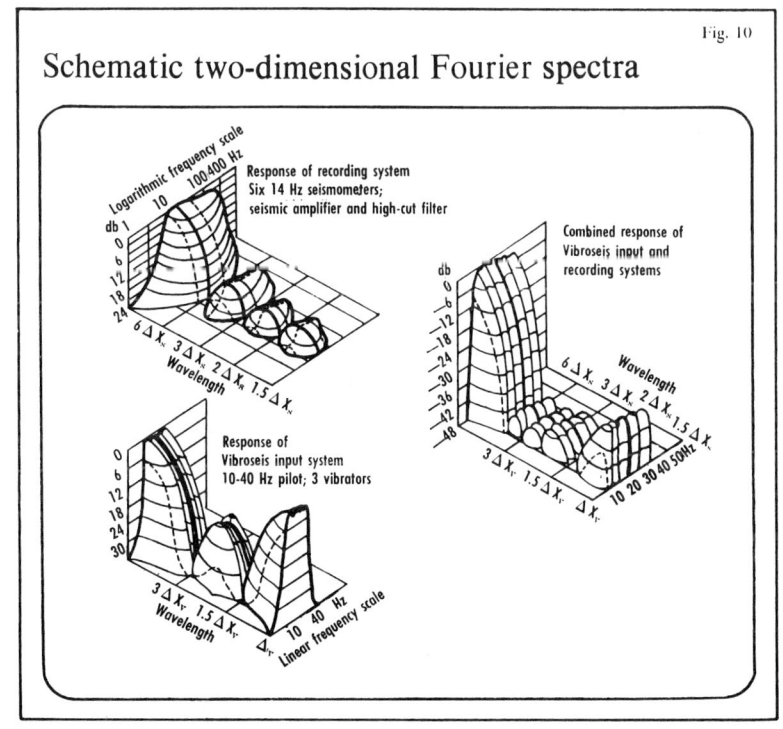

Fig. 10

Schematic two-dimensional Fourier spectra

Response of recording system
Six 14 Hz seismometers;
seismic amplifier and high-cut filter

Combined response of Vibroseis input and recording systems

Response of Vibroseis input system
10-40 Hz pilot; 3 vibrators

Rustler, i.e. 150 to 600 ft.

The Georgetown limestone, which caps the Edwards plateau, has a velocity of about 12 fpms and the wavelengths are about twice the values for the Cenozoic fill or two-thirds of the values for the Rustler.

They range between 300 and 1,200 ft. To use our vibrators to reduce the headwave interference in such a variety of geologic conditions, we should have to have large vibrator spacing.

It would be preferable to have more than just the three vibrators indicated in the Fig. 6 to do an adequate job.

On the other hand, if we look at the surface wave interference, we find that in the Rustler outcrop area the velocities range between 4.8 and 6 fpms. As a result, the wavelengths range between 120 to 600 ft.

In the Cenozoic fill the thicker the fill the lower the velocity and we may get 1.5 to 2.0 fpms for the Rayleigh type surface waves where the fill is thick. There the wavelengths range between 40 and 200 ft.

Along Georgetown-Edwards outcrop, velocity of the surface waves varies between about 3 and 4 fpms and the wavelengths renge from 80 to 400 ft.

Optimum source and receiver patterns should have the element (vibrator or seismometer) spacing less than half of the shortest wavelength and should use enough elements in the pattern to make the pattern as long as the longest wavelength.

Fig. 9 shows that if we were to use one spread configuration across this entire section we would need a spread that would accommodate wavelengths ranging from 40 to 1,800 ft.

In practice, of course, we could not stand an 1,800-ft pattern length because it would also cancel out many of the shallow reflections. Therefore, there must be a compromise in some parameter specifications to do the best job of attenuating the headwaves and surface concurrently without appreciably attenuating the reflections.

One way of looking at this problem is to use a two-dimensional Fourier transform diagram. Typical Vibroseis diagrams are shown in Fig. 10. They are shown in three-dimensions for a better amplitude.

The upper-left diagram represents the amplitude response of the recording system. The system includes six 14-hz seismometers in a linear pattern input to a seismic amplifier.

DESERT test site; vibrators in echelon pattern. Fig. 11.

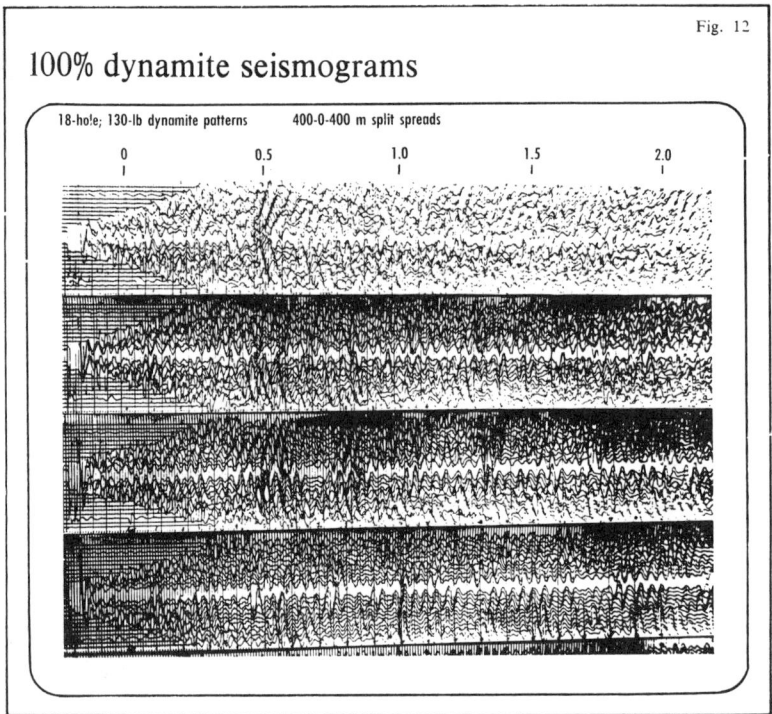

The low-frequency cut is the response of the 14-hz seismometers. An alias filter gives the high-frequency cut. The familiar logarithmic frequency scale is used for the amplitude spectrum on the upper-left diagram.

The attenuation by the six-element array is shown by the relative height of the first four lobes of the response pattern. T h e nonlinear wavelength scale is in multiples of the seismometer interval. (It is much easier to lay a spread out in feet than it is to lay it out in wave numbers.)

The response of three vibrators to a particular input signal (10-40 hz in the example) is shown in the lower left. The amplitude spectrum of a linear-swept frequency signal is symmetrical about the center frequency; therefore, we use a linear frequency scale to best illustrate Vibroseis signal and correlogram spectra. There are three vibrators and therefore three lobes to the vibrator response pattern.

We can find the combined recording system and vibrator input system responses by multiplying their amplitude responses.

This is done graphically, where amplitudes are recorded in db, simply by adding the attenuations for corresponding ground distances to get the combined response shown on the right.

This is the combined amplitude response of the Vibroseis input system, three vibrators using a 10-40 hz signal, and the recording system, using six seismometers in a linear array. The diagram shows the amount of attenuation that, theoretically, we would achieve by this combination for a range of wavelengths specified in terms of the spacing between the vibrators or between the seismometers.

The question now arises. This theory is all well and good, but does it improve the reflection quality on field data?

Fig. 11 shows the terrain of a test area in the desert where the three vibrators were able to give the pattern

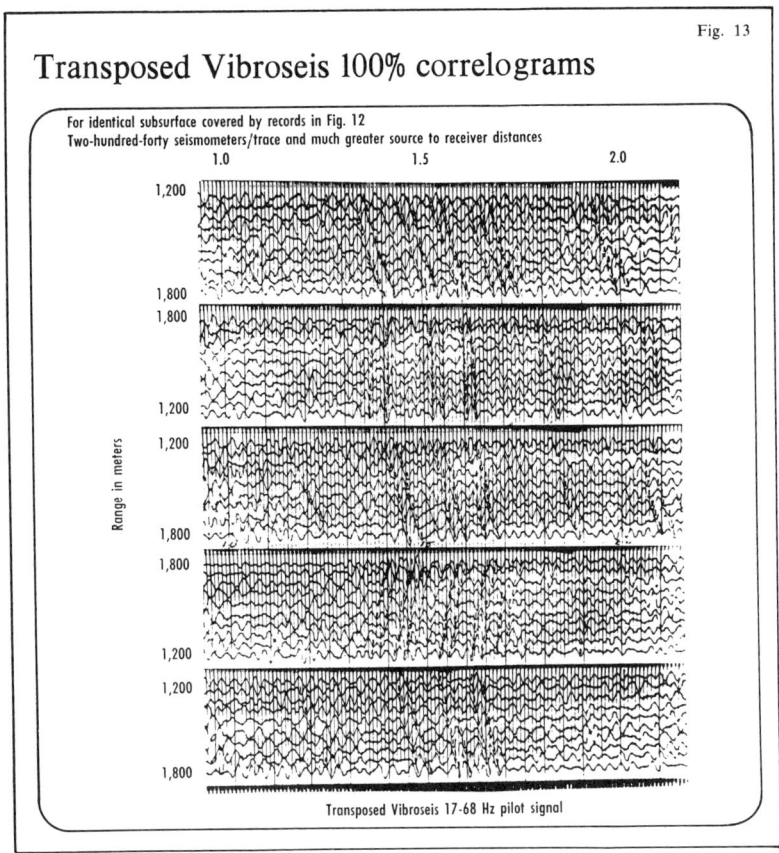

Fig. 13

Transposed Vibroseis 100% correlograms

For identical subsurface covered by records in Fig. 12
Two-hundred-forty seismometers/trace and much greater source to receiver distances

Transposed Vibroseis 17-68 Hz pilot signal

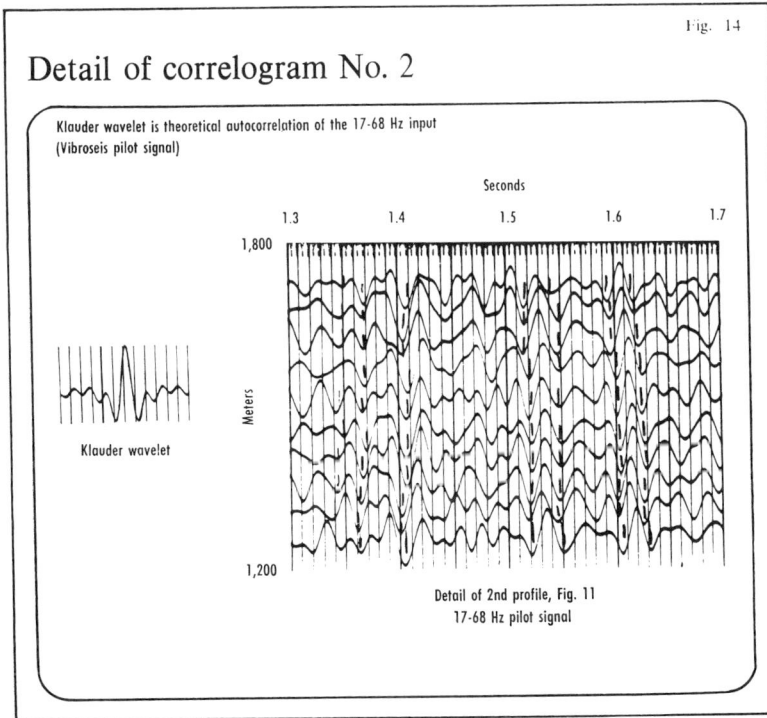

Fig. 14

Detail of correlogram No. 2

Klauder wavelet is theoretical autocorrelation of the 17-68 Hz input
(Vibroseis pilot signal)

Klauder wavelet

Detail of 2nd profile, Fig. 11
17-68 Hz pilot signal

breadth as well as length by using an echelon formation to attenuate surface-wave diffractions coming in from the side.

Fig. 12 shows a series of dynamite records that came from 18-hole patterns at the same location. Charges were about 3 m deep, the total charge weight was about 60 kg, and a 400-0-400 m split spread, with nine seismometers per traces, in line, recorded the shots.

There is an early strong event related to a very shallow lava flow and very little else in the way of pickable data. The Vibroseis data (Fig. 11) for the same subsurface converge was obtained using the transposed system.

In this mode the vibrators are moved to obtain successive depth points as the energy is recorded by two large groups of seismometers separated by five profile lengths.

Both the shot and Vibroseis records are 100% interlocking profiles. Notice, however, that the distance between the short and long traces on the Vibroseis records is 1,200 to 1,800 m from the vibrators rather than the 0-400 m from the shotpoint used in the dynamite spread.

The field techniques, therefore, are very different in many ways. The Vibroseis technique is different because it uses a long-swept frequency input signal, it is different because the input signal was chosen to give good wavelets, it is different because source-to-seismometer distances were chosen to give the best reflection quality, and it is different because the spacings of the seismometers and vibrators in the pattern were chosen to give the best attenuation of the interference waves.

To show how nearly like the symmetrical autocorrelation function the reflections are, Fig. 12 shows only part of profile No. 2. Notice especially the shape of some of the wavelets in the center and right reflection bands toward the middle of the figure and compare them with the ideal shape of the 17-68 hz Klauder wavelet on the left. The general similarity is good. Some attenuation of the high-frequencies in the input signal has broadened the reflection wavelets.

The Vibroseis system has a great potential, not only for the general record quality and economy that we are looking for but also for high resolution to get the high-quality data that we need to find much more obscure features than we can find with the blunderbuss methods of summing, CDP, shallow patterns, dynamite shots, etc. To realize this potential, we must work diligently to optimize the Vibroseis parameters.

References
1. Klauder, J. R., Price, A. S., Darlington, S., and Albersheim, W. J., 1960, "The theory and design of chirp radars," The Bell System Technical Journal, Vol. 39, No. 4, pp. 745-808.

SEISMIC DATA ENHANCEMENT—A CASE HISTORY*

R. J. GRAEBNER†

ABSTRACT

The theory relating to many methods—for example, multiple seismometer techniques—which the geophysicist may control to improve record quality is well known. However, its application has not been fully exploited. An example of the reduction of theory to practice in one area characterized by poor records is presented. It comprises a series of analytical tests designed to discover the cause of poor records, to examine the effect of each variable on the signal-to-noise ratio, and to evaluate the solutions predicted by theory. The tests showed that the poor record quality was attributable chiefly to relatively strong surface and near-surface waves propagating outward from the shot. Wave length filtering by means of suitable shot and seismometer patterns, and compositing through data processing methods, greatly improved record quality and permitted magnetic recording of reflected signals over a broad frequency range. The tests established, in the allotted time, that the quality of the data would meet clearly specified standards of performance.

Experience has shown that better seismic data can generally be obtained when the design of techniques is based on the special character of the signal and noise determined from simple tests rather than when the design is based on general assumptions.

INTRODUCTION

The increasing costs of finding oil compel geophysicists to evaluate more carefully than ever before the ways in which existing exploration techniques may best be utilized. The general subject of setting up standards of performance by which an exploration program can be judged has been explored in a recent paper by Agnich and Dunlap (1959). These authors suggest that means be established for evaluating the performance of each phase of an exploration program in terms of its predicted performance.

A prominent phase of many exploration endeavors consists of carrying out seismic surveys. Large portions of the world in which exploration activity is conducted can be characterized seismically as areas of poor record quality. Areas in which usable or even good seismic data were secured in the past may be considered poor in terms of contributing information on a specific or a new objective. Further, to realize fully the potential offered by modern data processing as means for solving difficult interpretive problems, it is necessary that the seismic data be recorded initially on the magnetic media in the best possible form. The task often arises, then, of determining the cause of inadequate record quality and of making practical estimates about what it takes to achieve a solution. This information is necessary in order to set up certain standards of performance for a particular situation.

A frequently used approach to the solution of a record quality problem, which was often adequate in the past, was the trial and error method. The chief disadvantages of this approach are four-fold. First, the rate of convergence toward a

* Manuscript received by the Editor October 5, 1959.
† Geophysical Service Inc., Dallas, Texas.

solution may be too slow. Second, recording techniques developed may be more elaborate and hence more costly than a simpler one which may be satisfactory. Third, if the problem to be solved is particularly difficult, a usable solution may be missed altogether. Fourth, the results of such procedures yield little information on how to modify techniques to meet changing conditions over the prospect.

An alternate approach consists of carefully controlled experimental tests designed to place in evidence essential features of the signal and noise characteristics. From a knowledge of these characteristics, it is possible to formulate the nature of the problem and, by combining this knowledge with the known behavior of the various recording methods, to estimate possible solutions. The methods or tools which the geophysicist may control to achieve an improvement in signal-to-noise ratio are the type of seismometer, method of planting, the number of seismometers in a group, the number of simultaneous shots, configuration of shot and seismometer arrays, shotpoint offset, charge size, shape of charge, charge depth, and filter-pass bands. The uses and functions of several of these recording parameters have been given careful theoretical treatment in the published literature. The wide range of capabilities of modern seismic instruments and instrument requirements for operations in difficult record areas have been discussed by Hammond and Hawkins (1958).

The theory and design of multiple seismometer arrays have been treated according to different points of emphasis by many authors, a partial listing of which would include Lombardi (1955), Parr and Mayne (1955), Smith (1956), Savit, Brustad, and Sider (1958), and White (1958). The conclusions of these authors about the applicability of multiple seismometer techniques vary somewhat according to the initial assumptions made about the problem, particularly about the noise characteristics. A noise analysis procedure designed to secure estimates of the properties of the reflected signals and the interfering noise as a basis for designing an optimum recording technique is described by Smith (1956).

Although theory has been treated in the literature, its application has not been fully exploited in practice. The need for its routine use in poor record areas is pointed up by the observations of Hammond and Hawkins (1958). These authors suggest that when a normal seismic record is obtained completely devoid of identifiable reflections, it is not possible to determine whether the signal-to-noise ratio is one to three or one to one hundred. In the first instance satisfactory records could probably be secured by several different combinations of recording devices. Here the concern would be the one of determining the most efficient and hence the most economical technique from several possible ones. For a situation in which the signal-to-noise ratio is in the vicinity of one to one hundred, it may or may not be possible to secure usable data by any known techniques. Obviously, in planning a seismic survey in a poor record area, it becomes vitally important to find where, in such a wide range of possible ratios of signal and noise, the actual signal lies.

The interfering noise can be defined in many ways. In seismic work, it is convenient to define the signal as any information-bearing wave or component of a wave which we desire to isolate. This information constitutes the data employed to develop the ultimate geologic interpretation. The noise then consists of all the remaining disturbances which appear on the records. Defined in this manner, the noise consists of (one) complex disturbances which are random in both amplitude and phase, and (two) coherent waves having identifiable characteristics. The coherent or systematic noise waves are commonly surface or near-surface waves initiated by the shot impulse. Under special circumstances, they may be multiple reflections, reflected refractions, refractions from shallow interfaces, or diffractions from sharp localized features. Results from careful studies of investigations into properties of shot-generated surface and near-surface waves have been reported by Howell, Neuenschwander, and Pierson III (1953) as well as by Dobrin, Lawrence, and Sengbush (1954). The results reported in the latter paper disclose a remarkable variety of coherent boundary waves. It is clear that the possible range in properties of noise is extremely broad but that in a specific situation, these properties may be bounded within narrow limits.

In Figure 1 are shown four pattern arrangements and their theoretical response curves to plane waves traveling in the direction of the seismic profile. The range of wave length distributions for two assumed noise conditions, both of which have been observed in actual tests, are shown. The noise locations refer to the dominant interfering noise occurring at ranges of record times at which it is desired to obtain usable reflection data. The illustration suggests two results, both of which have been observed in practice. One is that the behavior of the response curve can be critical. This is the situation for Noise I. Here the sharper cutoff needed to reject the noise wave lengths lying immediately adjacent to the signal is provided by the pattern most easily constructed, the linear array. The second result is that the response of several kinds of patterns, for example, patterns A and C may be nearly equivalent. Although the construction of one pattern is much more complicated than the other, each should yield a comparable improvement in record quality for the assumed conditions of signal and noise. A pattern of type B should be somewhat better than A and C if only Noise II exists. Pattern D is not optimum for any of the postulated noise conditions. This example suggests that knowledge of the problem is required to guide the proper selection of any array.

If complete knowledge of the signal and noise characteristics of an area were available, it would be possible to specify uniquely a best recording technique. Accepting the fact that field experimental work must be limited, one real task of practical geophysics, then, is the devising of simple tests to reveal the nature of specific problems and to adapt this knowledge to the design of recording techniques adequate to meet changing conditions around the prospect. Such tests must be simple enough to be performed reasonably fast with normal instrumentation and must yield data which can be interpreted by the normal staff concurrent with the shooting.

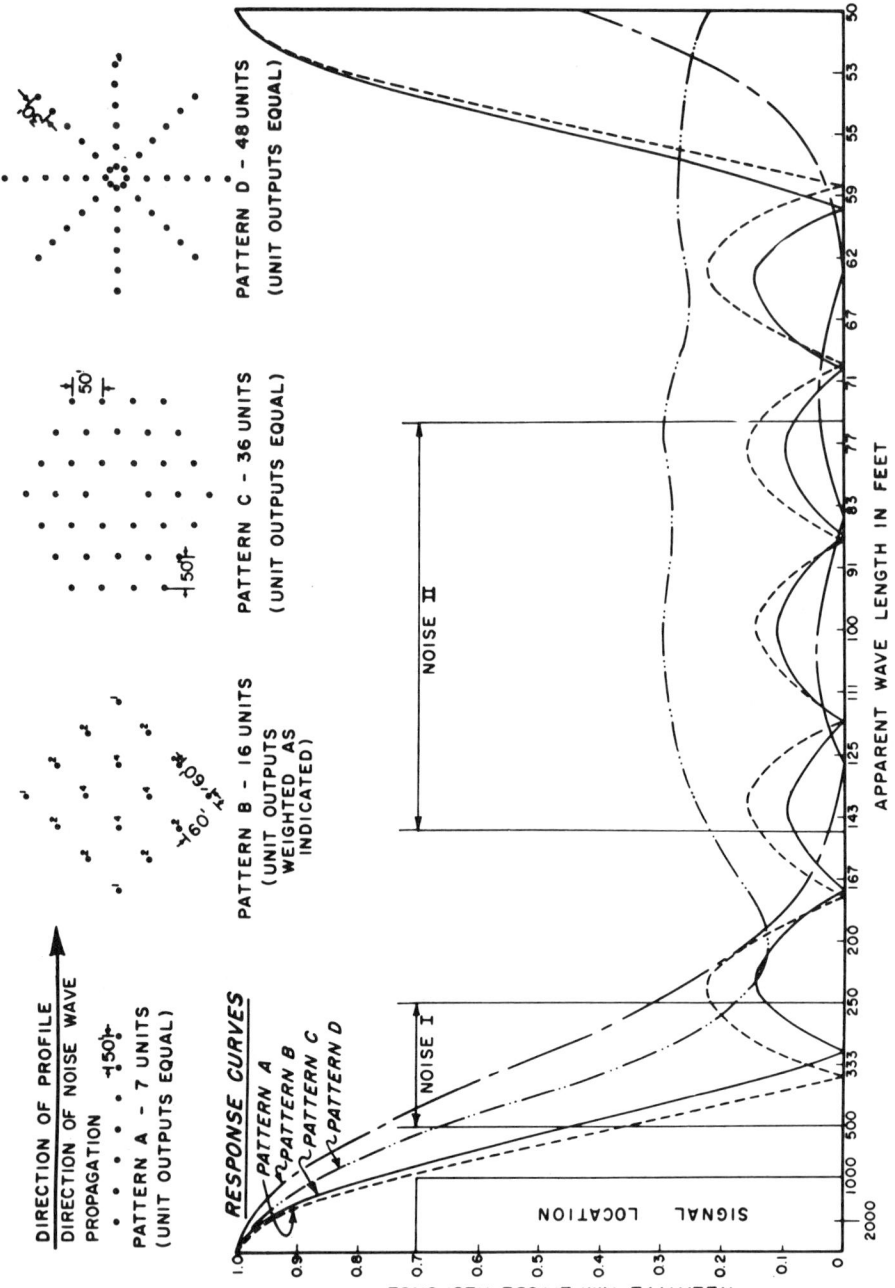

FIG. 1. Theoretical wave length response functions for plane waves traveling along direction of profile for four pattern arrangements. Applicability of array depends upon assumed noise condition.

The purpose of this paper is to describe one company's experience in investigating record quality problems in a routine fashion with an analytical type of testing procedure. This description is carried out by presenting one example in a case history form.

NATURE OF EXPERIMENTAL PROCEDURE

The testing process consists of the following four general phases: (one) the collection of data providing measures of chiefly those signal and noise characteristics useful in considering design of an optimum recording technique, (two) the interpretation of these data to identify the troubles which limit seismic record quality, (three) the prediction of solutions, and (four) the carrying out of simple tests to evaluate how well the problem responds to the predicted solutions.

A complete noise analysis is generally out of the question. The practicing geophysicist is confronted with the problem, on the one hand, of conducting an experiment with all factors held constant except the one whose effectiveness he wishes to isolate and, on the other hand, of attaining a satisfactory solution to a field problem with a practical amount of time and money. The conditions of the experiment often affect to some degree the evaluation of the parameter under investigation. It is apparent that a compromise must be struck between the detail of the data which can be obtained experimentally and operational considerations.

The wave form tests and interpretive procedures to be described represent a compromise which has been found to be a practical way of reducing theory to practice. The emphasis is on combining knowledge of the theoretical properties of recording devices with knowledge of the existing signal and noise conditions as revealed by experimental sampling to achieve useful data in an economical manner. The testing procedure follows the form outlined by Smith (1956).

CASE HISTORY

Test Site

The prospect is in an area where the oil operator had conducted earlier seismic surveys. Identification of this prospect and detailed structural information is omitted in this discussion at the request of the company owning the data.

A test site was selected in a part of the prospect in which the records were poor but not completely without evidence of reflected signals. The plan was to make the tests where measures of both the signal and noise could be secured and then to proceed into the "no reflection area" by performing abbreviated tests to determine whether the noise intensity or the signal level was changing. Through such observations, the operators sought to infer whether the change in record quality was attributable to variations in noise intensity or to the alteration of the signal owing to changing subsurface structural relationships.

A sample of the seismic data recorded in an earlier survey in the vicinity of the test site is shown in Figure 2. Although these data may have been useful in aiding the original exploration effort, seismic data of considerably improved quality were required to be consistent with the new exploration objectives. A

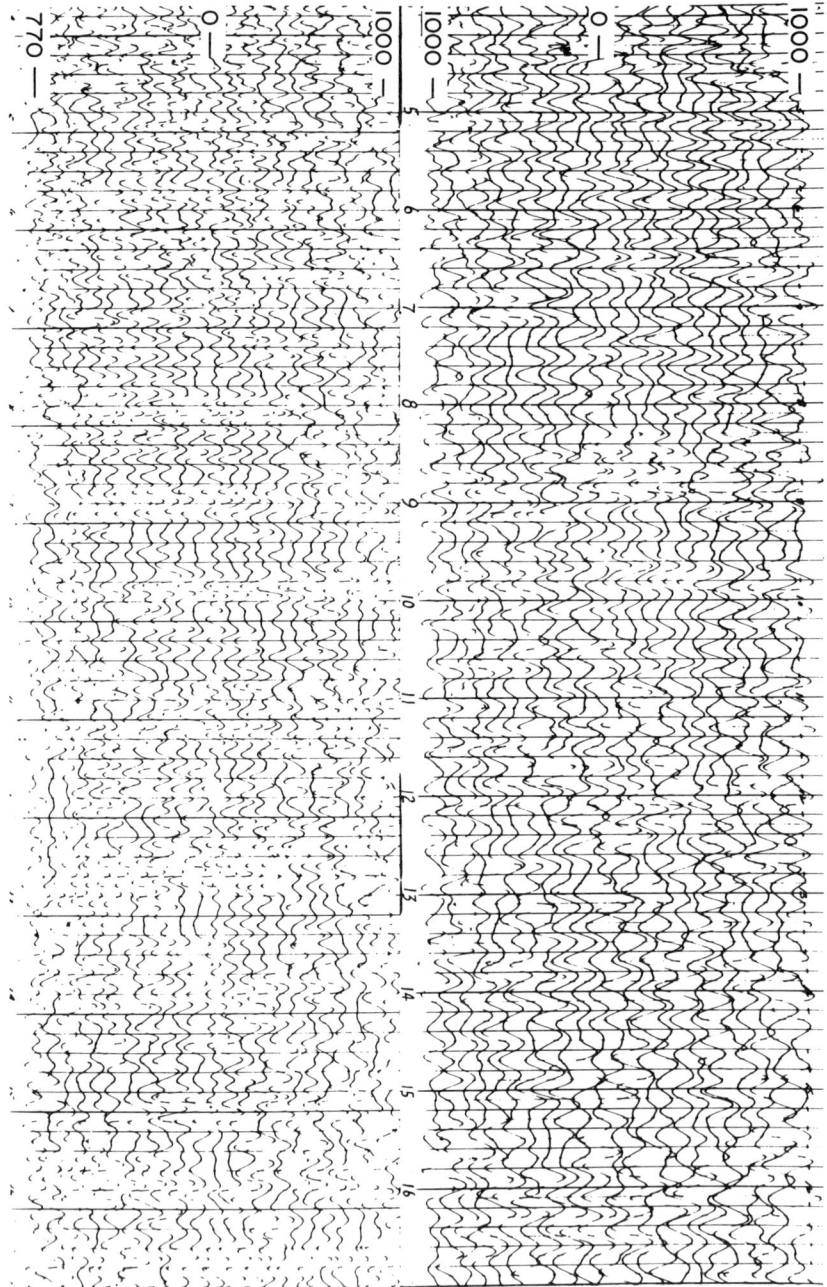

FIG. 2. Sample of quality of data secured in a prior seismic survey of prospect. Seismograms were recorded with one shothole, 2½ pound charges, and six seismometers per group.

study of these records does not reveal the specific causes of the poor record quality. It was the purpose of the tests to determine whether records adequate for the new exploration objectives could be obtained, and if so, to reveal the most efficient recording techniques. This program consisted of a sequence of tests designed to examine the effect of each variable on record quality insofar as seemed necessary. The test results shown in this paper, by way of illustration and discussion, are presented as typical samples characteristic of each test.

Test of Signal and Noise Characteristics

The basic data were recorded on a consecutive sequence of spreads ranging from 200 to 1,600 ft from a single shotpoint location. Charge sizes and hole depths were those thought to be typical for the area. The recording spreads consisted of single seismometers spaced at ten-ft intervals. Damped seismometers having a natural frequency of 14 cps were used for the most part; but on occasion, to investigate the filtering action of seismometers with respect to the observed signal and noise, both four-cycle and 27-cycle seismometers were also used.

The 14- and 27-cycle seismometers were S-32 models manufactured by Electro-Technical Laboratories, and the four-cycle seismometers were manufactured by Hall-Sears.

The recording equipment, manufactured by Texas Instruments Incorporated, consisted of a magneDISC recording unit and a 24-channel 7,000 amplifier system capable of being operated with either automatic gain control or constant gain. The frequency response of the instrumental system was flat within three decibels between 12 and 120 cps.

One constant gain recording was made at each spread setup. Replays made with automatic gain control facilitated the identification of the individual signal and noise events. The constant gain recordings allowed measurements of wave amplitudes. A sample of the data recorded in this test is shown in Figure 3. The data collected in this test were plotted on a time-distance graph shown in Figure 4, which displays the coherent events and their measurable characteristics.

The time-distance graph shows dominant sets of surface or near-surface waves between 0.6 and 2.0 seconds. Both the amplitude and frequency of these noise waves appear to decrease with increasing time and distance. A comparison of in-line and cross-spread data indicates that the strong surface or near-surface wave trains are developed by the shot and propagate outward from the shot. The dominant interfering noise appears to have a wave length distribution which peaks in the neighborhood of 45 ft. These waves have a definable set of properties extending over a limited range of values. They destroy substantially the continuity of the weaker reflected energy having arrival times up to two sec. Because of the dominance of the noise waves accurate signal measurements are difficult to make from data recorded with single seismometers. However, the most reliable measurements show that the signal-to-noise ratio varies roughly over a range from 1/1 to 1/4.

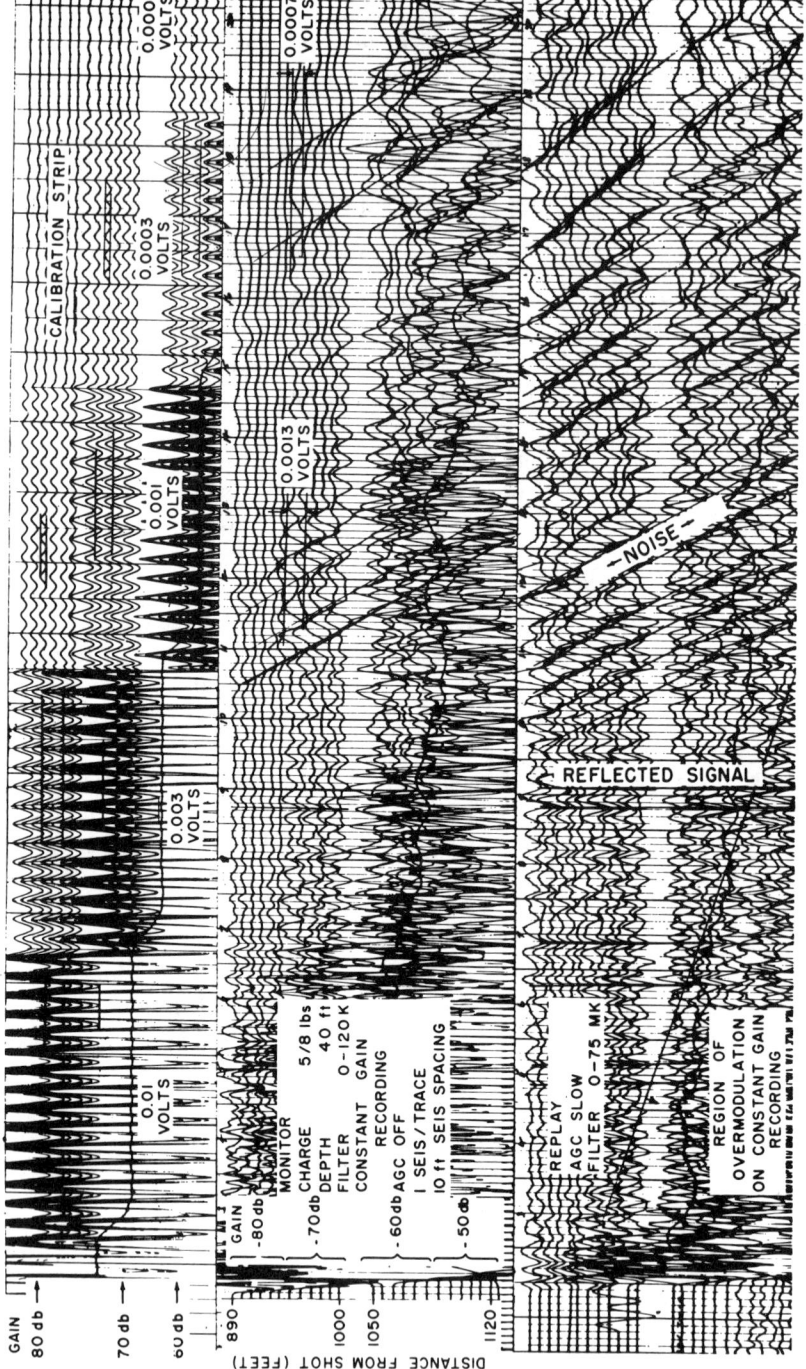

Fig. 3. A sample of the data recorded in the wave form tests. At the top is a signal calibration record. Next is shown a constant gain recording. The bottom record is a replay using automatic gain control.

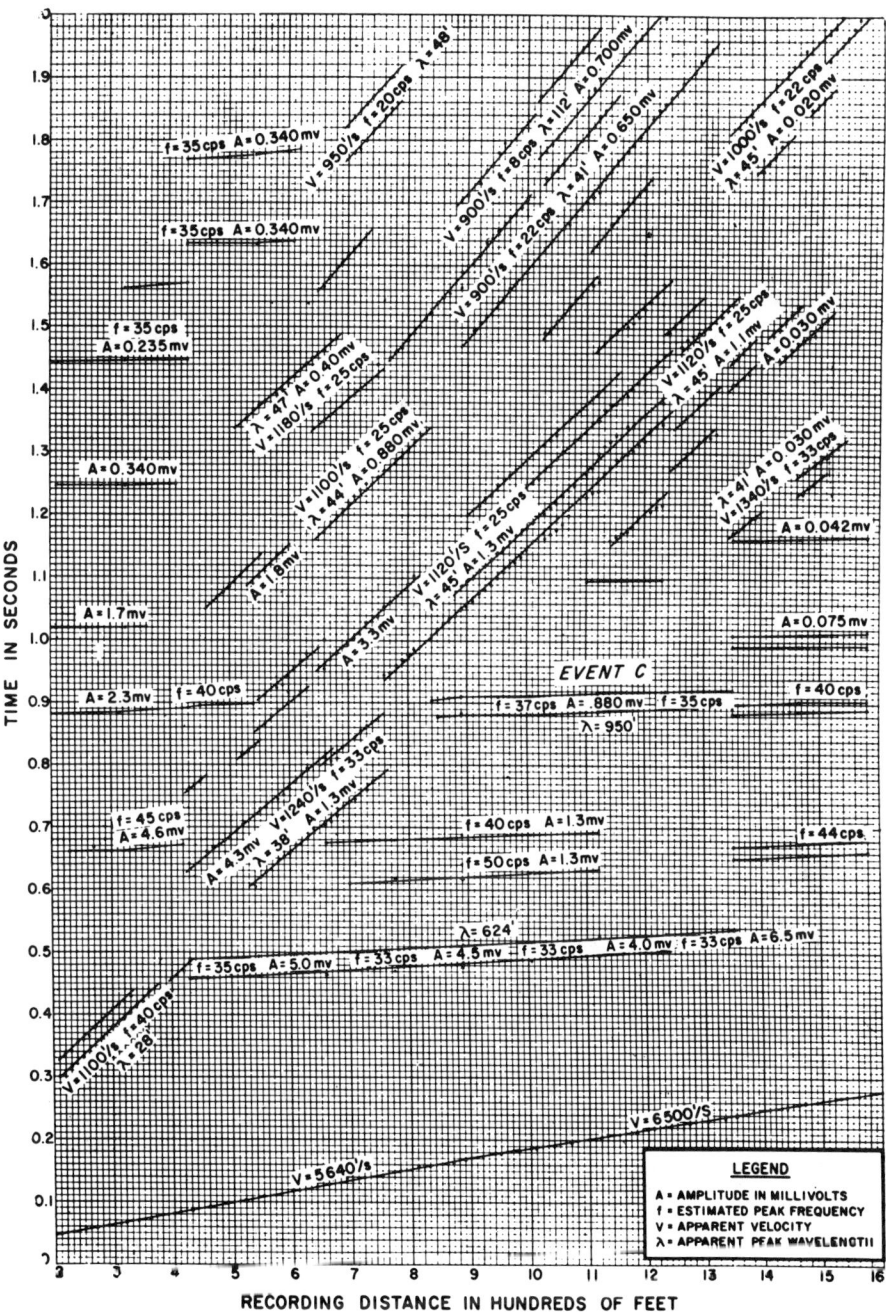

FIG. 4. Time-distance graph displaying the events recorded in the tests and their measurable characteristics.

Noise which did not exhibit coherence with the 10-ft sampling interval was judged to be random. These experiments showed that the random noise was much weaker than both the reflected signals and the coherent noise waves. Hence, it was not an important consideration in designing the recording geometry.

Effectiveness of Multiple Seismometer Arrays

The test results suggested the use of multiple shot and seismometer arrays designed to reject the interfering noise by wave length discrimination.

Basis of Array Design.—Although the properties of multiple seismometer patterns are well known, there is some speculation and difference of opinion about the application of this knowledge (Savit, Brustad, and Sider, 1958). Seismometer arrays can be treated as filtering devices whose response characteristics are a function of apparent horizontal wave lengths rather than frequencies. For example, the theoretical wave length response of a linear pattern consisting of six seismometers spaced uniformly at 15-foot intervals in line with the shot is shown in Figure 5. Since both the signal and noise waves are transient pulses, both are represented by broad frequency and wave length distributions rather than by single values. Commonly, the wave length distribution of the noise spans several of the zero and maximum points of such a theoretical seismometer curve. Hence, to predict the attenuation of a given noise relative to the signal achieved by any particular array, an average curve is more useful than the theoretical one. The

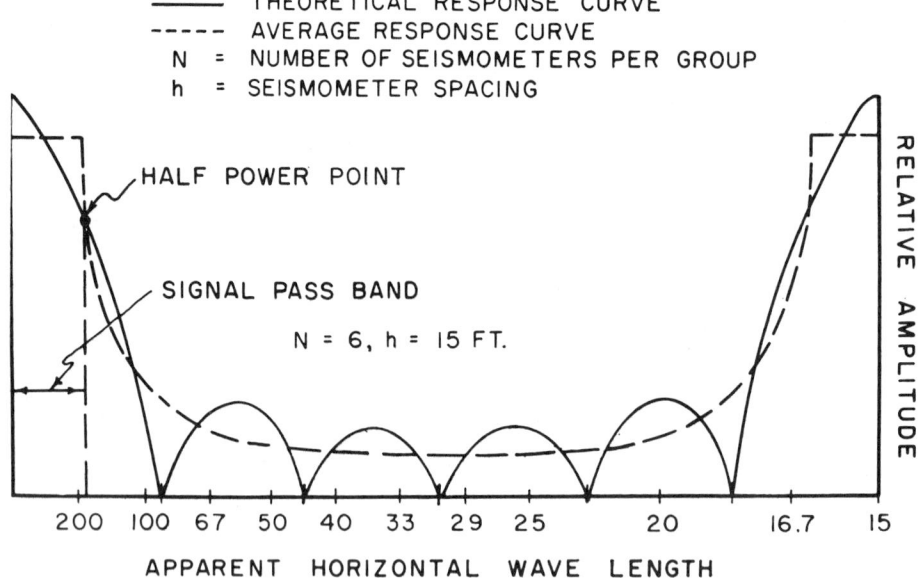

FIG. 5. Comparison of a theoretical and an average wave length response for a line of six seismometers spaced uniformly at 15-foot intervals.

method of averaging depends upon the definition of the band-pass and band-reject regions. The criterion for selection of a filter-pass band is based on the capacity of the filter to pass the signal within specified limits of distortion while rejecting the greatest amount of noise. The band-pass region of the multiple seismometer array was defined to be that portion of the region between the point of maximum response and the wave length for which the response is down 50 percent in power. This choice of cutoff point corresponds to that of the conventional highcut filter thought to be optimum for the reflection. To obtain the average response curve shown in Figure 5, the theoretical response curve was averaged at uniform intervals. These average amplitudes were then normalized with respect to the average amplitude in the band-pass region. This method of defining the band-pass and reject regions and of constructing averaged response curves is consistent with the criteria used for selection of electrical filters and is applicable to any type of array.

The maximum group length is limited by the requirement that the shortest signal wave length necessary to preserve reflection character falls within the signal pass band. The wave length response curves considered are periodic. For this reason the maximum seismometer spacing is governed by the shortest noise wave length passed through the electrical filter pass band. If we restrict the choice of arrays to a line of uniformly spaced seismometers having equal outputs, the above considerations set the following two criteria for selecting N, the total number of units in an array, and h, the seismometer spacing (Smith, 1956).

(1) Maximum group length is such that the product

$$Nh = 0.45\lambda_s, \quad \text{where} \quad \lambda_s = \frac{V_{s\,min}}{f_H}.$$

In these equations, $V_{s\,min}$ is the smallest value of the apparent horizontal velocity of the reflected signal, f_H is the highest frequency necessary to preserve reflection character, and λ_s is the corresponding shortest signal wave length. This f_H should coincide with the highcut filter setting. Values of $V_{s\,min}$ can be estimated from considerations of section velocity, depth to reflecting interface, recording distance, and dip. Use of this criterion assures that wave lengths of the reflected signals equal to or greater than λ_s will fall in the region between the maximum response and the half-power point.

(2) The limitation on the maximum seismometer spacing is given by

$$h_{max} < \frac{V_{n\,min}}{f_H}.$$

In this expression, $V_{n\,min}$ is the slowest noise velocity, a measure of which was obtained in the test results shown in Figure four.

Relation to Problem.—Gross estimates of the frequency distribution and the apparent wave length distribution of the signal and noise are shown in Figure 6.

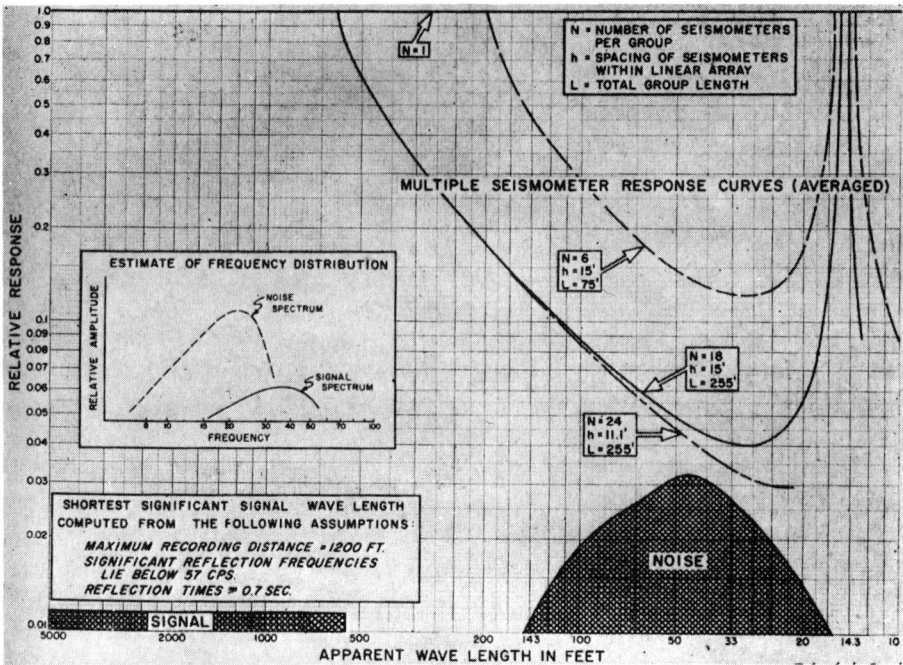

FIG. 6. Estimated frequency and wave length distributions of the signal and noise. The average response curves for three kinds of seismometer arrays are shown.

These estimates were made over a time range of 0.7 to 2.0 secs and a recording distance range of 100 to 1,600 ft. The frequency measurements are relatively crude in that they were made visually on paper records without benefit of accurate frequency analyzing equipment. Even after making allowances for errors in the estimated spectra, the measurements show that the signal and noise partially overlap in frequency but are considerably separated in wave length. The dominant noise appears to be localized in a particular region. It is evident that a complete separation of signal and noise cannot be achieved by electrical filtering alone. On the other hand, the substantial separation in the wave length distributions of the signal and coherent noise suggests that the enhancement of the desired signals by wave length filtering devices, such as arrays of seismometers, could be very effective.

Based on estimates of the signal and noise properties and the above mentioned criteria, the optimum seismometer pattern of linear density appeared to be 18 seismometers spaced uniformly at 15-ft intervals in line with the shot. Theoretically, use of this pattern should yield roughly a signal-to-noise ratio improvement of 20 to 1 over that provided by one seismometer and a 3 to 1 improvement over that provided by a pattern of six seismometers. The average ap-

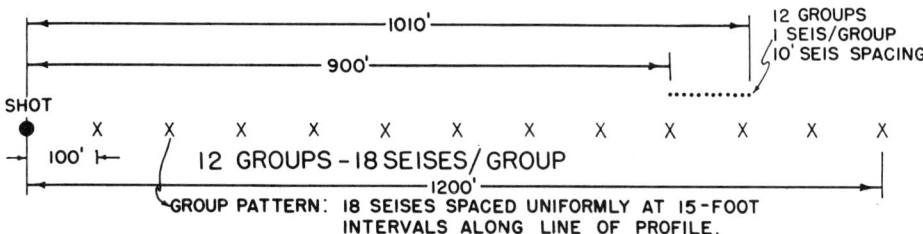

Fig. 7. Plan view of recording geometry used to record most of the experimental seismograms.

parent wave length response curves for these patterns are shown in Figure 6. The amplitude response curve for 24 seismometers per group indicates that there is essentially no additional benefit to be gained by adding six more seismometers to the pattern of 18 seismometers per group.

To test the validity of the analysis, brief experiments were conducted which allowed a quick comparison of actual results with those predicted by theory. The results of using the theoretically optimum pattern were compared with the results using arrays having appreciably different response characteristics in order to estimate the accuracy of the prediction.

The efficacy of the 18-seismometer pattern in rejecting the disturbing noise relative to a single seismometer per group was examined by using foldback spreads of the type shown in Figure 7. Recordings were taken with both fixed gain and automatic gain control. An example of a seismogram recorded by the latter method is shown as the top record in Figure 8. The spread, consisting of one seismometer per group, stretched between groups nine and ten of the multiple seismometer spread. Since no electrical filtering was done in the frequency range of interest and since a common shot was used, a comparison of the data from the two spreads should give a measure of the wave length filtering property of the pattern array. The diagonal line across the record notes the position at which the near-surface waves first hit each group. Before the noise waves arrive, reflections can be observed almost as well with single seismometers as with multiple seismometers. However, the noise waves almost completely obliterate all evidence of reflected events "D," "H," and "E" on the single seismometer traces whereas these reflections are well developed on the traces recorded with the pattern array. From previous considerations, one would expect such an improvement in appearance.

A comparison of the relative effectiveness of two different seismometer patterns is shown as the bottom record in Figure 8. In each instance the seismometers were spaced uniformly at 15-foot intervals in line with the shot. The data were recorded from a single shot on a foldback spread of the type illustrated beneath the seismogram. Again, before the noise waves arrive, reflections of the same quality were observed on each spread. In the region disturbed by the noise,

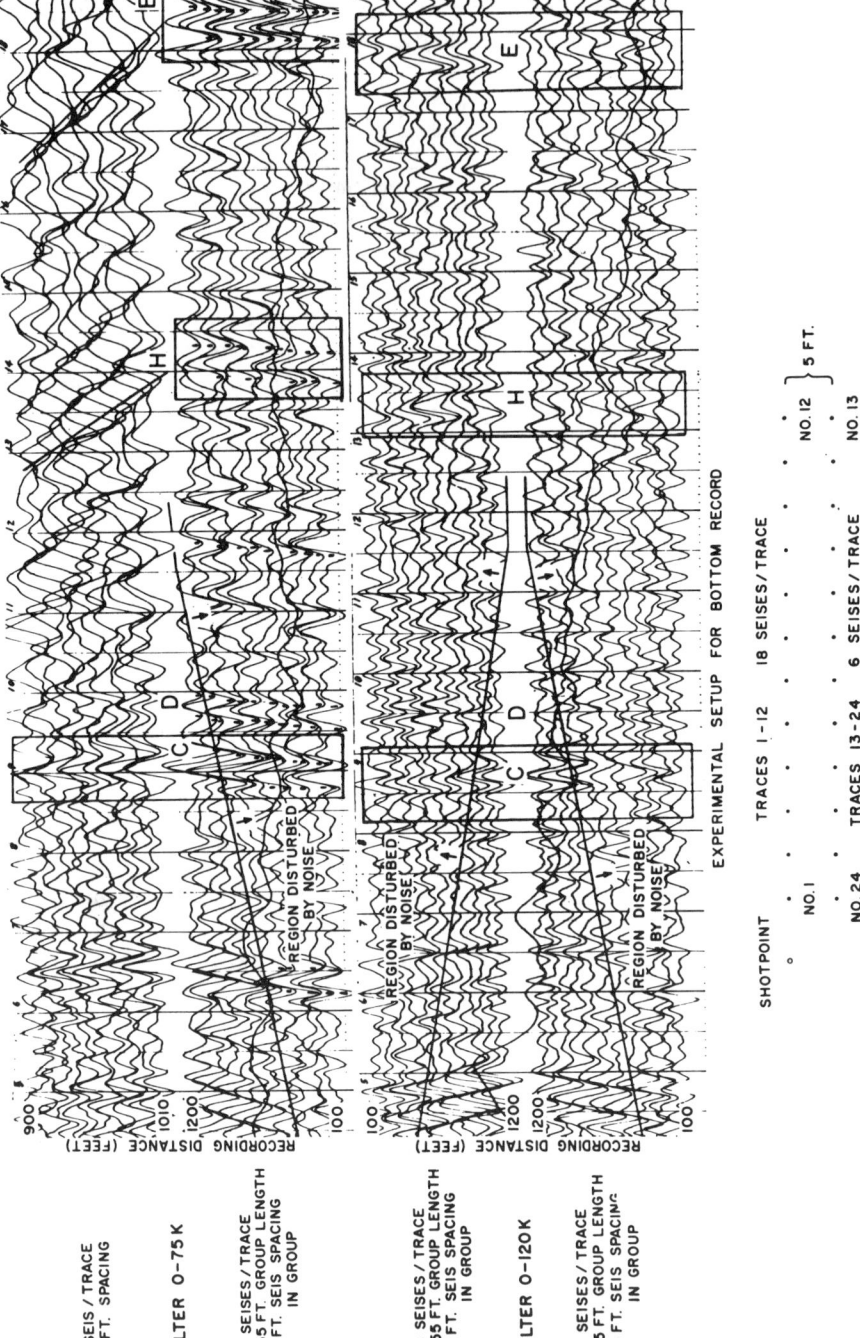

Fig. 8. Two comparisons of the effectiveness of multiple seismometer patterns. Experimental recording arrangement for top seismogram is shown in Figure 7. Seismometers in patterns are placed in a line along direction of profile at uniform intervals.

however, a significant improvement in record quality is provided by the pattern of 18 seismometers. The array of six seismometers per group was compared with the theoretically optimum linear array to establish whether the expected improvement in signal-to-noise ratio was observable.

The fact that these foldback tests provided results approximately as predicted, suggested that the use of multiple seismometers could be applied objectively over the prospect according to simple theory and that needed modifications in the array could be made in accordance with knowledge of changing conditions. The test suggested that the usefulness to be gained in compositing outputs could be predicted in a similar fashion.

Variations in elevations and weathering thickness decrease the effectiveness of the array by attenuating the higher frequencies in the signal spectrum. How large a variation in these variables can be accommodated in a particular situation can be estimated by calculating the changes produced in $V_{s\,min}$ and hence in λ_s. The lack of marked deviations of observation points from the time-distance curve of the first refraction arrivals shown in Figure 4 suggested that erratic time delays would not present a significant problem.

Effect of Hole Depth on Signal-to-Noise Ratio

In addition to conventional uphole velocity surveys, shots were recorded at several depths with constant gain amplifiers making use of an experimental arrangement similar to that shown in Figure 7. A sample of the test seismograms and a plot of the signal-to-noise ratio as a function of hole depth is shown in Figure 9. In this case, the spread of single seismometers recorded the input signals arriving at the eleventh and twelfth multiple seismometer groups. The particular type of spread geometry was chosen in order to secure a measure of both the desired reflected signal and noise from the same shot. Because of the dominant noise recorded with the single seismometers, amplitudes of the reflected signals were measured from the data recorded with patterns of seismometers. Noise amplitude measurements were made on seismograms recorded with single seismometers. For this reason, the signal-to-noise ratios indicated on the graph in Figure 9 represent the true ratio multiplied by a constant factor characteristic of the recording geometry.

Two marked changes in the properties of the reflected events occurred with respect to hole depth. These are the decrease in amplitude and the increase in frequency with increasing charge depth. In general the amplitude and frequency changes in the disturbing noise waves are considerably less than are those in the reflected events. Note that the noise amplitudes of the 100-ft shot are nearly the same as those of the 40-ft shot. The tests indicated that superior results were secured with 40-ft shots.

The reason for the rapid decrease in reflection amplitude and increase in frequency was not clearly understood. The drill logs showed very little change in the near-surface materials with hole depth. Part, but not all, of the progressive change

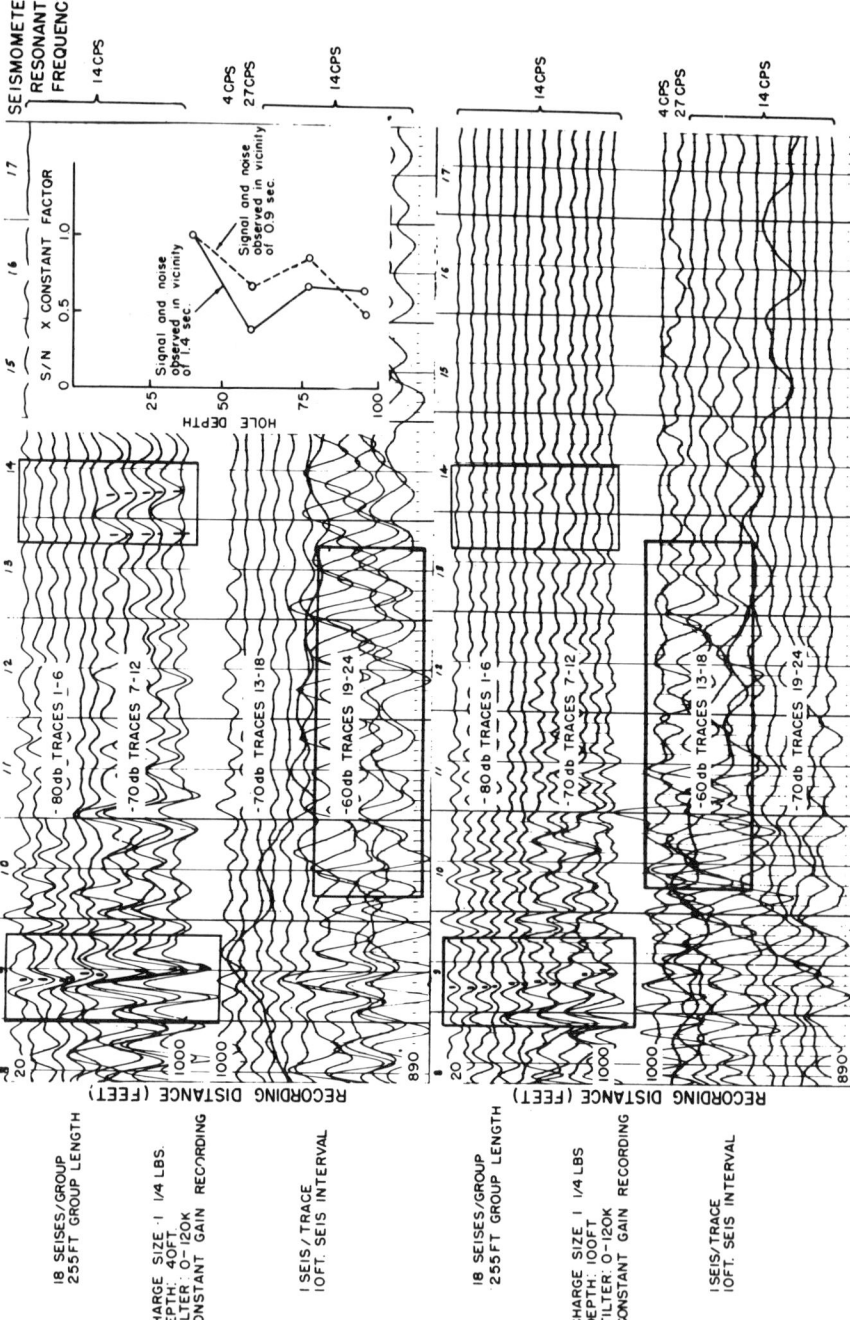

FIG. 9. Sample constant gain seismograms from charge depth test. Reflection character changes rapidly but noise remains nearly constant. For plot of signal-to-noise ratio vs hole depth, noise is measured on spread of single seismometers and signal on spread of pattern seismometers.

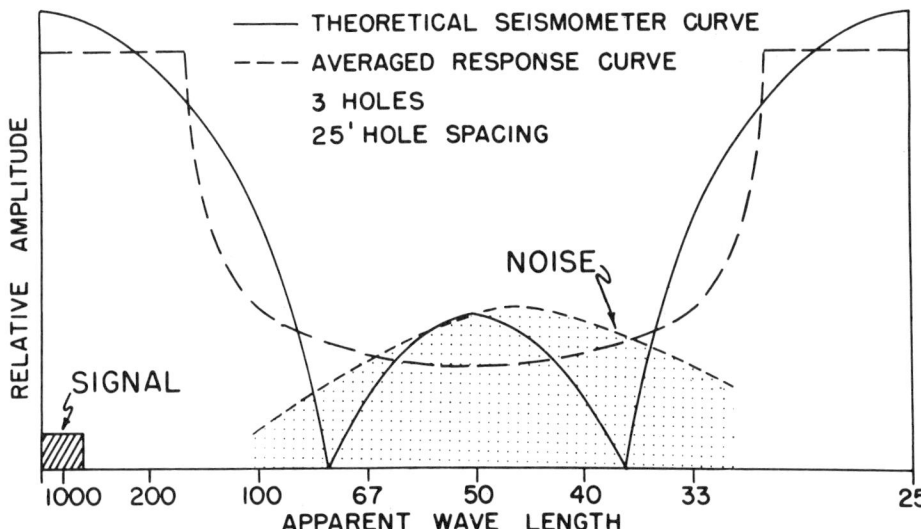

Fig. 10. Theoretical and average response curves of a pattern of three holes spaced at 25-foot intervals. Interfering noise falls in the general reject region. Tests of this shot pattern are shown in Figure 11.

in character with hole depth was attributable to the ghosting effect (Van Melle and Weatherburn, 1953, Musgrave, Ehlert, and Nash, 1958). It is possible that the shotpoint environment may change in such a manner that the pulse propagating outward from the deep shots may contain a much higher proportion of high frequencies than the pulse from shallow shots.

Because of the dependence of the design of pattern shots or seismometers on the properties of the noise, knowledge of whether the reflected signals or the type of interfering noise changes with hole depth becomes important in design considerations. For this reason the uphole tests were conducted as described. Such information would be partially obscured by conducting these tests with conventional reflection profiles recorded with automatic gain control.

Effectiveness of Multiple Holes

The fact that the multiple seismometers behaved approximately as postulated suggested the use of multiple holes as an additional wave length filter. A record shot with a three-hole pattern was compared with one shot with a single hole making use of the type of experimental setup shown in Figure 7. The three holes were spaced 25 ft apart along the profile line. The shot spacing was selected so that the dominant noise fell in the middle of the reject region of the wave length response curves of the shot pattern as shown in Figure 10. The total charge sizes for the two records were maintained essentially constant. Examples of both automatic gain control and constant gain recordings are shown in Figures

11 and 12. Note on the constant gain recording that the continuity of the noise waves observed with a single shot at the −80-db setting is considerably broken up by the three-hole pattern. Comparing the ratio of the signal-to-noise ratios measured from individual records, over the time range from 0.7 to 2.0 sec, reveals that the use of the three-hole pattern improves the signal-to-noise ratio three to four times. This improvement is approximately what one would predict. It suggested that should the signal-to-noise ratio become significantly worse, larger patterns of holes could be employed to advantage and that the advantage to be gained could be predicted according to design criteria similar to those employed for pattern seismometers. Alternatively, perhaps vertically distributed charges having equivalent wave length response characteristics could be employed (Musgrave, Ehlert, and Nash, 1958).

Study of Charge Size

The sizes of the explosive charges suitable for the prospect were characteristically small, and charges larger than two pounds were prohibited by permit restrictions over much of the prospect. Studies of the effect of charge size on the signal-to-noise ratio were conducted utilizing again the type of experimental setup illustrated in Figure 7. A sample of the test results is shown in Figure 13. In this example, the size of the charge used to record the top profile was approximately three times that for the bottom one. The recordings were made with constant gain and a 0–120 cps K-type filter pass band. Note that the noise waves between 1.0 and 1.3 secs at the −70db gain setting are essentially the same on both profiles, whereas the amplitudes of reflected events C, D, H, and E changed significantly. To remove the shot coupling variable insofar as possible, the ratio of the signal-to-noise ratios on one profile to the ratios on the other for corresponding events was determined. This calculation showed tripling the charge size increased the signal-to-noise ratio by a factor which varied between 1.5 and 2, over the time range between 0.7 and 2.0 sec. Even the smallest charge of $\frac{5}{8}$ pounds produced energy levels many times greater than the preshot background noise level. No reason was determined for the apparent non-linear behavior of surface waves and compressional waves with variation in charge size. The observation was taken as a characteristic of the signal and noise behavior, and this behavior was utilized to guide the design of the recording technique.

Effectiveness of Frequency Filtering

Attenuation of noise by frequency discrimination can be accomplished by electrical filtering or by choice of the resonant frequency of the seismometers.

Repeated playbacks of test records through the magneDISC were made to evaluate the effectiveness of various low-cut filters in rejecting noise waves while preserving the reflection character. In Figure 14 are displayed a suite of records in which all the recording parameters are constant except the low-cut filter setting. The four bottom profiles are constant gain replays of the top profile. The record-

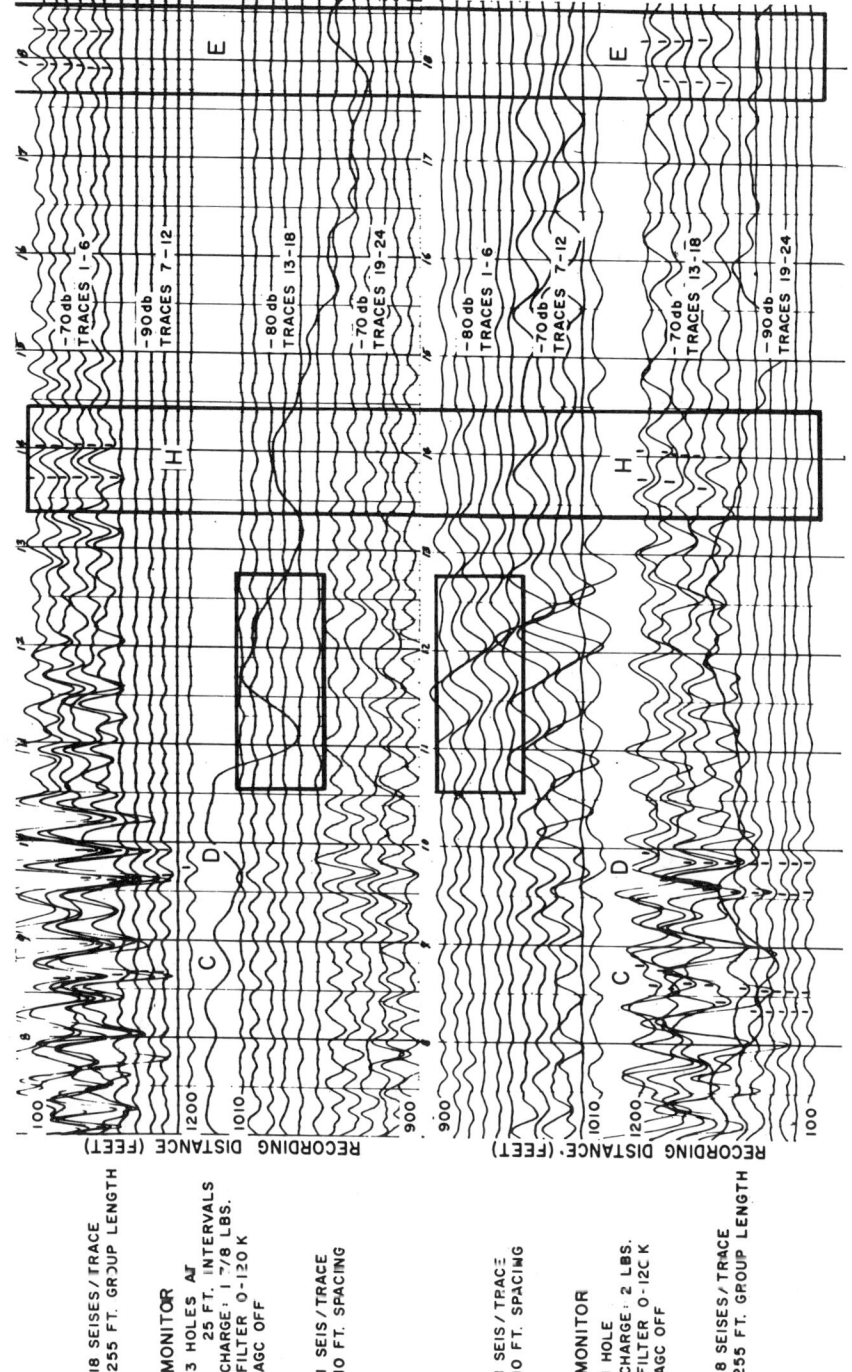

FIG. 11. Constant gain recordings to test effectiveness of multiple holes. The three-hole pattern yields an increase in signal-to-noise ratio three to four times that yielded by a single shot. Recording geometry is shown in Figure 7.

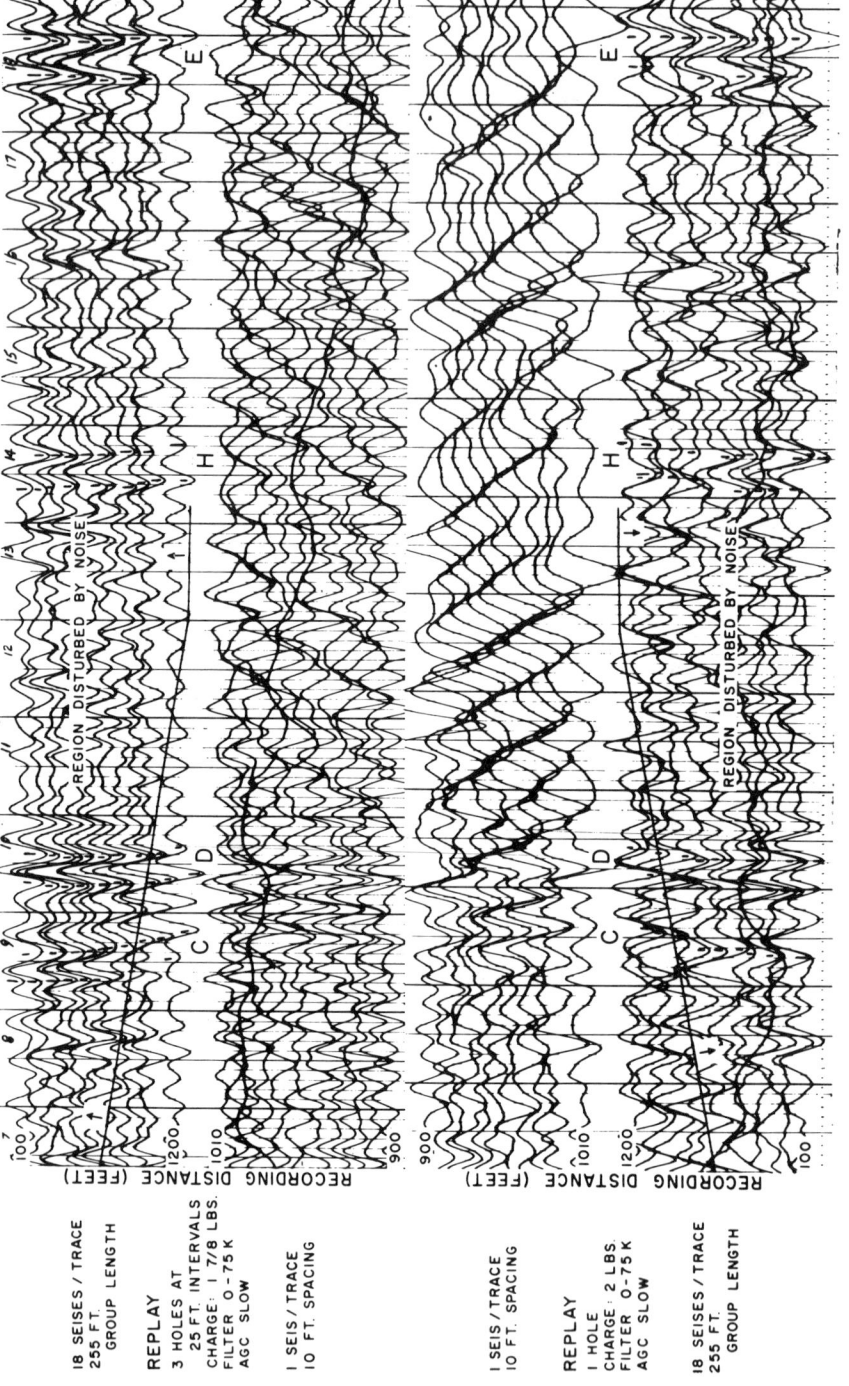

Fig. 12. Replays of seismograms shown in Figure 11 made with automatic gain control. Improvement in general record quality provided by the three-hole pattern is evident on these seismograms displayed in more conventional form.

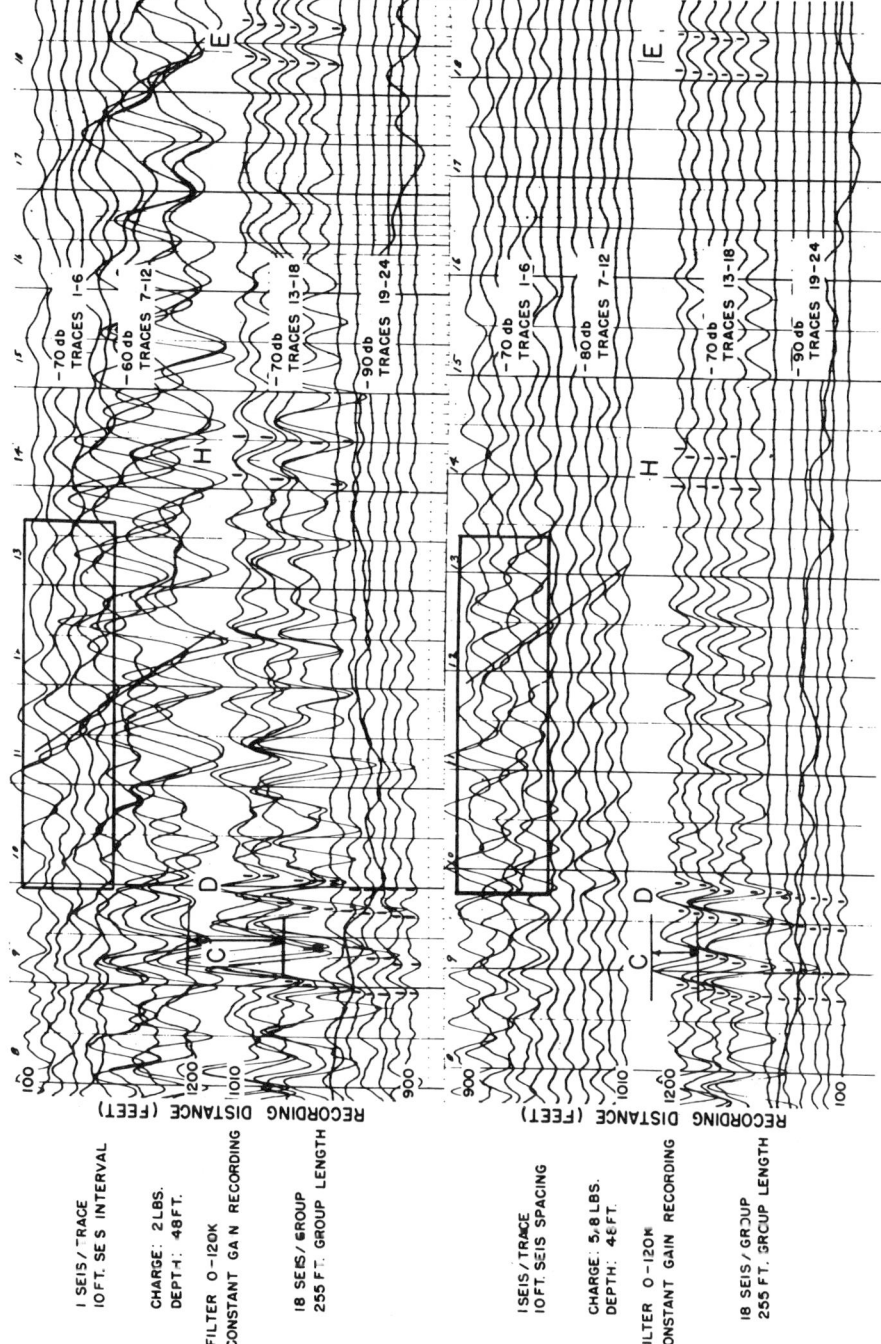

FIG. 13. Sample seismograms secured in study of effectiveness of charge size. Both charges are sufficiently large to yield energy levels greater than the preshot level. The larger charge yields an improved signal-to-coherent noise ratio. Recording geometry is shown in Figure 7.

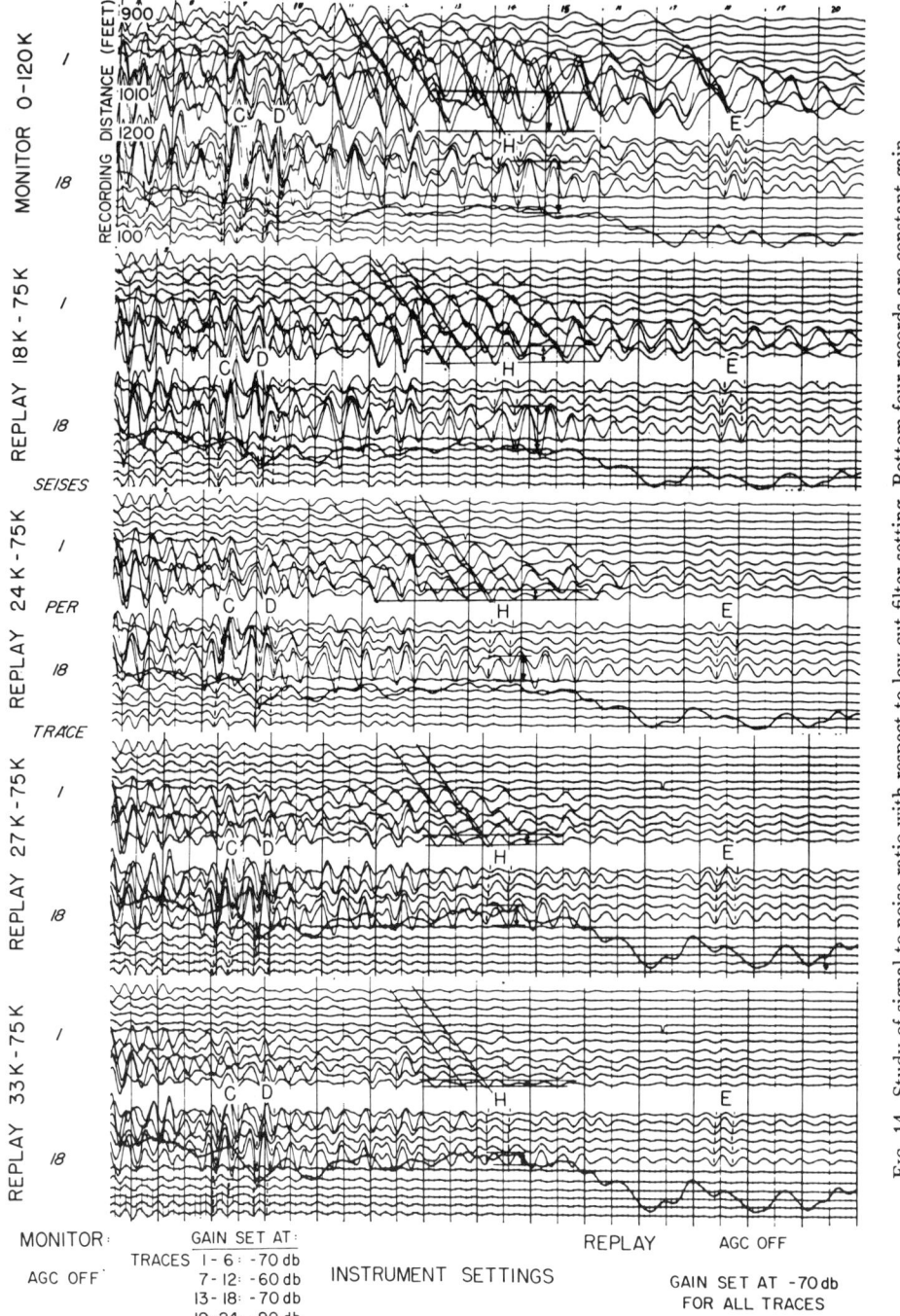

FIG. 14. Study of signal-to-noise ratio with respect to low-cut filter setting. Bottom four records are constant gain replays of top record. Recording geometry is shown in Figure 7.

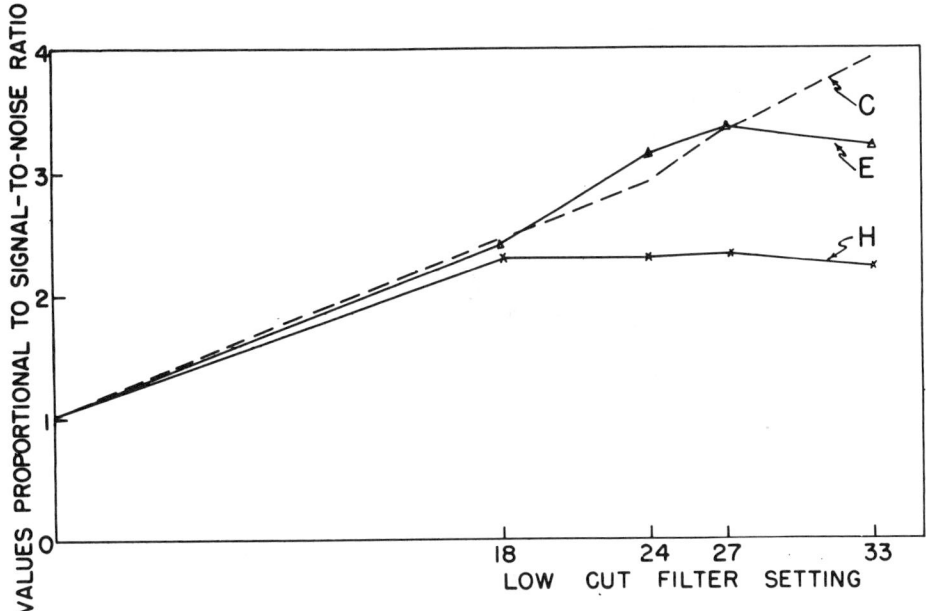

Fig. 15 Plot of signal-to-noise ratio as a function of low-cut filter setting for three reflections. Signal-to-noise ratio of event "H" does not improve by increasing the low-cut filter setting beyond 18 cps.

ing geometry used to obtain this profile was similar to the one shown in Figure 7. Signal-to-noise ratio measurements were made for several events on each record by measuring the noise on the traces representing single seismometers and the reflected signals on the traces representing multiple seismometers. Figure 15 shows the measured signal-to-noise ratio as a function of the low-cut filter for three events. The ratio values have been normalized with respect to the value existing on the open filter shot. This graph shows that a significant improvement is achieved by increasing the value of the low-cut filter to 18 cps on all events but that no additional relative rejection of low frequencies occurs for some events with additional filtering. This behavior suggests that, with low-cut filters higher than 18 cps, the signal is being filtered approximately as rapidly as the noise, with an attendant loss of reflection character. For this reason, a low-cut filter in the vicinity of 15 to 20 cps appeared to be an optimum choice.

Shown in Figure 16 is a filtered replay of the bottom record displayed in Figure 8. Note that in the region disturbed by the noise, the improvement provided by the pattern of 18 seismometers is approximately the same on the filtered record as on the unfiltered one shown in Figure 8. This particular test provides additional evidence of the overlapping of the signal and noise spectra and demonstrates that one cannot expect to achieve a solution to the record quality problem through electrical filtering alone.

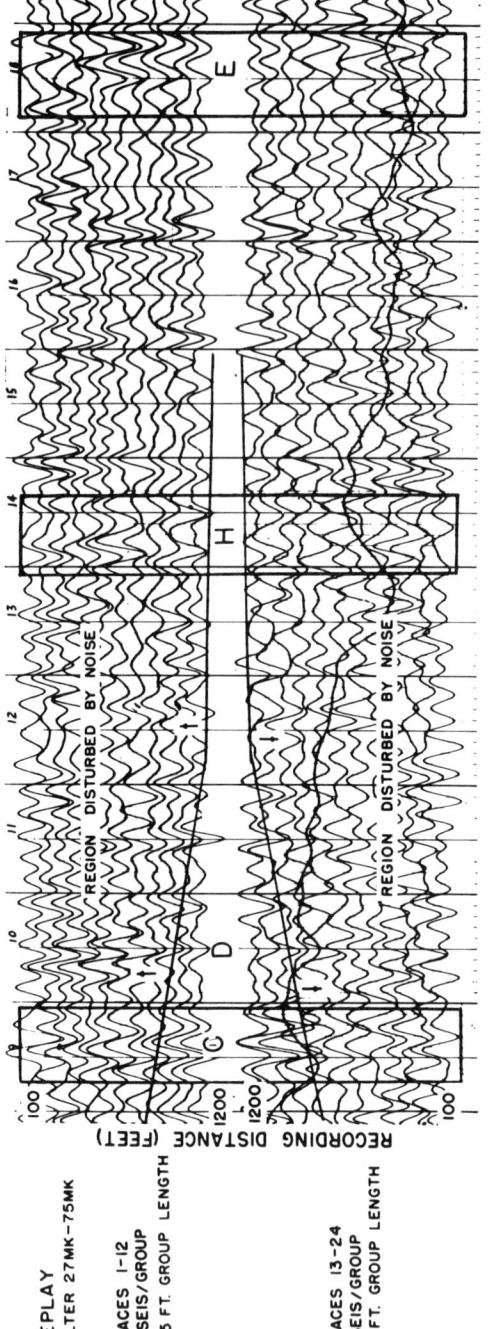

Fig. 16. Replay of bottom seismogram shown in Figure 8 with a 27 MK low-cut filter. Approximately the same degree of improvement is provided by the pattern of 18 seismometers on the filtered seismogram as on the unfiltered one shown in Figure 8.

Recordings were made at various stages in the experimental tests in which the filtering properties of 27-cycle, 14-cycle and 4-cycle seismometers were examined. One such comparison is shown in Figure 9. Although not demonstrated in Figure 9, the preponderance of evidence from tests of this nature showed that an improvement in signal-to-noise ratio could be achieved by substituting the 27-cycle seismometers for the 14-cycle type, but that this improvement was secured with some sacrifice in reflection character. For this reason, and because it was possible to attenuate severely the disturbing noise by wave length filtering, the 14-cycle seismometers were employed.

Utilization of the 14-cycle seismometers arrayed in patterns and a small amount of low-cut filtering permitted the use of the full dynamic range of the tape for recording primarily the desired reflected signals over a broad frequency range. Replays of the magnetically stored data were made through filter pass bands designed to emphasize particular reflection events when the need arose.

Extrapolation of Test Results over Prospect

The sequence of tests were conducted at several shotpoint locations distributed along a two-mile line. There is always a question of how applicable the results of tests conducted in a very localized area will be to the design of a suitable recording technique for the entire prospect. It is a commonplace observation that in a poor record area, the response will vary over wide ranges.

In this instance the tests served to provide a description of the problem and methods of solution in one location. Once accomplished, it was necessary only to keep track of the variation in noise characteristics to guide the modification of the initially-developed recording technique as needed. The changing conditions were followed by conducting very abbreviated tests with an experimental arrangement of the type illustrated in Figure 7. Such tests required very little production time. Operationally, the only additional work required was that involved in laying out a test cable alongside one-half of the spread arrangement being used and making the necessary recordings. The abbreviated tests were conducted when the technique adopted failed to produce the predicted result or when the driller's logs indicated a marked variation in near-surface conditions. Such tests could be interpreted in a meaningful fashion by relating the observations to the experiences developed in the initial and more comprehensive tests.

RESUME

To consider the nature of an exploration program designed to investigate a suspected anomalous feature, conferences attended by oil company geologists, geophysicists, and service company geophysicists were held. The objective of the proposed exploration program was clearly specified by the oil company. In studying the contribution toward the realization of the objective expected from seismic methods, it was observed that the data obtained in prior surveys in the prospect were inadequate to meet the new requirements. A review of these data and relevant geologic conditions indicated that by means of a carefully executed experimental program the cause of the poor records could probably be successfully discovered and suitable recording techniques could be developed. The quality of data needed was specified, and a time schedule was set for the investigation of the problem and the evaluation of proposed solutions. This entire procedure constituted a good example of the objective approach to exploration advanced by Agnich and Dunlap (1959).

Experimental tests were subsequently carried out, and the investigation and evaluation phase was completed in three production days. The test results indicated that the poor record quality in the area was attributable chiefly to the destruction of the reflection continuity by relatively strong surface and near-surface waves generated by and propagating outward from the exploding charge. Because of the time range over which it was desired to secure data, the device of offsetting the shot was not a possible technique for improving the signal-to-noise ratio. Owing to the apparent wave length separation between the reflected signals and interfering noise, wave length filtering by means of suitable seismometer and shot patterns and compositing through data processing methods provided a major increase in record quality. Use of wave length filtering permitted the recording of a broad frequency range of reflected signals on magnetic media.

Utilization of the techniques developed from the tests yielded results sufficiently improved from those obtained in an earlier survey to warrant a continued pursuit of the exploration objective. A sample of the reflection data recorded in the vicinity of the tests is shown in Figure 17. The top two seismograms displayed in Figure 17 and the seismograms displayed in Figure 2 were recorded at common surface positions.

COMMENTS ON GENERAL EXPERIENCES WITH PROCEDURE

The general analytical method of investigation, of which this case history is a specific example, has been used a number of years for studying record quality problems in many prospects around the world. The investigating procedure has been found to be a practical and generally useful means for discovering the nature of the problems and for estimating what improvement can be accomplished by practical means. On some occasions the answer to such a study has been that the problem probably cannot be solved by means of existing techniques or that the solution is too expensive in terms of anticipated rewards. At the other end of the scale of difficulty, problems were encountered which could be solved in several ways, and the task was to develop the most economic recording geometry from possible alternatives.

Experience has shown that the near-surface and surface waves commonly grouped under the general term of "ground roll" frequently constitute the dominant interfering noise and that these waves exhibit peaked wave length distributions characteristic of the near-surface materials. This latter condition would be expected since the characteristics of these waves are uniquely related to layering

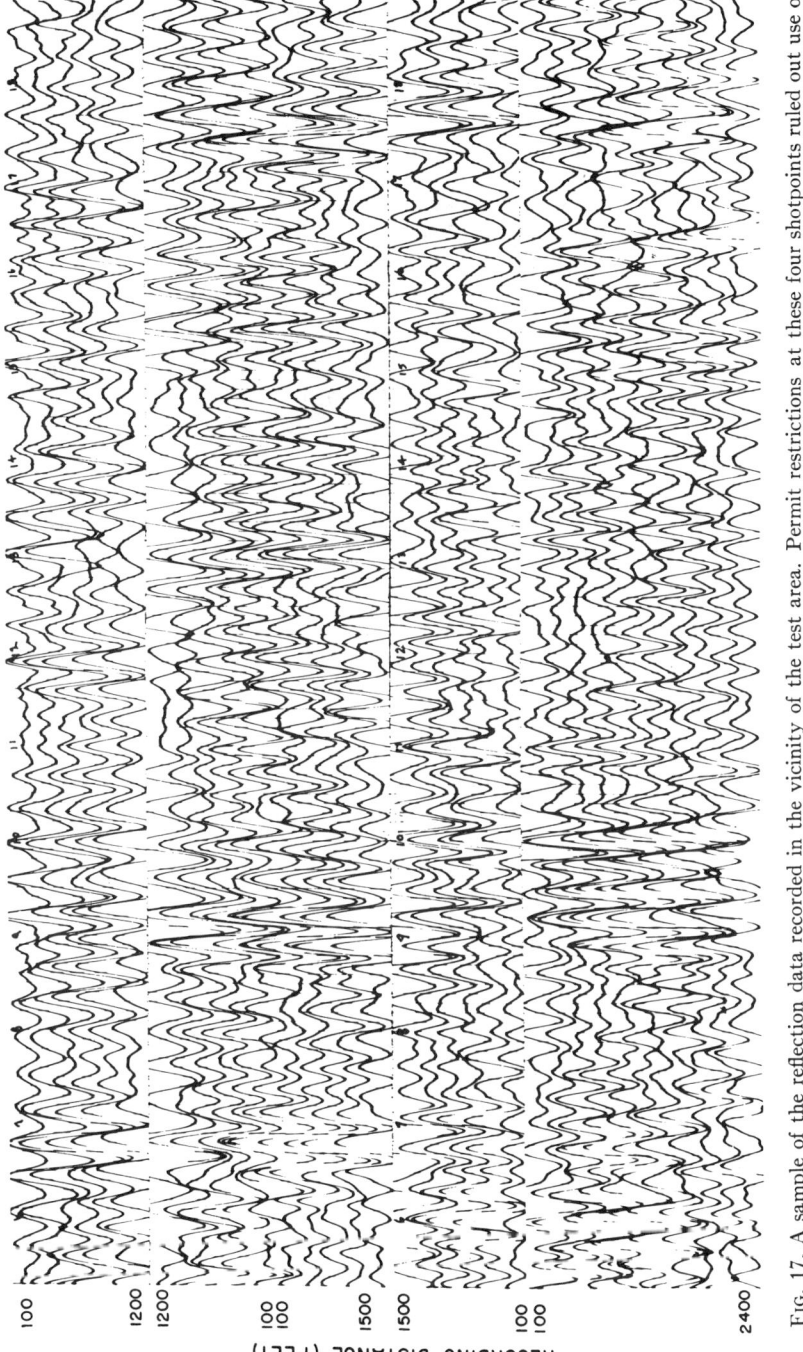

Fig. 17. A sample of the reflection data recorded in the vicinity of the test area. Permit restrictions at these four shotpoints ruled out use of optimum charge size and pattern holes. Seismograms were recorded with one hole, 1¼ pound charges, and 18 seismometers per group. The top two seismograms were obtained at same location as those displayed in Figure 2.

sequence, velocities, densities, and other physical properties of the shooting medium. Again, when coherent noise waves are the dominating interference, experience has shown that better seismic data can be obtained when the design of the recording technique is based on the special character of the noise determined from tests and sampling procedures than when the design is based on more general assumptions.

The amount of testing needed to keep track of changing conditions depends upon the difficulty of the problem and the rate of change in the physical properties of near-surface materials. A record quality problem characterized by high intensity noise coupled with complicated and rapidly changing subsurface structural conditions may require a sampling of the noise at every profile or at alternate profiles. Measures of noise intensities obtained at such close sampling intervals can be represented on a noise intensity contour map. The behavior of the contours on such a map provides a criterion for deciding whether disappearance of a reflected band of energy is attributable to increased noise intensity or to a rapidly changing structural condition. Whether or not such procedures are worth the effort depends specifically on the estimated value of achieving the exploration objective.

Frequently the noise consists of both coherent boundary waves and wave groups arriving in random directions. Such random waves may be largely disturbances from secondary sources set up by the impinging of the initial near-surface waves upon randomly located-near-surface inhomogeneities. In such situations it is important to establish which noise is dominant in the time range under investigation. In those cases in which both types of waves are sufficiently damaging, it may become necessary to construct two dimensional patterns having specified responses for particular directions of arrival. By the use of a suitably designed shot array, either horizontally or vertically distributed charges, it is often possible to reduce the energy traveling outward in the primary noise wave and hence reduce the strength of the potential secondary sources.

ACKNOWLEDGEMENTS

The author expresses his appreciation to the client company which generously permitted use of the data and to Geophysical Service Incorporated for permission to publish the paper. The author wishes also to give recognition to Mr. K. E. Burg and his staff for valuable help and suggestions.

REFERENCES

Agnich, F. J. and Dunlap, Jr., R. C., 1959, Standards of performance in petroleum exploration: Geophysics, v. 24, p. 916–924.
Dobrin, M. B., Lawrence, P. L., and Sengbush, R. L., 1954, Surface and near-surface waves in the Delaware Basin: Geophysics, v. 19, p. 695–715.
Hammond, J. W., and Hawkins, J. E., 1958, Getting the most out of present seismic instruments: Geophysics, v. 23, p. 795–822.

Howell, L. G., Neuenschwander, E. F., and Pierson III, A. L., 1953, Gulf Coast surface waves: Geophysics, v. 18, p. 41–53.
Lombardi, L. V., 1955, Notes on the use of multiple geophones: Geophysics, v. 20, p. 215–226.
Musgrave, A. W., Ehlert, G. W., and Nash, Jr., D. M., 1958, Directivity effect of elongated charges: Geophysics, v. 23, p. 81–96.
Parr, Jr., J. O., and Mayne, W. H., 1955, A new method of pattern shooting: Geophysics, v. 20, p. 539–564.
Savit, C. H., Brustad, J. T., and Sider, J., 1958, The moveout filter: Geophysics, v. 23, p. 1–25.
Smith, M. K., 1956, Noise analysis and multiple seismometer theory: Geophysics, v. 21, p. 337–360.
Van Melle, F. A., and Weatherburn, K. R., 1953, Ghost reflections caused by energy initially reflected above the level of the shot: Geophysics, v. 18, p. 793–804.
White, J. E., 1958, Transient behavior of patterns: Geophysics, v. 23, p. 26–43.

SEISMIC PROCESSING

COMMON REFLECTION POINT HORIZONTAL DATA STACKING TECHNIQUES*

W. HARRY MAYNE[†]

Techniques are described whereby multiple coverage of the subsurface is obtained. Detector spreads and shotpoints are arranged so that the channels representing common depth points are recorded with appreciably different horizontal distances between the shotpoints and detector stations. The channels which have a common reflection point are combined, or stacked, after appropriate corrections for angularity and travel time to datum have been applied. Reflections which follow the assumed travel paths are greatly enhanced, and other events are reduced. Methods for attenuating multiple reflections with respect to primaries are discussed in considerable detail.

Typical field comparisons between conventional and stacked traverses are shown to illustrate the degree of improvement which can be obtained in the signal-to-noise ratio.

General considerations applicable to field usage, and the geographic range of field experience are summarized.

INTRODUCTION

Improvements in the signal-to-noise ratio of seismic signals has been a continuing project with geophysicists for many years. As certain problems have yielded to solution, new ones have been encountered. From Kilipsch (1936), Rieber (1936), Poulter (1950), Woods (1953), Reynolds (1954), Parr (1955) to Graebner (1960), various techniques have been described to utilize the noise attenuation properties of multi-element arrays.

Applications of these and related techniques, either singly, or in combination, have produced solutions to many difficult problems. As the multiplicity of shotpoints and detectors is increased to cope with still more difficult situations, however, we are eventually confronted with an inherent limitation. As the arrays become larger and larger, the subsurface area which is averaged increases correspondingly. In practice the summation or integration of reflection arrivals from a subsurface area theoretically as great as ten acres may result from pattern dimensions of less than 700 ft. This, of course, tends to obscure the very detail which is being sought. The multiple-coverage, common-reflection point technique was devised to provide a practical means of increasing multiplicity without this limitation.

DESCRIPTION OF THE METHOD

Multiple coverage of the same subsurface with different shot and detector positions has been suggested for several purposes. Green (1938) advocated multiple paths centered about a common depth point to eliminate the effect of dip on velocity determinations.

This writer (1956) proposed that the information associated with a given reflection point, but recorded with a multiplicity of shotpoint and geophone locations, be combined algebraically after applying appropriate time corrections. Thus, if the reflected signals received along the several paths are adjusted for coincidence, their resultant sum will be proportional to the number of signals. Perturbations following other than the postulated ray paths will not be coincident, and hence will be degraded relative to the reflections. For random incidence the average theoretical enhancement will be proportional to the square root of the number of signals. This is analogous to pattern performance. Since, however, the source and receiving points have been selected so that the reflection point is common to all paths, the limitations of conventional pattern techniques no longer apply. The horizontal spacing between source and receiver is restricted only by the following considerations:

---

* Presented at the 31st Annual SEG Meeting, Denver, November 8, 1961. Manuscript received by the Editor July 23, 1962.

[†] Petty Geophysical Engineering Company, San Antonio, Texas.

1. The greatest distance which will permit coincidence adjustments of the requisite accuracy. The probable error in the postulated stepout increases with distance and must be kept small with respect to the reflection period.
2. The greatest distance over which the reflected signals persist with adequate similarity.

Magnetic tape recording equipment, Loper and Pittman (1954), has become readily available, and has made the necessary summation processes convenient and economical.

Figure 1 illustrates one simple field procedure for obtaining multiple coverage of the subsurface. This 24-detector station arrangement has the unique property of recording data from a specific reflection point on the same channel throughout a sequence of 12 shots. The multiplicity, i.e. the number of available paths which have a common reflection point, is 12 or one-half the number of stations in the spread. Note that the progressive variation in horizontal distance between the shotpoint and detector for the end stations is 22 intervals. Thus, summation of the number one channels of the 12 shots is equivalent to a 12 element array with a total length of more than one-half mile if typical station intervals are used. This is, of course, many times longer than would be permissible with conventional techniques, and thus provides effective attenuation of events with extremely great apparent wavelengths. Each of the shots recorded in this manner will require a different moveout correction program, but summation can be made without channel transposition.

ALTERNATE ARRANGEMENTS

It is possible to develop the desired multiple coverage in a variety of ways. In the following examples, 24-detector station spreads will be used. If a standard split-spread configuration is preferred, the arrangement of Figure 2 can be used. In this example both the spread and shotpoint are advanced two intervals so that a symmetrical setup is maintained. The depth points corresponding to each shot have been indicated by the dots underneath the setup. It is seen that channels from a total of 12 shotpoints must be

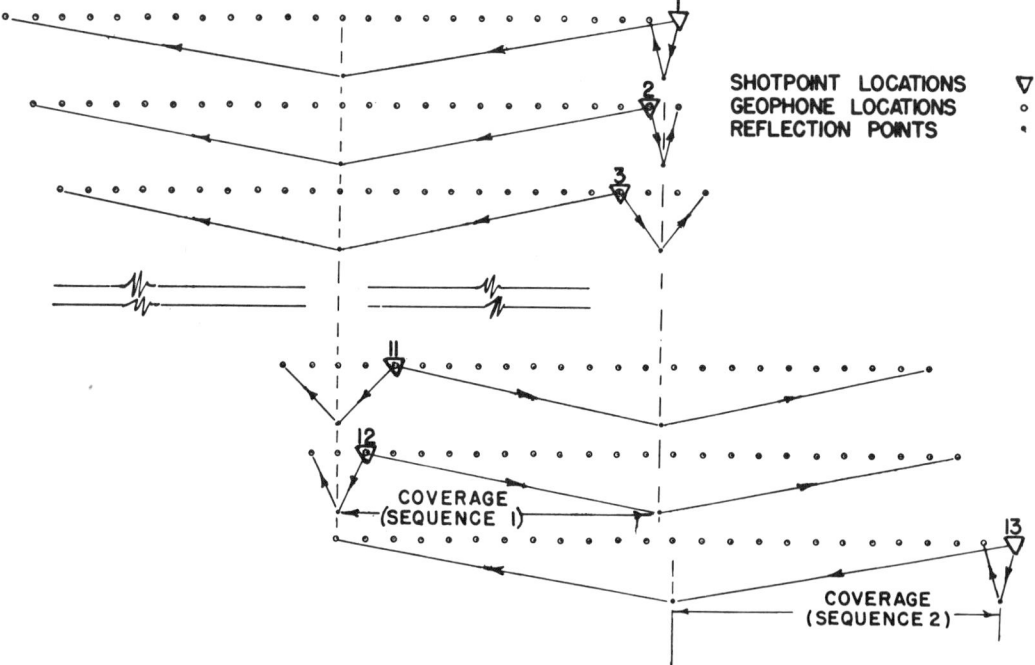

FIG. 1. Common-channel spread sequence giving 12-fold coverage of subsurface (multiplicity). Detector spread advances one station, shotpoint backs up one station for a total of twelve recordings. Common depth points are recorded on the same channel throughout each sequence of twelve recordings.

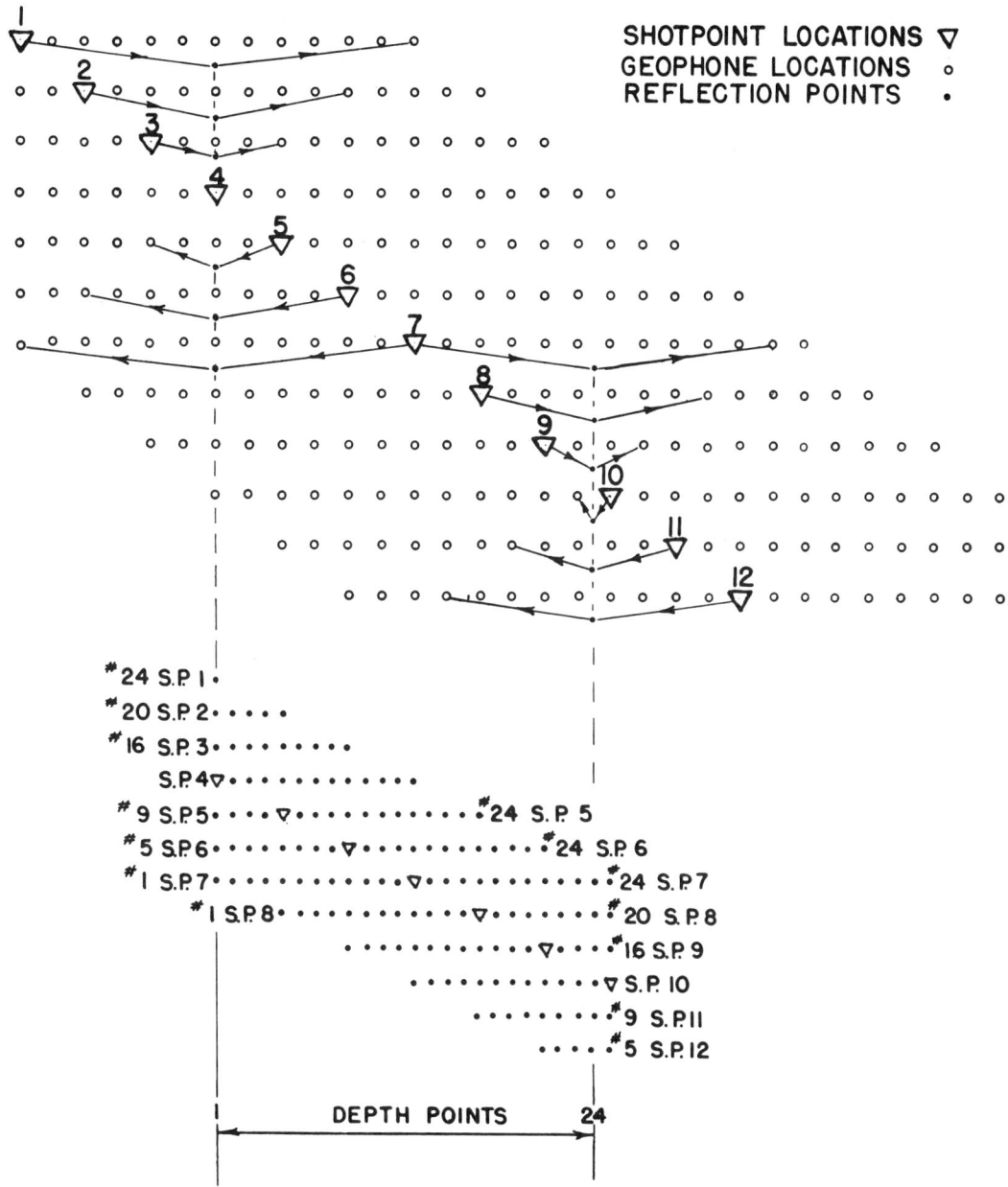

FIG. 2. Symmetrical straddle spread sequence giving six-fold multiplicity. Detector spread and shotpoint are advanced together by two stations for successive recordings.

used to obtain the full multiplicity of six.

Table 1 shows the channels recorded from each shotpoint which are used to compose the 24 depth points under Shotpoint 7.

In each of the previous examples only one shot is made into each instrument spread. Figure 13 illustrates a sequence which also yields six-fold multiplicity using three shots per spread, but allows the spread to be moved six stations ahead each time. Table 2 shows the corresponding channel composition schedule.

The preceding examples point the way to the

Table 1. Table illustrating the channel combination schedule for data recorded using the spread sequence of Figure 2. Depth points of the stacked resultant are composed of the channels included in each vertical column, as recorded from the indicated shotpoints. Depth points of the stacked resultant correspond to those recorded on channels one thru 12, respectively, of shotpoint 7. Stacked depth point 13 is under shotpoint 7, and stacked depth points 14 thru 24 correspond to those recorded on channels 13 thru 23 of shotpoint 7 respectively.

S. P. No.	Depth point No.																							
	1	2	3	4	5	6	7	8	9	10	11	12	13	14	15	16	17	18	19	20	21	22	23	24
1	24																							
2	20	21	22	23	24																			
3	16	17	18	19	20	21	22	23	24															
4	—	13	14	15	16	17	18	19	20	21	22	23	24											
5	9	10	11	12	—	13	14	15	16	17	18	19	20	21	22	23	24							
6	5	6	7	8	9	10	11	12	—	13	14	15	16	17	18	19	20	21	22	23	24			
7	1	2	3	4	5	6	7	8	9	10	11	12	—	13	14	15	16	17	18	19	20	21	22	23
8					1	2	3	4	5	6	7	8	9	10	11	12	—	13	14	15	16	17	18	19
9									1	2	3	4	5	6	7	8	9	10	—	13	14	15
10													1	2	3	4	5	6	7	8	9	10	11	12
11																	1	2	3	4	5	6	7	8
12																					1	2	3	4

general expression for the multiplicity of any particular progression. Thus if:

M = path multiplicity (as previously defined),
N = number of detector stations in the spread,
n = number of stations by which the spread is advanced,
S = number of shot positions for each spread,

the following expression will define the multiplicity for any sequence:

$$M = NS/2n.$$

Figure 6 illustrates a third arrangement which develops six-fold multiplicity. This is similar to the sequence of Figure 2 since the shot location and spread are both advanced two stations, and only a single shot per spread is taken. However, the shot is located at the end of the spread, which changes the geometry of the common-reflection-point paths. Table 3 shows the channel composition schedule. This type of arrangement is particularly efficient in attenuating multiple reflections.

MULTIPLE REFLECTIONS

Powerful evidence confirming the existence of multiple reflections was presented by Ellsworth et al (1948). Later experience and theoretical studies on synthetic seismograms have indicated that occurrence of this insidious form of "noise" is dangerously common.

The common-reflection point technique can be an excellent tool in reducing multiple reflections even though their source in many cases cannot be precisely determined. Figure 5 illustrates a rather typical example, assuming for simplicity that the shot and the near-surface velocity contrast generating the multiples are at the same level. A hypothetical velocity function of $6{,}000 + 0.2Z$, where Z = depth below the shot, has been as-

Table 2. Table illustrating the channel combination schedule for data recorded using the spread sequence of Figure 3. Depth points represented by the stacked data are the same as those recorded from shotpoint 2 C.

S. P. No.		Depth point No.																							
		1	2	3	4	5	6	7	8	9	10	11	12	13	14	15	16	17	18	19	20	21	22	23	24
1	A	17	18	19	20	21	22	23	24																
	B	15	16	17	18	19	20	21	22	23	24														
	C	13	14	15	16	17	18	19	20	21	22	23	24												
2	A	5	6	7	8	9	10	11	12	13	14	15	16	17	18	19	20	21	22	23	24				
	B	3	4	5	6	7	8	9	10	11	12	13	14	15	16	17	18	19	20	21	22	23	24		
	C	1	2	3	4	5	6	7	8	9	10	11	12	13	14	15	16	17	18	19	20	21	22	23	24
3	A									1	2	3	4	5	6	7	8	9	10	11	12	13	14	15	16
	B											1	2	3	4	5	6	7	8	9	10	11	12	13	14
	C													1	2	3	4	5	6	7	8	9	10	11	12
4	A																					1	2	3	4
	B																						1	2
	C																								

Table 3. Table illustrating the channel combination schedule for data recorded using the spread sequence of Figure 4. Depth points represented by the stacked data are the same as those recorded from shotpoint 6.

S. P. No.	Depth point No.																							
	1	2	3	4	5	6	7	8	9	10	11	12	13	14	15	16	17	18	19	20	21	22	23	24
1	21	22	23	24																				
2	17	18	19	20	21	22	23	24																
3	13	14	15	16	17	18	19	20	21	22	23	24												
4	9	10	11	12	13	14	15	16	17	18	19	20	21	22	23	24								
5	5	6	7	8	9	10	11	12	13	14	15	16	17	18	19	20	21	22	23	24				
6	1	2	3	4	5	6	7	8	9	10	11	12	13	14	15	16	17	18	19	20	21	22	23	24
7					1	2	3	4	5	6	7	8	9	10	11	12	13	14	15	16	17	18	19	20
8									1	2	3	4	5	6	7	8	9	10	11	12	13	14	15	16
9													1	2	3	4	5	6	7	8	9	10	11	12
10																	1	2	3	4	5	6	7	8
11																					1	2	3	4

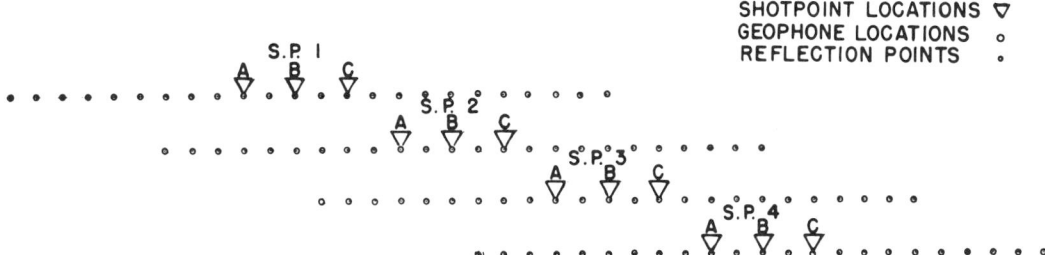

FIG. 3. Nonsymmetrical spread sequence which gives six-fold multiplicity using three shotpoint locations for each detector spread. The detector spread is advanced six stations forward after each sequence of three recordings. Shotpoints spaced two intervals apart.

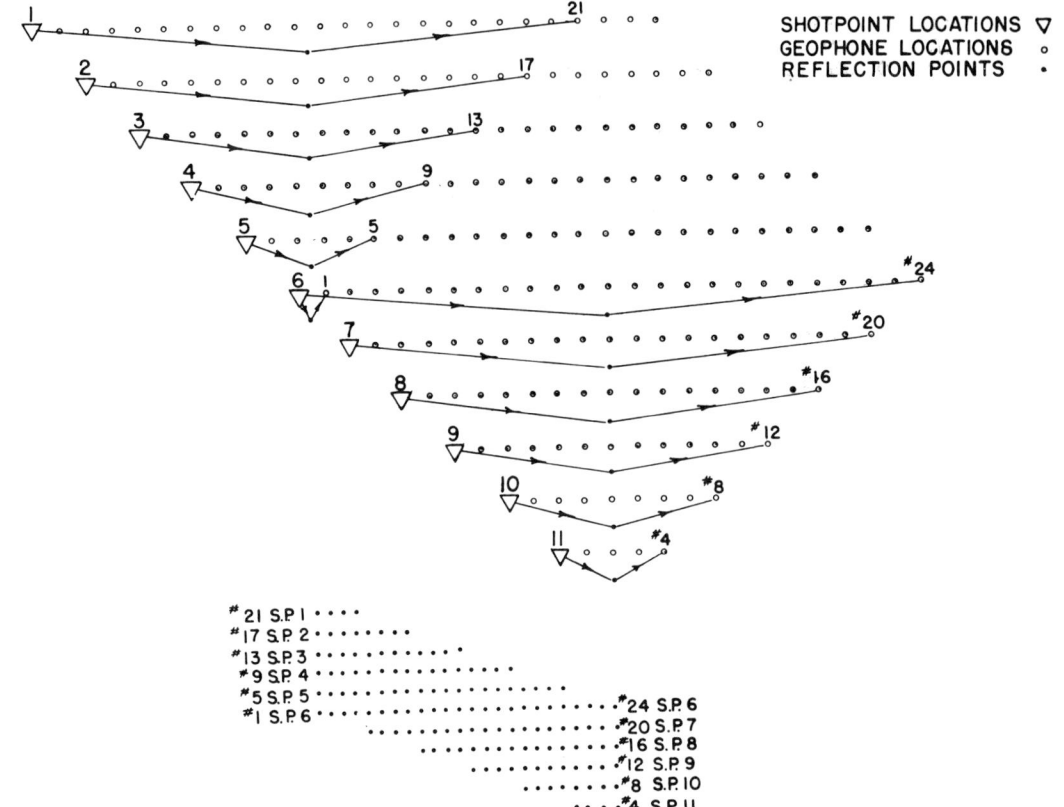

Fig. 4. Single-ended spread sequence giving six-fold multiplicity. Shotpoint and spread advanced together by two stations between successive recordings. Similar arrangements with the shotpoint at one end of the spread are preferred for use in attenuating multiple reflections.

sumed. An objective of major interest is postulated at a depth of 8,960 ft which corresponds to a vertical two-way time of 2.300 sec. The second-order reflection (simple multiple) from a shallow reflector at 3,898 ft would also have a total travel time of about 2.300 sec. The relative strength of the two reflections will, of course, depend on the reflection coefficients involved. Since the travel path of the multiple is confined to a lower velocity zone, it will exhibit the stepout shown by the upper curve of Figure 5, assuming straight-path geometry. The deep primary reflection will follow the stepout shown by the middle curve. The lower curve shows the difference in stepout between the two events. Hence if a number of channels with different shotpoint-to-detector distances are combined so that the primaries are in coincidence, the multiples will be out of phase as indicated by the lower curve. For example, suppose we combine five paths, and the multiple has a period of 0.035

Table 4. Table comparing the actual stepout differences obtained with the desired theoretical optimum values when the spread arrangement of Figure 4 is adapted to the example shown in Figure 5. The detector spacing is chosen to provide the desired overall difference in stepout between the nearest and the farthest common-reflection-point channels recorded. More attenuation can be obtained by using only five channels as shown, since there is so little difference in stepout between channels one and five.
Station interval is 215 ft.

Station No.	Distance	Actual Stepout	Stepout for Optimum Attenuation
1	215		0
5	1070		
9	1925	.006	.007
13	2785	.011	.014
17	3640	.019	.021
21	4500	.028	.028

Actual attenuation of 5 channel summation > 4.5 to 1.

sec. Attenuation behavior will be the same as for a five-unit pattern array. Hence we must have a minimum of 0.035/5 or 0.007 sec difference in stepout between the successive paths, or a total of 0.028 sec between the extremes if adequate attenuation is to be obtained. The estimated distances required to establish the necessary differences are indicated in Figure 5 by the circled points numbered one through five on the lower curve. Note that the distance to the fifth or farthest channel in this example is 4,500 ft. Shorter distances will seriously reduce the attenuation and should not be considered. Longer distances will not seriously affect the attenuation, but become increasingly unwieldy. Obviously paths with a common reflection point must be used if excessive subsurface averaging is to be avoided. Of equal importance is the fact that the amount and direction of dip of either reflector is not significant if common-reflection point geometry is applicable.

One problem is apparent from Figure 5. Since the differential-stepout-versus-distance curve is not linear, the desired station intervals are nonuniform. Suppose that data are to be recorded using the spread arrangement of Figure 4. If the required 0.028 sec stepout is developed between a station 1 and a station 21 (Table 3), our station interval should be 4,500/21 or approximately 215 ft. Table 4 shows the comparison between the theoretical stepout desired, and the actual values obtained. These approximations yield better than a 4.5-to-1 theoretical improvement in the primary-to-multiple ratio, assuming sinusoidal waves. There is so little stepout difference between the nearest channels (1 thru 4) and their next corresponding common-reflection-point channels (5 thru 8) that better attenuation is

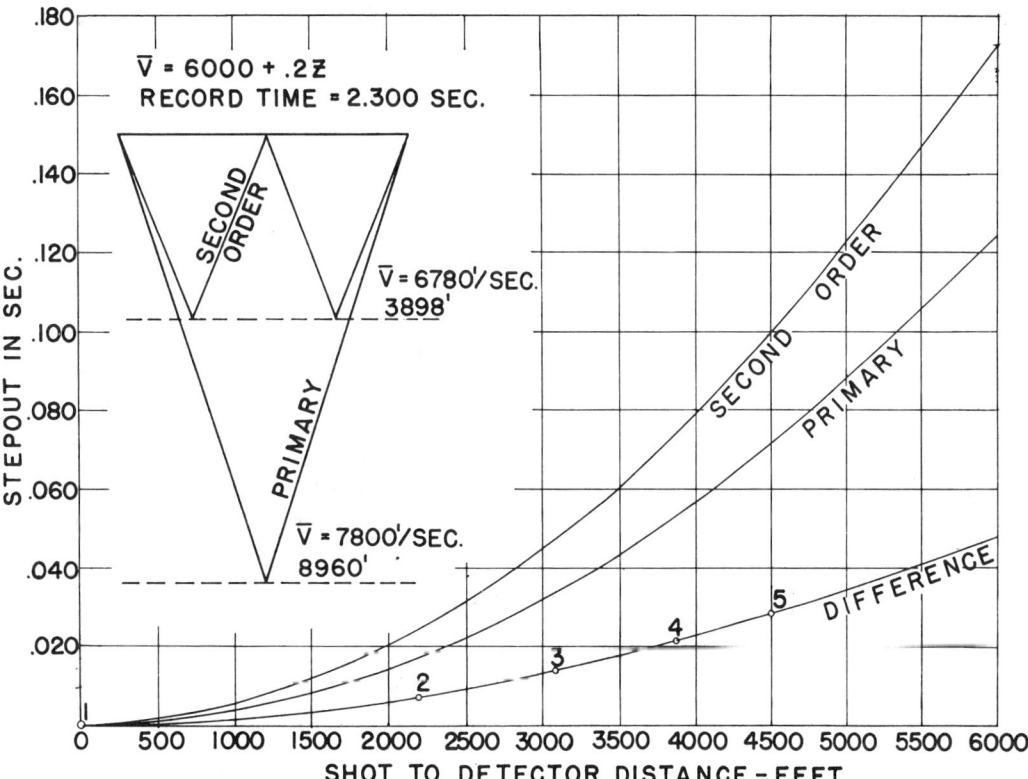

FIG. 5. Stepout as a function of shotpoint-to-detector distance for a primary reflection and a multiple with the same total travel time. Assumed velocity function is typical of the Gulf Coast of Texas. The lower curve represents the difference in stepout between the multiple and the primary at each distance. The circled points numbered 1 thru 5 on the lower curve show the distance at which channels should be recorded to obtain infinite theoretical attenuation of a sinusoidal multiple with a period of 0.035 sec. Corrections for the moveout of the primary would be applied, and the five channels combined to produce a single resultant.

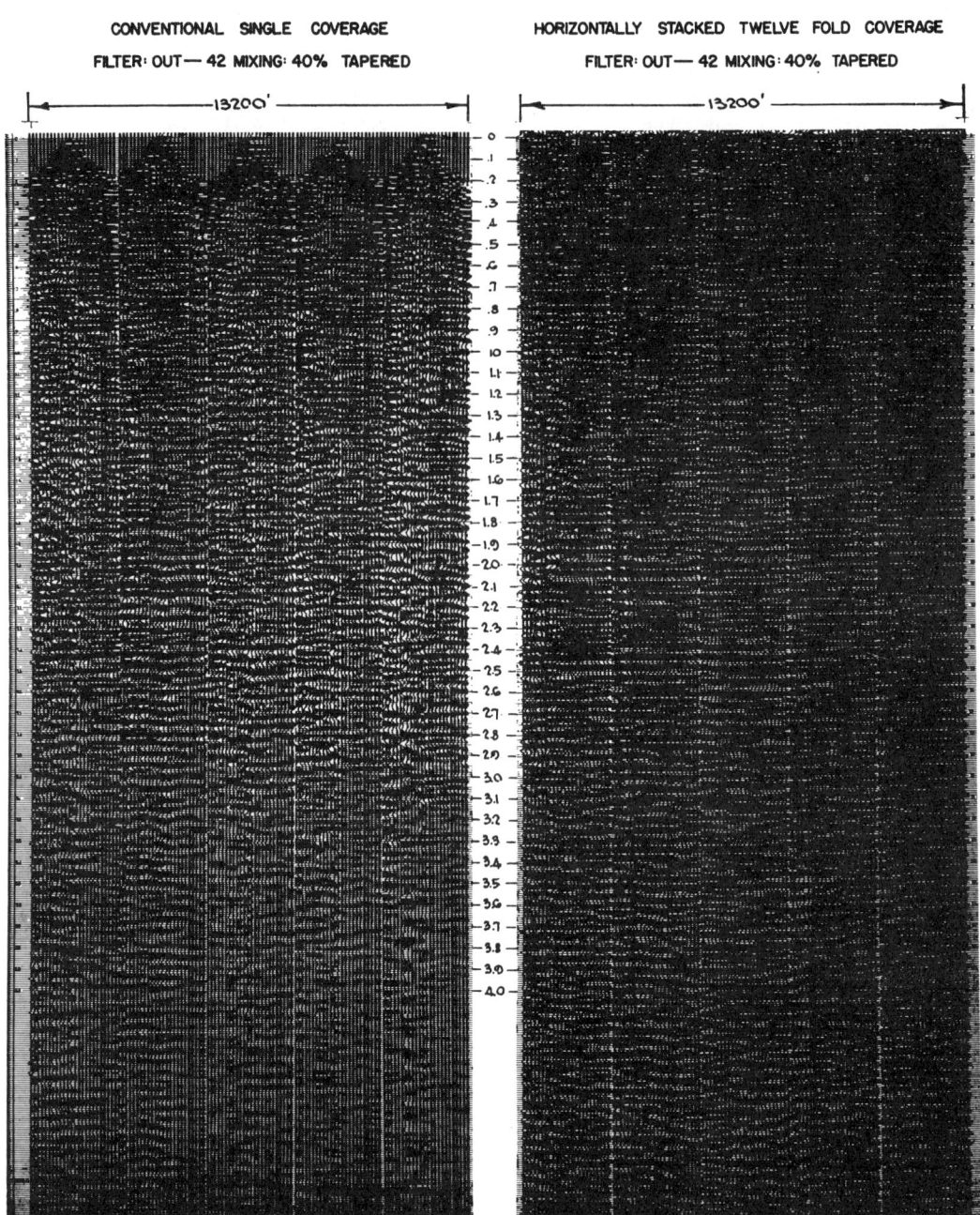

FIG. 6. Comparison of the same traverse recorded with conventional single coverage and stacked 12-fold multiplicity in the Powder River Basin of Wyoming. Shot and detector patterns, and playback settings were identical for both sections.

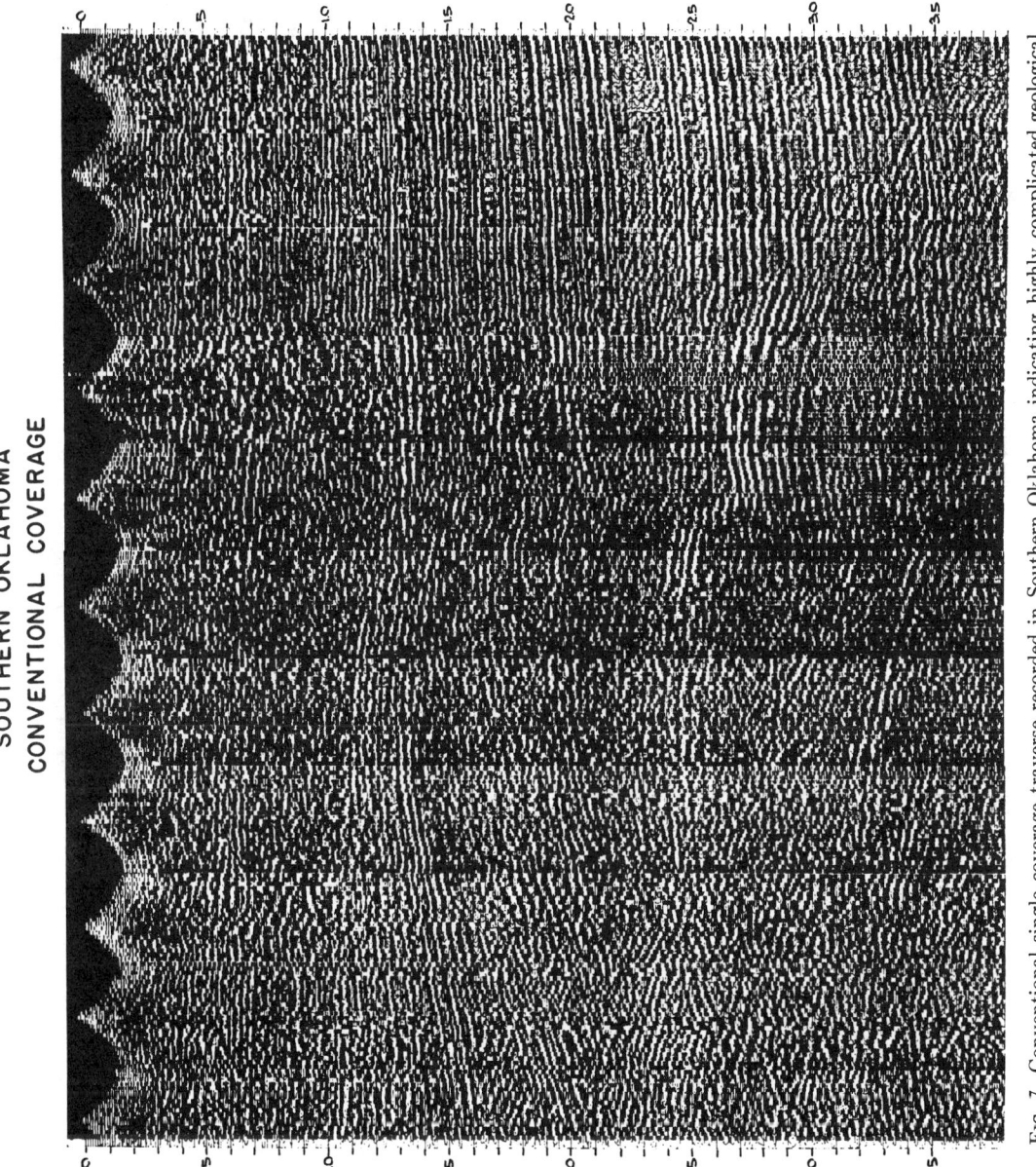

Fig. 7. Conventional single coverage traverse recorded in Southern Oklahoma indicating highly complicated geological conditions. Shot and detector patterns, and playback settings are identical with those used for Figure 8.

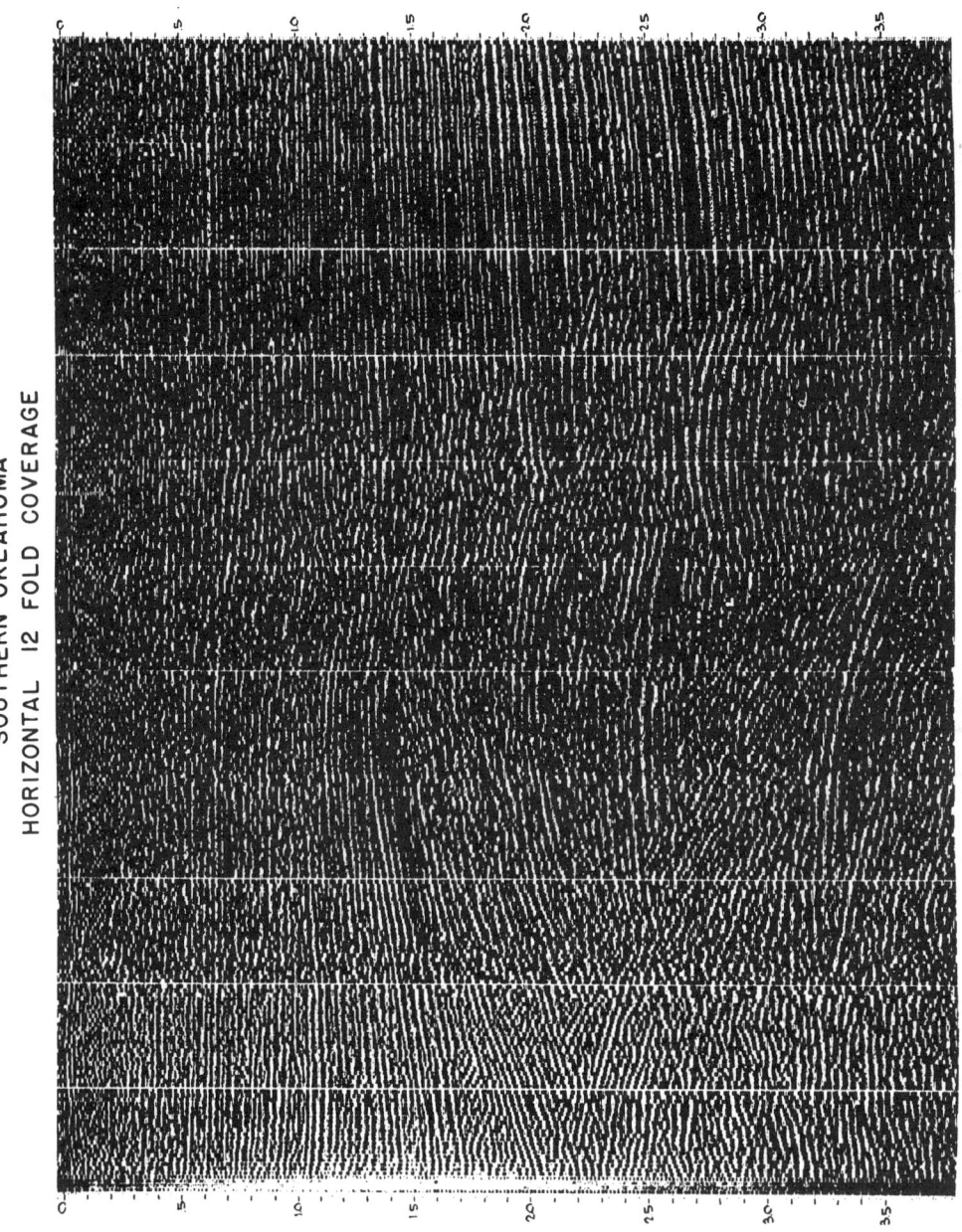

FIG. 8. Same traverse as Figure 7 using stacked 12-fold multiplicity. All field and playback parameters identical with those used in obtaining Figure 8.

obtained if channels 5 thru 8 are omitted. Use of all six channels reduces the corresponding ratio to 2.5 to 1.

The common-reflection-point technique is also advantageous when inverted-polarity mixing is used to attenuate multiple reflections. This form of skip mixing is effected by first adjusting the multiples to time coincidence. After adjustment the polarity of one channel is reversed, and it is then combined with any normal polarity channel having similar waveform and amplitude.

Attenuation of the multiple reflection will be limited only by the degree of waveform identity between the two signals, and the precision of the phase and amplitude adjustments.

Destructive interference will also attenuate the primary reflections unless they happen to be approximately one-half cycle out of phase on the channels which are combined. This means that the differential stepout between the multiple and the primary reflections must be one-half the period of the primary reflections, and the resultant primary reflection pulse will be broadened by one-half cycle. Thus, for a primary reflection period of 0.035 sec, Figure 5 indicates that channels separated by a spread distance of approximately 3,500 ft must be selected to avoid attenuation of the primary reflection. Although fortuitous dip relations might permit combination of channels closer together, this illustrates the order of subsurface averaging which will probably occur.

This excessive subsurface averaging will be avoided if common-reflection-point channels which possess the proper difference in stepout are used. The multiple reflection will be attenuated as before, but use of the common-reflection-point technique will eliminate subsurface averaging, and will also make the process independent of the dips involved.

DATA COMPARISONS

Figure 6 illustrates a conventional traverse in the Powder River Basin of Wyoming compared to the same traverse with 12-fold multiplicity.

Shot pattern, geophone arrangement and playback settings were identical.

Figures 7 and 8 show a similar comparison in Southern Oklahoma.

FIELD EXPERIENCE

Considerable field experience in a wide variety of terrain has been accumulated. While experience has varied somewhat from place to place, certain common considerations have become apparent.

1. Filter: Because of improved attenuation of "ground roll" and other extraneous low-frequency noise, wider-band filters can generally be used. They are preferable because reflection character and apparent damping are improved, and the required correction precision is somewhat reduced.
2. Spreads should be as long as practicable. Not only are they more efficient from a production standpoint, but data quality is enhanced. Detector stations should be at least 220 ft apart, and 440 ft is desirable if conditions permit. The longer spreads are much less vulnerable to multiple reflections, and complement the multiplicity obtainable with conventional pattern arrays.
3. The most effective shot and detector patterns permitted by economic considerations should be employed. The added multiplicity available with this technique should be used as a supplement to normal good practice and not in substitution thereof.
4. Preliminary traverses in an area should be recorded with greater multiplicity than may be necessary. Test processing can then be performed to select the most economical arrangement.
5. Corrections for moveout, weathering, and elevation must be accurately determined and precisely applied. Careful editing of corrections, and deletion of obviously poor data can be of great benefit.
6. Approximately the same daily production can be maintained using this method as would be attained with conventional operations.

CONCLUSION

The common-reflection-point, horizontal-data-stacking technique has added a new order of magnitude to the usable dimensions of multipath pattern array geometry. Signal-to-noise ratios have been enhanced well beyond the saturation point of conventional pattern methods. Field effectiveness has been demonstrated in Mississippi, South Louisiana, the deep Frio and Wilcox trends of the Texas Gulf Coast, the Delaware and Palo Duro Basins of Texas, the thrust areas of Southern Oklahoma, the Powder River Basin of

Wyoming and Montana, and Colombia, South America.

ACKNOWLEDGMENTS

The writer sincerely appreciates the confidence and cooperation of the many client companies who have accepted this development, and assisted in its application.

The wholehearted support of the Petty Company and all my co-workers is gratefully acknowledged.

REFERENCES

Ellsworth, T. P., Johnson, Curtis H., Sloat, John, Ittner, Frank, Waterman, Joseph D., Gutenberg, B., and Fu, C. Y., Dix, C. Hewitt, Lester, O. C. Jr., Walling, Dean, Dresbach, C. H., Hansen, Paul F., 1948, Multiple reflections, a symposium: Geophysics, v. 13, pp. 1–85.

Graebner, R. J., 1960, Seismic data enhancement—a case history: Geophysics, v. 25, p. 283.

Green, C. H., 1938, Velocity determination by means of reflection profiles: Geophysics, v. 3, p. 295.

Klipsch, P. W., 1936, Some aspects of multiple recording in seismic prospecting: Geophysics, v. 1, p. 365.

Loper, G. B., and Pittman, R. R., 1954, Seismic recording on magnetic tape: Geophysics, v. 19, p. 104.

Mayne, W. Harry, 1956, Seismic surveying: U. S. Patent 2,732,906 (application 1950), (Abstract: Geophysics, v. 21, p. 856).

Parr, J. O., Jr., 1955, Seismic surveying: U. S. Patent 2,698,927 (application 1953), (Abstract: Geophysics, v. 20, p. 689).

Poulter, Thomas C., 1950, The Poulter seismic method of geophysical exploration: Geophysics, v. 15, p. 181.

Reynolds, F. F., 1954, Design factors for multiple arrays of geophones and shot holes: Oil and Gas Journal, v. 52, no. 50, p. 145.

Rieber, F., 1936, A new reflection system with controlled directional sensitivity: Geophysics, v. 1, p. 97.

Woods, J. P., 1953, Shot hole array for eliminating horizontally traveling waves: U. S. Patent 2,642,146 (application 1950), (Abstract: Geophysics, v. 18, p. 954).

CORRELATION TECHNIQUES — A REVIEW[*]

BY

N. A. ANSTEY[**]

Abstract

Correlation techniques are in the process of passing from the research laboratory to the field. This paper seeks to aid the transition. The three main sections of the paper have these objects:

1. To state in words the basic principles on which correlation techniques are based, and the connection between these principles and more familiar concepts which are already among the tools of the exploration geophysicist.
2. To review briefly the published applications of correlation techniques in science generally.
3. To discuss some of the additional applications which are now emerging in exploration geophysics.

The treatment is graphical and illustrative, rather than rigorous. The paper is addressed to the practising exploration geoscientist and to the newcomer to correlation techniques. A list of about 150 references is given.

Introduction

Some things, like two peas in a pod, are similar; others, like chalk and cheese, are not.

Throughout science, however, we find instances where the situation is not as clear as this, and where it is desirable to establish a *measure* of the similarity between two quantities. For such applications we require the techniques of correlation.

The mathematical treatment of these techniques is given excellently in the existing literature (for example, by Lee, 1960), and the use of these techniques for research purposes has been established for more than a decade. Until recently, however, the actual operation of correlation has been a slow one, so that correlation techniques could hardly be considered for routine use. Today, correlation in *real time* is entirely practical, and there is little doubt that correlation techniques will soon take their place in the day-to-day operations of the practising exploration geophysicist. Consequently it seems appropriate, at this time, to present the basic propositions of correlation theory in physical

---

[*] Presented at the Twenty-fifth Meeting of the European Association of Exploration Geophysicists at Liège, 3rd-5th June 1964.
[**] Seismograph Service Limited, London, England.

or graphical terms, to review briefly the published applications of correlation techniques throughout the broad field of science, and to discuss the emergent applications in practical geophysics.

THE BASIC PROPOSITIONS

A good method of measuring the similarity between two waveforms is to multiply them together, ordinate by ordinate, and to add the products over the duration of the waveforms. Thus to assess the similarity between waveforms a and b of figure 1, we multiply ordinate p_1 by ordinate q_1, ordinate p_2 by

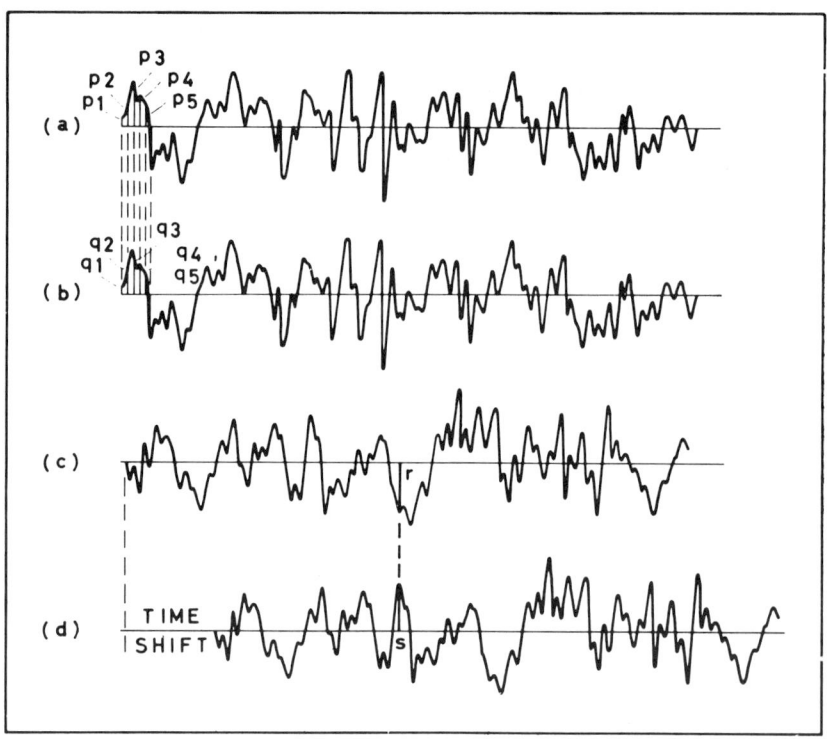

Fig. 1. Waveforms a and b are identical in shape and time. Waveforms c and d are identical in shape but d is shifted in time. The similarity between a and b is large, while that between b and c or between c and d is small.

q_2, p_3 by q_3, and so on, and finally we add these products to obtain a single number which is a measure of the similarity. In this example, waveforms a and b are identical, so that every ordinate, positive or negative, contributes a positive term to the sum; the sum is therefore large, and the waveforms are pronounced similar. If, however, we perform the same process on waveforms b and c, we find that each positive product is offset by another negative product, and the sum is small; the waveforms are pronounced dissimilar.

Now let us consider waveforms c and d. They are identical in shape, but one is displaced in time. If we perform the process of multiplying ordinates (of which r and s are typical), we find again that every positive product tends to be offset by a negative product, and that the sum is small. Thus if we were to plot the similarity between a waveform of the type of figure 1c and a time-shifted version of itself, we should expect the resulting curve to assume small values for large time shifts, and to rise to a large positive maximum when the time shift is zero. This curve is called the auto-correlation function.

The auto-correlation function of a waveform is a graph of the similarity between the waveform and a time-shifted version of itself, as a function of this time-shift.

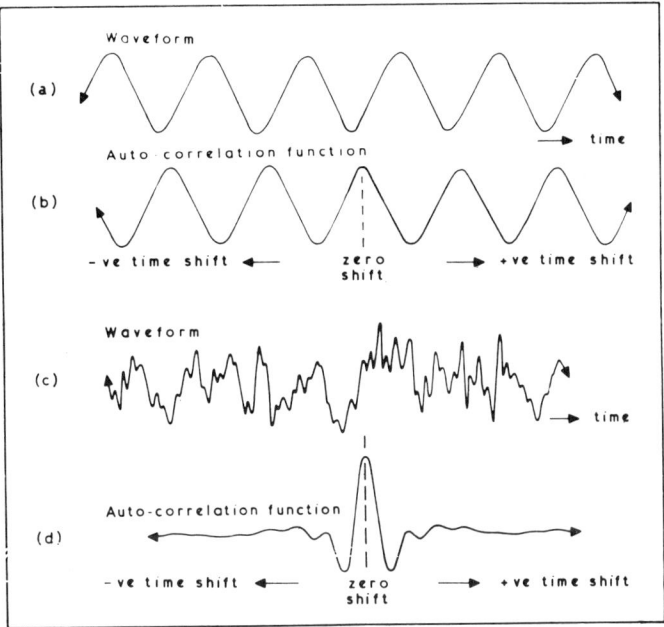

Fig. 2. Waveform *a* is a section of a sine wave. Waveform *c* is a section of continuous broad-band noise. Waveforms *b* and *d* are the corresponding auto-correlation functions. Note that the x-axis of an auto-correlation function represents time *shift*, not time.

We may obtain some feeling for the nature and merit of the auto-correlation function by considering the two examples of a sine wave and a sample of broad-band noise. The sine wave (figure 2) becomes exactly similar to itself whenever the time shift is an integral number of periods, and so the auto-correlation function must itself be periodic. In fact, the auto-correlation function of a sine wave is also sinusoidal, having the same frequency, and being symmetrical about the point which represents zero time shift (figure 2b).

The broad-band noise of figure 2c is quite different; a very small time shift is sufficient to destroy the similarity, and the similarity never recurs. The auto-correlation function (figure 2d) is therefore a sharp impulse, decaying quickly from the central maximum to very low values at large time shifts. Moreover, the same argument would apply if we took any other sample of noise having the same frequency content; the auto-correlation function would be the same. Evidently the auto-correlation function of a waveform does not depend on the actual waveform itself, but on its frequency content.

This is worth considering in Fourier terms. We know that the frequency content of a waveform is defined by the amplitude spectrum, and that a variety of waveforms can be synthesised by associating a particular amplitude spectrum with a variety of phase spectra. For example, if (as in figure 3) we consider cosinusoidal Fourier components and set them all to zero phase at the time origin, then the waveform represented by the sum of these components must have a maximum at the time origin (where all the components reinforce) and must be symmetrical about the time origin. The adoption of any other phase spectrum must decrease, in general, the degree of reinforcement at the time origin and the completeness of the cancellation away from the time origin; thus other phase spectra usually produce waveforms which have lesser peak amplitude and greater duration—which are *dispersed* over a longer time. We know also that we may derive the power spectrum of a waveform by squaring the absolute amplitude of the Fourier components; thus the power spectrum accents any departures from "flatness" in the amplitude spectrum, and discards completely the information in the phase spectrum. If now we associate these well-known facts with our previous conclusion that the auto-correlation function depends not on the actual waveform but only on its frequency content, we find it easy to accept this proposition:

> *The auto-correlation function contains the same information as the power spectrum of a waveform. However, this information is presented in the form of a function of time, rather than frequency. The function may be visualised by imagining that the Fourier components of the waveform are squared in amplitude, set into phase at the origin of a new time scale, and added together.*

Thus a *waveform* is synthesised by combining Fourier components with the amplitudes given by the amplitude spectrum and the phases given by the phase spectrum; the *auto-correlation function of a waveform* is synthesised by combining Fourier components with the amplitudes given by the power spectrum of the waveform, and with zero phase.

Now we know that if we inject an input waveform (having known amplitude and phase spectra) into a system (having known amplitude-frequency and phase-frequency responses) we obtain an output whose amplitude spectrum is

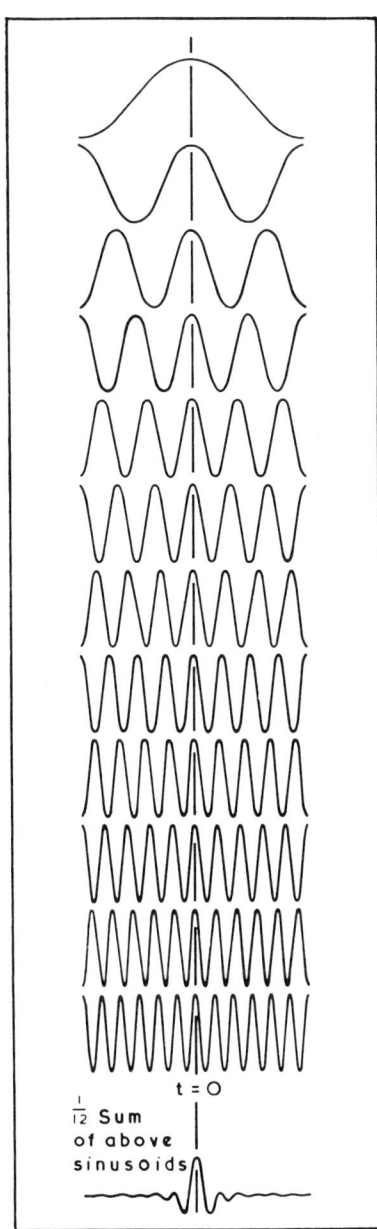

Fig. 3. Approximation to the synthesis of the waveform corresponding to a rectangular amplitude spectrum and a zero phase spectrum.

given by multiplying the input amplitude spectrum by the system amplitude-frequency response and whose phase spectrum is given by adding the input phase spectrum to the system phase-frequency response. For any waveform, therefore, we may define a system whose amplitude-frequency response has the same form as the input amplitude spectrum and whose phase-frequency response is exactly opposite to the input phase spectrum (that is, it offsets each phase lead with an equal lag, and each lag with a lead); then the amplitude spectrum of the output is equal to the power spectrum of the input, and the phase spectrum of the output is zero. Such a system is said to be a *matched filter* for its particular waveform. Thus we are led to this proposition:

The process of auto-correlation of a waveform is equivalent to passage of the waveform through its matched filter.

At this stage we have established two alternative approaches to the auto-correlation function: the approach through the concept of similarity as a function of time shift, and the approach through the power spectrum and the matched filter. In figures 4 and 5 we illustrate these two approaches, using as an example a waveform which has proved to be of particular utility in correlation techniques—the linearly swept-frequency waveform.

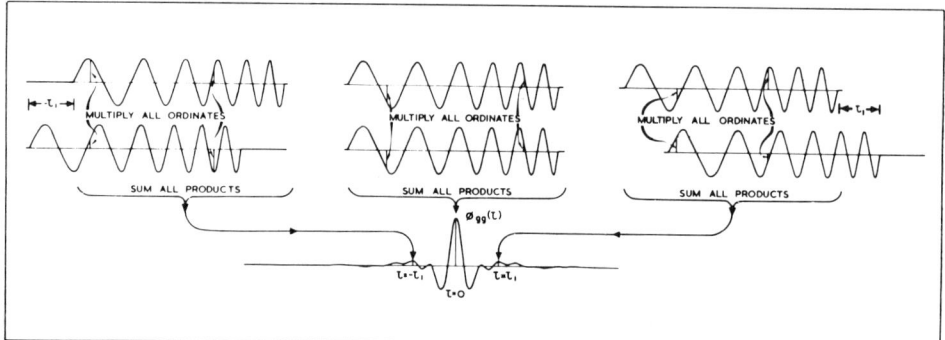

Fig. 4. Illustration of the process of constructing the auto-correlation function of a swept-frequency waveform.

In figure 4 we construct the auto-correlation function by the process of plotting the similarity between the waveform and a time-shifted version of itself. The centre section of the diagram illustrates the "overlay" position; for this the time shift is zero, the ordinate products are all positive and the sum is large. The left and right sections of the diagram illustrate the situation at a time τ_1 before and after the overlay position; some of the ordinate products are positive and some are negative, and their sum is small. Then our earlier propositions tell us that the variation of this product sum, as a function of the time shift, defines the form of the auto-correlation function shown on the lowest trace.

In figure 5 we illustrate the alternative approach from the power spectrum. Our intuition tells us, correctly, that if the sweep rate of the swept-frequency waveform is constant and small, then all frequencies are approximately equally represented; the amplitude spectrum is flat between the end frequencies of the sweep, and substantially zero beyond them. Consequently the power spectrum also has this rectangular form. If we now take a suitable set of Fourier components representing this power spectrum, draw them all in phase at the origin of a new time scale, and add them together (just as we did in figure 3), we obtain the auto-correlation function. For the particular case of a slow sweep, and for illustrative purposes only, we may visualise the swept-frequency waveform as the *sequential* arrangement of those parts of the Fourier components which make a positive contribution to its shape; loosely we may say that "the frequencies are strung out in line". Then the construction of the auto-correlation function (or the action of the matched filter, which is the same thing) may be viewed in terms of the phase shifts illustrated in figure 5.

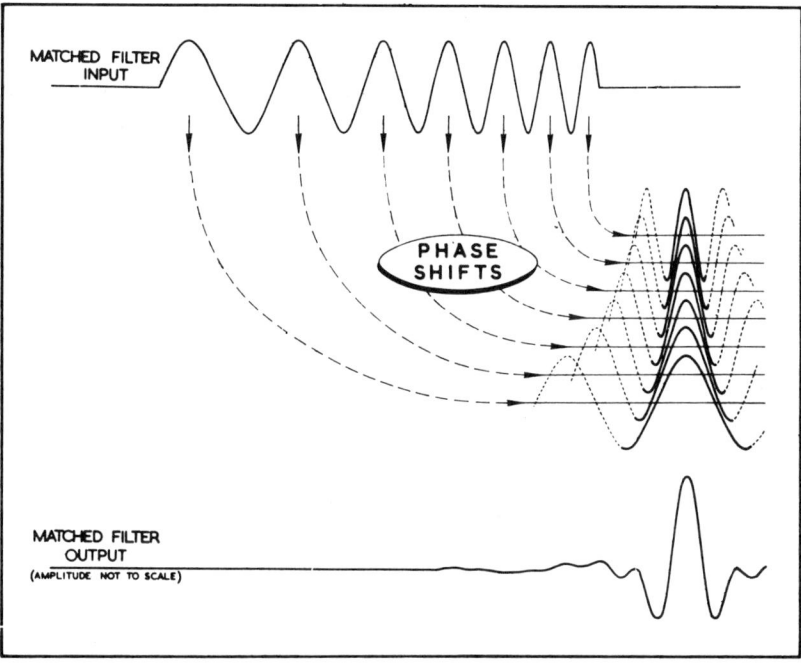

Fig. 5. Alternative approach to the process of constructing the auto-correlation function of a swept-frequency waveform.

We note that in the approach of figure 4 the central maximum of the auto-correlation function occurs because the similarity between the two waveforms is large when there is zero shift between them, and the parts of the auto-correlation function which correspond to large shifts are small because then the similarity between the waveforms is small. In the Fourier approach of figure 5, the maximum of the auto-correlation function occurs because all components are in phase at this point; away from the maximum the components interfere destructively to produce low amplitudes. The two approaches are, of course, interchangeable and equivalent. The merit of the Fourier approach is that it allows us to think of the auto-correlation function in terms which are very familiar to us, so that to the extent that we may look at a waveform and estimate its spectrum, we may also estimate its auto-correlation function. The merit of the similarity approach is that it remains valid for waveforms (for example, continuous noise-like waveforms) to which Fourier concepts may not be applied.

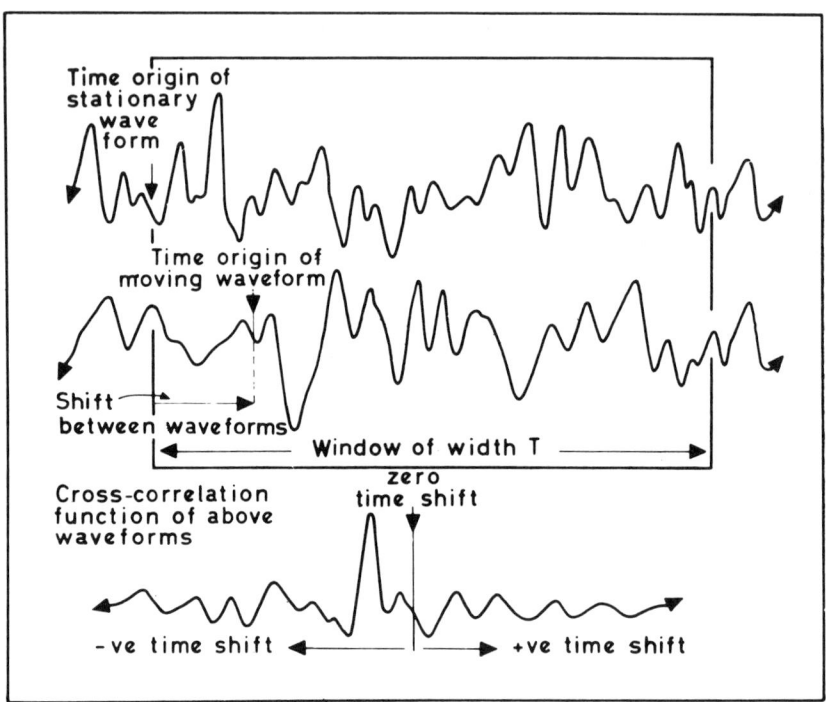

Fig. 6. The upper two traces represent the waveforms being cross-correlated. The bottom trace is the cross-correlation function; the maximum in it shows that for the indicated value of time shift there is a marked similarity between the waveforms, even though this is barely visible to the eye.

The auto-correlation function, then, is concerned with the similarity between a waveform and a time-shifted version of itself. As was implicit in our discussion of figure 1, we may apply the same measure of similarity to the case of two waveforms which are not identical. In figure 6 we visualise one waveform (the upper one) as stationary and the other as sliding bodily past it from right to left; then we view the two waveforms through a "window" of width T, and we assess the similarity of the two waveforms within this interval by our previous method of multiplying ordinates and summing products. The bottom trace plots this similarity as a function of the time shift between the waveforms.

The cross-correlation function of two waveforms is a graph of the similarity between the two waveforms as a function of the time shift between them.

The first example we should consider represents the basic problem of all communications and echo-ranging systems: a signal of known waveform is transmitted into a medium, and is received again unchanged in form but immersed in noise—what is the best way of detecting the signal? If we cross-correlate the received signal against the transmitted signal, the cross-correlation function must include two parts: a first part representing the auto-correlation function of the transmitted signal (which is common to both waveforms), and a second part representing the cross-correlation of the transmitted signal with the received noise. The form of the first part we know from the spectrum of

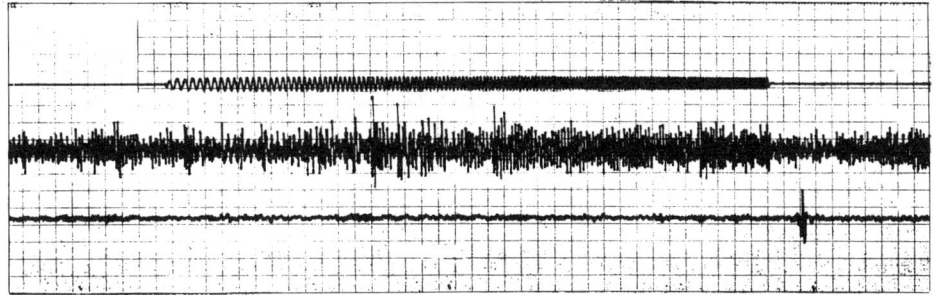

Fig. 7. The top trace represents the transmitted waveform. The centre trace shows this waveform buried in noise. The bottom trace is the cross-correlation of the other two; the signal (in the form of its auto-correlation function) is elevated well above the noise.

the transmitted signal. The second part, since it represents the cross-correlation of dissimilar waveforms, can contain only low amplitudes, and so it is reasonable to expect that the signal-to-noise ratio should be improved by the correlation process. We might say that during correlation the transmitted signal is *searching to find itself* in the received signal. The degree to which it can do so is indicated by figure 7, where the first trace represents the transmitted signal and the second trace the same signal immersed in noise to the extent that it cannot be positively identified by eye. The third trace represents the cross-correlation of the first two traces; the signal (in the form of its auto-correlation function) is now elevated far above the noise.

We have already seen that the auto-correlation function of a waveform can contain only those frequency components present in the waveform itself. By extension of this, we find it easy to accept this proposition:

The cross-correlation function of two waveforms contains only those frequencies common to both waveforms.

This suggests that it should be possible to use the correlation process as a general filter. For example, we may correlate a complicated waveform against a sample of a sine wave, and the output contains only those components of the complicated waveform which are very close to the frequency of the sine wave. Thus we have a sharply peaked filter, and it is not difficult to see that the sharpness of the peak can be increased without limit by taking longer and longer samples of the sine wave. This illustrates a general point:

Correlation techniques are most powerful when the waveforms being correlated consist of many cycles.

Correlation filters are not limited to sharply peaked responses. For example, we have said that the swept-frequency signal of the top trace of figure 7 has a spectrum which is flat within the sweep limits, and substantially zero beyond them. This means that the correlation process, in producing the third trace, does not see any of the noise frequencies of the second trace which are outside the sweep limits; the process of correlation against a signal having a rectangular spectrum is, as far as the noise is concerned, a filter having a near-ideal rectangular pass-band.

We may be even more general. For example, let us cross-correlate a signal —any signal—against the pulse-like waveform of the third trace of figure 4. We know from the foregoing material that this waveform represents a rectangular amplitude spectrum and a zero phase spectrum. The result of the cross-correlation, therefore, will be exactly the same as if we passed the original signal through a filter having a rectangular pass-band and zero phase. In completely general terms:

Cross-correlation against an arbitrary waveform is equivalent to passage through a filter having an amplitude-frequency response of the same form as the amplitude spectrum of the waveform, and a phase-frequency response of the same form as the phase spectrum of the waveform, but reversed in sign (that is, with a lead instead of a lag, a lag instead of a lead).

It is interesting to compare the cross-correlation process, which we see may be viewed as a *completely general linear operator*, with the process of convolution. This latter process, as we know, furnishes a general method of constructing the response of a system to an arbitrary input waveform when its response to a simple spike impulse is known. The input waveform is divided into a succession of impulses, a proportionally-scaled impulse response is drawn —reversed in time—for each one of these impulses, and the output waveform is obtained by adding all these impulse responses (figure 8). Although the graphical processes of correlation and convolution may appear to be different, the two processes have the same generality and are, in fact, equivalent.

Cross-correlation against a certain waveform is identical to convolution with the same waveform reversed in time.

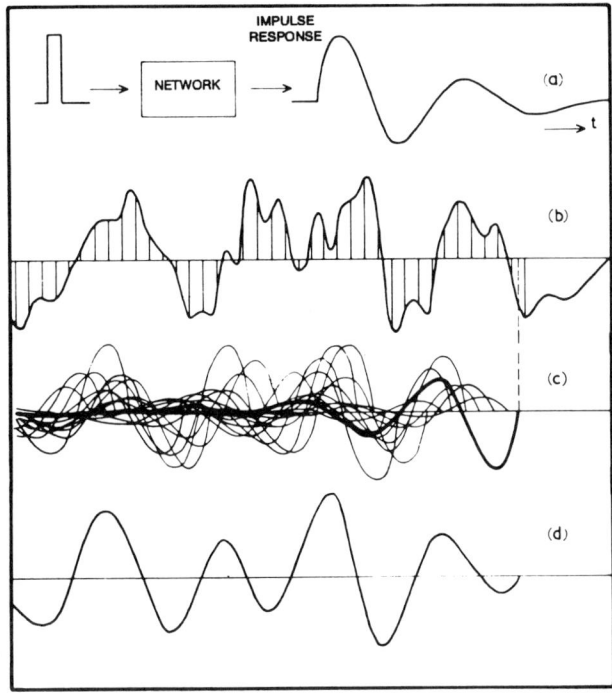

Fig. 8. The process of convolution. A spike injected into a certain network produces the impulse response, which is characteristic of the network. Any arbitrary input waveform (second line) may be represented as a succession of spikes. Each such spike generates an impulse response of appropriate amplitude and polarity (third line). The output waveform (fourth line) is the sum of these impulse responses.

We have previously stated that for every waveform there is a matched filter, whose output (when the waveform is injected at the input) is the autocorrelation function of the waveform, and we have specified the characteristics of the matched filter in the frequency domain. From the last proposition we may deduce an alternative specification:

The matched filter corresponding to a particular waveform has an impulse response which is the waveform itself, reversed in time.

We can illustrate this very simply in terms of a swept-frequency waveform (figure 9). Suppose that we have a "black box" such that the injection of a spike impulse at the input produces a swept-frequency waveform at the output, and let us say that the sweep starts at the low frequencies and rises to high frequencies. In other words, the black box delays the high frequencies more than the low frequencies. Then let us record this sweep, reverse it in time (for example by replay backwards), and inject it at the input of the black box. The high frequencies enter the box first but they suffer a long delay. The low frequencies come last, but they suffer a lesser delay. Consequently all frequencies emerge in their original relation, and form a simple impulse (which may be viewed as the auto-correlation function of the swept-frequency waveform, or as a band-limited version of the original spike).

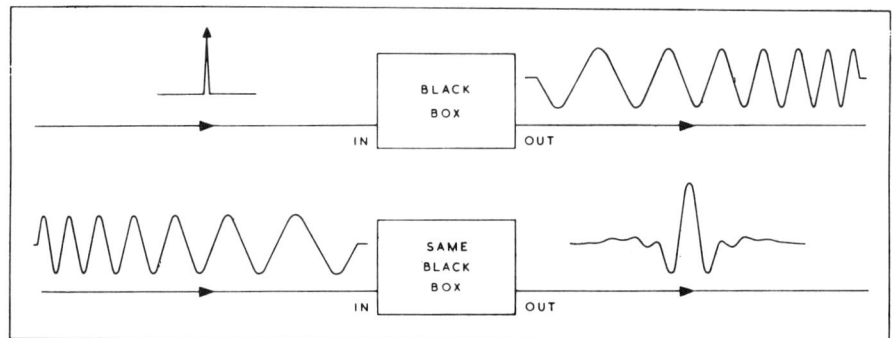

Fig. 9. A certain black box has an impulse-response of swept-frequency type (first line). The same black box is then the matched filter for this waveform reversed in time.

We have stated above that cross correlation is a linear operator. Perhaps we should form an explicit proposition from this:

Cross-correlation is a linear operation, so that when it is associated with other linear operations (filtering, attenuation, differentiation, and so on) the order in which these operations are performed does not affect the final result.

Our last proposition is concerned with the application of correlation techniques to methods of specifying the behaviour of a system. Thus any practical system modifies the waveform applied to it, producing an output which depends on the characteristics of the system. If we accept that the similarity between the output and the input is indicative of the characteristics of the system, and that cross-correlation is a method of measuring the similarity between two waveforms, we are not surprised to find that the cross-correlation function between input and output waveforms specifies the system behaviour over the bandwidth of the input signal. If we make the bandwidth of the input signal greater than the pass-band of the system, we have this useful result:

For a broad-band input, the cross-correlation function of the input and output of a system is the impulse response of the system.

This simplified discussion of the rudiments of correlation theory makes no pretence to be rigourous. More formal treatments may be found in the literature (for example Lytle, 1957; Lee, 1960; Turin, 1960; Barber, 1961). No mention has been made here of the approximation introduced when the waveforms considered are of finite duration; a formal treatment of this is given by Schroeder and Atal (1962).

Applications in Science Generally

1. Pulse-Compression Echo-Ranging Systems

The performance of all echo-ranging systems—radar, sonar, seismic, flaw-detection—is governed by the energy principle (Woodward, 1953; Stewart and Westerfield, 1959). In simplified terms, this states that, for given noise conditions, the detectability of an echo is prescribed by the signal *energy* (that is, by the product of its average power and its duration). Conventional echo-ranging systems cannot increase the duration of their search signals without sacrificing the resolution between reflectors, and so their approach to better penetration can be only through an increase in the power. Sooner or later, however, there comes a limit: electrical breakdown occurs in a radar waveguide, or cavitation occurs on the face of a sonar transducer, or extreme non-linearity occurs in the earth near a seismic charge. The solution to this difficulty is the transmission of a *long* search signal (so that high energy and long range are assured) and then the *compression* of the long signal to a short pulse (so that good resolution is preserved). This principle is known as pulse-compression.

The method of compressing the received signal is, as we might guess from all the foregoing material, the cross-correlation of the received signal against the transmitted signal (or—what is the same thing—the passage of the received signal through the filter matched to the transmitted signal). This process, under the conditions normally prevailing in echo-ranging systems, represents the theoretically optimum method of detecting an echo; as our above propositions show, it provides the virtual elimination of all noise frequencies outside the signal spectrum, it provides the shortest wavelet compatible with the spectrum employed, and it provides the highest peak wavelet amplitude compatible with the signal energy. Figure 7 may be taken as indicative of the advantages of a pulse-compression system.

One family of pulse-compression radar systems uses a swept-frequency search signal, and has come to be known as "chirp" radar (from the catch-phrase "Not with a bang, but with a chirp"). Relevant references include

Hüttman (1940, 1961), Dicke (1945), Darlington (1949), Cauer (1950), Krönert (1957), Chin and Cook (1959), Cook (1960, 1961), Klauder et al. (1960), Klauder (1960), May (1960, 1962), Meeker (1960), Meitzler (1960, 1961), Ohman (1960), O'Meara (1960), Fitch (1961a, b), Hoover (1961), Lange (1961), Lerner (1961), Merrihew (1961), Ramp and Wingrove (1961a, b), Tiberio (1961), Lohrmann (1962), Skolnik (1962), Bernfeld (1963), Mims (1963) and Milne (1964).

A second family of pulse-compression radar systems makes use of the fact that it is the spectrum of the search signal, rather than its actual waveform, which defines the form of the final compressed pulse. Thus various random and noise-like waveforms may be employed as search signals. Discussions relevant to these random-signal radars have been given by Guanella (1939), Lee and Wiesner (1950), Bussgang (1951), Faran and Hills (1952), Lytle (1957), Horton (1959), Turin (1960), Welti (1960), Fishbein and Rittenbach (1961), Katzman and Frost (1961), Miller (1961), Spilker and Magill (1961), Huffman (1962), Golomb (1963), Sakamoto et al. (1963), Spilker (1963), Zadeh et al. (1963) and Lockheed (1964). A remarkable early demonstration of a radar of this type was the registration of radar echoes from Venus, a distance of 28 million miles (Price et al., 1959). Pulse-compression is not peanuts.

When pulse-compression techniques are applied to sonar, the use of a noise-like search signal allows the detection of a submarine or surface vessel without there being any direct means for the latter to know that its presence has been detected. Pulse-compressive sonar systems have been discussed by Sproule and Hughes (1944), Stewart and Westerfield (1959), Parks and Downing (1960), Clay and Liang (1962), Federici (1962), Jones (1962), Middleton (1962), Rowlands (1962), Stewart and Allen (1962), Remley (1963), Wisotsky, Clay and Liang (1963), Allen and Westerfield (1964) and Parvulescu and Clay (1964).

A pulse-compressive method of seismic exploration, known commercially as the "VIBROSEIS"* system, is by now well known to members of this Association, having been described by Doty and Crawford (1953), Doty and Lee (1954), Crawford and Doty (1955), Crawford and Clynch (1955), Doty (1959), Crawford, Doty and Lee (1960, 1961) and Goupillaud and Lee (1963). This system, like its pulse-compressive counterparts in radar and sonar, promises to replace the conventional method for most applications.

A general summary of the theory and practice of pulse-compression systems has been given by Anstey (1963).

2. Communications

The problems of communication in a noisy medium have much in common with the problems of echo-ranging. The application of pulse-compression

---

* A Trademark of Continental Oil Co.

principles to the transmission of teletype and voice-frequency information can yield four major advantages: an improvement in signal-to-noise ratio, sophisticated measures to minimise fading, the provision of "secure" communications unintelligible to anyone without the appropriate matched filter, and conservation of radio bandwidth. Relevant references include Price and Green (1958), Sussman (1960), Wozencraft (1961), Winkler (1962), Aein and Hancock (1963), Anon (1964) and Chiles and Lafuse (1964).

3. The Detection of Periodic Signals in Noise

Two methods of recovering a periodic signal from noise emerge from our basic propositions above.

The first is applicable if the frequency of the periodic signal is known, and consists of cross-correlation of the noisy signal against a sinusoid. This is an extremely powerful method limited only by the purity and duration of the reference sinusoid; it is comparatively simple to recover a periodic signal immersed in 100 times its own amplitude of noise.

The second method is applicable if the frequency of the periodic signal is not known, and consists of the auto-correlation of the noisy signal. This is not as powerful as cross-correlation against a sinusoid—in effect we are correlating a dirty signal against a dirty signal instead of a dirty signal against a clean signal—but it is still capable of very large improvements in the signal-to-noise ratio. The auto-correlation function is the superposition of figures 2b and 2d, so that, provided the waveform sample considered has sufficient duration, the periodic signal is evident in the outlying parts of the auto-correlation function. These methods and their applications are discussed by Lee, Cheatham and Wiesner (1950), Lee and Wiesner (1950), Jones and Kelly (1952), Rudwick (1953), Brazier and Barlow (1956), Mercer (1958), Barlow (1959), Raemer and Reich (1959), Smith and Lambert (1960), the American Gas Association (1962), Bakewell (1963), Bass and Men' (1963) and Krauss (1963).

4. Determination of Spectra

Clearly the method of cross-correlating against samples of sinusoids may be used to establish the amplitude spectra of periodic signals (in the manner of a set of narrow-band filters), and the resolution with which the spectra may be established is limited only by the purity and duration of the reference sinusoids.

For non-periodic noise-like signals of the type found most commonly in nature, the best method for determining spectra is the construction of the auto-correlation function and the subsequent manipulation of this, by mathematical methods, to the power spectrum. A rapid approximate method of deriving the spectrum from the auto-correlation function is the cross-correlation of the latter against a slowly swept-frequency waveform, followed by rectification and smoothing.

A comparison of the determination of spectra by digital narrow-band filtering and by auto-correlation is given by Cartwright, Tucker and Catton (1961) in an application to oceanography. A modified correlation method is used by Akamatu (1956) for the determination of the spectra of microtremors; also relevant to this reference are Tomoda (1956), Melton and Karr (1957) and Ekre (1963). It begins to emerge that the ear, long thought to operate largely on a spectral basis, may include auto-correlation as one of its functions (Kraft, 1950; Stevens, 1950; Cherry, 1961).

5. General Filtering Operations

From our basic propositions, we may deduce that cross-correlation may be used as a completely general filter, with entirely independent control of amplitude and phase. Furthermore, a change from one filter characteristic to another involves only a change in the reference waveform against which the input is cross-correlated; no modification of the equipment itself is required. Thus cross-correlation represents an extremely flexible method of filtering (Jones and Morrison, 1954; Smith, 1958).

Since cross-correlation filtering is not an intuitively obvious process, an example may be helpful. Let us correlate an input waveform against a slowly and linearly swept-frequency signal. The output is small, because the waveforms are in general dissimilar; it is of a complicated nature, and longer than the input signal. Now let us correlate this output against the same swept-frequency signal reversed in time. The final output represents the input signal filtered through a near-ideal filter having a rectangular pass-band and zero phase. The same output would be obtained if, instead of the swept-frequency signal, we used *any* waveform of the same power spectrum (even a band-limited noise-like waveform). However, three advantages accrue from the use of a swept-frequency signal. First, the phase-compensation of any part of the main or ancillary equipment is a simple matter, achieved by shifting the phase of appropriate parts of the second swept-frequency waveform. Second, pass-bands of arbitrary shape may be obtained either by suitable amplitude modulation of the swept-frequency waveform or—more interestingly—by variations in the sweep rate. Third, after the first correlation the frequency components of the original signal may be regarded as separated in *time*, and this allows highly sophisticated data enhancement techniques in some applications.

6. Determination of Impulse-Response

One of our basic propositions showed that we could determine the impulse-response of a system by cross-correlating the output with a suitable broad-band input. But why should we wish to do so, when we can determine it much more easily by the simple injection of a spike?

The answer lies in figure 7. For an echo-ranging system may be said to determine the impulse-response of the transmission medium—a seismic reflection record is the impulse-response of the reflective earth—and so we may deduce that an impulse-response determined by cross-correlation is insensitive to the presence of noise or other signals which may be injected into the system between its input and its output. Thus the cross-correlation method is desirable whenever the determination must be made in the presence of a high noise level, or in the presence of other signals (such as the system's normal throughput), or on a system having a small dynamic range. A compelling illustration is in industrial process-control systems, where it is desirable to measure the characteristics of the system without stopping the plant (Goodman, 1955; Reswick, 1955; Balchen and Blandhol, 1960; Buland, Cooper and Margolis, 1960). Other examples occur in the testing of aircraft, buildings and acoustical systems (Lee and Wiesner, 1950; Bussgang, 1951; Goff, 1955; Mercer, 1958; Zverev and Kalachev, 1960; Gayford, 1961); further examples must surely emerge in neurophysiology.

7. The Identification or Location of Signal Sources

In our discussion of pulse-compressive echo-ranging systems, we visualised the transmission of a suitable known waveform, the identification of reflected versions of this signal by cross-correlation of the received and transmitted signals, and the consequent location of reflectors by study of the reflection travel times. For some purposes it is not necessary to generate a transmitted signal; if we observe a suitable broad-band signal occurring naturally at a first point, and if we correlate this against a signal observed at a second point, then we identify those disturbances which propagate between the two points, and we determine their apparent velocities.

Iyer (1958), Rykunov (1961) and Töksoz (1964) have applied this method to determine the velocities and directions of microseisms. Butterfield, Bryant and Dowsing (1960) have applied it to the measurement of the speed of moving steel strip in a rolling mill. Goff (1955) has applied it to the measurement of the acoustic absorption of panels. Burd (1964) has applied it to the identification of the path by which sound "leaks" out of a studio. Other applications include the determination of the contributions of each of many machines to the noise level in a control room, the determination of the relative contributions of inner and outer aircraft engines to the noise level in the cabin, the location of noise sources in mechanical, missile and sonar equipment, and the measurement of cross-talk in multichannel communication systems subject to intermodulation distortion (Lee and Wiesner, 1950; Goff, 1955; Gilford and Greenway, 1956;

Gilbrech and Binder, 1958; Bull, 1960; Keast and Kamperman, 1960; Lund and Urick, 1964). An inverted application is described by Denham (1963), who is concerned with determining the minimum spacing of the elements of an array which will ensure that the noise at these elements is substantially incoherent.

Clearly, many of the examples we have considered in this section may be taken as illustrating that during correlation each one of the waveforms is searching to find itself in the other. Thus correlation forms the basis of all systems of *pattern recognition*. All current attempts at the automatic recognition of speech, script and print represent some form of correlation technique. The same principles govern the matching of fingerprints and the identification of the after-shocks of an earthquake.

Emergent Applications in Exploration Geophysics

The most widespread and commercially important application of correlation techniques today is in the "VIBROSEIS" method of seismic reflection and refraction work. Indeed it was the commercial necessity of high routine production which furnished the incentive for the development, by S.S.L., of the real-time correlator now extensively used in "VIBROSEIS" exploration.

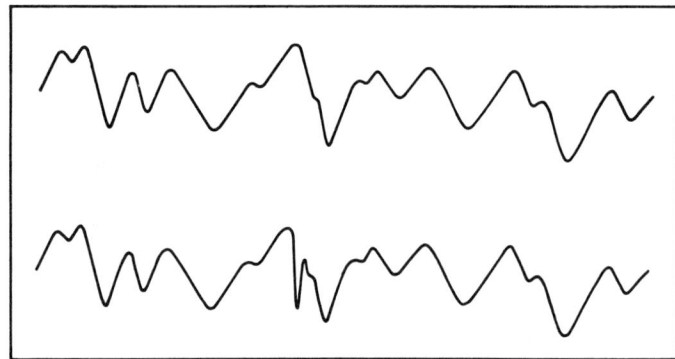

Fig. 10. Cross-correlation is not the correct process for distinguishing between these waveforms.

In this discussion, however, we are concerned with the whole field of exploration geophysics. We may expect a potential application of correlation techniques wherever some geophysical measurement may be expressed as a waveform, or as its two-dimensional equivalent—a pattern. Examples of waveforms are gravity and magnetic profiles, well logs and, of course, seismic traces. An example of a pattern is a multi-trace seismic reflection. Let us consider under what circumstances correlation techniques may usefully be applied to these examples.

First, we must reaffirm that correlation gives its clearest answers when the waveforms considered consist of many cycles.

Second, we must note that correlation gives an *average* measure of the similarity between two waveforms. Thus it is *quite unsuited* to the detection of one small pip on one of two otherwise-identical waveforms (figure 10). It is *perfectly suited* to the detection of an extended section of common content in two waveforms, particularly when this common content is obscured by noise or other irrelevant signals (figure 7).

Third, we must note that correlation is comparatively insensitive to local differences in the *amplitudes* of the two waveforms. Thus in figure 11 there would be only a minor difference between the auto-correlation function of waveform *a* and the cross-correlation function of waveforms *a* and *b*.

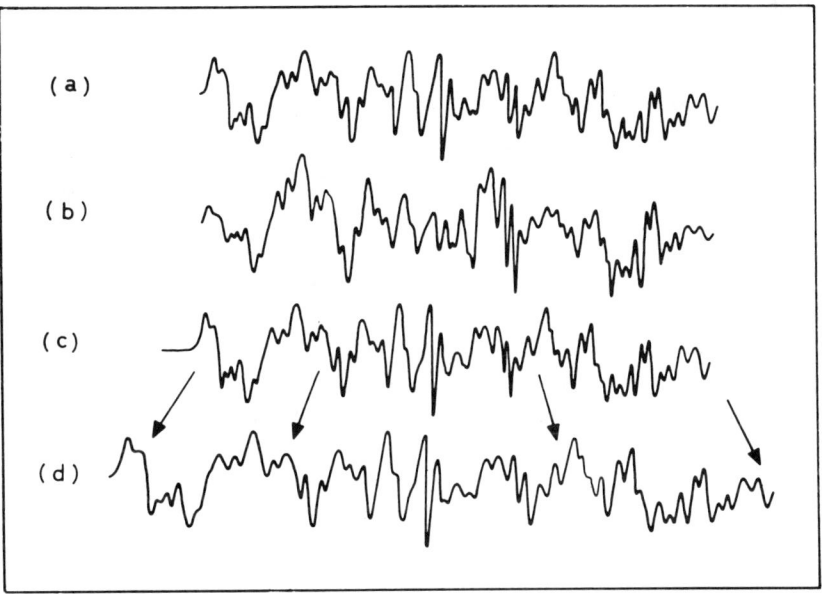

Figure 11. Traces *a* and *b* differ primarily in their local amplitudes; correlation is relatively insensitive to this type of difference. Trace *d* is a stretched version of trace *c*; correlation is very sensitive to this type of difference.

Fourth, we must note that correlation is extremely sensitive to the "stretching" of one of the waveforms. Thus the correlation process looks at waveforms *a* and *c* of figure 11, and pronounces them identical; it looks at waveforms *c* and *d*, and can see little similarity between them.

Let us now draw the conclusions relevant to the correlation of well logs. Suppose that special interest attaches to a particular segment of log in one hole, and that the problem is to find and explore this segment in a log from another hole. We cross-correlate the segment against the entire second log (or any lesser portion decreed by geological possibility). If a correlation is clearly indicated (by a sharp pulse in the cross-correlation function) the appropriate part of the second log may then be studied for significant differences in the balance of "amplitudes" and for significant differences in the detail of the logs. It is hardly necessary to stress that the correlation process can find the appropriate part of the second log even when these differences of amplitude or detail are sufficient to make the correlation invisible to the eye. The resolution with which the depth of the segment is determined is, of course, a function of the "bandwidth" of the waveform; if there is a chance periodicity in the beds of the segment the cross-correlation function is less sharp than if the spacing of the beds is random.

If the cross-correlation process produces several indications instead of one (and if this is geologically implausible), the segment which has been taken is too short. If it produces no indications of correlation (and if this is geologically implausible), the segment which has been taken is either geologically too long, or a thinning of the beds in the segment has occurred between the holes. In this case, we may try different degrees of compression or extension of the depth scale of the reference segment until a correlation is indicated (Haites, 1963). Shaub (1963) gives an alternative approach particularly applicable to pinch-outs.

By a straightforward change of terms, we may apply the same considerations to seismic reflection traces (Jones, 1962). Suppose that we have a seismic cross-section characterised by good quality data on both sides, but, for reasons associated with surface conditions or noise, poor data in the centre. Then if we can establish a reference trace in the good part of the section, we may use cross-correlation against this trace to push the continuity across the poor part of the section, even where this continuity is not visible to the eye.

Some features of this technique are worth exploring. First we note that there is an incompatability between the considerations which define the duration of the segment of reference trace employed. If we take a long segment (for example, 2 or 3 seconds of two-way time) we shall be able to detect a recurrence of its pattern, in another trace, even if it is buried deep in noise. However, this demands that the seismic section be conformable within very narrow limits (perhaps within 20 or 30 milliseconds in 2 or 3 seconds). Therefore, in using this technique, we must first decide how far across the section the continuity from a particular reference trace must be extended, and then, for this horizontal distance, we must decide for what duration of reference trace the section can be expected to remain approximately conformable. These decisions then define the maximum noise level in which the continuity may still be detected (Mut, 1959; Faust, 1963).

Clearly the selection of the reference trace is very important. We may use a synthetic trace (Peterson, 1955), or a single good field trace, or a trace compounded from several good ones. Alternatively, we may use a trace which represents the result of stacking many widely-separated reflection paths to a common depth point; in this way, of course, we reduce the contribution of the multiple reflections. Considerations like these suggest a method of field recording in which every device we know for improving record quality is applied just to one trace in every 20 (or 50, or 100), and then the other traces (obtained with a simpler and more economical technique) are used, in combination with a cross-correlation technique, to establish continuity between the special traces.

The above considerations are also very relevant to the whole problem of stacking. We have all seen examples where a number of apparently-useless records have been stacked to yield something useful. We have also seen examples where the stacked record has been no better than one or two of its components. Clearly what is required is a check whether a particular record will help or hinder the stacking; reinterpreted, this calls for a measure of the similarity between two traces, and therefore is a natural application of correlation techniques. Not only does the cross-correlation function decide whether the record is worth stacking or not; it also indicates the appropriate static correction with a certainty far beyond that obtainable (by the employment of doubtful velocities and an inadequate geometrical model) with a weathering correction. There is no better way of measuring a step-out than with an appropriate correlation technique.

We have spoken of giving the maximum possible enhancement technique to particular selected traces, and one of the techniques which we must consider is inverse filtering or deconvolution. By this we seek to compensate the filtering effect of the earth, off-setting frequency-selective attenuation by appropriate amplification. Correlation techniques are very relevant here. For example, it is well known that the implementation of inverse filtering requires a determination of the initial spectral content of the seismic trace, and that this is best done by correlation techniques (Robinson, 1957; Agard, 1961; Kunetz, 1961; Rice, 1962). It is also well known (Crawford and Doty, 1955) that in the "VIBROSEIS" technique it is feasible to effect inverse filtering by appropriate variation of the sweep *rate*, and that this allows any defined degree of inverse filtering with the minimum expenditure of energy.

Another process related to correlation is also relevant. In conventional work, the process of boosting the frequencies which are absorbed in the earth must clearly degrade the reflection signal relative to any broad-band noise which may be present, and thus the conventional inverse filtering process is constrained to exchange signal-to-noise ratio for resolution. If the signal-to-noise ratio is very poor, then we know that the optimum filter is the matched filter, whose amplitude-frequency response has the same form as the amplitude spectrum of the reflection signal. As we consider progressively improving signal-to-noise ratios, we pass through a ratio which allows detection of a reflection without any filtering at all, and then come to ratios which allow progressively increasing degrees of inverse filtering. Inverse filtering is appropriate only to very good records, and the possible degree of resolution improvement is defined rigorously by the signal-to-noise ratio. Therefore it is of the utmost importance to achieve signal-to-noise ratios as high as possible. This requires a modification of view for the exploration seismologist; for most of us a signal-to-noise ratio of unity makes a pleasant change, and we would not ordinarily see any point in increasing the ratio beyond a value of perhaps two or three. The attainment of much greater signal-to-noise ratios by methods which cannot be allowed to decrease the bandwidth of the signal must surely require help from the techniques of correlation.

We have said that inverse filtering affords a means of exchanging signal-to-noise ratio for resolution. Correlation techniques allow certain other exchanges which are of value for particular objectives. Thus multiplicative compounding and the use of "correlation arrays" allow various exchanges between signal-to-noise ratio, the directivity (or apparent-velocity resolution) of a geophone array, the physical dimensions of an array, and the clarity of indication of the onset of a signal (Dyk, 1956; Melton and Bailey, 1957; Welsby and Tucker, 1959; Linder, 1961; Tucker, 1963; Ksienski, 1964; Ryall, 1964; Shaw and Davies, 1964; Whiteway, 1964).

We have said that correlation is at its best when dealing with long waveforms. Consequently correlation against a waveform of the type of a single seismic wavelet is not a strong method. Correlation against many such wavelets simultaneously, however, can become a very strong method. For example, we may arrange a wavelet on each of many traces, align the wavelets as for a reflection having a particular step-out, and correlate the record against this pattern (Tullos and Cummings, 1961; Baranov and Picou, 1964). Similarly, we may search for the major apparent velocities present on a refraction record (McCamy and Meyer, 1964).

With a greater generality, we might say that two-dimensional correlation is appropriate whenever a geophysical result may be expressed as a number of separate short waveforms forming some sort of pattern. We should note that the correlation technique is very little disturbed by the fact that several such patterns may cross each other. A correlation method is adept at showing whether a waveform or a pattern is composed of several superposed components, and this fact should find application in many branches of exploration geophysics. It may be true that novel geophysical methods have proved unworkable in the past because of confusion from overlapping responses, and that such methods warrant review in the light of modern correlation techniques.

ACKNOWLEDGMENT

The author thanks the Directors of Seismograph Service Limited for their assent to the publication of this paper.

REFERENCES

AEIN, J. M. and HANCOCK, J. C., 1963, "Reducing the effects of intersymbol interference with correlation receivers", I.E.E.E. Transactions on Information Theory, July 1963, p. 167.

AGARD, J., 1961, "Analyse statistique et probabiliste des sismogrammes", D. Sc. thesis, University of Paris, April 1961.

AKAMATU, K., 1956, "Tomoda's method for calculating the correlation coefficients as applied to microtremor analysis", Journal of the Physics of the Earth, vol. 4, no. 2, 1956, p. 81.

ALLEN, W. B. and WESTERFIELD, E. C., 1964, "Digital compressed-time correlators and matched filters for active sonar", J.A.S.A., January 1964, p. 121.

AMERICAN GAS ASSOCIATION, 1962, "Sonic device pinpoints gas leaks from surface", Oil and Gas Journal, 17 December 1962, p. 71.

ANON, 1964, "RADA... New concepts in radio communications", International Electronics, May 1964, p. 19.

ANSTEY, N. A., 1963, "Pulse-compression: an application of correlation techniques in acoustics", M.Sc. dissertation, University of London, June 1963.

BAKEWELL, H. P., 1963, "Longitudinal space-time correlation function in turbulent airflow", J.A.S.A., June 1963, p. 936.

BALCHEN, J. G. and BLANDHOL, E., 1960, "On the experimental determination of statistical properties of signals and disturbances in automatic control systems", Automatic and Remote Control, (ed. J. F. Coales), the Proceedings of the First International Congress of the International Federation of Automatic Control, in Moscow, 1960, published by Butterworths 1961.

BARANOV, V. and PICOU, C. H., 1964, "Energy and vector record-sections", Geophysics, February 1964, p. 17.

BARBER, N. F., 1961, "Experimental Correlograms and Fourier Transforms", Pergamon Press.

BARLOW, J. S., 1959, "Autocorrelation and cross-correlation analysis in electroencephalography", I.R.E. Transactions on Medical Electronics, ME-6, p. 179.

BASS, F. G. and MEN', A. V., 1963, "Spatial correlation of travelling wave fluctuations in an unbounded turbulent medium", Soviet Physics—Acoustics, vol. 9, no. 3, p. 232.

BERNFELD, M., 1963, "Pulse compression techniques", Proc. I.E.E.E., 1963, p. 1261.

BRAZIER, M. B. and BARLOW, J. S., 1956, "Some applications of correlation analysis to clinical problems in electroencephalography", Electroenceph. Clin. Neurophysiol, vol. 8, p. 325.

BULAND, R. N., COOPER, G. R. and MARGOLIS, J., 1960, "Cross-correlation apparatus", U.S. Patent no 3,024,994, filed June 1960, issued March 1962.

BULL, M., 1960, "Space-time correlations of the boundary layer", Report no. 200, Department of Aeronautics and Astronautics, University of Southampton.

BURD, A. N., 1964, "Correlation techniques in studio testing", The Electronic and Radio Engineer, May 1964, p. 387.

BUSSGANG, J. J., 1951, "Correlation techniques in electro-acoustic measurements", M.Sc. thesis at Massachusetts Institute of Technology.

BUTTERFIELD, M. H., BRYANT, G. F. and DOWSING, J., 1960, "A new method of strip speed measurement using random waveform correlation", presented to the Society of Instrument Technology, October 1960.

CARTWRIGHT, D. E., TUCKER, M. J. and CATTON, D., 1961, "Digital techniques for the study of sea waves, ship motion and allied processes", presented to the Society of Instrument Technology, November 1961.

CAUER, W. A., 1950, "Verfahren zur Nachrichtenübermittlung durch", German patent no 892,772, filed December 1950, issued August 1953.

CHERRY, C., 1961, "Two ears—but one world", in "Sensory Communication" (ed. W. A. Rosenblith), M.I.T. and Wiley.

CHILES, W. H. and LAFUSE, H. G., 1964, "Sweeping carrier signals through interference", Electronics, 18 May 1964, p. 94.

CHIN, J. E. and COOK, C. E., 1959, "The mathematics of pulse compression", Sperry Engineering Review, October 1959, p. 11.

CLAY, C. S. and LIANG, W. L., 1962, "Continuous seismic profiling with matched filter detector", Geophysics, December 1962, p. 786.

COOK, C. E., 1960, "Pulse compression—key to more efficient radar transmission", Proc. I.R.E., March 1960, p. 310.

COOK, C. E., 1961, "General matched filter analysis of linear FM pulse compression", Proc. I.R.E., April 1961, p. 831.

CRAWFORD, J. M. and DOTY, W. E., 1955, "Method of and apparatus for determining the travel time of signals", U.S. Patent no 2,808,577, filed June 1955, issued October 1957.

CRAWFORD, J. M. and CLYNCH, F., 1955, "Method of coupling a vibrator to an elastic medium", U.S. Patent no. 2,910,134, filed June 1955, issued October 1959.

CRAWFORD, J. M., DOTY, W. E. and LEE, M. R., 1960, "Continuous signal seismograph", Geophysics, February 1960, p. 95.

CRAWFORD, J. M., DOTY, W. E., and LEE, M. R., 1961, "Further developments in the 'VIBROSEIS' system of exploration", presented to the European Association of Exploration Geophysicists, in The Hague, May 1961.

DARLINGTON, S., 1949, "Pulse transmission", U.S. Patent no. 2,678,997, filed December 1949, issued May 1954.

DENHAM, D., 1963, "The use of geophone groups to improve the signal-to-noise ratio of the first arrival in refraction shooting", Geophysical Prospecting, December 1963, p. 389.

DICKE, R. H., 1945, "Object detection system", U.S. Patent no. 2,678,997, filed September 1945, issued January 1953.

DOTY, W. E. and CRAWFORD, J. M., 1953, "Method of and apparatus for determining travel time of signals", U.S. Patent no. 2,688,124, filed February 1953, issued August 1954.

DOTY, W. E. and LEE, M. R., 1954, "Method and apparatus for determining the travel times of a signal propagated over a plurality of paths", U.S. Patent no. 2,874,795, filed October 1954, issued February 1959.

DOTY, W. E., 1959, "Compositing sequentially produced signals", U.S. Patent no. 3,065,453, filed January 1959, issued November 1962.

DYK, K., 1956, "Additive and multiplicative compounding", Geophysics, April 1956, p. 361.

EKRE, H., 1963, "Polarity coincidence correlation detection of a weak noise source", I.E.E.E. Transactions on Information Theory, January 1963, p. 18.

FARAN, J. J. and HILLS, R., 1952, "Correlators for signal reception", Technical Memo no. 27, U.S. Office of Naval Research, September 1952.

FAUST, L. Y., 1963, "Case history of Fargo Field", Geophysics, Dec. 1963, p. 990.

FEDERICI, M., 1962, "On the improvement of detection and precision capabilities of sonar systems", presented to the Symposium on Sonar Systems at the University of Birmingham, July 1962.

FISHBEIN, W. and RITTENBACH, O. E., 1961, "Correlation radar using pseudo-random modulation", presented to I.R.E. Convention, June 1961.

FITCH, A. H., 1961a, "A comparison of several dispersive ultrasonic delay lines using longitudinal and shear waves in strips and cylinders", Transactions I.R.E. International Convention Record, vol. 8, part 6, p. 284.

FITCH, A. H., 1961b, "Synthesis of dispersive delay characteristics by thickness tapering in ultrasonic strip delay lines", J.A.S.A., May 1963, p. 709.

GAYFORD, M. L., 1961, "Acoustic Techniques and Transducers", Macdonald and Evans.

GILBRECH, D. A. and BINDER, R. C., 1958, "Portable instrument for locating noise sources in mechanical equipment", J.A.S.A., vol. 30, September 1958, p. 842. See also Jacobsen, J.A.S.A., vol. 31 (1959), p. 1352 and vol. 32 (1960), p. 810.

GILFORD, C. L. and GREENWAY, M. W., 1956, "The application of phase coherent detection and correlation methods to room acoustics", B.B.C. Eng. Div. Monograph no. 9, November 1956.

GOFF, K. W., 1955, "An analog electronic correlator for acoustic measurements", and "The application of correlation techniques to some acoustic measurements", J.A.S.A., March 1955, vol. 27, no. 2, p. 223 and p. 236.

GOLOMB, S. W., 1963, "Ferreting signals out of noise", International Science and Technology, October 1963, p. 72.

GOODMAN, T. P., 1955, "Experimental determination of system characteristics from correlation measurements", Sc.D. thesis, M.I.T., June 1955.

GOUPILLAUD, P. L. and LEE, M. R., 1963, "Some theoretical aspects of the 'VIBROSEIS' system, and their practical implications", presented at the Annual Meeting of the Society of Exploration Geophysicists, New Orleans, October 1963.

GUANELLA, G., 1938, "Distance determining system", Swiss patent corresponding to U.S. Patent no. 2,253,975, filed May 1939, issued August 1941.

HAITES, T. B., 1963, "Perspective correlation", Bulletin of the American Association of Petroleum Geologists, April 1963, p. 553.

HORTON, B. M., 1959, "Noise-modulated distance measuring system", Proc. I.R.E., May 1959, p. 821.

HOOVER, C. W., 1961, "Applications of delay lines in correlation processing", abstract in J.A.S.A., November 1961, p. 1657.

HUFFMAN, D. A., 1962, "The generation of impulse-equivalent pulse trains", I.R.E. Transactions on Information Theory, volume IT-8, September 1962, p. 10.

HÜTTMAN, E., 1940, "Verfahren zur Entfermungsmessung", German Patent no. 768,068, filed March 1940, issued May 1955.

HÜTTMAN, E., 1961, "Impulskompression ohne dispersive Netzwerke", Nachrichtentechnik, September 1961, p. 422.

IYER, H. M., 1958, "A study of the direction of arrival of microseisms at Kew Observatory", Geophysical Journal, vol. 1, no. 1, p. 32.

JONES, H. J. and MORRISON, J. A., 1954, "Cross-correlation filtering", Geophysics, October 1954, p. 660; see also U.S. Patent no. 3,018,962.

JONES, H. J., 1962, "Applications of correlation analysis to seismic exploration", World Petroleum, March 1962, p. 50.

JONES, J. L. and KELLY, C. E., 1952, "Investigation of a simple auto-correlator", NAVORD report 2279, U.S. Naval Ordnance Laboratory, March, 1952.

JONES, J. L., 1962, "Theory and results of correlation measurements of acoustic reflections from lake, river and ocean bottoms", Report NOLTR 62-196 of the Naval Ordnance Laboratories, November 1962.

KATZMAN, M. and FROST, E., 1961, "Correlation optical radar", Proc. I.R.E., November 1961, p. 1684.

KEAST, D. N. and KAMPERMAN, G. W., 1960, "Acoustic instrumentation for measurements in the Minuteman missile silo", J.A.E.S., vol. 8, July 1960, p. 180.

KLAUDER, J. R., 1960, "The design of radar signals having both high range resolution and high velocity resolution", Bell System Technical Journal, July 1960, p. 809.

KLAUDER, J. R., PRICE, A. C., DARLINGTON, S. and ALBERSHEIM, W. J., 1960, "The theory and design of chirp radars", Bell System Technical Journal, July 1960, p. 745.

Kraft, L. G., 1950, "Correlation function analysis", J.A.S.A., November 1950, p. 762.

Krauss, J. B., 1963, "Computerized average response and autocorrelation methods as related to signal detection in noise", presented at the 1963 Rochester Conference on Data Acquisition and Processing in Biology and Medicine.

Krönert, R. von, 1957, "Impulsverdichtung", Nachrichtentechnik, April 1957, p. 148, and July 1957, p. 305.

Ksienski, A. A., 1964, "Multiplicative processing antennas for radar applications", presented to the Signal Processing Symposium at the University of Birmingham, July 1964.

Kunetz, G., 1961, "Essai d'analyse des traces sismique", Geophysical Prospecting, September 1961, p. 317. See also d'Hoeraene, J., "Deconvolution de traces réelles", Geophysical Prospecting, March 1962, p. 68.

Lange, F. H., 1961, "Entwicklungstendenzen der modernen Ortungstechnik: Korrelationsortungsverfahren", Nachrichtentechnik, vol. 11, no. 1, p. 2.

Lee, Y. W. and Wiesner, J. B., 1950, "Correlation functions and communication applications", Electronics, June 1950, p. 86.

Lee, Y. W., Cheatham, T. P. and Wiesner, J. B., 1950, "Application of correlation analysis to the detection of periodic signals in noise", Proc. I.R.E., October 1950, p. 1165.

Lee, Y. W., 1960, "Statistical Theory of Communication", Wiley.

Lerner, R. M., 1961, "Design of signals", chapter 11, "Lectures on Communication System Theory", (ed. Baghdady), McGraw Hill.

Linder, I. W., 1961, "Resolution characteristics of correlation arrays", Journal of Research of the National Bureau of Standards, May-June 1961, p. 245.

Lockheed Electronics Co., 1964, "New r-f radar will use optical pulse compressor", Editorial newsletter, Electronics, 28 February 1964, p. 17.

Lohrmann, D., 1962, "Ein Hibfssatz zur theoretischen und praktischen Behandlung des Cauer'schen Impulskompressionsverfahrens", Frequenz, April 1962, p. 156.

Lund, G. R. and Urick, R. J., 1964, "Vertical coherence of explosive reverberation", presented to the A.S.A., 8 May 1964 (abstract in J.A.S.A. May 1964, p. 1015).

Lytle, D. W., 1957, "On the properties of matched filters", Technical Report no. 17, Stanford Electronics Lab., Stanford University, June 1957.

May, J. E., 1960, "Wire-type dispersive ultrasonic delay lines", I.R.E. Transactions on Ultrasonics Engineering, June 1960, p. 44.

May, J. E., 1962, "Guided wave ultrasonic delay lines", presented at the 1962 Electronic Components Conference, Washington, D.C., May 1962.

McCamy, K. and Meyer, R. P., 1964, "A correlation method of apparent velocity measurement", Journal of Geophysical Research, 15 February 1964, p. 691.

Meeker, T. R., 1960, "Dispersive ultrasonic delay lines using the first longitudinal mode in a strip", I.R.E. Transactions on Ultrasonics Engineering, June 1960, p. 53.

Meitzler, A. H., 1960, "Ultrasonic delay lines using shear modes in strips", I.R.E. Transactions on Ultrasonics Engineering, June 1960, p. 35.

Meitzler, A. H., 1961, "Mode coupling occurring in the propagation of elastic pulses in wires", J.A.S.A., April 1961, p. 435.

Melton, B. S. and Bailey, L. F., 1957, "Multiple signal correlators", Geophysics, July 1957, p. 565.

Melton, B. S. and Karr, P. R., 1957, "Polarity coincidence scheme for revealing signal coherence", Geophysics, July 1957, p. 553.

Mercer, D. M., 1958, "The application of correlation techniques in noise analysis", Annals of Occupational Hygiene, vol. 1, p. 81.

Merrihew, H. W., 1961, "Pulse compression—a new look in radar", Tech-Rep Bulletin, July-August 1961, p. 2.

Middleton, D., 1962, "Acoustic signal detection by simple correlators in the presence of non-Gaussian noise", J.A.S.A., October 1962, p. 1598.

Miller, R. J., 1961, "Air and space navigation system using cross-correlation detection techniques", Electronics, 15 December 1961, p. 55.

MILNE, K., 1964, "The combination of pulse compression with frequency scanning for three-dimensional radars", presented to the Signal Processing Symposium at the University of Birmingham, July 1964.

MIMS, W. B., 1963, "The detection of chirped radar signals by means of electron spin echoes", Proc. I.E.E.E., August 1963, p. 1127.

MUT, S. C., 1959, "An evaluation of the long pilot cross-correlation technique", Geophysics, December 1959, p. 1143.

O'MEARA, T. R., 1960, "Linear-slope delay filters for compression", Proc. I.R.E., November 1960, p. 1916.

OHMAN, G. P., 1960, "Getting high range resolution with pulse compression radar", Electronics, 7 October 1960, p. 53.

PARKS, J. K. and DOWNING, J. J., 1960, "System considerations for random-signal sonars", Lockheed Technical Report LMSD-895060, December 1960.

PARVULESCU, A. and CLAY, C. S., 1964, "Reproducibility of signals", presented to the Signal Processing Symposium at the University of Birmingham, July 1964.

PETERSON, R. A., 1955, "Well logging system", U.S. Patent no. 3,011,582, filed July 1955, issued December 1961.

PRICE, R. and GREEN, P. E., 1958, "A communication technique for multipath channels", Proc. I.R.E., March 1958, p. 555.

PRICE, R., GREEN, P. E., GOBLICK, T. J., KINGSTON, R. H., KRAFT, L. G., PETTENGILL, G. H., SILVER, R. and SMITH, W. B., 1959, "Radar echoes from Venus", Science, 20 March 1959, p. 751.

RAEMAR, H. R. and REICH, A. B., 1959, "Correlation devices detect weak signals", Electronics, 22 May 1959, p. 58.

RAMP, H. O. and WINGROVE, E. R., 1961, "Principles of pulse compression", I.R.E. Transactions on Military Electronics, April 1961, vol. MIL-5, p. 109.

REMLEY, W. R., 1963, "Correlation of signals having a linear delay", J.A.S.A., January 1963, p. 65.

RESWICK, J. B., 1955, "Determine system dynamics—without upset", Control Engineering, June 1955, p. 50.

RICE, R. B., 1962, "Inverse convolution filters", Geophysics, February 1962, p. 4.

ROBINSON, E. A., 1957, "Predictive decomposition of seismic traces", Geophysics, October 1957, p. 767.

ROWLANDS, R. O., 1962, "Tapped delay-line detector for a modified chirp signal", presented to A.S.A., November 1962; abstract J.A.S.A. December 1962, p. 1972.

RUDWICK, P., 1953, "The detection of weak signals by correlation methods", Journal of Applied Physics, vol. 24, February 1953, p. 128.

RYALL, A., 1964, "Improvement of array seismic recordings by digital processing", Bulletin of the Seismological Society of America, February 1964, p. 277.

RYKUNOV, L. N., 1961, "A correlation method of determining the velocities of microseisms", Izvestiya, Academy of Sciences, USSR, July 1961, p. 1037; translated by American Geophysical Union, October 1961.

SAKAMOTO, T., TAKI, Y., MIYAKAWA, H., KOBAYASHI, H., SUZUKI, T., YOSHIDA, T., TAKEYA, T. and KOKUBU, M., 1963, "How coded-pulse techniques extend radar range", Electronics, 22 November 1963, p. 34.

SHAUB, YU. B., 1963, "The use of correlation analysis for the evaluation of geophysical data", Bulletin of the Academy of Sciences, USSR; translation in AGU series, April 1963, p. 358.

SHAW, E. and DAVIES, D. E., 1964, "Theoretical and experimental studies of the resolution performance of multiplicative and additive aerial arrays", presented to the Signal Processing Symposium at the University of Birmingham, July 1964.

SCHROEDER, M. R. and ATAL, B. S., 1962, "Generalised short-time power spectra and auto-correlation functions", J.A.S.A., November 1962, p. 1679.

SKOLNIK, M. I., 1962, "Introduction to Radar Systems", McGraw Hill.

SMITH, M. K., 1958, "A review of methods of filtering seismic data", Geophysics, January 1958, p. 44.

SMITH, M. W. and LAMBERT, R. F., 1960, "Acoustical signal detection in turbulent airflow", J.A.S.A., July 1960, p. 858.

SPILKER, J. J. and MAGILL, D. T., 1961, "The delay-lock discriminator—an optimum tracking device", Proc. I.R.E., September 1961, p. 1403.

SPILKER, J. J., 1963, "Delay-lock tracking of binary signals", I.E.E.E. Transactions on Space Electronics and Telemetry, March 1963, p. 1.

SPROULE, D. O. and HUGHES, A. J., 1944, "Improvements in and relating to systems operating by means of wave trains", British Patent no. 604,429, filed June 1944, issued July 1948.

STEVENS, K. N., 1950, "Autocorrelation analysis of speech sounds", J.A.S.A., November 1950, p. 769.

STEWART, J. L. and WESTERFIELD, E. C., 1959, "A theory of active sonar detection", Proc. I.R.E., May 1959, p. 872.

STEWART, J. L. and ALLEN, W. B., 1962, "Pseudonoise-correlation techniques in underwater acoustic studies", presented to Fourth International Congress on Acoustics, Copenhagen, August 1962.

SUSSMAN, S. M., 1960, "A matched filter communication system for multipath channels", I.R.E. Transactions on Information Theory, June 1960, p. 367.

TIBERIO, U., 1961, "Sul rapporto eco-rumore che e possibile consequire nei radar mediate la compressione di impulso", Alta Frequenza, September 1961, p. 665.

TOKSÖZ, M. N., 1964, "Microseisms and an attempted application to exploration", Geophysics, April 1964, p. 154.

TOMODA, Y., 1956, "A simple method for calculating the correlation coefficients", Journal of the Physics of the Earth, vol. 4, no. 2, p. 67.

TUCKER, D. G., 1962, "Multiplicative arrays in radio-astronomy and sonar systems", presented to the Symposium on Sonar Systems, University of Birmingham, 9-12 July 1962.

TULLOS, F. N. and CUMMINGS, L. C., 1961, "An analog seismic correlator", Geophysics, June 1961, p. 298.

TURIN, G. L., 1960, "An introduction to matched filters", I.R.E. Transactions on Information Theory, June 1960, p. 311.

WELSBY, V. G. and TUCKER, D. G., 1959, "Multiplicative receiving arrays", J. Brit. I.R.E., vol. 19, p. 369.

WELTI, G. R., 1960, "Quaternary codes for pulsed radar", I.R.E. Transactions on Information Theory, June 1960, p. 400.

WHITEWAY, F. E., 1964, "Theory and application of seismometer arrays", presented to I.E.R.E. symposium on "Modern Techniques for Recording and Processing Seismic Signals", May 1964.

WINKLER, M. R., 1962, "Chirp signals for communications", paper 14/2, Communications Systems, WESCON 1962. Summary in Electronics, 10 August 1962.

WISOTSKY, S., CLAY, C. S. and LIANG, W. L., 1963, "Use of a hydro-acoustic transducer in continuous seismic profiling", presented at the 66th meeting of the A.S.A., University of Michigan, November 1963.

WOZENCRAFT, J. M., 1961, "Sequential reception of time-variant dispersive transmissions", chapter 12 of "Lectures on Communication System Theory" (ed. Baghdady), McGraw Hill.

ZADEH, L. A., ABRAMSON, N., BALAKRISHNAN, A. V., BRAVERMAN, D., EDEN, M., FEIGENBAUM, E. A., KAILATH, T., LERNER, R. M., MASSEY, J., MUELLER, G. E., PETERSON, W. W., PRICE, R., SEBESTYEN, G., SLEPIAN, D., THOMASIAN, A. J. and TURIN, G. L., 1963, "Report on progress in information theory in the U.S.A., 1960-1963", I.E.E.E. Transactions on Information Theory, vol. IT-9, October 1963, p. 221. (Note: a substantial part of this periodical is now concerned with matter relevant to pulse-compression.)

ZVEREV, V. A. and KALACHEV, A. I., 1960, "Application of frequency modulation to acoustical measurements", Akusticheskii Zhurnal, April 1960, p. 205; translated by American Institute of Physics.

THE DIGITAL PROCESSING OF SEISMIC DATA[†]

DANIEL SILVERMAN[*]

The paper discusses the background of the problem of signal and noise in the seismic process, and the application of the principles of communication theory to this problem.

The limitations of the seismic process are discussed along with the types of noises involved, the methods of rejecting noise, the use of filters to reduce noise, characteristics of filters, and the relationships between frequency domain, time domain, mathematical, and digital filters.

In the discussion of the electronic data processing of seismic information, the characteristics of an ideal seismic digital computer system are developed in relation to the characteristics of seismic data. The choice between digital and analog field recording is discussed in relation to the needs of the seismic process and the quality of the seismic data.

Among the mathematical processes discussed are velocity filtering and a number of types of Wiener filtering, including horizontal stacking, deghosting, deconvolution, and multitrace digital filtering.

EDITOR'S NOTE:

The following paper was prepared to accompany the June issue, which was devoted entirely to the basic theory of digital filtering. However, before publication, it was decided that the nonmathematical elements of our membership, for whom the paper could well reveal an adequate understanding, might not see it at all, because of the formidable nature of the rest of the volume. The decision was then made to delay publication so that, included in an ordinary issue, it could serve its real purpose, namely, to give those who have only to hear the associated words from time-to-time, an adequate understanding of what the words mean.

INTRODUCTION

In recent years, there has been a growing interest in the subject of the digital processing of seismic data. Unfortunately, the accent has been on the "digital" part of the subject, whereas, the real value of the new system lies in the "processing" part, and what it can do with the seismic data. The real breakthrough has been in the success of an extensive research and development program in the application to seismic data of the mathematical processes of communication theory. The use of digital computers is purely incidental. However, without the fast computers available today this processing would not be economically feasible.

LIMITATIONS OF THE SEISMIC PROCESS

Oil and gas exploration, a vital oil industry activity, depends largely on the seismic process to provide the basic data, from which the shapes, attitudes, and structural relations of subsurface rock strata are determined. However, today, as throughout its active life, the seismic method is working against, what appear to be, irreducible difficulties.

The difficulties which beset the seismic process can be expressed by one word: "Noise". Noise is everything that is not desired signal. That is (a) everything that is generated by the seismic source travelling over other than the desired paths, (b) everything generated by nonseismic sources, and (c) everything that is added to the seismic signal during the detecting, recording, and processing of the seismic data. Noise is everything in the seismic data that is not related to the subsurface structure. The desired reflection signals, with noise superimposed, are detected and recorded. The problem is to identify the signal in the presence of the noise. Seismic operations must maximize the signal-to-noise (S/N) ratio.

There are two principal ways to do this: (a) to conduct the field operations so that a minimum of noise is generated, or the noise that is generated is cancelled during the recording process, and (b) to process the recorded data so as to reduce or eliminate the noise that is recorded.

[†] Manuscript received by the Editor 31 March 1965; revised manuscript received 26 May 1967.
[*] Pan American Research Corporation, Tulsa, Oklahoma.

This paper is concerned primarily with the latter method.

Over the years, much has been done to improve the signal-to-noise ratio in seismic operations. Since the dominant frequency of some of the noise is appreciably different from that of the signal, early in the history of the seismic method, frequency filters were used in the recording system to diminish wind noise, ground roll, and similar noise before the amplified signals were recorded. Later, arrays of detectors and shots were used to cancel some of the noise waves.

With the advent of magnetic recording, it has been possible to reproduce the recorded seismic signals so as to achieve greater flexibility in processing. Processing may include frequency filtering, stacking, common-depth-point compositing, multiple visual displays, etc. All of these techniques have helped the interpreter to obtain more information from the seismic data.

COMMUNICATION ASPECTS OF THE SEISMIC PROCESS

However, it is possible to go only so far with these techniques, and a new approach is required if the seismic process is to be effective for the increasingly more difficult exploration problems being encountered today. Fortunately, some twenty years ago mathematicians and electrical engineers were developing this new approach which they called communication theory. Out of their work have come some new ideas applicable to the seismic process.

Communication theory was developed to solve some of the problems involved in telephone, radio, radar, and sonar communication. Each of these fields has a classic communication system. Each has a source of information, a transmission channel, and a receiver. In a similar way the seismic method may be considered as a communication process. The telephone system has a wire channel, radio and radar have air or vacuum channels, and sonar a water channel, while the seismic process uses the earth as its channel. A signal created by a dynamite explosion is transmitted through the channel, over many paths, with many changes in amplitude and phase, and is finally detected by a receiver. In the telephone and radio systems the information transmitted is the signal itself, while in the radar, sonar, and seismic systems what is mainly important is the time of travel of the signal. However, before the traveltime can be determined, the signal must be identified in the noise, and in this respect, all of the systems are alike.

One of the advantages that the communications engineer often has over the geophysicist is that he puts a known signal into his channel and normally knows what to look for in the received signal-plus-noise. The geophysicist, on the other hand, puts into the earth (in the conventional process)[1] a sharp pulse-like wide frequency band signal. Because of its broad frequency band and the absorption of high frequencies in the earth, the received signal generally bears very little obvious resemblance to the input signal. Thus, since the geophysicist does not know in advance what signal to look for in the received signal-plus-noise, he must find new criteria for separating the signal from the noise.

Because of the special conditions involved in signalling through the earth, the communications theory of the electrical engineer was not directly applicable to the seismic process. A small group of scientists working at the Massachusetts Institute of Technology became interested in applying communications theory to the seismic communication problem. This work was initiated by Professor G. P. Wadsworth and Dr. J. G. Bryan of the Mathematics Department and carried out, under the supervision of Professors R. R. Shrock and P. M. Hurley of the Department of Geology and Geophysics, by a small group of graduate students under the leadership of Enders A. Robinson and S. M. Simpson, Jr. They described their ideas of how this might be done to a group of exploration geophysicists who, in view of the promise of this program, assisted in organizing a group of sponsoring companies to support the work. Starting with about twenty participating members, this work continued for four years. While there was no major result that could be applied at that time to the seismic process, the group, known as the Geophysical Analysis Group (GAG), wrote a number of reports, doctoral theses, and papers describing their work.

The salient aspects of this study were the subject of the June 1967 number of GEOPHYSICS.

Unfortunately, at that time there were few

---

[1] Within the past few years a new seismic process has been developed, called the VibroSeis, which uses a particular kind of vibratory signal called a swept frequency signal as input to the earth. The received signal is converted to conventional form by the process of correlation with the input signal.

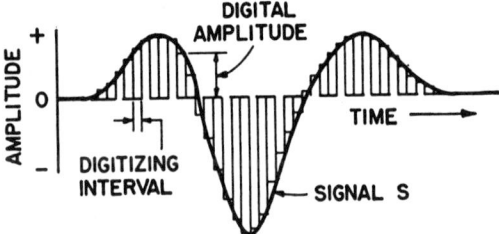

Fig. 1. Digitization of a time function.

scientists among the exploration geophysicists who could capitalize on these publications, and the reports lay dormant in company files for a number of years. However, the work continued at MIT on a reduced basis, and a number of the graduate students received their Ph.D.'s and took positions in the oil industry. Here, they formed nuclei, about whom others gathered to work in this field. Today, some twelve years after the first Ph.D graduated from the GAG program, the oil industry is applying the results of their efforts and subsequent developments in an extensive program that holds great promise.

WHAT IS DIGITAL PROCESSING?

Digital processing is technically the mathematical processing of data by means of a digital computer. However, from the point of view of this paper, digital processing is considered to be the application of certain new mathematical operations to seismic data processing. These are "filtering" operations which are more effective in enhancing signal-to-noise ratio than are the conventional seismic analog filters. However, before discussing these new filtering processes, it might be well to discuss what filtering is and does.

Digital versus analog signals

There are two ways to express a signal: (a) as a continuous curve or time function representing the amplitude of the signal as a function of time (this is the analog form); or (b) as a series of numbers, each of which represents the amplitude of the signal at a specific value of time (this is the digital form. Generally, the values of time correspond to a series of equal intervals of time called the "digitizing interval." This is shown schematically in Figure 1 in which the continuous signal S is an analog signal. The area under the signal S is divided into narrow rectangles. The height of each rectangle corresponds to the amplitude of the curve at the time corresponding to the midpoint of the rectangle. The width of each rectangle is equal to the digitizing interval.

Frequency Domain versus Time Domain

An analog signal is most often expressed as a continuous time function of varying amplitude (and polarity). In this form, in which it is expressed as an amplitude varying with time, it is said to be expressed in the time domain. This time-varying signal can be considered as made up of a group of monofrequency signals of different amplitudes and phases. When the signal is expressed in terms of the amplitudes of its component signals versus frequency and phases of its component signals versus frequency, this signal is said to be expressed in the frequency domain. There are mathematical processes for converting a signal in the time domain to its equivalent (or transform) in the frequency domain, and vice versa. The amplitude and phase characteristics of a time function are called its amplitude and phase spectra.

Filter characteristics

A filter is a device that will change a given input signal I, passing into a filter F, (Figure 2) into an output signal which is the filtered signal O. The filter F is shown bounded by a dashed line. This is to indicate that it can be considered either as a physical electronic apparatus (analog filter), or as a mathematical program or operator (digital filter), which, acting upon the input signal in digital form, will provide a digital output representing the filtered signal.

Every filter has a "characteristic" which describes its mode of operation. For example, an ideal "low-pass filter" has a characteristic (Figure 3) which indicates that all frequencies below a particular value, the cutoff frequency, are passed (ideally) without attenuation, while those frequencies above the cutoff frequency are highly attenuated. Figure 3 shows what is called the

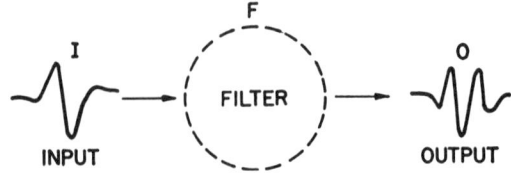

Fig. 2. Analog filtering.

amplitude-frequency characteristic of the filter. Figure 4 shows a phase-frequency characteristic of an ideal filter. This displays the phase shift as a function of the frequency of the signals passing through the filter. These two characteristics, amplitude and phase, together fully describe the action of the filter in terms of the frequencies of the signals which might be applied to it. These characteristics are said to be expressed in the "frequency domain" since they express amplitude and phase effects in terms of frequency.

It is possible also to express the characteristic of a filter in the "time domain." This is illustrated in Figure 5 in which the filter F has an input signal I which comprises a "unit-impulse" or "spike." The output O is the signal output of the filter corresponding to the unit impulse input. This output (Figure 6) is called the "impulse response" of the filter. It is a time function, the amplitude of which, expressed as a function of time, is the time-domain characteristic of the filter. The impulse response of the filter is often called the filter "operator."

Frequency Filters

A filter may be considered as a device for partially separating components of a mixture, commonly a mixture of signal and noise. The greater the differences between the components, the better they can be separated by filtering and, conversely, the more alike the components, the poorer the separation.

In the seismic process, vibrations are recorded which arrive from many different sources and over many different paths in the earth. Some of these vibrations are considered to be signal and some are noise. The problem is to determine which are which, and to separate the signal from the noise. This is done by filtering.

There are two principal categories of filters, a) frequency filters and b) optimum filters.

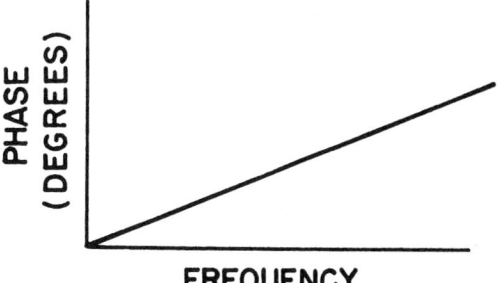

FIG. 4. Phase-frequency characteristic of an analog filter.

Frequency filters are designed on an arbitrary basis for passing or rejecting certain frequency components of the signal. At the same time, the phase characteristic must follow certain requirements. Frequency filters may be used in analog or digital form. The conventional analog filters, which are constructed of inductances and capacitances are frequency filters. On the other hand, digital frequency filters can be designed with the same or improved characteristics.

In the seismic process, some of the noises recorded, such as ground roll and wind noise, have pronounced frequency components. These can be strongly rejected on the basis of frequency content. Also, there may be noises of specific frequencies, such as 60-cycle power line interference, which can be filtered by narrow-band frequency rejection filters. However, frequency filters, or those which operate by passing a selected band of frequencies are of relatively limited value where the noise and the signal have substantially the same frequency content. In this case, separation must be made on the basis of other characteristics. This is what communication theory can do. From the nature and composition of both the signal and the noise it can provide a filter to enhance the S/N ratio.

Optimum filters

Frequency filters are designed on an arbitrary basis without reference to the signal or the noise,

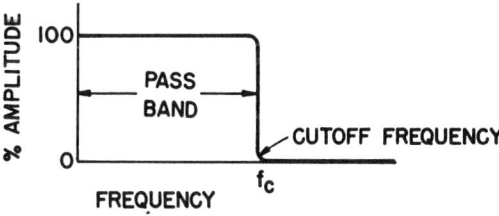

FIG. 3. Amplitude-frequency characteristic of an analog filter.

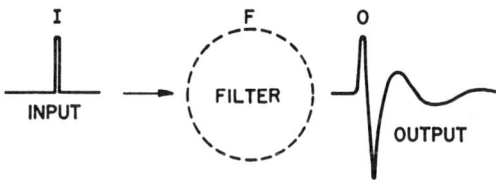

FIG. 5. Time-domain filtering.

and without reference to the effectiveness of the filter. There is another type of filter, generally called an "optimum" filter, that is designed on the basis of a specific character of signal and noise. The optimum filter is designed on some basis of optimality. For example, one type of optimum filter, the Wiener filter, is designed on the basis of the *actual input* signal and *desired output* signal to have an optimum fit between the *actual output* signal and *desired output* signal, as defined by the least squared error criterion.

The late Professor Norbert Wiener of MIT and Dr. C. E. Shannon of the Bell Telephone Laboratories were the early leaders in the development of communication theory. Since their early work, others such as Professor Y. W. Lee of MIT have made important contributions. However, much of the present results are derived from the early work of Wiener, whose name is applied to a general type of mathematical filter. In the Wiener filter an *input* and a *desired output* are specified, and a filter operator is calculated such that, when applied to the input signal, it will give an actual output which is the optimum fit to the desired output, as defined by the least-square error criterion.

There is one major difference between the kind of filtering that is involved in the new optimum filter processing and the conventional analog frequency filtering. In analog processing, the electronic assembly, or filter, is designed on some arbitrary basis, and is applied separately to each seismic trace. Normally its characteristics are designed for an average seismic trace, and the same filter is generally used for all traces in a record and for many records.

In the new optimum filters, the filter operators can be designed in accordance with the character of the signal and noise on each of the traces in the record to be processed. In other words, before a trace is filtered, the character of the noise can be determined, and the optimum filter can be designed to remove noise of that particular character from that particular trace.

Convolution filtering

Having a signal and a filter operator, how is the "filtering" accomplished in a digital computer? There are several principal ways: a) in the time domain, by a mathematical process called convolution, b) by a modified convolution called recursive filtering, and c) by multiplying ampli-

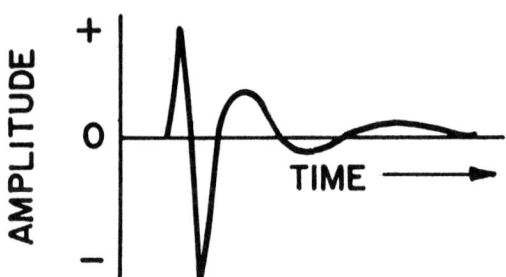

FIG. 6. Impulse response characteristic of a filter.

tude spectra and adding phase spectra of the signal and the operator to obtain the amplitude and phase spectra of the filtered output.

The physical process of convolution is illustrated by Figure 7. Consider that the wavelet of Figure 1 is to be convolved with the impulse response operator of Figure 6. The impulse response is the result of passing a unit impulse through the filter. But the input wavelet A can be considered as being made up of a large number of impulses, or spikes, similar to the narrow rectangles of Figure 1. Each of these rectangles, in passing through the filter, acts like a separate impulse and creates an impulse response signal of

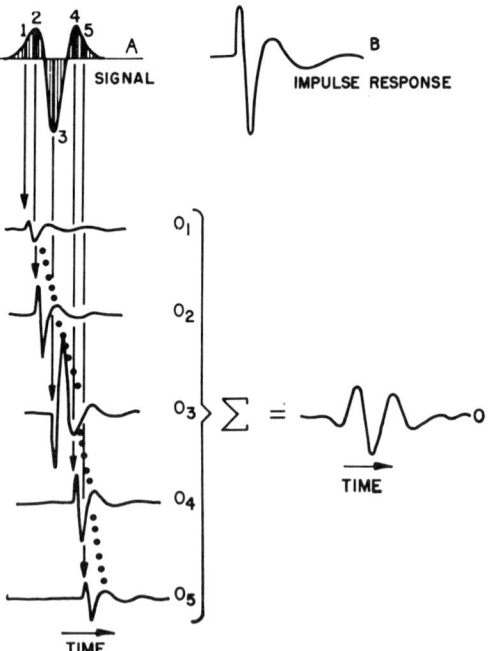

FIG. 7. Convolution illustrated on the basis of impulse response.

its own. However, the amplitude of an impulse response signal is proportional to the amplitude of the impulse causing it. Also the impulse response signal starts at the time of the impulse. The output of the filter corresponding to the passage of the complete input signal A through the filter is the sum, *in proper time relation*, of all of these separate impulse response signals. This is shown schematically in Figure 7 in which the input signal A is considered (for convenience) to be represented by five ordinates (or impulses), 1, 2, 3, 4 and 5. Each of these impulses creates its own impulse response signal drawn below in approximate amplitude, polarity, and time, and numbered O_1, O_2, O_3, O_4, and O_1. The sum of these five signals is the output signal O, which is the results of the convolution of signal A with the operator B.

The digital process of convolution is illustrated schematically in Figure 8, in which an input signal A is convolved with an operator B. The signal A is illustrated in trace (a). It is made up of digital values $A_1, A_2, \ldots A_N$. Time runs to the left. The filter operator is illustrated in trace (b). The operator digital values are $B_1, B_2, \ldots B_M$. Time runs to the right. The digitizing interval t is the same for both functions. Trace (c) represents the filtered output resulting from the convolution process. It comprises a number of digital values $C_1, C_2, \ldots C_Z$, computed in accordance with the relations listed at the bottom of the figure.

C_1 is equal to the product of A_1 and B_1. That is, the two functions (a) and (b) are positioned so that their initial values A_1 and B_1 are in time coincidence. These values are multiplied together to give C_1. Next, as in (c), function (a) is shifted one interval t to the right. Then A_1 is lined up with B_2 and A_2 with B_1. These pairs are multiplied, and $C_2 = A_1B_2 + A_2B_1$. Then (a) is shifted another interval t to the right and the three pairs of values are multiplied and added to give C_3, and so on. Finally the last value of C which is indicated as C_Z is equal to the product of the last values A_N times B_M.

The equations at the bottom of Figure 8 indicate that the output signal $F = (C_1 + C_2 + \ldots C_Z)$ can be considered as the sum (*in proper time relation*) of

$$A_1(B_1 + B_2 + \cdots B_M)$$
$$+ A_2(B_1 + B_2 + \cdots B_M)$$

and so on, where $(B_1, +B_2 + \ldots B_M)$ is the impulse response operator. This can be expressed as the sum, in proper time relation

$$F = \sum (A_1 + A_2 + \cdots A_N)$$
$$(B_1 + B_2 + \cdots B_M)$$

or

$$F = \sum (A_1 + A_2 + \cdots A_N)$$

(Impulse response),

which is the same as shown in Figure 7.

The output function F has Z terms, where $Z = N + M - 1$. The duration of the filtered signal is one unit less than the time duration of the input signal A plus the operator B. The filtered signal is always longer than the input signal.

Recursive filtering

In electrical analog filtering, it is well known that, to process a signal more than once through a given filter, that is, to filter the filtered output, will in effect sharpen (increase the Q of) the filter. For example, multiple cascade stages of a low Q filter will act like a high Q filter. In the same way, a feedback filter, in which the output is fed back

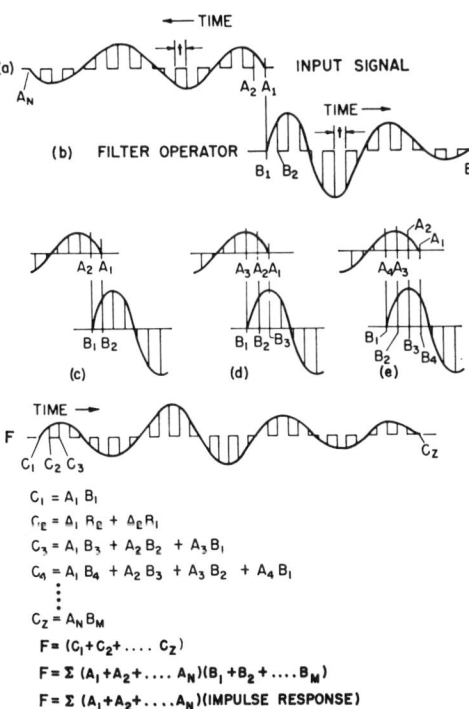

FIG. 8. Convolution illustrated on the basis of digital computation.

to pass again through the filter, will approximate the effect of a multistage, and therefore a sharper filter.

This same procedure of increasing the Q of a filter by appropriate feedback, can be accomplished by means of a class of digital filters called recursive filters (Shanks, 1967). In recursive filters, the input signal is filtered with a short simple operator to form a first filtered signal that goes to the output circuit. Part of this filtered signal is delayed and fed back through a second filter with a short simple operator, to form a second filtered signal. The two filtered signals are added and the combined filtered signal is sent to the output circuit. Again, part of this combined filtered signal is delayed, fed back through the second filter and added to the first filtered signal, and so on. Thus, by the use of two short operators, a signal can be filtered in recursive form, to provide the equivalent filtering of a very much longer convolution filter operator, with a consequent saving in time and cost.

Fourier transform filtering

Digital filtering in the frequency domain is accomplished in terms of the amplitude and phase characteristics or spectra of the signal to be filtered and of the filter operator. The amplitude spectrum of the *filter output* is obtained by *multiplying* the amplitude spectra of the *input signal* and of the *filter operator*. The phase spectrum of the *filter output* is obtained by *adding* the phase spectra of the *input signal* and of the *filter operator*.

Having the signal and the operator in time domain form, it is, of course, necessary to convert them to the frequency domain to obtain their amplitude and phase characteristics. Then after obtaining the filter output in terms of its amplitude and phase characteristics, it is necessary to convert back to the time domain.

The conversion from time to frequency domain and vice versa involve the Fourier transforms. Recently there has been developed a new procedure (or algorithm) for performing these transformations. This is known as the Cooley-Tukey algorithm (Cooley and Tukey, 1965). This algorithm effectively speeds up the transform process.

Digital processing

While there are three principal methods of digital filtering, that is, convolution, recursive filtering, and Fourier transform filtering, there are several ways in which the processing can be carried out on a digital computer:

1) Convolution involving multiplication in the normal operating cycle of the computer.

2) Convolution involving the use of a fast multiply (F.M.) box. This is a small, fast, wired-program computer to perform the multiply-and-add steps of the convolution process. This F.M. box can speed the convolution by a factor of 10 or more.

3) Convolution by recursive filters operating in the normal machine cycle.

4) Convolution by recursive filters with a special recursive F.M. box. The recursive process does not lend itself well to the use of the standard F.M. box. However, a special design of recursive F.M. box will permit faster convolution than in (2).

5) Filtering by Fourier transform using the Cooley-Tukey algorithm.

In the convolution process, the number of multiplications goes up as the product of the lengths (that is, the number of digital values) of the signal and the operator. For short and medium length operators, process 2 offers the greatest speed. For extremely long operators, process 5 may offer the greatest speed because of the computing short cuts of the Cooley-Tukey algorithm. If and when a specially designed F.M. box is available for use with the recursive filters, process 4 will probably be the fastest filtering process.

At the present time, the vast majority of the filter processing in seismic operations is by means of the convolution process with the fast multiply or convolution box.

The ease with which a signal can be digitally filtered, irrespective of the particular shape of the operator, provides a filtering system of great flexibility and power. The digital computer therefore provides a tool to do all present-day seismic processing plus the new mathematical processing.

In discussing analog filters and frequency-domain processes, on the one hand, and digital filters and time-domain processes on the other, there is a tendency to think of them as two separate and distinct classes. However, the actual situation is not this definite.

At one extreme, the frequency domain involves continuous or analog signals, which can be filtered by inductance-capacitance filters. However, it is possible to construct a type of time domain filter that will accept analog signals. This is known as a

"Delay Line Filter." It is constructed with a multielement delay line, with potentiometers at each unit delay tap, and a circuit to add the outputs, in proper phase, of all the potentiometers. The multiplying settings of the potentiometers correspond to the digital values of the convolving operator. This type of filter accepts input of the signal in analog form and the operator in digital form.

At the other extreme, with the digital computer, both the time function and the operator are in digital form, and the filtering takes place in the time domain.

However, it is possible to filter a time function with an operator, both of which are expressed in the frequency domain in the form of amplitude and phase spectra. The amplitude spectrum of the filtered signal is the product of the amplitude spectra of the signal and the operator. The phase spectrum of the filtered signal is the sum of the phase spectra of the signal and the operator. These multiplications and additions can of course be done on the digital computer.

Thus it is possible, with special analog apparatus, to do time-domain filtering and, with digital apparatus, do frequency-domain filtering.

PROCESSING EQUIPMENT REQUIRED

Characteristics of seismic data

The digital computing problem involved in the processing of seismic data is a sizable one. Consider that a 5-sec, 24-trace record, digitized at 2 ms intervals, would have 60,000 discrete amplitude values, or 60,000 numbers to represent it. With a filter operator of 200 ms duration (digitized at 2 ms intervals) the convolution process would require about 6 million multiplications and additions. In the conventional high-speed general purpose computer each multiplication cycle might take about 20 μsec, and it would require 120 sec to convolve or filter this record. The seismic data problem is therefore one which requires a special digital computer system, particularly one that will multiply rapidly. Manufacturers have looked into this problem and have designed special, fast-multiply units (F.M.U.) with which it is possible to make a multiply-plus-add step in a time of the order of 1 μsec or less. This will speed the seismic computation process by a factor of 10 to 20 or more.

This aspect of the seismic data problem, that is, the great volume of data, is reflected also in the need for large temporary storage in the computer. Since it is desirable to store many records simultaneously for joint processing, the internal storage of the computer should preferably be able to handle up to 10 or 20 million characters. This requires a large disc or drum storage.

It is also desirable to have a fast, sequential, single-trace plotter adapted to plot each trace as it is computed. An ideal computer should therefore have the ability to carry on two or more programs concurrently, such as computing, plotting, reading input, and writing output. This feature of the computer is called multiprocessing or time sharing.

In normal digital computer operations, data are recorded on the tape in a series of short blocks separated by gaps. As the tape is transported past the reading heads, the data in a block is read and passed into core storage. The maximum size of the block is normally dependent on the size of the core storage available. At the end of each block a gap is provided in which no data are recorded. This gap signals the tape drive to stop until the computer can utilize the data from the previous block. The computer then signals the tape to start and the second block is read into the core. Normally these data blocks contain up to 5000 "words." As indicated above, a 24-trace seismic record might have as many as 60,000 or more digital values or words, which is much more than can be handled at one time in the core memory of the conventional computer.

As the seismic record is digitized in the field, it is recorded in real time, that is, while it is happening. To gap the data would require stopping the recording and storing the incoming signals during the short interval while the "gap" is passing the record heads. This involves expensive core storage and logic circuits in the field recording instruments. Conversely, it is possible to record the original data continuously on the tape (gapless format) without interrupting the signal to form gaps. This requires modification of the computer operation to read this gapless tape. Both systems are possible, and are in use.

The above aspects of the seismic data problem require an increase in the normal complement of apparatus (and cost) of the digital computer. A third aspect sometimes serves to decrease the cost. This results from the fact that the range of seismic signal amplitudes and the number of binary digits required to express this range (14 or 15 bits) is appreciably less than the size range of numbers that most high speed digital computers can handle (36 or 48 bits or more). Thus smaller and cheaper processing units might be considered for use.

Characteristics of the seismic digital computer system

A digital computer system designed for seismic data processing can therefore be characterized as having:

1) High-speed processor;
2) Concurrent programming—multiprocessing;
3) Ability to handle ungapped tape;
4) Medium precision (15 bits);
5) Special fast multiply unit for high-speed convolution;
6) Large volume drum and/or disc storage;
7) Fast sequential single-trace plotter;
8) Analog magnetic tape input and output;
9) Analog-to-digital and digital-to-analog conversion apparatus.

One of the immediate problems in this seismic data system is concerned with the lack of a suitable standard of data record format. At the present time, there are a number of digital tape formats in general use in seismic work, and the possibility is that there may be many others in the future. Some of these are:

1) GSI format using one-inch tape with 21 tracks, without record gaps,
2) IBM format with 7 tracks on one-half inch tape with and without record gaps,
3) IBM format with 9 tracks on one-half inch tape with and without record gaps.

This lack of standardization makes for inflexibility, inefficiency, and higher cost since each processing center must add special equipment to read all tape formats, or hire a contractor to make tape transcriptions.

Another problem in the design of a digital processing system is the matter of display of the computed traces. Today, in most large analog playback centers, all traces on a record are processed in parallel and plotted in parallel, in real time. The digital computer is a fast single-trace serial processor. It can process a single trace in a small fraction of real time. As the traces are computed, they are in core memory. They can be transferred to the drum or disc memory and stored until the full complement of 24 traces are complete. They can then be read out of drum or disc, multiplexed, converted to separate analog traces, and plotted on a multitrace plotter. However, all of these steps are expensive in computer time. The ideal way to display the computed traces would be to have a very fast single-trace sequential plotter that could plot a 5,000-word trace in 1/20 to 1/50 sec. The conventional photographic seismic plotter is much slower than this, recording a 5000 word trace in 5 sec. What is required is a plotter possibly 50 to 100 times faster than present seismic plotters. In this way, as soon as a single trace is computed, it could be read directly out of core to the plotter. With this fast plotter, the simultaneous processing and plotting of different trace treatments can be carried on at very little additional cost.

Digital versus analog field recording

The seismic signals, as generated and detected, are in analog form. These must be amplified and recorded in reproducible form. Both the amplifiers and recorders have a limited recording range. That is, there is a relatively limited range of signal amplitude that can be recorded without distortion. In view of the very wide range in amplitude of the seismic signals as a function of time, it is necessary to insert a variable gain control between the seismometer and the recorder to hold the signal level between limits set by the recorder.

The variation of gain with time should be known so that, in the processing step, the original signal amplitude can be recovered. This can be done with a so-called "programmed gain control" (PGC). The gain of the system as a function of time is set by the PGC prior to initiation of the shot. Because of the difficulty of judging in advance the response of the earth, it is difficult to preset the PGC with sufficient accuracy.

The best system available today for this purpose is the "binary gain ranging" amplifier. This is an amplifier designed with fixed gain elements with gain adjustable in ratios of two to one or 6 db steps. In operation, the amplifier adjusts its gain to suit the signal amplitude. By recording which elements of the amplifier are working, the gain can be determined precisely. This amplifier operates like an automatic volume control except that it operates in fixed gain ratio increments.

The amplified-gain-controlled signals are conventionally recorded on analog magnetic tapes. These tapes can be taken to the computer center, digitized, and entered into the computer. Conversely, these signals can be digitized in the field and recorded on digital magnetic tapes. One of the important questions which must be considered concerns the necessity or the advisability of digital recording versus analog recording in the field.

If all that is desired is to process the records digitally in the same way that they are now done in analog processing, it is doubtful whether field digitization is necessary. In other words, the digitization, processing, and analog display of data originally recorded on an analog magnetic tape can be done as precisely (or more so) on a digital computer as the analog processing of this same tape.

However, the real advantage of digital processing lies in the new optimum filters that are not readily adaptable to analog processing. Thus, the need for field digital recording (with possibly greater cost) in comparison with field analog recording, must be considered in relation to the need for the high quality of digital field recording in the new optimum processes. Also the digital processing of seismic data with these new filters, particularly with multitrace filters, will undoubtedly cost more than the present analog processing. These added costs must be weighed in relation to the improvement in the processed seismic data resulting from the digital processing.

The quality of seismic data

Discussion of the new mathematical processes and field digital recording cannot be separated from the very important question of the "quality" of seismic data, irrespective of the form of recording. The word "quality" here refers to the high fidelity or lack of distortion in the signal, the dynamic range of the recording, the precision of amplitude control between traces and as a function of time along the trace, and so on. There are many details of present field instruments, recorders, and playback equipment that remain to be improved. It is possible, of course, to "get by" by using present field instruments, but this may involve expensive processing of poor data, and, possibly, failure to make the best interpretation by not using the best possible data. Points to which every geophysicist must address himself, if he is to get the most out of his exploration program, are (1) the close inspection of the equipment specifications and maintenance of field instruments, (2) field operations and quality of the recorded data, and (3) the requirements of the processing steps.

One of the advantages of magnetic recording has been the ability to reproduce the original signals so that new processes, as they are developed, can be applied to the old data. However, it is impossible to get more out of a reproducible recording than went into it. So, whether or not it is desired today to use a process that requires better input data, a new process may be used in the future that may require higher quality data. It makes sense to start now to record the best possible data.

Part of this problem involves recording in as wide a frequency band as possible, consistent with the noise conditions. The time interval at which the analog signals are digitized (whether at 1 ms, 2 ms, 4 ms, etc.) inherently sets the upper frequency limit of the recorded data, and is a form of prefiltering. For example, digitizing at 4 ms intervals sets an effective upper frequency limit of the order of 100 cps. Careful consideration should be given to the choice of the digitization interval in the original field recording.

Having decided on the proper field recording instruments, the manner of recording, and the ideal digital computer, what can be done today, and what are the prospects for tomorrow?

TYPES OF MATHEMATICAL PROCESSES AVAILABLE

Digital processing can handle all of the record processing that is now being done with analog instruments, such as

A) Static and normal-moveout corrections;
B) Record edit;
C) Trace-amplitude equalization;
D) Notch rejection and band-pass filtering;
E) Harmonic analysis;
F) Correlation;
G) Convolution; and so on.

However, the main advantages of the new digital processing lie in the use of the new mathematical programs, such as

A) Velocity filtering (also called fan filtering, pie slice);
B) Wiener filtering;
 a) Horizontal stacking,
 b) Deghosting,
 c) Deconvolution,
 d) Multitrace digital filtering.

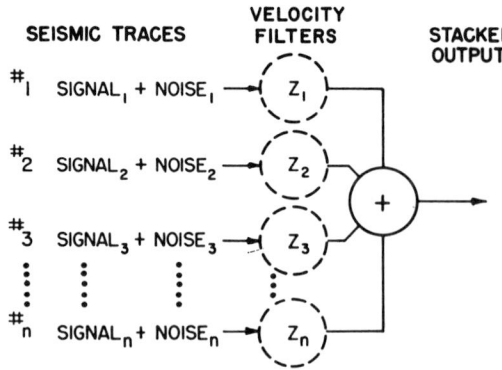

FIG. 9. Schematic illustration of velocity filtering.

Some of these processes have been described in the literature. Others have been in confidential use by geophysical service organizations. There are, undoubtedly, still others that are being used, at least on an experimental basis.

Velocity filtering

Velocity filtering, otherwise known as pie slice or fan filtering, is an optimum filtering process that can be described as a wide-frequency band multichannel filter with which coherent events on the record, which have a desired apparent moveout, are passed without appreciable alteration over a wide-frequency band, while all other events which have other moveouts are severely attenuated. (Embree, et al., 1963; Fail and Graw, 1963, 1964).

With the advent of magnetic recording and the resultant capability of playback, the adding of records (stacking or compositing) has provided a useful means to enhance signal and attenuate noise. For this purpose, the noise can be random, or coherent, but at a different moveout than that of the signal. This stacking process can be improved greatly by filtering each of the traces to be stacked (each with an appropriate different optimum filter) before adding. Each of the filters is designed for the particular spread geometry, the particular trace in the spread, and the desired range in moveout of the events on the record. This process is called velocity filtering.

This is shown schematically in Figure 9 where traces 1, 2, 3 . . . n represent the series of traces of the record to be velocity filtered, each composed of a signal component and a noise component. Each of the traces is filtered with an appropriate filter $Z_1, Z_2, Z_3, \cdots, Z_n$ and then all traces are added to provide a resultant trace. This is the first trace of the velocity filtered record. Next, traces 2 to $(n+1)$ are selected and the same series of filters Z_1 to Z_n used to provide the second filtered trace, and so on.

In designing the filters, a certain range of moveout (such as $+1$ ms to -1 ms per trace or $+2$ ms to -2 ms per trace) is chosen. All events on the record within this range of moveout are passed with nominal attenuation, while events with greater moveout are attenuated by a factor of up to 20 db.

Velocity filtering is useful in complex areas where there are many interfering events with different moveouts. If the record contains multiples of shallow primaries which have different moveouts than the deep primaries, velocity filtering may permit attenuation of the multiples without alteration of the deep primaries. It serves to minimize noise events which could not be removed by conventional methods without serious modification of signal character.

Velocity filtering is a process of compositing of adjacent depth points, and in effect is an overlap or heavy mix of adjacent traces. If any trace or group of traces have high bursts of noise, this noise energy will carry through on adjacent traces of the filtered record, and will have the moveout of the desired passband. In this case the filtered noise may have the appearance of and may be misinterpreted as signal.

Wiener filtering

Wiener filters form a broad class of digital filters which are designed on the basis of the specific characteristics of the signal and noise, in the record to be filtered. Theoretically, the Wiener filter is an optimum filter, and processing with this filter should give an output record which has the best fit to the desired record, based on the least-squared error criterion.

Under this general class of Wiener filters there are those designed for specific processes which are often identified by special names, such as filters for horizontal stacking, deghosting, deconvolution, etc.

Horizontal stacking.—The addition or stacking of signals is a good method of reducing noise. This is the basis for the use of increasingly large arrays of seismometers. However, as the size of the arrays increase, there may be partial cancellation of signal, which places a limit on the useful size of such arrays.

This limit, based on destructive interference of reflection signal, can be extended by adding or compositing signals from *common* reflecting points, rather than from *adjacent* reflecting points. This process is called common-depth-point compositing. In this process, the range from shotpoint to seismometer spread is varied, keeping the reflecting point the same. Optimistically, the reflection signal will be the same on each record while initiation and reception environments will be different for each shot and thus much of the noise on the records will be different; stacking of these records will tend to reduce the noise and thus be beneficial.

This process permits stacking records shot with widely different range without deterioration of primary reflection signals. On the other hand, as the range changes, the normal moveout of the multiple reflections changes and becomes different from that of the primaries. So this method permits stacking of records to add primaries in-phase, while (because of different normal moveout) the multiples will be added out-of-phase, and will be partially cancelled.

While stacking of the unfiltered traces gives some multiple attenuation, it is possible to design Wiener filters for each of the traces such that, when the traces are filtered before compositing, a greater attenuation of the multiples will result.

This is shown schematically in Figure 10 in which the three traces to be stacked are X_1, X_2, and X_3. The primary events P are in alignment while the multiple events M of X_2 and X_3 are delayed by 15 and 17 ms, respectively, from event M of X_1. Their unfiltered sum is shown in the trace below, labelled Simple Stack. In this trace the amplitude of the multiple event M is reduced by about 6 db. However, by filtering each trace with an appropriate Wiener filter Y, the output of the filtered traces labelled Optimum Stack shows almost 20 db attenuation of the multiple event M.

The filters Y are designed on a least-mean-square-error basis for the particular spread geometry, multiple velocity function, the particular traces used, and the record time. Claims are made that this processing (a) offers greater multiple attenuation with fewer stacked traces, (b) more effective utilization of excess multiple normal moveout, and (c) preservation of high-frequency signal content.

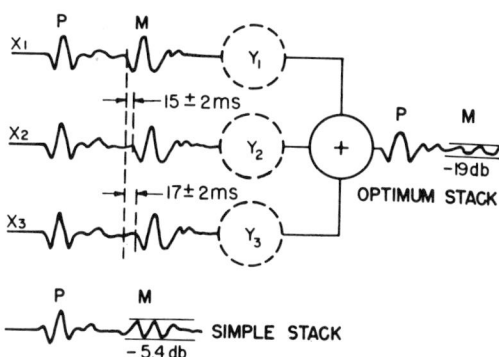

FIG. 10. Schematic illustration of optimum horizontal stacking.

Optimum horizontal stacking differs from velocity filtering in several important respects.

1) Optimum horizontal stacking composites traces having the same (common) reflecting point, while velocity filtering composites traces having different (adjacent) reflecting points.
2) The optimum horizontal stacking operators are derived on the Wiener least mean square error basis, while the velocity filter operators are generally obtained by means of a two-dimensional Fourier transformation of an assumed pass-characteristic in the frequency-wavenumber plane.

Deghosting filters.—In seismic operations where shots are fired below the surface, energy that travels upward from the shot may be reflected downward at the base of weathering or at the surface, and form a down-traveling, reflected, seismic wave. Like the original down-traveling, direct wave from the shot, this down-traveling, reflected wave will produce a series of primary and multiple reflections, similar to those formed by the direct wave, but delayed in time by approximately twice the traveltime from the shot to the upper reflectors. These delayed events are called ghosts, and complicate the interpretation of the record by (a) the increased number of events, (b) lack of information as to which events are primaries and which are ghosts, (c) interference between ghost reflection events from a shallower horizon and true primary reflections from a deeper horizon, and so on.

Ghosting may be minimized by shooting with a "distributed charge" in the field recording process, which minimizes the amplitude of the

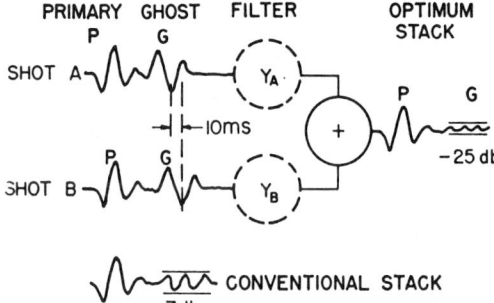

FIG. 11. Schematic illustration of deghosting.

seismic wave that travels upward from the shot. However, by taking two conventional shots at different depths in the shothole, two traces will be obtained on which the delay of the ghost events behind the primary events will be different. If these traces are placed with the primary events in time coincidence, the ghost events will be out of coincidence. Then if the traces are added, the primary events will be added in-phase while the ghost events will be added out-of-phase, and so reduced in amplitude. By designing Wiener operators for these traces and filtering them, optimum cancellation of the ghost events can be accomplished (Schneider et al, 1964).

This is illustrated with synthetic signals in Figure 11, corresponding to two shots with a difference in uphole time of about 10 ms. Simple stacking of the two traces gives the resulting trace shown below labelled Conventional Stack. For the timing and wavelets shown, the ghost event is reduced about 7 db. However, if the traces are filtered with an optimum Wiener filter Y before stacking, the resulting optimum stack shows about 25 db attenuation of the ghost event.

Deconvolution filtering.—Convolution, as discussed above, is a mathematical process in which a time function and an operator are combined in a process of multiplications and additions. The result of this process provides an output time function which is the result of filtering the input time function by the operator. This process is precisely specified. Convolution of a given time function and a given operator will always give the same answer. Furthermore, it is always possible to compute the filtered (convolved) record given the time function and the operator.

Since seismic wave signals are often filtered or convolved in an undesirable way by passage through the earth, it is desirable to be able to "de-filter" or "de-convolve" the received signals, to obtain the unconvolved signals before passage through the earth. This reverse process, or deconvolution, is of great interest to geophysicists (D'Hoeraene, 1962; Graebner et al, 1964; Rice, 1962). It is also one that is poorly understood. There is no simple, straightforward, precisely specified process for performing deconvolution of a seismic trace. Since the convolution process which selectively modifies the time function is unknown, there is no obvious way in which the deconvolution process can exactly recover the original time function. However, there are several multistep, complex, mathematical processes that may *approximate* the deconvolution.

In most of these deconvolution processes, the signal and noise content of the time function is examined by means of the autocorrelation function. From this a deconvolution operator is calculated. The deconvolution is then performed by convolving this deconvolution operator with the time function. It might seem strange to deconvolve a time function, the purpose of which is to shorten a signal wavelet, by convolving it with an operator, where convolution always lengthens a signal. While the deconvolved signal is always longer than the original signal, the deconvolution operators are designed to make the deconvolved signal appear to be shorter by distributing the energy so that the early portion of the signal contains most of the energy. Thus the later portion of the deconvolved signal is often of such a low amplitude as to be almost negligible.

In the seismic process, the geologic structure of the earth can be approximated by a series of horizontal strata each having a characteristic density and seismic wave velocity. Each of the interfaces between these strata will act as reflectors of seismic waves with reflection coefficients which depend upon the density and velocity of the strata above and below the interface. The set of reflection coefficients, spaced in time, in accordance with the traveltime through each of the strata, becomes the reflectivity function of the earth for primary reflections (Sengbush et al., 1961). If the reflection coefficients are assumed to be sharp impulses, the reflectivity function is a series of irregularly time-spaced spikes of varying magnitudes and polarities. If the shot pulse or wavelet is assumed to be time invariant, a synthetic record, with primary reflections only, can be provided by convolving the shot pulse with the reflectivity function.

The objective in seismic exploration is to obtain this reflectivity function of the earth. But what is recorded, is something that approximates the reflectivity function *convolved* with the shot pulse (plus a number of additional events such as multiple reflections). The advantage of deconvolution (if it could be carried out), would be to substitute, for the field record with its conventional reflection wavelets, representing the presence of reflection interfaces, a record substituting shorter, simpler events or spikes for the conventional reflection wavelets. This type of record will permit higher resolution interpretations than are possible with longer wavelets. Even an *approximate* deconvolution may shorten the reflection wavelet, and be helpful in resolving closely spaced reflections. This is particularly important in exploration for stratigraphic traps.

Deconvolution is important in another area also, where the normal reflection wavelets are lengthened by ringing or reverberation. This is very common in offshore operations where the water acts as a reverberation layer, trapping the seismic energy. Long reflection reverberation wave trains may form and, by overlapping each other, make it difficult to detect the separate reflections.

The deconvolution process inherently broadens the frequency content of the signal by selectively amplifying the low amplitude frequencies. This requires that the record be recorded originally in broad frequency band, with adequate dynamic range and low noise content.

Multitrace digital filtering—One of the real advantages of the seismic process is the ability to use receiver arrays, and to record the signals on each of the detector groups in the array. These multiple detector (multiple trace) records can be used to determine the direction of arrival of waves, and to discriminate against unwanted waves arriving from certain directions. This is done by the geophysicist in the conventional visual analysis of record sections. It is also made use of in the velocity filtering process, to reject certain waves based on direction of arrival, as represented by the moveout across a group of traces.

All of the recorded traces have information regarding both the signal and the noise, and it is important to utilize all of this information.

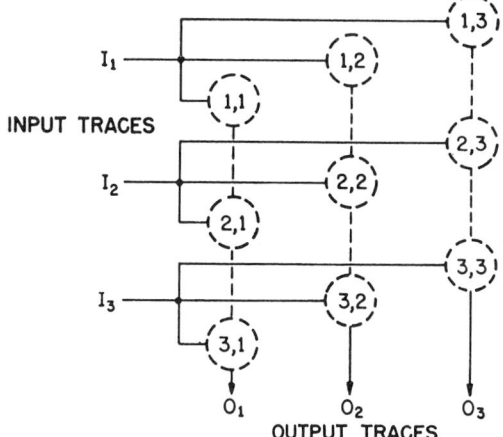

Fig. 12. Schematic illustration of multitrace wiener filtering.

Multitrace Wiener filters are designed to do this. (Foster et al., 1964.) Consider that there are n traces. In the most general case, each trace is filtered separately through n different filters, say $(1, 1)$ $(1, 2)$, $\cdots$, $(1, n)$ for trace 1; $(2, 1)$, $(2, 2)$, $\cdots$, $(2, n)$, for trace 2, and so on. Then when the outputs of the n^2 different filters are combined, the filtered output of each trace will have contributions from each of the other traces. This is illustrated in Figure 12 for the case of three traces. Here the input traces, identified as I_1, I_2, I_3, are each filtered with three filters, and the outputs of the nine filters are combined into three groups of three each, so as to include in each output trace O_1, O_2, O_3 contributions from each of the three input traces.

This type of mutlitrace filter in its most general form is expensive in computation time. It requires that n^2 individual filter operators be calculated. And, each of these operators must be convolved with an appropriate trace. However, it is possible to use fewer filters (one for each trace instead of n) as illustrated in Figure 10. While poorer than the more general form, this multitrace filter is more effective than single trace filters. Theoretically, multitrace filters can provide the optimum filtering possible. Since it costs so much to gather field data, serious consideration should be given to using these multitrace filters in spite of their cost.

CONCLUSIONS

In conclusion, it is well to point out that the real value of digital processing lies in the processes and not in the hardware. While their usefulness has not been fully evaluated, there is good probability that these new processes, such as Wiener filtering, deconvolution, and multitrace filtering, will provide better information about subsurface structure than are now available. Work will undoubtedly continue on the development of new criteria for designing filter operators which may be more effective than those now in use. Also new techniques will be provided such as recursive filtering, the Cooley-Tukey Algorithm, and others, that will permit more rapid (and less costly) data processing.

However, it is well to sound a note of caution! These new processes are no panacea! Statistical processes require great amounts of data, and in seismic operations, data are difficult and expensive to collect. For the same data, however, the new processes should provide more information than the conventional analog processing.

Nothing has been said about other attractive possibilities of using the computer in seismic processing. One of the main advantages of the digital computer is that it can be completely programmed to the most minute detail. Thus, once the programming is complete and correct, the subsequent processing always follows the program without error. On the other hand, in manual operations, there is always the possibility of error in setting of switches, insertion of records, and so on.

As more knowledge is gained by geophysicists about the interpretive processes, they may be able to specify the steps by which decisions are made, and so design an ideal model of the interpretive process. It may then become possible to use the computer to make interpretive decisions which can be automated and the process of interpretation speeded. Thus, several interpretations can be made, each on the basis of different assumed criteria. In this way, the best interpretation can be found. Another use of the computer involves the storage of subsurface data in the computer memory and the plotting and contouring of maps based on different assumed conditions.

Even without these longer range developments, the present mathematical processing will undoubtedly help the geophysicist by providing more and better subsurface information from which he can derive a better picture of the subsurface geology.

REFERENCES

Backus, M. M., Burg, J. P., Baldwin, D., and Bryan E., 1964, Wide band extraction of mantle P waves from ambient noise: Geophysics, v. 29, p. 672–692.
Cooley, J. W., and Tukey, J. W., 1965, An algorithm for the machine calculation of complex Fourier series: Mathematics of Computation, v. 19, p. 297, 301.
D'Hoeraene, J., 1962, Deconvolution de traces realles: Geophys. Prosp. v. 10, p. 68–83.
Embree, P., Burg, J. P., and Backus, M. M., 1963, Wide band velocity filtering—The pie slice process: Geophysics, v. 28, p. 948–974.
Fail, J. P., and Grau, G., 1963, Les filtres en eventail: Geophys. Prosp., v. 11, p. 131.
Fail, J. P., Grau, G., and Layotte, P. C., 1964, Amelioration du rapport signal/bruit a l'aide du filtrage en eventail: Geophys. Prosp., v. 12, p. 258–282.
Foster, M. R., Sengbush, R. L., and Watson, R. J., 1964, Design of suboptimum filters for multitrace seismic data processing: Geophys. Prosp., v. 12, p. 173–191.
Graebner, R. J., and Prince, E. R., 1964, Marine seismic systems: Paper presented at the 34th Annual International SEG Meeting, Los Angeles, California.
Kunetz, G., 1961, Essai d'analyse de traces sismiques: Geophys. Prosp., v. 9, p. 317–341.
Rice, R. B., 1962, Inverse convolution filters: Geophysics, v. 27, p. 4–18.
Robinson, E. A., and Treitel, Sven, 1964, Principles of digital filtering: Geophysics, v. 29, p. 395–404.
Schneider, W. A., Larner, K. L., Burg, J. P., and Backus, M. M., 1964, A new data processing technique for the elimination of ghost arrivals on reflection seismograms: Geophysics, v. 29, p. 783–805.
Sengbush, R. L., Lawrence, P. L., and McDonal, F. J., 1961, Interpretation of synthetic seismograms: Geophysics, v. 26, p. 138–157.
Shanks, John L., 1967, Recursion filters for digital processing: Geophysics, v. 32, p. 33–51.

Seismic data display and reflection perceptibility

Frank J. Feagin*

ABSTRACT

Relatively little attention has been paid to the final output of today's sophisticated seismic data processing procedures—the seismic section display. We first examine significant factors relating to those displays and then describe a series of experiments that, by varying those factors, let us specify displays that maximize interpreters' abilities to detect reflections buried in random noise.

The study.—From psychology of perception and image enhancement literature and from our own research, these conclusions were reached: (1) Seismic reflection perceptibility is best for time scales in the neighborhood of 1.875 inches/sec because, for common seismic frequencies, the eye-brain spatial frequency response is a maximum near that value. (2) An optimized gray scale for variable density sections is nonlinearly related to digital data values on a plot tape. The nonlinearity is composed of two parts (a) that which compensates for nonlinearity inherent in human perception, and (b) the nonlinearity required to produce *histogram equalization*, a modern image enhancement technique.

The experiments.—The experiments involved 37 synthetic seismic sections composed of simple reflections embedded in filtered random noise. Reflection signal-to-noise (S/N) ratio was varied over a wide range, as were other display parameters, such as scale, plot mode, photographic density contrast, gray scale, and reflection dip angle. Twenty-nine interpreters took part in the experiments. The sections were presented, one at a time, to each interpreter; the interpreter then proceeded to mark all recognizable events.

Marked events were checked against known data and errors recorded. Detectability thresholds in terms of S/N ratios were measured as a function of the various display parameters. Some of the more important conclusions are: (1) With our usual types of displays, interpreters can pick reflections about 6 or 7 dB below noise with a 50 percent probability. (2) Perceptibility varies from one person to another by 2.5 to 3.0 dB. (3) For displays with a 3.75 inch/sec scale and low contrast photographic paper (a common situation), variable density (VD) and variable area-wiggly trace (VA-WT) sections are about equally effective from a perceptibility standpoint. (4) However, for displays with small scales and for displays with higher contrast, variable density is significantly superior. A VD section with all parameters optimized shows about 8 dB perceptibility advantage over an optimized VA-WT section. (5) Detectability drops as dip angle increases. VD is slightly superior to VA-WT, even at large scales, for steep dip angles. (6) An interpreter gains typically about 2 dB by foreshortening, although there is a wide variation from one individual to another.

INTRODUCTION

Seismic data display has evolved over the years from a few galvanometer traces on a strip of photographic paper, produced in the field in the early 1930s, to today's corrected and processed section presentation produced from digital tape. Advanced data gathering equipment and techniques and sophisticated data processing must be credited for much of the improvement we see in our data, but another major advance from the interpreter's point of view has been the form of the display itself. The potency of our present section displays lies in the way that individual time functions are laid out spatially so that relationships or coherence between display elements can be readily perceived by the interpreter.

There was a quantum jump in seismic display effectiveness when the section presentation was introduced in the 1950s, but it is the author's opinion that we have not paid as much attention to our displays since that time as we should have[1]. This is not to say that there has not been progress, but usually the improvements have been measured in terms of increased plotting speed, lower cost, or the ease with which a display could be duplicated. We have not made the research effort required to understand the psychophysics of displays. Some understanding is necessary before logical steps can be taken toward enhancing the human-perceived effectiveness of seismic sections.

There are reasons why we have not aggressively pursued display research. It is difficult to quantify an improvement in display quality. Most of our display appraisals have been qualitative (and rather subjective) in the past. How can we know for sure that one type of display is superior to another—and under what conditions? We are dealing with a system which includes not only the display

---
[1] Some early reflection detectability work was reported by Dyk and Eisler (1951) and by Kim and Dyk (1965).

Manuscript received by the Editor January 20, 1980.
*Formerly Exxon Production Research Co., Houston; presently 7 Beavertail Point, Houston, TX 77024.
0016-8033/81/0201—106$03.00. © 1981 Society of Exploration Geophysicists. All rights reserved.

but the human eye and brain as well, a perceptual system which must be analyzed and evaluated as a whole. We have been reluctant to do research in this area because meaningful perception experiments are difficult to design, carry out, and interpret. We have not felt qualified.

Another reason we have avoided display research is that plotters with enough flexibility and control for doing research have not been available. Many of our seismic displays are relatively large, and at the same time, very high spatial resolution is essential—a combination not required for most other display applications and a combination demanding a very precise (and expensive) plotter. For display experimentation, a plotter must also be capable of a variety of modes, scales, and transfer functions.

The time now seems appropriate to commence some research on seismic displays. For one thing, several precision plotters have recently become available which can produce small or large plots with the desired resolution and other attributes needed for seismic display research. Also, largely because of research in other areas, there has evolved in recent years new knowledge in image enhancement, psychology of perception, pattern recognition, and human engineering—all of which deal with ideas having some relevance to our seismic display system (section presentation plus eye-brain). It seems reasonable to expect at least some useful fallout from these developments.

The work reported here concerns itself with effectiveness of displays from the standpoint of reflection detectability only. Seismic sections are used by interpreters in many ways. Sometimes *seeing* the reflections is not the problem; putting the reflections into geologic context is often the greater challenge. A display optimized from a detectability standpoint may not, in all cases, be the best for some other interpretation requirements. This limitation should be kept in mind when considering conclusions drawn from this research. But the importance of detectability cannot be denied; if a reflector is not seen, it is not likely that much will be done with it.

The research can be divided into two parts: (1) An initial study aimed at gaining some understanding of the significant perceptual factors involved in our presently used seismic displays and optimizing them accordingly; and (2) a series of experiments designed to measure quantitatively the actual performance of seismic interpreters, using synthetic sections of various kinds, as a means of evaluating displays and display parameters on a practical and statistical basis.

INITIAL STUDY

One basic factor to be considered in connection with a seismic section is the matter of its size, i.e., its time scale in inches/sec and its horizontal scale in traces/inch. Should there be, from a technical standpoint, any perceptual advantage of one scale over another?

The curve shown in Figure 1 can be found in a number of psychology textbooks. It shows the spatial frequency response of a normal human eye-brain at levels of illumination commonly used in interpretation offices. The horizontal axis is marked in cycles per degree. The cycles per degree measure is the number of spatial cycles subtended by a 1-degree angle at the viewer's eye, from whatever distance the display is viewed.

A typical 25-Hz seismic reflection on a 1.875 inch/sec scale, held at 20 inches from the eye, corresponds to about 4.6 cycles/degree. The curve shows near maximum response at that spatial frequency. At a 3.75 inch/sec scale, the response is slightly down on the low-frequency side of the peak.

The curve, therefore, predicts that these commonly used scales should be near optimum from the standpoint of perceptibility and

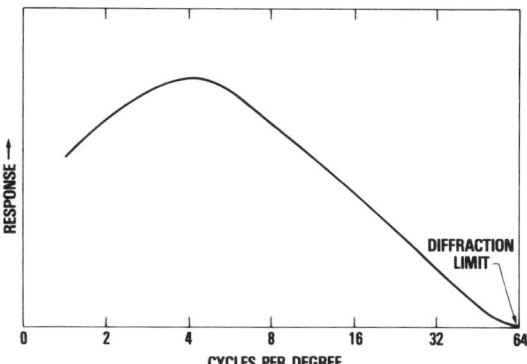

FIG. 1. Spatial frequency response of the human eye-brain.

suggests that there might be some preference for the smaller scale. Larger scales, of course, are more useful for detailed study of already detected reflections. Certainly we need more than one or two scales available. It is interesting to note that the shape of the curve in Figure 1 is similar to the spatial frequency response curve of an ordinary spherical lens. This is not too surprising, since part of the human eye *is* a lens. But an ordinary lens does not have a peak with response dropping at low spatial frequencies. A lens response increases all the way to zero frequency. The drop seen in Figure 1 at low frequencies is not the result of deficiency in the human eye lens; the curve shows the combined response of the eye *and* the brain. According to modern psychology literature, the very low spatial frequencies undergo a loss somewhere in the communication link between the eye and the brain.

At the high-frequency end, the curve of Figure 1 can be seen to go to zero at about 60 cycles per degree. This is the diffraction limit for a normal eye. Section scales, of course, must never be made so small that details of significance in the data, such as higher frequency components, approach the diffraction limit.

Section size is also determined partially by the horizontal scale, usually expressed in traces per inch. The important spatial bandwidth needed for good perception in the horizontal dimension extends from the trace repetition frequency at the upper end (perhaps a bit higher to accommodate a few harmonics) down to as low frequency as is consistent with the high-frequency requirement. The decrease in eye-brain response at the extreme low frequencies is undesirable but is not seriously detrimental to perception in the case of seismic data.

The horizontal scale should be related to the time scale of a section. The optimum aspect ratio appears to result when individual trace width is about equal to a half-wavelength of the predominant frequency in the data. At typical seismic frequencies, this criterion results in a product of time scale and horizontal scale of about 95 (inches per second multiplied by traces per inch). The relationship is not critical, however, because deviation by ±30 percent has little effect on perceptibility.

Another display parameter considered worthy of study is the gray scale used on variable density sections. Just what should be the relationship between instantaneous amplitudes in the seismic time function and the spatially distributed shades of gray on the section?

In the past, we tended to accept the inherent transfer characteristic of the section plotter being used, modifying it only to the limited extent possible through adjustment of plotter gain or manipulation of photographic variables. Gain and photography were

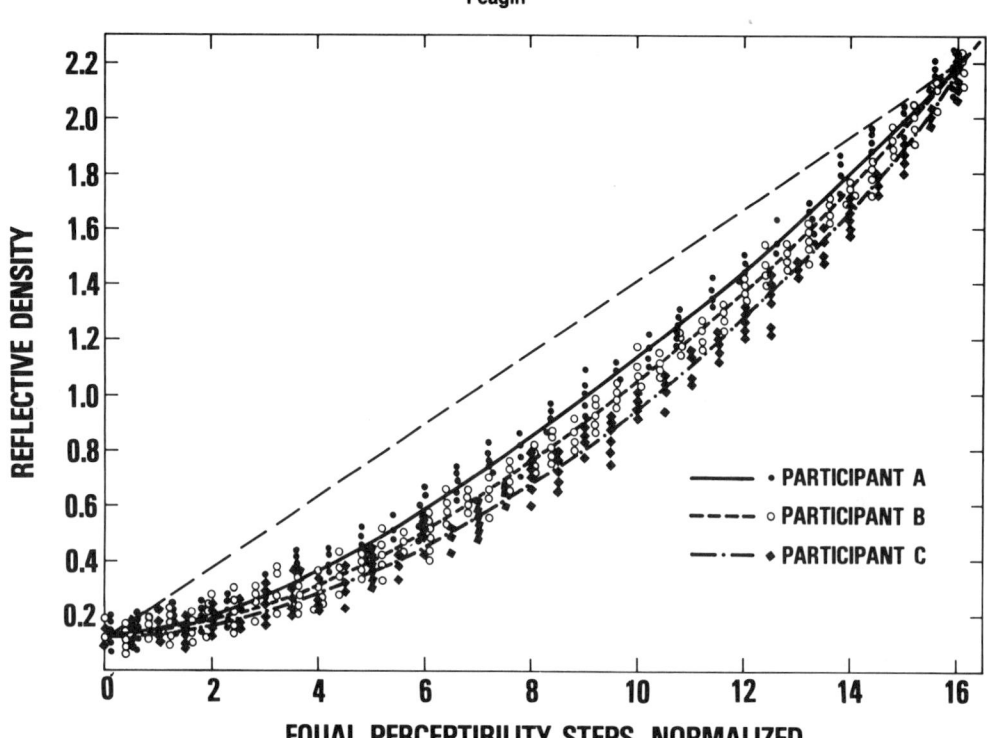

FIG. 2. Gray scale perceptibility curve.

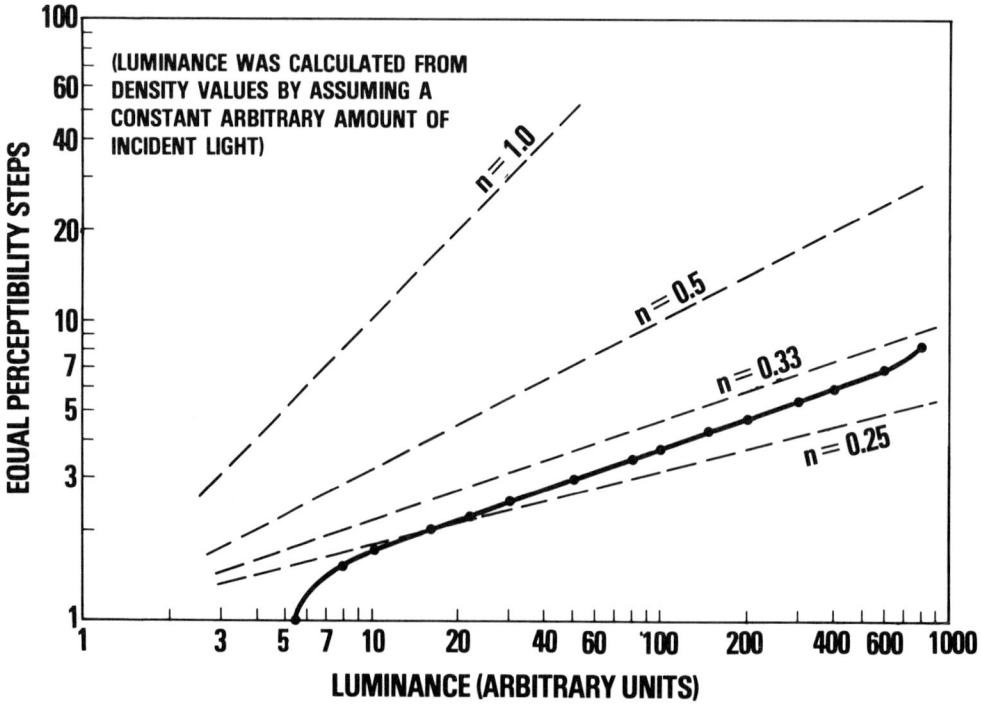

FIG. 3. Fit of experimental gray scale data to Steven's law ($R = kS^{0.33}$).

usually adjusted to produce displays most pleasing to the eye or set in such a manner that interpreters felt the display to be most interpretable. This was a practical and effective, if not very scientific, procedure.

Measurements with a microdensitometer on a number of our variable density sections showed that the overall transfer function which we typically use is a very nonlinear one. We then set about to identify the several kinds of nonlinearities present and distinguish hardware-imposed nonlinearities from those which might have some human perceptual basis. We would like to develop an optimum transfer function. One might well ask, "Why shouldn't the transfer function be linear?" It is a good question, since we try at every point in the seismic system to preserve linearity. But since the human perceptual apparatus is an essential part of the display system, we must ask, "What does the word linear mean in this context? Just what does a human perceive as being a linear or equal-step gray scale?"

Ernst Heinrich Weber, physiologist, and Gustave Theodor Fechner, physicist (both of Germany) concluded as a result of psychophysical research in the mid-19th century that the various human senses respond approximately proportionally to the logarithm (log) of their respective stimuli. It would seem logical, therefore, to make reflective density a linear function of signal amplitude, since density is the logarithm of the reciprocal of the reflection coefficient. The conclusions of Weber and Fechner have been challenged by Stevens (1951), Beardslee and Wertheimer (1958), and Hurvich and Jameson (1966). Stevens' work shows that human responses are more nearly proportional to a power function of the stimulus

$$R = kS^n$$

and that in the case of luminance, exponent n is equal to $1/3$. (The k is an assumed constant amount of incident light in arbitrary units.)

We carried out an experiment of our own to determine a linear gray scale as perceived by humans. A large number of shades of gray were photographically produced on small cards. A range of reflective density from 0.1 to 2.2 was covered in very small steps with some intentional duplicates. Eight participants were enlisted, and each was asked to arrange the cards in order of increasing density. Whenever differences were undetectable, the participant was asked to group those cards together. A background (table top) of intermediate density was used, and the experiments were carried out under average office illumination.

Results of the experiments are shown in Figure 2. Curves for only three participants are shown in order to keep the graphs readable. The three curves are typical. One concludes that although there are measurable differences between individuals, there is also much similarity in their responses. It is obvious, too, that a linearly perceived gray scale is *not* made up of equal density steps.

An average of the curves in Figure 2 was replotted in Figure 3 on log-log coordinates and with density converted to luminance as follows:

$$L = K \times 10^{-D}.$$

The K is an assumed constant amount of incident light in arbitrary units; it has no effect on the shape or slope of the curve in Figure 3. The graph in Figure 3 can be seen to have a slope of approximately $1/3$ (0.33) for most of its length, in good agreement with Stevens' (1951) findings.

Several synthetic sections were plotted using an overall transfer characteristic such that signal amplitudes were linearly converted to the equal perceptibility steps of Figure 2. The sections were "washed-out" in appearance, and detection of synthetic reflections

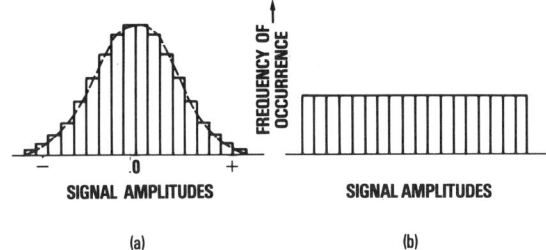

FIG. 4. Histogram equalization.

in random noise was very poor. This result led us to try to explain why we have a natural tendency to plot variable density sections with gain advanced to the point that some degree of overdrive, saturation, or peak clipping results. Our explanation follows.

One rather effective image enhancement technique that has been used to process pictures transmitted for great distances through space is called *histogram equalization*. Histogram equalization is a process whereby each picture element (pixel) value in an image is nonlinearly transformed to a different specific value in such a way that the frequency of occurrence of each incremental range of transformed pixel values is the same for all ranges in the processed image.

To carry out histogram equalization on a seismic section, of course, one must know the distribution of signal amplitudes in the data to be plotted in order to convert to shades of gray on a section having a flat distribution. Most seismic digital data when linearly recorded and processed will have a distribution of digital values (or histogram) which is bell-shaped, approximately normal, or Gaussian. Figure 4a shows such a curve. By means of a nonlinear transfer function, distribution of digital values on the tape can be transformed into digital values on another tape so that the new distribution is boxcar-shaped as in Figure 4b, thus equalizing the histogram. The transfer function required to histogram-equalize data from a Gaussian distribution to a boxcar distribution is shown in Figure 5. (The transfer function can be shown to have the shape of the integral of the Gaussian curve.) When the equalized tape is plotted, the plotter transfer function must be designed so that digital

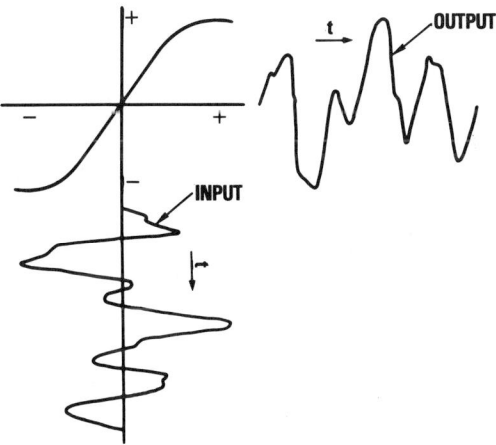

FIG. 5. Transfer function for histogram equalization.

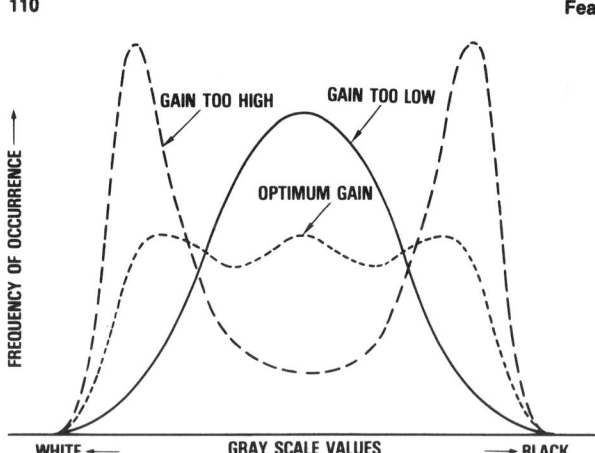

FIG. 6. Effect of plotter gain upon VD.

data values are plotted linearly proportional to equal-perceptibility gray scale steps.

But one might well ask, "Why should a redistribution of pixel values on a display result in any enhancement? Why would it make a weak reflection any easier to see?" Our seismic data have the majority of data samples falling in the range near zero as shown by the Gaussian histogram, with fewer samples near the maximum negative and maximum positive values. It seems logical, therefore, to stretch the display gray scale in the midrange, thus providing increased amplitude resolution (or dynamic range) for the large number of data values near zero at the cost of less amplitude resolution for the less frequently occurring large values. Admittedly we gain something and we lose something when we histogram equalize, but the net effect is a gain; the available gray scale steps are more efficiently used. A more technically accurate treatment of histogram equalization from an information theory viewpoint can be found in Rosenfeld and Kak (1976, sections 5.4.1 and 6.2.3).

Several tapes of synthetic data were made up with reflections of various strengths embedded in filtered random noise. A computer program was written for performing histogram equalization. It was applied to the synthetic data, and the sections were plotted. Before plotting, a nonlinear transfer function was set into the plotter to take into account the nonlinear perceptibility curve of Figure 2. Other nonlinearities in the plotter hardware and photographic parameters were taken out. The plotter which was used is convenient for this kind of experimentation; any transfer function can be set in which it can be adequately represented by 16 points, or shades of gray.

The histogram-equalized sections were definitely more effective than the ones plotted with perceptually linear transfer functions. Tests with interpreters showed them to be about 10 dB more effective.

FIG. 7. Typical synthetic section (VA-WT).

Similar sections were then plotted without using the histogram equalization program but with the plotter transfer function set to approximate a typical plotter characteristic, where the shape of the curve is determined largely by the density-log-exposure curve of the film or photopaper. Several different settings of plotter gain were used. Figure 6 shows the effect of various plotter gains upon the gray scale distribution histogram. It can be seen that if gain is kept low enough, the transfer characteristic is fairly linear; the distribution comes out essentially as it goes in—it is Gaussian. If gain is set very high, a double-peaked distribution results, which means that most of the pixels are either black or white with relatively few in the mid-grays. If gain is carefully adjusted, however, a fairly flat distribution results as shown in Figure 6. This is a reasonable approximation to histogram equalization.

We made tests using computer-produced histogram equalization and using controlled-gain overdrive. We found no measurable difference in weak signal detectability. It should be emphasized, however, that for controlled-gain overdrive to be effective, the gain must be properly set. The average absolute value (AABV) for the data in the section is a convenient guide. We have found that, if gain is set so that only those negative and positive data values greater than twice the AABV value are overdriven, the distribution is quite good. The AABV criterion is only a general guide; there are instances, of course, where more or less gain is desirable in order to enhance some special part of a section.

As a result of these tests, we concluded that an optimum transfer function for a variable density (VD) section is one which deliberately inserts two kinds of nonlinearities: (1) that needed to convert the usual Gaussian amplitude distribution into a histogram-equalized one, and (2) that required to force the plotter to plot in terms of equally perceptible gray scale steps. Any other nonlinearities that are inherent in the plotter hardware and photography should be taken out or taken into account when combining the other nonlinearities.

It is difficult to be as analytic with VA-WT sections as with variable density sections. Two parameters in addition to scale were investigated for VA-WT sections. Tests were made to determine what portion of the seismic wiggle should be blacked-in and to determine optimum plotter gain. Consider an arbitrary scale of from -100 through 0 to $+100$ to indicate the blacked-in part of the signal, with zero being the point where half of the waveform is black. With this scale we find a broad optimum extending from -20 to $+20$. On the matter of plotting gain, we found by experiment that the gain is optimized when amplitudes larger than about 2.5 times the AABV value for the section were overdriven.

PERCEPTIBILITY EXPERIMENTS

The second part of the display research consisted of a series of experiments, involving synthetic sections, designed to allow an evaluation of various display parameters as they affect an interpreter's ability to detect weak reflections. The objective was to get answers through experimentation to the following questions per-

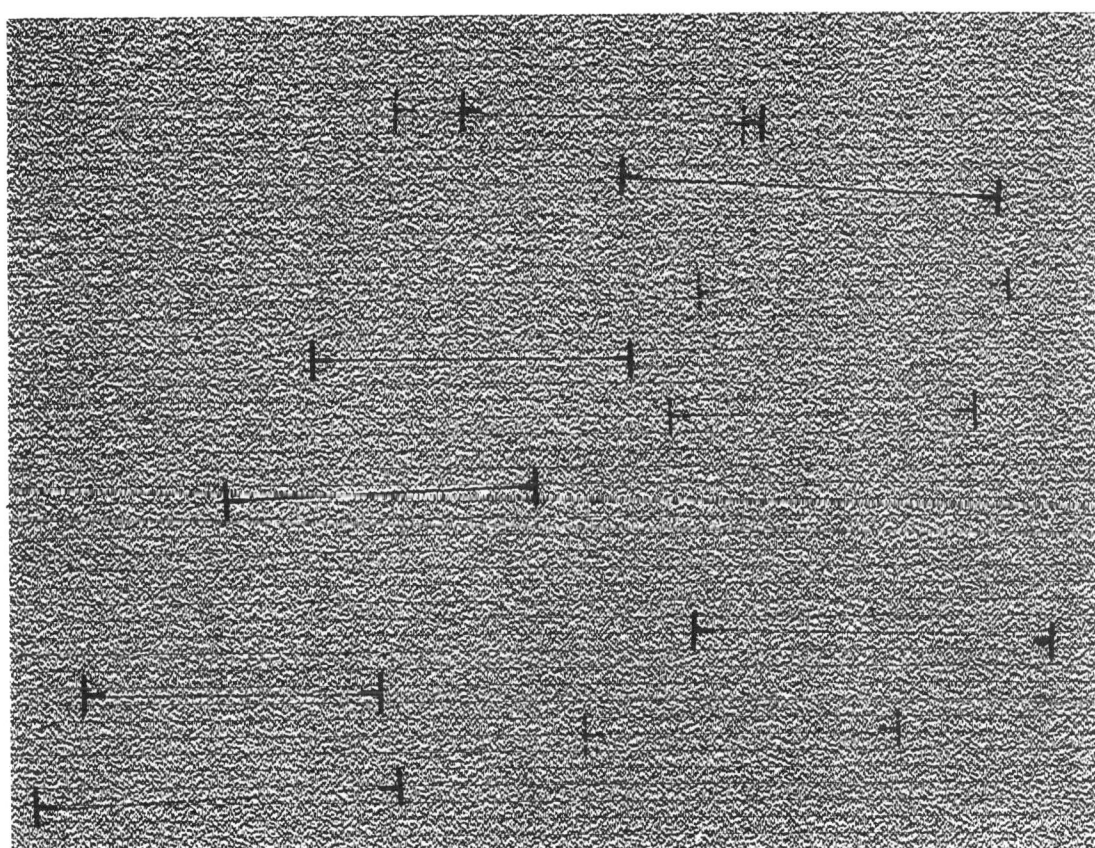

FIG. 8. Section of Figure 7 as marked by an interpreter. The dashed marks indicate foreshortening.

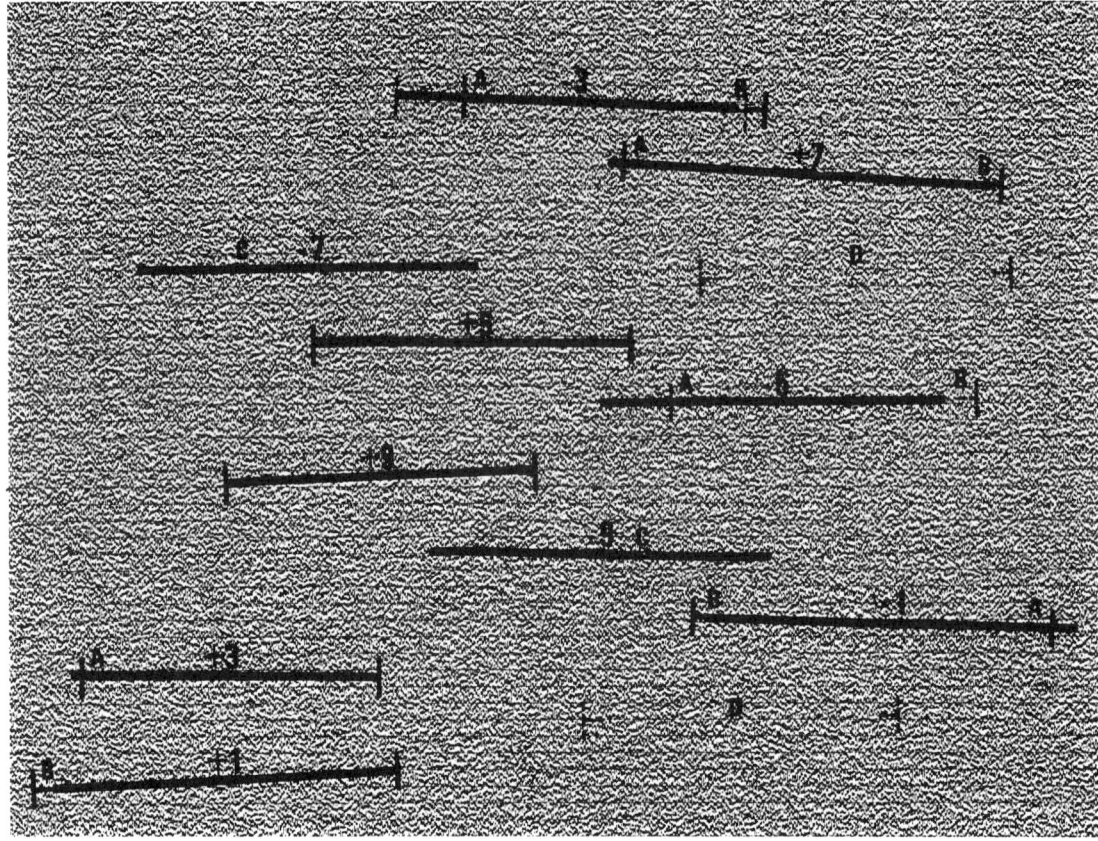

FIG. 9. Marked events compared with actual reflections.

taining to seismic data displays on a statistically meaningful scale:

(1) How weak can a reflection be and still be detectable? That is, how far down into noise can an interpreter be expected to see when using one of our common types of displays? A related question is: How much must the S/N ratio of a reflection be improved to change from a low to high probability of detection?

(2) How much perceptibility variation is there between interpreters?

(3) From a reflection perceptibility standpoint, which is the better display mode, variable density or VA-WT? This is an old question and, at one time, a rather hotly debated one among interpreters.

(4) From a practical standpoint, how does section scale affect detectability?

(5) Does density contrast on a display matter? Does it matter how wide the range of reflective densities is—how black the blacks are, how white the whites?

(6) It is common practice for an interpreter to view a seismic section horizontally at a sharp angle in order to enhance the detectability of weak events. How much does this technique, referred to as foreshortening, help?

(7) What is the effect of reflection dip angle upon detectability? Are steeply dipping events more difficult to see? If so, how much more difficult?

(8) Do time lines interfere with detectability?

(9) Does signal polarity make any difference?

(10) Some interpreters prefer sections plotted with a vertical white trace placed every 24 traces on the section so that the data can be more closely related to time lines. Does the practice of gapping interfere with detectability?

We spent some time designing a suite of synthetic sections which would yield a maximum of information with a minimum of each participant's time. A total of 37 sections were made up of simple (second-derivative Gaussian) reflection wavelets embedded in filtered, random additive noise. The frequency spectra of the wavelets and of the noise were made the same. Straight-line reflections were randomly placed on the sections, varying in length and dip, with each section being different except for two sections that were made alike, as a check on each interpreter's consistency. Various section scales, modes, and density contrasts were used. The sections typically contained 10 reflections each, with S/N ratios varying over a wide range.

A total of 29 interpreters participated in the experiments. Each was asked to mark on a clear film overlay all of the coherent events he or she could see on each of the 37 sections. The participant was asked first to mark without foreshortening, then, having done so, to mark additional events with a different colored pencil making use of foreshortening. No time constraints were imposed. The interpreters took from 4 to 6 hours each to mark all of the sections.

The synthetic sections were not very realistic in a number of ways. There were no curved or discontinuous reflections, no

FIG. 10. Synthetic VA section with steep dips with the S/N ratio held constant.

simulated geology, no coherent noise (such as multiple reflections), and no variation in frequency spectra as a function of seismic time. Considerable thought was given to creation of these data before deciding upon the rather simplified sections. We felt that it was very important that we be able to put numbers on some of the display parameters. Sections made to be more realistic seemed likely also to yield more qualitative results.

Figure 7 shows a typical synthetic section. This happens to be a VA-WT section. There are 10 reflections on the section. The display was plotted to a scale of 3.75 inches/sec and 25 traces/inch. This particular section has reflections ranging in S/N ratio from −9 to +9 dB. Some can be easily picked, some cannot. Signal-to-noise ratio is defined, for purposes of these experiments, to be the ratio of the maximum peak amplitude of the wavelet to the root-mean-square (rms) value of the noise, expressed in decibels.

One interpreter marked on an overlay the events on this section[2] as shown in Figure 8. He was asked to mark only the ends of the reflections. He was given the information that all reflections were straight lines but not necessarily flat or of the same length. The solid marks show which reflections were seen without foreshortening. The dashed marks show events marked when the participant made use of foreshortening.

[2] The reader will understand that Figures 9 through 11 are intended only to illustrate the kind of synthetic sections used in the experiments. Because of the usual deterioration caused by reproduction and reduction in size, one cannot expect to make meaningful tests of perceptibility from these printed figures.

When the data were brought back to the lab for analysis, another overlay was superimposed which showed exactly where the reflections were located and showed the strength of each in terms of S/N ratio (see Figure 9). All errors were measured and taken into account in the analysis of the data. It should be noted that four different kinds of errors were made. Errors marked A are examples where the interpreter marked reflection terminations too short, i.e., he did not perceive the coherence extending as far as it really did. Errors marked B were made when reflections were marked too long. Errors marked C are instances where the interpreter missed events altogether. Those marked D are cases where the interpreter indicated a reflection that was not there, or more correctly, that was not intentionally put into the data. There are instances, of course, even in filtered random noise (no reflections present) where straight-line coherence will persist for short distances on a section.

Figure 10 shows an example section that was used to measure the effect of steep dip angles on detectability. There were also other steep dip sections where reflections were randomly placed rather than with dip angle increasing in an orderly way. All of the reflections in Figure 10 have the same S/N ratio, although it does not appear that way because of the rapid decrease in perceptibility with increasing dip angle. Figure 11 shows the reflection picks made by one interpreter. The straight lines on the section indicate where all of the reflections are located. The lines have been displaced slightly to the left in order not to cover the reflections.

Figure 12 shows an example of one kind of graph made for each set of display parameters and for each interpreter. The graph shows

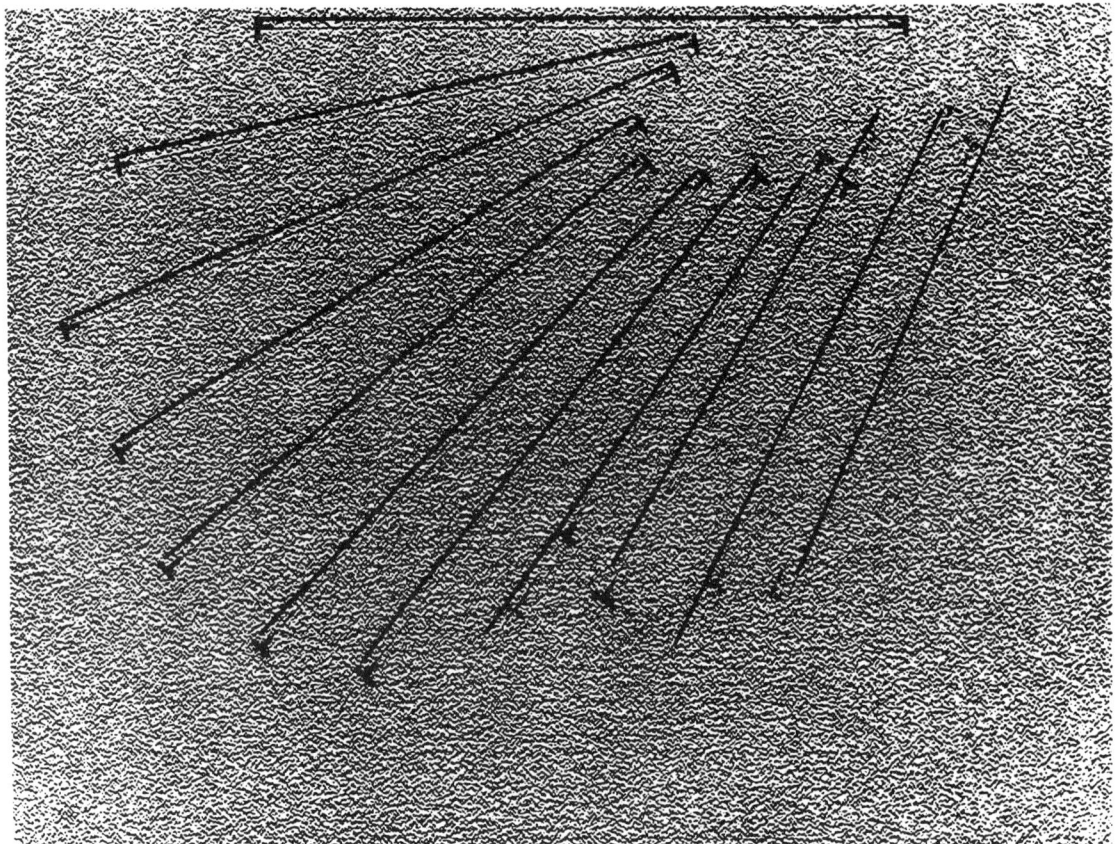

FIG. 11. Section of Figure 10 showing interpreter picks compared with actual reflections.

the increase in magnitude of errors made by the participant as he progressed from strong reflections toward weaker and weaker ones. Errors were simply measured in inches. The absolute magnitudes of errors were plotted, i.e., an error was considered an error regardless of whether the end of a reflection was marked too long or too short. For this set of parameters and for this interpreter, the threshold of detection can be seen to be about −6 dB.

Three things seem to happen almost simultaneously as the interpreter approaches the rather sharply defined threshold of Figure 12. All types of errors described previously become evident, i.e., errors (both positive and negative) abruptly get very large. Reflections weaker than the threshold are not perceived, and although it does not show in Figure 12, the interpreter also begins to mark events that are not there.

For each interpreter and for each set of display parameters, a graph such as Figure 13 can also be made which takes into account all types of errors and shows the interrelationships between the S/N ratio of a reflection and the probability of picking it with various degrees of accuracy. One significant conclusion that can be drawn from Figure 13 is that for reasonably high probabilities of detection, a small improvement in S/N ratio can improve accuracy tremendously. This points up the importance of even small improvements in S/N ratios through data processing or other techniques.

The threshold figure that we derive from this graph is the result of a combination of perceptual factors that are in operation. One factor is the effectiveness of the display. This, of course, is the factor in which we are primarily interested. A second factor is the individual perception capability of the particular interpreter and varies from person to person. A third factor is the aggressiveness of the interpreter. This last statement requires some explanation.

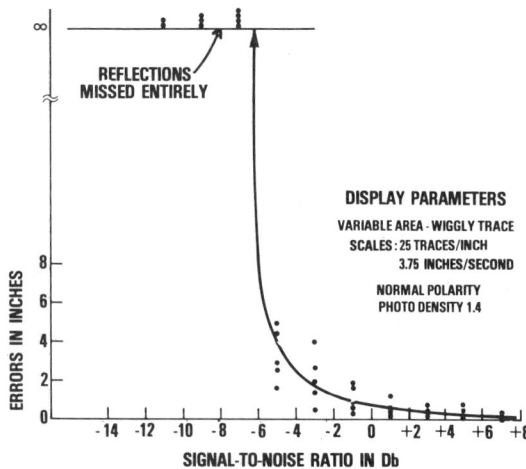

FIG. 12. Typical error curve.

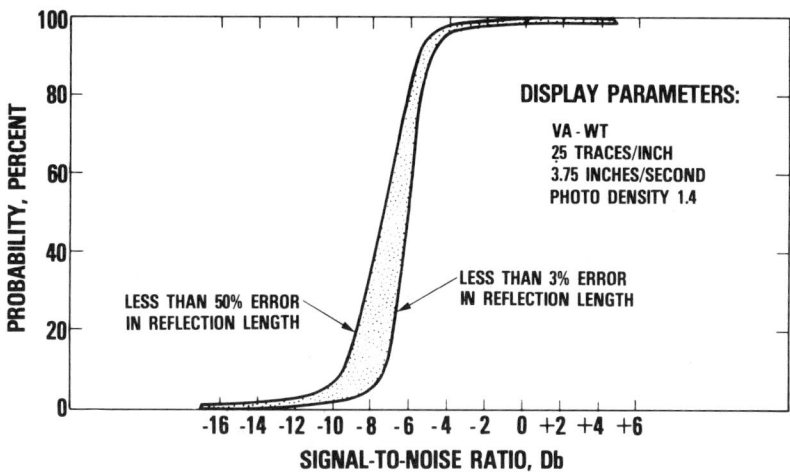

FIG. 13. Probability of reflection detection (typical performance).

We have found that each interpreter has a different criterion or standard as to when a weak event should or should not be marked as a reflection. Some people are conservative; when the judgment is left to them, they will not mark anything that is doubtful. Others are more aggressive—perhaps optimistic is a better word —and will mark a number of events that are questionable, perhaps even in their own minds. In a real life exploration situation, of course, the degree of aggressiveness may depend upon a number of external conditions or pressures which bear on the interpreter such as the quality of reflections in the area or the economic importance of the prospect. Since we are interested primarily in display effectiveness and not as much interested in individual capability or aggressiveness, we looked for a way to remove the effects of individual variables or to normalize the data.

Figure 14 illustrates one of two normalizing procedures that was used. This graph was used to normalize aggressiveness. The number of events marked is plotted against the number of correctly marked ones for a particular set of display parameters and for each interpreter. Since there is a range of S/N ratios on each group of sections, the larger the number of events picked correctly, the better is that person's threshold. The relationship between the number picked correctly and the S/N ratio is shown on the vertical axis.

The ideal interpreter would follow the heavy dashed line, with every reflection being picked correctly until he ran out of reflections. Any additional picks would necessarily lie on the horizontal dashed line. An actual interpreter typically follows something like the solid curve. At a point near the knee of the curve, the interpreter is getting into very weak reflections, and he starts missing some and marking some incorrectly. Therefore his per-

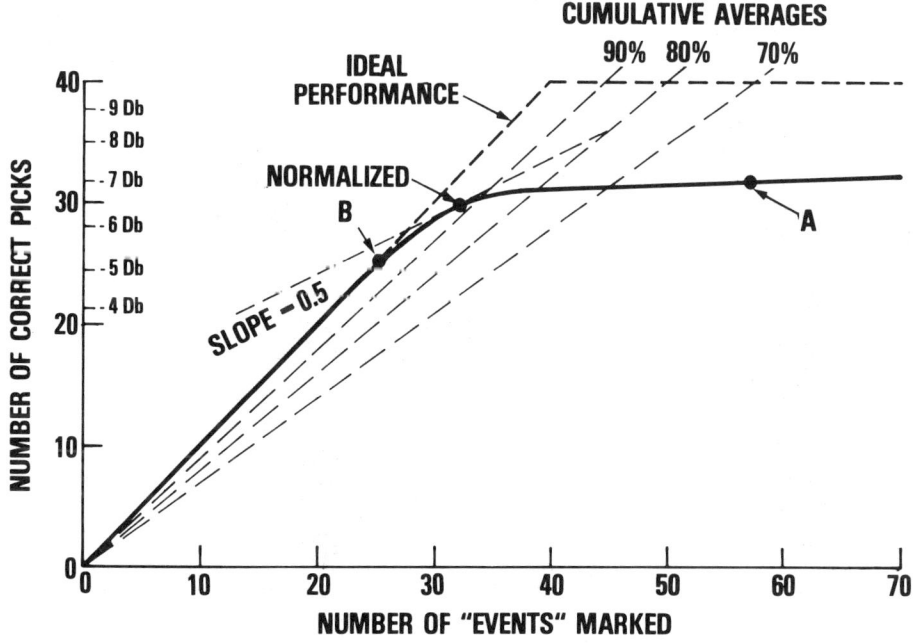

FIG. 14. Threshold normalization for a particular set of display parameters.

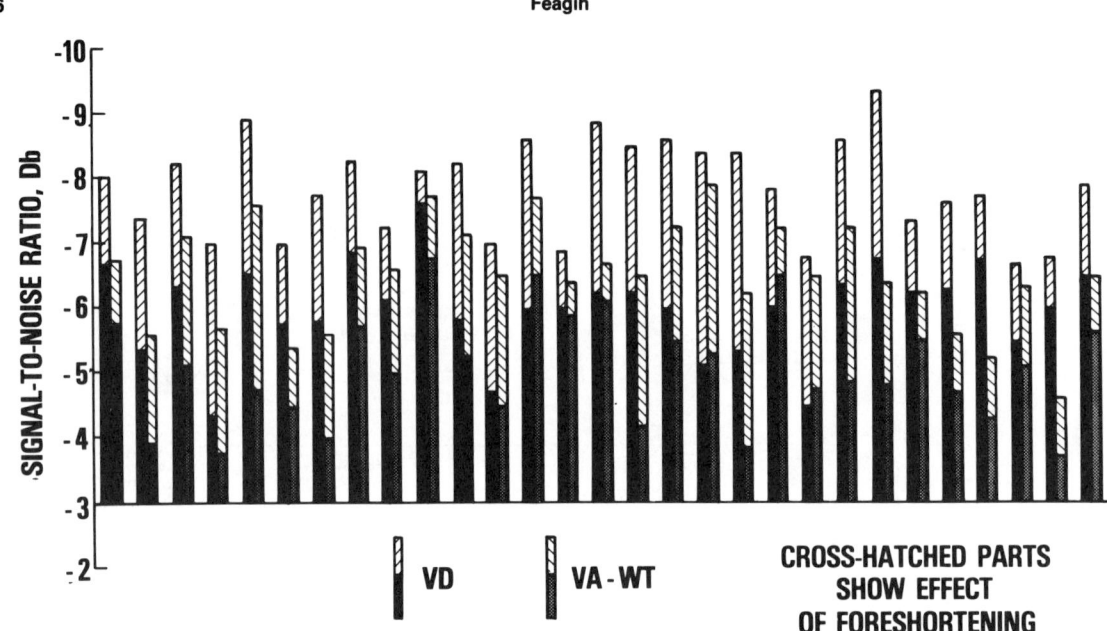

FIG. 15. Individual interpreter perceptibility averages for 29 participants.

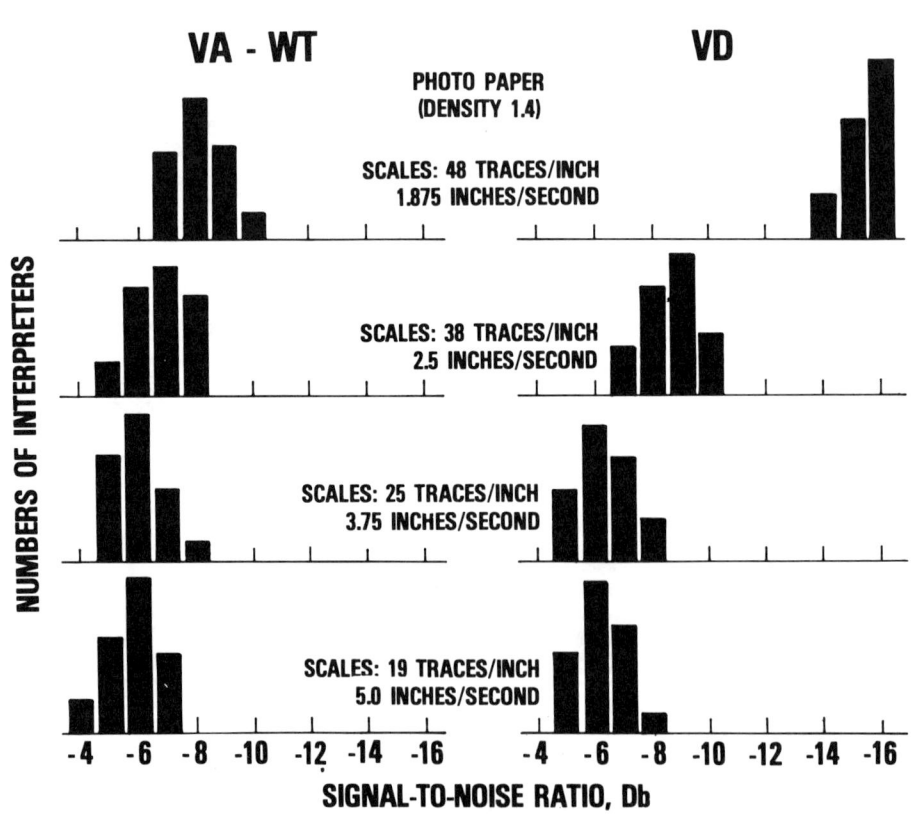

FIG. 16. Effect of scale on perceptibility.

centage of accurate picks starts to drop. An aggressive interpreter might not stop making picks until he was well out on the curve, say at point A. One more conservative might stop much sooner at point B where he was still "batting" 1000.

It is interesting to note that the performance curve of Figure 14 continues to have a slight upward trend instead of leveling to the horizontal at some S/N threshold value. This means that an aggressive participant continues to improve his threshold a little as he makes more and more picks. His "batting average" drops very low, but he still manages to get one right now and then, and each time he does he has succeeded in picking a still weaker reflection. Some perceptual psychologists argue that there is really no threshold for any of the human senses but rather that the probability of detection gets lower and lower as S/N ratios are decreased (Swets, 1964; Green and Swets, 1966). Be that as it may, for our purposes, there *is* a practical threshold. Our interpreters could not consider pushing aggressiveness to the point that only one reflection was picked correctly for every 10 or more incorrectly picked.

It is obvious that we just cannot take as a threshold value the S/N ratio where each interpreter chose to stop on his performance curve. The S/N figure where each individual stopped was projected forward or backward along his curve to the point where he was, or could have been, picking events with 50 percent probability of being correct. That is where the slope of the curve is 0.5. This does not mean that half of all reflections marked up to that point are wrong. As can be seen in Figure 14, at the 50 percent probability point the participant has a cumulative average of about 93 percent.

Another normalizing procedure was used to take out the effect of variation in perceptual ability from one individual to another. The S/N threshold figures (for VD and VA-WT separately) averaged over all 29 participants were compared with the corresponding average threshold figures for each person. A kind of handicap figure was thus derived (either positive or negative) for

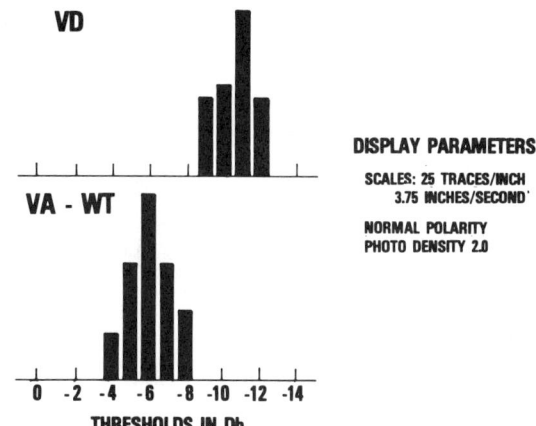

FIG. 17. Threshold comparison of high-contrast displays.

each interpreter which was used to raise or lower all of his threshold figures.

By means of these two normalizing procedures, we were more nearly able to measure display effectiveness alone as a function of display parameters, the principal objective of this research. However, we do have some interest in the amount of variation in perceptibility between individual interpreters. Figure 15 shows that variation. It also shows several other things. The bars represent the S/N thresholds for each interpreter normalized for aggressiveness but not normalized for individual perceptibility variations. The higher the bars, the greater is that person's ability to pick weak reflections in a noisy background. The bars are paired for each person; the solid bars are for VD, and the dotted bars are for

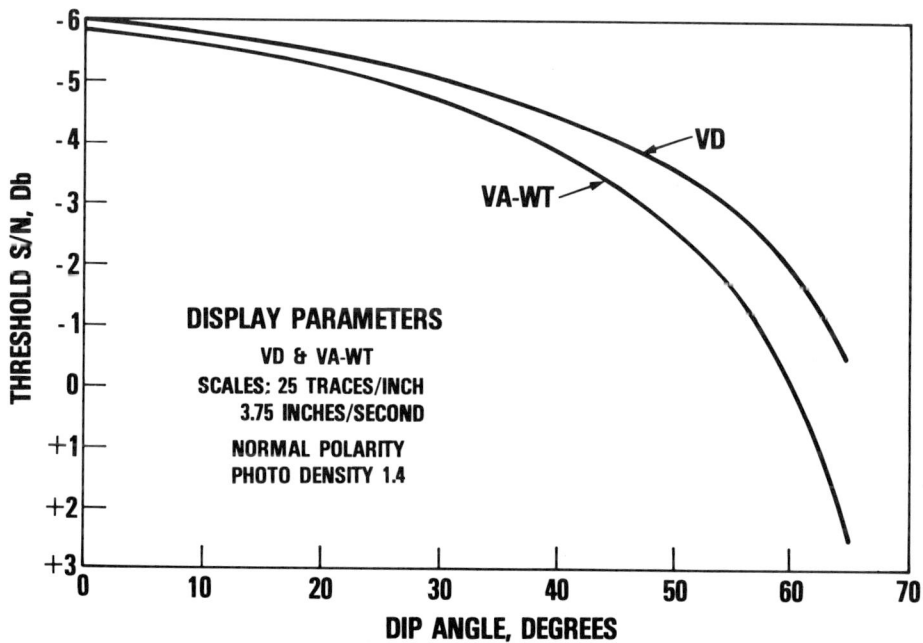

FIG. 18. Effect of dip angle upon perceptibility.

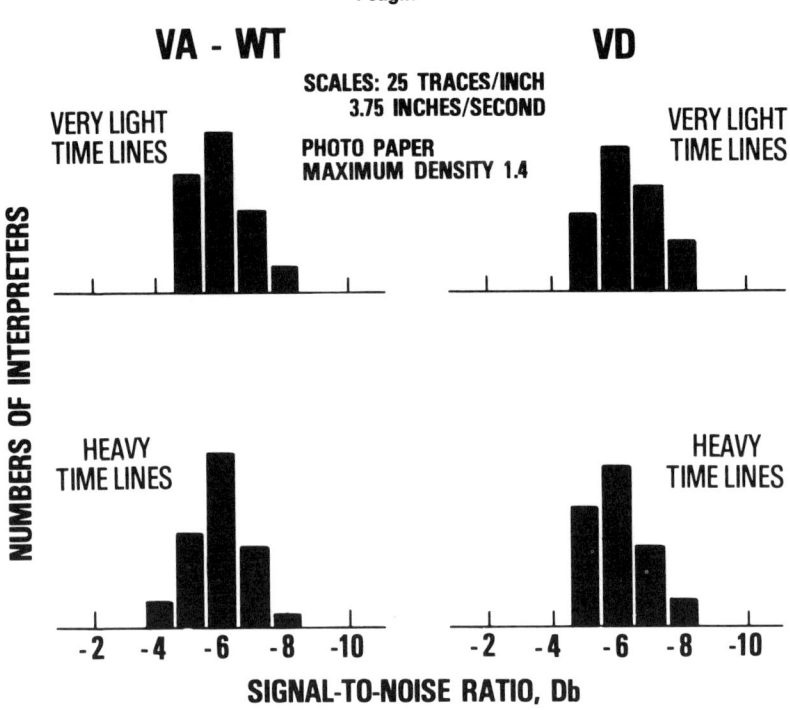

FIG. 19. Effect of weight of time lines upon detectability.

VA-WT. Each bar represents an average over all VD or VA-WT sections, respectively. The enhancement in detectability from the use of foreshortening is indicated by cross-hatching for each person.

Conclusions that one can draw from Figure 15 are as follows:

(1) There is about 2.5 to 3.0 dB maximum variation between individual perceptibility averages.

(2) Foreshortening helps an average of about 2.0 dB, but there is a rather wide variation between individuals.

(3) Each interpreter, without exception, has a better overall average with VD than with VA-WT. It is important to note, however, that these figures are averaged for each person over all types and sizes of displays. As will be seen in the figures to follow, the advantage of VD over VA-WT does not show up for some types of displays but appears rather strikingly for others.

Figure 16 shows histograms of normalized threshold figures for various display scales, for both VA and VA-WT. There is a definite advantage (about 8 dB) for VD over VA-WT at small scales, but at the popular 3.75 inch/sec scale and with usual photographic paper contrast, the two modes appear about equally effective. Small scales show an advantage for both VD and VA-WT, but the advantage is much greater for VD. Such small scales, of course, are not suitable when one is studying details in the data, but from the standpoint of reflection recognition, they are superior. Interpreters often make good use of this superiority when they use small sections for regional studies.

Another situation where VD is superior is where photographic contrast is high and the VD gray scale is properly controlled. Figure 17 shows the VD advantage even with a 3.75 inch/sec scale when a high-contrast display is used.

Figure 18 shows the effect of dip angle on perceptibility. We see that when dip angles are large, perceptibility suffers. The curve drops a little more rapidly for VA-WT displays than for VD.

The histograms of Figure 19 show that even with moderately heavy time lines the detectability thresholds of the participants are about the same as with sections with very light (normal) time lines.

Figure 20 indicates that signal polarity has no significant effect upon a reflection's detectability. We have called polarity normal when the large single peak on the wavelet is plotted toward black on VD sections and in the direction of increasing black area on VA-WT sections.

It can also be seen in Figure 20 that section gapping does not, on the average, impair detectability. It should be noted, however, that the gapped sections which were used in the tests made use of synthetic reflections with only small dips. It is possible that the break in reflection continuity caused by the gaps might have some effect for much steeper dips.

CONCLUSIONS FROM EXPERIMENTS

We have reached the following conclusions with respect to the questions posed at the beginning of the experiments:

(1) With our usual displays, interpreters can pick reflections 6 or 7 dB below noise with 50 percent probability. An improvement in S/N ratio of 1 dB can be very significant.

(2) Average perceptibility varies from one interpreter to another by about 2.5 or 3.0 dB.

(3) For our most common type of displays (3.75 inch/sec time scale and low contrast photographic paper), VD and VA-WT are about equally effective. But a VD section with all parameters

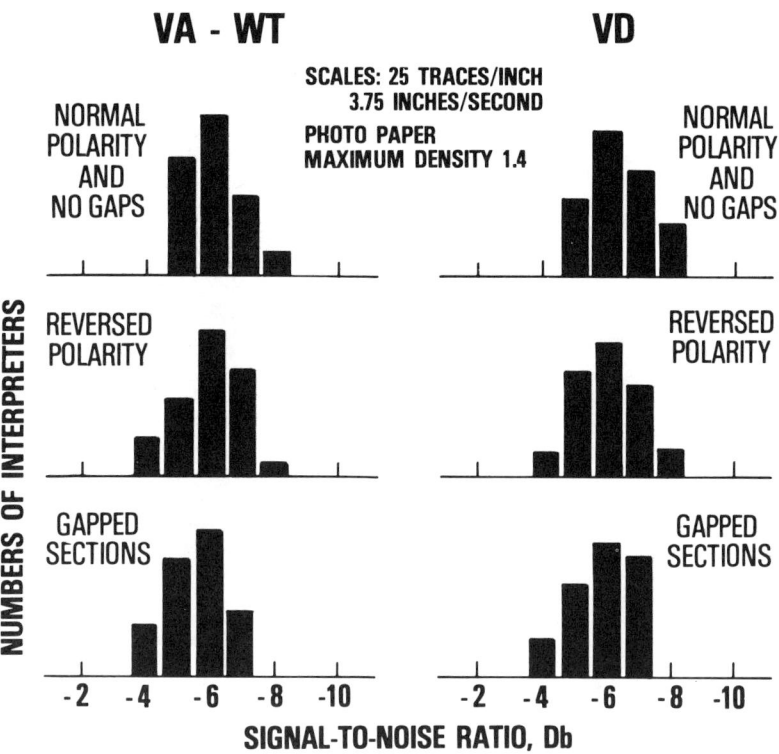

FIG. 20. Effect of reflection polarity and section gapping.

optimized (small scale and high density contrast) shows about 8 or 9 dB advantage over an optimized VA-WT section.

(4) From a perceptibility standpoint, small scales are preferable.

(5) High-density contrast is desirable, especially for VD sections, but gray scale must be properly controlled.

(6) The interpreter gains an average of about 2 dB by foreshortening, although there is a large variation about that average from interpreter to interpreter.

(7) Detectability drops rapidly as dip angle increases. VD is slightly better for steep dips.

(8) Time lines, unless extremely heavy, do not affect perceptibility.

(9) Signal polarity makes no difference.

(10) Section gapping has little effect, at least for small dips.

Although not a part of the original set of questions posed, several other interesting observations were made during the analysis of the data:

(1) When detectability thresholds were plotted against years of interpretation experience, there was no observable correlation, indicating that reflection perception skills do not in general improve with experience. Experience ranged from 6 months to more than 41 years. As pointed out earlier, however, perception is only one of the tasks required of an interpreter. *There can be no question that other important interpreter skills do improve with experience.*

(2) Each participant was asked before marking the displays whether he felt he generally could see reflections better on a VD or on a VA-WT section. The preferences were about evenly divided. As shown by Figure 15 however, no one did better with VA-WT. Experimental psychologists often report this kind of result; a priori participant preference very often is not a reliable indicator of optimality.

(3) It was found that errors made in picking reflection terminations were made by marking reflections too short about four times as often as picking them too long.

(4) The participants were remarkably consistent in marking the reflections on the synthetic sections. It was mentioned earlier that there were two identical sections submitted to each person. These were submitted several hours apart so that there could be no chance that the person would recognize the sections. In each case, the interpreter marked the two sections almost exactly alike. Most of the interpreters even made the same errors on both sections.

PRACTICAL MEANING OF RESULTS

Just what meaning do the results of this research have for practical geophysical operations? The following suggestions represent further interpretation and summarization of the results.

(1) If there is a critical problem in detecting seismic reflections in an area because of poor S/N ratios, then a small-scale, high-contrast variable density section with controlled gray scale should be of considerable advantage to an interpreter. Other display scales and modes may be required in addition for reasons other than detectability.

(2) It is important to make sure that all VD sections are plotted with a correct transfer function, i.e., one which takes into account

desirable nonlinearities. The superiority of variable density sections deteriorates rapidly if the transfer function is not reasonably near optimum. One good way to check this is to measure reflective densities on a section at about 100 randomly disposed points with a microdensitometer. The densities should then be converted into linearly perceived gray scale steps in accordance with Figure 2. A histogram of the frequency of occurrence of the gray scale values should approximate a boxcar shape for optimality. If this is not the case, then the overall transfer function should be modified. This can often be done by changing photographic parameters and/or the amount of gain overdrive. If this is not feasible, then a special plot tape can be made using a digital computer to make the changes in transfer characteristic. The book by Rosenfeld and Kak (1976, section 6.2.3) should be helpful in designing a transfer function for converting from one histogram to another.

Variable density displays have lost popularity in recent years for a number of practical reasons. First, it is much easier to control the quality of a VA or VA-WT section because photographic variables have less effect on results. Also, it is easier to make reproductions of an original VA or VA-WT plot. Almost any town of even medium size has facilities for making such copies. Finally, the VA-WT section has to a large extent become the common medium of exchange between companies when data are being traded or purchased.

These are reasons which cannot be minimized, and where reflections are reasonably good, one cannot take exception to them. But when *seeing* reflections is a problem and when it is important that we do our best job, variable density sections can have a significant advantage provided we spend the time and money to make them correctly. But then, that is a price we pay for superior results at *every* point in the seismic system.

(3) If signal-to-noise ratios are very poor (near the threshold of detectability), an interpreter might want to consider deliberately marking those reflections a bit longer than he would normally mark them because of the demonstrated general tendency to mark them short. At a 3.75 inch/sec scale, the extension should be about 0.25 inch. At smaller scales, the extension should be proportionately less.

(4) If the geology is complex and dips are steep, variable density sections are preferable even for good reflections. This is true for larger as well as smaller scales.

(5) Extremely heavy time lines should be avoided, but average weight or even moderately heavy lines cause no problems as far as detectability is concerned.

(6) Foreshortening is a very useful technique, and we should continue to employ it. It might be well to point out, however, that in those cases where a reflection can *only* be seen by making use of foreshortening, it is being picked with somewhat higher probability of error than events which can be seen without foreshortening. During perception experiments, participants were asked not to foreshorten until all events had been marked which could be seen without foreshortening. This, of course, was done in order to be able to evaluate the effectiveness of foreshortening. An interpreter in his regular daily work, however, often will start foreshortening almost as soon as he encounters a section. No inference here is intended that there is anything wrong with this practice, only that those reflections which can be seen without foreshortening are picked with a lower probability of error.

In conclusion, it is worth restating that the models (synthetic displays) used in this study are unrealistic in several ways; consequently, one must be careful about drawing sweeping and generalized conclusions based on this experiment. We do believe, however, that the stated conclusions are legitimate where S/N ratios are poor, where the noise tends to be random, where reflections are straight (or with low-order curvature), and where detection of reflections is the interpreter's primary problem rather than some other interpretational consideration.

ACKNOWLEDGMENTS

This paper is based on work done by the author prior to retirement from Exxon Production Research Co. The author is indebted to the company for releasing the paper for publication.

Thanks are due a number of people for assistance in carrying out this research—to L. L. Lenz for his help in producing the synthetic data tapes, from which the sections were produced, to Z. J. Nikolic for helpful consultation from time to time, and to Mona D. Guiler for her help in analysis and plotting of the large quantity of data generated by the experiment.

Special thanks go to the 29 seismic interpreters for their interest and encouragement and for their considerable time and effort expended in marking the synthetic sections.

REFERENCES

Beardslee, D. C., and Wertheimer, M., 1958, Readings in perception: Princeton, N.J., D. Van Nostrand Company, Inc., p. 45.
Dyk, K., and Eisler, J. D., 1951, A study of the influence of background noise on reflection picking: Geophysics, v. 16, p. 450.
Green, D. M., and Swets, J. A., 1966, Signal detection theory and psychophysics: New York, John Wiley and Sons, Inc., chap. 5.
Hurvich, L. M., and Jameson, D., 1966, The perception of brightness and darkness: Boston, Allyn and Bacon, Inc., p. 6.
Kim, D. Y., and Dyk, K., 1965, The effect of signal-to-noise ratio on seismic signal picking: Presented at the 35th Annual International SEG Meeting November 16, in Dallas; abstr. Geophysics, v. 30, p. 1238.
Rosenfeld, A., and Kak, A. C., 1976, Digital picture processing: New York, Academic Press.
Stevens, S. S., 1951, Handbook of experimental psychology: New York, John Wiley and Sons, Inc., chap. 1.
Swets, J. A., 1964, Signal detection and recognition by human observers: New York, John Wiley and Sons, Inc., p. 122.

SEMBLANCE AND OTHER COHERENCY MEASURES FOR MULTICHANNEL DATA†

N. S. NEIDELL* AND M. TURHAN TANER*

The concept of semblance is introduced, along with a descriptive review of several of the more common likeness or coherence measures. Measures are considered from three points of view: the domain in which they are applied, the philosophy of their design, and the manner in which they are used.

Crosscorrelation, the most familiar of the likeness criteria, is examined in detail. Differences of design philosophy are noted as expressing themselves by a change in the normalization. Semblance is shown to be related to an energy-normalized crosscorrelation and to share certain features of the summation method or stack which has been used recently as a coherence measure.

Several coherence measures, including semblance, are considered in a problem environment —the determination of stacking velocities from multiple ground coverage seismic data. A noise-free synthetic example is studied in order to compare discrimination thresholds of the various methods. Semblance, when properly interpreted, proves to have the greatest power of discrimination among the candidates examined for the particular application.

INTRODUCTION

Multichannel data are used in a wide variety of contexts, most often in order to take advantage of any information redundancy which may be present. Common signal in the channels can be measured, enhanced, extracted, shaped, or even ignored. Each of these operations, however, requires the use of some coherence measure.

Coherence techniques for multichannel data are too broad a subject for comprehensive discussion within the confines of this paper. Our presentation singles out for study a few of the more common coherence measures. Using these as a framework, the concept of semblance is introduced and examined.

Three points of view are taken in evaluating each coherence measure: First, we note the domain in which a measure applies; second, we clarify the philosophy underlying the measure's design, including such considerations as normalization; third, we specify the objectives toward which the measure aims. Using these descriptive criteria, we see that semblance is a time domain coherence measure relating directly to an appropriately defined output/input energy ratio. The similarities and differences between semblance and some of the other coherence measures are explored. Most noteworthy among the properties of semblance is its sensitivity to amplitude diversity or variance.

Examples illustrate some of the ideas discussed. These are taken from problems associated with determining optimal stacking velocities in multiple ground coverage reflection seismology, an area in which coherence techniques have received much attention. The examples thus do double duty: they lend credence to our conclusions and they contribute directly to the solution of an important problem in exploration seismology.

GENERAL REMARKS CONCERNING COHERENCE MEASURES

Coherence measures express in quantitative form the likeness of data content among data

† Presented at the 39th Annual International SEG Meeting in Calgary, Alberta, September 16, 1969. Manuscript received by the Editor November 24, 1969, revised manuscript received December 16, 1970.
* Seismic Computing Corporation, Houston, Texas 77036.
© 1971 by the Society of Exploration Geophysicists. All rights reserved.

channels. What are considered data and the uses toward which the resulting measurements are put vary greatly and are tailored to the specific problem. This section attempts to define in a broad sense three views appropriate to any coherence measure: the domain of application, philosophy of design, and specific utilization. Once these ideas and concepts have been adequately explained, we are in a good position to look at various measures individually.

Domains of application

A basic question before undertaking any data analysis is the following: Is there any transformation which will simplify the task? When a transformation is employed, we say that we are working in another domain.

Two domains are most commonly employed, the time domain or original coordinate space of the independent variables and the Fourier frequency domain. Frequency domain measures are favored most often in filtering applications. We shall mention briefly certain other domains of importance and the transformation sequences through which they may be addressed.

Principal-component-type transformations lead to domains where the input data covariance attains its extremal values. The review by Rao (1964) gives the statistical reasoning which underlies implementation of principal component techniques. Linear combinations of the input channels are taken to be a new data domain in which the data channels are uncorrelated in a mathematical sense.

The Hadamard domain (Pratt, Kane, and Andrews, 1969) has served valuably in certain problems requiring the compression of information. Here data are decomposed into orthogonal square-wave components. This technique is rapidly gaining popularity owing to the existence of a fast algorithm akin to the Fast Fourier Transform (Cooley and Tukey, 1965 and Shanks and Cairns, 1968) but one requiring no multiplications whatsoever.

New domains often result from the adoption of novel views and the application of the corresponding transforms to existing problems. The domain used will depend upon and complement the design philosophy of the particular coherence measure and the specific problem at hand.

In closing, we note that many coherence measures can operate in any one domain, and, conversely, there are many domains in which any one measure may be applied. The first part of the preceding statement is obvious; after all, we all are cognizant of more than one coherence measure which can be applied in the time domain. We can illustrate the validity of the second tenent with ease as well. For example, the definition of a normalized correlation coefficient $\tilde{R}_{xy}(0)$ is simply

$$\tilde{R}_{xy}^2(0) = \frac{R_{xy}^2(0)}{R_{xx}(0)R_{yy}(0)},$$

where $R_{xy}(0)$ is the unnormalized correlation coefficient of a data series x with a data series y. If we now apply the same definition to a single frequency band in the cross-spectral domain, we derive the coherence coefficient $\gamma_{xy}(\omega)$

$$\gamma_{xy}(\omega) = \frac{|G_{xy}(\omega)|^2}{G_{xx}(\omega)G_{yy}(\omega)}.$$

Here $G_{xy}(\omega)$ is the ω frequency band of the cross-spectrum of data series x and data series y. (See Bendat and Piersol, 1965.) Of course, the interpretation of the measures in the two domains is quite different.

Design philosophy

Coherence measures are most easily designed by starting with the problem objectives. Some common objectives for multichannel studies include signal prediction, detection of shifts, extraction of common signal, signal classification, and process control. Clearly, the coherence measure should convey information relevant to the objective. This information may be either the objective itself or some rating of the goal fulfillment. Scales for rating goal fulfillment can be based upon energy, entropy, statistics, and a host of other quantities or disciplines.

In prediction problems, for example, the mean square prediction error is the most usual coherence—or, more appropriately, lack-of-coherence—measure. This quantity may be considered to be either an energy measure or a statistical one. A Wiener solution to the prediction problem (Wiener, 1949) leads to the use of crosscorrelation functions, which might have been employed initially if a statistical viewpoint had been adopted in attacking the problem. The Wiener solution to the shift-detection problem leads directly to identification of the lag value for which crosscorrelation attains its maximum. Here the lag value is itself the objective, i.e., the shift.

There are many design philosophies worthy of consideration. Some of the more important ones are based upon principles of statistics, classification, energy, pattern recognition, and mathematical modeling. Specifically, we might think about using crosscorrelation, moment functions (Kendall, 1948), likelihood concepts (Green, Kelly, and Levin, 1966), ambiguity/similarity functions (Neidell, 1969), noise/signal measures (Simpson, 1955, 1967), energy, or entropy (Pelto, 1954).

It is evident that most of the coherence measures bear intimate relation to one another and that each can be interpreted from a number of points of view. We have already commented upon the duality existing between variance, a statistical concept, and energy, a physically grounded principle. Normalized correlation coefficients are also related to regression coefficients and thus can be interpreted in terms of mathematical modeling.

We can move on now, recognizing that the general viewpoint we have adopted includes the entire field of curve fitting in hyperspaces using generalized norms or distance measures.

Coherence utilization

We may choose to derive an operator (linear or otherwise) which maximizes or minimizes a coherence measure. We might devise an algorithm by which the coherence measure itself attains the objective. Alternatively, a systematic search for suitable ranges of the parameters might be undertaken until an appropriate extremum of the coherence measure is located. Iterative processes might be initiated which converge upon the desired results. All the ingenuity and mathematical tools available should be brought to bear in the use of our coherence measure. As for the preceding considerations, the multiplicity of choice for a coherence measure is great.

Let us now discuss crosscorrelation as a coherence measure and introduce the concept of semblance. After that, we shall consider these two, together with some other coherence measures, in a problem environment.

Crosscorrelation as a coherence measure.—Crosscorrelation is a much used, but often little appreciated, coherence measure. We shall adopt a deterministic (nonstatistical) viewpoint and examine what crosscorrelation does and does not measure.

Suppose that there are M independent data channels across which a coherent signal follows a lag trajectory $k(i)$. Let us take a time gate of $N+1$ samples from each trace, symmetrically disposed about sample $k(i)$; this choice makes the gate boundaries parallel the trajectory $k(i)$. We wish to establish for the signal which crosses this gate a coherence measure, which uses crosscorrelation. We will work directly with untransformed data, hence remaining in the time domain. (See Figure 1.)

For the moment, let us choose

$$k(i) = k + (i - 1)\alpha, \quad (1)$$

where α is a constant equal to an integral multiple of the sampling interval. Now, the unnormalized crosscorrelation between two channels p channel-index units apart, for the lag following trajectory $k(i)$, is just

$$\frac{1}{N+1} \sum_{j=k-(N/2)}^{k+(N/2)} f_{i,j} f_{i+p,j(i+p)}$$

$$= \frac{1}{N+1} \sum_{j=k-(N/2)}^{k+(N/2)} f_{i,j} f_{i+p,j+(i+p-1)\alpha}, \quad (2)$$

where $f_{i,j}$ is the amplitude of the jth sample in the ith channel.[1] The usual normalization of the crosscorrelation function is given by the expression

$$\frac{\sum_{j=k-(N/2)}^{k+(N/2)} f_{i,j} f_{i+p,j+(i+p-1)\alpha}}{\sqrt{\sum_{j=k-(N/2)}^{k+(N/2)} f_{i,j}^2 \sum_{j=k-(N/2)}^{k+(N/2)} f_{i+p,j+(i+p-1)\alpha}^2}} \quad (3)$$

and is justified by appealing to statistical philosophy based on regression.

An interesting observation may now be made. The denominator of the normalized crosscorrelation function is also the geometric mean of the energy in the two channels over the time gate chosen. The geometric mean is usually used for averaging ratios. It would seem that energy—a physical quantity—ought to be treated by arithmetic averaging and so a normalized crosscorrelation should be

---

[1] A notation has been chosen which facilitates computer programming and emphasizes that the signal trajectories need not lie along paths through the samples.

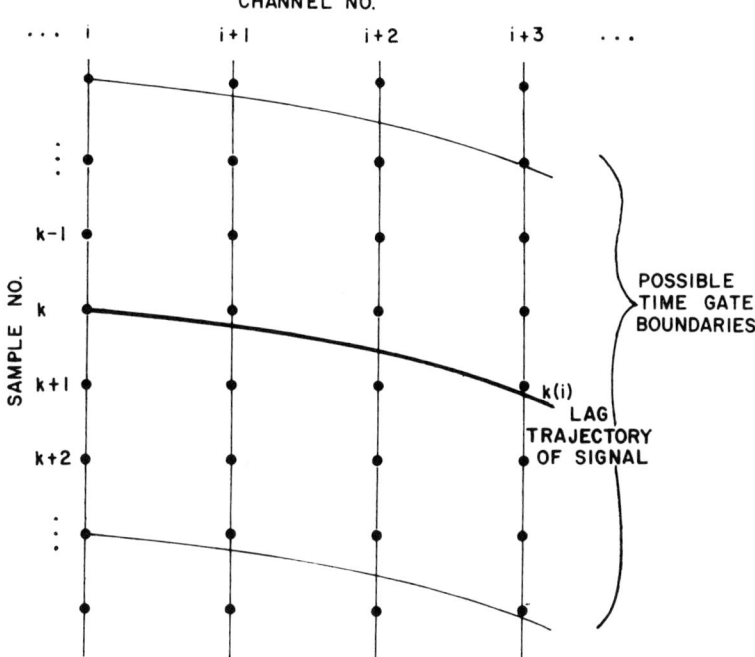

Fig. 1. Sampled multichannel data.

$$\frac{2\sum_{j=k-(N/2)}^{k+(N/2)} f_{i,j}f_{i+p,j+(i+p-1)\alpha}}{\sum_{j=k-(N/2)}^{k+(N/2)} f_{i,j}^2 + \sum_{j=k-(N/2)}^{k+(N/2)} f_{i+p,j+(i+p-1)\alpha}^2}. \quad (4)$$

In fact, the preceding expression is a legitimate normalization and varies between ± 1 according to the likeness and phase of the signal in the channels.

One outstanding difference between the two schemes of normalization is that if the rms signal amplitudes in the two channels are different and in the ratio γ one to the other and if the phase and shapes of the signal are identical, geometric normalization will give unit correlation, while the arithmetic normalization will give a correlation of

$$\frac{2\gamma}{1+\gamma^2}, \quad (\gamma \leq 1). \quad (5)$$

Hence, arithmetic normalization embodies a sensitivity to the scaling of the data channels which, depending on the intended applications, may or may not be desirable. In this case, the difference between a physical and a statistical design philosophy expresses itself in terms of a normalization difference.

For M data channels, a sum of all the possible unnormalized crosscorrelations should be a more reliable coherence measure than any single member of that sum. Starting with expression (2), we may write the unnormalized crosscorrelation sum as

$$\sum_{p=1}^{M-1}\sum_{i=1}^{M-p}\frac{1}{N+1}\sum_{j=k-(N/2)}^{k+(N/2)} f_{ij(i)}f_{i+p,j(i+p)}, \quad (6)$$

where we have returned to the most general notation for the signal trajectory $j(i)$. Expression (6) may be normalized in the usual statistical manner; and, after reordering summations and introducing constants to maintain unit maximum amplitude, we have

$$\frac{2}{(M-1)M}\sum_{j=k-(N/2)}^{k+(N/2)}\sum_{p=1}^{M-1}\sum_{i=1}^{M-p} \frac{f_{ij}f_{i+p,j(i+p)}}{\sqrt{\sum_{j=k-(N/2)}^{k+(N/2)} f_{ij}^2 \sum_{j=k-(N/2)}^{k+(N/2)} f_{i+p,j(i+p)}^2}}. \quad (7)$$

In statistical notation, expression (7) becomes

$$\frac{2}{(M-1)M} \sum_{i=1}^{M} \sum_{i'>i} \frac{R_{ii'}(0)}{\sqrt{R_{ii}(0)R_{i'i'}(0)}},$$

where

$$R_{ii'}(0) = \sum_{j=k-(N/2)}^{k+(N/2)} f_{ij(i)} f_{i'j(i')}.$$

While developing the alternative normalization for crosscorrelation given by expression (4), we talked about energy. Expression (6) can be viewed analogously. Let us start with the well-known and easily verified identity

$$\sum_{j=k-(N/2)}^{k+(N/2)} \left\{ \sum_{i=1}^{M} f_{ij(i)} \right\}^2$$
$$= \sum_{j=k-(N/2)}^{k+(N/2)} \left\{ 2 \sum_{p=1}^{M-1} \sum_{i=1}^{M-p} f_{i,j} f_{i+p,j(i+p)} \right. \quad (8)$$
$$\left. + \sum_{i=1}^{M} f_{i,j}^2 \right\}.$$

It is now evident that the unnormalized crosscorrelation sum of expression (6) is exactly equivalent to $\frac{1}{2}$ of an energy difference, or

$$\sum_{j=k-(N/2)}^{k+(N/2)} \sum_{p=1}^{M-1} \sum_{i=1}^{M-p} f_{i,j} f_{i+p,j(i+p)}$$
$$= \frac{1}{2} \sum_{j=k-(N/2)}^{k+(N/2)} \left\{ \left\{ \sum_{i=1}^{M} f_{i,j(i)} \right\}^2 \right. \quad (9)$$
$$\left. - \sum_{i=1}^{M} f_{i,j}^2 \right\}.$$

In statistical notation, we may write equation (9) as

$$\sum_{i=1}^{M} \sum_{i'>1} R_{ii'}(0)$$
$$= \frac{1}{2} \sum_{i=1}^{M} \left(\sum_{i'=1}^{M} R_{ii'}(0) - R_{ii}(0) \right).$$

Equation (9) shows the unnormalized crosscorrelation sum to be equal to half the difference between an output energy and the gate input energy. The output trace is taken to be an unnormalized sum of the inputs along the trajectory of interest—in the parlance of exploration seismology, a stack.

If all the $f_{i,j}$ were equal, it follows from equation (9) that the unnormalized crosscorrelation sum would be equal to $M(M-1)/2$ times the common trace input energy. Hence, we can normalize the crosscorrelation sum of expression (6) to unit maximum amplitude by using the average trace energy and the preceding constant. We would then have as a counterpart to expression (7)

$$\frac{\dfrac{1}{N+1} \dfrac{2}{M(M-1)} \dfrac{1}{2} \sum_{j=k-(N/2)}^{k+(N/2)} \left\{ \left\{ \sum_{i=1}^{M} f_{ij(i)} \right\}^2 - \sum_{i=1}^{M} f_{ij}^2 \right\}}{\dfrac{1}{N+1} \dfrac{1}{M} \sum_{j=k-(N/2)}^{k+(N/2)} \sum_{i=1}^{M} f_{ij}^2} \quad (10)$$

$$= \frac{\sum_{j=k-(N/2)}^{k+(N/2)} \left\{ \left\{ \sum_{i=1}^{M} f_{ij(i)} \right\}^2 - \sum_{i=1}^{M} f_{ij}^2 \right\}}{(M-1) \sum_{j=k-(N/2)}^{k+(N/2)} \sum_{i=1}^{M} f_{ij}^2}.$$

In statistical notation, the right-hand side of equation (10) is

$$\frac{\sum_{i=1}^{M} \sum_{i' \neq i} R_{ii'}(0)}{(M-1) \sum_{i=1}^{M} R_{ii}(0)}.$$

Expression (10) is an energy-normalized crosscorrelation sum.

A comparison of expressions (7) and (10) reveals some interesting facts. The statistically normalized function varies in the range $-1 \leq S \leq 1$, while the energy normalized function varies in the range

$$\frac{-1}{M-1} \leq S \leq 1,$$

where S represents the functions of expressions (7) and (10). In our discussion of the semblance coefficient, the meaning of this biased range will be made clear. Computationally, expression (7) requires $M(M+1)(N+1)/2$ multiplications, while expression (10) needs only $(M+1)(N+1)$ multiplications to operate over the same gate. Both figures are exclusive of multiplications used in interpolating values to follow given trajectories. Again, expression (7) has no sensitivity to changes in rms amplitude from trace to trace, while expression (10) seriously penalizes such variation.

We now understand that crosscorrelation is a time domain coherence measure whose properties depend upon the design philosophy which is employed. Although only two viewpoints, energy and variance, have been discussed, it is evident that there must also be others. In the following section which introduces the semblance coefficient, we shall see crosscorrelation considered again, but this time from the direction of coherence use. Crosscorrelation will further be related to a coherence (rather, lack-of-coherence) measure but from the viewpoint of signal/noise philosophy.

The semblance coefficient

Suppose that after use of a multichannel coherence measure an optimal signal trajectory has been identified. In many applications the problem to be faced next is the extraction or enhancement of signal following the trajectory of interest. Simple compositing or stacking along this trajectory is one accepted and commonly employed enhancement technique, which is optimal under some fairly realistic and robust assumptions about signal and noise. At ths point, one may ask how to go about measuring the effectiveness of the enhancement or extraction procedure itself.

A physically meaningful quantity of relevance to this goal is the output/input energy ratio. The semblance coefficient is defined to be the normalized output/input energy ratio, where the output trace is a simple compositing or sum of the input traces. In mathematical terms, S_c, the semblance coefficient, is just

$$S_c = \frac{\sum_{j=k-(N/2)}^{k+(N/2)} \left\{ \sum_{i=1}^{M} f_{i,j(i)} \right\}^2}{M \sum_{j=k-(N/2)}^{k+(N/2)} \sum_{i=1}^{M} f_{i,j(i)}^2}. \quad (11)$$

In statistical notation,

$$S_c = \frac{\sum_{i=1}^{M} \sum_{i'=1}^{M} R_{ii'}(0)}{M \sum_{i=1}^{M} R_{ii}(0)}$$

$$= \frac{\sum_{i=1}^{M} \left\{ 2 \sum_{i'>i} R_{ii'}(0) + R_{ii}(0) \right\}}{M \sum_{i=1}^{M} R_{ii}(0)}.$$

Hence, the semblance coefficient is the normalized output/input energy ratio for one possible multichannel compositing which we have singled out for study owing to its special properties.

It should be clear from the identity (8) that the semblance coefficient can be used as a coherence measure as well as a compositing performance measure. We should point out, however, that as a coherence measure the semblance coefficient is biased and includes terms not containing crosschannel information. Now we can explain why the energy-normalized crosscorrelation sum had a biased range of values. This quantity, which was given by expression (10), may be rewritten in terms of the semblance coefficient and takes the form

$$\frac{1}{M-1}(MS_c - 1). \quad (12)$$

The semblance coefficient is thus proportional to the output/input energy ratio. At the same time, a biased semblance coefficient is precisely equivalent to a multichannel coherence measure as defined by expression (10).

For a simple linear signal-noise model, the semblance coefficient can also be shown to be equal to a signal energy to total energy ratio, under the assumption that the noise sum over all channels at any time is zero. Specifically, for the jth time increment in the ith channel, the model is

$$f_{i,j(i)} = S_{j(i)} + n_{i,j(i)}, \quad (13)$$

where $S_{j(i)}$, $n_{i,j(i)}$ stand for signal and noise, respectively. Now, in terms of the model (13) definition (11) becomes

$$S_c = \frac{\sum_{j=k-(N/2)}^{k+(N/2)} \left[M^2 S_{j(i)}^2 + 2S_{j(i)} \sum_{i=1}^{M} n_{i,j(i)} + \left(\sum_{i=1}^{M} n_{i,j(i)} \right)^2 \right]}{M \sum_{j=k-(N/2)}^{k+(N/2)} \left(M S_{j(i)}^2 + 2S_{j(i)} \sum_{i=1}^{M} n_i + \sum_{i=1}^{M} n_{i,j(i)}^2 \right)}$$

$$= \frac{M \sum_{j=k-(N/2)}^{k+(N/2)} S_{j(i)}^2}{M \sum_{j=k-(N/2)}^{k+(N/2)} S_{j(i)}^2 + \sum_{j=k-(N/2)}^{k+(N/2)} \sum_{i=1}^{M} n_{i,j(i)}^2}, \quad (14)$$

since the second and third terms in the numerator and second term in the denominator are each equal to zero. Equation (14) is precisely the ratio of signal energy to total energy over the selected window.

To complete this discussion, we call attention once again to Simpson's similarity measure (really coherence measure—see Simpson, 1955; 1967). His ratio λ' was developed from energy considerations and used linear data channel combinations, though only taken in channel pairs. The same linear signal-noise model as that of equation (13) was enlisted. It follows then that λ' must be simply related to the semblance coefficient, and we shall expicitly develop this relationship as an illustration of the generality with which we may now treat coherence measures.

Simpson's ratio λ' (1967, p. 486) in our notation is just

$$\lambda' = \frac{\sum_{j=k-(N/2)}^{k+(N/2)} \sum_{p=1}^{M-1} \sum_{i=1}^{M-p} \{f_{i+p,j(i+p)} - f_{i,j(i)}\}^2}{\sum_{j=k-(N/2)}^{k+(N/2)} \sum_{p=1}^{M-1} \sum_{i=1}^{M-p} \{f_{i+p,j(i+p)} + f_{i,j(i)}\}^2}. \quad (15)$$

If we consider the easily verified identity

$$\sum_{j=k-(N/2)}^{k+(N/2)} \sum_{p=1}^{M-1} \sum_{i=1}^{M-p} \{f_{i+p,j(i+p)}^2 + f_{i,j(i)}^2\}$$

$$= \sum_{j=k-(N/2)}^{k+(N/2)} \left\{ \sum_{i=1}^{M-1} (M-i) f_{i,j(i)}^2 + \sum_{i=2}^{M} (i-1) f_{i,j(i)}^2 \right\} \quad (16)$$

$$= \sum_{j=k-(N/2)}^{k+(N/2)} (M-1) \sum_{i=1}^{M} f_{i,j(i)}^2$$

and use also identity (9), along with the definition (11) of the semblance coefficient, we find that

$$\lambda' = \frac{1 - S_c}{\left(1 - \frac{2}{M}\right) + S_c}. \quad (17)$$

λ' was intended to be a measure of dissimilarity and we can now show that it has such properties. If M is large enough that $2/M$ may be ignored, we see that

S_c	0	1/4	1/2	3/4	1
λ'	1	3/5	1/3	1/7	0

where the values of λ' have been listed under the corresponding values of S_c.

Let us turn our attention toward applying some of these coherence measures to a practical multi-channel problem, the determination of multiple ground coverage stacking velocities. One of our goals will be to establish the relative sensitivity of the various measures in this problem context.

Coherence measures in stacking velocity determination

Reflection seismic exploration has been a multi-channel discipline for many years. Reiber's work in Volume I of GEOPHYSICS (1936) illustrates that the concept of multichannel coherence or sensitivity predates acceptance by the petroleum industry of multiple ground coverage methods (see Mayne, 1962; 1967 and Courtier and Mendenhall, 1967) substantially. Six basic subheadings encompass the majority of multichannel methods in reflection seismology:

I. multiple reflection elimination,
II. static trace shift corrections,
III. stacking velocity and interval velocity determinations,
IV. stacking,
V. space-time frequency filtering,
VI. reflection picking.

All of these subject areas deserve extensive comment as to their relevance to coherence measures. Once again, however, we must restrict our comments because of the breadth of the subject matter.

Current interest in coherence measures stems from their relevance to the determination of stacking velocities (topic III). In fact, stacking velocity determination probably employs a greater variety of coherence measures than any other single subject area in exploration seismology. The coherence measures range from subjective visual discernment of continuity as employed in trial stack displays to the semblance concept as used in computing the Velocity Spectrum.[2] In velocity determination via methods akin to the latter, the objective is to establish the relative amount of coherent signal arriving at any normal incidence time as a function of the stacking (or rms) velocity.

Four principal coherence measures are routinely employed and have gained acceptance within the industry: semblance, unnormalized correlation, statistically normalized correlation, and mean amplitudes. Semblance was introduced by Koehler and Taner (1967); unnormalized correlation was employed by Schneider and Backus (1968); while the use of the statistically normalized correlation was reported simultaneously by several authors at the Pacific Coast SEG velocity symposium (1968). Mean amplitudes were first discussed by Garotta and Michon (1967). Most recently, Robinson (1969) developed a frequency-wavenumber method which uses the transformed counterpart of unnormalized correlation.

All of these stacking velocity determination methods work with the common depth (or more precisely ground) point trace gathers. They begin with the familiar hyperbolic move-out equation

$$T_x^2 = T_0^2 + \frac{x^2}{\overline{V}^2}, \qquad (18)$$

which dates from Dix (1955).

In equation (18), T_x is two-way arrival time; T_0 is two-way normal incidence time; X is offset distance; and $\overline{V}$ is dip-weighted rms velocity or stacking velocity. From this point, two alternate approaches dominate: equation (18) may be used directly, leading to an analysis of coherence along hyperbolic trajectories; or a preliminary moveout correction is applied, resulting in an analysis of coherence along nearly linear alignments. It is worth mentioning that the latter technique requires correlations at nonzero lags of traces already distorted by a preliminary moveout; the distortion may be responsible for those disadvantages of the linear alignment technique relative to those that use mean amplitudes reported by Herman (1968) and Garotta and Michon (1967).

Schneider and Backus employed a linearized analysis of residual moveout first differences called dynamic correlation analysis. The crosscorrelation combinations used depended upon the multiplicity of ground coverage. Correlations among widely separated traces were not computed, thus minimizing violation of the linear trajectories assumed for the correlation maxima. At the same time, some information redundancy was sacrificed. These authors compensated for such loss by averaging over adjacent ground points. Merdler, Backus, Schneider, and King (1968) extended this principle of using spatial continuity to improve signal/noise ratio through the vehicle of a moveout scan map, but the perils and risks of the approach are clear and need little elaboration beyond that given by Taner, Cook, and Neidell (1970). Velocity filtering (Fail and Grau, 1963) defined the most coherent moveout residual from the crosscorrelations which had first been appropriately arranged.

In simple terms, and in the notation of our earlier sections, the velocity filtering method (for 600 percent coverage) essentially searches an

---

[2] Service name of the Seismic Computing Corporation.

expression like

$$\sum_{i=1}^{M-4} \frac{1}{N+1} \sum_{j=k-(N/2)}^{k+(N/2)} \cdot f_{i,j+(i-1)\alpha} f_{i+4,j+(i+3)\alpha} \quad (19)$$

for the α value of maximum coherence. Expression (19) may be an average of this quantity taken over several common ground points. The crosscorrelation calculation is unnormalized and unbiased by information not relating to more than one data channel.

Koehler and Taner's method is also rather straightforward. They search the hyperbolic trajectories (18) directly, using expression (12) as a coherence measure. Other workers search the same suite of curves using measures given by relations (6)—unnormalized crosscorrelation—or (7)—statistically normalized crosscorrelation.

Garotta and Michon proposed an analysis somewhat like that of Schneider and Backus but substituted a summing operation for the mean product. When a time gate is used, their coherence measure is

$$\sum_{i=1}^{M} \frac{1}{N+1} \sum_{j=k-(N/2)}^{k+(N/2)} f_{i,j+(i-1)\alpha}, \quad (20)$$

which replaces expression (19). An obvious extension of their approach would be to substitute a mean amplitude criterion for the ones commonly considered by methods like those of Koehler and Taner. We shall examine this possibility later.

Robinson's (1969) frequency domain approach, although novel, is basically a dynamic correlation analysis with many disadvantages. Since the transformed coherence measure considers the equivalent of all possible crosscorrelations, nonlinear residual trajectory effects may operate to the detriment of the analysis. At the same time, zero-trace-lag correlation information, or better its transformed equivalent, is also employed, biasing the calculation and reducing its possible power of resolution. The zero-phase spectral whitening technique suggested for normalizing the data would tend to make all events look alike, thus degrading the resolving power and discrimination of the technique further yet. Robinson's technique lacks flexibility and cannot conveniently be applied directly to hyperbolic trajectories; it must be a residual method. In short, there appear to be no advantages either of theoretical or computational nature for performing analysis in the frequency-wavenumber domain.

Comparisons of the various techniques are few in number and often lacking in objectivity; however, one point emerges which was verified by more than one investigator. Both Garotta and Michon and Herman found the mean amplitude coherence measure to have advantages. As Herman concluded, "It is evident that in the area of low S/N (signal over noise) ratios the summation method gives more precise results than crosscorrelation (unnormalized), assuming equal stacking redundancy . . . ."

The discussion introducing the semblance coefficient makes clear the intimate relation between that quantity and the summation result which was referred to as an output. More specifically, in terms of the simple linear signal-noise model of equation (13), the semblance coefficient is the ratio of signal energy to total energy, while the summation represents an estimate of total signal amplitude. Since numerous techniques of exploration seismology are evaluated by simple graphic displays, it is worth noting that the dynamic range required to represent total signal amplitude is greater than that required by the semblance measure when signal/noise ratios remain greater than unity.

Semblance also is connected with crosscorrelation. For the same signal-noise model [equation (13) again], unnormalized crosscorrelation will simply estimate the signal energy. In practical situations, this quantity will require an even greater dynamic range than either summation or the semblance coefficient. Statistically normalized crosscorrelation [equation (7)], in the absence of amplitude diversity or phase distortion, will behave like the energy normalized crosscorrelation; and we already appreciate how that quantity is linked to the semblance coefficient [see equation (12)]. Where amplitude or shape differences exist, the semblance-based measures will apply greater penalties than the statistically derived scale, tending again to require a smaller dynamic range for display.

Noise present on the data channels affects semblance primarily through the apparent amplitude and shape diversity it creates. The precise character of the effects depends on the noise statistics and the signal-noise interactions. It would thus be instructive to perform an experi-

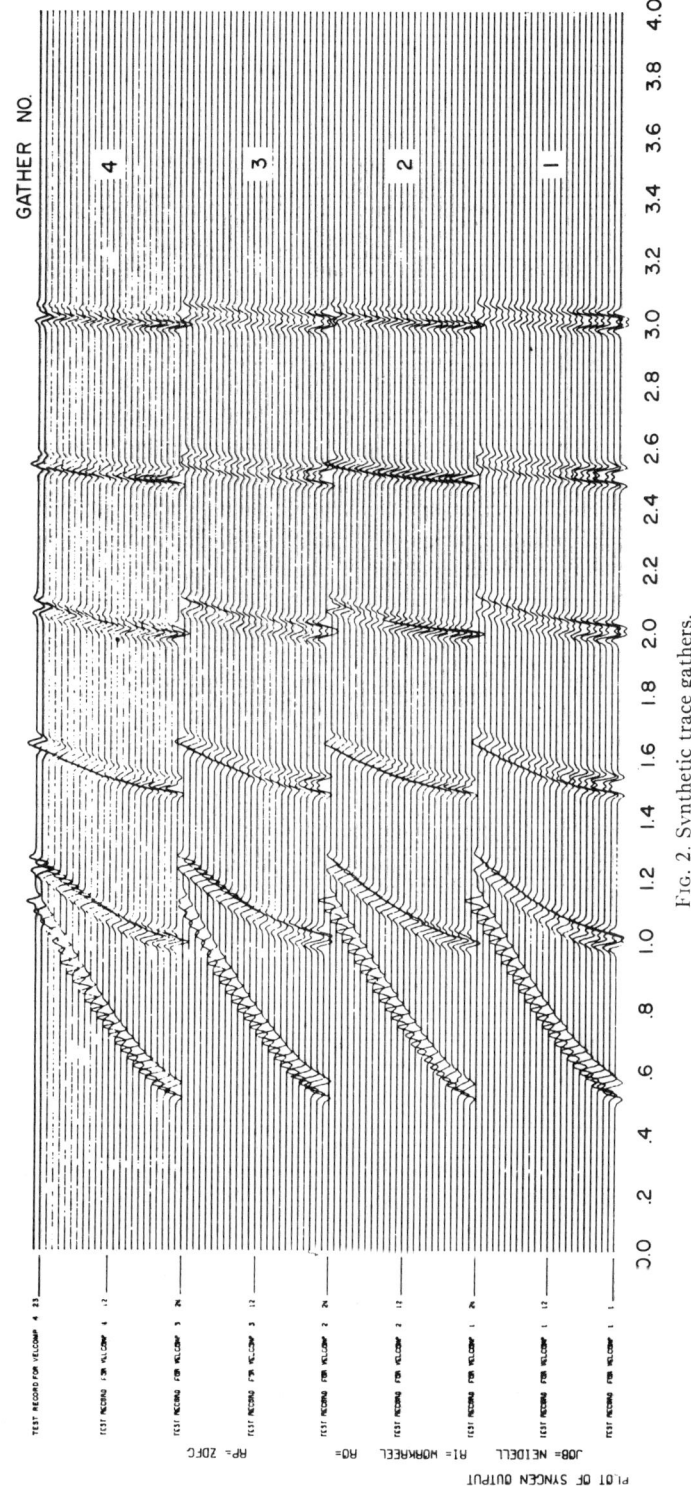

FIG. 2. Synthetic trace gathers.

ment on noise-free data to establish in some sense the discrimination or resolving power of several candidate coherence measures. We can in this way also gain a further appreciation of the properties of semblance.

A comparison of coherence measure threshold discrimination.—Figure 2 depicts four gathers of synthetic common ground point data. These data were designed specifically to test the resolving power and discrimination thresholds of candidate coherence measures. Each event indicated in the data is a doublet, the pulses being separated by 20 msec in two-way normal incidence time. In gathers 1 and 2, the rms velocities of the doublets are 50 ft/sec apart, while in gathers 3 and 4, the rms velocities differ by 200 ft/sec. In the odd numbered gathers, all events are Ricker wavelets of 40 msec width and unit amplitude. In the even gathers, the ghost or second event of each doublet is a 30 msec Ricker wavelet of half unit amplitude. Table 1 summarizes the characteristics of the synthetic data.

Since the objective of this computation is a resolution test rather than the modeling of physical reality, the time separation and velocity increments for the doublets were chosen so that the trajectories tend to cross. It is clear from Figure 2 that some of the doublets cannot easily be distinguished by visual examination. Hence, we have set up a demanding discrimination task for the several coherence measures.

Our procedure is direct and begins with discrete searches of the hyperbolic trajectories given by equation (18) over the two parameter space T_0, $\overline{V}$; we use a fixed time gate, time step, and velocity increment but different coherence measures. Contoured results will be presented for final comparison.

Since the data are noise free, any coherence measure if employed to sufficient accuracy will accomplish the desired task of event discrimination. Finite accuracy computer arithmetic and evaluation via contour or trace plots lead to a more demanding criterion. The coherence measures must make the distinctions and do so within a limited dynamic range of outputs. The practical state of routine seismic data processing parallels this type of discrimination.

With the preceding understanding of what the proposed experiment involved, three measures were examined: unnormalized crosscorrelation (called cross-power in the computations), semblance (energy normalized crosscorrelation), and summing. We conjecture that the results for a statistically normalized crosscorrelation will share many of the gross features of semblance but will exhibit less resolving power within the given dynamic range of display, particularly in the case where the doublet events differ in shape and amplitude. Computation of this coherence measure was deleted because of the additional programming effort required and the order of magnitude of additional arithmetic calculations needed (which are economic handicaps for its routine implementation).

All the calculations used a 12 msec time step,

Table 1. A summary of synthetic data characteristics

Normal incidence times—seconds	Gather Number											
	1			2			3			4		
	rms vel. ft/sec	Ampl.	Ricker* width	rms vel. ft/sec	Ampl.	Ricker* width	rms vel. ft/sec	Ampl.	Ricker* width	rms vel. ft/sec	Ampl.	Ricker* width
.500	6000	1	40	6000	1	40	6000	1	40	6000	1	40
.520	6050	1	40	6050	.5	30	6200	1	40	6200	−.5	30
1.000	8000	−1	40	8000	−1	40	8000	−1	40	8000	−1	40
1.020	8050	−1	40	8050	−.5	30	8200	−1	40	8200	−.5	30
1.500	9000	1	40	9000	1	40	9000	1	40	9000	1	40
1.520	9050	1	40	9050	.5	30	9200	1	40	9200	.5	30
2.000	10000	−1	40	10000	−1	40	10000	−1	40	10000	−1	40
2.020	10050	−1	40	10050	−.5	30	10200	−1	40	10200	−.5	30
2.500	12000	1	40	12000	1	40	12000	1	40	12000	1	40
2.520	12050	1	40	12050	.5	30	12200	1	40	12200	.5	30
3.000	14000	−1	40	14000	−1	40	14000	−1	40	14000	−1	40
3.020	14050	−1	40	14050	−.5	30	14200	−1	40	14200	−.5	30

* Widths are in milliseconds.

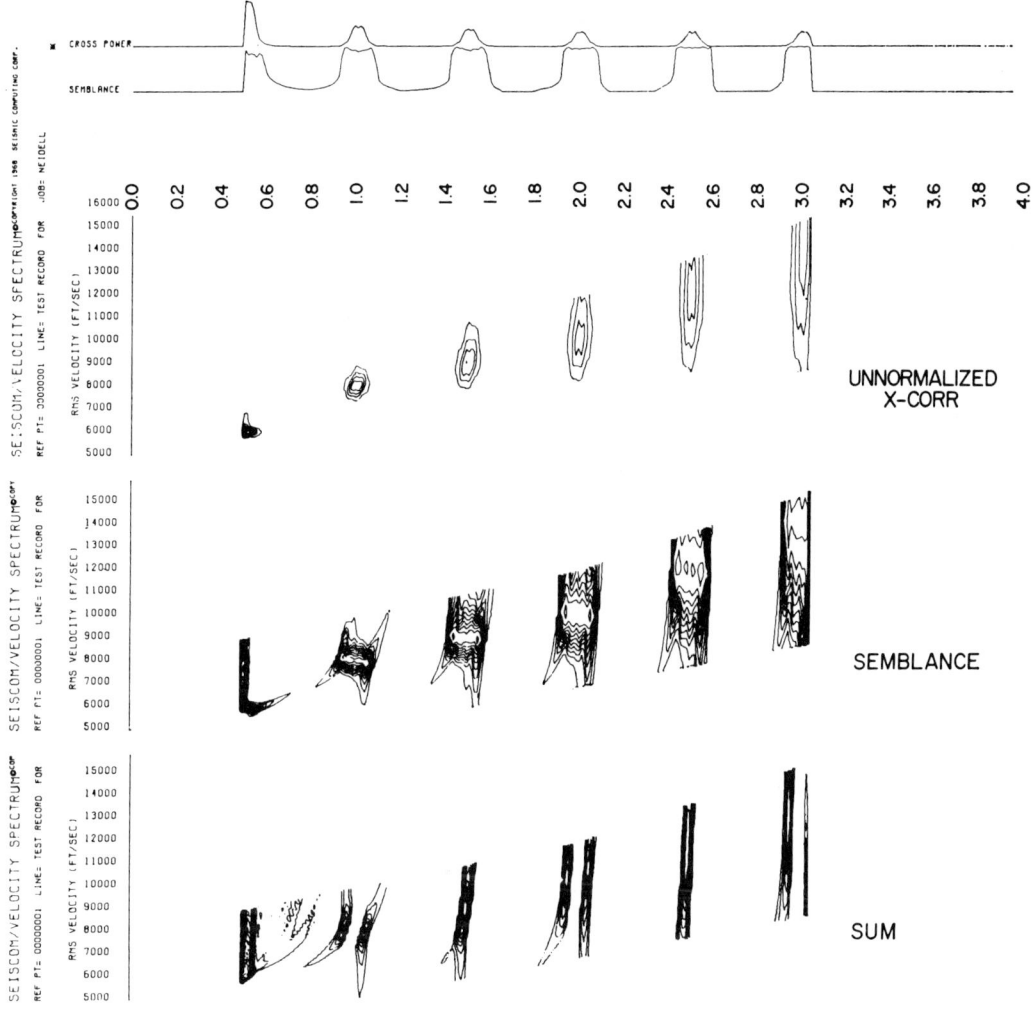

Fig. 3. Discrimination test. Gather 1.

a 50 ft/sec velocity increment, and a 48 msec time gate. The data were sampled at a 4 msec rate. Hence, for gathers 1 and 2, our doublet search is operating right at the size of the velocity step and at less than twice the time step size. For gathers 3 and 4, the stringent requirement on velocity is somewhat relaxed because the target velocities are now four times the search velocity step size. The contoured output results which we shall compare have the highest 70 percent of the data contoured at 10 levels.

Figures 3, 4, 5, and 6 show the computed results. In each, the contoured unnormalized correlation, semblance, and sum are compared. The generic relationship of the three methods is remarkably clear and is probably the first observation we might make. The second observation would be that semblance has recognized two events in almost every case, whereas the other two methods do not. We should also point out that semblance appears to increase its discrimination properties when the two events of the doublet are dissimilar. In this circumstance, the unnormalized crosscorrelation results deteriorated because they were unduly influenced by the larger event. The summing approach achieves good event discrimination when the trajectory crossover does not occur near the center of the gather; when the cross-over takes place near the center, this technique can identify only one event.

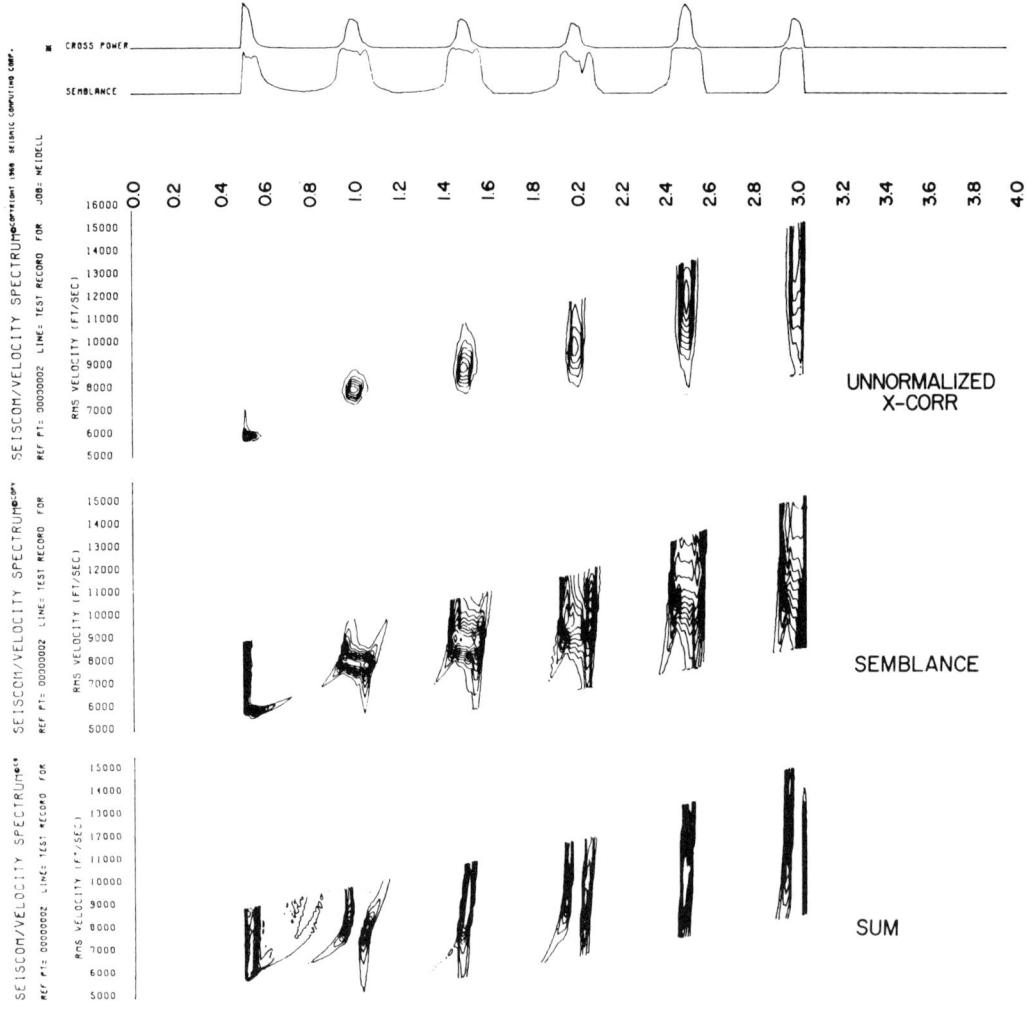

Fig. 4. Discrimination test. Gather 2.

Before we talk about ability to define event parameters, we should be reminded that this experiment was designed as a test of resolving power within the confines of a common form of display; hence, we should not expect any of the methods to perform well for the identification of event parameters. Semblance, nevertheless, performs best at this task as well, providing that we compensate for a bias effect which we know to be present in noise-free semblance calculations. We point out that elements of the bias effect are also present in the summing calculation, so it is worth discussing.

When an event first appears in a noise-free gate and is subjected to a semblance calculation, large semblance values will be computed for gates of similar makeup because the zero numerical values will be considered to look alike. Where noise is present, much of this tendency is effectively suppressed. Hence in our noise-free case, we should expect to see the semblance maximum broadened by the width of the search time gate at its onset and tail end. Also, a velocity aliasing effect comes into play, causing the maximum semblance contour to resemble a football with its center at the correct velocity but indicating too high a velocity too early and too low a velocity later on.

The effect just described operates on each event of the doublet pair and on the pair taken jointly. It is evident from the computed results that, de-

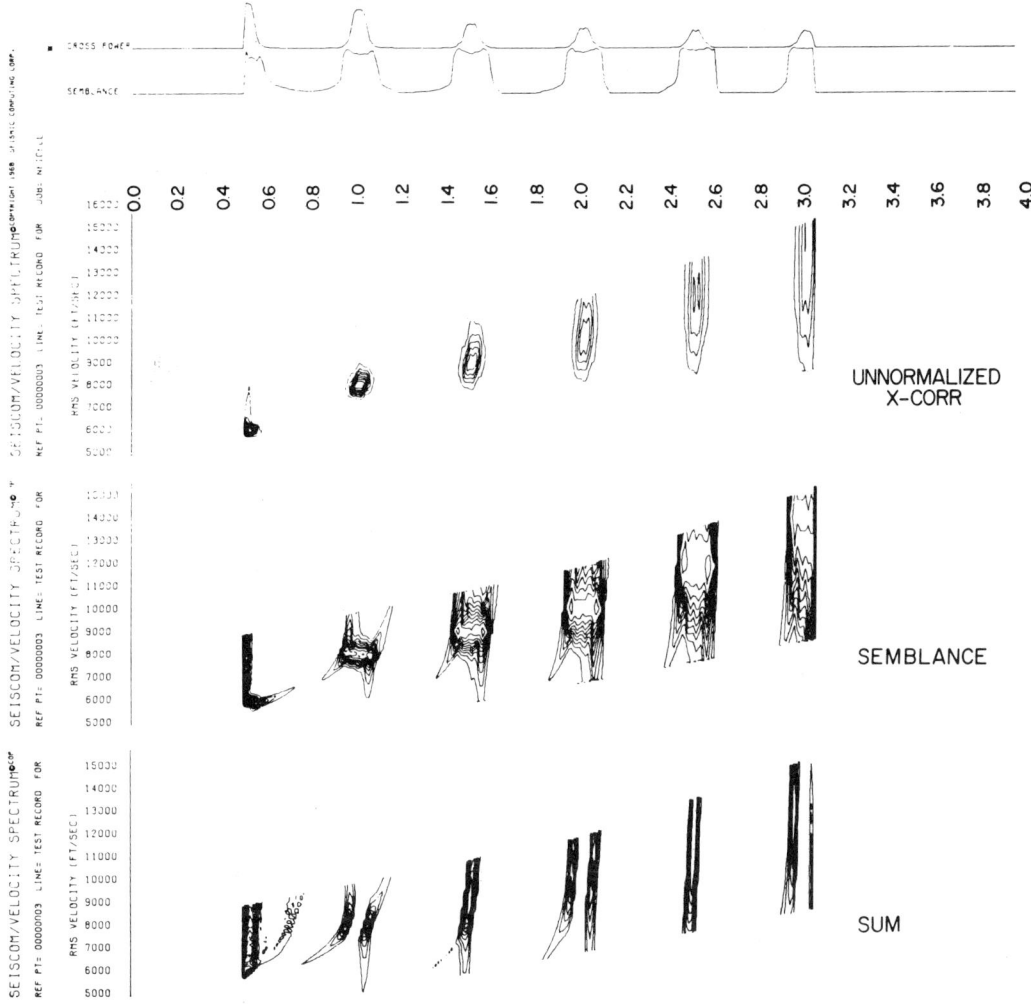

Fig. 5. Discrimination test. Gather 3.

spite our understanding, there are not many instances where we may unambiguously pick the correct parameters for a doublet. Nevertheless, the semblance results are better defined and closer to being correct in more cases than those derived from either of the alternative methods. Thus, we may also infer that semblance has an edge in parameter identification, as well as in power of resolution.

In summarizing the particular computational experiment we have just made, we can say that semblance has better properties of resolution and parameter identification than either of the two other methods examined. We add, however, that the other coherence methods will undoubtedly perform better than semblance in different problem environments and with different criteria for evaluation.

CONCLUDING REMARKS

This presentation has examined and discussed as a group coherence measures used with multichannel data. The concepts of semblance and of the semblance coefficient have been introduced and put into their proper place within this group. An attempt has been made to provide a unified treatment for this oft neglected subject, so that the relations among the members of the coherence measure family might be clarified. In working toward this goal, we considered each measure

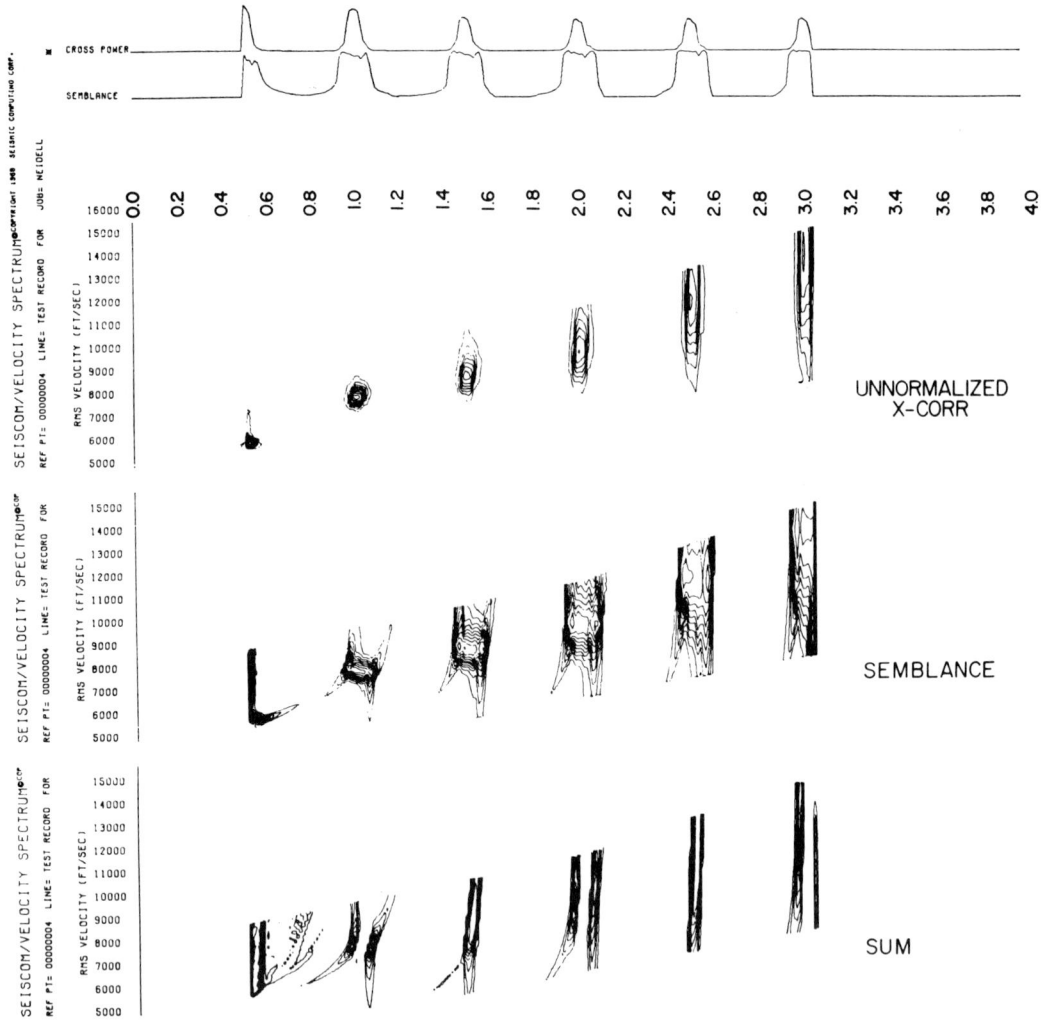

Fig. 6. Discrimination test. Gather 4.

from three points of view: its domain, its design philosophy, and its manner of employment. Many interesting insights into matters such as normalization arose during the investigations. As a consequence of our adopting a more general view toward coherence measures, they may be designed and applied according to scientific principles rather than tradition.

Computational examples explored the relative resolving power of several coherence methods in a problem situation drawn from exploration geophysics. This paper introduced semblance; hence, the particular problem chosen deliberately highlights the merit of this calculation. We should not lose sight of the worth of other coherence methods in other problem contexts.

ACKNOWLEDGMENTS AND HISTORICAL FOOTNOTES

Semblance is an outgrowth of the "Rho" function coherence measure suggested by Dr. F. Koehler several years earlier. M. T. Taner then attached energy interpretations to the concept. Since that time a number of people in the Seiscom organization have contributed to the understanding of and formalization of the semblance concept.

The author wishes to thank Mr. Robert Veazey

and Miss Kathleen Ward for assisting with the computations.

REFERENCES

Bendat, J. S., and Piersol, A. G., 1965, Measurement and analysis of random data for engineering applications: Lecture notes, UCLA extension course, August 30–September 3, 1965.

Cooley, J. W., and Tukey, J. W., 1965, An algorithm for the machine calculation of complex Fourier series: Math. Comput., v. 19, p. 297–301.

Courtier, W. H., and Mendenhall, H. L., 1967, Experiences with multiple coverage seismic methods: Geophysics, v. 32, p. 230–258.

Dix, C. H., 1955, Seismic velocities from surface measurements: Geophysics, v. 20, p. 68–86.

Fail, M. P., and Grau, G., 1963, Les filters en eventail: Geophys. Prosp., v. 11, p. 131–163.

Garotta, R., and Michon, D., 1967, Continuous analysis of the velocity function and of the move-out corrections: Geophys. Prosp., v. 15, p. 584–597.

Green, P. E., Kelly, E. J., and Levin, M. J., 1966, A comparison of seismic array processing methods: Geophys. J. of R.A.S., v. 11, p. 67–87.

Herman, A. S., 1968, Summary of velocity determination techniques: Pacific Coast Symposium on Velocity, Los Angeles.

Kendall, M. G., 1948, The advanced theory of statistics, v. 1 and 2: London, Griffin Co. Ltd.,

Koehler, F., and Taner, M. T., 1967, Velocity spectra—Digital computer derivation of velocity function: Preprint, 37th SEG Meeting, Oklahoma City.

Mayne, W. H., 1962, Common reflection point horizontal stacking techniques: Geophysics, v. 27, p. 927–938.

——— 1967, Practical considerations in the use of common reflection point techniques: Geophysics, v. 32, p. 225–229.

Merdler, S., Backus, M., Schneider, W., and King, T., 1968, Estimation of continuous reflection parameters from seismic data: 38th SEG Meeting, Denver.

Neidell, N. S., 1969, Ambiguity functions and the concept of geological correlation: Preprint, Computer Applications Symposium, AAPG, Dallas.

Pacific Coast SEG, 1968, Symposium on velocity: Los Angeles.

Pelto, C. R., 1954, Mapping of multicomponent systems: J. of Geol., v. 62, p. 501–511.

Pratt, W. K., Kane, J., and Andrews, H. C., 1969, Hadamard transform image coding: Proc. of the IEEE, v. 57, p. 55–68.

Rao, R. C., 1964, The use and interpretation of principal component analysis in applied research: Sankhya, Series A., v. 26, p. 329–358.

Reiber, F., 1936, A new reflection system with controlled directional sensitivity: Geophysics, v. 1, p. 97–106.

Robinson, J. C., 1969, HRVA—A velocity analysis technique applied to seismic data: Geophysics, v. 34, p. 330–356.

Schneider, W. A., and Backus, M. M., 1968, Dynamic correlation analysis: Geophysics, v. 33, p. 105–126.

Shanks, J. L., and Cairns, T. W., 1968, Use of a digital convolution device to perform recursive filtering and the Cooley-Tukey algorithm: IEEE Trans. of Comp., v. C-17, p. 943–949.

Simpson, S. M., Jr., 1955, Similarity of output traces as a seismic operator criterion: Geophysics, v. 20, p. 254–269.

——— 1967, Traveling signal-to-noise ratio and signal power estimates: Geophysics, v. 32, p. 485–493.

Taner, M. T., Cook, E. E., and Neidell, N. S., 1970, Limitations of the reflection seismic method; Lessons from computer simulations: Geophysics, v. 35, p. 551–573.

Weiner, N., 1949, Time series: Cambridge, MIT Press.

PREDICTIVE DECONVOLUTION: THEORY AND PRACTICE†

K. L. PEACOCK* AND SVEN TREITEL*

Least-squares inverse filters have found widespread use in the deconvolution of seismograms. The least-squares prediction filter with unit prediction distance is equivalent within a scale factor to the least-squares, zero-lag inverse filter. The use of least-squares prediction filters with prediction distances greater than unity leads to the method of predictive deconvolution which represents a more generalized approach to this subject.

The predictive technique allows one to control the length of the desired output wavelet, and hence to specify the desired degree of resolution. Events which are periodic within given repetition ranges can be attenuated selectively. The method is thus effective in the suppression of rather complex reverberation patterns.

INTRODUCTION

The Wiener filter is one of the most effective tools for the digital reduction of seismic traces. It constitutes the keystone of many current deconvolution methods. In one realization this filter is used to deconvolve a reverberating pulse train into an approximation of a zero-delay unit impulse. More generally it is possible to arrive at Wiener filters which remove repetitive events having specified periodicities. In this context the Wiener filter is better viewed as a predictor of coherent energy than merely as a spiker of "leggy" wave trains.

The prediction filter used in this treatment gives rise to the method of *predictive deconvolution*. We remark that Robinson's Ph.D. thesis (1954), if written today, would be entitled "*Predictive Deconvolution* of Time Series with Applications to Seismic Exploration," since the older term *decomposition* has given way to the newer term *deconvolution*. The method of predictive deconvolution has been described in a paper by Robinson (1966), in which the author advocates a prediction distance greater than unity. A discussion of the general properties of the digital Wiener filter has been given by Robinson and Treitel (1967).

BASIC CONCEPTS

The digital filtering process is described by the discrete convolution formula

$$y_\tau = \Delta t \sum_t x_t a_{\tau-t},$$

where x_t is the input, a_t is the filter, y_τ is the output, and Δt is the sampling increment. No loss of generality will result if we assume Δt to be unity. In the sequel t and τ are discrete time variables and $\Delta t = 1$ unless otherwise specified.

If a_t is a *prediction* operator with prediction distance α, the output y_τ will be an estimate of the input x_t at some future time $t+\alpha$. We thus write

$$y_\tau = \sum_t x_t a_{\tau-t} = \hat{x}_{t+\alpha}, \qquad (1)$$

where $\hat{x}_{t+\alpha}$ is an estimate of $x_{t+\alpha}$.

An error series may be defined as the difference between the true value $x_{t+\alpha}$ and the estimated or predicted value $\hat{x}_{t+\alpha}$,

$$\epsilon_{t+\alpha} = x_{t+\alpha} - \hat{x}_{t+\alpha}. \qquad (2)$$

Thus ϵ_t is an output series which represents the nonpredictable part of x_t.

Replacement of the term $\hat{x}_{t+\alpha}$ in equation (2) with its equivalent as defined by equation (1) results in

---

† Presented at the 38th Annual International SEG Meeting in Denver, Colorado, October 1, 1968. Manuscript received by the Editor October 10, 1968.

* Pan American Petroleum Corp., Research Center, Tulsa, Okla.

$$\epsilon_{t+\alpha} = x_{t+\alpha} - \sum_t x_t a_{\tau-t}. \quad (3)$$

The z-transform of equation (3) is

$$z^{-\alpha} E(z) = z^{-\alpha} X(z) - X(z) A(z). \quad (4)$$

Multiplication of both sides of equation (4) by z^α yields

$$E(z) = X(z) - z^\alpha X(z) A(z)$$
$$= X(z)[1 - z^\alpha A(z)]. \quad (5)$$

The quantity $[1-z^\alpha A(z)]$ is the z-transform of the so-called *prediction error operator*. It is seen to be the difference between the zero-delay unit spike and the prediction operator $A(z)$ delayed by the prediction distance α. Thus one may calculate the error series ϵ_t by computing $\hat{x}_{t+\alpha}$ from equation (1), and follow with a subtraction as defined by equation (2). Alternatively, one may compute the error series in a single step by use of the prediction error operator. Let us assume that a seismic trace is represented by the convolution of an uncorrelated reflection coefficient series with a reverberating pulse train, which by its nature is rich in repetitive energy (see Appendix A). The prediction error filter will then remove the predictable portion of such a trace, which to a good approximation will be given by the repetitive energy in the reverberations. The output of this filtering operation is the error series ϵ_t, which within the framework of this model constitutes the estimate of the reflection coefficient series of the layered subsurface.

Suppose that the prediction operator is given by the n-length series

$$a_t = a_0, a_1, \cdots, a_{n-1}.$$

Then the corresponding prediction error operator with prediction distance α is

$$f_t = 1, \overbrace{0, 0, \cdots, 0}^{\alpha - 1 \text{ zeros}}, -a_0, -a_1, \cdots, -a_{n-1}.$$

We must now deal with the explicit design of the prediction operator, a task which will be accomplished in the next section.

THE LEAST-SQUARES PREDICTIVE FILTERING MODEL

A general least-squares filter model involves the three signals illustrated in Figure 1, namely (1) the input signal x_t, (2) the desired output signal z_t, and (3) the actual output signal y_t. Minimization of the energy existing in the difference between the desired output z_t and the actual output y_t, i.e., minimization of the expression

$$I = \sum_t (z_t - y_t)^2,$$

results in the least-squares, or Wiener filter described by Robinson and Treitel (1967).

The n-length Wiener filter results from the solution of the normal equations with matrix representation,

$$\begin{bmatrix} r_0 & r_1 & \cdots & r_{n-1} \\ r_1 & r_0 & \cdots & r_{n-2} \\ \vdots & & & \vdots \\ r_{n-1} & r_{n-2} & \cdots & r_0 \end{bmatrix} \begin{bmatrix} f_0 \\ f_1 \\ \vdots \\ f_{n-1} \end{bmatrix} = \begin{bmatrix} g_0 \\ g_1 \\ \vdots \\ g_{n-1} \end{bmatrix}, \quad (6)$$

where r_t is the autocorrelation of the input, g_t is the crosscorrelation between the desired output and the input, and f_t is the Wiener filter.

We have seen in the previous section that the prediction operator is that filter which acts on an input trace up to time t and estimates the trace amplitude at some future time $t+\alpha$. Thus it is reasonable to define the desired output for the predictive filter as a time-advanced version of the input x_t. We can now express the prediction filter in terms of a particular Wiener filter, namely the one for which the desired output trace is simply a time-advanced version of the input trace.

In order to solve equation (6) we must know

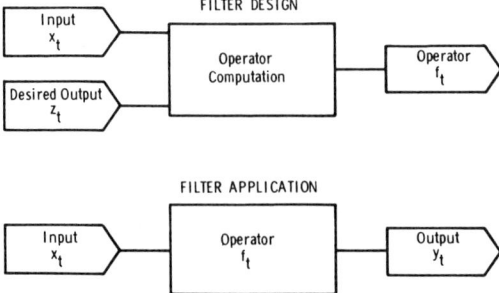

FIG. 1. A general model which illustrates the design and application of the Wiener filter.

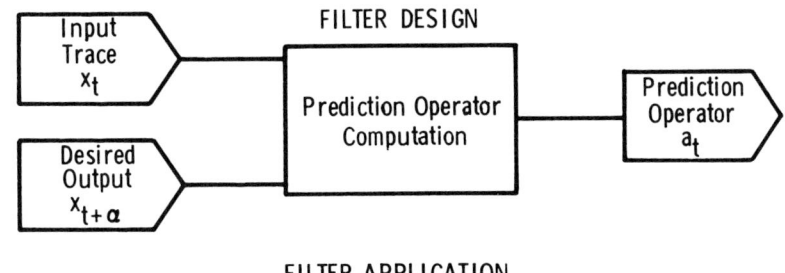
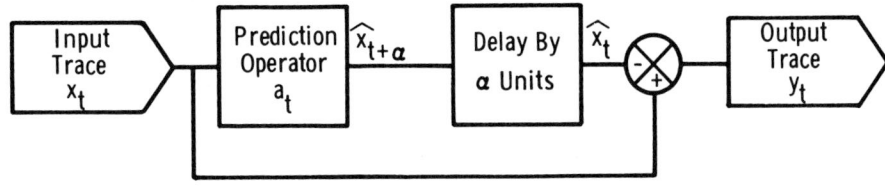
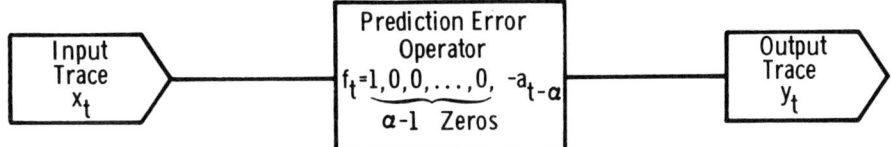

FIG. 2. The predictive filter model which illustrates prediction operator design (upper diagram) and application (center diagram). Alternately, the prediction error operator may be formed and utilized as illustrated in the lower diagram.

the autocorrelation of the input and the positive lag coefficients of the crosscorrelation between the desired output and the input. The autocorrelation of the input trace is given by

$$r_\tau = \sum_t x_t x_{t-\tau},$$

while the crosscorrelation between the desired output and input traces is

$$g_\tau = \sum_t z_t x_{t-\tau}. \quad (7)$$

Since the desired output for the prediction operator is a time-advanced version of the input, i.e., since

$$z_t = x_{t+\alpha},$$

equation (7) becomes

$$g_\tau = \sum_t x_{t+\alpha} x_{t-\tau} = \sum_t x_t x_{t-(\tau+\alpha)} = r_{\tau+\alpha}.$$

Thus, the crosscorrelation between the desired output and the input is by definition equal to the autocorrelation of the input for lags $\geq \alpha$. The normal equations (6) become

$$\begin{bmatrix} r_0 & r_1 & \cdots & r_{n-1} \\ r_1 & r_0 & \cdots & r_{n-2} \\ \vdots & & & \vdots \\ r_{n-1} & r_{n-2} & \cdots & r_0 \end{bmatrix} \begin{bmatrix} a_0 \\ a_1 \\ \vdots \\ a_{n-1} \end{bmatrix} = \begin{bmatrix} r_\alpha \\ r_{\alpha+1} \\ \vdots \\ r_{\alpha+n-1} \end{bmatrix}. \quad (8)$$

The solution to the above matrix equation yields the prediction operator illustrated in the upper diagram of Figure 2. This prediction operator can be utilized as indicated in the center diagram of Figure 2. Alternately, the corresponding prediction error operator can be formed from the prediction operator and utilized as indicated in the lower diagram of Figure 2.

We have thus shown how one can use the Wiener least squares error criterion to generate the least squares prediction operator, and how the prediction error operator is derived from its corresponding prediction operator. Our aim in the next section is to indicate how the prediction *error* operator can also be expressed in the form of a particular Wiener filter.

Let us modify the above system by first subtracting the coefficient r_i from both sides of the ith row of each equation such that the right-hand side vanishes (the original system is shown within the rectangle),

$$
\begin{array}{rl}
-r_1 + & \boxed{r_0 a_0 + r_1 a_1 + \cdots + r_{n-1} a_{n-1} = r_1} \quad -r_1 \\
-r_2 + & \boxed{r_1 a_0 + r_0 a_1 + \cdots + r_{n-2} a_{n-1} = r_2} \quad -r_2 \\
& \vdots \\
-r_n + & \boxed{r_{n-1} a_0 + r_{n-2} a_1 + \cdots + r_0 a_{n-1} = r_n} \quad -r_n.
\end{array}
$$

Let us next augment the above system in the form,

$$-r_0 + \quad r_1 a_0 + r_2 a_1 + \cdots + r_n a_{n-1} = -\beta$$

$$
\begin{array}{rl}
-r_1 + & \boxed{r_0 a_0 + r_1 a_1 + \cdots + r_{n-1} a_{n-1} = r_1} \quad -r_1 \\
-r_2 + & \boxed{r_1 a_0 + r_0 a_1 + \cdots + r_{n-2} a_{n-1} = r_2} \quad -r_2 \\
& \vdots \\
-r_n + & \boxed{r_{n-1} a_0 + r_{n-2} a_1 + \cdots + r_0 a_{n-1} = r_n} \quad -r_n.
\end{array}
$$

PREDICTIVE FILTERING AND DECONVOLUTION

We shall demonstrate that the least-squares deconvolution filter which ideally transforms an unknown signal to an impulse at zero delay is equivalent to the prediction *error* filter for which the prediction distance α is unity. The matrix relation for the prediction operator a_t with prediction distance unity ($\alpha = 1$) and length n is obtained by setting $\alpha = 1$ in equation (8),

$$\begin{bmatrix} r_0 & r_1 & \cdots & r_{n-1} \\ r_1 & r_0 & \cdots & r_{n-2} \\ & & \vdots & \\ r_{n-1} & r_{n-2} & \cdots & r_0 \end{bmatrix} \begin{bmatrix} a_0 \\ a_1 \\ \vdots \\ a_{n-1} \end{bmatrix} = \begin{bmatrix} r_1 \\ r_2 \\ \vdots \\ r_n \end{bmatrix}. \quad (9)$$

The above system may be written in the form of the n simultaneous linear equations,

$$
\begin{aligned}
r_0 a_0 + r_1 a_1 + \cdots + r_{n-1} a_{n-1} &= r_1 \\
r_1 a_0 + r_0 a_1 + \cdots + r_{n-2} a_{n-1} &= r_2 \\
&\vdots \\
r_{n-1} a_0 + r_{n-2} a_1 + \cdots + r_0 a_{n-1} &= r_n.
\end{aligned}
$$

This system may be written

$$
\begin{aligned}
r_0 - r_1 a_0 - r_2 a_1 - \cdots - r_n a_{n-1} &= \beta \\
r_1 - r_0 a_0 - r_1 a_1 - \cdots - r_{n-1} a_{n-1} &= 0 \\
&\vdots \\
r_n - r_{n-1} a_0 - r_{n-2} a_1 - \cdots - r_0 a_{n-1} &= 0,
\end{aligned}
$$

for which the associated matrix equation is

$$\begin{bmatrix} r_0 & r_1 & \cdots & r_n \\ r_1 & r_0 & \cdots & r_{n-1} \\ & & \vdots & \\ r_n & r_{n-1} & \cdots & r_0 \end{bmatrix} \begin{bmatrix} 1 \\ -a_0 \\ \vdots \\ -a_{n-1} \end{bmatrix} = \begin{bmatrix} \beta \\ 0 \\ \vdots \\ 0 \end{bmatrix}. \quad (10)$$

We now see that the Wiener filter of equation (10) can be identified as the unit prediction *error* operator associated with the prediction operator of equation (9). Let us rewrite equation (10) in the form

$$\begin{bmatrix} r_0 & r_1 & \cdots & r_n \\ r_1 & r_0 & \cdots & r_{n-1} \\ & & \vdots & \\ r_n & r_{n-1} & \cdots & r_0 \end{bmatrix} \begin{bmatrix} b_0 \\ b_1 \\ \vdots \\ b_n \end{bmatrix} = \begin{bmatrix} \beta \\ 0 \\ \vdots \\ 0 \end{bmatrix}, \quad (11)$$

where

$$b_0 = 1$$
$$b_i = -a_{i-1}, \quad i = 1, \cdots, n$$
$$\beta = \sum_{i=0}^{n} b_i r_i.$$

In Appendix B we describe the standard deconvolution method, which is based on the use of the least-squares, zero-delay inverse filter. We note that the system of normal equations for the inverse filter given by equation (B-1) is identical to the system of equations (11), except for a scale factor β. Thus the $(n+1)$-length prediction error operator with prediction distance unity is identical to the zero-delay inverse filter of length $(n+1)$, except for a scale factor.

We shall now show that the predictive filter with prediction distance greater than unity can also serve as a deconvolution operator, and thus it turns out that the predictive filtering technique constitutes a more generalized approach to deconvolution. We remark that under certain assumptions described in Appendix A, the autocorrelation of an input seismic signal can be identified with the autocorrelation of the source wavelet[1].

The inverse filter described in Appendix B shapes the unknown source wavelet to an impulse at zero lag time. We will show here that the predictive filter shapes the unknown source wavelet of length $\alpha+n$ to another unknown wavelet of length α. Thus, by having control of the desired output wavelet length, one may specify the desired degree of resolution.

The predictive filter matrix equation for filter length n and prediction distance α is given by equation (8),

$$\begin{bmatrix} r_0 & r_1 & \cdots & r_{n-1} \\ r_1 & r_0 & \cdots & r_{n-2} \\ & & \vdots & \\ r_{n-1} & r_{n-2} & \cdots & r_0 \end{bmatrix} \begin{bmatrix} a_0 \\ a_1 \\ \vdots \\ a_{n-1} \end{bmatrix} = \begin{bmatrix} r_\alpha \\ r_{\alpha+1} \\ \vdots \\ r_{\alpha+n-1} \end{bmatrix},$$

or

$$r_0 a_0 + r_1 a_1 + \cdots + r_{n-1} a_{n-1} = r_\alpha$$
$$r_1 a_0 + r_0 a_1 + \cdots + r_{n-2} a_{n-1} = r_{\alpha+1}$$
$$\vdots \qquad (12)$$
$$r_{n-1} a_0 + r_{n-2} a_1 + \cdots + r_0 a_{n-1} = r_{\alpha+n-1}.$$

The above system can be augmented in such a way that the prediction operator is converted into its corresponding prediction error operator. This is accomplished by the addition of suitable terms to both sides of the equations (12). Proceeding as in the case of the unit prediction error filter (equations (9) et seq.), one obtains,

$$-r_0 1 - r_1 0 - \cdots - r_{\alpha-1} 0 + r_\alpha a_0 + r_{\alpha+1} a_1 + \cdots + r_{\alpha+n-1} a_{n-1} = -\rho_0$$
$$-r_1 1 - r_0 0 - \cdots - r_{\alpha-2} 0 + r_{\alpha-1} a_0 + r_\alpha a_1 + \cdots + r_{\alpha+n-2} a_{n-1} = -\rho_1$$
$$\vdots$$
$$-r_{\alpha-1} 1 - r_{\alpha-2} 0 - \cdots - r_0 0 + r_1 a_0 + r_2 a_1 + \cdots + r_n a_{n-1} = -\rho_{\alpha-1}$$
$$\vdots$$
$$-r_\alpha 1 - r_{\alpha-1} 0 - \cdots - r_1 0 + \boxed{r_0 a_0 + r_1 a_1 + \cdots + r_{n-1} a_{n-1} = r_\alpha} \quad -r_\alpha$$
$$-r_{\alpha+1} 1 - r_\alpha 0 - \cdots - r_2 0 + \phantom{\boxed{}} r_1 a_0 + r_0 a_1 + \cdots + r_{n-2} a_{n-1} = r_{\alpha+1} \quad -r_{\alpha+1}$$
$$\vdots$$
$$-r_{\alpha+n-1} 1 - r_{\alpha+n-2} 0 - \cdots - r_n 0 + \phantom{\boxed{}} r_{n-1} a_0 + r_{n-2} a_1 + \cdots + r_0 a_{n-1} = r_{\alpha+n-1} \quad -r_{\alpha+n-1}$$

where the original set (12) is enclosed by the rectangle. The associated matrix equation is

---

[1] The source wavelet is here meant to be the shot pulse modified by near-surface reverberations.

$$\begin{bmatrix} r_0 & r_1 & \cdots & r_{\alpha+n-1} \\ r_1 & r_0 & \cdots & r_{\alpha+n-2} \\ & & \vdots & \\ r_{\alpha-1} & r_{\alpha-2} & \cdots & r_n \\ r_\alpha & r_{\alpha-1} & \cdots & r_{n-1} \\ & & \vdots & \\ r_{\alpha+n-1} & r_{\alpha+n-2} & \cdots & r_0 \end{bmatrix} \begin{bmatrix} 1 \\ 0 \\ \vdots \\ 0 \\ -a_0 \\ \vdots \\ -a_{n-1} \end{bmatrix}$$

$$= \begin{bmatrix} \rho_0 \\ \rho_1 \\ \vdots \\ \rho_{\alpha-1} \\ 0 \\ \vdots \\ 0 \end{bmatrix}, \quad (13)$$

where

$$\rho_0 = r_0 - (r_\alpha a_0 + r_{\alpha+1} a_1 + \cdots + r_{\alpha+n-1} a_{n-1})$$
$$\rho_1 = r_1 - (r_{\alpha-1} a_0 + r_\alpha a_1 + \cdots + r_{\alpha+n-2} a_{n-1})$$
$$\vdots$$
$$\rho_{\alpha-1} = r_{\alpha-1} - (r_1 a_0 + r_2 a_1 + \cdots + r_n a_{n-1}).$$

The solution of the above matrix equation yields the prediction error operator with prediction distance α. Let us interpret this equation in terms of the Wiener filter model, where the left-hand matrix is the input autocorrelation matrix, and where the elements of the right-hand column vector constitute the positive lag values of the crosscorrelation between the desired output and the input. Subject to the assumptions given in Appendix A, the autocorrelation function $r_0, r_1, \cdots, r_{\alpha+n-1}$ can be identified with the autocorrelation of a source wavelet of length $\alpha+n$. However, we still require an interpretation of the crosscorrelation,

$$g_\tau = \underbrace{\rho_0, \rho_1, \cdots, \rho_{\alpha-1},}_{\alpha \text{ terms}} \underbrace{0, \cdots, 0.}_{n \text{ zeros}}$$

Although the crosscorrelation function is complicated, we can make one important observation. Since the crosscorrelation vanishes for lags greater than $\alpha-1$, the length of the implied desired output wavelet cannot be greater than α. In other words, the input wavelet is of length $\alpha+n$, while the implied desired output wavelet is of length α, and hence the prediction error operator shortens an input wavelet of length $\alpha+n$ to an output wavelet of length α. Since α is an independent variable, we are free to select whatever length we choose for the desired output wavelet. We conclude that the predictive filter leads to a more generalized approach to deconvolution, in which one may control the desired degree of resolution or wavelet contraction.

We have shown earlier in this section that the zero-delay least-squares, inverse filter is equal within a scale factor to the prediction error filter with prediction distance unity ($\alpha=1$). Experience has taught us that the output from these filters cannot in general be interpreted with ease. This is due to the presence of high-frequency components in the deconvolved trace, which result from the fact that this kind of deconvolution makes use of inverse, or "spiking" filters. One improves this condition by passing the raw deconvolved trace through suitable low-pass filters, by smoothing the autocorrelation function, or by other related means. We suggest that the use of prediction error filters with arbitrary prediction distance α leads to a deconvolution method in which one has more effective control on the desired degree of resolution. It is also significant to note that the inverse filter deconvolution method requires the insertion of an arbitrary scaling factor into the right side of the normal equations (i.e., the element β of equation (11) is arbitrary). No such scaling factor is needed in the predictive filter model, and hence the trace-to-trace amplitude variation which occurs in the input data can be preserved if so desired. In addition, no time need be spent by the computer in analyzing the output data to determine the scaling factor.

It is of some interest to establish how the predictive filters presented in this section perform on an idealized, noise-free reverberation model. These matters are discussed in Appendix C, where we also show that under appropriate simplifying conditions the prediction error filter becomes identical to the 3-point filter of Backus (1959).

APPLICATIONS OF PREDICTIVE DECONVOLUTION

The concept which permits resolution control by means of the prediction distance parameter has been introduced in the previous section. We have seen that the crosscorrelation between the desired output and the input is zero between lag positions α and $\alpha+n-1$, and we were thus able to deduce that the implied desired output pulse cannot be of length greater than α. We note from the autocorrelation matrix of equation (13) that the predictive filter does not utilize any autocorrelation coefficient beyond lag position $\alpha+n-1$. Our model thus implies that the source wavelet is of length $\alpha+n$.

Since this filter attempts to shape the input into some desired output, we can argue that the autocorrelation of the actual output data will tend to vanish between lag positions α and $\alpha+n-1$. This is because the autocorrelation of the implied desired output wavelet vanishes for lags greater than $\alpha-1$. The predictive filter will thus modify the input in such a way that the autocorrelation of the actual output will tend to vanish between α and $\alpha+n-1$. We cannot expect the autocorrelation to be 0 everywhere beyond lag$=\alpha+n-1$, since the filter computation makes use of no autocorrelation coefficients for lags greater than $\alpha+n-1$.

Anstey (1966) describes how the autocorrelation can be an interpretative aid in the analysis of reverberatory problems. When we consider that the predictive filter is designed only from knowledge of the input autocorrelation and that the magnitude of this input autocorrelation at a particular lag is an indication of the degree of predictability at that lag, we see that the autocorrelogram is a very important entity to gauge the effectiveness of dereverberation by means of predictive deconvolution.

Thus we set our parameters α (the prediction distance) and n (the prediction operator length) such that predictable (i.e., repetitive) energy having periods between α and $\alpha+n-1$ time units will tend to be removed, and hence the autocorrelation of the output will tend to vanish between lags α and $\alpha+n-1$.

Another means to measure the effectiveness of the predictive deconvolution process has been given by Wadsworth et al (1953). These authors point out that the reduction in energy content of the output trace relative to the input trace gives a measure of the predictable energy removed by the filtering operation in the range $t=\alpha$ to $t=\alpha+n-1$.

A given reverberation may be characterized as either "short-period" or "long-period." Long-period reverberations appear on a correlogram as distinct waveforms which are separated by quiet

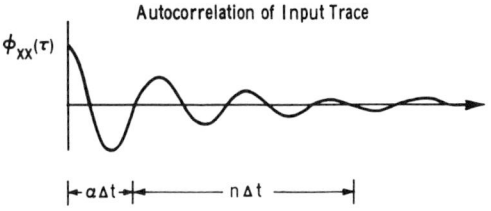

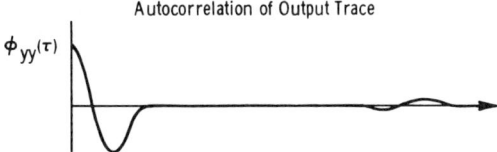

Fig. 3. A typical autocorrelation of an input trace with a moderate amount of short-period reverberation is illustrated by the upper diagram. The lower diagram illustrates the appearance of the output autocorrelation after application of the prediction error operator with prediction distance α and length n, as established in the upper diagram. If the reverberating pattern on the autocorrelation is highly regular, one need set n such that only the first full cycle is spanned. If the pattern is irregular, the significant portion of the reverberation should be spanned.

zones. Short-period reverberations appear on a correlogram in the form of decaying waveforms which are not separated by any noticeable quiet intervals.

The upper diagram of Figure 3 illustrates the autocorrelation of a typical trace which exhibits a moderate degree of short-period reverberation. The prediction distance α is chosen to specify the degree of wavelet contraction desired. As α approaches unity, more contraction and consequently more high-frequency noise is introduced. Thus we choose α so that we may obtain a compromise between wavelet contraction and signal-to-noise ratio in the output trace. Preliminary studies indicate that α should be set roughly equal to the lag that corresponds to the second zero crossing of the autocorrelation function. The lower diagram of Figure 3 illustrates the fact that the autocorrelation of the output signal trace tends to zero between lags α and $\alpha+n-1$.

Figure 4 shows three different predictive deconvolution runs on offshore traces having reverberations with characteristics somewhat in-between our definitions of the "short-period" and "long-period" types. The product $\alpha\Delta t$ has been given the values 32, 16, and 4 ms. Since this data has a sampling interval of 4 ms, the third run actually corresponds to deconvolution by the

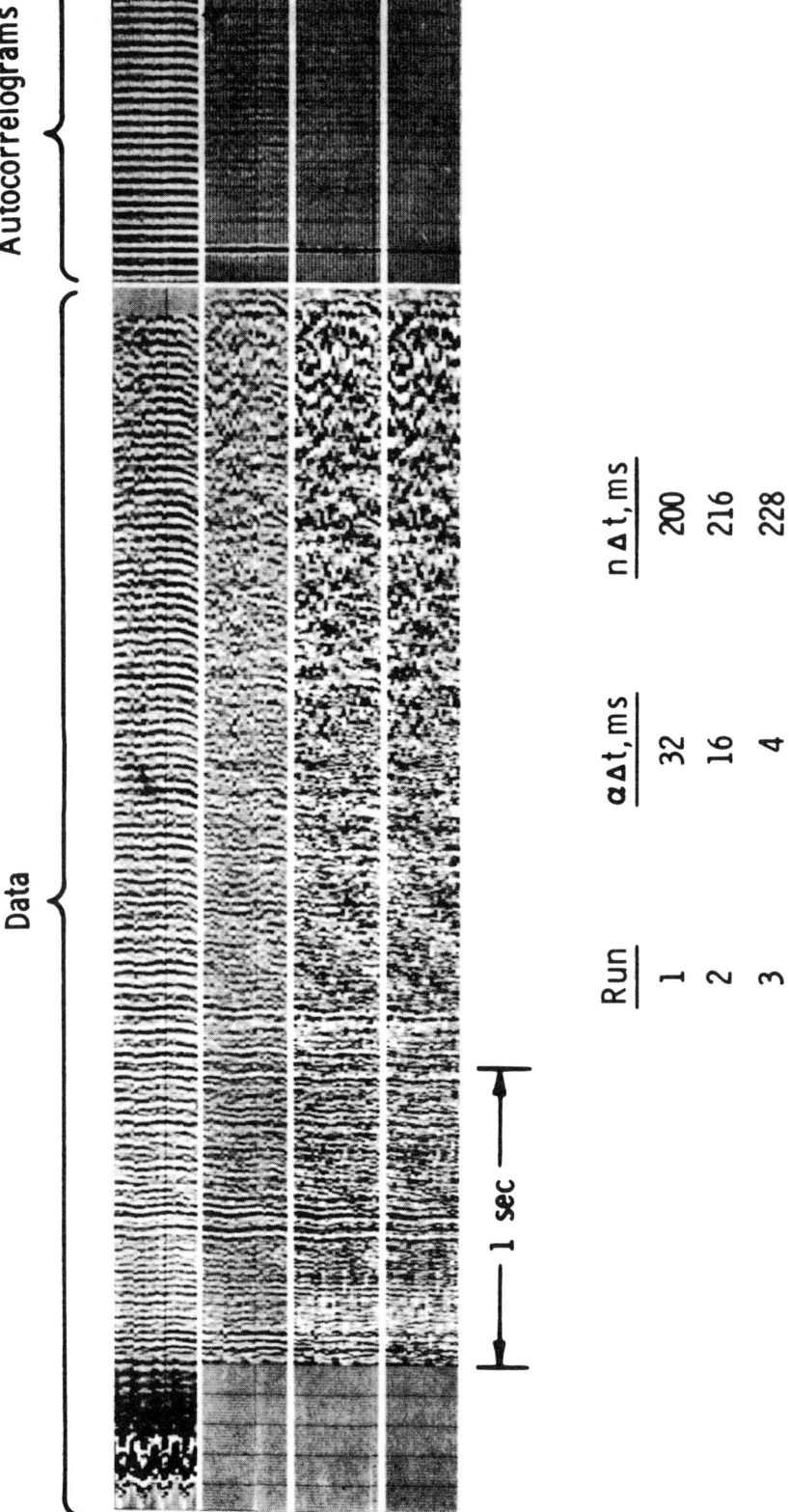

FIG. 4. An illustration of the effect which a variation in α, the prediction distance, has upon the output data. There appears to be some value of α which gives the best compromise between deringing and signal-to-noise ratio. Seismograms—top to bottom. Input; Outputs 1, 2, 3.

Run	$\alpha \Delta t$, ms	$n \Delta t$, ms
1	32	200
2	16	216
3	4	228

zero-delay least-squares inverse filter. We see that for $\alpha \Delta t = 32$ ms ($\alpha = 8$), we obtain a good dereverberation which does not exhibit the noise build-up associated with the smaller prediction distances.

The upper diagram of Figure 5 illustrates the autocorrelation of a typical trace with long-period reverberations. We define the appropriate prediction distance α such that the window to be deleted on the autocorrelogram begins just before the onset of the first multiple indication. Depending upon the nature and period of the reverberation, we define the filter length n such that the window to be deleted spans one, two, or more orders of the multiple pattern. The autocorrelation of the resulting output trace will show very little energy between lags α and $\alpha + n - 1$. In addition, further repetitions of the waveform centered at multiples of $\alpha + n/2$ will be attenuated.

We note that the predictive filter enables us to suppress selected waveform portions of the autocorrelation function. This is highly advantageous since some waveforms on the autocorrelation might be due to accidentally strong correlations between certain primary reflections, and in this case we would choose not to suppress them. We may indeed avoid their suppression by the proper selection of the prediction distance α, and we then concentrate on those waveforms associated with reverberations. We remark that we can often successfully remove the long-period multiple indication from the autocorrelation by means of the predictive filter. Even so, the net change on the section itself is not always significant. Perhaps more study will reveal better ways of selecting the parameters such that long-period reverberations will be better attenuated by the predictive filter.

Figure 6 illustrates a predictive deconvolution run on a record with short-period reverberations. The data and associated autocorrelograms show the pulse compression which has been obtained.

Figure 7 depicts a predictive deconvolution run on marine data which exhibits long-period reverberations. We note that the prediction distance in this case is 150 ms and that the filter length is only 60 ms. If one were to deconvolve these traces with the unit prediction error filter, it would be necessary to make the filter length at least equal to 210 ms. This would require much more time to process the data. Kunetz and Fourmann (1968) have reached similar conclusions by a somewhat different line of reasoning. We note

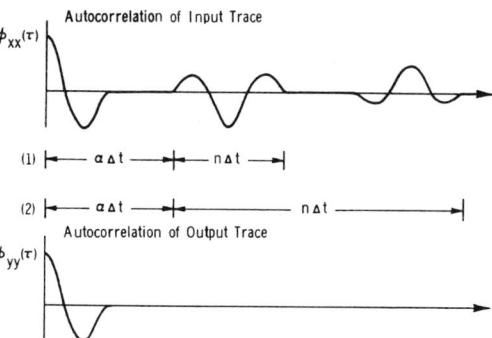

FIG. 5. A typical autocorrelation of an input trace with a moderate amount of long-period reverberation is illustrated by the upper diagram. The lower diagram illustrates the appearance of the output autocorrelation after application of the prediction error operator with prediction distance α and length n as established in the upper diagram. The parameters should be defined as indicated in (1) if the ringing is of a first-order nature, or as indicated in (2) if the ringing is of a second-order nature.

that we have achieved a successful attenuation of the reverberations on the autocorrelograms and a moderately successful dereverberation of the data itself.

CONCLUSIONS

The predictive filter is a very flexible tool for the deconvolution of seismic traces. The ability to specify the prediction distance implies the ability to control output resolution, and this means that a broad range of complex reverberatory problems can be successfully attacked with the present methods. Repetitive waveforms of a particular period can be selectively attenuated, and this is accomplishable without any significant disturbance of waveforms which one may wish to retain. The autocorrelogram is a valuable interpretative device for reverberation analysis and should be used on a routine basis. It would be desirable to have still better criteria for the determination of optimum values of such filter parameters as the prediction distance α and the filter length n. More research and evaluation of the methods presented in this paper are therefore in order.

ACKNOWLEDGMENTS

The authors wish to express their thanks to Mr. C. W. Frasier for the use of some of his unpublished results and to the Pan American Petroleum Corporation for permission to publish this paper.

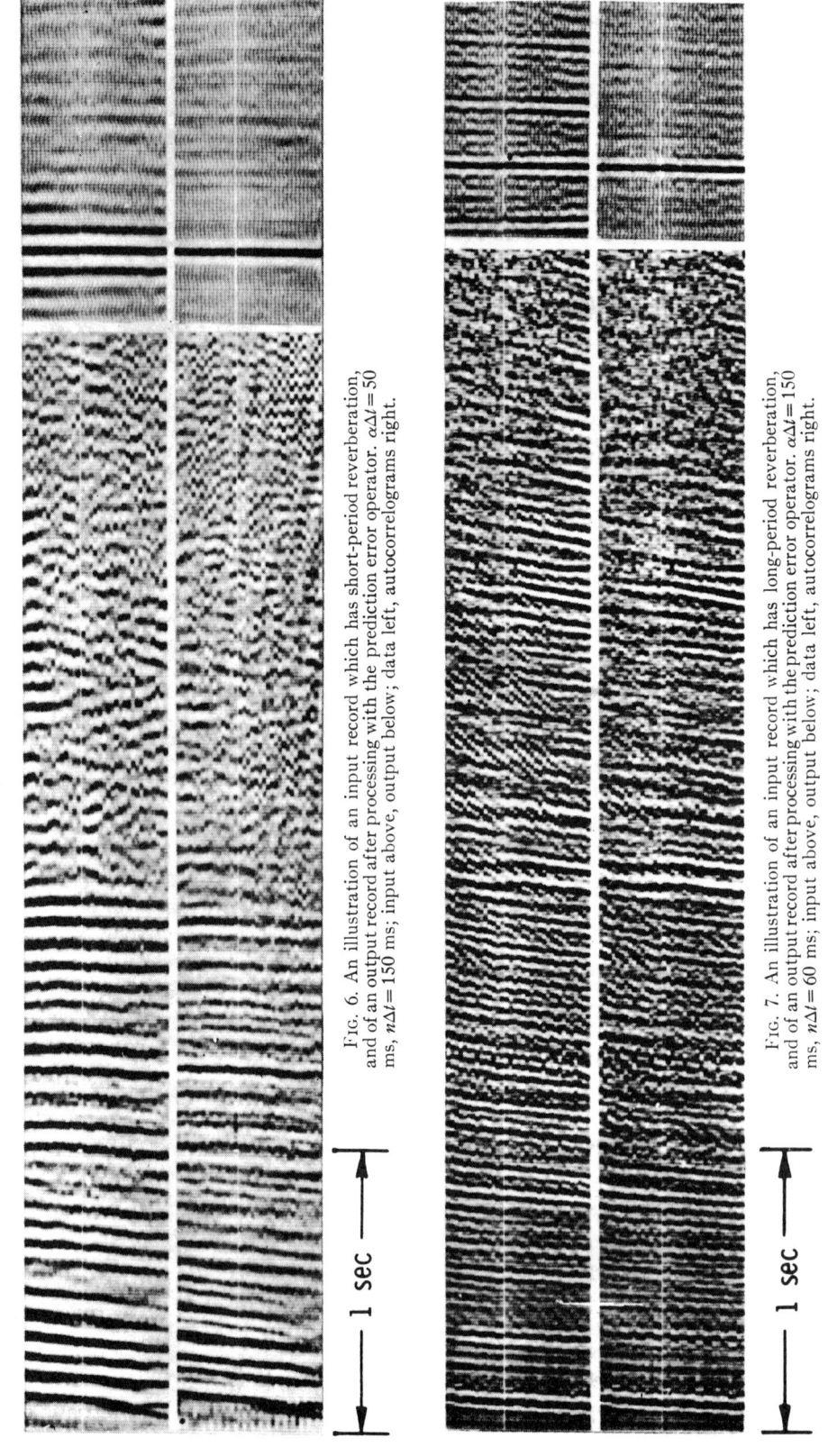

FIG. 6. An illustration of an input record which has short-period reverberation, and of an output record after processing with the prediction error operator. $\alpha\Delta t = 50$ ms, $n\Delta t = 150$ ms; input above, output below; data left, autocorrelograms right.

FIG. 7. An illustration of an input record which has long-period reverberation, and of an output record after processing with the prediction error operator. $\alpha\Delta t = 150$ ms, $n\Delta t = 60$ ms; input above, output below; data left, autocorrelograms right.

REFERENCES

Anstey, N. A., 1966, The sectional autocorrelogram and the sectional retrocorrelogram: Geophys. Prosp., v. 14, p. 389–426.

Backus, M. M., 1959, Water reverberations—Their nature and elimination: Geophysics, v. 24, p. 233–261.

Jenkins, G. M., 1961, General considerations in the analysis of spectra: Technometrics, v. 3, no. 2, p. 133–166.

Kunetz, G., and Fourmann, J. M., 1968, Efficient deconvolution of marine seismic records: Geophysics, v. 33, p. 412–423.

Robinson, E. A., 1954, Predictive decomposition of time series with applications to seismic exploration: Ph.D. thesis, MIT, Cambridge, Mass.

——— 1966, Multichannel z-transforms and minimum-delay: Geophysics, v. 31, p. 482–500.

——— and Treitel, S., 1967, Principles of digital Wiener filtering: Geophys. Prosp., v. 15, p. 311–333.

Wadsworth, G. P., Robinson, E. A., Bryan, J. G., and Hurley, P. M., 1953, Detection of reflections on seismic records by linear operators: Geophysics, v. 18, p. 539–586.

APPENDIX A

THE AUTOCORRELATION OF A SEISMIC TRACE

Under the proper assumptions the autocorrelation of a seismic trace is an estimate of the autocorrelation of the "basic" seismic wavelet.[2] The derivation presented here is similar to one given by Robinson and Treitel (1967).

Suppose we have a signal x_t which results from the convolution of a basic wavelet p_t with an uncorrelated series n_t, where we assume that n_t can be identified with the reflection coefficient series of a layered medium (Robinson, 1954), that is,

$$x_t = p_t * n_t.$$

The z-transform of the autocorrelation of x_t is given by

$$\Phi_{xx}(z) = [P(z)N(z)][P(1/z)N(1/z)].$$

The above equation can be rewritten in the form,

$$\Phi_{xx}(z) = [P(z)P(1/z)][N(z)N(1/z)],$$

which is the z-transform of

$$\phi_{xx}(\tau) = \phi_{pp}(\tau) * \phi_{nn}(\tau). \qquad (A-1)$$

Therefore the autocorrelation of x_t is equal to the convolution of the autocorrelation of p_t with the autocorrelation of n_t. Since n_t is an uncorrelated series, we obtain

---

[2] The basic seismic wavelet is assumed to be either the initial shot pulse, or the initial shot pulse modified by near-surface reverberations.

$$\phi_{nn}(\tau) = E_n \quad \text{for } \tau = 0$$

and

$$\phi_{nn}(\tau) = 0 \quad \text{for } \tau \neq 0,$$

where E_n is the energy in n_t. Thus equation (A-1) reduces to

$$\phi_{xx}(\tau) = \sum_t \phi_{nn}(t)\phi_{pp}(\tau - t) = E_n \phi_{pp}(\tau),$$

and we see that the autocorrelation of x_t is simply a scaled version of the autocorrelation of p_t. This means that subject to the above assumptions, we can obtain an estimate of ϕ_{pp} even though we do not know p_t itself. The consistency of the autocorrelation estimates can be improved through use of suitable weighting functions. A good discussion of these matters is given by Jenkins (1961).

APPENDIX B

THE INVERSE FILTER MODEL

The Wiener filter model requires that the autocorrelation of the input and the positive lag values of crosscorrelation between the desired output and the input be known. The basic seismic wavelet is generally unknown; however, we can calculate its autocorrelation and the required crosscorrelation if we make the proper assumptions.

Appendix A shows that an estimate of the basic wavelet autocorrelation can be obtained from the input trace. If we assume the desired output to be an impulse at zero lag time, the crosscorrelation between desired output and input also becomes an impulse at zero lag time. In other words, since the crosscorrelation is given by

$$\phi_{dp}(\tau) = \sum_t d_t p_{t-\tau} \quad \text{for } \tau = 0, 1, \cdots, n-1,$$

where $d_t = 1, 0, 0, \cdots$ is the desired output signal and $p_t = p_0, p_1, p_2, \cdots$ is the basic wavelet or input signal, we see that

$$\phi_{dp}(\tau) = p_0, 0, 0, \cdots ;$$
$$\tau = 0, 1, \cdots, n-1,$$

which can be scaled in the form,

$$\phi_{dp}(\tau) = 1, 0, 0, \cdots .$$

The matrix equation for the Wiener filter (Robinson and Treitel, 1967) then becomes,

$$\begin{bmatrix} r_0 & r_1 & \cdots & r_{n-1} \\ r_1 & r_0 & \cdots & r_{n-2} \\ & & \vdots & \\ r_{n-1} & r_{n-2} & \cdots & r_0 \end{bmatrix} \begin{bmatrix} f_0 \\ f_1 \\ \vdots \\ f_{n-1} \end{bmatrix} = \begin{bmatrix} 1 \\ 0 \\ \vdots \\ 0 \end{bmatrix}, \quad \text{(B-1)}$$

where the f_t are the n coefficients which shape the basic wavelet p_t to an approximation of the impulse at zero lag time.

We have thus assumed a model of the form

$$x_t = p_t * n_t,$$

where x_t is the signal trace and n_t is an uncorrelated series which represents the reflection coefficients of the layered subsurface. Since the desired output is an impulse at zero lag time, we see that the model requires a filter f_t such that

$$f_t * p_t \doteq 1$$

or

$$f_t \doteq p_t^{-1},$$

where the symbol $\doteq$ means "approximately equal to." The filter f_t then deconvolves the input trace as follows:

$$y_t \doteq x_t * f_t \doteq p_t * p_t^{-1} * n_t$$

$$y_t \doteq \delta_t * n_t = n_t,$$

where δ_t is the unit impulse function, and in this sense the output trace tends to approximate the subsurface reflection coefficient series.

APPENDIX C

A STUDY OF A TWO-LAYER MARINE REVERBERATION MODEL

Impulse response of first-order component

Figure C-1 represents an idealized noise-free model of an offshore seismic situation. Reflector 1 is the water surface, reflector 2 is the water bottom, reflector 3 is some strong interface beneath the water bottom, and S is the source location just beneath the water surface. The associated normal incidence reflection coefficients are 1, c_1 and c_2, respectively, while the transmission coefficient across reflector 2 is t_1. If c is the downward reflection coefficient, the corresponding upward reflection coefficient is $-c$. From physical considerations, we know that the magnitudes of all re-

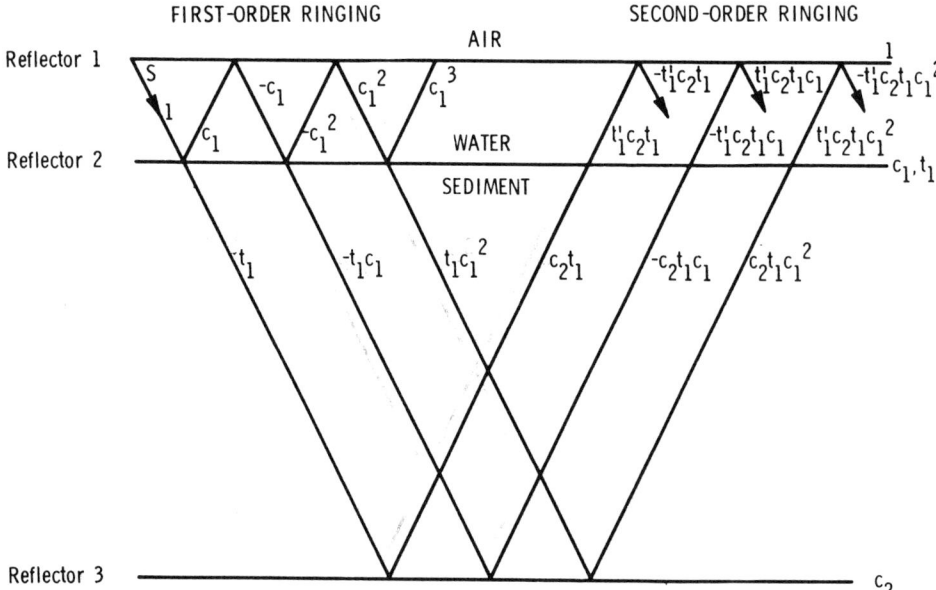

Fig. C-1. First- and second-order ringing in a 2-layer marine model.

flection coefficients are less than unity. The two-way traveltime through the water layer is τ_1.

Let us compute the first-order reverberation portion of the two-layer impulse response as indicated in Figure C-1. This response can be expressed in terms of the following z-transform:

$$R_1(z) = 1 - c_1 z^{\tau_1} + c_1^2 z^{2\tau_1} + \cdots . \quad \text{(C-1)}$$

Multiplication of equation (C-1) by $c_1 z^{\tau_1}$ produces

$$c_1 z^{\tau_1} R_1(z) = c_1 z^{\tau_1} - c_1^2 z^{2\tau_1} + c_1^3 z^{3\tau_1} + \cdots , \quad \text{(C-2)}$$

while addition of equations (C-1) and (C-2) yields the expression

$$R_1(z) = \frac{1}{1 + c_1 z^{\tau_1}} . \quad \text{(C-3)}$$

We have thus obtained the z-transform of the first-order reverberation portion of the two-layer impulse response.

(C.1) *Impulse response of second-order component*

Figure C-1 indicates the pattern of all raypaths which contribute to second-order reverberations. A component from the original impulse is reflected from reflector 3 and re-enters the water layer to be reflected from the surface. At this point its amplitude is $-t_1' c_2 t_1$, where t_1' is the upward transmission coefficient across reflector 2. Hence an impulse of amplitude $-t_1' c_2 t_1$ is introduced into the first layer, which again will generate the associated first-order ringing already given by equation (C-3). However, the onset of this ringing occurs with a time delay of $\tau_1 + \tau_2$, where τ_2 is the two-way traveltime in the second layer. The z-transform of this component is

$$R_{2,1}(z) = z^{\tau_1+\tau_2}(-t_1' c_2 t_1) \frac{1}{1 + c_1 z^{\tau_1}},$$

where the subscripts of $R(z)$ denote response order and associated components, respectively. Likewise, the next pulse entering the water layer in the above manner generates the second component of the second-order response,

$$R_{2,2}(z) = z^{2\tau_1+\tau_2}(t_1' c_2 t_1 c_1) \frac{1}{1 + c_1 z^{\tau_1}} .$$

The third component of the second-order response is

$$R_{2,3}(z) = z^{3\tau_1+\tau_2}(-t_1' c_2 t_1 c_1^2) \frac{1}{1 + c_1 z^{\tau_1}} .$$

One can continue this analysis up to any number of additional components. Summation of the above series of equations produces the complete second-order response,

$$R_2(z) = R_{2,1}(z) + R_{2,2}(z) + R_{2,3}(z) + \cdots$$

$$= z^{\tau_1+\tau_2}(-t_1' c_2 t_1) \frac{1}{1 + c_1 z^{\tau_1}}$$

$$\cdot [1 - c_1 z^{\tau_1} + c_1^2 z^{2\tau_1} + \cdots]$$

$$= z^{\tau_1+\tau_2}(-t_1' c_2 t_1) \frac{1}{(1 + c_1 z^{\tau_1})^2}$$

where we recall that $|c_1| < 1$. Since $z^{\tau_1+\tau_2}$ is simply a delay factor and $-t_1' c_2 t_1$ is a constant, we may shift the time origin and normalize the second-order response. This yields

$$R_2(z) = \frac{1}{(1 + c_1 z^{\tau_1})^2}, \quad \text{(C-4)}$$

an expression which we see to be the square of the first-order response given by equation (C-3).

Removal of first-order ringing

Let us incorporate the first-order impulse response into the predictive deconvolution model. We will assume that the two-way traveltime through the water layer is τ_1 sample units. Thus our impulse response becomes,

$$x_1(t) = 1, \underbrace{0, 0, \cdots, 0,}_{\tau_1 - 1 \text{ zeros}}$$
$$-c_1, \underbrace{0, 0, \cdots, 0,}_{\tau_1 - 1 \text{ zeros}} c_1^2, \cdots .$$

In order to compute the predictive filter, we require the autocorrelation of $x_1(t)$, which is

$$r_\tau = 1 + c_1^2 + c_1^4 + \cdots , \quad \tau = 0.$$
$$r_\tau = 0, \quad 0 < \tau < \tau_1.$$
$$r_\tau = -c_1(1 + c_1^2 + c_1^4 + \cdots)$$
$$= -c_1 r_0, \quad \tau = \tau_1,$$

and so on. Thus the autocorrelation of $x_1(t)$ can

be written

$$r_\tau = E_x, \underbrace{0, 0, \cdots, 0}_{\tau_1 - 1 \text{ zeros}}, -cE_x, \cdots,$$

where E_x is the energy in $x_1(t)$. Let the filter length n be *less* than τ_1, and let the prediction distance be $\alpha = \tau_1$. Then the normal equations become

$$\begin{bmatrix} r_0 & 0 & \cdots & 0 \\ 0 & r_0 & \cdots & 0 \\ & & \vdots & \\ 0 & 0 & \cdots & r_0 \end{bmatrix} \begin{bmatrix} a_0 \\ a_1 \\ \vdots \\ a_{n-1} \end{bmatrix} = \begin{bmatrix} r_{\tau_1} \\ 0 \\ \vdots \\ 0 \end{bmatrix}.$$

The only member of this system whose right side does not vanish is,

$$r_0 a_0 = r_{\tau_1},$$

and thus,

$$a_0 = r_{\tau_1}/r_0 = -c_1 r_0/r_0 = -c_1.$$

The associated prediction error operator is

$$f_2(t) = 1, \underbrace{0, 0, \cdots, 0}_{\tau_1 - 1 \text{ zeros}}, c_1. \quad \text{(C-5)}$$

In practice it is not necessary to set the prediction distance α exactly to τ_1. The present model permits α to take on any value as long as it is less than or equal to τ_1. Furthermore, the filter length must be such that the inequality $\alpha + n > \tau_1$ holds true.

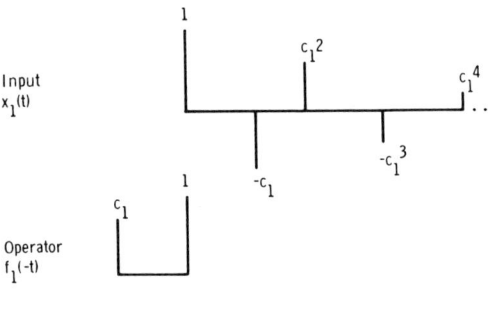

FIG. C-2. Deconvolution of a first-order ringing system. The operator is shown in time-reversed form.

We note that the z-transform of the prediction error operator of equation (C-5) is $1 + c_1 z^{\tau_1}$ which is the inverse of the first-order impulse response given by equation (C-3). If this first-order impulse response is convolved with the above prediction error filter, the output will be 1; in other words, we will have deconvolved the ringing signal. Figure C-2 illustrates the input signal $x_1(t)$, the prediction error operator $f_1(t)$, and the output signal $y_1(t)$ for this situation.

Removal of second-order ringing

Let us use the predictive deconvolution method on the second-order portion of the two-layer impulse response given by equation (C-4). This response can be written,

$$x_2(t) = 1, \underbrace{0, 0, \cdots, 0}_{\tau_1 - 1 \text{ zeros}}, -2c_1, \underbrace{0, 0, \cdots, 0}_{\tau_1 - 1 \text{ zeros}},$$

$$3c_1^2, \underbrace{0, 0, \cdots, 0}_{\tau_1 - 1 \text{ zeros}}, -4c_1^3, \cdots.$$

The autocorrelation of $x_2(t)$ is

$$r_\tau = 1 + 4c_1^2 + 9c_1^4 + 16c_1^9 + \cdots$$

$$= \frac{1 + c_1^2}{(1 - c_1^2)^3}, \quad \tau = 0.$$

$$r_\tau = 0, \quad 0 < \tau < \tau_1.$$

$$r_\tau = -2c_1 - 6c_1^3 - 12c_1^5 - 20c_1^7 + \cdots$$

$$= \frac{-2c_1}{(1 - c_1^2)^3}, \quad \tau = \tau_1.$$

$$r_\tau = 0, \quad \tau_1 < \tau < 2\tau_1.$$

$$r_\tau = 3c_1^2 + 8c_1^4 + 15c_1^6 + 24c_1^8 + \cdots$$

$$= \frac{-c_1^4 + 3c_1^2}{(1 - c_1^2)^3}, \quad \tau = 2\tau_1,$$

and so on. Thus the normalized autocorrelation of x_t becomes

$$r_\tau = 1 + c_1^2, \underbrace{0, 0, \cdots, 0}_{\tau_1 - 1 \text{ zeros}},$$

$$-2c_1, \underbrace{0, 0, \cdots, 0}_{\tau_1 - 1 \text{ zeros}}, 3c_1^2 - c_1^4, \cdots.$$

If the filter length is $n=\tau_1+1$ and the prediction distance is $\alpha=\tau_1$, the normal equations become,

$$\tau_1-1 \text{ rows} \left\{ \begin{bmatrix} 1+c_1^2 & \overbrace{0 \quad 0 \cdots 0}^{\tau_1-1 \text{ columns}} & -2c_1 \\ 0 & 0 \quad 0 \cdots 0 & 0 \\ \vdots & & \vdots \\ 0 & 0 \quad 0 \cdots 0 & 0 \\ -2c_1 & 0 \quad 0 \cdots 0 & 1+c_1^2 \end{bmatrix} \begin{bmatrix} a_0 \\ a_1 \\ \vdots \\ a_{\tau_1-1} \\ a_{\tau_1} \end{bmatrix} = \begin{bmatrix} -2c_1 \\ 0 \\ \vdots \\ 0 \\ 3c_1^2 - c_1^4 \end{bmatrix} \right.$$

The two nonvanishing equations of this system yield the solution

$$a_0 = -2c_1$$

and

$$a_{\tau_1} = -c_1^2.$$

Hence the associated prediction error operator is

$$f_2(t) = 1, \underbrace{0, 0, \cdots, 0,}_{\tau_1 - 1 \text{ zeros}}$$

$$2c_1, \underbrace{0, 0, \cdots, 0,}_{\tau_1 - 1 \text{ zeros}} c_1^2. \qquad (C\text{-}6)$$

This particular prediction error operator is identical to the three-point filter of Backus (1959). We thus see that in the noise-free case the present predictive deconvolution model yields the classical results obtained on the basis of strictly deterministic considerations. The predictive deconvolution scheme allows a more general attack on the dereverberation problem, as the present treatment has sought to demonstrate.

It is not necessary to set the prediction distance exactly equal to τ_1, nor is it necessary to set the filter length exactly equal to τ_1+1. However, these parameters must be set such that $\alpha \leq \tau_1$, $n \geq \tau_1$, and $\alpha+n \geq 2\tau_1$.

The z-transform of equation (C-6) is

$$F_2(z) = 1 + 2c_1 z^{\tau_1} + c_1^2 z^{2\tau_1} = (1 + c_1 z^{\tau_1})^2,$$

which is the inverse of the z-transform of the second-order impulse response given by equation (C-4). Thus, convolution of the second-order impulse response $x_2(t)$ with the prediction error operator $f_2(t)$ produces a zero delay spike (Figure C-3). In other words, the second-order ringing system has been deconvolved by means of the prediction error operator.

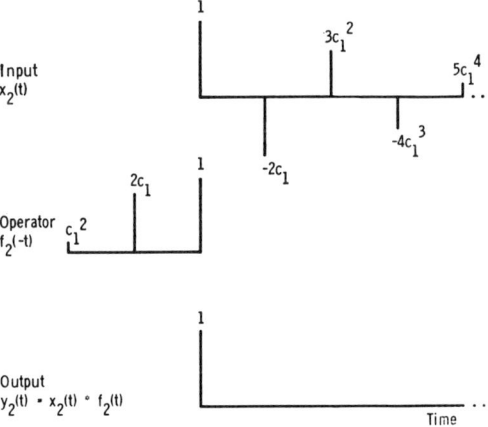

Fig. C-3. Deconvolution of a second-order ringing system. The operator is shown in time-reversed form.

ESTIMATION AND CORRECTION OF NEAR-SURFACE TIME ANOMALIES

M. TURHAN TANER,* F. KOEHLER,* AND K. A. ALHILALI§

The problem of computing static corrections for CDP seismic reflection data is discussed. A new approach is presented and it is related to various existing approaches. The approach consists of using crosscorrelation computations to find time shifts which appear to align the traces of each common-depth-point. These shifts are expressed in terms of surface corrections, one for each source and receiver position; a residual NMO correction for each common-depth-point; and a fixed correction for each common-depth-point.

These simultaneous equations form an overdetermined set which can be solved for the unknown static and NMO corrections. The least-square-error solution to these equations has an important indeterminancy which is discussed. Methods for its resolution are proposed.

Application of the technique to real data is illustrated by several examples. Validity of the corrections is demonstrated by velocity analyses before and after correction of the traces.

INTRODUCTION

Sheriff (1969) defines "statics" as "corrections applied to seismic data to eliminate the effect of variations in elevation and of weathering thickness or velocity." Estimation of these near-surface time anomalies has been the subject of numerous papers since the earliest days of seismic exploration. On single-coverage conventional seismic records, the accuracy of these estimates was limited by the experience of the interpreter in visually smoothing key reflections. Increased coverage and increased volume of data made it difficult to check by eye the consistency of these statics.

It became evident, however, that stacking velocity computations and the quality of sections obtained by the common-depth-point technique depended very much on the accuracy and consistency of these time corrections. The use of digital computers, making possible the economical handling of many time consuming tasks, aided in producing a geologically plausible section. Papers by Hileman et al (1968), Garotta and Michon (1968), Irvine and Worley (1969), Disher and Naquin (1970), and Sherwood and Donaldson (1970) describe various approaches to the computation of statics by the use of digital computers. We will try to relate some of these approaches to one another and present an additional one of some generality.

BASIC ASSUMPTIONS

In automatic statics computation, the aim is to estimate the anomalous variations of arrival time due to near-surface effects. The effects of variation of the surface elevation, or the change of weathering thickness or weathering velocity, result in phase and amplitude distortion of the propagating seismic wavefront. One would expect the near-surface layers to behave like a filter whose response depends on the path of the seismic wave. In order to eliminate these near-surface

---

Presented at the 24th Annual SEG Midwestern Regional Meeting, April 2, 1971, Oklahoma City, Okla. Manuscript received by the Editor November 26, 1973; revised manuscript received January 23, 1974.
* Seiscom Delta, Inc., Houston, Tex. 77036.
§ Formerly Seiscom Delta, Inc., Houston; presently, Sun Oil Co., Dallas, Tex. 75221.
© 1974 Society of Exploration Geophysicists. All rights reserved.

effects, we must estimate the response of this filter. This computation is not practically feasible at this time, and we proceed by making some simplifying assumptions while keeping the objectives of the solution in mind.

The generally-used assumption is the simplified one that the effect of the near surface is to introduce pure time delays. These are furthermore specified as "surface-consistent," i.e., a given location on the surface is associated with a constant time delay regardless of the wave path. Experience over the past few years has shown that the assumptions of pure time delay and of surface-consistency of the delay have resulted in reasonably good processed sections.

When one considers the low velocity of the weathered layer relative to the higher velocities of the subweathering layers, it becomes evident that surfaceconsistency is an acceptable assumption; because of refraction, waves travel along paths which are close to the vertical in the low-velocity layer. Therefore the variation of time delay along various paths remains reasonably small. This assumption, however, becomes weaker when the near-surface layers producing the time anomalies are of higher velocity than the layers below. Permafrost regions are good examples of this case; waves entering the permafrost zone refract away from the normal, and this can produce different delays for different ray paths at the same receiver or source position (Figure 1).

Accepting surface-consistency of time delays as a good initial assumption, we may nevertheless be prepared to relax it somewhat in the course of solution computations.

Our next assumption is that the near-surface effects do not vary with reflection time. If statics may be considered surface-consistent, it follows that they should be considered time-consistent (or time-constant), also. This is not true for excessively large and abrupt variations of elevation or weathering, and these must be handled by different techniques. However, the name "statics" is derived from the time-constant effect of the near-surface layers, and this is the assumption we shall use here.

In many instances we do not have any information about the total thickness of the weathered layer, so it is not possible to compute absolute corrections. Since our objective is to obtain a seismic section that represents deeper horizons in their correct relative positions without the influence of the shallow layers, we are content to consider the desired statics as relative time shifts with respect to some datum plane. Thus, we are

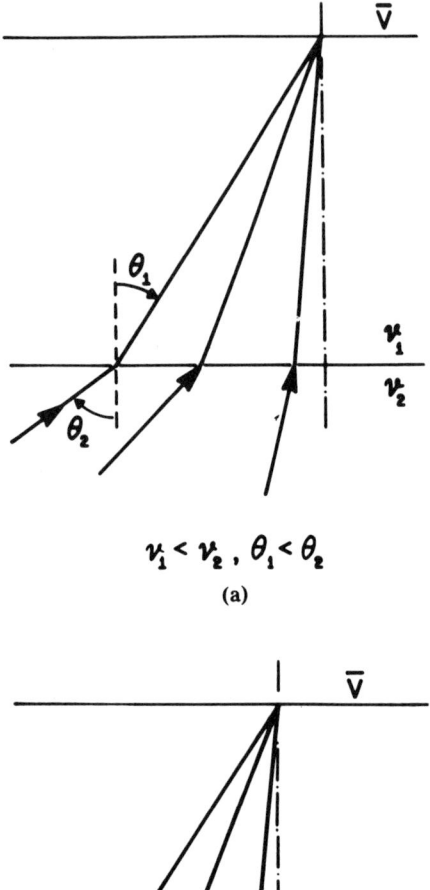

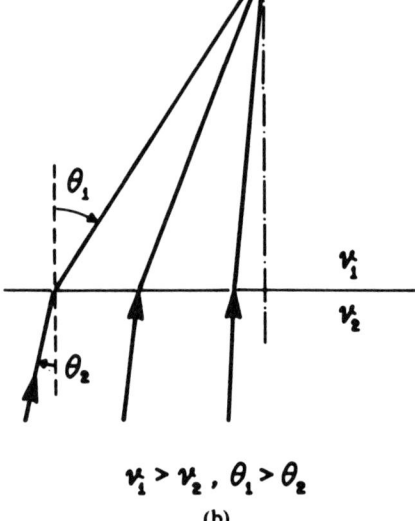

FIG. 1. (a) Low-velocity weathered layer over high-velocity layer; since $v_1 < v_2$, $\theta_1 < \theta_2$. (b) High-velocity surface layer over low-velocity layer (permafrost regions); since $v_1 > v_2$, $\theta_1 > \theta_2$.

concerned with measuring the differences of the surface effects between different surface points; we wish to compute the time shifts between traces belonging to different source and/or receiver positions.

One of the most convenient tools for estimation of relative time shifts between traces is the crosscorrelation function. Figure 2 shows a crosscorrelation function computed between two traces $f_1(t)$ and $f_2(t)$; the amplitude of the crosscorrelation function being given by

$$C(\tau) = \frac{\sum f_2(t)f_1(t+\tau)}{\sqrt{\sum f_2^2(t) \cdot \sum f_1^2(t)}},$$

where τ is the time shift between the traces.

The shift τ_m at which the largest positive peak of the crosscorrelation occurs, represents the time shift we seek. This means that if we shift trace $f_2(t)$ by a lag of τ_m, it will be best aligned with trace $f_1(t)$. Or we could say that the trace $f_1(t)$ is delayed by a lag of τ_m with respect to trace $f_2(t)$. If we consider that trace $f_2(t)$ is in its "proper" position, we are led to apply a static correction of amount $-\tau_m$ to trace $f_1(t)$.

Since $C(\tau)$ is normalized by the power of the two traces, the maximum amplitude C_m represents a good statistical measure of the similarity between them. If $\tau_m \neq 0$ and $C_m = 1.0$, the two traces are identical except for a time shift. If $0 < C_m < 1.0$, there is less-than-perfect similarity; perhaps different noise patterns or some phase distortion may exist. The closer to zero C_m becomes, the more unreliable the lag τ_m becomes. For this reason C_m suggests itself as a reliability factor in static computations, allowing the rejection of τ_m lags picked for C_m values less than some preselected threshold value.

PRINCIPAL TRACE PLANES

The trace-to-trace variations of the arrival time of reflections depend on a combination of static and dynamic effects. In order to separate these effects, we need to consider the traces in a suitably organized form. Such a form is the familiar "stacking diagram." Since we are concerned mainly with near-surface effects, it is more convenient to use the "surface diagram" form. This type of trace classification was used by Morgan (1970) in his wavelet maps. Figure 3 shows the surface diagram for 12-fold common-depth-point coverage with a 24-trace cable. The horizontal axis represents "receiver position," and the vertical axis represents "source position." We could think of each seismic trace, representing the amplitude of the recorded seismic wave, as appearing perpendicular to this plane at a point which represents its source and receiver coordinates. Traces at points aligned vertically lie in a plane which we will call the "common-receiver plane." All the traces on this plane are received at the same surface position. Traces at points aligned horizontally lie in the "common-source plane"; all

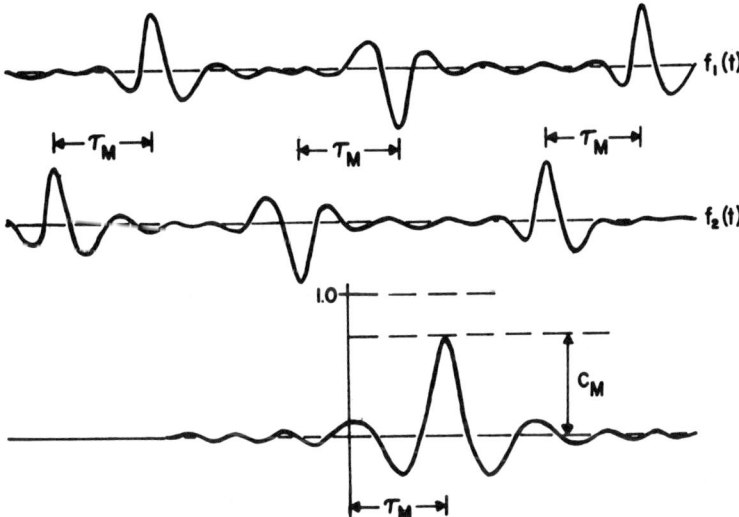

FIG. 2. Crosscorrelation function between two traces.

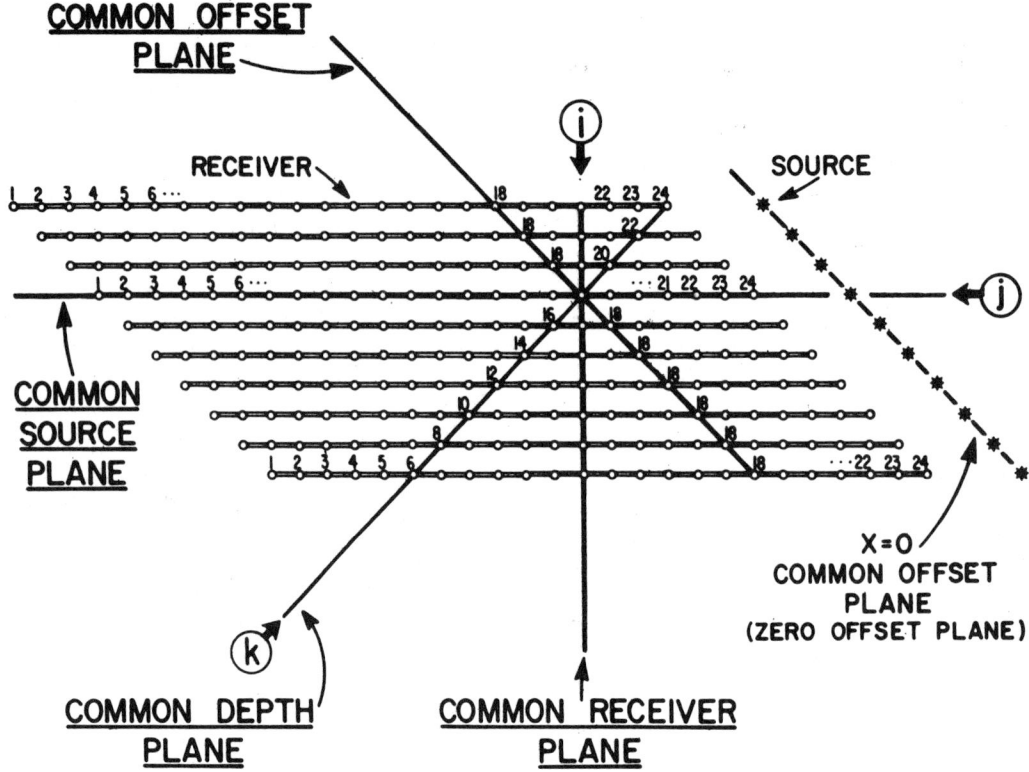

FIG. 3. Surface and receiver positions diagram showing four principal trace planes.

the traces on this plane are generated from the same source position.

There are two other principal planes which are at 45 degrees to the common-receiver and common-source planes. One of these represents the familiar CDP or common-depth-point traces, whose center points between source and receiver positions are common. The other is the common-offset-plane, which represents the traces having the same source-to-receiver distance. As we can see in Figure 3, common-depth-point and common-offset planes are normal to each other.

One of the most interesting common-offset planes is the one that passes through the source positions where offset is equal to zero; this is the "CDP-stack" plane. The intersection point of this plane and a particular common-depth-point plane represents the actual surface position of that CDP-stacked trace. Therefore we could interpret the application of NMO corrections as the transformation of a trace from its original position to the CDP-stack plane.

Even for surface sources, differences in the characteristics of source and receiver arrays make it desirable to separate static corrections into two components:

a) Source statics: time delays introduced into the downgoing wave in the vicinity of a source.
b) Receiver statics: time delays introduced into the upgoing wave in the vicinity of a receiver.

Let us now look at the influence of these effects on the differences of arrival time along the four principal planes. We shall use Figure 3 and Table 1.

Let us consider the differential arrival time picked from a crosscorrelation function computed between two adjacent traces on the common-receiver plane. The differential arrival time contains differential source statics. These two traces are also on two different common-depth-point planes so that the influence of dip is present, and they are on two different common-offset planes so that the influence of velocity is present.

Similarly, we see that the differential arrival time between two adjacent traces on the common-source plane also contains the dip and velocity components and differential receiver statics. Traces along the common-offset and common-depth-point planes belong to different common-source and common-receiver planes; therefore, differential arrival times between such traces contain both the differential source and receiver statics. The major influence on the common-offset plane is dip (and faults, if present); lateral variation of velocity can also be included here. The major influence on the common-depth-point plane is, of course, velocity. Morgan (1970) has shown that study of the wavelet character along these four principal planes yields useful lithological information.

GENERAL APPROACH

Looking again at Table 1, we see that by the use of crosscorrelations (computed over small time gates taken continuously along the time axis) on all the principal trace planes it is possible to compute dip, velocity, and statics simultaneously. However, for the restricted objective of statics computation, we are able to simplify some of the procedures.

Delays caused by velocity are generally much larger than those caused by the other factors. Therefore, if we first apply normal-moveout correction according to the best estimated velocity function, we obtain two benefits. First, it reduces the size of dynamic delays to the scale of residual normal moveout. Second, since all time windows are more or less aligned in a similar manner, it allows the use of fewer (one or two) crosscorrelation windows. Furthermore, if we are able to estimate the source and receiver statics from independent field observations, we can apply these to the data and thus reduce the problem to one of residual statics. As a further simplification, Hileman et al (1968) suggest the assumption that residual statics have zero mean over a spread length. We shall see that this last assumption is not always necessary.

The next compromise we could make is to reduce the number of principal planes over which time differences are computed. It is mainly in this respect that various approaches to automatic static computations differ.

Disher and Naquin (1970) have chosen the separation provided by the common-source and common-receiver planes. Crosscorrelations along the common-source plane are computed for adjacent receiver positions. These are averaged to give the estimate of differential receiver statics.

Table 1. Factors influencing differential arrival times on four principal trace planes.

	DYNAMIC		STATIC	
	DIP	VELOCITY	SOURCE	RECEIVER
COMMON RECEIVER PLANE	✓	✓	✓	O
COMMON SOURCE PLANE	✓	✓	O	✓
COMMON OFFSET PLANE	✓	O	✓	✓
COMMON DEPTH POINT PLANE	O	✓	✓	✓

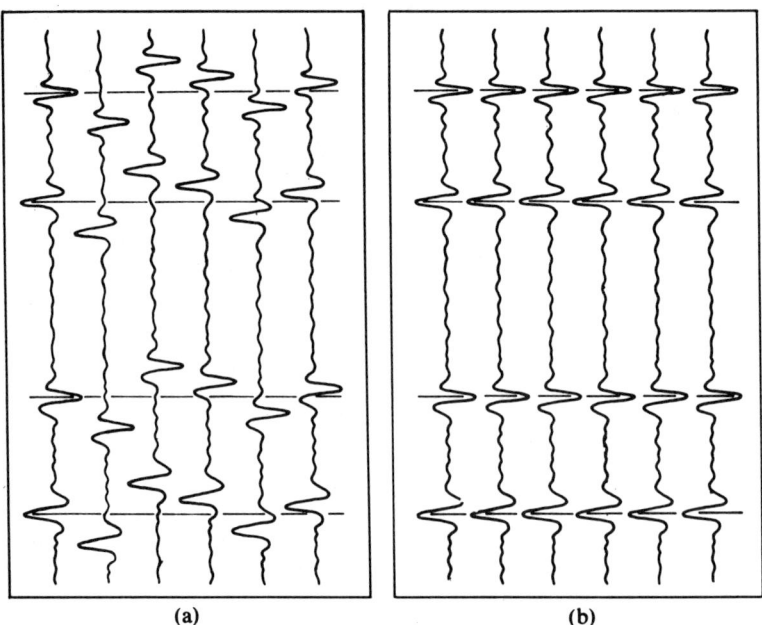

FIG. 4. (a) CDP-gather traces before alignment of primaries (with statics), left. (b) CDP-gathers traces after alignment of primaries (after statics correction for optimum stacking), right.

Similarly, differential source statics are estimated from crosscorrelations computed on the common-receiver planes. Looking back to Table 1, we see that these lags contain the influence of velocity (this time the residual NMO) and dip. The authors show that the redundancy of differential source and receiver statics allows one to eliminate noisy and inconsistent crosscorrelations from the computations. The statics themselves are then computed by integration (or by cumulative total) of the differential statics, while the influence of dip and residual NMO is estimated and separated by low-pass wavenumber filtering.

Hileman et al (1968) and Irvine and Worley (1969) have chosen to work on the common-depth-point plane. The advantages of this plane are that its residual NMO is small and it is insensitive to dip; the dominant contribution to the measured time shifts is the sum of differential source and receiver statics.

It is feasible to use one long time-window on this plane, in contrast to the several short windows used by Disher and Naquin (1970). The authors indicate that residual NMO, varying systematically with offset, produces erroneous results and should be removed before application of the averaging technique. This can be done either by estimating (Irvine and Worley, 1969) or by recomputing velocity (Hileman et al, 1970). However, as shown by Hileman et al, the averaging technique leaves small-amplitude undulations longer than one or two spread-lengths on the stacked data.

Irvine and Worley (1969) describe a method in which the time differentials are computed by a "summation correlation" technique that is superior to pair-wise crosscorrelations. Source and receiver statics are then separated by averaging along common-source and common-receiver planes, as in the previous method. The effect of residual NMO is investigated, and approximately compensated, by scatter analysis of time differentials over a number of adjacent CDP gathers. The authors point out that unsatisfactory results would be obtained for extremely large statics and for statics whose spatial period is longer than the cable length.

THE PRESENT TECHNIQUE

The objective of the present technique is to compute surface-consistent time shifts that yield a seismic section in which (a) each CDP gather is aligned to produce an optimally-stacked trace, and (b) the stacked section displays a degree of lateral signal continuity representative of real geology.

The first operation, then, is to obtain for each trace a time shift which aligns all primary reflections within a CDP gather. Figure 4a shows schematically a gather with arbitrary time shifts, while Figure 4b shows the same traces after alignment of the reflections. We could accomplish this by using one of several techniques: (1) Computing all pair-wise crosscorrelations and picking a consistent set of delays. (2) Shifting one trace at a time until we maximize the sum of weighted crosscorrelations or the semblance (see Neidell and Taner, 1971).

In the following discussion we assume that the traces are NMO-corrected with an initial stacking velocity. Each trace is identified by its surface coordinates; index i represents its receiver position, index j represents its source position, and index k denotes the CDP gather of which it forms a part. Then from the surface diagram,

$$k = \tfrac{1}{2}(i + j). \qquad (1)$$

As shown in Figure 5, all traces with the same i index belong to the same common-receiver plane, and all traces with the same j index to the same common-source plane.

Selection of correlation windows

In the selection of an appropriate approach, the influences of signal-to-noise ratio and of faulted reflections must be considered in addition to those of residual NMO and dip. Most of the problems can be overcome by proper choice of the initial processing parameters (such as deconvolution and band-pass filtering) and by care in selecting the correlation windows. The latter should be chosen so that (a) the window covers a time zone where primary events are dominant (thus minimizing multiple interference); (b) the window is sufficiently long to cover a number of primary reflections (thus minimizing multiple and noise interference); and (c) the window is reasonably deep (thus minimizing differential residual NMO between the primaries).

The basic equations

Let us consider an individual trace with surface coordinates i and j for which a time shift has been computed. This time shift T_{ijk} is a sum of several effects:

$$T_{ijk} = R_i + S_j + C_k + M_k(j - i)^2, \qquad (2)$$

where

R_i = receiver static at ith receiver position,
S_j = source static at jth source position,

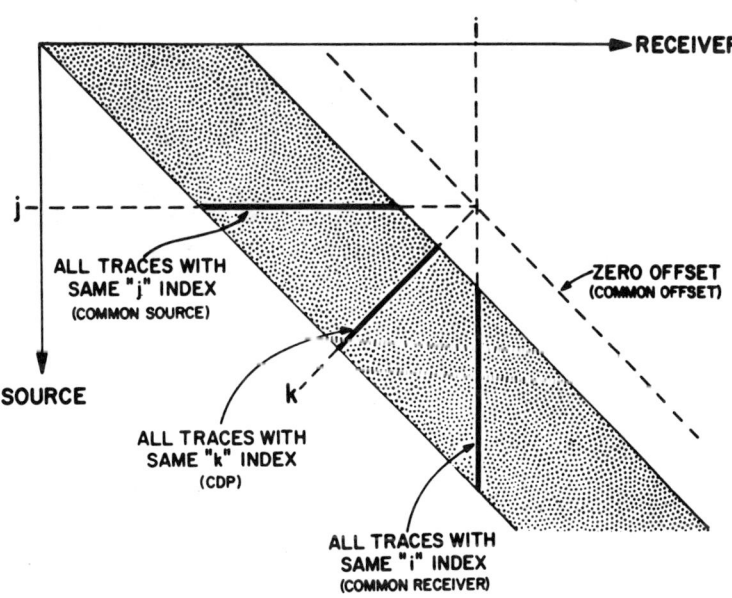

FIG. 5. Surface diagram showing trace coordinate relationship.

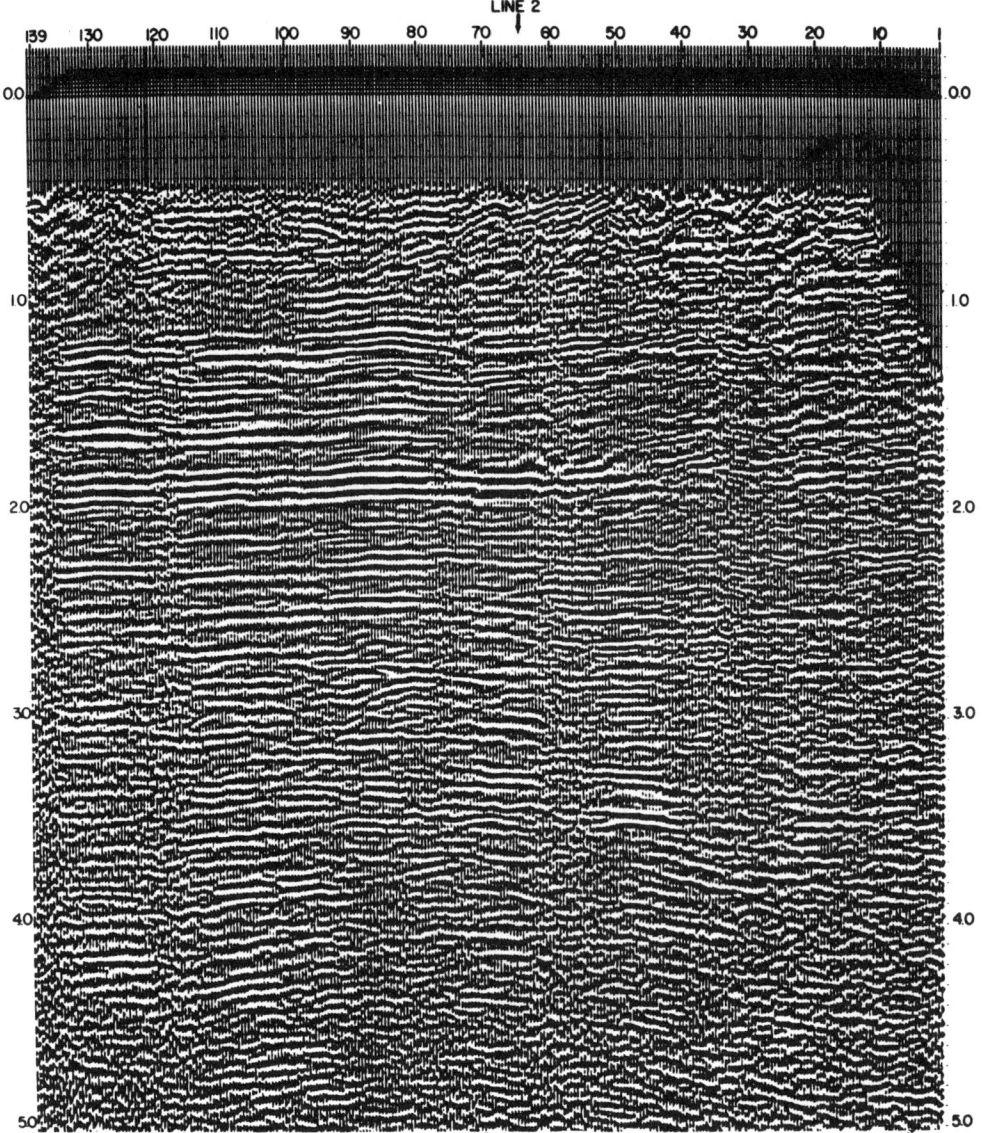

FIG. 6. Marine Line 1. Preliminary stacked section.

C_k = arbitrary time shift for kth CDP gather,
M_k = residual NMO component at kth CDP gather, and
$(j-i)$ = source-to-receiver distance.

C_k (the arbitrary shift at the kth CDP gather) has to be considered an unknown quantity, because the application of a constant shift to all traces in the gather does not disturb the reflection alignment within it.

There is one equation [in the form of equation (2) above] for each trace in the seismic line, and there are four sets of unknowns. Therefor we have more equations than unknowns. These equations may be solved by setting up the normal equations which yield the least-mean-square error and then either by iteration or by normal matrix operations.

We stress here, however, that the homogeneous system of equations obtained by setting all T_{ijk} equal to zero,

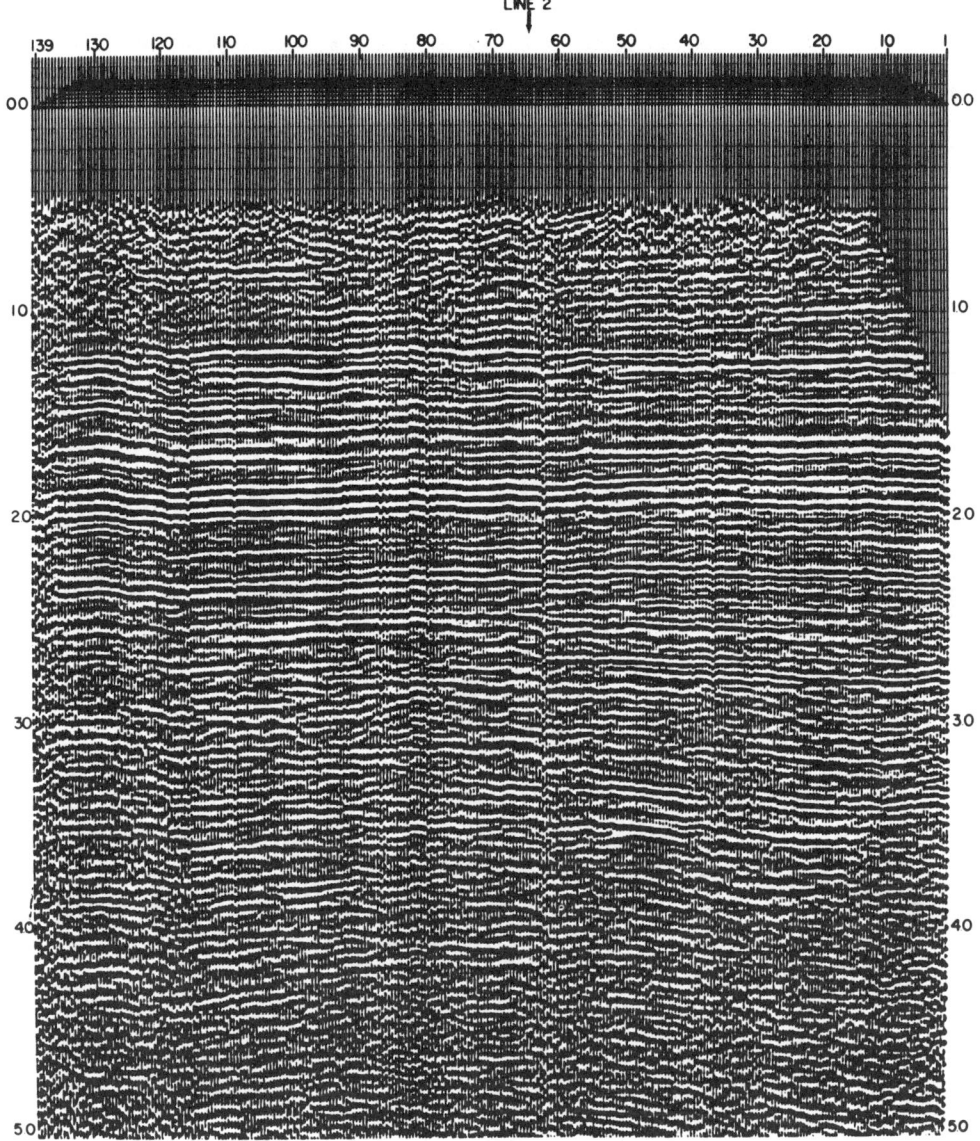

FIG. 7. Marine Line 1. Monitor section phase 1.

$$R_i + S_j + C_k + M_k(j - i)^2 = 0, \quad (3)$$

has nontrivial solutions of the form:

$$R_i = a_0 + a_1 i + a_2 i^2 + a_3 i^3$$
$$S_j = b_0 + b_1 j + b_2 j^2 + b_3 j^3, \quad (4)$$
$$C_k = c_0 + c_1 k + c_2 k^2 + c_3 k^3,$$

and

$$M_k = d_0 + d_1 k.$$

This indicates that the computed statics are indeterminate by a cubic polynomial (see Appendix.)

This difficulty is overcome if we can define the constants for any one of the equations. If we consider that the original stacked section (with field statics applied) is our most reliable source of information, then we could require that the final stacked section (with residual statics also applied) should be an optimum general match to

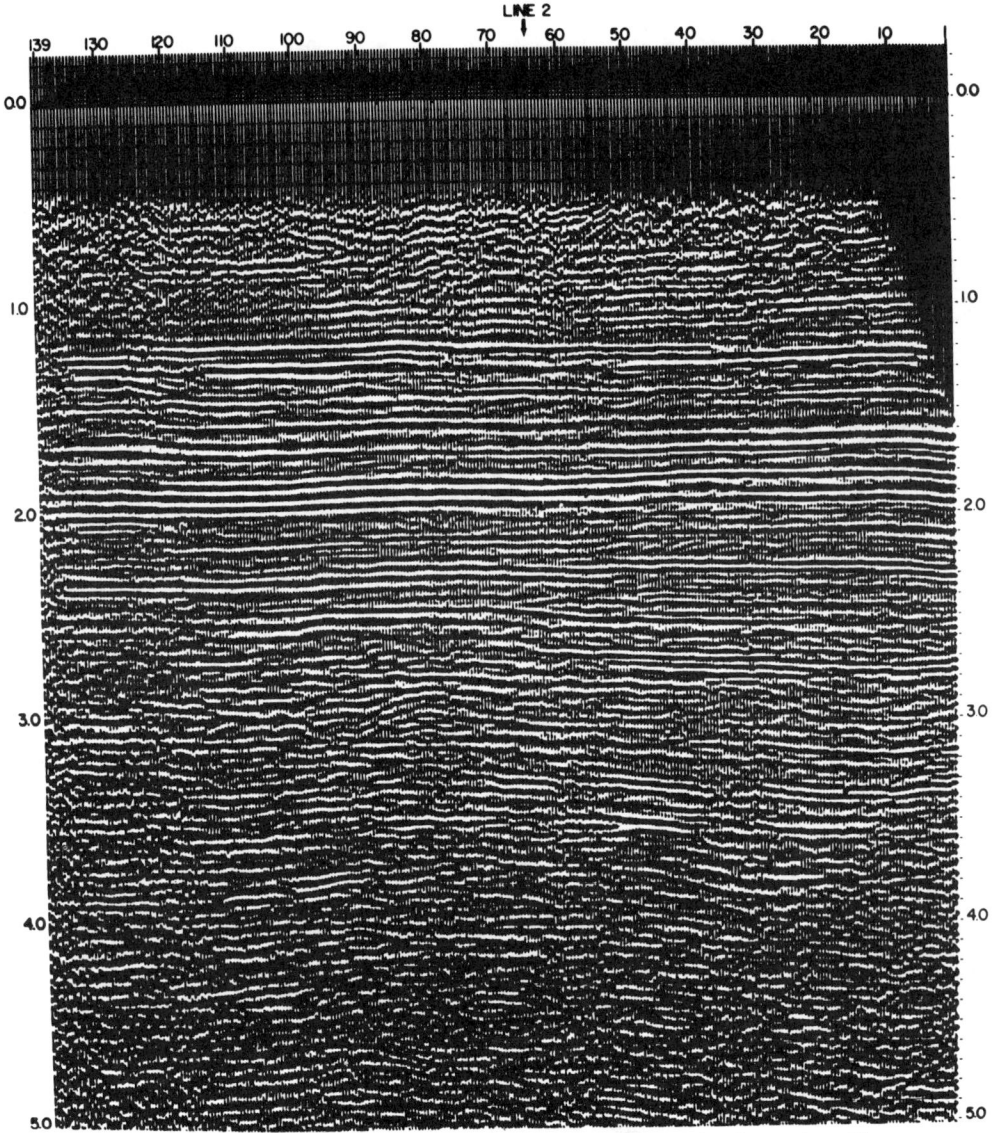

FIG. 8. Marine Line 1. Final statics-corrected monitor section from phase II.

the original stacked section. This means that the coefficients for C_k in equation (4) are no longer arbitrary; they are all equal to zero (see Appendix).

The method described here is divided into two parts (representing the two stated objectives): Phase 1, time shifts are computed for all traces in each CDP gather, in order to optimize the stacked trace. Phase 2, a surface-consistent solution of the statics is computed and a monitor section is generated for the geophysicist's review. The computed statics are then applied to the original data for normal seismic processing.

Application to field data

As our first example, we have taken an offshore line from the Gulf Coast. Figure 6 shows the preliminary stacked section of line 1. Figure 7 shows the output section from the computations of phase 1; this is shown here solely for a full

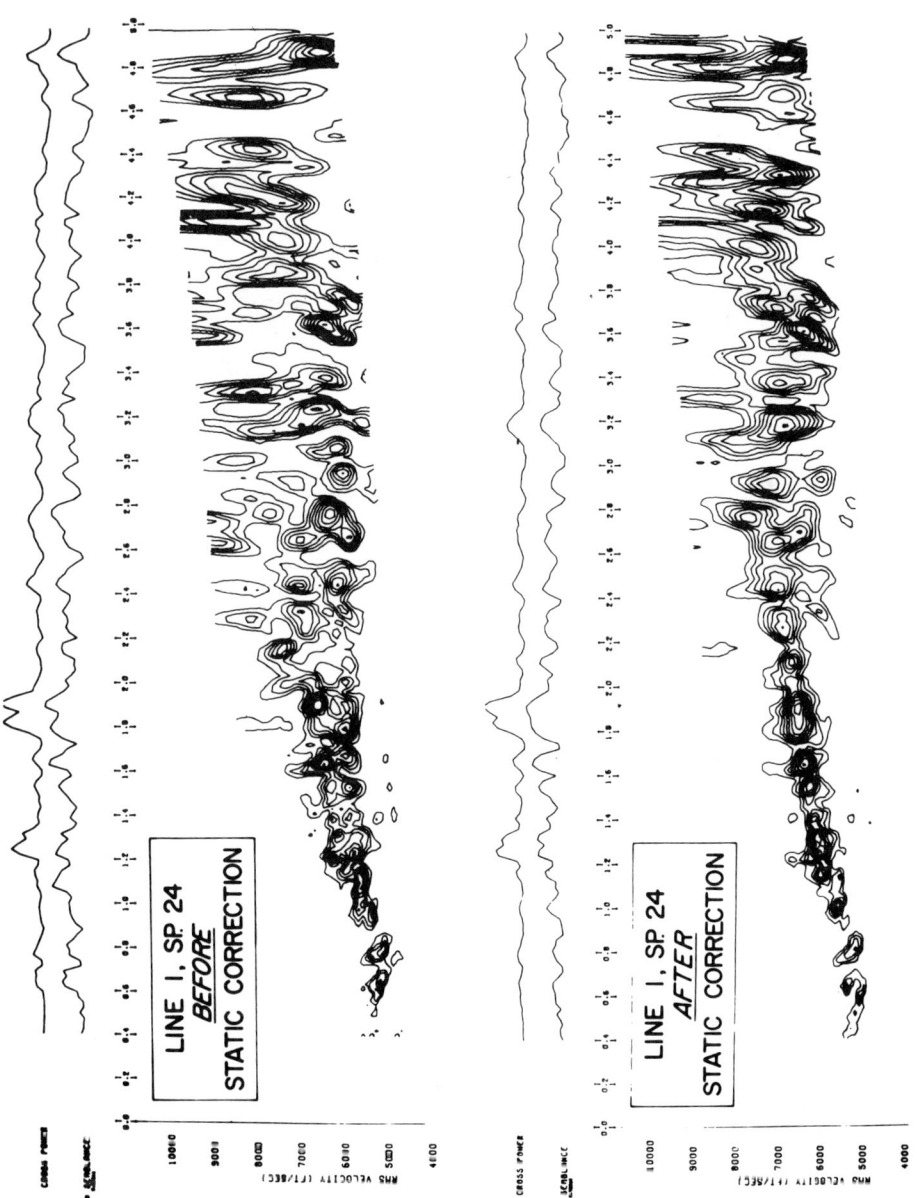

FIG. 9. Marine Line 1, velocity spectra at SP 24 before and after static corrections.

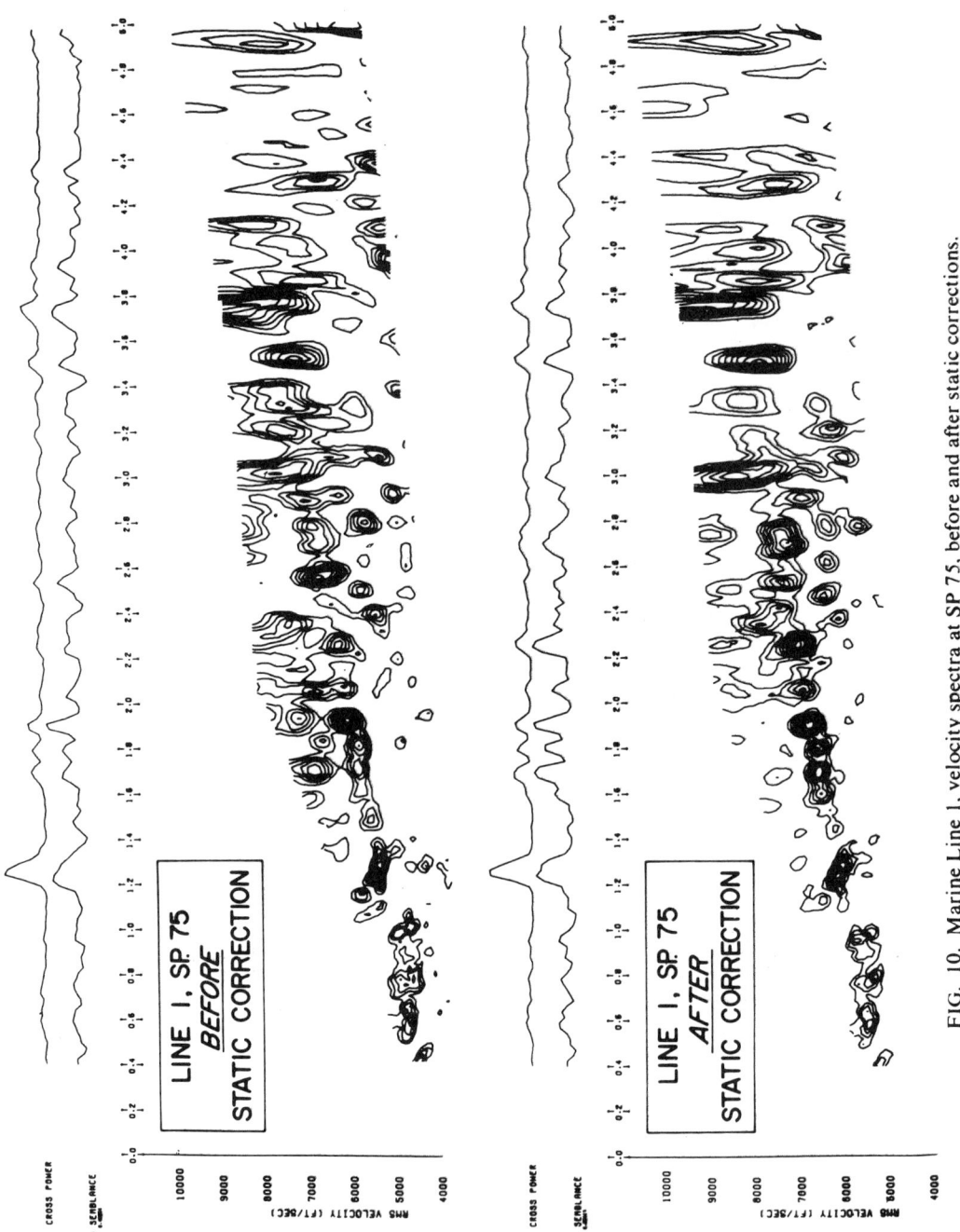

FIG. 10. Marine Line 1, velocity spectra at SP 75, before and after static corrections.

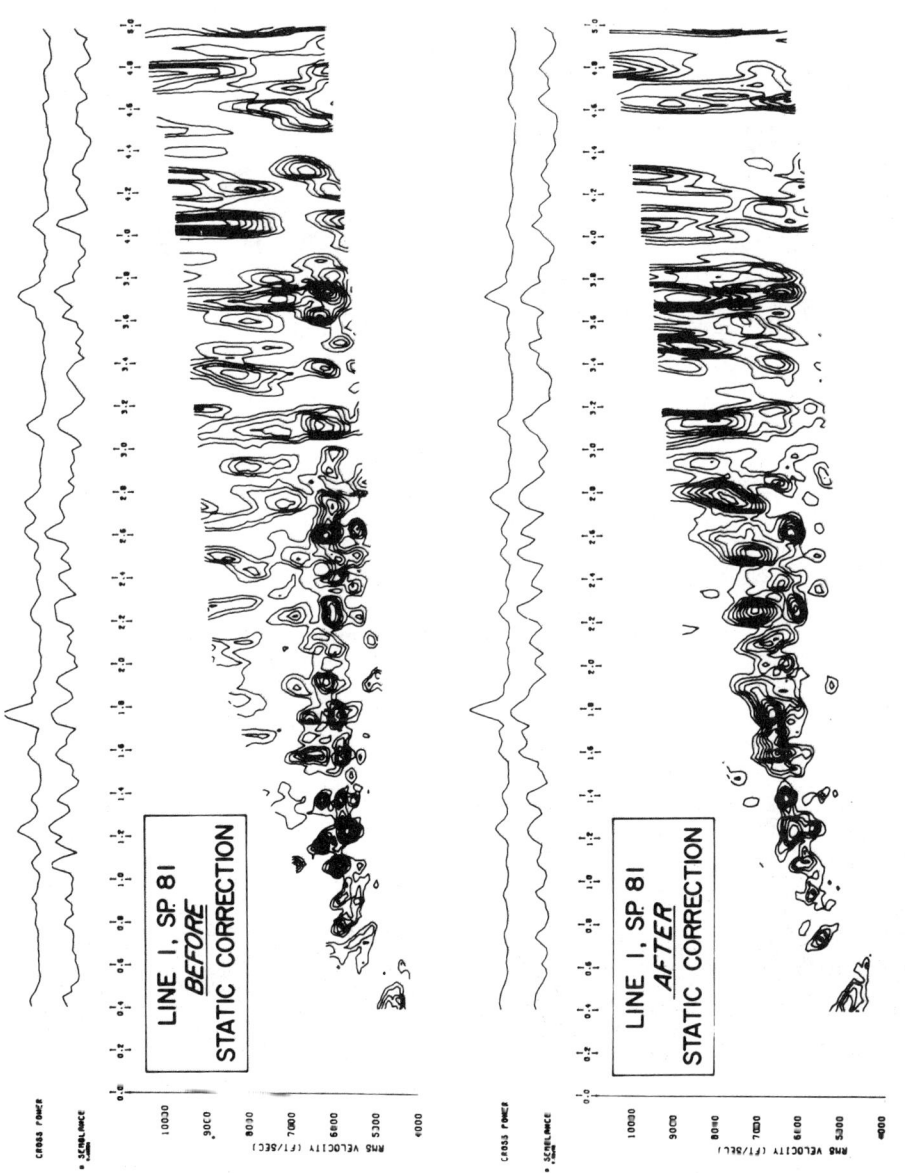

FIG. 11. Marine Line 1, velocity spectra at SP 81, before and after static corrections.

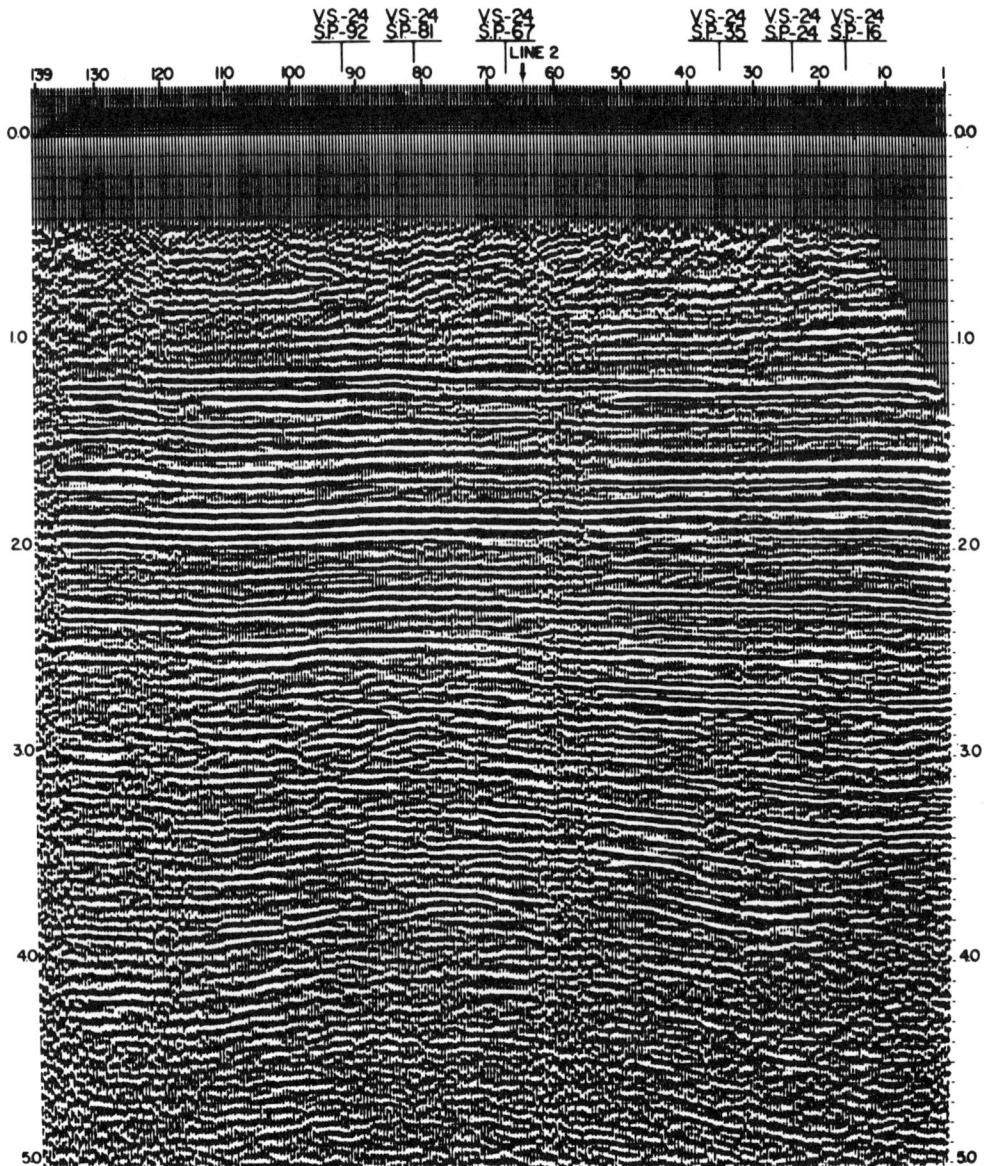

FIG. 12. Final stacked section with static corrections and final stacking velocities.

demonstration of the technique. Comparing this figure with the preliminary stacked section, we see that each individual stacked trace contains well-defined reflections. However, Figure 7 lacks spatial continuity, because of the arbitrary time shifts C_k.

Figure 8 shows the monitor output from the program where the values of C_k are computed and corrected. In routine practice this is compared with the preliminary stacked section shown in Figure 6, in order to establish the degree of success of the computation of automatic statics. The processing then reverts to the original field data, to which the computed statics are applied before the final analysis of stacking velocities and further normal processing.

One of the most convincing demonstrations of the value of the statics correction is obtained by

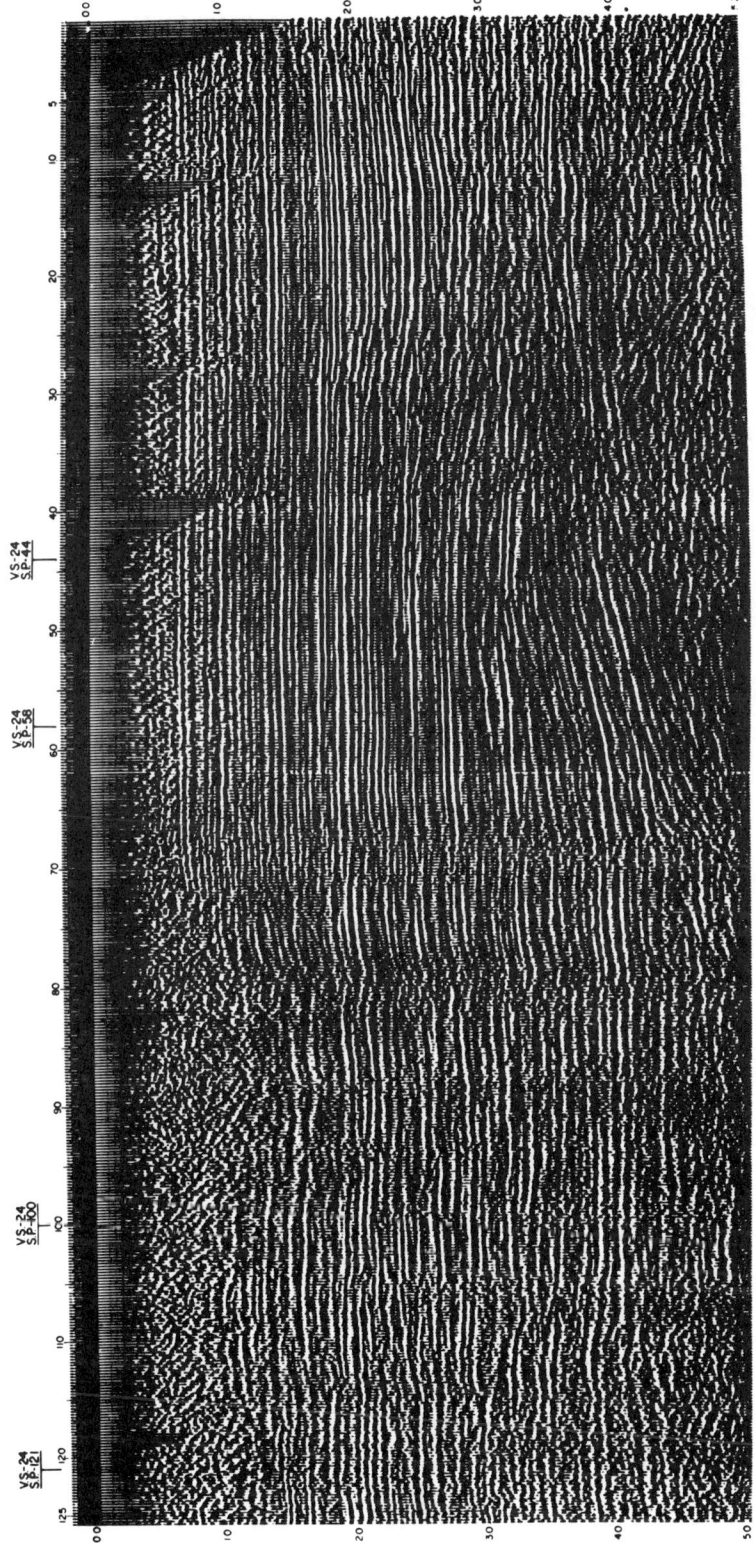

FIG. 13. Land Line 3. Preliminary 600 percent CDP stacked section.

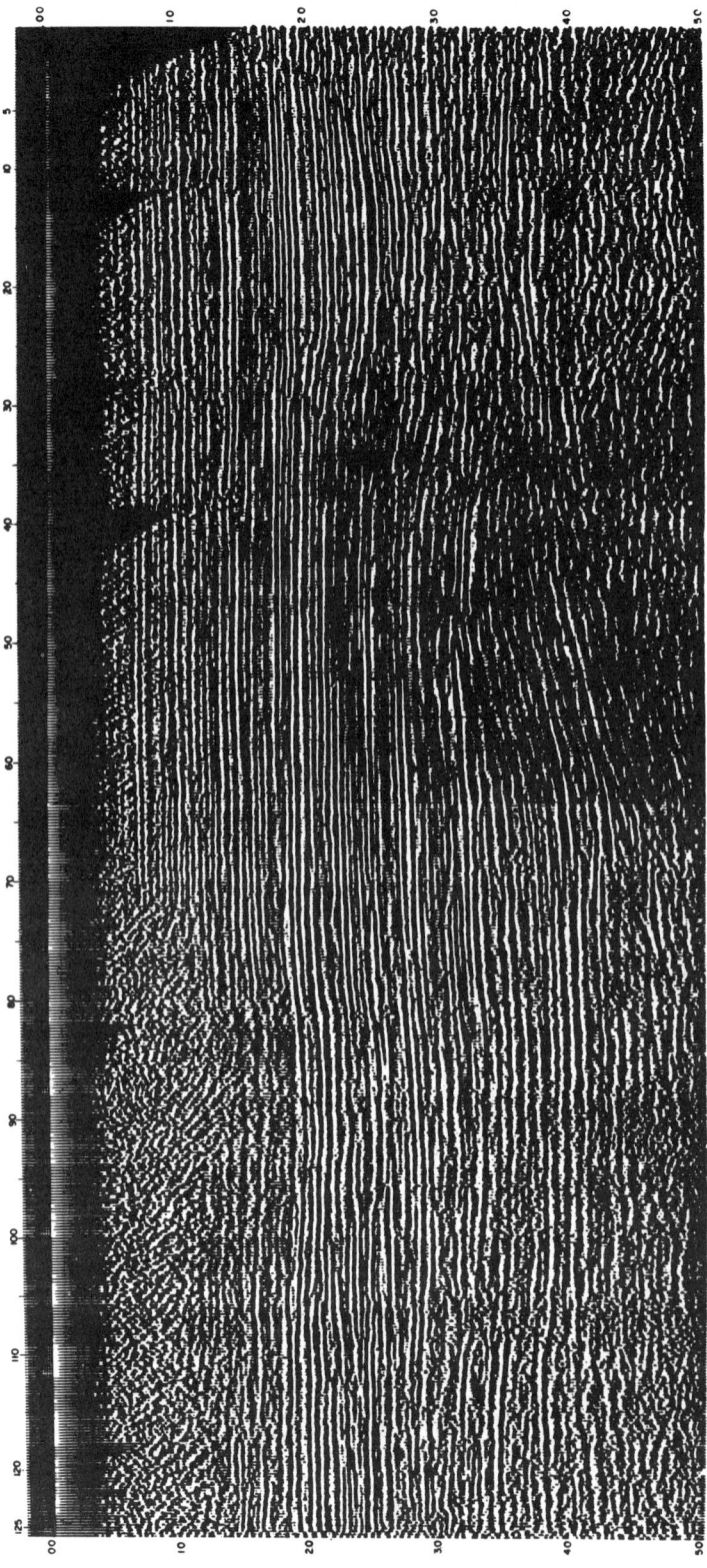

FIG. 14. Land Line 3. Statics corrected monitor section from phase II.

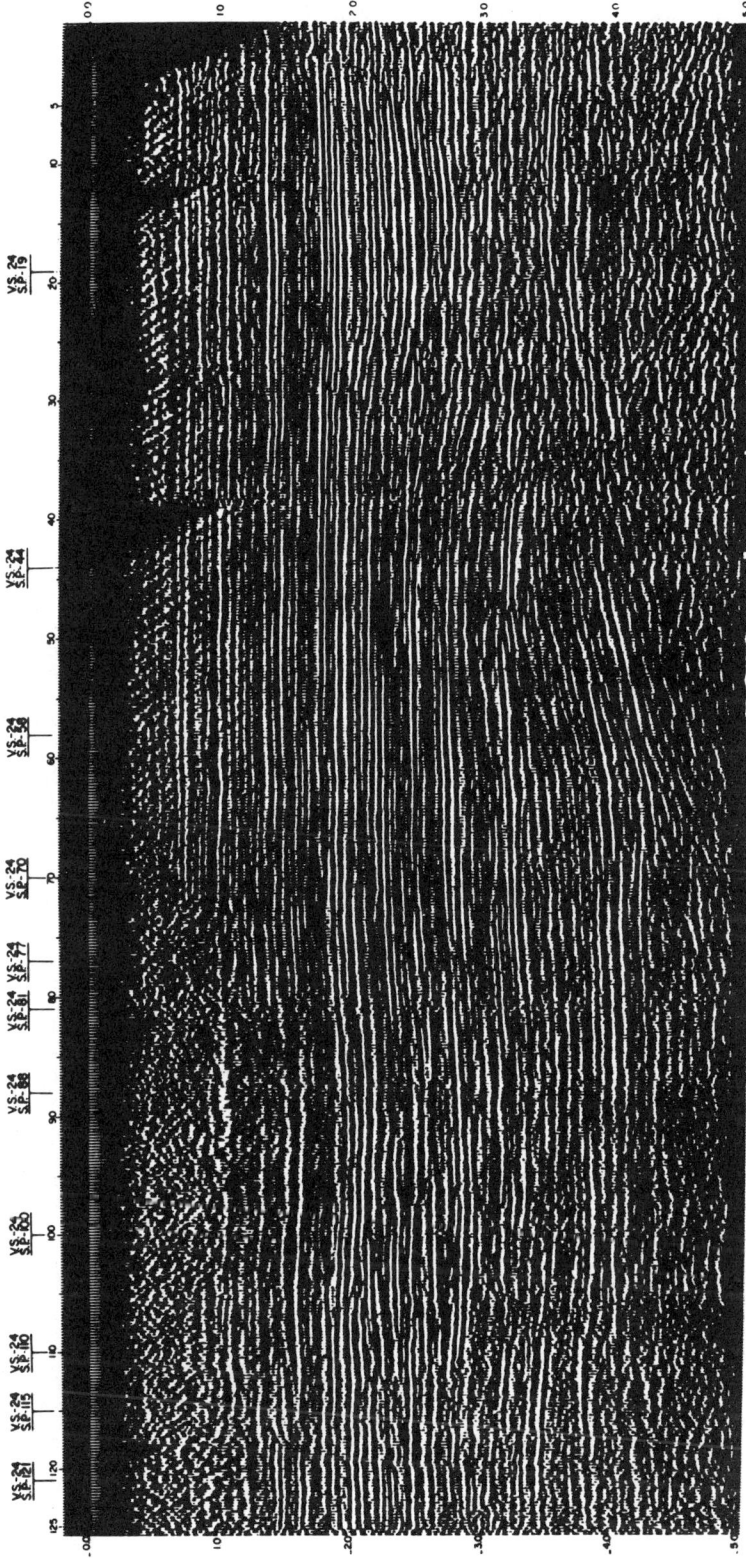

FIG. 15. Land Line 3. Final stacked section with static corrections and new velocities.

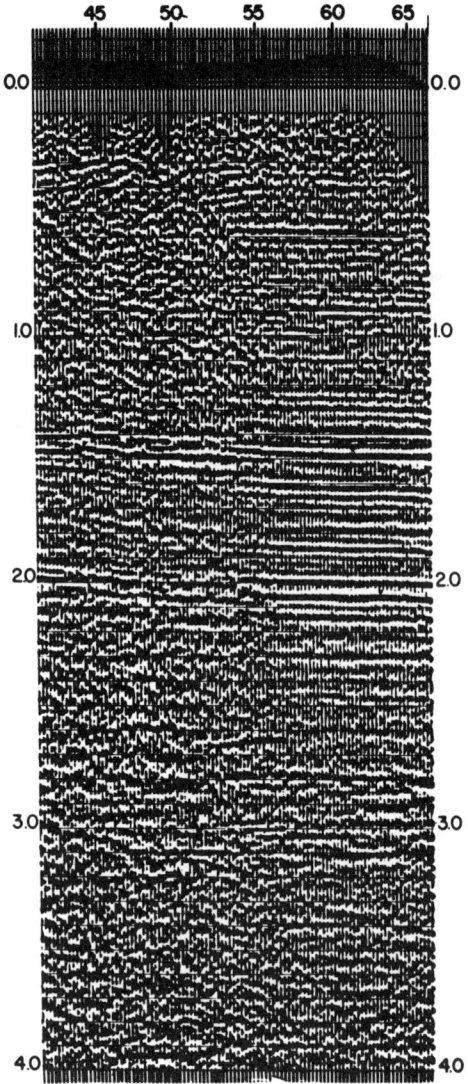

FIG. 16. Land Line 4. Preliminary stack section.

12 shows the final CDP stacked section generated after static and NMO correction with final stacking velocities.

Earlier in the paper we discussed the benefits of working in the common-depth-point plane; in this plane we are not seriously affected by dips or by faults. The next example is a land line (Figure 13) disturbed at the left by a local thickening of the low-velocity near-surface layer. Comparing this to the phase 2 output shown in Figure 14, we see that all dips and faults are preserved on the

comparing velocity analyses before and after such corrections. It should be noted that since the equations include residual NMO as one of the unknowns, a reasonable regional velocity before static computation is sufficient; the program computes and removes the influence of residual NMO from the static delays. Figures 9, 10, and 11 show, for comparison, velocity spectra at three different locations on line 1, before and after static corrections. The velocities after static corrections are more consistent, and in all cases they show a considerable increase of maximum semblance. Figure

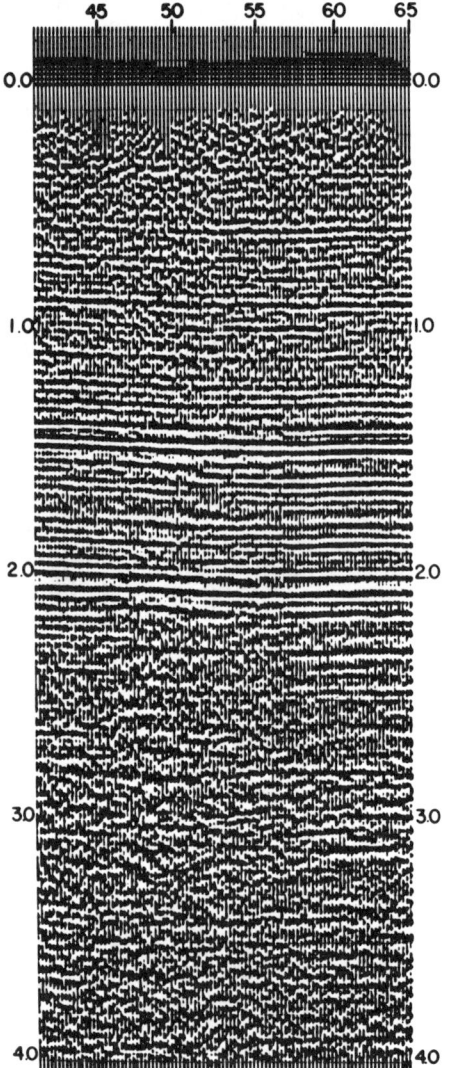

FIG. 17. Land Line 4. Statics corrected monitor section from phase II.

right-hand side. The disturbed area at the left now appears more consistent with the rest of the section. The final stacked section is shown in Figure 15.

Sections representing the preliminary stack and phase 2 monitor are given in Figures 16 and 17, for another land line. This example demonstrates that if statics are properly corrected, both the final seismic section and the final velocity analyses maintain a reasonable consistency along the line. Figure 18 shows a velocity spectrum computed at SP 57, which is in the area of better data. The next two velocity spectra are taken in the disturbed zone and are shown in Figures 19 and 20. The velocity spectra computed after static corrections are much more consistent than the ones before static corrections, and they compare well with the spectrum at SP 57.

CONCLUSIONS

Organization of seismic traces into four principal planes on the surface diagram has been discussed. On the basis of this coordinate system a surface-consistent statics computation procedure can be simply formulated. One must note, however, the inherent indeterminancy present in such computations. Statics computed here can be indeterminate by a polynomial of cubic order. Different orders of indeterminancies may be present, depending upon the type of problem formulation for which the surface diagram is used.

The validity of computed statics must be

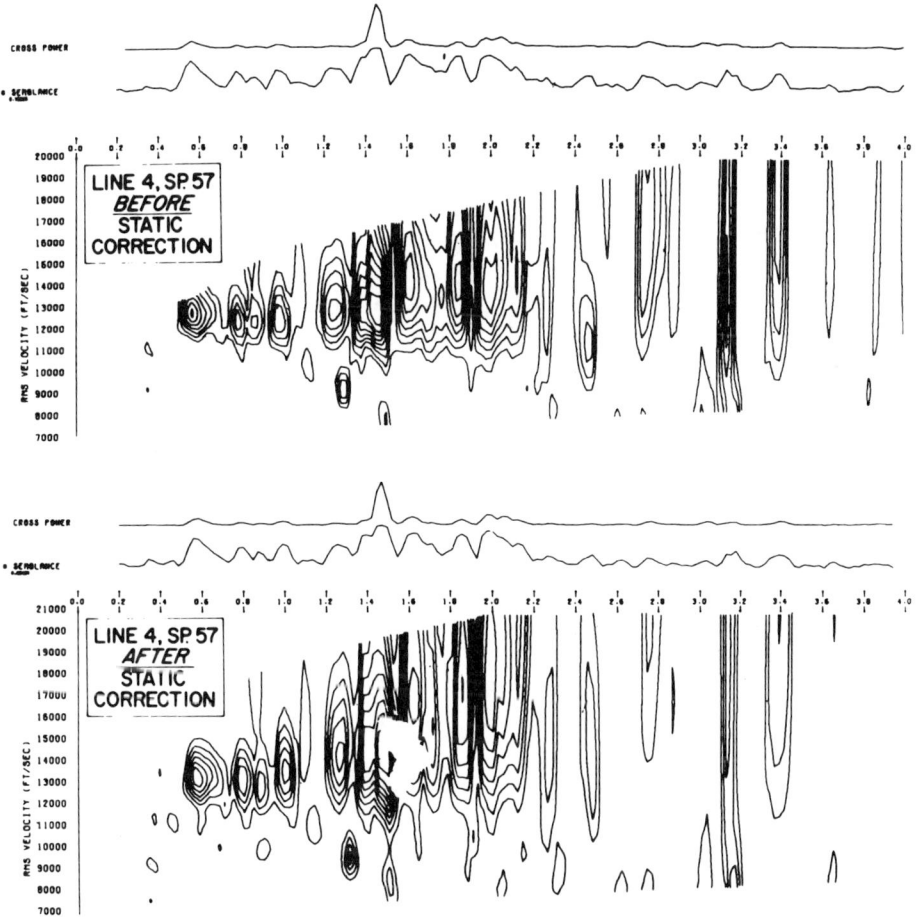

FIG. 18. Land Line 4, velocity spectra at SP 57 before and after static corrections.

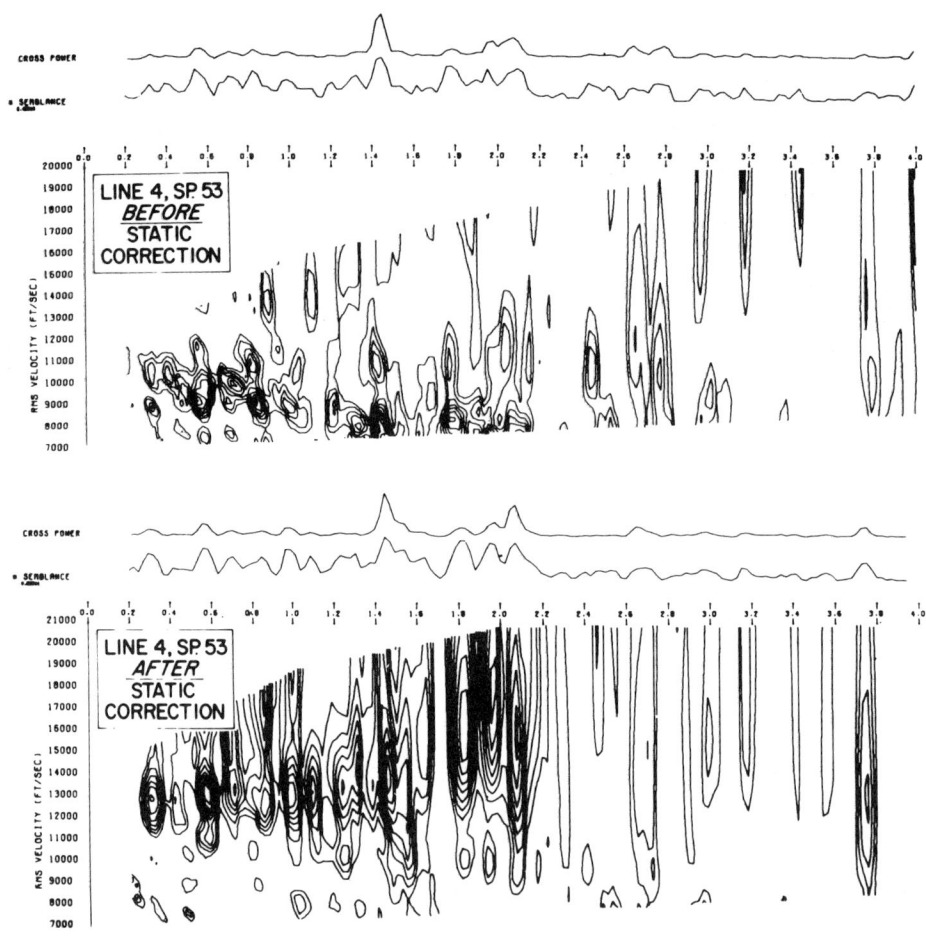

FIG. 19. Land Line 4, velocity spectra at SP 53 before and after static corrections.

checked before they are applied to the original data, prior to further processing. The monitor section from phase 2 compared to the preliminary processed section gives a good idea of the general structural match and degree of sectional data improvement. We feel, however, the best validity check is provided by comparing the velocity spectra before and after statics application. Correct statics give the appearance of focusing on the spectra. The scattered events appearing on the velocity spectra before static corrections will form larger peaks with less scattering after static corrections are applied. The resulting velocity functions will also exhibit a reasonable degree of spatial consistency.

ACKNOWLEDGMENTS

The authors would like to take this opportunity to extend their appreciation to Mr. Nigel A. Anstey and Dr. Franklyn K. Levin for their helpful comments and encouragement in preparation of this paper.

REFERENCES

Disher, D. A., and Naquin, P. J., 1969, Statistical automatic statics analysis: Presented at the 39th Annual International SEG Meeting, September 17, 1969, Calgary, Alta.

Garotta, R., and Michon, D., 1968, Static corrections in reflection seismology: Presented at the 13th meeting of the EAEG, Salzburg, Austria.

Hileman, J. A., Embree, P., and Pflueger, J. C., 1968, Automated static corrections: Geophys. Prosp., v. 16, p. 326–358.

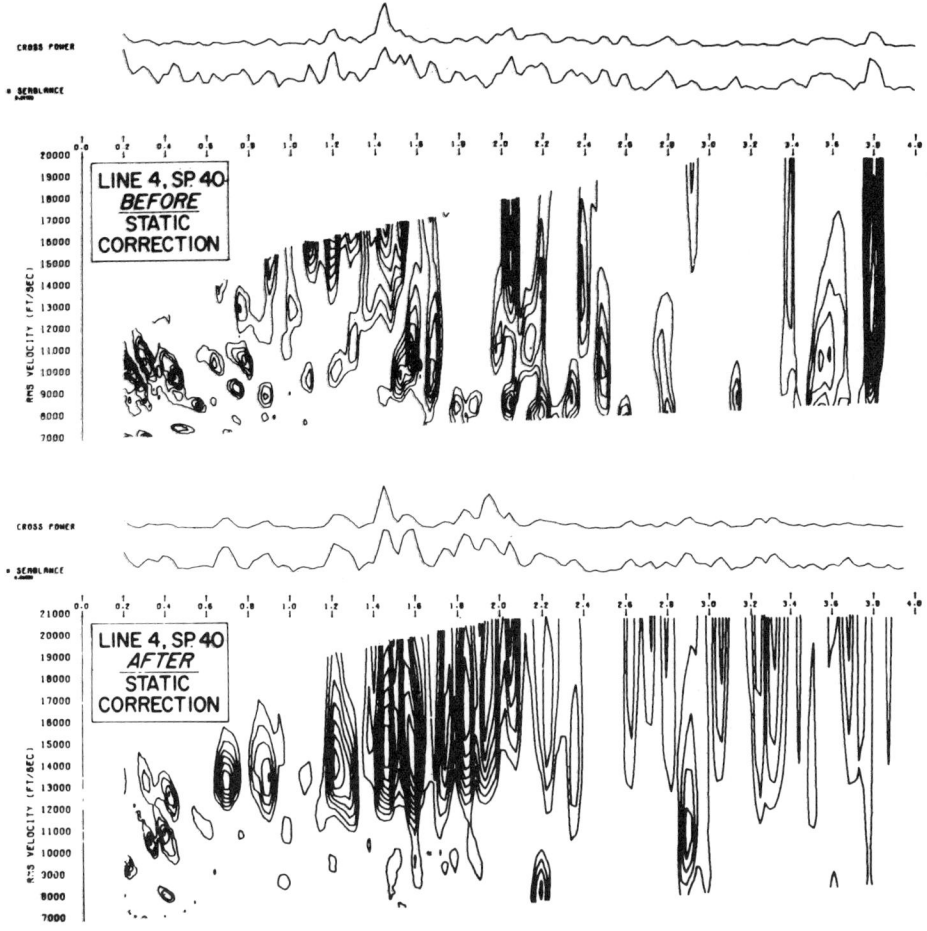

FIG. 20. Land Line 4, velocity spectra at SP 40 before and after static corrections.

Irvine, B. M., and Worley, J. K., 1969, The application and limitations of automatic residual static correction techniques: Presented at the 39th Annual International SEG Meeting, September 17, 1969, Calgary, Alta.

Morgan, N. A., 1970, Wavelet maps: A new analysis tool for reflection seismograms: Geophysics, v. 35, p. 447–460.

Neidell, N. S., and Taner, M. T., 1971, Semblance and other coherency measures for multichannel data: Geophysics, v. 36, p. 482–497.

Sheriff, R. E., 1969, Glossary of terms used in geophysical exploration: Geophysics, v. 33, p. 183–228.

Sherwood, J. W. C., and Donaldson, John, 1970, Automatic residual static corrections: Presented at the 40th Annual International SEG Meeting, November 8, 1970, New Orleans, La.

APPENDIX

The general static equation is given by

$$T_{ij} = R_i + S_j + C_k + M_k(j-i)^2, \quad (1A)$$

where

T_{ij} = total time shift for trace (ij),
R_i = receiver static at ith receiver position,
S_j = source static at jth source position,
C_k = arbitrary CDP group shift at kth CDP position,
M_k = time-averaged residual NMO coefficient at kth CDP position,
i = receiver position index,
j = source position index,
k = CDP position index, where $k = \frac{1}{2}(i+j)$, and
$(j-i)$ = offset position index.

Indices (i, j) range over a given set of discrete points. The values of T_{ij} are given data, and we wish to solve for R_i, S_j, C_k and M_k.

The first problem which must be solved is the problem of indeterminancy of the solution of equation (1A). This can be done by finding all possible solutions of the homogeneous equation related to (1A);

$$R_i + S_j + C_k + M_k(j - i)^2 = 0. \quad (2A)$$

We find the general solution of (2A) by the method of polynomial, or power series, representation of the variables. We look for solutions of (2A) in the form:

$$\begin{aligned} R_i &= a_0 + a_1 i + a_2 i^2 + \cdots \\ S_j &= b_0 + b_1 j + b_2 j^2 + \cdots \\ C_k &= c_0 + c_1(i+j) + c_2(i+j)^2 + \cdots \\ M_k &= d_0 + d_1(i+j) + d_2(i+j)^2 + \cdots \end{aligned} \quad (3A)$$

We determine the coefficients in (3A) so that it will be a solution of (2A). If we substitute (3A) into the left side of (2A) and equate to zero the coefficients of $i^n j^m$ for $n+m \leq 3$, we get

$$a_0 + b_0 + c_0 = 0 \quad (n = m = 0), \quad (4A)$$

$$\begin{aligned} a_1 + c_1 &= 0 \quad (n = 1, m = 0) \\ b_1 + c_1 &= 0 \quad (n = 0, m = 1), \end{aligned} \quad (4A1)$$

$$\begin{aligned} a_2 + c_2 + d_0 &= 0 \quad (n = 2, m = 0) \\ b_2 + c_2 + d_0 &= 0 \quad (n = 0, m = 2) \\ -2c_2 + 2d_0 &= 0 \quad (n = 1, m = 1), \end{aligned} \quad (4A2)$$

$$\begin{aligned} a_3 + c_3 + d_1 &= 0 \quad (n = 3, m = 0) \\ b_3 + c_3 + d_1 &= 0 \quad (n = 0, m = 3) \\ 3c_3 - d_1 &= 0 \quad (n = 2, m = 1) \\ 3c_3 - d_1 &= 0 \quad (n = 1, m = 2). \end{aligned} \quad (4A3)$$

The general solution of equations (4A) through (4A3) can be expressed as follows:

$$c_0 = -(a_0 + b_0), \quad (5A)$$

$$\begin{aligned} b_1 &= a_1 \\ c_1 &= -a_1, \end{aligned} \quad (5A1)$$

$$\begin{aligned} b_2 &= a_2 \\ d_0 &= c_2 = -\tfrac{1}{2} a_2, \end{aligned} \quad (5A2)$$

$$\begin{aligned} b_3 &= a_3 \\ c_3 &= -\tfrac{1}{4} a_3 \\ d_1 &= -\tfrac{3}{4} a_3. \end{aligned} \quad (5A3)$$

$b_0, a_0, a_1, a_2, a_3 =$ arbitrary constants. The fourth degree terms in (2A) are

$$a_4 i^4 + b_4 j^4 + c_4(i+j)^4 + d_2(i+j)^2 (j-i)^2,$$

and this will be identically zero, only if

$$a_4 = b_4 = c_4 = d_2 = 0.$$

In the same way, we must have

$$a_n = b_n = c_n = d_{n-2} = 0 \quad \text{for } n \geq 4.$$

The general solution of (2A) is therefore of the form:

$$\begin{aligned} R_i &= a_0 + a_1 i + a_2 i^2 + a_3 i^3, \\ S_j &= b_0 + a_1 j + a_2 j^2 + a_3 j^3, \\ C_k &= -(a_0 + b_0) - a_1(i+j) \\ &\quad - \frac{a_2}{2}(i+j)^2 - \frac{a_3}{4}(i+j)^3, \quad (6A) \\ M_k &= -\frac{a_2}{2} - \frac{3a_3}{4}(i+j), \end{aligned}$$

$b_0, a_0, a_1, a_2, a_3 =$ arbitrary constants.

The general solution of the least-square-error system derived from (1A) is equal to any particular solution plus the general solution of (2A). We derive the normal least-square-error system of equations corresponding to (1A) as follows: Let,

$E_i =$ set of index pairs (i, j) in which the first index has a given value i;

$F_j =$ set of index pairs (i, j) in which the second index has a given value j; and

$G_k =$ set of index pairs (i, j) such that $\tfrac{1}{2}(i+j)$ has a given value k.

Let $N(E_i), N(F_j), N(G_k)$ be the number of members of each set E_i, F_j, G_k, respectively. The least-square-error system corresponding to (1A) is:

$$N(E_i)R_i = \sum_{E_i} [T_{ij} - S_j - C_k - M_k(j-i)^2],$$

$$N(F_j)S_j = \sum_{F_j} [T_{ij} - R_i - C_k - M_k(j-i)^2],$$

$$N(G_k)C_k = \sum_{G_k} [T_{ij} - R_i - S_j - M_k(j-i)^2], \quad (7A)$$

$$\left[\sum_{G_k}(j-i)^4\right]M_k = \sum_{G_k}[j-i]^2[T_{ij} - R_i - S_j - C_k].$$

The system of equations (7A) gives the condition for the error quantity,

$$\sum_{(i,j)}[T_{ij}-R_i-S_j-C_k-M_k(j-i)^2]^2, \quad (8A)$$

to be a minimum.

In order to have a determinate problem, we replace (8A) by the condition:

$$\sum_{(i,j)}[T_{ij}-R_i-S_j-C_k-M_k(j-i)^2]^2 \quad (9A)$$

$$+\lambda\left[\sum_i R_i^2 + \sum_j S_j^2 + \sum_k C_k^2 + \sum_k M_k^2\right]$$

$$= \min,$$

where $\lambda > 0$. The normal equations for (9A) are obtained from (7A) by the addition of λ to the coefficients on the left-hand side. In matrix notation this amounts to adding λ to all elements on the main diagonal.

The new system is:

$$[N(E_i) + \lambda]R_i = \sum_{E_i}[T_{ij} - S_j - C_k - M_k(j-i)^2],$$

$$[N(F_j) + \lambda]S_j = \sum_{F_j}[T_{ij} - R_i - C_k - M_k(j-i)^2], \quad (10A)$$

$$[N(G_k) + \lambda]C_k = \sum_{G_k}[T_{ij} - R_i - S_j - M_k(j-i)^2],$$

$$\left[\sum_{G_k}(j-i)^4 + \lambda\right]M_k = \sum_{G_k}(j-i)^2[T_{ij} - R_i - S_j - C_k].$$

System (10A) has a unique solution for any $\lambda > 0$; this solution can be found to any degree of accuracy by iterative methods.

Seismic Signal Processing

LAWRENCE C. WOOD AND SVEN TREITEL

Invited Paper

Abstract—Seismic prospecting for oil and gas has undergone a digital revolution during the past decade. Most stages of the exploration process have been affected: the acquisition of data, the reduction of this data in preparation for signal processing, the design of digital filters to detect primary echoes (reflections) from buried interfaces, and the development of technology to extract from these detected signals information on the geometry and physical properties of the subsurface. The seismic reflection is generally weak, and it must be strengthened by the use of signal summing (stacking) procedures. The determination of depths to a target horizon requires knowledge of the propagational velocities of seismic stress waves, and a wealth of technology has evolved for this purpose. More recently, it has been possible to relate signal amplitude to the physical properties of the medium traversed and, in particular, to make inferences about the oil and gas content of the buried rocks. Much of the exploration effort occurs in offshore areas, where reverberations in the water layer mask reflections from below. The method of predictive deconvolution has been most effective in its ability to attenuate these reverberations, making it possible to detect reflections from structures at depth. Seismic signal processing is neither pure science nor pure art, and offers a continuing challenge to the practitioners of both cultures.

I. Introduction

MASSIVE AMOUNTS of seismic data are recorded and processed on a routine basis by the oil industry. Seismic surveys are carried out on a surface grid in order to build up a three-dimensional picture of the subsurface geology in a region, and each survey mile contains around 50 million bits of information, making modern data processing impossible without high-speed digital computers. Domestic oil companies acquired and processed around 390 000 line miles of data in the United States alone during 1973, at a total cost of about $370 million [1]. These figures do not include a similar international exploration effort.

Seismic signal processing can be divided into three categories: 1) data acquisition, 2) data processing, and 3) data interpretation. Though this paper will deal mainly with data processing, data acquisition and interpretation will be treated where necessary. Only modern reflection seismology methods are discussed as they are presently used to explore for hydrocarbon reserves. We shall not deal with processing techniques used in earthquake seismology, nuclear detection, earth crustal studies, and architectural engineering. General exploration objectives include the mapping of subsurface geological structures, the detection of hydrocarbon accumulations, and the estimation of total energy reserves in an area. As shown in Fig. 1, most reservoirs [2] are associated with geological formations having convex upward structures (anticlines) and linear displacements (faults) rather than with concave upward structures (synclines). Many deposits also relate to lateral changes in composition (stratigraphic traps). Time differences between reflected seismic signals map structural deformation, whereas amplitude changes

Manuscript received September 26, 1974; revised November 11, 1974.
The authors are with the Research Center, Amoco Production Company, Tulsa, Okla. 74102.

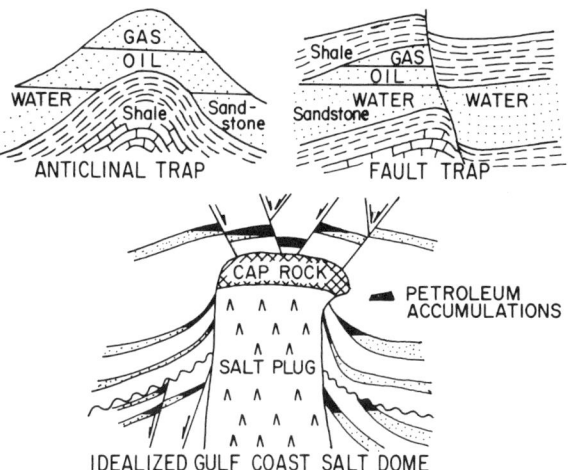

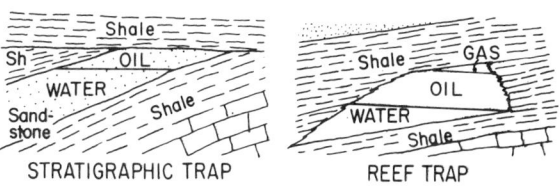

Fig. 1. Typical structural configurations for trapping hydrocarbons [2], [18].

of reflected signals may indicate the presence of hydrocarbons. The first part of this paper describes some basic time and amplitude adjustments, while the latter part deals with design of digital filters.

II. Data Acquisition

Readers unfamiliar with exploration methods must acquire a rudimentary knowledge of seismic field procedures [3], energy sources [4], and recording instruments [5], before seismic signal processing can be fully understood. Only those procedures currently employed in common-depth-point (CDP) surveying [6] will be reviewed in this paper. "Split-spread" methods [7]–[9] are not treated. A brief review of data acquisition practices precedes discussion of seismic processing methods.

Fig. 2 shows the essence of a seismic data gathering system. Elastic disturbances created by seismic energy sources propagate through the earth, where interfaces between geological strata reflect spreading wavefronts. Arrival times of single-bounce echoes (primary reflections) at surface receivers permit the determination of depths and inclination angles of reflectors when subsurface velocities are known. The receivers shown in Fig. 2 actually represent a composite array (group) of

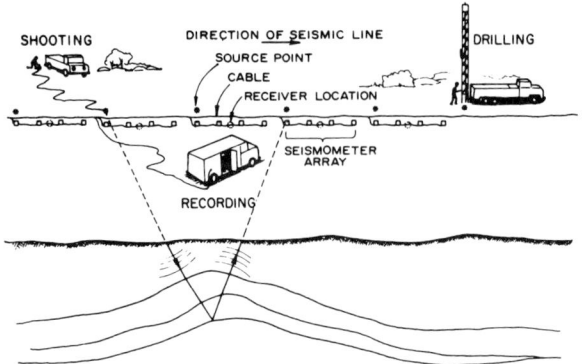

Fig. 2. Configuration of source points and seismometer arrays for common-depth-point (CDP) surveying.

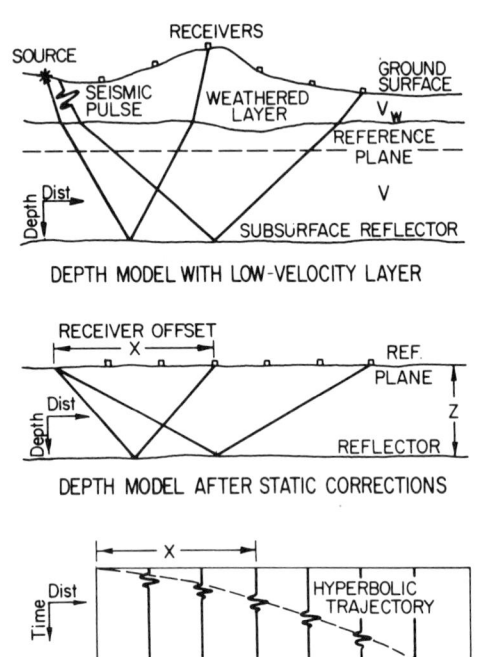

Fig. 3. Basic models of a reflection seismograph system.

transducers (seismometers) as illustrated in Fig. 3. These groups may consist of up to 100 individual geophones laid out in various linear and spatial patterns with group intervals (distance between groups) ranging from 50 to 900 ft. Each time a source is activated it is common practice to record either 24, 48, or 96 group arrays (traces) on digital tape simultaneously as a single recording. The seismic "master" cable joining these groups typically ranges from 1 to 3 mi in length. Seismic surveys are conducted along parallel straight lines and individual lines may extend for distances of 1 to 100 mi for total survey distances of 1 to 1000 mi or more.

Many different types of energy sources are used to generate seismic waves. Dynamite and other high energy explosive sources provide the simplest and most efficient means of releasing energy, but environmental considerations have led to the development of many alternate sources: implosive air guns, electrical sparkers, vibrating chirp systems [10], [11], etc.

A seismic source must provide good reflection signal-to-noise ratios at all times of interest. Weak sources are, therefore, laid out in arrays similar to receiver arrays, and signals generated by multiple source arrays are summed (stacked) together in a process called "vertical stacking." Vertical stacking should not be confused with "horizontal stacking" that sums traces lying in a common-depth point plane. Source and receiver arrays cancel unwanted ambient noise, attenuate surface waves, smooth time variations caused by surface irregularities, and discriminate against scattered energy. Many authors [12]–[14], [20] have studied optimum design of arrays, and an important problem in acquiring data in a new prospect is to relate array configurations and vertical stacks with reflection quality.

Modern instruments and data acquisition systems record fairly accurate facsimiles of the ground's response to incoming reflected energy. Many references [5], [15], [16] cover this subject in good detail; consequently, we review only those aspects most relevant to signal processing. Seismometers convert particle velocity to electrical voltages for land surveys, whereas in marine work they convert pressure variations to voltage. As mentioned in the preceding, an array may have up to 100 transducers with the array signal recorded on a single channel. A total of 24, 48, or 96 array signals are multiplexed and recorded digitally on magnetic tape (a single seismic record) by an instrument truck as shown in Fig. 2. Seismic traces seldom exceed 6 s in length because hydrocarbon reservoirs rarely occur below 30 000 ft in geologic basins, where velocities average around 15 000 ft/s. Special chirp systems [10], [11], however, may record 15 s of data. Reflected signals contain frequencies from a few hertz to a few hundred hertz, and field data is usually sampled at 1, 2, or 4 ms rates with alias frequencies (half the sampling frequencies) of 500, 250, or 125 Hz, respectively.

Digital seismic-recording systems have dynamic ranges around 80 dB. Exploration geophysicists define a decibel as $20 \log_{10} (A/A_0)$, where A/A_0 is the amplitude ratio. Signals, however, may rise 100 dB above ambient noise levels. Digital processing is able to recover another 20 or 30 dB of signal lying within the noise. Reflection amplitudes decay about 100 dB in the first 4 s of recording, primarily due to attenuation losses along the travel path. Consequently, amplifier gain levels change many times during recording in order to preserve signal amplitude for subsequent processing. Modern gain systems include instantaneous-floating-point and binary-gain control. Binary-gain amplifiers record the times of gain changes to allow recovery of signal amplitude. Field instruments record 16 bits. Most processing programs require only 12 bits, while final plotter output displays use the most significant 8 bits. A 24-trace seismic record contains around 1 million bits of information, and a typical marine crew may acquire 200 such records a day. Having discussed the rudiments of seismic data acquisition, we now consider processing these gigantic data sets on a computer.

III. PRELIMINARY CORRECTIONS

Signal processing begins with the demultiplexing of field records. This results in a work tape with signal traces in sequential order. Trace data is preceded on tape by header information giving elevations, seismometer group intervals, sampling rate, word size, trace length, and similar information. The creation of a work tape in a format compatible with central computing center requirements is one of the largest and most frustrating processing tasks, despite industry attempts [17] to standardize tape formats. Adjustment of times and amplitudes

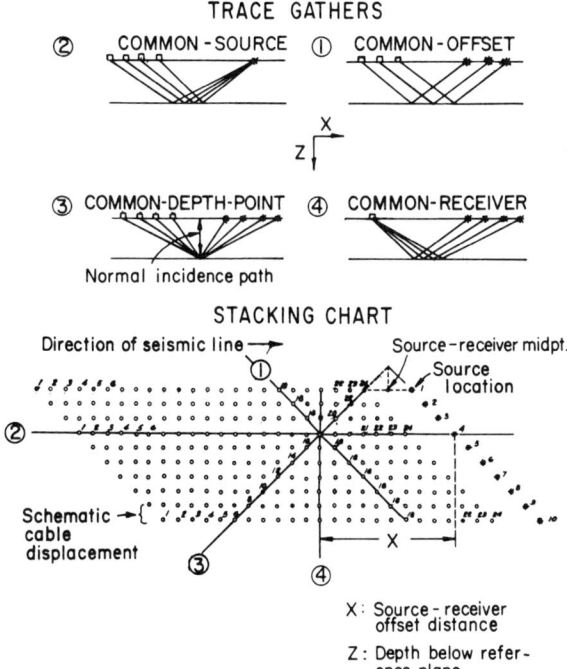

Fig. 4. Source and receiver positions corresponding to four principal planes used for sorting seismic traces [19].

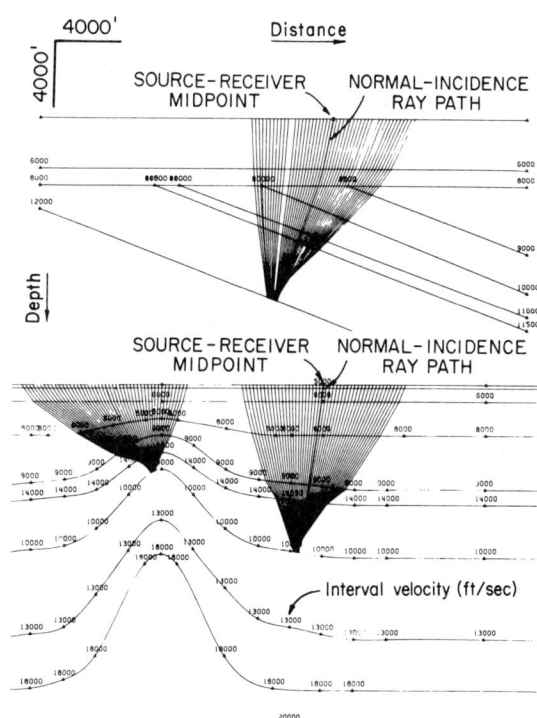

Fig. 5. Ray-path diagrams for CDP traces. Normal-incidence ray paths correspond to ideal CDP compositing.

to correct for various physical phenomena follows demultiplexing and reformatting. We briefly discuss time adjustments before proceeding to the important topic of relative amplitude preservation.

Geophysicists divide time corrections into static and dynamic categories [18]. A "static correction" consists of the application of a time shift, or translation, to an entire trace. In other words, a constant time correction term is added or subtracted from all reflection times, regardless of record time or reflector depth. "Dynamic corrections," on the other hand, vary with record time and therefore depend on reflector depth. Elevation changes and near-surface inhomogeneities severely degrade trace-to-trace continuity and the purpose of automatic static computations is to remove time variations caused by anomalous conditions at the earth's surface.

A region of very low velocity extends from the earth's surface to a depth of several tens to hundreds of feet; at this point velocities change either gradationally or abruptly from values near 2000 to 5000 ft/s or more. Time delays associated with this "weathered layer" disrupt reflection continuity (i.e., trace-to-trace alignment) and pose a major obstacle to successful processing of data acquired on land. Seismic lines recorded at sea, however, do not usually require static adjustments because of the uniform water layer of constant elevation.

An assumption underlying all automatic correction programs [19] is that a simple translation of a trace converts it into a model trace that would have been recorded had sources and receivers been vertically displaced downward to a reference plane with no weathering material present (see Fig. 3). This time delay is assumed to be "surface consistent," that is, to be a sum of an "initiation" or source related component and a contribution characteristic of a given surface or "receiver" position. The validity of these two assumptions is confirmed by the great success of modern static correction programs [19], although near-surface layers behave as complicated filters whose impulse responses distort amplitude and phase characteristics of seismic wavefronts. Further discussion of static and dynamic corrections requires knowledge of seismic trace sorting procedures.

Traces are usually collected into one of four kinds of data sets [19] or "gathers," depending on different objectives. For this purpose a diagram called a "stacking chart" is used (see Fig. 4). Sources are activated sequentially in the field (Fig. 2) with each initiation creating a "common-initiation" record. Surface positions of sources and receivers are displaced vertically on a stacking chart for clarity. Fig. 4 illustrates common-initiation gathers of 24 traces. These diagrams define four principal trace gathers called "common-initiation," "common-receiver," "common-offset," and "common-depth-point." Common-initiation gathers consist of traces having the same source, common-receiver gathers consist of traces having identical surface locations and common-offset gathers are traces with the same source-receiver distance ("offset").

IV. HORIZONTAL (CDP) STACKING

Traces with a CDP have a common midpoint [6] between source and receiver (Fig. 4); in the case of horizontal interfaces they also have common points on reflecting interfaces called "common-reflection-points" (CRP). Otherwise, common subsurface reflection points, however, migrate laterally and spread apart as structures become more complicated; nevertheless, complex geometries frequently cause small dispersion [21] of CRP locations (Fig. 5), so that CDP trace compositing is still successful. The last two decades have shown that horizontal (CDP) stacking is the single most important step in seismic signal processing. Stacking consists of the simple sum of traces contained in a CDP plane to produce a single composited trace. The latter's surface position is then equated with the common source-receiver midpoint. The summing of CDP traces succeeds because primary CRP reflections are in phase and add constructively, whereas ambient noise and other seismic signals not in phase cancel. Compositing increases reflection signal-to-noise ratios by factors approaching $\sqrt{N}$, where N is the number ("fold") of CDP traces summed. Horizontal stacks of 12, 24, and 48 fold are routinely produced.

CDP traces must be corrected for travel-time differences caused by varying ray path distances prior to stacking. This latter correction [18], called "normal-move-out" (NMO), depends on depth (record time) to the reflecting horizon and is, therefore, classified as a dynamic correction. NMO is defined as the increase in reflection time due to an increase in distance from source to receiver for a horizontal reflecting interface in a homogeneous medium of constant velocity. A simple expression for an NMO time increment can be derived from Fig. 3:

$$\Delta T_{\text{NMO}} = T_x - T_0 = \frac{1}{V}\sqrt{4Z^2 + X^2} - T_0 \tag{1a}$$

$$\Delta T_{\text{NMO}} \sim \frac{X^2}{4VZ} \tag{1b}$$

where $T_0 = 2Z/V$ is the two-way reflection time for the zero-offset trace, T_x is the two-way reflection time for a trace of offset distance X, V is the velocity of the medium, Z is the depth to the reflecting horizon, and X is the source-receiver separation (offset).

NMO correction involves the subtraction to a time increment ΔT_{NMO} from each record time T_x, with interpolation as necessary. This correction converts a trace of offset distance X into a zero-offset trace that would have been initiated and recorded at a common source-receiver midpoint (Fig. 4). Equations (1) show the dynamic nature of the NMO correction, because even in this most elementary case it is a function of depth, velocity, and offset.

Two facts contribute greatly to the success of CDP stacking. First, reflection time-distance curves (T_x versus X) from complicated structures are approximated well by a hyperbolic relationship [21] of the form,

$$T_x^2 = (T_0 + \Delta T_{\text{NMO}})^2 = T_0^2 + \frac{X^2}{V^2}. \tag{2}$$

Second, the purpose of NMO corrections is to align single-bounce ("primary") reflections prior to summing. Multiple-bounce ("multiple") reflections travel at lower average velocities than primary reflections with the same arrival times, since velocity usually increases with depth. Therefore, multiple reflections having greater NMO are misaligned and attenuated in CDP stacking.

NMO corrections and CDP compositing create new traces [21] called "normal-incidence traces" (NIT). These correspond to identical source and receiver positions (i.e., zero offset). The zero-offset traces also have identical incident and reflected ray path segments, as shown in Fig. 5. The NIT ray paths form right angles with reflecting horizons at points of reflection called "normal-incident points" (NIP). Thus CDP stacking produces a suite of NIT traces with reflection travel paths normal to subsurface horizons (Fig. 5).

V. Velocity Analysis

The most important variable in seismic prospecting is velocity, because distances to subsurface reflectors are calculated from observed travel times and known velocities. Seismic waves propagate with the velocity of sound in rock, which depends on chemical composition and local geology. Velocities increase with depth as a general rule, and vary from speeds of 1100 ft/s in air up to values approaching 21 000 ft/s in deep sedimentary basins. This information is obtained either through direct measurements in wells, or derived indirectly from seismic reflections with the aid of NMO relationships. We will not elaborate on well surveys where seismometers and sources are placed at varying depths in the well [18], because most velocity determinations make use of redundancy inherent in CDP surveys. Well surveys, however, are always used when they are available. The objective of CDP compositing is to increase signal-to-noise ratios to a level sufficient to insure reliable identification of primary events. Velocity as a function of time, however, must be known very accurately in order to apply proper NMO corrections prior to summing traces.

Hyperbolic characteristics of reflection time–distance curves provide a means for establishing the necessary velocity–time relationships by scanning CDP ensembles along hyperbolic trajectories for signal coherence. These scans establish a function to use in calculating NMO corrections. Reflection times as a function of distance do not satisfy hyperbolic relationships when more than one subsurface layer exists; nevertheless, this second-order approximation works very well even in areas of complex structural geology provided parameters are determined correctly [21]. Only reflections from a single interface in a homogeneous medium have a truly hyperbolic time–distance curve (see (1)).

Reflection times for a horizontal reflector below a sequence of N horizontal layers with constant interval velocities can be described by an infinite power series of the form [21],

$$T_{X,N}^2 = C_1 + C_2 X^2 + C_3 X^3 \cdots . \tag{3}$$

A hyperbolic approximation analogous to the single-layer case results from retention of the first two terms,

$$T_{X,N}^2 = T_{0,N}^2 + \frac{X^2}{V_{\text{rms}}^2} \tag{4a}$$

where

$$T_{0,N} = \sum_{k=1}^{N} \frac{2Z_k}{v_k} \tag{4b}$$

$$V_{\text{rms}} = \left(\frac{1}{T_{0,N}} \sum_{k=1}^{N} v_k^2 t_k\right)^{1/2} \tag{4c}$$

and X is the offset distance, N is the number of layers overlying the reflecting horizon, Z_k is the thickness of the kth layer, v_k is the interval velocity of the kth layer, t_k is the two-way travel-time in kth layer, $T_{0,N}$ is the two-way traveltime to the bottom of the Nth layer for the NITth trace, and V_{rms} is the rms velocity. In the limit, (4) reduce correctly to describe a single-layer, and then there is no difference between rms and interval velocity.

An expression can be derived [22] from (4c) for calculating interval velocities v_N in the multilayered situation when rms velocities are known. Velocity spectra described below are one way of measuring rms velocities. These average velocities $\bar{V}_N$ are calculated in succession beginning from $\bar{V}_1 = v_1$ using the following relationship:

$$v_N = \left(\frac{\bar{V}_N^2 T_{0,N} - \bar{V}_{N-1}^2 T_{0,N-1}}{T_{0,N} - T_{0,N-1}}\right)^{1/2} \tag{5}$$

where $\bar{V}_N$ is the rms velocity to the bottom of the Nth layer, and $\bar{V}_{N-1}$ is the rms velocity to the top of the Nth layer. Hyperbolic approximations (4) are accurate within 2 to 5 percent in geologic areas of simple structural deformations, that is where inclination angles of interfaces do not exceed about 15°; and interval velocities (5) are often estimated with accuracies between 5 and 10 percent for use in stratigraphic studies and for detection of hydrocarbon accumulations.

A velocity versus time display called a "velocity spectrum" [21] is generally used to determine hyperbolic parameters, which are calculated from CDP traces assuming that traveltimes of reflections from a CRP lie along a hyperbola. The determination of velocity becomes a matter of scanning various hyperbolic trajectories for maximum reflection coherency. Spectra are generated by incrementing normal incidence traveltimes $T_{0,N}$ and keeping them constant while incrementing $V_{\rm rms}$ at regular intervals between some minimum and maximum value. Each $(T_{0,N}, V_{\rm rms})$ pair defines a hyperbola, and coherency of data contained in a gate about this curve (Fig. 6) is measured. Traces are scanned with various hyperbolas whose apexes are fixed at the origin (i.e., $X = 0$ and $T_{0,N}$ = constant). A velocity spectrum consists of a three-dimensional surface of coherency as a function of normal incidence time $T_{0,N}$ and rms velocity $V_{\rm rms}$.

This spectrum may be displayed as contour lines which represent the intersection of level planes of constant coherency with the coherency surface. Interpretation of velocity spectra requires skill and experience, because multiple reflections and other seismic events in addition to primary reflections tend to align themselves along hyperbolic trajectories. A spectral interpretation consists of the location of peaks on the coherency surface that correspond to primary reflections. These peaks are then suitably joined to obtain an average stacking velocity ($V_{\rm rms}$) versus time ($T_{0,N}$) function display.

Coherence measurements are a crucial part of the determination of effective stacking velocities from multifold seismic data, and they have received considerable attention in [22]. The basic problem is to establish the similarity that exists between various time gates centered about hyperbolic trajectories (see Fig. 6). The main task is to measure alignment. Cross correlation and semblance are two commonly used statistical measures.

Cross correlation functions may or may not be sensitive to amplitude changes between time gates, this sensitivity depending on normalization procedures. The following normalized coherency function [23] employing zero-lag values of autocorrelation and cross correlation functions is not sensitive to rms signal amplitude variations between channels:

$$S = \frac{2}{M(M-1)} \sum_{i=1}^{M} \sum_{i>i'} \frac{R_{ii'}(0)}{\sqrt{R_{ii}(0) R_{i'i'}(0)}} \quad (6)$$

where M is the number of CDP traces, $R_{ii}(0)$ is the zero-lag value of the autocorrelation function of the ith trace, and $R_{ii'}(0)$ is the zero-lag value of the cross correlation function between the ith and i'th traces. This cross correlation measure varies between -1 and 1, where 1 corresponds to perfect signal coherency.

Another useful quantity for measuring multichannel coherence is semblance, which is defined [23] as the normalized output/input energy ratio. Output energy is measured on a composited time gate obtained by summing input time gates. The semblance coefficient S_c is sensitive to channel amplitude differences and varies between 0 and 1, with 1 denoting identical signals. It, too, can be expressed in terms of zero-lag

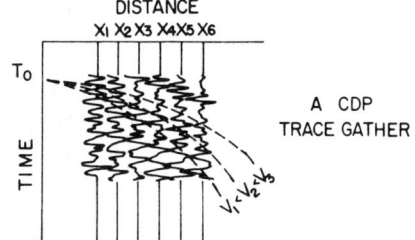
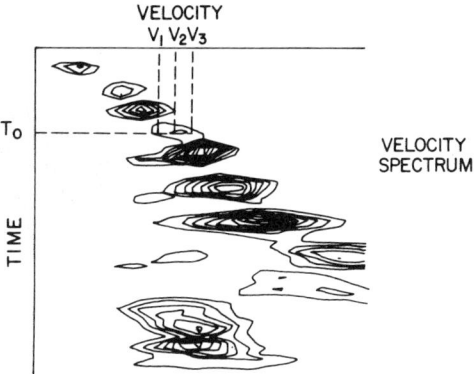

Fig. 6. A velocity spectrum displaying coherency as a function of reflection time and rms velocity.

values of correlation functions,

$$S_c = \frac{\sum_{i=1}^{M} \sum_{i'=1}^{M} R_{ii'}(0)}{M \sum_{i=1}^{M} R_{ii}(0)}. \quad (7)$$

These coherence measures are closely related, and the optimum statistic for extracting velocity information from CDP trace gathers may not yet have been found.

VI. RELATIVE AMPLITUDE PRESERVATION

Prior to 1970, seismic amplitudes were used almost exclusively as a qualitative tool for identifying seismic events. A recent development [24]–[26] relating large amplitude anomalies, called "bright spots," with the possible presence of hydrocarbon accumulations has added a new and significant dimension to the search for oil and gas deposits. Exploration has been generally restricted to the location of structural features such as anticlines, faults, and salt domes (Fig. 1) delineable with trace-to-trace differences in reflection time arrivals. Structural traps favoring the accumulation of hydrocarbons are drilled successfully about 20 percent of the time; however, amplitude information increases these percentages by helping to pinpoint changes in rock composition, layer thickness and stratigraphic conditions. The approach is successful [26] about 70 percent of the time in locating gas accumulations in young, unconsolidated Pleistocene (not over 12 million years in age) deposits in offshore Nigeria, Indonesia, and the U.S. Gulf Coast [29]. The "bright-spot" technology is being refined in the hope that it can treat amplitude anomalies observed in older, petroliferous sediments around the world in areas such as Alaska, the North Sea, California, and the continental United States.

Porous rocks at depth are usually filled with salt water, but may contain oil or gas. The new technology works best in lo-

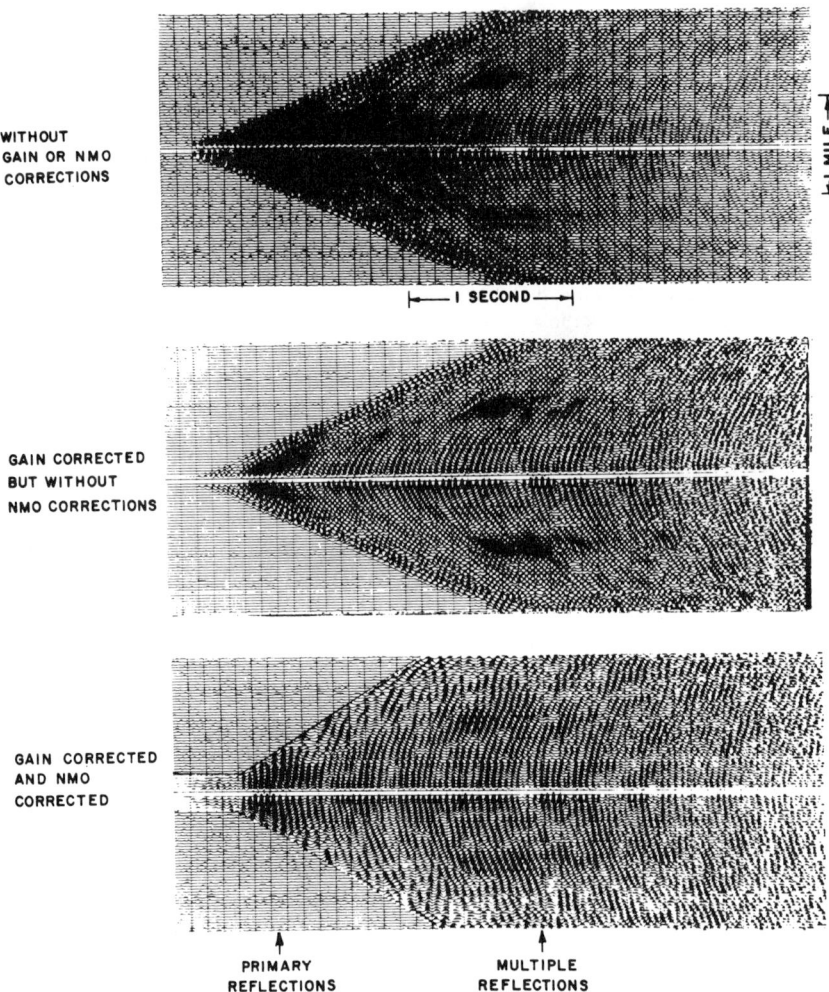

Fig. 7. Two CDP trace gathers showing the effects of gain and NMO corrections.

cating gas reservoirs because they cause a greater variation of reflection amplitudes. The amount of energy reflected at an elastic interface depends on the change in acoustic impedance (i.e., velocity-density product) across the interface. Gas-filled rocks have much lower velocities and, therefore, greater acoustic contrasts than either oil or brine saturated rocks, thereby reflecting a greater percentage of incident energy.

A simple expression [27] relates reflected and refracted amplitudes across an interface for the special case of plane waves incident on plane interfaces at normal incidence. In general, compressional (P) and shear (S) wave modes propagate in an elastic medium. Snell's law and Fermat's principle of minimum-time paths govern refraction at an interface, where incident energy splits into reflected and refracted P and S modes. No mode conversion occurs for normal incidence, and a simple normal-incidence reflection coefficient relates incident and reflected amplitudes,

$$A_r = \frac{\rho_2 v_2 - \rho_1 v_1}{\rho_2 v_2 + \rho_1 v_1} \tag{8}$$

where 1 is the medium containing the incident wave, 2 is the medium containing the transmitted wave, ρ is the density, and v is the interval velocity. A corresponding transmission coefficient relates incident and transmitted amplitudes,

$$A_t = \frac{2\rho_1 v_1}{\rho_1 v_1 + \rho_2 v_2}.$$

Appropriate equations for oblique incidence are much more complicated [28], because mode conversion must be taken into account. Nevertheless, normal incidence coefficients are very useful and quite accurate for stacked traces in areas having simple geological structures.

Gas has a much larger effect on reflected amplitudes than oil, so that amplitude anomalies associated with gas/brine contacts are greater than those related to oil/brine and gas/oil interfaces. Normal incidence reflection coefficients for gas-filled sandstones encased in slow-velocity shales may approach 40 percent, as compared to 10 percent or less for brine-charged sandstones. Coefficients for oil-bearing sands exhibit intermediate values. Thus hydrocarbons may produce amplitude anomalies around 12 dB. Polarity reversals also characterize hydrocarbon accumulations, because acoustic impedances across the upper interface of gas and, to a lesser extent, oil reservoir sands encased in shale decrease (negative $\rho_2 v_2 - \rho_1 v_1$ differences). In contrast, velocities and densities associated across the upper interface of a brine-filled reservoir sand encased in shale usually increase with depth (positive $\rho_2 v_2 - \rho_1 v_1$ differences). Thus rapid lateral increases in amplitude and sudden changes in polarity as shown in Fig. 8 may indicate a hydrocarbon accumulation at depth.

Trace amplitudes may vary 100 dB during the first four seconds of recording. Hence current digital instruments with dynamic ranges around 80 dB may be insufficient, but nevertheless they represent a significant improvement over the 40-dB analog systems in use several decades ago. The industry neglected amplitudes prior to the discovery of the "bright-spot" technique, and tended to destroy relative reflection amplitude relationships by improper use of automatic gain control (AGC) and trace average amplitude equalization procedures. Reflection amplitudes require some kind of time-dependent adjustment after corrections for gain recording functions have been made. This is because the human eye cannot assimilate dynamic ranges of 80 dB. The main factors contributing to reflection amplitude decay include attenuation caused by reflection and transmission coefficients, diverging wavefronts and frequency-selective absorption.

Techniques for the design of inverse gain functions to preserve relative reflection amplitude variations within and between traces fall into statistical and deterministic categories. Both approaches attempt to correct traces for average attenuation rates while preserving instantaneous variations caused by changes in subsurface acoustical impedances. Deterministic approaches define general models to describe many of the possible factors affecting amplitudes such as diverging wavefronts, frequency-selective absorption, reflection and transmission losses, source and receiver array effects and so on. Statistical approaches, on the other hand, produce average gain functions based on collections of traces sorted by common range, source, receiver, etc.

These average gain curves may be exponential functions of the form $a_0 \exp(a_1 t)$ and $(a_0/t) \exp(a_1 t)$, or polynomials of the form, $a_0 + a_1 t + a_2 t^2 + \cdots + a_N t^N$.

Arbitrary constants $a_i (i = 0, N)$ are determined by statistical regression. Trace amplitudes are then corrected by multiplication with an inverse gain function $g^{-1}(t)$.

$$G(t) = \frac{a(t)\bar{g}}{g(t)}$$

where $a(t)$ is the instantaneous trace amplitude (including polarity) corrected for recording gain, $g(t)$ is the gain function consisting of an average instantaneous trace amplitude (e.g., absolute value) obtained through statistical regression, and $\bar{g}$ is some desired average absolute-value of trace amplitude (e.g., 307 for 12-bit data, where 2047 is the maximum possible value).

In this manner, all traces have similar absolute amplitudes over all time gates, and they can be displayed conveniently (Fig. 7). Relative trace-to-trace reflection amplitude variations caused by changing subsurface conditions are thus preserved. The statistical approach is used most often, but sometimes gives poor results in areas of low signal-to-noise ratio, where the regression coefficients $(a_0, a_1, \cdots, a_N)$ tend to be affected by noise. Deterministic models often yield better results in such noisy areas.

The "bright-spot" technique is used to locate hydrocarbons, to determine reservoir dimensions [29], and to establish fluid content, either oil or gas. These estimates ultimately provide reserve figures used in economic evaluations of prospects prior

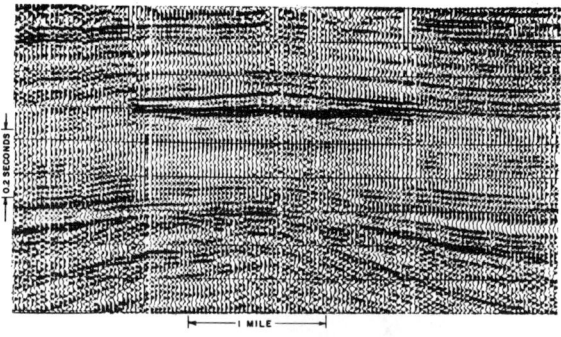

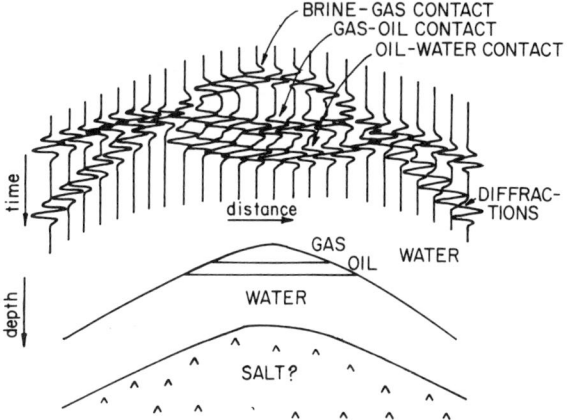

Fig. 8. An amplitude anomaly exhibiting many seismic features of an idealized "bright-spot" associated with a hydrocarbon reservoir.

to lease sales. Bright spots possess diagnostic features in addition to their large reflection amplitudes and to their polarity reversals, all of which serve to pinpoint hydrocarbon indicators (HCI) as well as lithologic change indicators (LCI). Large amplitude events of limited lateral extent having no inclination or dip on a stacked section sometimes correspond to reflections from gas/brine, gas/oil, or oil/brine interfaces (Fig. 8). These "contact events" constitute an important HCI. They are essentially horizontal on a stacked section because fluids tend to align themselves along gravitational equipotential surfaces regardless of the complexity of geological structures. Contact events may indicate the presence of hydrocarbons and help to define reservoir dimensions. Slow velocities also characterize hydrocarbon accumulations. Their effect is to delay reflections from fluid contacts. Contact events from thick reservoirs often have a convex downward appearance ("velocity pulldown") on a stacked section. Consequently, many bright spots have a "fisheye" appearance, as shown in Fig. 8. This effect occurs because reflections from the top of the reservoir are convex upward in accordance with the geological structure, whereas slower reservoir velocities cause contact events to be concave downward. Diffracted wavefronts from edges of reservoirs where hydrocarbons terminate add to this "fisheye" effect, and provide an additional HCI. Another criterion is the marked attenuation of amplitudes of reflections originating from horizons beneath reservoirs. Large transmission losses and strong reverberations associated with shallow accumulations attenuate or "mask" reflections from underlying strata and deeper reservoirs.

Modeling is still another important aspect of bright spot interpretation [29]. Here the objective is to assist geophysical

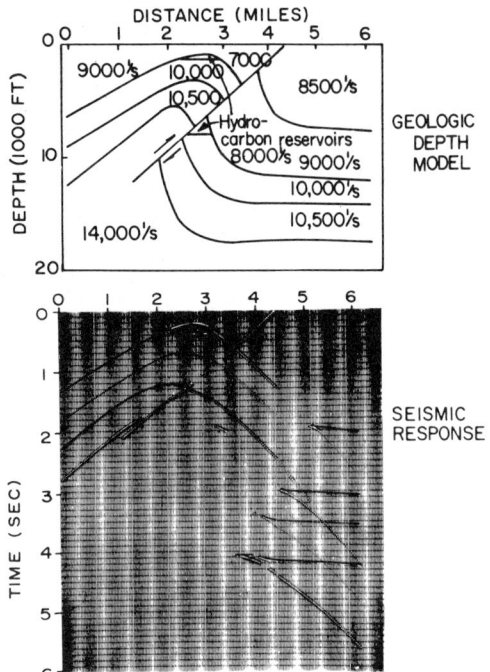

Fig. 9. A depth model of a faulted anticline structure with hydrocarbon reservoirs. Idealized normal-incidence traces show the corresponding seismic reflection times and amplitudes. (Courtesy of Dr. B. T. May and Dr. F. Hron.)

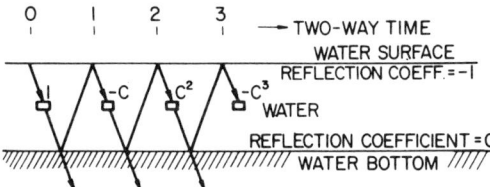

Fig. 10. Reverberations in the water layer where ray paths have been drawn as slanted lines in order to illustrate time dependence [33].

interpretation by means of computer simulated reflection amplitude anomaly patterns. This is done with synthetic traces computed from geological depth models (Fig. 9). Sophisticated modeling procedures produce synthetic records, and model parameters such as layer thicknesses and velocities are varied iteratively until times and amplitudes match observations within specified tolerances. Success depends on the ability to record as many high-frequency components as possible in the field, so that subsequent deconvolution (see Section VII) and source pulse-compression techniques can improve the resolution of thin layers.

Amplitude anomalies do not always indicate hydrocarbon accumulations. Reflected signal strengths depend on subsurface impedance contrasts, and many factors other than hydrocarbon accumulation can cause large impedance contrasts. Thin lenses of lava and tightly cemented layers of silt, lime and lignite known as "hard streaks" give rise to bright spots similar to those associated with hydrocarbon reservoirs [30]. Low-saturation gas sands [31] and rocks deposited in shallow water environments also produce large amplitude reflections. Furthermore, by no means are all hydrocarbon deposits commercial, and careful interpretations must be made to establish thicknesses, fluid content, saturation levels, areal extent and other similar variables.

The bright-spot method works best in outlining gas reservoirs in unconsolidated sand reservoirs at depths not exceeding 6000 ft. Amplitude anomalies associated with older rocks at greater depths are exceedingly difficult to interpret because rocks are more indurated, have less pore space and, therefore, smaller impedance changes across elastic interfaces. Multiple reverberations as well as geological complexities tend to become more bothersome with depth. Onshore surface conditions further complicate interpretations because of changes in the shallow layers, topography, and variations in source and receiver coupling. Despite these many difficulties, amplitude anomalies have defined many new oil and gas fields throughout the world, and many unexpected benefits have resulted from attempts to extract meaningful information from seismic amplitudes.

VII. THE METHOD OF PREDICTIVE DECONVOLUTION

A substantial fraction of the globe's deposits of oil and gas is buried in subsurface rocks overlain by water. Typically, a seismic source imparts a pulse of energy into the water just a few feet below the surface. This source pulse travels from the water into the rock formations below it, where it is split into a large number of waves traveling along various paths determined by the material properties of the medium. Whenever such a wave encounters a change in acoustic impedance (which is the product of rock density and rock propagation velocity), a certain fraction of the incident wave is reflected upwards. Seismic detectors situated on the water surface record the continual motion of the water under the impact of seismic waves impinging from below. This recording is performed digitally at a fixed sampling increment. The resultant set of discrete observations is called a "marine seismic trace", and constitutes a sample of a time series.

The interpreter of such marine recordings is faced with the task of extracting the direct reflections which give him information about the subsurface geometry from a recording which contains a wealth of background interference and noise. One of his several problems is the presence of the so-called "multiple reflections" or "reverberations." These slowly decaying wave trains usually arise in the water layer, which tends to act as a strong waveguide because it is bounded above and below by media of radically differing acoustic impedances. The water reverberation phenomenon came to light when it was observed that seismic traces recorded in water depths greater than 10 ft or so exhibit a marked sinusoidal, or "ringing" appearance.

During the past two decades very significant strides have been made in a continuing effort to remove reverberations from marine data. One of the more successful approaches is based on a rather simple theoretical model of a reverberating trace. The treatment given below is an abbreviated version of a discussion by Robinson [33].

Consider an ideal source located on the water surface emitting a unit spike (or unit pulse) at time $t = 0$, and assume that a detector just below the water surface responds only to downward motion (see Fig. 10). Both the air water and water rock interfaces are strong reflectors. We restrict ourselves to plane wave fronts whose raypaths are perpendicular to the interfaces, although for the sake of clarity these paths have been drawn as slanting lines in Fig. 10. Under such so-called "normal incidence" conditions, we may associate a reflection coefficient of -1 with the lower surface of the air-water inter-

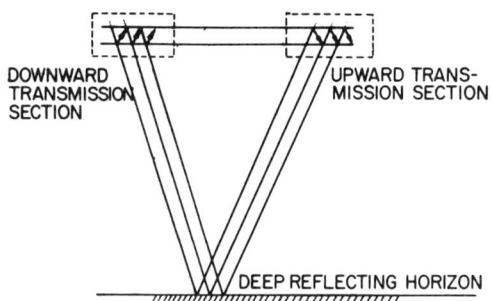

Fig. 11. The cascading effect of the water layer on a reflection from great depth [33].

face, while we let the water-bottom reflection coefficient be c, whose magnitude must be less than unity from physical considerations. A source pulse generated in the water layer will reverberate between these two strong reflectors, although part of the energy will be propagated into the underlying rocks. Let the integer n represent one roundtrip, or two-way traveltime in the water layer. Then the downgoing unit spike, which occurs at time $t = 0$, is followed at intervals of n time units by successive downgoing spikes whose values are $-c$, $+c^2$, $-c^3$, and so on. The z transform[1] of such a water-confined reverberation spike train is

$$C(z) = 1 - cz^n + c^2 z^{2n} - c^3 z^{3n} + \cdots. \quad (9)$$

Since $|c| < 1$, this convergent geometric series can be summed to yield

$$C(z) = \frac{1}{1 + cz^n}.$$

An inverse filter to remove the water reverberations is therefore

$$A(z) = \frac{1}{C(z)} = 1 + cz^n.$$

Because $|c| < 1$, it follows that both the water reverberation spike train $C(z)$ as well as the corresponding "dereverberation" filter $A(z)$ are minimum-delay [32].

In actuality, a part of the energy originally present in the downgoing unit spike travels into the medium below the water layer, in which it continues to propagate until it encounters a deep reflector. At this point, some of the incident energy is reflected upwards, and when this reflected pulse enters the water layer from below, it in turn becomes partially trapped and causes reverberations (see Fig. 11). In other words, the water layer affects the deep reflection returns twice—once on the way down, and once on the way up. To a good approximation, the z transform of the resulting spike train can be obtained by merely cascading the response (9) with itself [32] and thus

$$C(z) = (1 - cz^n + c^2 z^{2n} - c^3 z^{3n} + \cdots)^2$$
$$= \frac{1}{(1 + cz^n)^2}. \quad (10)$$

This reverberation spike train is also minimum delay. An inverse filter to remove the cascaded set of reverberations is

[1] Geophysicists define the z transform as $C(z) = \sum_{n=-\infty}^{+\infty} c_n z^n$ rather than as $C(z) = \sum_{n=-\infty}^{+\infty} c_n z^{-n}$, as electrical engineers do.

now given by

$$A(z) = \frac{1}{C(z)} = (1 + cz^n)^2$$
$$= 1 + 2cz^n + c^2 z^{2n} \quad (11)$$

or

$$A(z)C(z) = (1 + 2cz^n + c^2 z^{2n})(1 - cz^n + c^2 z^{2n} - c^3 z^{3n} + \cdots)^2$$
$$= 1.$$

We see that the filter $A(z)$ *deconvolves* the reverberation spike train $C(z)$ to the unit spike at zero delay.

The deconvolution method just described is of slight practical value because the reverberation patterns encountered in petroleum exploration are far more involved. Now it turns out that the minimum-delay property of the reverberation spike train $C(z)$ is quite general in the sense that the unit spike response of an arbitrary system of horizontally stratified layers also is minimum delay [35]. In practice, the source pulse is a broader time function, say b_t, rather than merely a unit spike. If this source pulse is reasonably sharp, as will be the case for an explosion of dynamite, then we may expect b_t to have most of its energy concentrated near its front end, i.e., to be "front loaded." Front-loaded time functions tend to be approximately minimum delay, and for the moment we assert that the source pulse b_t does in fact have this property.

We assume that the reverberation *pulse* train, say r_t, is the convolution of the reverberation spike train c_t of (10) with the pulse b_t,

$$r_t = c_t * b_t$$

where the asterisk denotes convolution. We imagine that the marine seismic trace x_t arises from the linear superposition of a large number of deep reflections, each of which has the characteristic shape of the pulse train r_t. Let ϵ_t be a series of spikes whose amplitudes represent the values of the deep reflection coefficients, and whose times represent the two-way travel time to these reflectors. Our model of the marine seismic trace x_t is, therefore,

$$x_t = c_t * b_t * \epsilon_t$$
$$= r_t * \epsilon_t.$$

Next, we assume that the series ϵ_t is uncorrelated and random. In particular, this means that the series ϵ_t is totally unpredictable, in the sense that knowledge of the amplitudes and arrival times of the first k deep reflections does not permit us to make any deterministic statement about the amplitude and arrival time of the $(k + 1)$th reflection. Of course, we cannot prove that the actual earth has this property (and there are some demonstrable cases for which it does not), but the practical success of a deconvolution approach based on this model suggests that the random and uncorrelated representation of the series ϵ_t is generally reasonable.

On the other hand, the reverberation pulse train r_t is predictable if we assume, as we do here, that both the source pulse b_t as well as the reverberation spike train c_t are minimum delay [36]. Let ϕ_τ be the autocorrelation of the marine trace x_t. Then we have,

$$\phi_\tau = E\{x_t x_{t+\tau}\} = E\{r_t r_{t+\tau}\} * E\{\epsilon_t \epsilon_{t+\tau}\}$$

where E is the expectation operator. But since ϵ_t is random and uncorrelated,

$$E\{\epsilon_t \epsilon_{t+\tau}\} = E\{\epsilon_t^2\} = P\delta_{\tau 0}$$

where P is the power in the series ϵ_t, and where $\delta_{\tau 0}$ is the Kronecker delta. Therefore,

$$\phi_\tau = P E\{r_t r_{t+\tau}\}$$

and P is a scale factor which does not affect the final result and will thus be neglected. We conclude that the trace autocorrelation ϕ_τ is equal to the autocorrelation of the reverberation pulse train r_t within an arbitrary scale factor. Furthermore, the minimum-delay property of r_t enables us to predict its reverberation component c_t if we compute a prediction operator for prediction distance n, where we recall that n is two-way travel time in the water layer. If we delay the output of such a prediction operator by n time units and subtract it from r_t, we obtain the nonreverberatory component of r_t, namely, the source pulse b_t. The linearity of the prediction operator allows us to apply it to the entire trace x_t, suppressing from the data the reverberatory components c_t.

Let a_t be such a prediction operator. For the simplest case, this operator is given by (11), but in practice a far more general approach results from the use of Wiener theory [34]. Minimization of the mean square error between a desired output and an actual output yields a set of normal equations involving the trace autocorrelation coefficients ϕ_t. If we identify the desired output with an input advanced by n time units, the $(m+1)$ length least squares prediction operator a_t is the solution of the system,

$$\begin{bmatrix} \phi_0 & \phi_1 & \cdots & \phi_m \\ \phi_1 & \phi_0 & & \\ \vdots & & & \\ \phi_m & \phi_{m-1} & \cdots & \phi_0 \end{bmatrix} \begin{bmatrix} a_0 \\ a_1 \\ \vdots \\ a_m \end{bmatrix} = \begin{bmatrix} \phi_n \\ \phi_{n+1} \\ \vdots \\ \phi_{n+m} \end{bmatrix} \quad (12)$$

The autocorrelation matrix of this system contains only the $(m+1)$ independent elements $\phi_0, \phi_1, \cdots, \phi_m$, and these are arranged in such a manner that all elements on the main diagonal as well as any super- or subdiagonals are equal. This so-called Toeplitz structure enabled Levinson [37] to obtain an efficient recursive algorithm for the solution of the normal equations. It is of interest to note that the case $n = 1$ leads to a set of normal equations arising in the linear prediction approach to speech compression [38].

The prediction operator coefficients $a_0, a_1, \cdots, a_m$ can be used to construct the prediction error operator for prediction distance n,

$$1, \underbrace{0, 0, \cdots, 0}_{n-1 \text{ zeroes}}, -a_0, -a_1, \cdots, -a_m.$$

This prediction error operator is then convolved with the marine trace x_t to yield,

$$z_t = x_t - a_0 x_{t-n} - a_1 x_{t-n-1} - \cdots - a_m x_{t-n-m}.$$

The series z_t, therefore, represents the deconvolved marine trace, from which the reverberation spike train c_t has been removed. Alternatively, z_t is the prediction error series associated with the prediction error operator for prediction distance n, where n is two-way travel time in the water layer.

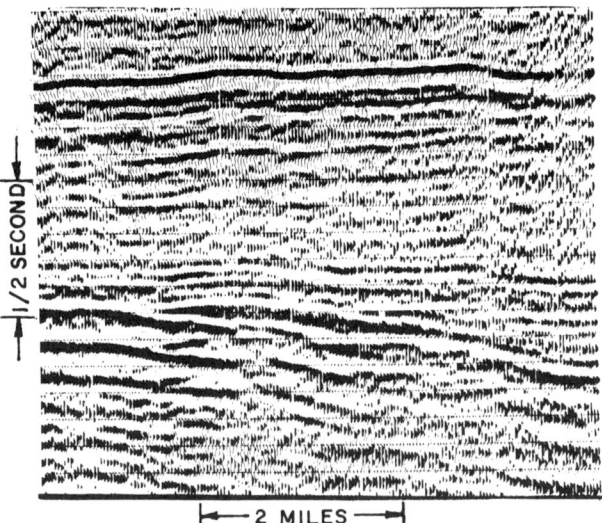

Fig. 12. An example of composited (CDP) marine data.

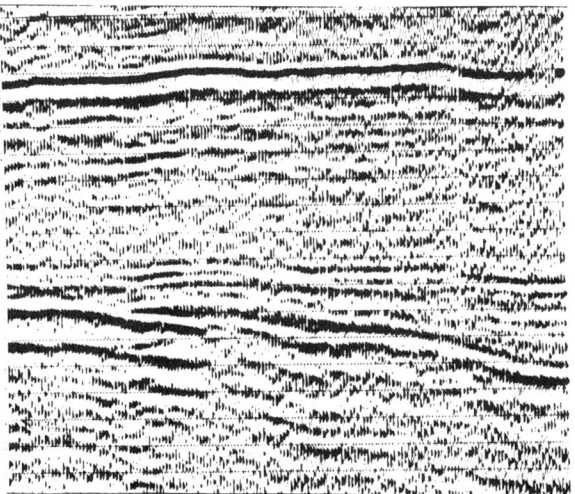

Fig. 13. The seismic data shown in Fig. 12 after application of predictive deconvolution.

The approach we have described is called the method of predictive deconvolution, which dates back to the work of Geophysical Analysis Group (GAG) at Massachusetts Institute of Technology between 1952 and 1957 [39]. Fig. 12 shows a selected portion of a marine seismic line which has been stacked. The vertical scale is two-way travel time, while the horizontal scale represents distance. In Fig. 13 we may observe the output after every trace has been filtered with a predictive deconvolution operator. We note that a significant amount of reverberating energy has been removed from the input data. It is customary to follow the dereverberation procedure with a number of further digital filter applications designed to compress the source pulse and to provide greater emphasis to the deeper reflections. This goal is accomplished with Wiener shaping filters (see Section VIII), which are designed for a selected number of gates on each trace. The variations in source pulse shape with travel time can be accounted for and, in effect, the Wiener shaping filters are applied in a time-varying manner (see Fig. 14).

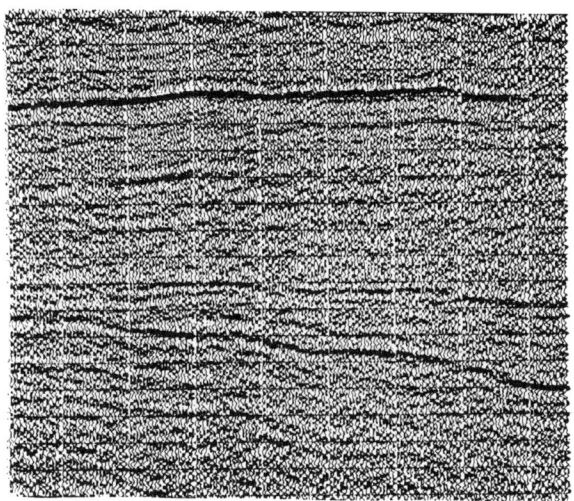

Fig. 14. The seismic data shown in Fig. 12 after Wiener filtering and time-varying pulse compression.

Query: Very nice indeed, but how often does real data respond to such treatment?

Answer: To the extent that it obeys our model's assumptions, namely that:
1) the layered earth is a linear system;
2) the reverberation spike train and the source pulse are minimum delay;
3) the deep reflector reflection coefficient series is random and uncorrelated.

In actuality, of course, these assumptions may or may not be upheld. All we can say is that widespread application of the predictive deconvolution technique has demonstrated its ability to remove reverberations, and thereby to permit the identification of reflections from depth. In instances for which assumption 2) is inappropriate, viable alternatives exist, of which the homomorphic deconvolution approach [40], [41] is one.

Our treatment of the method of predictive deconvolution has necessarily been quite brief. Unmentioned here is a wealth of implementational detail, such as the choice of proper lag windows for the autocorrelation coefficients ϕ_t, the determination of "good" values for the operator length, $(m + 1)$ and the two-way travel time in the water layer (n), etc. The interested reader must be referred to the cited references for further discussion of these by no means negligible points.

VIII. Wiener Shaping Filters

We have already seen how the method of predictive deconvolution is based on Wiener's least mean-square error criterion. In geophysical work the need to alter the shape of a given signal pulse often arises. The problem can be attacked with zero-phase bandpass filters in the frequency domain. However, the amount of control one has on the shape of the output pulse is limited, and we have found it more expedient to design such shaping filters in the time domain.

We consider the problem of finding an $(m + 1)$-length filter $f_t = (f_0, f_1, \cdots, f_m)$, which shapes an $(n + 1)$-length input pulse $b_t = (b_0, b_1, \cdots, b_n)$ into an $(m + n + 1)$-length desired output pulse $d_t = (d_0, d_1, \cdots, d_{m+n})$ in such a way that the error energy between the desired output d_t and the actual $(m + n + 1)$-length output $c_t = (c_0, c_1, \cdots, c_{m+n})$ is minimized.

The actual output is the convolution of the filter with the input

$$c_t = \sum_{s=0}^{m} f_s b_{t-s}.$$

The error energy I is

$$I = \sum_{t=0}^{m+n} (d_t - c_t)^2 = \sum_{t=0}^{m+n} \left(d_t - \sum_{s=0}^{m} f_s b_{t-s}\right)^2. \quad (13)$$

The preceding error energy is at its minimum value if its partial derivatives with respect to each of the filter weighting coefficients $f_0, f_1, \cdots, f_m$ equal zero. We have,

$$\frac{\partial I}{\partial f_j} = \sum_{t=0}^{m+n} 2\left(d_t - \sum_{s=0}^{m} f_s b_{t-s}\right)(-b_{t-j}) = 0$$

which gives

$$-\sum_{t=0}^{m+n} d_t b_{t-j} + \sum_{t=0}^{m+n} \left(\sum_{s=0}^{m} f_s b_{t-s}\right) b_{t-j} = 0$$

or

$$\sum_{s=0}^{m} f_s \sum_{t=0}^{m+n} b_{t-s} b_{t-j} = \sum_{t=0}^{m+n} d_t b_{t-j}, \quad j = 0, 1, \cdots, m.$$

Now

$$\sum_{t=0}^{m+n} b_{t-s} b_{t-j} = \phi_{j-s}$$

and

$$\sum_{t=0}^{m+n} d_t b_{t-j} = g_j$$

where r_j is the autocorrelation of the input pulse b_t, g_j is the cross correlation between the input pulse b_t and the desired output pulse d_t. We thus obtain,

$$\sum_{s=0}^{m} f_s \phi_{j-s} = g_j, \quad j = 0, 1, \cdots, m. \quad (14)$$

This system of $(m + 1)$ linear simultaneous equations in the unknowns $f_0, f_1, \cdots, f_m$ can also be written in the matrix form

$$\begin{bmatrix} \phi_0 & \phi_1 & \cdots & \phi_m \\ \phi_1 & \phi_0 & & \\ & & \cdot & \\ & & \cdot & \\ & & \cdot & \\ \phi_m & \phi_{m-1} & \cdots & \phi_0 \end{bmatrix} \begin{bmatrix} f_0 \\ f_1 \\ \cdot \\ \cdot \\ \cdot \\ f_m \end{bmatrix} = \begin{bmatrix} g_0 \\ g_1 \\ \cdot \\ \cdot \\ \cdot \\ g_m \end{bmatrix} \quad (15)$$

where $\phi_{-j} = \phi_j$ because b_t is real valued. We note that the normal equations (15) for the Wiener shaping filter reduce to the normal equations (12) for the predictive deconvolution filter if we identify the crosscorrelation vector $(g_0, g_1, \cdots, g_m)$ with the vector $(\phi_n, \phi_{n+1}, \cdots, \phi_{n+m})$. This is so because in the case of predictive deconvolution the desired output $d_t =$

r_{t+n}, where r_t is reverberation pulse train. Hence,

$$g_j = \sum_t d_t r_{t-j} = \sum_t r_{t+n} r_{t-j}$$
$$= \sum_t r_{t+(n+j)} r_t = \phi_{n+j}, \quad j = 0, 1, \cdots, m.$$

The method of predictive deconvolution is accordingly seen to constitute a particular realization of the Wiener shaping filter. Solutions of the more general system (14) are again readily obtained with Levinson's algorithm.

An expression for the normalized minimum square error, E_N, results when the normal equations (14) are substituted into the error energy relation (13),

$$E_N = 1 - \sum_{t=0}^{m} f_t g'_t$$

where g'_t is the normalized cross correlation coefficient [42],

$$g'_t = \frac{g_t}{\sum_{t=0}^{m+n} d_t^2}.$$

It follows that

$$0 \leq E_N \leq 1$$

and the extreme cases $E_N = 0$ and $E_N = 1$ correspond, respectively, to perfect agreement and to no agreement between the actual output c_t and the desired output d_t.

Space unfortunately precludes the presentation of numerical examples illustrating these principles, and the reader is referred to the afore cited references for further details.

IX. Conclusions

In this paper, we have outlined how seismic data are acquired, interpreted, and processed. One of the major problems we have omitted, however, is the processing sequence, that is, the order in which the corrections are made and the filters applied. For example: Should static corrections be estimated before or after NMO? Should predictive deconvolution be applied before or after CDP stacking? Should Wiener filters be applied before or after velocity analysis? Unique answers to such questions do not appear to be available.

Processing sequences depend on geological conditions and vary from area to area. Geophysical objectives tend to determine a particular sequence, while geological factors cannot always be considered. Land data require static corrections to adjust for surface irregularities, and therefore tend to undergo more complicated processing sequences than marine data. As a general rule, however, both land and marine records are sorted into CDP trace gathers, subjected to velocity analysis, NMO corrected and composited. Scaling, static corrections and digital filtering alter this basic flow. A typical marine sequence might consist of the following steps:

demultiplexing
reformatting
gain recovery
sorting for relative amplitude scaling
bandpass filtering
predictive deconvolution
Wiener filtering
CDP sorting
velocity analysis
NMO correction
CDP stacking
Wiener filtering
modeling, migration, and interpretation.

Static corrections complicate the processing of data acquired on land. The NMO correction procedure requires the use of an initial average velocity function in order to reduce trace-to-trace time variations of reflections. Automatic static correction programs cannot handle large time increments between traces, and this fact necessitates the use of an initial gross, average velocity function. Once static corrections have been determined, they can be applied before the NMO corrections to produce a final processed seismic line. If an average velocity function is used for preliminary NMO estimation, a typical land processing sequence might consist of the following steps:

demultiplexing
reformatting
gain recovery
sorting for relative amplitude scaling
bandpass filtering
predictive deconvolution
Wiener filtering
NMO correction
automatic static correction.

A final seismic line on land might consist of the following steps after demultiplexing, reformatting, and amplitude scaling:

static corrections
bandpass filtering
predictive deconvolution
Wiener filtering
CDP sorting
velocity analysis
NMO correction
CDP stacking
Wiener filtering
modeling, migration, and interpretation.

New technology continually alters and modifies the flow of these sequences. Such innovations present a never-ending challenge to the ingenuity of geophysicists involved in the processing of exploration seismic data.

Acknowledgment

The authors wish to thank S. N. Domenico for helpful discussions, G. Bard for preparing the illustrations, and Amoco Production Company for permission to publish. Appreciation is also extended to K. L. Peacock, R. C. Heiser, P. W. Johnson, B. T. May, F. Hron, R. D. Bjerstedt, and P. F. Barron for help with the data processing required for the illustrations.

References

[1] "Outlay for U.S. seismic work hits record high, SEG reports" *Oil and Gas J.*, vol. 72, no. 32, p. 64, Aug. 1974.
[2] A. I. Levorsen, *Geology of Petroleum*. San Francisco, Calif.: Freeman, 1958.
[3] N. A. Anstey, "Seismic prospecting instruments: Signal Characteristics and instrument specifications," *Geoexploration Monog.*, ser. 1, vol. 1, no. 3 (Gebrüder Borntraeger, 1 Berlin 38), 1970.
[4] F. S. Kramer, R. A. Peterson, and W. C. Walter, Eds., *Seismic Energy Sources 1968 handbook*. United Geophysical Corp. (Bendix Corporation subsidiary), 1968.
[5] B. S. Evenden and D. R. Stone, "Seismic prospecting instruments: Instrument performance and testing," *Geoexploration Monog.*, ser. 1, vol. 2, no. 3 (Gebrüder Borntraeger, 1 Berlin 38), 1971.
[6] W. H. Mayne, "Common reflection point horizontal data stacking techniques," *Geophysics*, vol. 27, no. 6, pp. 927-938, Dec. 1962.

[7] L. L. Nettleton, *Geophysical Prospecting for Oil.* New York: McGraw-Hill, 1940.
[8] C. A. Heiland, *Geophysical Exploration.* Englewood Cliffs, N.J.: Prentice-Hall, 1940.
[9] J. J. Jakosky, "Exploration geophysics," Trija, 2nd edition, 1950.
[10] J. M. Crawford, W. E. N. Doty, and M. R. Lee, "Continuous signal seismograph," *Geophysics*, vol. 25, no. 1, pp. 95–105, Feb. 1960.
[11] R. L. Geyer, "The VIBROSEIS system of seismic mapping," *J. Canadian Soc. Explor. Geophys.*, vol. 6, no. 1, pp. 39–57, Dec. 1970.
[12] L. V. Lombardi, "Notes on the use of multiple geophones," *Geophysics*, vol. 20, no. 2, pp. 215–226, Apr. 1955.
[13] M. K. Smith, "Noise analysis and multiple seismometer theory," *Geophysics*, vol. 21, no. 2, pp. 337–360, Apr. 1956.
[14] M. Holzman, "Chebyshev optimized geophone arrays," *Geophysics*, vol. 28, no. 2, pp. 145–155, Apr. 1963.
[15] A. J. Hermont, "Design principles for seismic reflection amplifiers," *Geophys. Prospect.*, vol. 4, no. 3, pp. 279–293, Sept. 1956.
[16] R. L. Gray, J. H. Leitinger, and J. C. Hollister, "Determination of seismic system distortion and its compensation using digital filters," *Geophysics*, vol. 33, no. 2, pp. 285–301, Apr. 1968.
[17] E. J. Northwood, R. C. Weisinger, and J. J. Bradley, "Recommended standards for digital tape formats," *Geophysics*, vol. 32, no. 6, pp. 1073–1084, Dec. 1967.
[18] C. H. Dix, *Seismic Prospecting for Oil.* New York: Harper & Brothers, 1952.
[19] M. T. Taner, F. Koehler, and K. A. Alhilali, "Estimation and correction of near-surface time anomalies," *Geophysics*, vol. 39, no. 4, pp. 441–463, Aug. 1974.
[20] J. O. Parr, and W. H. Mayne, "A new method of pattern shooting," *Geophysics*, vol. 20, no. 3, pp. 539–564, July 1955.
[21] M. T. Taner, E. E. Cook, and N. S. Neidell, "Limitations of the reflection seismic method; lessons from computer simulations," *Geophysics*, vol. 35, no. 4, pp. 551–573, Aug. 1970.
[22] M. T. Taner, and F. Koehler, "Velocity spectra-digital computer derivation and applications of velocity functions," *Geophysics*, vol. 34, no. 6, pp. 859–881, Dec. 1969.
[23] N. S. Neidell, and M. T. Taner, "Semblance and other coherency measures for multichannel data," *Geophysics*, vol. 36, no. 3, pp. 482–497, June 1971.
[24] C. I. Craft, "Detecting hydrocarbons—for years the goal of exploration geophysics," *Oil and Gas J.*, vol. 71, no. 8, pp. 122–125, Feb. 1973.
[25] "Lithology and direct detection of hydrocarbons using geophysical methods," in *Symp. Rec. Geophys. Soc. Houston*, Oct. 1973.
[26] A. L. Hammond, "Bright spot: Better seismological indicators of gas and oil," *Science*, vol. 185, no. 4150, pp. 515–517, Aug. 1974.
[27] F. S. Grant, and G. F. West, *Interpretation Theory in Applied Geophysics.* New York: McGraw-Hill, 1965.
[28] W. M. Ewing, W. S. Jardetsky, and F. Press, *Elastic Waves in Layered Media.* New York: McGraw-Hill, 1957.
[29] J. P. Lindsey, and C. I. Craft, "How hydrocarbon reserves are estimated from seismic data," *World Oil*, vol. 177, no. 2 pp. 23–25, Aug. 1973.
[30] D. McNabb, "Bright-spot warning: it's not infallible," *Oil and Gas J.*, vol. 72, no. 34, pp. 50–51, Aug. 26, 1974.
[31] S. N. Domenico, "Effect of water saturation on seismic reflectivity of sand reservoirs encased in shale," *Geophysics*, vol. 39, no. 6, pp. 759–769, Dec. 1974.
[32] S. Treitel and E. A. Robinson, "The stability of digital filters," *IEEE Trans. Geosci. Electron.*, vol. GE-2, pp. 6–18, Nov. 1964.
[33] E. A. Robinson, "Multichannel z-transforms and minimum-delay," *Geophysics*, vol. 31, pp. 482–500, June 1966.
[34] K. L. Peacock and S. Treitel, "Predictive deconvolution: theory and practice," *Geophysics*, vol. 34, pp. 155–169, Apr. 1969.
[35] E. A. Robinson, *Multichannel Time Series Analysis with Digital Computer Programs.* San Francisco, Calif.: Holden-Day, 1967.
[36] E. A. Robinson, *Random Wavelets and Cybernetic Systems.* London, England: Charles Griffin, 1962.
[37] N. Levinson, "The Wiener rms error criterion in filter design and prediction," *J. Math. Phys.*, vol. 25, pp. 261–278, Jan. 1947.
[38] J. Makhoul, "Linear prediction: a tutorial review," this issue, pp. 561–580.
[39] Issue on the MIT Geophysical Analysis Group reports, *Geophysics*, vol. 32, pp. 441–525, June 1967.
[40] A. V. Oppenheim, R. W. Schafer, and T. G. Stockham, "Nonlinear filtering of multiplied and convolved signals," *Proc. IEEE*, vol. 56, pp. 1264–1291, Aug. 1968.
[41] T. J. Ulrych, "Application of homomorphic deconvolution to seismology," *Geophysics*, vol. 36, pp. 650–660, Aug. 1971.
[42] S. Treitel and E. A. Robinson, "The design of high-resolution digital filters," *IEEE Trans. Geosci. Electron.*, vol. GE-4, pp. 25–38, June 1966.

SEISMIC VELOCITIES

GEOPHYSICS, VOL. XX, NO. 1 (JANUARY, 1955), PP. 68-86, 10 FIGS.

SEISMIC VELOCITIES FROM SURFACE MEASUREMENTS*

C. HEWITT DIX[†]

ABSTRACT

The purpose of this paper is to discuss field and interpretive techniques which permit, in favorable cases, the quite accurate determination of seismic interval velocities prior to drilling. A simple but accurate formula is developed for the quick calculation of interval velocities from "average velocities" determined by the known x^2-T^2 technique. To secure accuracy a careful study of multiple reflections is necessary and this is discussed.

Although the principal objective in determining velocities is to allow an accurate structural interpretation to be made from seismic reflection data, an important secondary objective is to get some lithological information. This is obtained through a correlation of velocities with rock type and depth.

INTRODUCTION

Reflection measurements of seismic velocities, using only data taken near the surface of the ground, have been made for many years (Green, 1938). Continued study of this problem has led to several notable publications (Steele, 1941; Gardner, 1947; Hansen, 1947; Hansen, 1948; Savit, 1951; Widess, 1952; Brustad, 1953; Pflueger, 1954). I presented the central formula (equation (12) below) at a meeting of the *Deutsche Geophysikalische Gesellschaft* in Hannover in October, 1953, and was informed there by Dr. H. Dürbaum that this same formula was a special case of a more general one presented by him at the May, 1953 meeting of the European Association of Exploration Geophysicists in Paris. I am grateful to Dr. Dürbaum for making his manuscript available to me prior to its publication (1954). Since my treatment is very much more elementary than his and is a little closer to the practical exploration problems, I decided to proceed with its publication.

I became aware of the theoretical possibilities of determining seismic velocities from reflection records twenty years ago. But it was Curtis H. Johnson and N. R. Shade who in 1941 called my attention to the remarkable accuracy it was possible to obtain if the field work was carefully arranged. The field work was done by Western Geophysical Company, which had been doing this for some years. Early work (1938) was also done by G.S.I., under Frederick E. Romberg's initiation and direction.

A sound, but rather laborious, procedure of interpretation by successively determining interval velocities from the top layer downward was developed and used by me in 1941. From 1941 onward to 1948, I received much help from conversations with Curtis H. Johnson, N. R. Shade, Wylie R. Price, Jr., Robert Woods, John A. Legge, Jr., John J. Rupnik, Robert H. Mansfield, Maynard W. Harding, John Woolson, and Ethel W. McLemore. The idea of working in the

---

* Manuscript received by the Editor April 19, 1954.
† California Institute of Technology, Pasadena, California.

68

neighborhood of zero horizontal distance was developed and used by John A. Legge, Jr. about ten years ago.

It would be much better if I could present an example to accompany the material that follows. This may be possible in the future. An example would have the great merit that a clear distinction could be made between simple idealized theory and the practical application of this theory. I am presuming that many of my readers will have available data that can serve for an example. The theory to follow gives most of the organizing framework which can guide practice.

FIELD TECHNIQUE

The field technique has been described by Hansen (1947). This technique has been used several years prior to 1941 by Western Geophysical Company and was used by Romberg in 1938. It is a natural extension of continuous profiling.

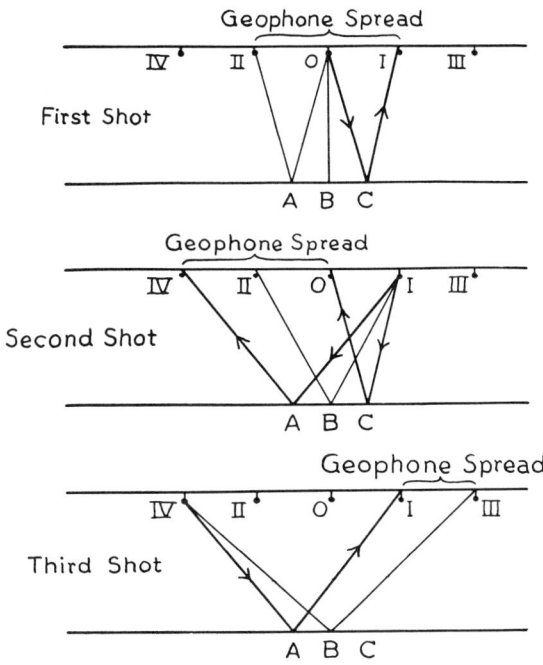

FIG. 1. Three steps in field procedure of velocity reflection profiling, showing time ties due to ray-path reversal.

The objective of the technique is to secure a large horizontal distance from source to receiver so that the slope of the x^2-T_x^2 graph may be determined with some accuracy (Dix, 1952, p. 124–126). This is done by the following combination, illustrated in Figure 1.

The first shot is a regular symmetric split. The second shot is linked to the first via the time from S.P. 0 to C to S.P. I and the reverse. The third shot is linked to the second via the time from S.P. I to A to S.P. IV and the reverse. This gives a continuously-tied range of distance, x, from o to the separation between S.P. IV and S.P. III.

By exchanging positions about the central line, OB, we obtain another set of continuously tied records. But each new such record is also tied to one of the former set with respect to a ray reflected at the center point, B.

This procedure can clearly be generalized to 7, 9, or any odd number of shot holes.

For more complete information, especially as regards dips and curvatures of reflectors, a regular continuous profile from S.P. IV to S.P. III would be required.

Clearly every means possible should be used to make the datum correction for every trace as accurate as possible. For this, up-hole (or down-hole) shooting, and possibly, also, shallow refraction surveying, should be used, together with first break information, on all records.

It is helpful for the interpreter if the instrument operator orients his records in such a way that all of them can be laid out on a table with bottom traces tying to top traces. This requires, for example, reversal of the orientation on the second shot (Fig. 1) with respect to the first. Then the orientation is again reversed for the third shot relative to the second.

It is most essential to shoot so that the above-described time-ties can be used and also to obtain good data for the datum correction.

PLANE REFLECTOR, ZERO-DIP CASE

First Layer

This case has been well understood for many years (Green, 1938). Here we find

$$T_x^2 = T_0^2 + (1/V^2)x^2, \quad (1)$$

where T_x is the reflection time at distance x and V is the velocity, so that if we plot x^2 as abscissa and T_x^2 as ordinate we get a straight line of slope $1/V^2$ with an intercept T_0^2. The depth (or thickness) of the first layer is

$$\Delta z_1 = VT_0/2. \quad (2)$$

Second Layer

Let us look at this problem first in its simplest form. Suppose we have two layers of constant velocity V_1 and V_2. Suppose we have already determined V_1 and Δz_1 by equations (1) and (2) above. Suppose further that we have x^2, T_x^2 graphed for the second reflector. This will not plot as an exact straight line (unless $V_2 = V_1$) but will instead be a curve very slightly concave toward small values of T_x^2. Reference to Figures 2 and 3 will show the true situation very clearly. If we were always to calculate the time over straight paths, as typified by $SABAS$ and $SCDEF$, then the corresponding x^2, T_x^2 graph would be an exact straight line, say JK in Figure 3. But, for a geophone at F the least time path is say $SGDHF$ and this corresponds to a shorter time than the path $SCDEF$. So the least time plots at L below K. The effect is smaller for smaller x's so that the line JK and the curve JL are tangent at J.

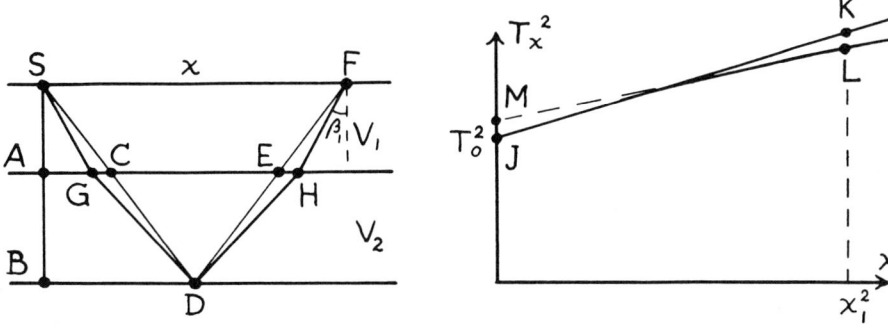

FIG. 2. Two-layer case showing straight ray paths through C and E and minimum time paths through G and H.

FIG. 3. x^2, T_x^2 graph curve, JL, with tangent line at J, JK, and tangent line at L, ML.

We shall solve the problem by removing the effect of the upper layer. First we need the angle β_1 between FH and the vertical (Fig. 2). We know that

$$\sin \beta_1 / V_1 = dT_x/dx. \tag{3}$$

If we could determine V_1 (which we now assume determined) and dT_x/dx, we could use equation (3) to determine β_1. A direct determination of dT_x/dx would be to replace the derivative by its approximating difference quotient $\Delta T_x/\Delta x$. But as the T_x vs. x graph is curved we must keep ΔT_x down to, say, about 0.010 second or the chord will not be a good approximation to the tangent. ΔT_x requires two time measurements, each with an estimated probable error of ± 0.001 second in favorable cases, which would correspond to a probable error for ΔT_x of ± 0.0014 seconds or about 14%. Even under favorable circumstances, this would lead to at least a 14% probable uncertainty range for $\sin \beta_1$ if (3) is used directly. This leads to far too great an uncertainty for β_1 if we are to use it to calculate V_2 with good accuracy.

To avoid this difficulty we draw the tangent to the x^2, T_x^2 curve at x_1 (which is easy to do because this curve is so nearly straight). We get the dashed straight line LM, which has the equation

$$T_x^2 = M + (1/V_{A_2}^2(x_1))x^2. \tag{4}$$

Differentiation of equation (4) with respect to x gives, at $x = x_1$,

$$T_x dT_x/dx = x_1/V_{A_2}^2(x_1). \tag{5}$$

Using equations (3) and (5) we get

$$\sin \beta_1 = x_1 V_1 / (T_x V_{A_2}^2(x_1)). \qquad (6)$$

Equation (6) gives β_1 in terms of quantities all of which can be measured with very high precision compared with that obtained by calculating dT_x/dx from $\Delta T_x/\Delta x$.

From β_1, SG and FH can be computed. So we can compute the time to be removed from T_x to give the time from G to D to H. This gives $(T_x)_R$ (or T_x reduced). Also from β_1, $2AG$ can be computed, and this can be subtracted from x to give $(x)_R$ (or x reduced).

If we plot $(T_x)_R^2$ against $(x)_R^2$ we get a straight line whose equation is

$$(T_x)_R^2 = (T_0)_R^2 + (1/V_2^2)(x)_R^2. \qquad (7)$$

The slope gives us V_2 and the intercept gives us $(T_0)_R$ and the thickness of the second layer is

$$\Delta z_2 = V_2 (T_0)_R / 2. \qquad (8)$$

The above calculation can be replaced by one which is somewhat simpler and much easier to apply. The above calculation does however show the principles involved in a very clear way; a generalization of this is readily extendable to more complicated cases. We now consider the simpler calculation.

Note that the x^2, T_x^2 graph (Fig. 3) may be carried right down to the immediate neighborhood of J. Indeed the tangent at J is easier to find than the tangent at L because the curve is straighter at J than at L.

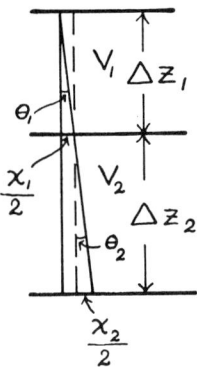

Fig. 4. Geometry of the down-traveling ray for the case of arbitrarily small x and no dip.

Refer to Figure 4. Let $\Delta z_1 = \Delta T_1 V_1 / 2$ and $\Delta z_2 = \Delta T_2 V_2 / 2$ define ΔT_1 and ΔT_2 respectively. Then

$$\begin{aligned} x_1 + x_2 &= \Delta T_1 V_1 \tan \theta_1 + \Delta T_2 V_2 \tan \theta_2 \\ &\approx \Delta T_1 V_1 \sin \theta_1 + \Delta T_2 V_2^2 \sin \theta_1 / V_1 \\ &\approx (V_1^2 \Delta T_1 + V_2^2 \Delta T_2)(x_1 + x_2)/(V_{A_2}^2(0) T_x). \end{aligned}$$

Divide through by x_1+x_2. Then take the limit as $x_1+x_2 \to 0$ and the above becomes exactly (after multiplication by $V_{A_2}^2(0)(\Delta T_1 + \Delta T_2)$)

$$V_{A_2}^2(\Delta T_1 + \Delta T_2) = V_1^2 \Delta T_1 + V_2^2 \Delta T_2. \tag{9}$$

This shows precisely what kind of average $V_{A_2}(0)$ represents. Also (9) can be solved for V_2, since all other quantities in it are known.

Note that equation (9) is only valid near $x=0$. If the reflections are such that we cannot work near $x=0$ we must not use (9) as it is not valid for large x. If we are forced to use large x's exclusively, the best procedure I know is to revert to the rather painful process leading to equation (7). I have never encountered this need in practise excepting in the case of an occasional transverse-wave-type reflection.

n Layers

This result may be readily generalized to n layers. In fact the same procedures show that, exactly,

$$V_{A_n}^2 \sum_1^n \Delta T_i = \sum_1^n V_i^2 \Delta T_i, \tag{10}$$

and

$$V_{A_{n-1}}^2 \sum_1^{n-1} \Delta T_i = \sum_1^{n-1} V_i^2 \Delta T_i. \tag{11}$$

Subtract equation (11) from (10) to get

$$V_n^2 = \left(V_{A_n}^2 \sum_1^n \Delta T_i - V_{A_{n-1}}^2 \sum_1^{n-1} \Delta T_i \right) / \Delta T_n. \tag{12}$$

This result shows that errors made on earlier and later interfaces have no effect on the nth interval velocity. If, for example, $V_{A(n-2)}^2$ corresponded to a multiple but was included by mistake, this mistake would have no effect on the calculation of V_n. Such errors are not cumulative. One may note also in passing that the necessary and sufficient condition for $V_n = V_{A_n}$ is that $V_{A_n} = V_{A_{n-1}}$.

Correction to Beginnings of Reflections

We have tacitly assumed up to this point that reflection times are minimum travel times from datum level at the source to reflector to datum level at the geophone. This is not correct, as, in practice, we are not able to pick the beginnings of reflections. So we try to make a correction. This correction is clearly a weak point in an otherwise remarkably accurate procedure. However, as we shall see, depths are not very sensitive to errors in timing and interval velocities may in some cases be found with very good accuracy in spite of this difficulty. For other sources of error, see the discussion by Widess (1952). For very careful work the correction for x at the datum ought to be included.

Suppose we work with a good reflection. For each trace we pick each peak and trough. The corresponding T_x's are squared and plotted against x^2. We then have a series of V_A^2's for this reflection. We can get an idea of the variation of V_A^2 with T_0. Our interest is in limiting values of V_A^2 and T_0 as we approach the beginning of the reflection.

Now we may look at the problem from another viewpoint, making the hypothesis that the shape (or character) of the pulse does not change with x. Then if the first reflection is being picked at T_x, we estimate this time is δ too large. Then the corrected V_A^2 will be

$$(V_A)_\delta^2 = x^2 / \{(T_x - \delta)^2 - (T_0 + \delta)^2\} \approx V_A^2 (1 + \delta/T_0). \tag{13}$$

So the hypothesis of unchanging shape is equivalent to the supposition that V_A^2 depends linearly on δ (measured say near $x=0$). Whether or not this hypothesis is correct can be determined by picking all peaks and troughs and plotting as outlined in the preceding paragraph. Usually the deviations from the hypothesis of uniformity of shape do not appear to be significant, so we use this hypothesis. Its use should be restricted to a limited range of angles of incidence near zero, as otherwise it simply is not correct.

Notice in equation (13) that the effect of the δ-correction is always to increase $(V_A)_\delta$. We find then that, for example,

$$(\Delta z_1)_\delta = (V_A)_\delta (\Delta T_1 - \delta)/2 = V_A (1 + \delta/\Delta T_1)^{1/2} (1 - \delta/\Delta T_1) \Delta T_1/2$$
$$\approx \Delta z_1 (1 - \delta/2T_0). \tag{14}$$

Thus the corrected depth differs from the uncorrected depth by half as much as it would if the V_A^2 had not been influenced by the correction. If we extract the square root of both sides of (13) we find that the velocity also is changed by the same percentage but in the opposite direction. The factor $1/2$ which occurs in both corrections removes some of the damage done by the uncertainty of δ but not enough in case the reflections are very good.

In case the reflections are very good it may be worth while to measure the variation of V_A^2 by the direct method outlined above, since deviations from the hypothesis of uniformity of shape with x usually appear to be in the direction of minimizing still further any errors in Δz due to uncertainties in δ.

Relative Corrections and Interval Velocities

The situation here is not nearly so bad as it was in the last section. Here we have again to deal with corrections to the beginning of reflections. But now our corrections must be applied to two successive reflections and our primary interest is to minimize the relative errors in these corrections.

Referring to (12), if the actual correction made is δ_i and the true correction δ_i^T and $\Delta_i = \delta_i^T - \delta_i$ for the ith reflection then let us compare the true squared interval velocity, $(V_n^T)^2$ with the corrected squared interval velocity, $(V_n)^2$.

After some reductions we find

$$(V_n{}^T)^2 \approx (V_n)^2(1 + (\Delta_n - \Delta_{n-1})/\Delta T_n). \tag{15}$$

Thus the percentage error is measured by $(\Delta_n - \Delta_{n-1})/\Delta T_n$. ΔT_n is the vertical interval reflection time between reflectors and is usually not very large, perhaps of the order of 0.1 second. When the $(n-1)$st and nth reflections are good enough so that corresponding phases of them can be accurately matched, then $\Delta_n - \Delta_{n-1}$ can be reduced to something like 0.005 second. Then in such a case the error in measuring $V_n{}^2$ would be about 5% and that in measuring V_n would be about 2.5%. The reader will remember that this estimate of accuracy is made assuming ideal conditions of plane non-dipping reflectors bounding homogeneous strata. The accuracy could not be as good as 2.5% in an actual case.

But one is easily led by these considerations to hope that even when $\Delta T_n \approx 0.025$ second, if $\Delta_n - \Delta_{n-1}$ can be kept down to 0.005 second, the error in V_n will be ideally 10% and practically less than, say, 20%. However this hope must not prevent a strictly objective assessment of the range of uncertainty at the end of the work.

It may come as a shock to some that 20% accuracy is regarded as good. Remember, however, that 0.025 seconds time interval is very small—it corresponds to an interval of the order of 100 feet thick. In ordinary well shooting for velocities, comparable accuracy can be achieved only with comparable care. With more recent sonic logging techniques much greater accuracy can be expected. Remember, we make the assumption that no well is available.

In the struggle to minimize $\Delta_n - \Delta_{n-1}$, every device that will improve accuracy should be used. Basically what is wanted is to determine the shape of the recorded signal pulse. I usually try to do this by tracing this shape as best I can on a piece of transparent plastic. I try to average out noise effects as much as possible. This somewhat unsatisfactory operation can be much improved by adjusting times for shallow irregularities and also for regular changes due to variation of angle of incidence and to dip (if present) and then averaging (or mixing). This is the type of operation that has been so fruitful in studying daily variations of the earth's magnetic field. A reproducible recording mechanism is clearly very useful in such an operation. However, if the interpreter has some understanding of the characteristics of his instruments and the reflections are reasonably good, with very few oscillations (obtained by using a minimum of filtering), the tracing procedure is usually sufficient to permit one to identify positive and negative (relatively speaking only) reflections and so make a fairly accurate correlation of common phase positions on adjacent reflections. In this one must try to extract from the data a maximum of extractable information.

When one has made such a study, one must then confront the positive, negative sequence with the interval velocity sequence. If the densities do not change appreciably in successive layers, then a negative reflection must correspond to a decrease of velocity and a positive to an increase of velocity with increasing depth.

From this one can usually determine the absolute signs of the reflections. When an inconsistency remains, one then has to appeal to the fact that the sign of the reflection depends, not on V, but on ρV, the product of the density and velocity. Reasonable variations of density must of course be permitted.

Multiple Reflections

Any multiple reflections which are present will of course be first picked and interpreted as if they were not multiples. This will often lead to an absurd result when equation (12) is applied blindly. When an absurd result is obtained, the first order of business is to search for the cause. If care has been used to eliminate various disturbing effects, then the two most probable causes of trouble are usually (1) mistakes in calculation and (2) multiples. Clearly (1) must be first eliminated as a possibility before (2) can be very seriously considered.

Although every reflector produces multiples, except in very unusual cases of focusing, the multiples that are easily picked are those involving a reflection from either the ground-to-air interface and/or the interface between the low velocity weathering and the higher velocity layer below. The ground-to-air interface reflects nearly 100% of the energy coming up to it. It is not unusual for the base of the low velocity to reflect 50% of the pulse coming up to it. Therefore we would expect one or both of these shallow interfaces to be involved.

Some of the more probable multiple possibilities are indicated in Figure 5 Cases (c), (d), (g) and (k) correspond to what we shall refer to as *simple multiples*. More specifically, these cases correspond to *simple doubles*. Among the simple doubles (c), (d), and (k) involve two extra paths through the low-velocity weathered layer. Cases (a), (b), and (l) involve an initial upward propagation. But (a) and (b) differ by the time from shot up to base of the low-velocity layer counted only once.

It will be clear that among the doubles one may expect types (a) or (b) to be stronger than types (c) or (g). Types (b) and (i) are to be classed along with (a). Types (e) and (f) are equivalent as far as times are concerned but may be different with respect to pulse shape details. Types (e) and (f) should be intermediate in strength between (c) and (d).

In cases (j) through (m) a shallow reflector not detected as a single is shown at (S) to (S'). Such a failure to detect a shallow reflector occurs quite frequently because these are not usually regarded as having any interest. However, for velocity profiling it is very necessary to try to detect such reflectors. This requires shooting a small charge at the center shot—often only a cap. Sometimes a hidden shallow good reflector can be inferred from the velocity profile. The great danger to the interpretation due to such multiples is that (k), since it comes in so early on the record may not even be suspected of being a multiple. It is difficult to guard against this kind of error except by getting the single from (S) (S').

In interpreting multiples the shape or character plays a major part. For any reasonably small range of times, about all we can hope to be able to detect is the

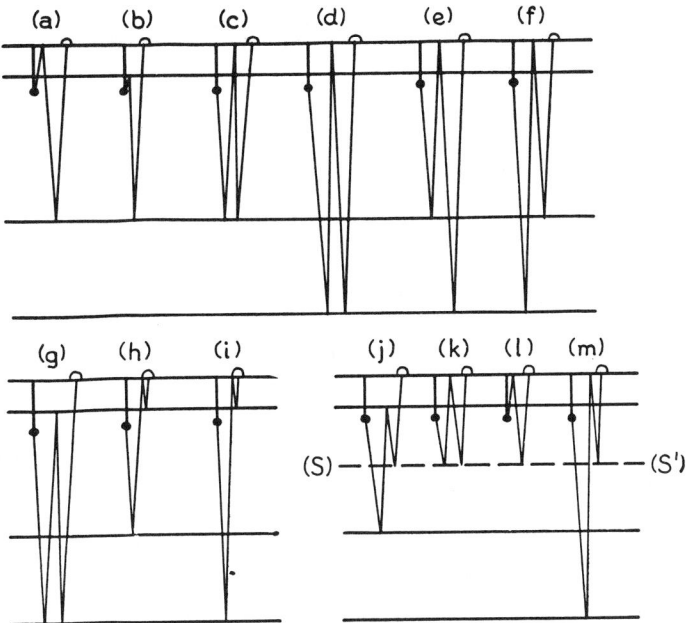

Fig. 5. Various multiple-reflection paths.

sign + or −. But we must be able to make this analysis—otherwise we have to make a series of alternative hypotheses and select the hypothesis least in conflict with the other related interpretive results. Such an elimination process must be carried out where weak reflections are involved, as is quite often the case with multiples.

One characteristic of simple doubles is that they are always negative. This useful fact may help in making an analysis where the sign is otherwise difficult to fix.

A characteristic of multiples involving initial upward propagation (as (a), (b), and (l)) is that they are sensitive to shot depth and are usually most reliably detected by means of this. An otherwise difficult separation of (a) and (b) may be made on this basis.

If the multiple is made by a reflection at the ground-air interface, two passages through the low-velocity weathering are involved. Thus double the layer above datum is added and this influences V_A^2. A time correction along rays must be made. But this correction supposes the validity of geometrical optical principles which are known to be only very roughly valid. Such a correction has to be smoothed or the weathering irregularities, being doubled, will introduce more irregularity than can be allowed to exist. These irregularities are certainly introduced in the double passage through the weathering but they are to a great extent smoothed out physically in the subsequent downward and upward propagation processes.

Now let us consider how some cases appear in the interpretation.

Cases (a), (b), (h), and (i) are most likely to be uncovered when the corresponding single is regarded as the $(n-1)$st reflection and one of these as the nth, using equation (12) to find the "interval velocity." In case (a), this interval velocity will be approximately the average velocity from the shot level to the ground surface. In case (b) this will be approximately the average velocity from the shot level to the base of the low-velocity layer. In cases (h) and (i) this will be approximately the average low velocity of the shallow layer.

Cases like (c) correspond to such low $V_A{}^2$'s that when used in (12) they may lead to negative $V_n{}^2$'s and imaginary V_n's! But, in addition, one should be able to verify the fact that they are multiples by showing, with some precision, just what singles generate them (the exception is given by cases like (j) through (m)). Thus we should be able to interpret (c) in terms of double the corresponding single section plus double the layer from shot level to the surface of the ground. The principal uncertainty in this calculation is δ_1 and δ_2, the two corrections to the beginning of the single and the double respectively. Sometimes the relative corrections can be accurately made (say within ± 0.005 seconds) but their absolute values are not known. It is then possible to add to δ_1 and to δ_2 the same correction term δ_c and solve for δ_c on the hypothesis that our interpretation of the relation of single to multiple is correct. If we get a reasonably small (say less than ± 0.020 seconds) value for δ_c in this way, we may regard $\delta_1+\delta_c$ and $\delta_2+\delta_c$ as probably closer than δ_1 and δ_2 as the absolute (approximate) corrections to the reflection beginning.

The study of the multiples can thus be turned to give more information about the singles and so instead of being a nuisance may be converted into a useful tool. This optimistic statement cannot be made in many cases because disturbing factors are too strong to permit such a sharp view of the situation. However, this fact need not prevent very accurate interpretations where the records do permit such results.

DISTURBING EFFECTS

Thus far we have supposed no dip, no curvature, homogeneous layers—that is, the simplest ideal situations. Some disturbing factors can be taken into account and their effects reduced. Methods of attacking such problems are now considered.

Dip

A correction for dip has been given by Gardner (1947) which is directly applicable to our case also. He does not give the derivation so I supply it below. In triangle $S_I S_{II} S_I'$ (Figure 6) note that

$$(S_{II}S_I')^2 = V^2 T_x^2 = x^2 + (S_I S_I')^2 + 2x(S_I S_I')\sin\theta$$
$$= x^2 + V^2 T_0^2 + x^2 \sin^2\theta - 2VT_0 x \sin\theta + 2VT_0 x \sin\theta - 2x^2 \sin^2\theta$$
$$= V^2 T_0^2 + x^2 \cos^2\theta$$

or
$$T_x^2 = T_0^2 + (\cos^2 \theta/V^2)x^2. \tag{16}$$

So if we plot T_x^2 against $x^2 \cos^2 \theta$ the slope will give us $1/V^2$.

Or if we already have the x^2, T_x^2 plot then the slope gives $\cos^2 \theta/V^2$ and if we can find θ we can then find V. M. B. Favre (1953, personal communication) has shown how θ may be computed without knowing V, from time data alone. This result one gets as follows. From the split with shot point at S_0 with geophones from S_I to S_{II} one gets

$$\sin \theta = (V/x)(T_{0II}^2 - T_{0I}^2)/2T_0 \tag{17}$$

and from equation (16),

$$\cos \theta = (V/x)(T_x^2 - T_0^2)^{1/2}. \tag{18}$$

Dividing equation (17) by (18) one obtains Favre's relation

$$\tan \theta = (T_{0II}^2 - T_{0I}^2)(T_x^2 - T_0^2)^{-1/2}/2T_0. \tag{19}$$

I had formerly made the calculation by a successive approximations procedure. For the first approximation let $\theta = 0$ and calculate $(V)_1$ from the slope of the x^2-T_x^2 graph. Then using $(V)_1$ calculate $(\theta)_2$ (the second approximation to θ) from the approximation to (17)

$$\sin (\theta)_2 \approx (V)_1 \Delta T/x. \tag{20}$$

Then $(V)_2 = (V)_1 \cos (\theta)_2$ gives the second approximation to V. And so on. Usually the second approximation is good enough. But Favre's formula or its approximate expression

$$\tan \theta \approx \Delta T(T_x^2 - T_0^2)^{-1/2}, \tag{21}$$

is better.

The above considerations apply strictly only to one layer. For the second layer we have the ΔT of the second reflection at S.P. 0 which gives, in the plane of the survey, the direction of the ray issuing from and returning to S_0. The generalization to three dimensions can also be carried through (Dix and Lawlor, 1943; Dürbaum, 1953). The calculation made with reference to Figure 4 above is really a calculation to determine the limiting radius of curvature of the symmetric x-T_x graph at $x=0$ ("symmetric" refers here to the symmetry of S_I and S_{II} with respect to S_0 in Figure 6 and the symmetry is not exact but only is true to first order in x). We can thus compute the removal of the first layer and the problem is reduced to that of computing the second layer. For the finite case it is good to have regular continuous-profile data along the line of the survey because this permits shifting of the center time to its proper position of symmetry in the second layer.

The above outline applies to any number of layers. I shall not clutter the paper with the slightly painful details.

However, the special case in which the reflecting interfaces are all parallel is especially simple and also somewhat important so we look at it briefly here.

Refer to Figures 7 and 4. To first order in $x(=x_1+x_2)$, if we add to T_{xR} (means T_x shot toward right) the time $x \sin \theta / V_1$ and use $x \cos \theta$ in place of x we reduce the problem to the zero dip problem excepting that we have lost our symmetry. Using the same x and shooting toward the left (not shown), we have to replace x by $x \cos \theta$ and T_{xL} (means T_x shot toward left) by $T_{xL} - x \sin \theta / V_1$. We again get a reduction to the case of Figure 4 but without symmetry. By averaging these two

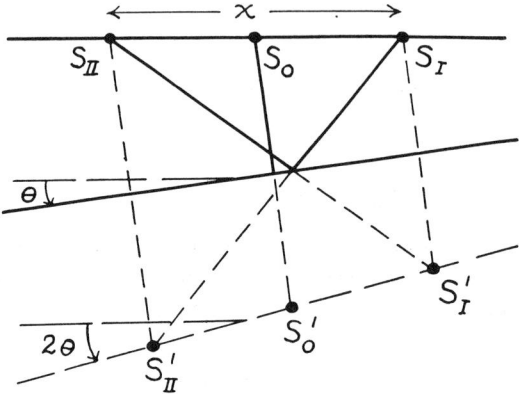

Fig. 6. Geometrical arrangement used for calculation of the dip correction when the distance from shot to geophone increases symmetrically about a center point, S_0.

cases we again get symmetry. In this way we see that equation (16) may be applied to all reflections. But x does not appear in equation (9), so $x \cos \theta$ does not appear there also (actually the effect of θ is only hidden in V_1). Thus if V_1 is properly computed, (9) and (12) may be used for the other layers just as they stand if the other reflectors are parallel to the first.

The process I have used in the general case (starting with the second layer problem) is to assume parallelism of the first and second reflectors as a first approximation. Using this first approximation to the velocity (in the second layer) and the split ΔT for the second reflector, then calculate the first approximation to the angle between the first and second reflector. With this angle the reduced problem for the second layer can be made symmetric so that Gardner's relation (16) may be applied to compute a second approximation to the velocity. This process is continued until further changes, with successive steps, of V_2 and the angle between the first and second reflectors become negligible.

The procedure outlined in the above paragraph is clearly applicable to any number of layers.

It may be interesting to note that in a regular continuous profile a type of cor-

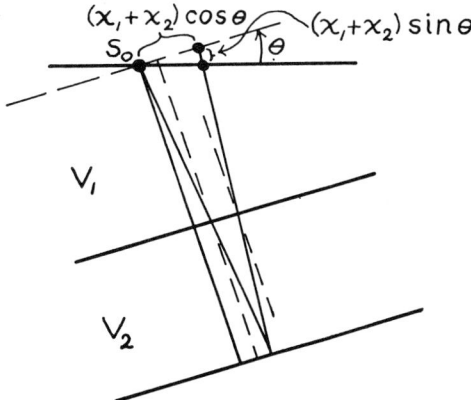

FIG. 7. Geometry needed to handle case of parallel dip with arbitrarily small distance from shot to geophone, x_1+x_2.

rection for shallow velocity variations is often useful. For example, in parts of Canada it is useful to use an especially deep datum because important velocity variations exist in the relatively shallow layers that are below the ordinary low-velocity layer. The field set-up is shown in Figure 8. From the figure one may

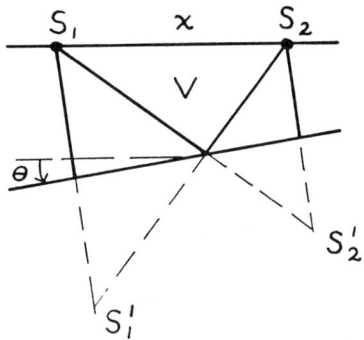

FIG. 8. Shot-to-shot tie over a plane dipping reflector.

show that if T_{01} and T_{02} are the times from shot points S_1 and S_2 for zero geophone distance (corrected to a shallow datum) and T_x is the time from S_1 to S_2 a distance x apart, then

$$V^2 = x^2/(T_x^2 - T_{01} T_{02}) \tag{22}$$

exactly (Pflueger, 1954, p. 339–340). But this formula, which is correct for plane reflectors, is very sensitive to curvature, which we now proceed to consider.

Curvature

The effect of curvature in the general case is most difficult to correct. In practice graphical methods have been used.

Two cases may be distinguished, namely (a) the case where the reflector is concave upward and (b) where it is concave downward. If there are no buried foci the calculation for the first layer by (1) using the field set-up of Figure 1 is not affected by curvature. Similar remarks apply to deeper layers also.

For the field arrangement discussed at the end of the preceding sub-section (Fig. 8) the effect of curvature is very important. The situation may be understood by reference to Figure 9 in which S_1 and S_2 are two shot points, C is the center of

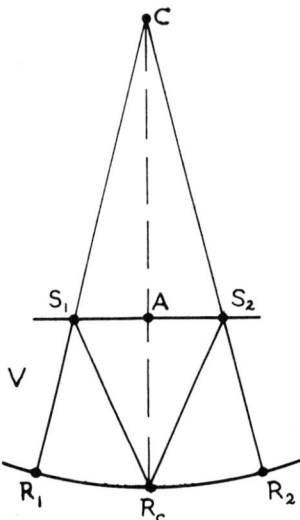

FIG. 9. Shot-to-shot tie over a curved non-dipping reflector.

curvature of a circular cylindrical reflector $R_1R_cR_2$ of radius of curvature $CR_1 = \rho$. Let V be the true velocity so that $S_1R_1 = VT_{01}/2$, $S_2R_2 = VT_{02}/2$ and $S_1R_c = VT_x/2$. So in triangle S_1AC we have

$$(\rho - VT_{01}/2)^2 = x^2/4 + (\rho - AR_c)^2 \tag{23}$$

and in AR_cS_1

$$(AR_c)^2 = (VT_x)^2/4 - x^2/4. \tag{24}$$

If we eliminate AR_c and solve for ρ we get

$$\rho = (V/4)(T_x^2 - T_{01}^2)/[(T_x^2 - x^2/V^2)^{1/2} - T_{01}]. \tag{25}$$

If we write

$$\overline{V}^2 = x^2/(T_x^2 - T_{01}^2) \tag{26}$$

we can rewrite (25) as

$$\rho = (Vx^2/4\overline{V}^2)/[\{T_{01}^2 + x^2(1/\overline{V}^2 - 1/V^2)\}^{1/2} - T_{01}]. \tag{27}$$

The same relation is obtained if the curvature is reversed. The sign of the curvature is automatically taken care of by equation (27). This result is quite useful for making corrections which are relatively shallow but still below the ordinary shallow weathering datum. In Figure 10, $0-S$ is a line of shot points. For each adjacent pair of shot points, $1/\overline{V}^2$ is measured according to equation (26). Between each such pair we plot the value of $1/\overline{V}^2$ as ordinate. Then we draw a smooth curve (possibly a straight line as indicated) and let the smoothed value represent $1/\overline{V}^2$. From T_{01} and T_{02} and V, the center of curvature is found and ρ is found, so one can easily draw the circular arc representing the average curved reflector for each pair of shot points. If we use reasonably good shallow reflections with large enough x values these circular arc segments will fit together reasonably well and we shall have a basis for making a deep datum correction.

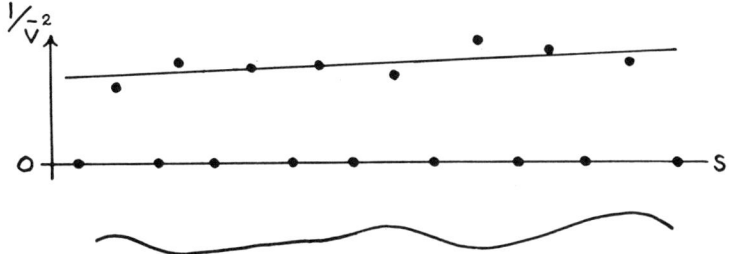

Fig. 10. Cross-section showing continuous profile line with shallow curved reflector and the plot of $1/\overline{V}^2$ points used for curvature calculations.

There are dangers in the blind application of this method. If the shallow reflection is too deep the method will probably make a serious over-correction because of a tendency for diffraction effects to smooth out irregularities that originate at appreciable depth. Clearly also the correction will vary with dip from reflection to reflection on the same spread if there is an appreciable variation of dip.

The objectionable hypothesis has been made that we are dealing with a velocity, V, that is (a) smoothly varying laterally and (b) not variable vertically. Significant failure of either of these conditions may seriously distort the interpretation. My attitude is that, to be very serious, these failures should show in the data. For example, a sudden change in V (laterally) should show in the plot of $1/\overline{V}^2$ as shown in Figure 10. A serious failure of (b) should be expressed in terms of serious incoherence between the deep and shallow interpretations. But if neither of these failures is detectable, then one may suppose that actual deviations are not important. As always, one must look for weak spots in the interpretation and seek an understanding of their causes.

Now we go to a short discussion of the effect of buried foci.

First, how do we detect such effects? The best way seems to be to note first that for a fixed x (distance from shot point to geophone) there are two times, one corresponding to shooting toward the left and the other corresponding to shooting toward the right. Let these be respectively T_{xL} and T_{xR}. Then we may write $T_x^2 = T_{xL} \cdot T_{xR}$ for each trace distance. As buried foci are more likely to be encountered at larger x's (Dix, 1952, p. 363), we watch the plotted points on the x^2-T_x^2 plot for rather abrupt systematic deviations from an almost straight line. The main reason to plot every trace on the x^2-T_x^2 graph is to detect such effects. Physically, the effect of a buried focal line (Dix, 1952, p. 361–362) is to shift the phase forward for all spectral components of the pulse by 1/4 wave length. But the foci are associated with points on the reflecting surface other than the point of least reflection time and these other points may not be where the curvature of the reflector has buried foci at all. The effect is only rarely clear-cut in field practice, so it may not appear as abruptly as the simple idealized view presented above might indicate.

Our purpose is not to correct for focusing effects but to avoid being led astray by them. This is usually possible by a careful use of the x^2-T_x^2 plot, using every trace and referring to the records. When such influences appear we simply avoid using the corresponding points.

An irregular trend of x^2-T_x^2 points will usually not be due to buried foci shooting in both left and right directions but only in one of these, at least for the smaller x-values. This is due to the fact that, except for the center traces, the points of reflection shooting toward the right and toward the left are different for the same x-value.

Thus far we have referred to curvature effects as if only one layer were involved. If one wishes to correct for effects in deeper layers, then a careful graphical interpretation seems almost necessary. However, an aquaintance with the main tools of geometrical optics has some value. The matter is best studied by regarding the central point of reflection as a source and studying the upward propagation from this point in an axially symmetrical corresponding optical system. These matters are the subject of many treatises and will not be discussed here.

Miscellaneous Disturbing Effects

One of the most important of these is background noise. This may be due to surface waves, interface waves, many small reflections, wind, etc. This noise generally means that times are readable with an accuracy of approximately ± 0.002 second, to cite a fairly frequent practical situation. Thus if $T_x = T_c \pm 0.002$, then $T_x^2 = T_c^2 \pm 0.004 T_c$ so that the scatter on the plot is proportional to T_c. Thus the uncertainty in both the slope and the intercept increases with T_x. The situation is not quite this bad, actually, since for the deeper reflections larger x ranges can usually be used, the main limitation here being associated with the angle of reflection.

Disturbances due to drilling fluids in the weathering, due to "hole fatigue," due to "geophone plant" variation, due to instrumental variations may all be serious. The first three effects are difficult to account for but may be reduced with special care. The last is a question of good engineering in the instruments. This is usually cared for better in the high frequency range than in the low range. This fact may cause trouble in handling the deeper reflections. One of the best ways to detect this effect is to make an indicial record (Dix, 1952, p. 374–379) and make sure that the indicial responses match on the lower frequency rear end of the recorded pulses as well as in the beginnings. This places a severe requirement on the geophones but not an unduly severe one.

Perhaps the worst disturbing feature is lack of simplicity of geological structure. Clearly a fault or a sudden change of dip or almost any but the most simple undisturbed bedding will cause much trouble. Usually such trouble is easy to see—if at all serious it is likely to render the survey useless.

ACCURACY

We have already considered certain inaccuracies or ranges of uncertainty which are necessarily involved in any measurement system. For each result we have to ask ourselves "How well do we know this—within what range, with what degree of probability?" And whenever we have answered such questions it is necessary to ask further "how reliable are the ideal simplifying hypotheses in the background—the necessary assumptions without which we cannot proceed at all?"

The question of accuracy may be handled directly. Any assemblage of data, e.g., a collection of x^2-T_x^2 points on a graph, may be duplicated, say, by making n prints from an original. Each of these n prints may be handed to each of n competent persons to measure, say, T_0 and V_A^2. We will then get n values for each of these quantities. After eliminating gross mistakes, the extreme range of these quantities can be divided into a not very large number of intervals and the total number of measurements in each interval can be counted and divided by n. These ratios are the frequencies for the various ranges of say T_0 and V_A^2. Each frequency may be plotted. As the measurements tend to cluster, the frequency plot may be approximated by the Gaussian normal error curve given by

$$p(x) = (1/\sigma(2\pi)^{1/2}) \exp(-x^2/2\sigma^2) \qquad (28)$$

where σ is the standard error. 2σ is the distance between inflection points on the bell shaped curve. The range from -2σ to $+2\sigma$ is for many of our purposes a good practical measure of certainty—there is about one chance in 20 that a measure will fall outside of this range. So if the n measures of V_A^2 lead to $(V_A)^2 = (9.120 \times 10^6 \pm 0.011 \times 10^6)(m/s)^2$ then we will suppose that V_A^2 has 19 chances out of 20 of being within the range 9.142×10^6 to 9.098×10^6.

The above procedure is very long and cumbersome. A shorter procedure is to apply the method of least squares (Jeffreys, 1939, p. 121–126; Whittaker and Robinson, 1940, chapter 9) to calculate the "best fit" values *and* the range of variation associated with each.

When the standard errors have been computed, one finds the errors of derived quantities by using the so-called "law of propagation of errors" that is discussed in all books on the subject. Or if one wishes, he can use the direct procedure involving repeated calculations throughout the entire study. The great advantage of this latter type of procedure is that in it every quantity that is measured has a clear meaning—the shorter least squares process is sometimes misunderstood.

But when we have made such calculations, what of the validity of the background hypotheses? Is it legitimate to use a straight line to fit the x^2-T_x^2 data? Is it a good enough approximation to assume layers of constant velocity? And there are many other questions of this kind always. Many detailed considerations of the main possible kinds of measurement uncertainties are well presented in the paper by Widess (1952).

The answers to all of these questions must always depend on experience. Even if we can't give precise answers, it is good to ask these questions as it keeps us on guard against the clearly erroneous supposition that our simplifying hypotheses are absolutely correct.

If we are to make geophysical investigations at all we must take chances. The correct language for understanding all such matters is the language of the theory of probability.

REFERENCES

Brustad, J. T., 1953, Curved path delta-T analysis: Geophysics, v. 18, p. 738 (Abstract).

Dix, C. H., 1952, Seismic prospecting for oil: New York, Harper and Brothers.

———, and Lawlor, R. C., 1943, Computation of seismic dips below an unconformity: Geophysics, v. 8, p. 105–118.

Dürbaum, H., 1953, Possibilities of constructing true ray paths in reflection seismic interpretation: Geophysical Prospecting, v. 1, p. 125–139.

———, 1954, Zur Bestimmung von Wellengeschwindigkeiten aus reflexionsseismischen Messungen: Geophysical Prospecting, v. 2, p. 151–167.

Gardner, L. W., 1947, Vertical velocities from reflection shooting: Geophysics, v. 12, p. 221–228.

Green, C. H., 1938, Velocity determinations by means of reflection profiles: Geophysics, v. 3, p. 295–305.

Hansen, R. F., 1947, A new system of seismic reflection profiles: Boletin Inform. Petroleras, Buenos Aires, v. 24, p. 237–247.

———, 1948, Multiple reflections of seismic energy: Geophysics, v. 13, p. 58–85.

Jeffreys, H., 1939, Theory of probability: Oxford, Oxford University Press.

Pflueger, J. C., 1954, Delta-T formula for obtaining average seismic velocity to a dipping reflecting bed: Geophysics, v. 19, p. 339–341.

Savit, C. H., 1951, Routine delta-T analysis for field use: Geophysics, v. 16, p. 562 (Abstract).

Steele, W. E., Jr., 1941, Comparison of well survey and reflection "time-delta-time" velocities: Geophysics, v. 6, p. 370–377.

Whittaker, E. T., and Robinson, G., 1940, The calculus of observations, London, Blackie and Son, Ltd.

Widess, M. B., 1952, Salt solution, a seismic velocity problem in western Anadarko Basin, Kansas-Oklahoma-Texas: Geophysics, v. 17, p. 481–504.

Reprinted by permission of Blackwell Scientific Publications
Ltd. from M. Al-Chalabi, 1974, An analysis of stacking, rms,
average, and interval velocities over a horizontally layered
ground: *Geophysical Prospecting*, v. 22, no. 3, p. 458-475.

AN ANALYSIS OF STACKING, RMS, AVERAGE, AND INTERVAL VELOCITIES OVER A HORIZONTALLY LAYERED GROUND*

BY

M. AL-CHALABI**

Abstract

AL-CHALABI, M., 1974, An Analysis of Stacking, rms, Average, and Interval Velocities of a Horizontally Layered Ground, Geophysical Prospecting 22, 458-475.

A correct derivation of rms, average and interval velocities from one another and from common depth point stacking velocities requires a clear understanding of the relationships between these velocities. We relate the average velocity to the rms velocity through a "heterogeneity factor" which is a quantity that gives a measure of the degree of velocity heterogeneity in the ground. The interval velocity is a quantity which varies according to the method of its derivation. The difference between rms and stacking velocities depends on the heterogeneity factor and on the length of the spread. Unless allowed for, this difference can reverse the advantages of long spreads and cause large errors in interval velocity determinations. It may be removed through a number of techniques. The accuracy of stacking velocities in the presence of random "noise" is independent of the heterogeneity factor. Relevant expressions can be broken down into simple formulae which give the accuracy quickly and with good precision.

Introduction

Much of our information about the velocity distribution in the ground is derived from stacking velocities obtained from common depth point stacks. These stacking velocities are used as bases for estimating the rms velocities and are often treated as being synonymous with them. The rms velocities may be used to estimate interval and average velocities. A correct derivation of one velocity from another requires a clear understanding of the relationships between these velocities. In the present work, we deal quantitatively with these relationships and look into the accuracy of stacking and rms velocities. We pay a particular attention to the role played by the velocity heterogeneity of the ground in determining these relationships. To keep the treatment as simple as possible we restrict the investigation to a horizontally layered ground and deal with basic principles only.

---

* Paper presented at the Thirty-fifth Meeting of the European Association of Exploration Geophysicists in Brighton, June 1973.
** Exploration and Production Research Division, BP Research Centre, Sunbury-on-Thames, Middlesex, England.

The Relation between Average and rms Velocities

On a horizontally stratified ground, the average velocity to the nth interface is given by

$$V_a = \frac{1}{T_0} \sum_{k=1}^{n} v_k t_k \tag{1}$$

where v_k is the velocity of the kth layer, t_k is the two way traveltime within the kth layer and T_0 is the two way normal incidence time defined by

$$T_0 = \sum_{k=1}^{n} t_k = 2 \sum_{k=1}^{n} h_k/v_k,$$

h_k being the thickness of the kth layer.

The corresponding rms velocity is given by

$$V_{rms} = \left(\frac{1}{T_0} \sum_{k=1}^{n} v_k^2 t_k \right)^{1/2} \tag{2}$$

In appendix 1 we show that

$$V_{rms}^2 - V_a^2 = \frac{V_a^2}{D^2} \sum_{k=1}^{n-1} h_k \sum_{j=k+1}^{n} h_j \frac{(v_k - v_j)^2}{v_k v_j} \tag{3}$$

where D is the depth to the nth interface. The quantity

$$g = \frac{1}{D^2} \sum_{k=1}^{n-1} h_k \sum_{j=k+1}^{n} h_j \frac{(v_k - v_j)^2}{v_k v_j} \tag{4}$$

gives a measure of the degree of velocity heterogeneity in the ground. We shall call this quantity the *heterogeneity factor*. It is a positive quantity, being equal to zero only when all of the layers have the same velocity (a homogeneous ground). Its value is independent of the order of layering.

From equation (3) we get

$$\frac{V_{rms}^2 - V_a^2}{V_a^2} = g \tag{5}$$

i.e.,

$$V_{rms}/V_a = (1 + g)^{1/2} \tag{6}$$

Equations (3-6) illustrate quantitatively the observation that the rms velocity equals the average velocity when the ground is homogeneous and progressively exceeds it as the ground becomes more heterogeneous.

The Interval Velocity

It is well known from work with continuous velocity logs and synthetic seismograms that the velocity of an interval does not remain constant over any significant thickness of rock. A truer picture is the squiggly form of a velocity log. Let us then look into the meaning of the quantity which we are calling interval velocity.

The interval velocity v_e can be calculated from

$$v_e^2 = (V_b^2 \tau_b - V_t^2 \tau_t) / (\tau_b - \tau_t) \qquad (7)$$

where V_t and V_b are the rms velocities at the top and bottom of the interval and τ_t and τ_b are the corresponding normal incidence traveltimes (Dix 1955). Consider an interval consisting of n segments (or layers) each one having a uniform velocity w (figure 1). The interval velocity W of this interval is then given by

$$W^2 = (V_{n+1}^2 \tau_{n+1} - V_1^2 \tau_1)/(\tau_{n+1} - \tau_1) \qquad \text{by equation (7)}$$

$$= (\sum_{j=1}^{n} V_{j+1}^2 \tau_{j+1} - V_j^2 \tau_j)/(\tau_{n+1} - \tau_1).$$

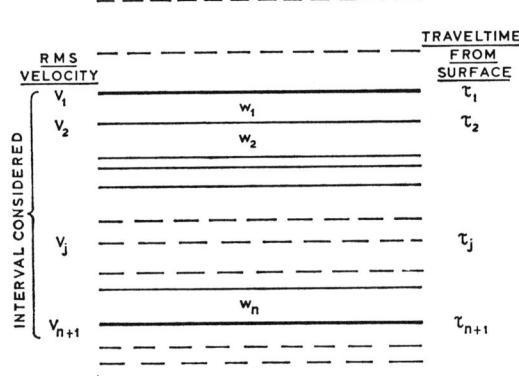

Fig. 1. A sketch of an interval consisting of n segments.

At the jth segment, equation (7) gives

$$V_{j+1}^2 \tau_{j+1} - V_j^2 \tau_j = w_j^2 (\tau_{j+1} - \tau_j).$$

Therefore,

$$W^2 = \sum_{j=1}^{n} w_j^2 (\tau_{j+1} - \tau_j)/(\tau_{n+1} - \tau_1) \qquad (8)$$

which is the formula for the rms velocity of the interval under consideration. Therefore, we conclude that *an interval velocity calculated from two rms velocities is the rms velocity of the interval.*

The interval velocity may also be calculated from the two average velocities at the top and bottom of the interval. In this case, a procedure similar to that used in deriving equation (8) gives

$$W = \sum_{j=1}^{n} w_j (\tau_{j+1} - \tau_j)/(\tau_{n+1} - \tau_n) \qquad (9)$$

which is the formula for the average velocity of the interval. Therefore, *an interval velocity calculated from two average velocities is the average velocity of the interval*.

Hence, the interval velocity is a quantity that depends on the method by which it is calculated. Suppose we had rms velocities obtained from a CDP stack and average velocities obtained from measurements at a well nearby. In this case, an interval velocity calculated using the rms velocities will be larger than the velocity of the same interval calculated using the average velocities. The difference between the two results depends on the heterogeneity of the interval in accordance with the difference between rms and average velocities discussed in the previous section.

THE STACKING VELOCITY AS AN ESTIMATE OF THE RMS VELOCITY

The rms velocity is usually estimated from the stacking velocity. In the absence of errors, the quantity

$$B = V_s - V_{rms} \qquad (10)$$

represents the *bias* in this estimate. It can be shown (Al-Chalabi 1973) that the bias is a non-negative quantity which increases as the quantity

$$\sum_{k=1}^{\alpha} F_k (v, h) \, p^{2k}$$

increases. $F_k(v, h)$ is a complicated non-negative function of the velocities and thicknesses of the layers, increasing in magnitude as the heterogeneity factor g increases. $p(= \sin \theta_i/v_i)$ is the ray parameter, θ_i being the angle of incidence at the ith layer.

Over the same ground, the ray parameter increases with increasing offset and decreases with increasing depth. Therefore, there is always a rapid increase in the bias B as the maximum offset increases. However, for the same spread geometry, increasing depth does not necessarily reduce the bias; when there is a large increase in g with depth, the consequent increase in $F_k(v, h)$ can swamp the decrease in p and cause B to increase. The fact that the discrepancy between the stacking velocity and the true rms velocity could increase with depth has generally been overlooked in the geophysical literature.

These considerations are illustrated in figures 2 and 3. We based our data

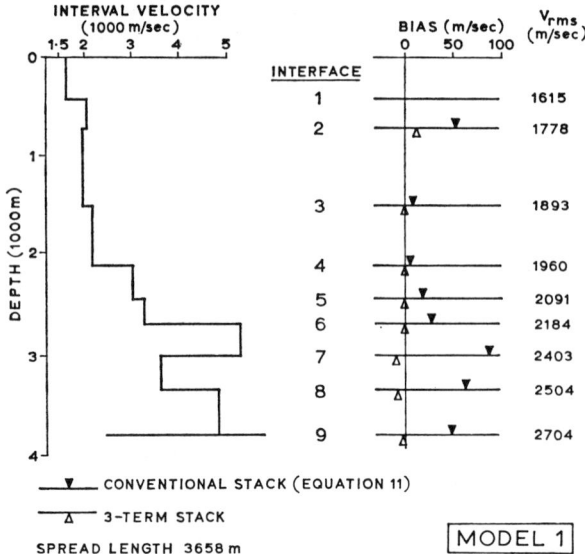

Fig. 2. A simplified model from the North Sea showing the bias in the estimate of V_{rms}.

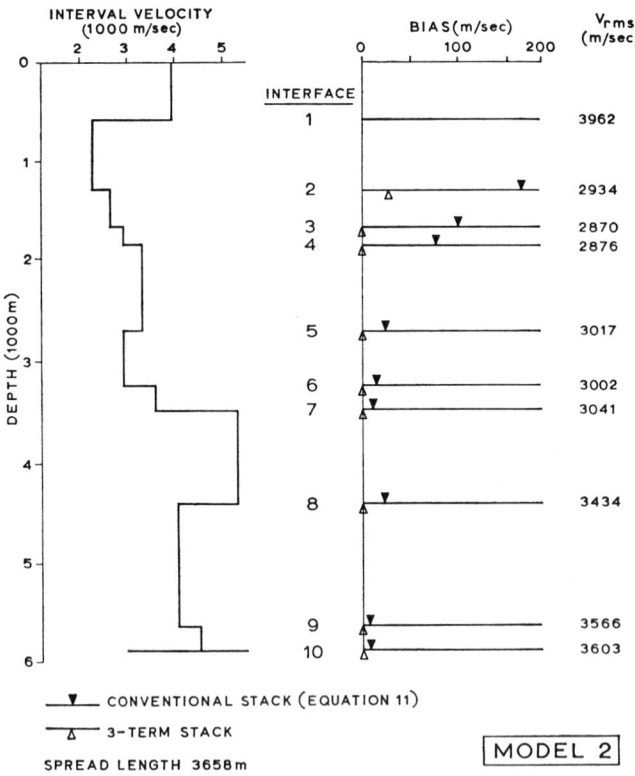

Fig. 3. A simplified model from Alaska showing the bias in the estimate of V_{rms}.

on a 3658 m (12 000 ft) spread length, which is somewhat longer than is normally used in practice. The stacking velocity corresponding to each interface was calculated from

$$V_s^2 = [m \sum_{i=1}^{m} X_i^4 - (\sum_{i=1}^{m} X_i^2)^2]/[m \sum_{i=1}^{m} T_i^2 X_i^2 - \sum_{i=1}^{m} T_i^2 \sum_{i=1}^{m} X_i^2] \quad (11)$$

where m is the number of stacked traces and X_i and T_i are respectively the offset and total traveltime corresponding to the ith trace. This value of V_s^2 corresponds to the slope of a least squares fit through the traveltimes in the T^2-X^2 plot. For our purpose, this is an adequate simulation of the process of obtaining stacking velocities in practice. The bias value is indicated by a black triangle above each line.

Model 1 (figure 2) is simplified from an actual model from the North Sea. There are only small variations in interval velocity in the top four layers. Consequently, the heterogeneity factor is low and the bias decreases steadily with depth as p gets smaller. The increase in interval velocity below the fourth interface and the larger increase below the sixth interface produce large increases in the heterogeneity factor. Thus, at a depth of 3000 m, the bias is much larger than at shallower horizons, say at 1500 m. This example stresses the possibility of significant increases in the bias with depth despite the decrease in the spread length/depth ratio.

Model 2 (figure 3) is simplified from Alaskan well data. The direct waves travel with a high velocity through the fast permafrost layer allowing the use of the full record from the second interface downwards. Hence, a large bias exists at shallow levels caused by a large spread length/depth ratio. The large bias is further emphasized by the high heterogeneity factor arising from the sudden decrease in velocity below the permafrost layer. In this example, the bias decreases steadily with depth, except for an insignificant rise at the 8th interface.

In the previous two examples, the stacking velocity was calculated from exact traveltimes. These times correspond to the onset times of the reflection wavelets on the traces of a CDP gather. In practice, random variations in the quality of the reflection data generally cause random time shifts in the wavelets. These variations arise from several factors such as signal-to-noise ratio, imperfect static corrections, irregular spread geometry, etc. We shall refer to the net effect of these factors as random "noise". For our purpose, it was convenient to simulate this noise by random time errors (jitter) superimposed on the exact reflection times of the horizon being considered. Bodoky and Szeidovitz (1972) make a statistical investigation of these time errors. They accept the hypothesis that these errors are normally distributed.

Figure 4 shows a histogram of stacking velocities corresponding to the seventh interface of model 1 (figure 2). This histogram represents a population of 200 stacking velocities. In calculating each stacking velocity, a different time jitter from the same random number distribution (of 2.3 ms standard deviation) was superimposed on the exact reflection times for interface 7.

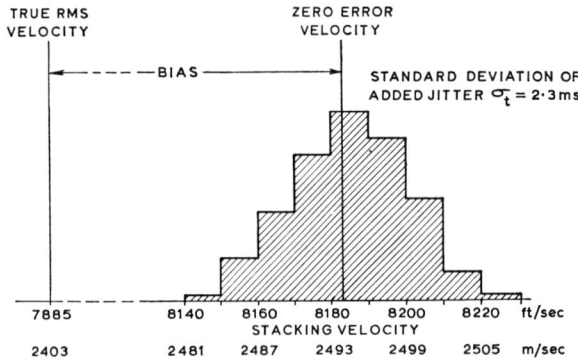

Fig. 4. A histogram of stacking velocities for interface 7 of model 1.

The stacking velocity was then calculated by equation (11). Figure 4 shows that the peak of the histogram is located at the zero error velocity, i.e. the stacking velocity which would have been obtained had the data been jitter free as in the data of figure 2. This histogram is typical of a large number of histograms corresponding to other tested models. In these histograms, χ^2—tests show that the stacking velocities are normally distributed at the 5% level of significance, the mean falling close to the zero error velocity. When the time errors are large, some skewness in the distribution should be expected. This is because a negative time error increases the velocity by a greater amount than an equal positive error would decrease it.

Thus, in our model, the difference between a given stacking velocity and the true rms velocity is made up of the sum of a random error caused by random noise and a systematic component which is the bias. Methods of estimating the accuracy of stacking velocities in the presence of random noise are given in the last section. Methods for removing the bias are given in appendix 2. Semi-systematic factors such as a laterally variable weathering velocity (see for example Schneider 1971, figure 26) are not covered by the present work.

Errors Caused by the Bias

In many practical applications the stacking velocity can be safely treated as the rms velocity. However, when the bias is large, serious errors could arise unless allowance is made for it. The determination of interval velocity is one example. Stacking velocities are generally used for V_t and V_b in equation (7) to calculate the velocity of a given interval. When the bias values in V_t and V_b are of comparable magnitudes they tend to cancel out, leaving the interval velocity relatively unaffected. In many cases, however, a large bias exists on one interface of the interval only. In such cases, the bias in V_t will not cancel that in V_b and the calculated interval velocity will contain a large error. For example, if no correction is made for the bias in calculating the velocity of the interval between the sixth and seventh interface of model 1 (figure 2) the interval velocity will be overestimated by 14%. In the same way, the velocity of the interval between the 4th and 5th interface of model 2 (figure 3) will be underestimated by 9%. These errors are too large for most purposes. They indicate the need to take into account any significant bias when calculating the interval velocity. If the bias on one interface is negligible, the error in interval velocity is roughly of the order of $B_e D/h$ where B_e is the larger bias and h and D are respectively the thickness and average depth of the interval. This result can be verified from equation (7).

The use of long spreads is another example where the improved accuracy of the stacking velocity (gained by using a long spread) could be swamped by the increased bias. An example is shown in figure 5. The data correspond to interface 7 of model 1. Random noise is simulated by a random jitter of

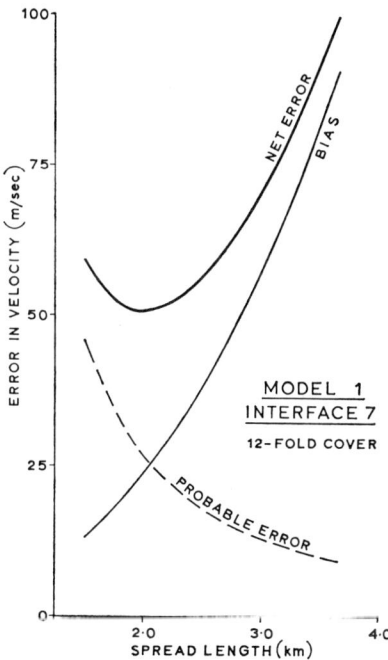

Fig. 5. A plot showing the increase in net error with increasing spread length when the bias is not allowed for.

5 ms standard deviation superimposed on the true traveltimes. The stacking velocity is calculated by equation (11). The probable error plot refers to the magnitude of 0.675 σ_s where σ_s is the standard error in the stacking velocity calculated by equation (12) below. The net error is the total magnitude of the probable error and the bias. As the spread gets longer the rate of increase of the bias increases while the rate of decrease of the probable error decreases. Therefore, after a certain optimum value of spread length (2,000 m in this case) the net error starts to increase. This example is representative of a large number of practical cases. It shows that large bias values should be corrected. Otherwise, increasing the spread length becomes an expensive way of accumulating more errors in what is treated as the rms velocity.

The Accuracy of Stacking Velocities

We now consider the accuracy of stacking velocities in the presence of random effects which we are collectively calling random noise. We disregard the bias and treat V_s and V_{rms} as equivalent. We again simulate the random noise by a random time jitter superimposed on the traveltime data.

Equations for Estimating the Accuracy

Tests have shown that for all practical purposes the accuracy of stacking velocities is independent of the heterogeneity factor, i.e. of the type of velocity variations in the ground. Figure 6 summarizes this fact. The coefficient of variation is plotted against the heterogeneity factor for a large number of models. The stacking velocity corresponds to the bottom interface in each model. The models were generated in such a way that the reflection times of the bottom interface always produced the same NMO in the outermost trace.

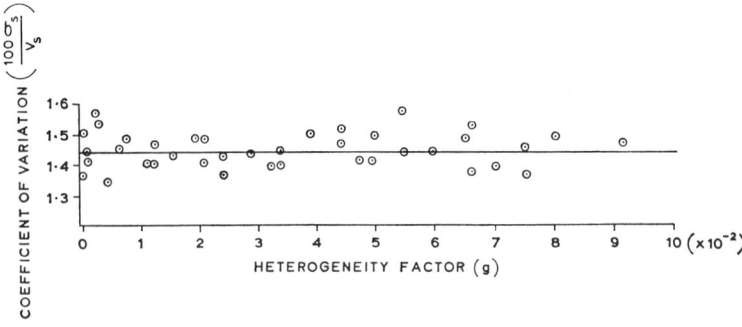

Fig. 6. A plot of the coefficient of variation of stacking velocity versus the heterogeneity factor showing the scattering of points about a horizontal line.

The variance σ_s^2 of each stacking velocity was obtained numerically. It corresponded to the variance of a population of 200 stacking velocities. These velocities were generated in a way similar to that used for the velocities of figure 4. The standard deviation of the time jitter was 5.8 ms. The points in figure 6 plot about a horizontal line showing no correlation with the heterogeneity factor. The horizontal line corresponds to the theoretical value of the coefficient of variation which is calculated by equation (12) below.

These results suggest that the accuracy of stacking velocities can be estimated from the traveltime data, without taking the actual velocity distribution into account. If the stacking velocity is obtained through equation (11), or through an algorithm equivalent to it, then the stacking velocity variance is given by

$$\sigma_s^2 = \sigma_t^2 V_s^2 \left(\sum_{i=1}^{m} T_i^2 x_i^2 \right) / \left(\sum_{i=1}^{m} T_i^2 x_i \right)^2 \tag{12}$$

where

$$x_i = m X_i^2 - \sum_{j=1}^{m} X_j^2,$$

σ_t^2 is the variance of the time jitter, m is the number of stacked traces, and X_i and T_i are respectively the offset and traveltime corresponding to the ith trace. Replacing T_i by $\Delta T_i + T_0$, where ΔT_i is the NMO of the ith trace and T_0 is the two-way normal incidence traveltime, equation (12) reduces to

$$\sigma_s^2 \backsim 0.25\, \sigma_t^2 V_s^2 \left(\sum_{i=1}^{m} x_i^2 \right) / \left(\sum_{i=1}^{m} \Delta T_i x_i \right)^2 \tag{13}$$

A Method for a Quick Estimation of the Accuracy of Stacking Velocities

Suppose that we replaced the ratio of the two sums in equation (13) by the quantity $K'/\Delta T_e^2$ where K' is a constant and ΔT_e is an effective NMO, the exact expression for which is to be determined by experiment. In this case an approximately linear relationship will exist between this quantity and σ_s^2.

Thus, we would write (13) as

$$\sigma_s^2 \backsim K\, \sigma_t^2\, V_s^2 / \Delta T_e^2 \tag{14}$$

where $K = 0.25\, K'$. The form of equation (14) is quite useful. A large number of experiments was conducted to establish whether the use of this empirical relation was justified. The experiments consisted of searching for an expression for ΔT_e which would produce consistent values of K over a velocity range of

1524-7620 m/sec (5000-25 000 ft/sec), a time range of 1.0-5.0 sec and a spread length of up to 3658 m (12 000 ft). Among the expressions tested were

$$(\sum_{i=1}^{m} \Delta T_i^2/m)^{1/2}, \ (\sum_{i=1}^{m} \Delta T_i X_i / \sum_{i=1}^{m} X_i), \ (\sum_{i=1}^{m} \Delta T_i^2 X_i^2 / \sum_{i=1}^{m} X_i^2)^{1/2}, \text{ and } \Delta T_{max},$$

the latter being the NMO corresponding to the largest offset. In all of these expressions it was assumed that the offsets of the traces in the CDP gather increased at equal increments. The expression

$$\Delta T_e = \Delta T_{max}$$

produced the most consistent values of K. Assuming a 12-fold cover, a value of K was determined in the above ranges on a least squares basis. This value was 0.196. Accordingly, we may write (14) as

$$\sigma_s^2 \backsim 0.196 \ \sigma_t^2 \ V_s^2 \ \frac{12}{N} \ \frac{1}{\Delta T_{max}^2}$$
$$\backsim 2.4 \ \sigma_t^2 \ V_s^2 / (N \ \Delta T_{max}^2) \qquad (15)$$

where N is the multiplicity of cover. The standard error of the stacking velocity is

$$\sigma_s \backsim 1.54 \ \sigma_t \ V_s / (\sqrt{N} \ \Delta T_{max}) \qquad (16)$$

The discrepancy between the exact value of σ_s of equation (12) and that calculated by equation (16) decreases as time and velocity increase and as the spread length decreases. Over ranges of practical interest, the discrepancy is usually less than 5%. In specific problems where the ranges V_s, N, T_0, and X are narrower than those quoted above, a different value for K may be determined, if required, to obtain σ_s with even greater precision. It was therefore concluded that the use of equations (15) and (16) was justified for gathers in which the offsets of successive traces increase at a fairly regular interval and the offset of the innermost trace is small relative to the total spread length. When the offset of the innermost trace is not small (say greater than six times the increment between the offsets of successive traces) the discrepancy could become quite large (15% or more). In this case, and in cases where the offsets of successive traces do not increase at regular intervals, the accuracy estimate may be obtained from equations (12) or (13).

From equation (16), the coefficient of variation is

$$100 \ \sigma_s / V_s \backsim 154 \ \sigma_t / (\sqrt{N} \ \Delta T_{max}) \qquad (17)$$

Equations (15-17) can be used to estimate the accuracy of stacking velocities directly. A slide-rule is adequate for the purpose. Equation (17) was used to construct a chart of the coefficient of variation as a function of ΔT_{max} for various values of N (figure 7). This chart simplifies further the estimation of the accuracy of stacking velocities. As an example, suppose that we obtained a velocity from a 2 × 12 stack (N = 24). Suppose also that the NMO of the outer traces was about 95 ms and that σ_t was estimated to be 5 ms. For ΔT_{max} = 95 and $N = 24$, we read 0.33 on the vertical axis. Therefore, the estimated accuracy of the stacking velocity is 0.33 × 5 ≃ 1.7 per cent.

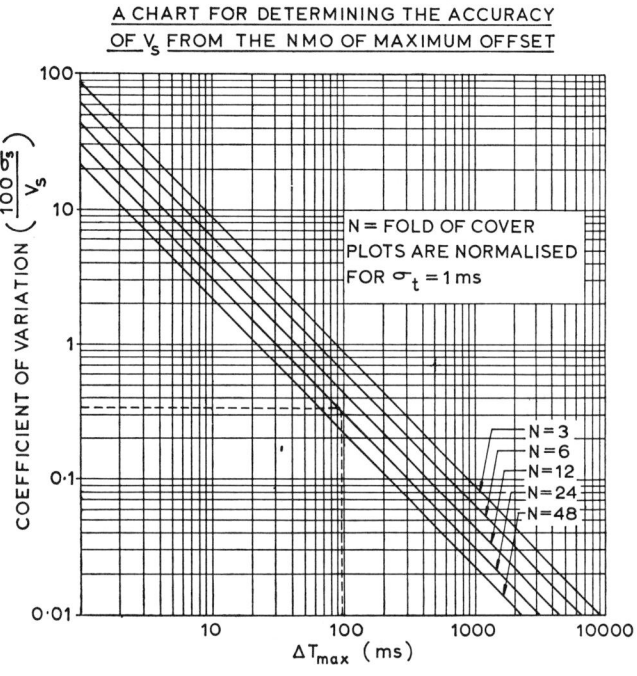

Fig. 7. Chart for determining the accuracy of V_s from the NMO of maximum offset.

In a strict sense the above considerations apply only to stacking velocities obtained by an algorithm totally equivalent to equation (11). However, the principle of equation (11) is in many ways very similar to the measure of coherence used in other methods. Therefore, equation (12)-(17) should provide a good estimate of the accuracy of stacking velocities obtained by other methods. The value of σ_t is the quantity most prone to uncertainty. This value is a function of the signal-to-noise ratio, of the dominant signal and noise frequencies, of poor statics, etc. A fairly reasonable estimate of σ_t may be obtained by inspecting the record of the corrected CDP gather. Methods of estimating σ_t more rigorously are not central to the present work. A useful discussion of the estimation of σ_t has been given by Bodoky and Szeidovitz (1972).

Appendix 1

Derivation of Equation (3)

From equations (1) and (2) we get

$$V_{rms}^2 - V_a^2 = [(\sum_{k=1}^{n} t_k)(\sum_{k=1}^{n} v_k^2 t_k) - (\sum_{k=1}^{n} v_k t_k)^2] / T_0^2$$

$$= [\sum_{i=1}^{n} t_i^2 v_i^2 + \sum_{k=1}^{n}\sum_{j=1}^{n} t_k t_j v_j^2 - \sum_{i=1}^{n} t_i^2 v_i^2$$

$$- \sum_{k=1}^{n}\sum_{j=1}^{n} t_k t_j v_k v_j] / T_0^2 \qquad (j \neq k)$$

$$= [\sum_{k=1}^{n-1}\sum_{j=k+1}^{n} t_k t_j (v_k^2 + v_j^2) - 2\sum_{k=1}^{n-1}\sum_{j=k+1}^{n} t_k t_j v_k v_j] / T_0^2$$

$$= \frac{1}{T_0^2} \sum_{k=1}^{n-1} t_k \sum_{j=k+1}^{n} t_j (v_k - v_j)^2.$$

Therefore,

$$V_{rms}^2 - V_a^2 = \frac{V_a^2}{D^2} \sum_{k=1}^{n-1} h_k \sum_{j=k+1}^{n} h_j \frac{(v_k - v_j)^2}{v_k v_j}. \tag{3}$$

Appendix 2

Methods of Removing the Bias

We present below three methods for estimating the true rms velocity from the stack data. Other methods (e.g. Brown 1969) might also be useful.

A. *The Model Simulation Method*

In this method, the stacking velocities corresponding to all major reflectors are picked out. An approximate model of the ground is then constructed and used as a basis for estimating the bias values. The procedure comprises the following steps:

1) The stacking velocities are assumed to represent the true rms velocities of the ground. The interval velocities are computed in equation (7). The thickness of each layer is then worked out.

2) For each interface in the resulting model, reflection traveltimes corresponding to all offsets used in the stack are computed by a ray-tracing method.

3) A new stacking velocity corresponding to each interface is computed from these traveltimes. Equation (11) may be used for this purpose. An estimate of the bias at each interface is obtained by subtracting the rms velocity (the original stacking velocity) from the newly calculated stacking velocity.

4) The estimated bias values are subtracted from the original stacking

velocities. The resulting velocities will be usually fairly close to the true rms velocities. They are regarded as the rms velocities of another approximate model. Steps 1-3 are repeated.

5) Steps 1-4 are repeated 3 or 4 times. This is usually sufficient for the estimated rms velocities to converge to their correct values. Once the final rms velocities have been obtained, the determination of the thickness and velocity of each layer becomes a trivial matter.

In some cases, the estimated rms velocities oscillate at successive stages without converging. In such cases, the last two steps can be replaced by:

4) The model defined by the original stacking velocities is regarded as a first stage approximation. The set of stacking velocities obtained in step 3 are again treated as the rms velocities of the ground (without subtracting the bias values). Steps 1-3 are repeated twice to generate a second stage and a third stage approximation models. In general, the discrepancy between the estimated and the correct bias continues to increase at successive stages.

5) For each interface, the bias may be regarded as a function of the stage of approximation. This bias is then treated as ordinate, the stages being equally spaced along the abscissa (figure B1). A quadratic fitted to the bias values at the three stages of approximation may be extrapolated back to the zeroth stage to obtain an estimate of the true bias. The quadratic extrapolation (or some variations of it) works consistently well although the principle is arbitrary. An example of the quadratic extrapolation is shown graphically in figure B1. The estimated bias corresponds to interface 7 in model 1.

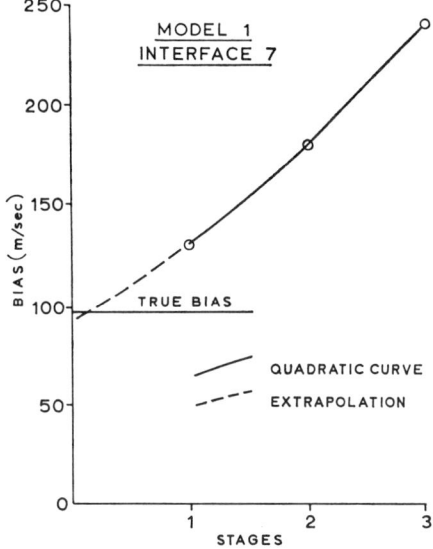

Fig. B1. An example of quadratic extrapolation in model simulation.

The method of model simulation is simple and requires very limited computer time. The bias estimates are relatively insensitive to small errors in the stacking velocity (up to about 50 m/sec). Such errors are carried through to the estimated rms velocity. The main difficulty with this method lies in picking out the major reflectors (interfaces) correctly.

B. *The Method of Shifting Stack*

The decrease in the bias with decreasing offset (decreasing ray parameter p) was discussed in the text. At zero offset ($p = 0$) the bias vanishes. The principle of the method is to estimate the velocity at zero offset by extrapolating back from a set of stacking velocities. Each of these stacking velocities is obtained from a different set of traces from the trace gather. Suppose we had data for a 24-fold stack. We can stack the inner 12 traces (traces 1-12) and obtain a stacking velocity, then stack the inner 14 traces (traces 1-14) and obtain another stacking velocity, then stack the inner 16 traces and so on until all 24 traces had been stacked. We should then end up with 7 different stacking velocities, increasing as the maximum offset increases. The chosen traces could have also been varied in a different manner, e.g. we could have stacked traces 1-14, then 3-16, then 5-18, etc. Let us define an effective offset as being a single quantity that represents the offsets of all of the stacked traces. A convenient but arbitrary form of the effective offset is

$$X_e = (\sum_{i=1}^{m} X_i^2/m)^{1/2}$$

where X_i is the offset of the ith trace and m is the number of stacked traces. We can plot the seven velocities versus their effective offset and extrapolate back to the zero offset velocity. A low order polynomial fitted to the points may be used in the extrapolation. In fact, on a V_s versus X_e^2 plot, the points would fall almost on a straight line. The near-linear relationship between V_s and X_e^2 is not an obvious one. It was deduced from a large number of numerical experiments. It can also be justified analytically. The V_s versus X_e^2 plot is recommended for the purpose of this method. The $V_s^2 - X_e^2$ relationship was also found to be approximately linear, but to a slightly lesser extent than the $V_s - X_e^2$ relationship. Figure B2 shows an example of $V_s - X_e^2$ and $V_s - X_e$ plots. The example corresponds to interface 7 of model 1.

In practice, the $V_s - X_e^2$ points will generally scatter about a straight line because of the presence of noise in the stacked data. The rms velocity may be estimated from the intercept of a straight line fitting the points on a weighted least squares basis, the weights being proportional to the inverse of the variance of each velocity.

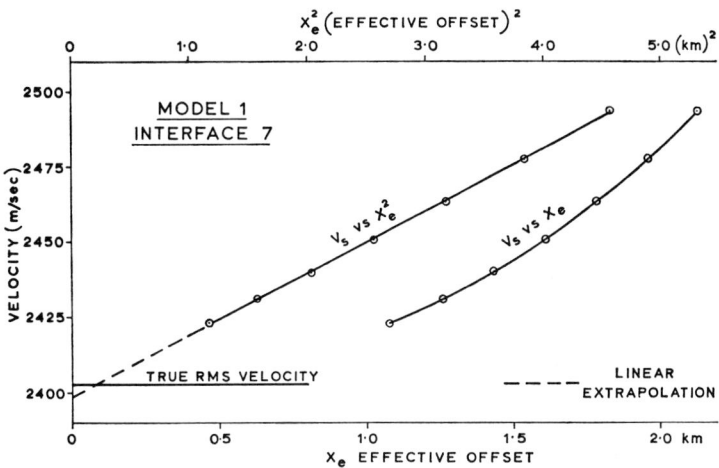

Fig. B2. Estimation of the true *rms* velocity by a linear extrapolation from the V_s versus X_e^2 plot.

An important feature of this method is that the estimate of V_{rms} at each interface is independent of errors made at other interfaces. The principle of the method can be extended to other problems in which the stacking velocity varies systematically with offset such as problems involving dipping layers.

C. *The Three-Term Series Truncation*

Methods of obtaining stacking velocities are essentially based on a two-term truncation of the series

$$T_i^2 = C_1 + C_2 X_i^2 + C_3 X_i^4 + C_4 X_i^6 + \ldots \tag{B1}$$

where X_i and T_i are respectively the offset and true reflection time corresponding to the ith trace, $C_j (j = 1, 2, \ldots, \infty)$ is the coefficient of the jth term and $C_2 = 1/V_{rms}^2$ (Taner and Koehler 1969). The bias in the estimate of V_{rms} arises as a result of this truncation. One possibility for reducing the bias would be to base the stack on a larger number of terms. The three-term truncation of equation (B1) is particularly useful because it usually gives a very good approximation to the true traveltime (Al-Chalabi 1973). If a stack is carried out according to this three-term truncation the stacking velocity obtained from

$$V_s^2 = 1/C_2$$

should be very close to the true rms velocity. This approach was tested on a number of models including those of figures 2 and 3. Stacking velocities were obtained by using a least squares criterion similar to that used in deriving equation (11). Results for models 1 and 2 are shown in figures 2 and 3. The bias values corresponding to these three-term stacking velocities are indicated by white triangles below each line. They demonstrate the drastic reduction

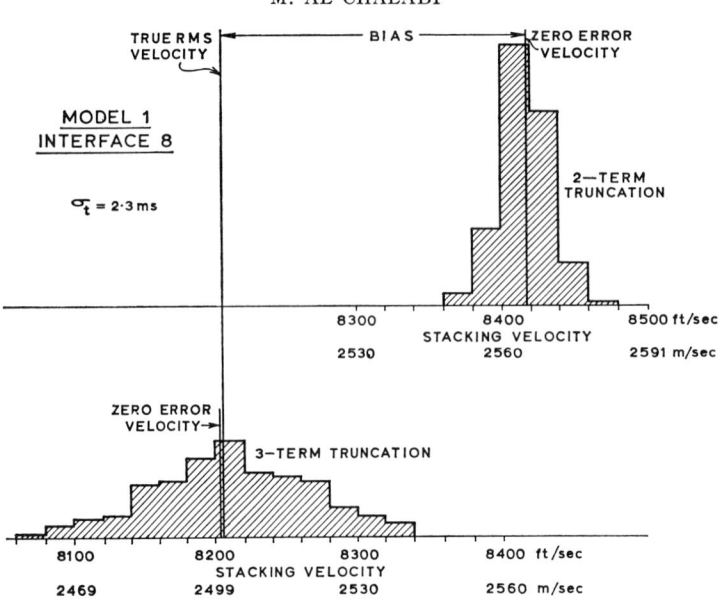

Fig. B3. A comparison between the histograms of two-term and three-term stacking velocities. Despite the high variance of the three-term velocities there is here a clear advantage in stacking according to three terms.

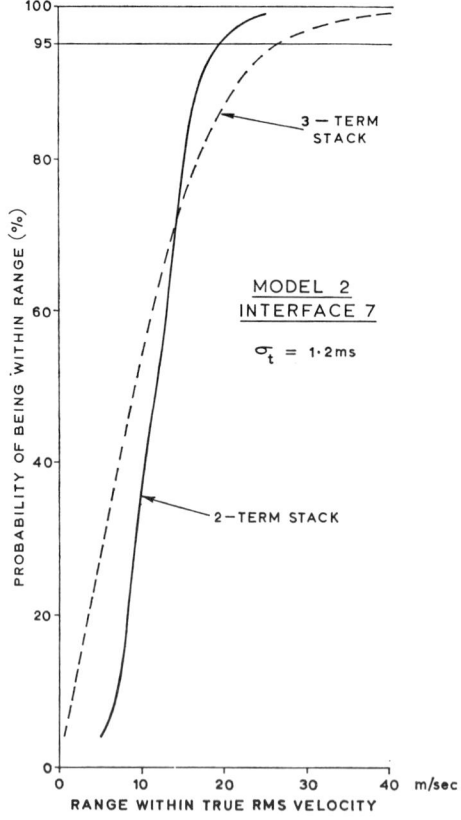

Fig. B4. Plots of the probability of being within a given range of the true *rms* velocity for a two-term and a three-term stack.

in the bias achieved by a three-term stack. However, numerical experiments in the presence of simulated random noise indicate that the variance of a velocity obtained from a three-term stack is generally about 10 times greater than the variance of the corresponding velocity obtained by the conventional two-term stack. Figure B3 shows a comparison between the histograms of stacking velocities produced by a two-term truncation and a three-term truncation. The histograms correspond to the eighth interface in model 1. They were generated in the same way as the histogram of figure 4. The standard error of the added time jitter was 2.3 ms. As in all of the tested models, the three-term stacking velocities are centred about the true rms velocity but their variance is much larger than in the case of the two-term velocities.

In this particular example, there is some advantage in stacking according to three-terms despite the large variance; the three term stack produces a closer estimate of the true rms velocity than a two term stack. Usually, however, the improvement gained by stacking according to three terms is offset by the loss of accuracy. An example is shown in figure B4. In this example, the bias and the random errors are combined to give the probability of being within a certain range of the true rms velocity. The plots correspond to the seventh interface of model 2. The standard deviation of the added jitter is 1.2 ms. Although the jitter is low, the probability of being within a close range of the true rms velocity reverses in favour of a two-term stack at a range of only 14 m/sec.

In conclusion, the method of three-term stack is not recommended for general use. It is presented here mainly because it arises as an interesting possibility when considering the time-distance relationship of equation (B1).

Acknowledgements

I am grateful to Dr Peter O'Brien for reading the manuscript and to him and Dr Andrew Lucas for some interesting discussions. I thank the Chairman and Board of Directors of the British Petroleum Company Ltd. for their permission to publish this work.

References

AL-CHALABI, M., 1973, Series approximation in velocity and traveltime computation. Geophysical Prospecting 21, 783-795.

BODOKY, T., and SZEIDOVITZ, Zs., 1972, The effect of normal correction errors on the stacking of common-depth point traces. Geophysical Transactions of the Hungarian Geophysical Institute Roland Eötvös, 20 (3-4), 47-57.

BROWN, R. J. S., 1969, Normal-moveout and velocity relations for flat and dipping beds and for long offsets. Geophysics 34, 180-195.

DIX, C. H., 1955, Seismic velocities from surface measurements. Geophysics 20, 68-86.

SCHNEIDER, W. A., 1971, Developments in seismic data processing and analysis (1968-1970). Geophysics 36, 1043-1073.

TANER, M. T., and KOEHLER, F., 1969, Velocity spectra—Digital computer derivation of velocity functions. Geophysics 34, 859-881.

TIME-DEPTH AND VELOCITY-DEPTH RELATIONS IN WESTERN CANADA*

C. H. ACHESON†

Every time-depth curve studied so far in western Canada can be divided into at most a few segments, precisely represented by equations of the type $t = az^n + b$. The particular values of a, b, and n that define each segment are of fundamental importance. Best values of these quantities have been determined by statistical methods at several places in the basin, and their areal variation is significant in terms of lithologic changes and basin position.

INTRODUCTION

Detailed analysis of several hundred time-depth curves, carried out a few years ago (Acheson, 1959), shows that remarkably simple time-depth and velocity-depth relations hold throughout the sedimentary section in western Canada.

Each time-depth curve consists of a few major segments (corresponding to geologic layers) having time-depth and velocity-depth equations of the form:

$$t = az^n + b, \quad (1)$$

$$V = \frac{1000}{an} z^{1-n}. \quad (2)$$

The first relation was found empirically and the second (representing secular[1] variation only) was obtained from it by differentiation. The quantities z, t, and V represent depth in feet, time in milliseconds, and velocity in feet per second; and the velocity relation contains the scale factor 1,000, to express velocity in the proper units.

It is now shown that relationships (1) and (2) are applicable to formations of all types and ages. The specific numerical values of a, b, and n at various wells are geologically significant in terms of lithology, pressure, and basin position. The quantity $1-n$, in particular, is found to vary between 0 and 1/6. This indicates that velocity is proportional to depth raised to a power less than or equal to one-sixth; this links (2) to important theoretical one-sixth-power relations whose applications are worldwide.

[1] Long-term; i.e., unaffected by purely local change.

Under these circumstances, velocity changes may be related to compaction; and a working hypothesis is evolved to suggest an explanation for the anomalous increase in surface velocities westward in this basin, in the direction of increasingly younger rocks.

ANALOGOUS RELATIONS

Theoretical one-sixth-power laws

Theoretical one-sixth-power relations:

$$V = cz^{1/6}, \quad (3)$$

$$V = c(\Delta P)^{1/6}, \quad (4)$$

representing the velocity of sound in porous media in terms of either depth z or pressure difference ΔP, have been derived mathematically by Hara (1935), Iida (1939), Gassmann (1950, 1951, 1953), White and Sengbush (1953), and Brandt (1955). They were developed by applying Hertz' theory of elasticity to model granular aggregates designed to simulate porous media (Graton and Fraser, 1935; Love, 1941; Timoshenko and Goodier, 1951).

Relations (3) and (4) hold for a number of models. For each, however, the coefficient c represents a different, rather complicated function of rock properties. Iida and Gassmann, independently, used the hexagonal packing of equal spheres (Figure 1) to demonstrate (3). To demonstrate (4), Brandt used a more involved model consisting of a liquid-saturated, unconsolidated, randomly stacked aggregate of nonspherical grains of different sizes.

---

* Presented at the 32nd Annual SEG Meeting, Calgary, Alberta, September 20, 1962. Manuscript received by the Editor April 4, 1963.

† Imperial Oil Enterprises, Calgary, Alberta, Canada.

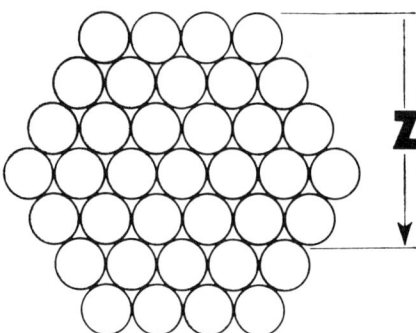

Fig. 1. Model granular aggregate, used by Iida and Gassmann independently, in mathematical demonstrations of the formula for the velocity of sound in porous media: $V = Cz^{1/6}$.

Experimental verification of relations (3) and (4) in ideal model aggregates was obtained by Iida (1939), Paterson (1954, 1956), and Berry (1959).

Empirical relations

Several analogous empirical formulas for longitudinal seismic velocities have been developed in various parts of the world.

Extensive statistical studies by Faust (1951) in the United States and Canada showed that velocity varies as the one-sixth power of the product of depth and geologic age in an average section of sand and shale. The formula was later modified (Faust, 1953) for use with a more complete geologic section, as follows:

$$V = 1948(ZTL)^{1/6}, \qquad (5)$$

where Z, T, and L represent depth in feet, geologic age in years, and a factor L dependent on lithology.

Investigation of velocity survey data in the Triassic, Permian, and Carboniferous at 46 wells in England (Wyrobek, 1959) indicates a relationship similar to (3) of the form

$$V = cz^{1-n}. \qquad (6)$$

Laboratory measurements by Berry (1959) in sandstone cores from the Tertiary in Colombia, and the Pennsylvanian in Texas indicate that velocity depends on the pressure difference ΔP at depth z, according to the formula

$$V = c(\Delta P)^{1-n}. \qquad (7)$$

As in western Canada, the quantity $1-n$ in equations (6) and (7) is not precisely one-sixth. Apparent variation from 1/17 to 1/4 is reported by Wyrobek; from 1/26 to 1/11 by Berry. Its significance is discussed below.

RELATIONSHIPS IN WESTERN CANADA: BEST VALUES FOR a, b, AND n

The western Canada basin contains a fairly complete geologic section from Cambrian to Tertiary (Webb, 1954), in which continuous-velocity information is available at hundreds of wells.

All time-depth curves are influenced by a kind of "noise," caused by velocity fluctuations due to purely local changes in lithology. This noise usually seems to be randomly distributed with respect to depth, and does not affect the general shape of the longer segments of time-depth curves in this region, but it may distort the shape of some of the shorter segments. Best-fit equations can thus be directly determined for the longer segments only. Reasonable inferences can often be made concerning equations for the shorter segments, however, as will be shown below.

There are probably several ways to determine best-fit time-depth equations. We might try to proceed graphically by plotting a set of curves between t and z^n for various powers of n, assuming that the best n could be found from inspection of the curve nearest to a straight line. Unfortunately, the variation in n is too subtle for this procedure, and results would be somewhat subjective and unreliable. To avoid this, an objective statistical method has been developed and adapted to machine computation for this study.

Sixty-two wells were chosen for investigation, as shown in Figure 2. Computations for wells A and B are presented in more detail to illustrate this method and to point out the significance of certain results; the remainder are discussed in more general terms.

TIME-DEPTH RELATIONS—WELL A

There are four significant segments in the time-depth curve for well A, as indicated in Figure 3 by the graph between t and $z^{.91}$. This particular graphical representation was used on the basis of

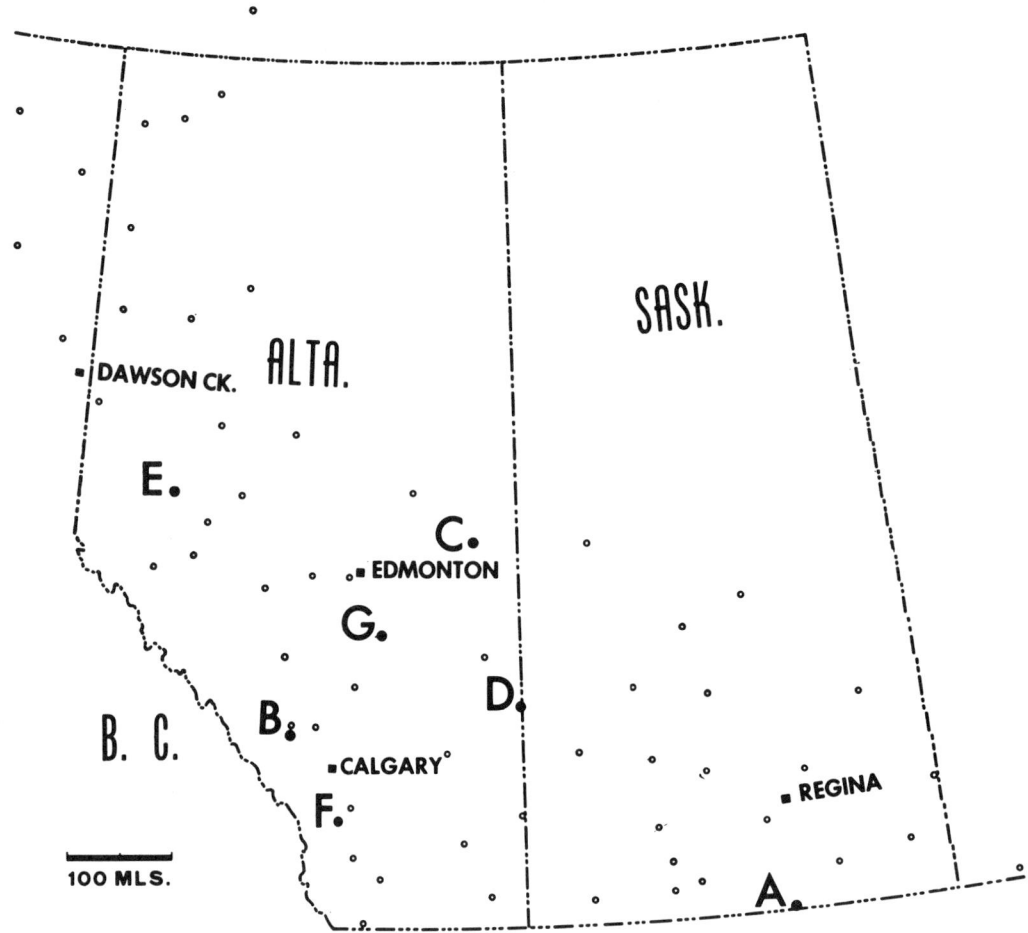

Fig. 2. Location of 62 wells chosen for precise determination of the parameters a, b, and n in the time-depth relation $t = az^n + b$.

an analysis of velocity data in western Canada, presented in a previous paper (Acheson, 1959); it delineates the segments somewhat more clearly than would a standard plot of t versus z.

The first segment—well A

We shall now compute the time-depth equation for the first layer, using data from the continuous-velocity survey, as shown in Table 1. Here, z represents the depth in feet below the ground (elevation 2,276 ft) and t represents vertical time

Table 1. Time-depth data—well A

Depth z below surface (ft)	Time t below time-datum (ms)
625	58.8
800	84.9
1,000	114.2
1,200	143.2
1,400	170.7
1,600	199.1
1,800	224.6
2,000	248.5
2,200	275.4
2,400	302.7

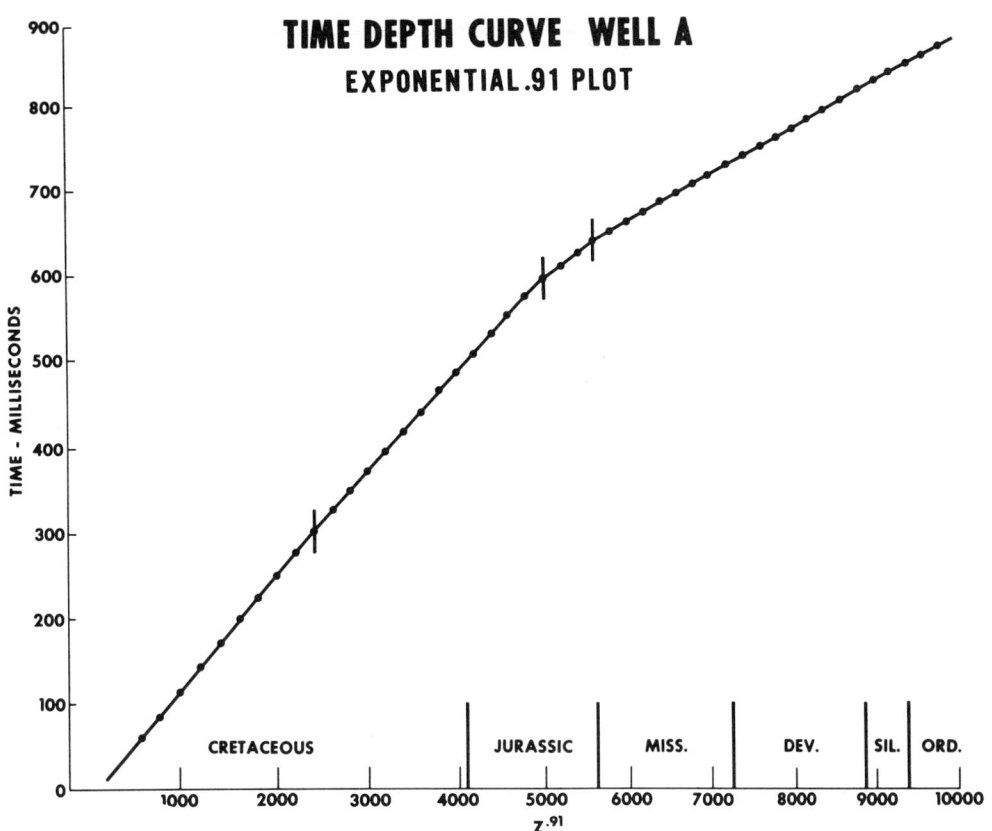

FIG. 3. Preliminary plot to delineate the segments of the time-depth curve at well A, prior to determining their best-fit equations more precisely.

below a datum at the base of the lower-velocity layer (elevation 2,013 ft).

Since the best value of $1-n$ is likely to be somewhere between 0 and 1/6, n probably varies between 1 and 5/6. This range may be restated for convenience in the reverse order, in decimals, as 0.83 to 1.00.

Only for the best value of n is the graph between t and z^n a truly straight line. Straight-line equations between t and z^n should thus fit the data better for this value than for any other. Such equations were calculated by least squares for fourteen values of n, from 0.78 to 0.91. Twelve significant figures were carried in the computations, as tests showed that it was advisable to do so. Correlation coefficients and standard deviations of observed minus calculated values were compiled in each case. Correlation coefficients of 0.999 were obtained for all these equations, indicating that they are significant within 99-percent confidence limits (Levy and Preidel, 1957).

The standard deviation s is influenced by both noise in the data and the quality of fit. Since the part of the standard deviation due to random noise is a constant, the change in the standard deviation is a measure of the quality of fit among these equations. Plotted against n (Figure 4), s is a minimum at 0.85. This is the best value of n.

Corresponding values of a and b are 0.477 and -54.8, and the best-fit equation is thus

$$t = 0.477z^{.85} - 54.8. \qquad (8)$$

The maximum discrepancy between times computed from equation (8) and the observed values in Table 1 is less than two ms. Provided the noise in the data is random, the estimate of a is within $\pm$ one percent at the 95-percent confidence level,

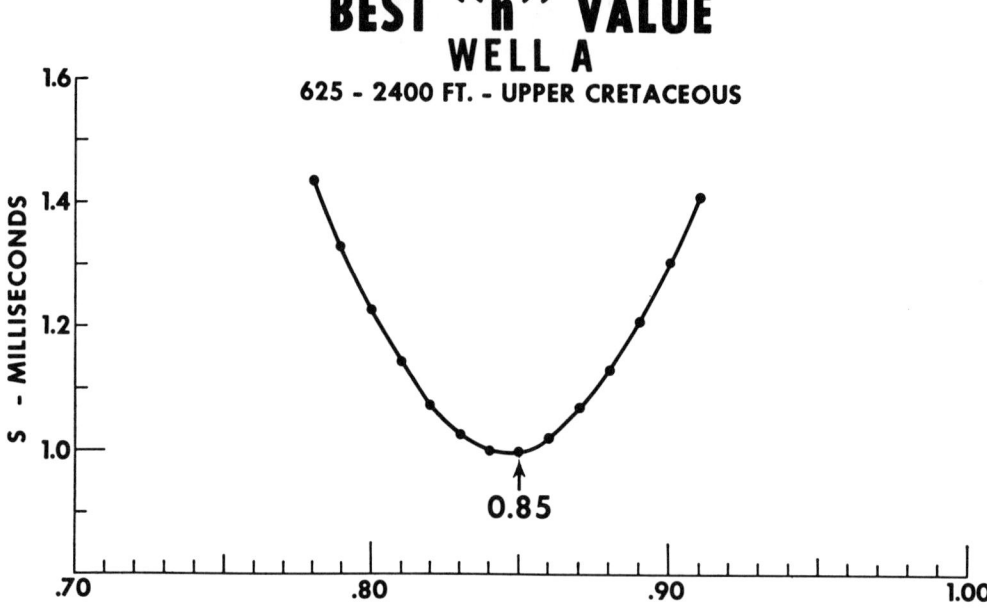

Fig. 4. The standard deviation between observed and calculated times for a number of equations $t = az^n + b$ having different values of n. The best n is indicated by the minimum.

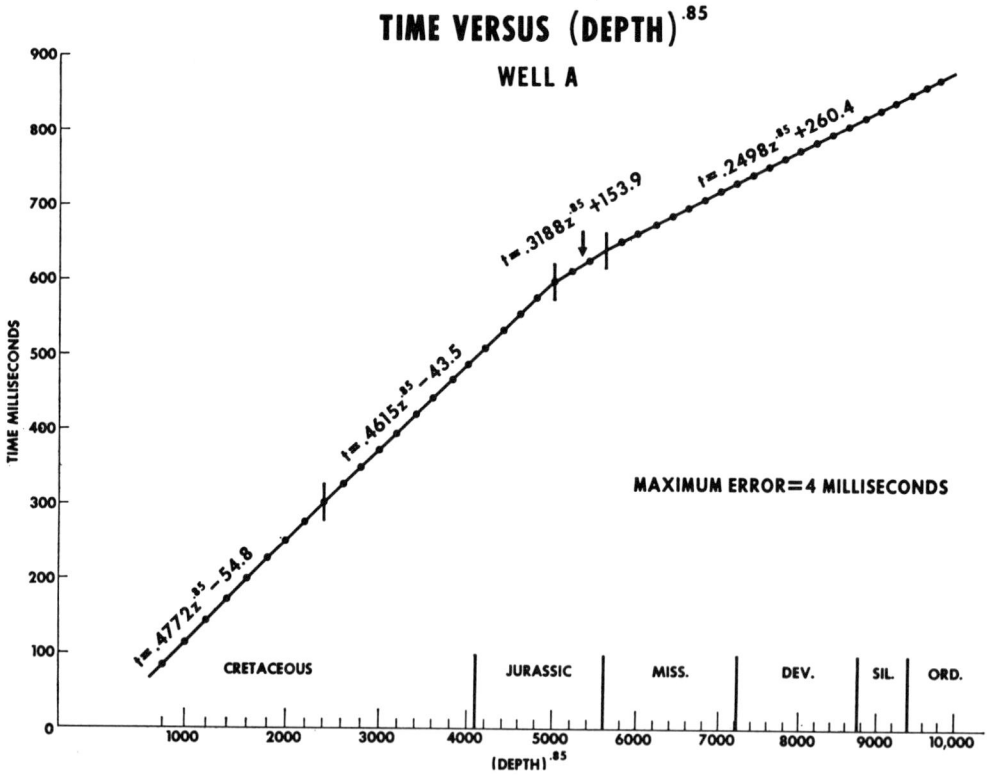

Fig. 5. Best-fit time-depth curves for well A, represented as four straight lines between t and $z^{.85}$.

and corresponding small changes in b tend to compensate the error. Thus, equation (8) represents a significant and reliable expression for time-depth variation (Dixon and Massey, 1957).

Time-depth curves for all segments of well A

Values of n obtained for layers 2 and 4 were found to be within limits of experimental error of this same value. A minimum standard deviation was not obtained for layer 3 within the above range of n values, probably because the true shape of this thin (400-ft) layer is distorted by noise (see below, Sources of Error in Measuring n).

Assuming that $n=0.85$ is a reasonable value for well A, the equations of the four layers are as given in Table 2.

The plot of t versus $z^{.85}$ (Figure 5) indicates four straight-line segments that fit the data within a maximum discrepancy of 4 msec in 10,000 ft of section.

In each of these, a is the slope and b the intercept. The slope varies from segment to segment in accordance with changes in the gross lithology, but it does not seem to be affected by change in age, for the same segment transgresses several time-system boundaries.

Velocity-depth relations—well A

Velocity-depth relations corresponding to the time-depth relations in Table 2 are given in Table 3.

In comparison to the continuous-velocity log for well A (Figure 6), these equations indicate only secular variation in velocity with depth. They do not show variations due to purely local changes in porosity or lithology, but they form an excellent basis for separating local from secular change if it is desired to do so.

Table 2. Time-depth equations for $n=0.85$—well A

Layer	Equation	Standard deviation (ms)
1	$t = .477z^{.85} - 54.8$	0.9987
2	$t = .462z^{.85} - 43.5$	1.4141
3	$t = .319z^{.85} + 153.9$	0.4220
4	$t = .250z^{.85} + 260.4$	0.6300

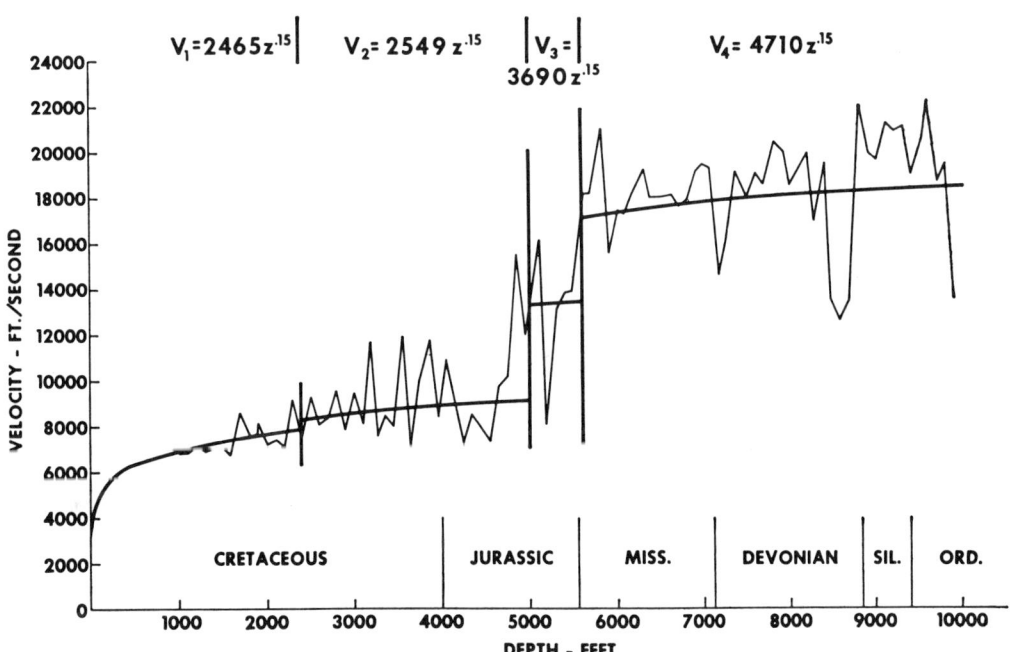

FIG. 6. Best-fit velocity-depth curves for well A, compared to the continuous-velocity log.

Table 3. Velocity-depth relations—well A

Layer	Velocity-depth relation
1	$V_1 = 2465 z^{.15}$
2	$V_2 = 2549 z^{.15}$
3	$V_3 = 3690 z^{.15}$
4	$V_4 = 4710 z^{.15}$

Time-depth relation—well B

The best value of n at well A is in excellent agreement with the theoretical value 5/6, or 0.83, predicted by one-sixth-power laws. In order to show that this agreement does not hold everywhere, reference is made now to results from a second well, B (Figure 7).

Well B lies several hundred miles west of A in the geologically disturbed belt of the Alberta foothills. The section is entirely Cretaceous shale in the first 10,000 ft below casing (1,000–10,000 ft), and the time-depth curve is one continuous segment. The surface elevation is 5,575 ft, and the elevation of the time datum is 5,450 ft. A list of time values is supplied in Table 4.

By the same methods as before, the best value of n was found to be 0.97 (Figure 7) and the best-fit time-depth equation is

$$t = .0894 z^{.97} - 3.7. \quad (9)$$

The correlation coefficient is 0.999 and the residual standard deviation is 1.04 ms, indicating significance of a high order, and a very close fit. The maximum discrepancy between times observed and times calculated from equation (9) in the entire 10,000 ft of section is 2.3 ms.

Values of n in the subintervals 1,000–5,000 ft and 1,000–7,000 ft were found to be 0.98 and 0.96. Thus, $n = 0.97$ is a reasonable best-value for the entire section, and relation (9) is significant and reliable.

Comparison of wells A and B

Velocity-depth equations for the first layer in wells A and B are respectively:

$$V = 2,469 z^{.15}, \quad (10)$$

$$V = 11,532 z^{.03}. \quad (11)$$

These equations are shown graphically in Figure 8.

In the same way, particular equations may be obtained for the rate of change of velocity with depth,

$$R = \left(\frac{1-n}{an}\right) z^{-n}.$$

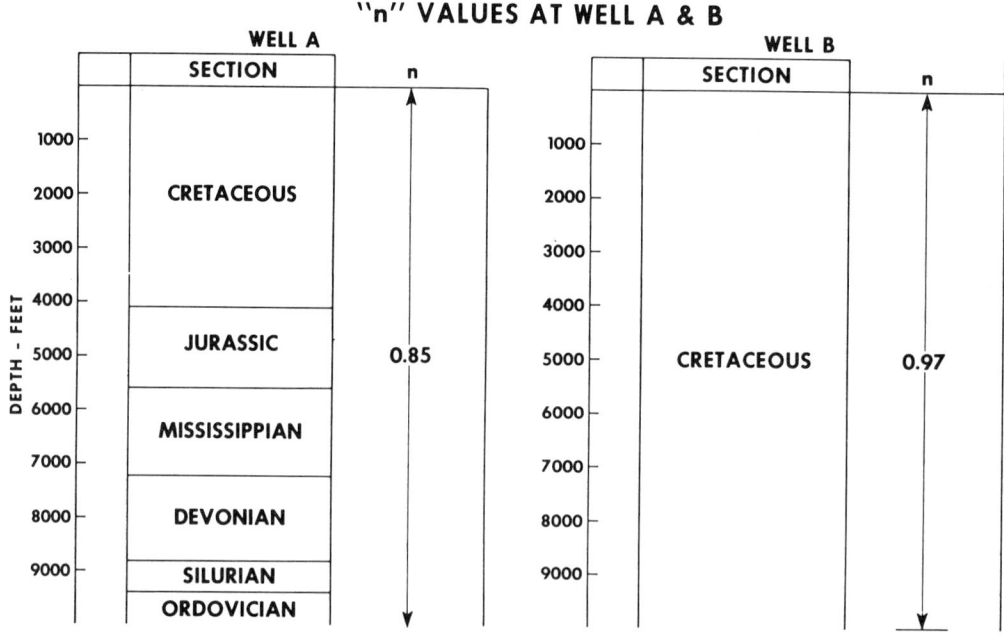

FIG. 7. Comparison of n at wells A and B to illustrate the dependence of this parameter on basin position.

Table 4. Time-depth data for well B

Depth from surface (ft)	Time from datum (ms)
1,000	68.9
1,250	86.8
1,500	103.2
1,750	120.9
2,000	140.1
2,250	157.9
2,500	174.0
2,750	189.4
3,000	206.4
3,250	224.0
3,500	240.6
3,750	258.0
4,000	275.6
4,250	292.4
4,500	309.2
4,750	325.9
5,000	343.8
5,250	360.6
5,500	377.7
5,750	393.0
6,000	409.4
6,250	425.0
6,500	441.7
6,750	457.4
7,000	474.9
7,250	492.6
7,500	510.0
7,750	526.5
8,000	543.7
8,250	557.9
8,500	575.0
8,750	592.0
9,000	608.9
9,250	624.8
9,500	641.3
9,750	660.0
10,000	676.8

If R is divided by the corresponding expression for the velocity at a depth of one foot, the result is equal to $(1-n)/z^{-n}$. This may be interpreted to mean that, at any given depth, the rate of change of velocity relative to the velocity at one foot is a function of n only. It is larger at A where n is smaller than it is at B where n is nearly unity and the velocity is almost constant.

It is surprising that the near-surface velocities in B (Figure 8) are so much larger than velocities at much greater depth in A, since each section is almost entirely shale. A marked difference is suggested in the present compaction and also in the degree that it may be possible to reach.

VELOCITY, LITHOLOGY, AND PRESSURE

Relationships among velocity, lithology, and pressure have been studied intensively in the past few years. Recent papers by Birch (1960, 1961) contain a discussion of the effects of pressure, porosity, alteration, anisotropy, and composition. They also provide an excellent bibliography of nearly a hundred references to the work that has been carried on (including important papers by Hughes and associates) that are especially pertinent to this discussion.

Local and secular velocity variations are evident in all continuous-velocity logs; and Wylie, Gregory, and Gardiner (1956), Hicks and Berry (1956), Pickett (1960), Brandt (1960), and Sarmiento (1961) have developed ways, from theory and/or experiment, to relate such changes to variations in porosity, lithology, and pressure.

Relation (2) is not concerned in this way with purely local variations, but can be used to interpret the geologic importance of large-scale velocity changes.

Relation between "a" and lithology

Changes in a may indicate either a change in gross lithology at any given location or a change in basin position. For example, the four values of a in Table 2 imply velocity changes among layers whose gross lithologies are shale, sandstone, calcareous sandstone, and limestone (Figures 5 and 6). The difference in a between equations (8) and (9), on the other hand, is associated with a change in basin position.

Relation between "n" and pressure

The quantity n is more directly related to overburden pressure. Restating relation (1) in the equivalent form:

$$t - b = az^n,$$

and taking logarithms of both sides, we obtain

$$\log(t - b) = n \log z + \log a.$$

Differentiating with respect to z, setting z/t and dz/dt equal to $\overline{V}$ and V respectively (where $\overline{V}$ and V represent average and interval velocity respectively), and rearranging terms, we have

$$\frac{\overline{V}}{V} = n\left(1 - \frac{b}{t}\right). \quad (12)$$

In the special case of the first layer, b may be reduced to zero by correcting for the low-velocity layer and the equation becomes

$$\overline{V} = nV. \quad (13)$$

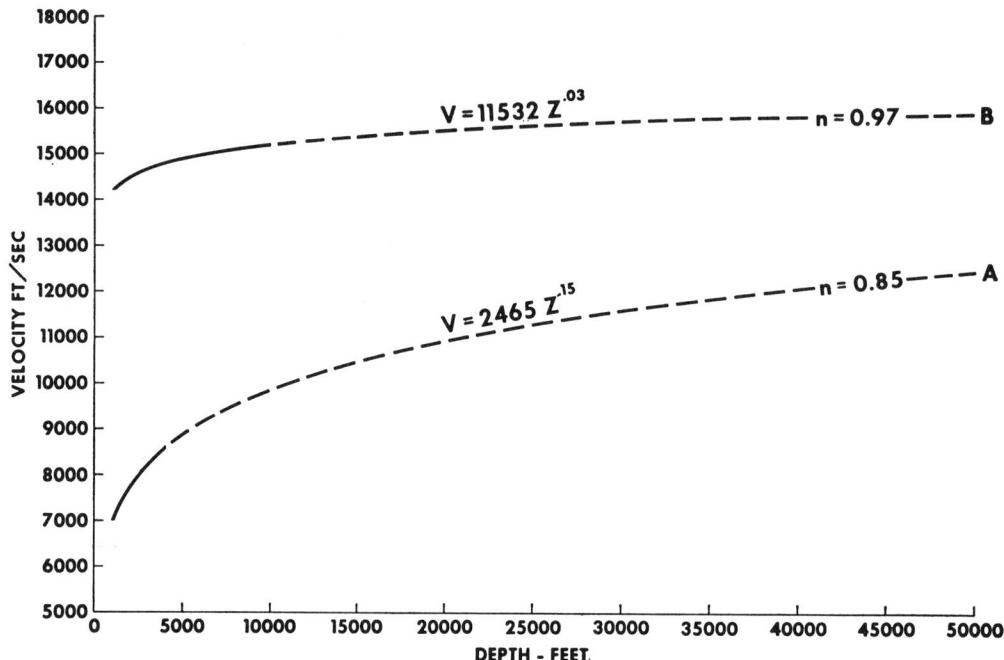

Fig. 8. Comparison of velocity-depth curves for the first segments of wells A and B. Note that as n approaches unity, the velocity tends to become constant.

Equation (13) holds also for the deeper layers, provided the intercepts of their time-depth curves are reduced to zero by correcting suitably for delay.

It seems reasonable from (13) to suppose that n increases as the section becomes more compact. In relatively uncompacted sections, the average velocity through the section to a given depth should be less than the instantaneous velocity at that depth and n considerably less than one. In more compact sections, the average and interval velocities should be more nearly equal, and n should be larger. In very compact sections, n should be nearly one, and the velocity should be almost constant.

SOURCES OF ERROR IN MEASURING n

Reliable n values can usually be obtained from single segments of the time-depth curve for sections not less than 1,500 ft thick. In such cases (as in well B) values apply not only to the whole depth range but also to subintervals. At well G, as a further example, values of n for the intervals 600–3,000 ft, 600–2,200 ft, and 600–1,500 ft were 0.89, 0.88, and 0.88 respectively.

Spurious n values are obtained, on the other hand, when two segments separated by a discontinuity are inadvertently used together in a calculation. At well E, for example, an n value of 0.77 was obtained for the Cretaceous shale section between 1,500 and 8,000 ft. Examination of the time-depth graph indicated a distinct break at 3,750 ft, and values of n computed for the two intervals 1,500–3,750 ft and 3,750–8,000 ft were 0.84 and 0.86. The true value of n, therefore, is not 0.77, but some value in the neighborhood of 0.84.

Spurious n values may also be obtained where very short, single segments are distorted by "noise." Velocity discontinuities that always

WESTERN CANADA BASIN
n VALUES - UPPER CRETACEOUS

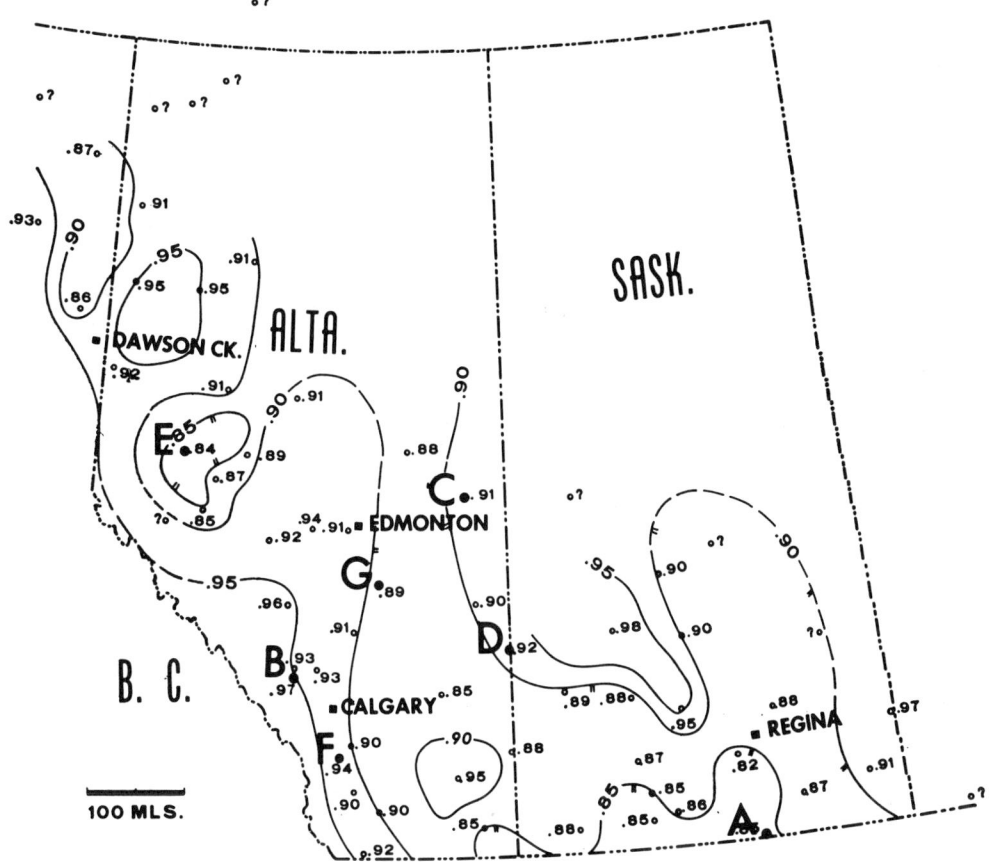

Fig. 9. Regional variation of n in the Colorado group of the Upper Cretaceous, or its equivalent.

occur every few feet in the section introduce subdued saw-toothed irregularities in the time-depth curve. When, as in layer 3 of well A, the segment is so short that these serrations affect the true shape of the curve, it is possible to obtain a spurious value of n either much greater than one if the local shape of the curve is concave upward, equal to one if it is straight, or less than 0.83 if it is unusually concave downward.

BEST VALUES OF a, b AND n—WESTERN CANADA BASIN

Determination of "n"—Upper Cretaceous section

The Upper Cretaceous section is a group of clastic sediments, predominantly shale, whose thickness increases from a few hundred feet in the eastern part of the basin to more than 10,000 ft in the west. The time-depth curve of the Colorado group within this section is a single segment that may be used to determine best values of a, b, and n across the basin.

At 52 of the 62 wells under investigation, the Colorado group was thick enough to provide reliable direct measurements of n and the associated parameters, the distribution of thicknesses being roughly as shown in Table 5.

Table 5. Thicknesses of Colorado group for n determination

Thickness (ft)	Percentage of 52 wells
1,500–2,500	50
2,500–5,000	35
over 5,000	15

In the computations for *n*, it was found that correlation coefficients were almost invariably 0.999, indicating highly significant relations, and standard deviations were commonly of the order of one to three ms. Thus, the time-depth relations were roughly of the same order of significance and reliability as at wells A and B.

The distribution of *n*, shown in Figure 9, indicates a northwest-trending trough, with high values to both the east and west. This systematic variation between approximately 0.83 and 1 implies that $1-n$ lies between 0 and 1/6. Consequently, velocity is directly proportional to depth raised to a power equal to or less than one-sixth.

Significance of "n" in the complete section

The Tertiary section was investigated at two wells only, for its occurrence is limited, but in these locations, values of *n* did not appear to differ significantly from those of the Colorado group

The Lower Cretaceous and Jurassic sections were encountered in many wells and appear to have essentially similar values of *n* to those in the Upper Cretaceous at the same locations.

Reliable *n* determinations were obtained for various layers in the predominantly carbonate section of the pre-Jurassic in only about 25 percent of the 52 wells. These varied between 0.81

FIG. 10. Theoretical velocity that would be predicted at 100-ft depth if the sediments of the Colorado group were to continue upward to surface in all areas.

WESTERN CANADA BASIN
$R \equiv \left(\frac{dv}{dz}\right)_{100}$
UPPER CRETACEOUS

FIG. 11. Theoretical rate of velocity change predicted at 100-ft depth if the sediments of the Colorado group were to continue upward to surface in all areas.

and 0.95 and generally agreed fairly closely with values obtained for the Cretaceous in the same wells. Thus, n seems to be largely independent of age and lithology, but is predominantly a function of pressure and basin position.

REGIONAL VELOCITY VARIATION—UPPER CRETACEOUS SHALES

Velocity behavior in the near-surface

The best-fit equations for time and velocity that have been determined for the Colorado group of the Upper Cretaceous will now be used to investigate regional variation of near-surface velocity.

The following relations represent velocity and rate of change of velocity at 100 ft respectively:

$$V_{100} = \frac{1,000}{an} (100)^{1-n}, \quad (14)$$

$$R_{100} = \frac{V_{100}}{100} (1 - n), \quad (15)$$

where it can easily be verified that R_{100} is equivalent to dv/dz at 100 ft.

Although, if the weathering is ignored, the Colorado group extends to the surface in many parts of the region, it is covered by younger formations in some places. The quantities in (14) and (15) are, therefore, not true geologic measures but rather predictions of the behavior these velocities would exhibit if the section continued to the surface throughout the region.

Velocity change as a measure of compaction

The map of V_{100} (Figure 10) indicates a velocity increase toward the mountain front in the direction of increasingly younger rocks. This anomalous trend, which has been amply verified in practice by many direct near-surface measurements, requires explanation.

Since greater change results in velocity when pressure is applied to unconsolidated material than to consolidated, rate of change of velocity is probably a measure of compaction. In the map of R_{100} (Figure 11) the smaller numerical values correspond to a relatively slow increase in velocity and probably represent areas where compaction is approaching a maximum for conditions prevailing in that part of the region.

If the velocity in shales is a function of their compaction, it must also be a function of the pressure that was applied. But if pressure has had an effect, some account must be taken of former pressure systems. Great horizontal and vertical pressures must have existed during the formation of the Rocky Mountains. In the plains area, there was a considerable load over the entire region in the form of now-eroded Tertiary (Webb, 1954) and/or, perhaps, thousands of feet of glacial ice, and the velocity that we measure today was probably largely set at the time when this maximum load was present. At that time, however, the surface of the ground was in a far different position from that which it is in today (Figure 12) and it is probably this surface, or something approximating it, that we should use as origin of coordinates.

Changes of origin of this sort result in interesting and significant effects. The best value of n at well F for the section between 650 and 6,400 ft referred to ground surface was found to be 0.94. Changing the origin by adding 3,000 ft to all depth values resulted in a new best value of n of 0.83. Similarly at well G, changing the origin by adding 1,000 ft to all depths reduced the best value of n from 0.89 to 0.83. Moreover, the new time-depth equations were found to fit the data better, for decreases of more than 5 percent in standard deviation resulted at both wells.

A tentative relationship between n and ΔZ indicated by these results is shown in Figure 13.

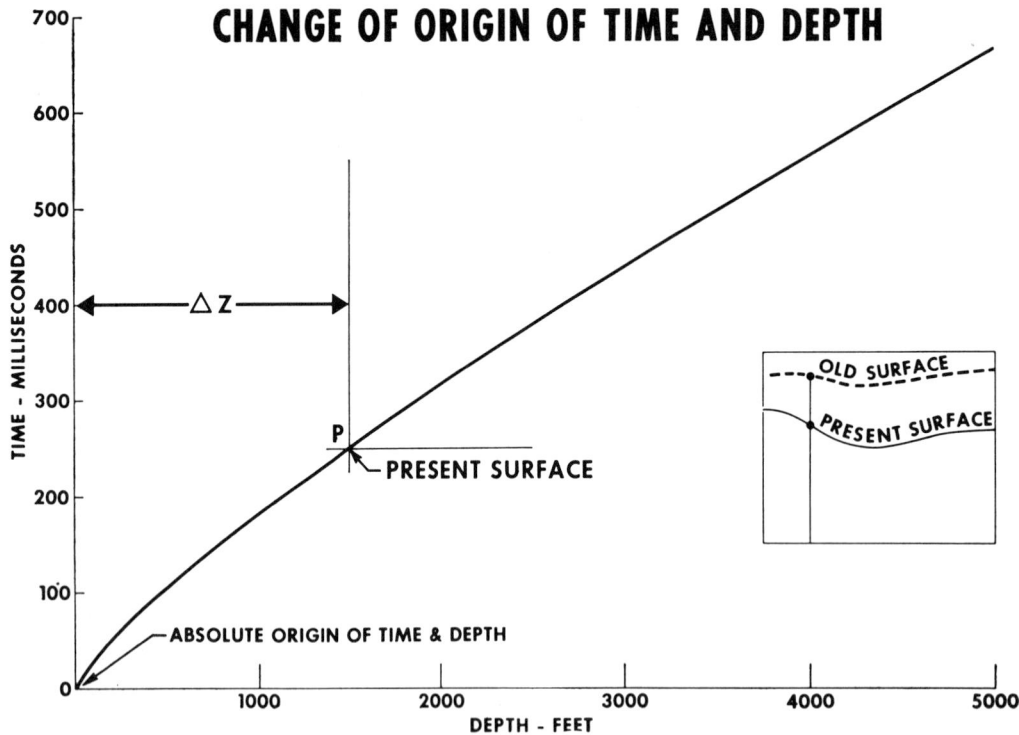

FIG. 12. Change of origin upward to allow for the former load of the eroded section and/or glacial ice.

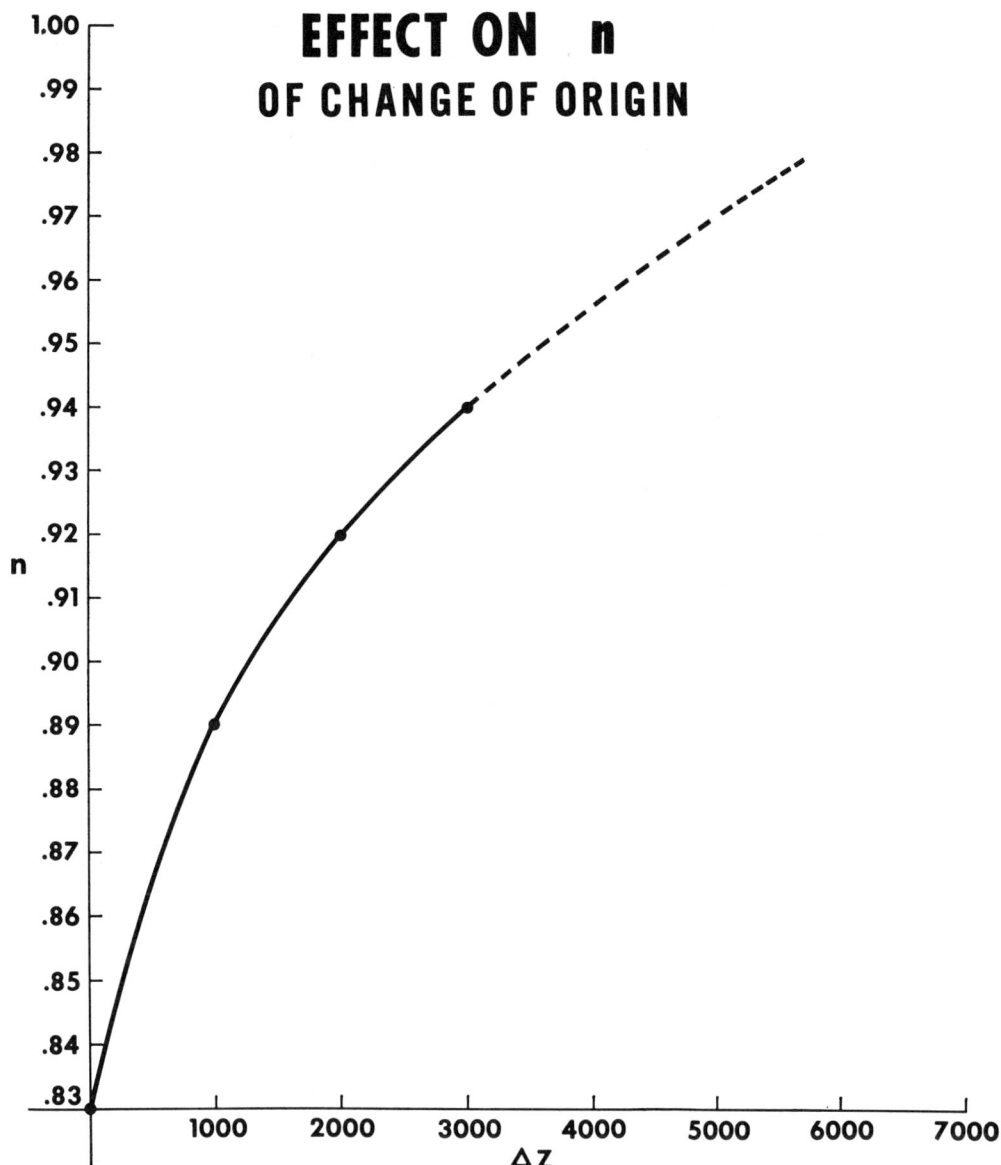

Fig. 13. Plot of n versus ΔZ, which is the change in the origin of coordinates.

This is incomplete for several reasons. Considerably more work would be needed to establish the relationship firmly. The standard deviation was not found to be an absolute minimum at 0.83, although it was changing very slowly. Complete investigation would involve further statistical study and critical evaluation of the theoretical value 0.83 that is beyond the scope of this paper. With these restrictions in mind, however, it is possible to discuss in a qualitative way the effect of shifting the origin of coordinates at wells along the profile BC.

The anomalous increase in near-surface velocities in the direction CB (Figure 14) of increasingly younger rocks may be noted along a surface which is an arbitrary small distance (say 100 ft) below the ground. Ground surface itself cannot be used, for here the velocity predicted is constantly zero. A hundred feet below the former geologic surface defined by the new origins of coordinates, on the other hand, the velocity change is very considerably less. It seems reasonable to suppose that near-surface velocities in this region have been affected more by depositional history than by age itself, and that rocks of approximately the same age along this profile have been subjected to greater pressure in the west than in the east. This points out an interesting difference between the relations discussed in this paper and formulas published by Faust. In this region we are able to suggest that geologic age has not been a primary influence in velocity variation so much as what has happened to the rocks, and that what has happened here is not simply related to geologic age.

SUMMARY AND CONCLUSIONS

It has been demonstrated that the simple time-depth equation $t = az^n + b$ holds very exactly in the western Canada basin for all types of lithology and for all ages of rock.

Changes in the parameter a reflect mainly changes in the gross lithology at any one location or changes in basin position for the same type of lithology. Although the effect of age may be present, it is probably obscured in this region by lithologic change.

Changes in the parameter n occur in a regular pattern across the basin and seem to be largely independent of both lithology and age. Combined with corresponding changes in the parameter a, they are probably closely related to variations in pressure and compaction.

Similarity that has been noted between the velocity-depth equation and theoretical one-sixth-power laws for the velocity of sound in porous media is used to suggest an explanation for the anomalous basinward increase of surface velocity in the direction of increasingly younger rocks. Although the theory is far from complete and more work is needed to establish it firmly, better agreement can be reached between the

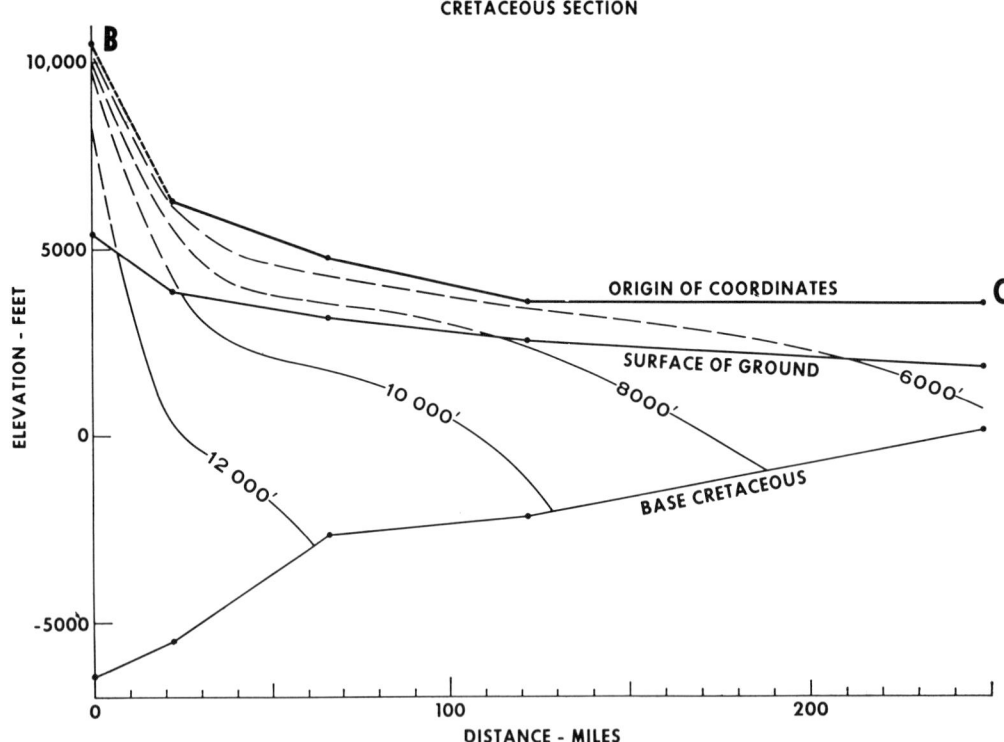

Fig. 14. The anomalous velocity change westward at ground surface along profile BC is greatly attenuated along the locus of the new origin of coordinates.

empirical formulas for western Canada, the formulas derived by Faust, and theoretical one-sixth-power relations by moving the origin of coordinates upward from the present surface. Under these conditions, a former geologic surface is defined along which velocities change less rapidly, and it is suggested that anomalous increase in velocities toward the mountains is mainly an effect of pressure.

ACKNOWLEGMENTS

The author wishes to thank Mr. C. J. Chapman of the Exploration Research Department for interest and helpful suggestions, the Systems and Computer Services Department for cooperation in writing the machine program and handling the data, and the management of Imperial Oil, Limited for permission to publish this material.

REFERENCES

Acheson, C. H., 1959, The correction of seismic time maps for lateral variation in velocity beneath the low-velocity layer: Geophysics, v. 24, pp. 706–724.
Berry, James E., 1959, Acoustic velocity in porous media: Petroleum Transactions A.I.M.E., v. 216, pp. 262–270.
Birch, Francis, 1960, The velocity of compressional waves in rocks to 10 kilobars, Part 1: Jour. of Geoph. Res. v. 65, pp. 1083–1102.
——, 1961, The velocity of compressional waves in rocks to 10 kilobars, Part 2: Jour. of Geoph. Res., v. 66, pp. 2199–2224.
Brandt, H., 1955, A study of the speed of sound in porous granular media: Jour. of App. Mech., v. 22, pp. 479–486.
——, 1960, Factors affecting compressional wave velocity in unconsolidated marine sand sediments: The Jour. of the Acoust. Soc. of America, v. 32, pp. 171–179.
Dixon and Massey, 1957, Introduction to the theory of statistical analysis: 2nd ed., New York, McGraw-Hill Book Co., Inc.
Faust, L. Y., 1951, Seismic velocity as a function of depth and geologic time: Geophysics, v. 16, pp. 192–206.
——, 1953, A velocity function including lithologic variation: Geophysics, v. 18, pp. 271–288.
Gassmann, Fritz, 1950, Uber Die Elastizitat Poroser Medien. Vierteljahrsschrift der Naturforschenden Gessellschaft in Zurich, v. 96, pp. 1–23.
——, 1951, Elastic waves through a packing of spheres: Geophysics, v. 16, pp. 271–288.
——, 1953, Note: Geophysics, v. 18, p. 269.
Graton, L. C., and Fraser, H. J., 1935, Systematic packing of spheres with particular relation to porosity and permeability: The Jour. of Geol., v. 43, pp. 785–909.
Hara, G., 1935, Theorie der Schwingungs-Austreitung in Gekornten Substanzen und Experimentelle Untersuchungen in Kohlepulvei: Elektrische Nachrochten Techvik, v. 12, pp. 191–200.
Hayakawa, M., and Balakrishna, S., 1961, An explanation for the high ultrasonic velocity in Indian Rocks: Geophysical Prospecting, v. 9, pp. 74–85.
Hicks, W. G., and Berry, J. E., 1956, Application of continuous-velocity logs to determination of fluid saturation of reservoir rocks: Geophysics, v. 21, pp. 739–754.
Hughes, D. S., and Cross, J. H., 1951, Elastic wave velocities at high temperatures and pressures: Geophysics, v. 16, pp. 577–593.
——, and Kelly, J. L., 1952, Variation of elastic velocity with saturation in sandstone: Geophysics, v. 17, pp. 739–752.
Iida, K., 1939, Velocity of elastic waves in granular substance: Bull. of the Earthquake Res. Inst., v. 17, pp. 783–807.
Levy, H., and Preidel, E., 1957, Elementary statistics: New York, Thomas Nelson and Sons.
Love, A. E. H., 1944, A treatise on the mathematical theory of elasticity: 4th ed., New York, Dover Publications.
Parasnis, D. S., 1960, The compaction of sediments and its bearing on some geophysical problems: The Geoph. Jour. of the Royal Astronomical Soc., v. 3, pp. 1–28.
Paterson, N. R., 1954, A theoretical approach to the calculation of seismic wave-velocity in sedimentary formations: Trans. Roy. Soc. Canada, Ser. III, Sec. 4, v. 48, pp. 60–64.
——, 1956, Seismic wave propagation in porous granular media: Geophysics, v. 21, pp. 691–714.
Pickett, G. R., 1960, The use of acoustic logs in the evaluation of sandstone reservoirs: Geophysics, v. 25, pp. 250–274.
Sarmiento, R., 1961, Geologic factors influencing porosity estimates from velocity logs: Bull. A.A.P.G., v. 45, pp. 633–644.
Timoshenko, S., and Goodier, J. N., 1951, Theory of elasticity: 2nd ed., New York, McGraw-Hill Book Co., Inc.
Webb, J. B., 1951, Geological history of the plains of western Canada: Bull. A.A.P.G., v. 35, pp. 2291–2315. 1954, Western Canada Sedimentary Basin Sympos., Rutherford Memorial Volume, A.A.P.G., pp. 3–28.
White, J. E., and Sengbush, R. L., 1953, Velocity measurements in near-surface formations: Geophysics, v. 53, pp. 54–69.
Wylie, M. R. J., Gregory, A. R., and Gardner, L. W., 1956, Elastic wave velocities in heterogeneous and porous media: Geophysics, v. 21, pp. 41–70.
Wyrobek, S. M., 1959, Well velocity determinations in the English Triassic, Permian, and Carboniferous: Geophysical Prospecting, v. 7, pp. 218–230.

APPARENT VELOCITY FROM DIPPING INTERFACE REFLECTIONS[†]

F. K. LEVIN[*]

When a seismic reflector is a dipping plane and the subsurface can be approximated by a single bed, the velocity needed to stack CDP data is higher than the velocity with which the energy travels in the subsurface. The ratio of the stacking velocity to the energy-travel velocity increases from unity for strike lines to the secant of the dip angle for dip lines. Although the reflection points are not common, reflections stack even for steep dips. Stacking velocities for multiple reflections increase with the order of the multiple, but multiple reflections may also be stacked. To convert from reflection time to depth requires knowledge not given by the stacking velocity. In theory, a split-spread procedure can furnish the needed information.

INTRODUCTION

The effect of a dipping reflector on common depth point (CDP) processing has been considered by several investigators. Cressman (1968) pointed out that because of the way the normal moveout (NMO) correction is applied, CDP stacking works as well for dipping beds as it does for horizontal beds; although in the former case the reflections being summed do not have common reflection points. Brown (1969) examined in detail the NMO correction and velocity relations for dipping beds. Both Cressman and Brown dealt with realistic subsurfaces, subsurfaces characterized by velocities varying with depth. In a recent paper, Taner and Koehler (1969) avoided the difficulties associated with dipping bed calculations by tracing rays with a digital computer.

A physical understanding of the dependence of CDP-determined velocities on dip can be obtained if we restrict ourselves to an extremely simple subsurface; namely, one with a single reflecting plane. Although stemming from an unreasonably oversimplified picture, the expressions derived from such a subsurface are those actually used in much CDP data processing. Hence, there is some reason for presenting our results in spite of their obvious limitations.

THE PRIMARY REFLECTION

The subsurface we shall consider is shown in Figure 1. Because we have only one bed, the distinction between rms velocity, interval velocity, and average velocity that must be made when the subsurface is layered has no meaning here; we can assume there is a constant velocity V between the surface of the ground and the reflector. To keep our deviation general, we write the equation of the reflecting plane as

$$x \cos \alpha + y \cos \beta + z \cos \phi = d, \quad (1)$$

where α is the angle between the normal to the plane and the profile direction, taken here to be the x axis, and ϕ is the dip angle of the plane. α, β, and ϕ are direction cosines of the normal to the plane. We place our source at the origin $(0, 0, 0)$ and the geophone at $(X, 0, 0)$. d is the distance from the source to the reflecting plane, measured along the perpendicular to the plane. D is the corresponding distance from a point along the profile halfway between the source and geophone. Notice that the line along which D is measured *does not* intersect the reflector at the reflection point. The difference between single-fold and CDP geometries can be summarized as follows: in the

[†] Manuscript received by the Editor September 29, 1970; revised manuscript received January 19, 1971.
[*] Esso Production Research Company, Houston, Texas 77001.
© 1971 by the Society of Exploration Geophysicists. All rights reserved.

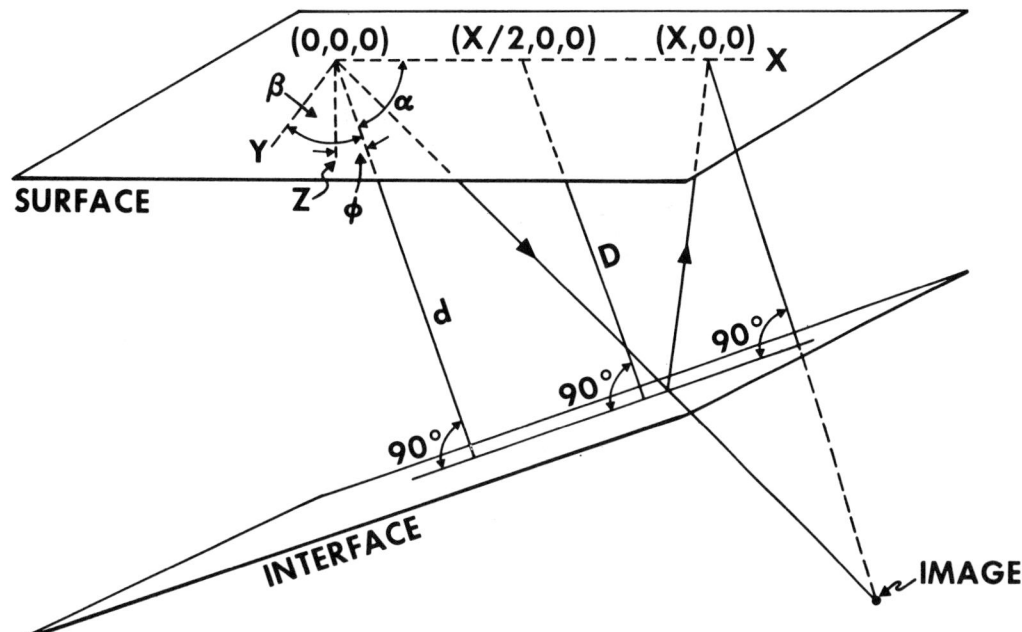

Fig. 1. Drawing of a layer bounded by the surface and a dipping, plane reflector. The source is at (0, 0, 0); the geophone, at (X, 0, 0).

CDP method, D is constant; for single-fold exploration, d is constant.

The mathematics of computing reflection traveltime involves nothing more complicated than setting up an image geophone (or an image source). As shown by Slotnick (1959), the traveltime to and from the reflector is given by

$$V^2 t^2 = 4d^2 + X^2 - 4dX \cos \alpha. \qquad (2)$$

Rewriting equation (2) in terms of D, we have

$$t^2 = \left(\frac{2D}{V}\right)^2 + X^2 \cdot \left(\frac{\sin \alpha}{V}\right)^2.$$

We define V_{NMO} as

$$V_{\text{NMO}} = V/\sin \alpha \qquad (3)$$

to get

$$t^2 = \left(\frac{2D}{V}\right)^2 + \frac{X^2}{V_{\text{NMO}}^2}. \qquad (4)$$

Hence, there is an apparent velocity V_{NMO} which will flatten the reflection; but this apparent velocity is not the same as V.

We can draw several conclusions from equation (4). First, to the degree that a constant velocity to a reflector is a good approximation, the use of V_{NMO} lets us stack CDP data for steep or for small dips. The steepness of the dip doesn't affect our ability to stack. Second, the time intercept for zero spread length is the two-way distance at the halfway point divided by the constant velocity V. As long as the source and geophone stations are arranged symmetrically around a surface point, the data may be stacked; although clearly we do not stack data from the same subsurface points. Third, the stacking velocity will always be the same as or higher than V.

Let us look more closely at the way V_{NMO} depends on the direction of the profile line relative to the direction of the dip of the reflector. The angle α is not one that is available to an exploration geophysicist. We normally think in terms of the dip angle ϕ and the angle between the profile line and a dip line. We'll call this latter angle θ (Figure 2). In terms of θ and ϕ, equation (3) is

$$V_{\text{NMO}}/V = (1 - \sin^2 \phi \cos^2 \theta)^{-1/2}. \qquad (5)$$

In Figure 3, we plot the ratio V_{NMO}/V as a function of θ for several dip angles. For profiles in the strike direction, V_{NMO} and V are the same. For

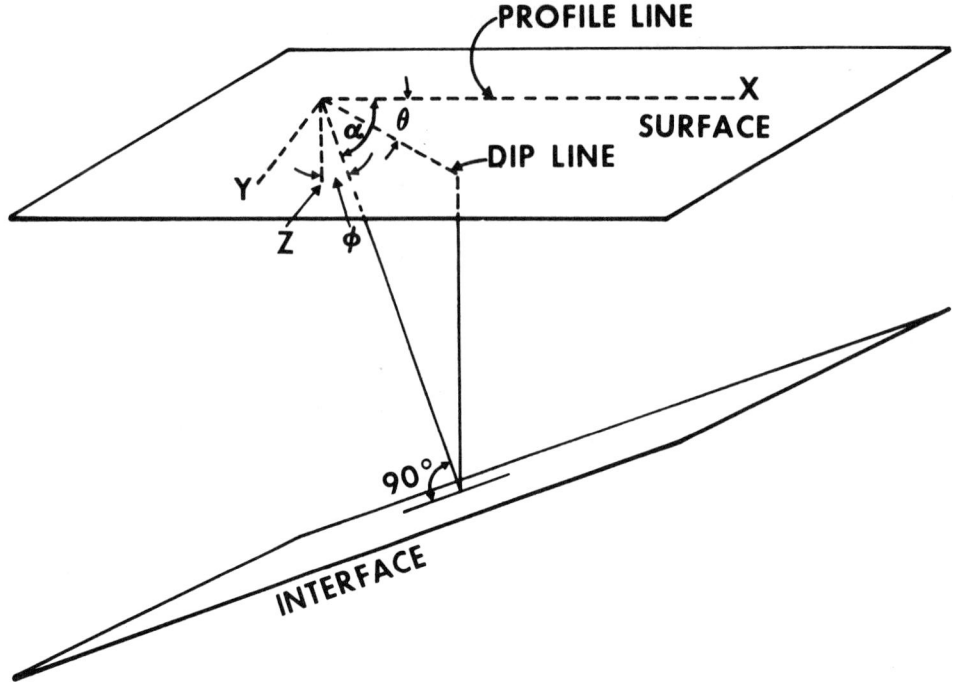

FIG. 2. Drawing showing the relation between the angles α, ϕ, and θ.

profiles in any other direction, V_{NMO} is greater than V. The greatest difference occurs for profiles in the dip direction. For dip lines,

$$V_{NMO_{dip}}/V = \sec \phi. \qquad (6)$$

A plot of this expression (Figure 4) shows that for dips of less than 10 degrees, the velocity difference can be ignored. It is obvious that if we average velocities found for lines in different directions, the average will be greater than V but less than $V_{NMO_{dip}}$.

Conversion of traveltime to depth for a layered subsurface is a complicated, sequential process.

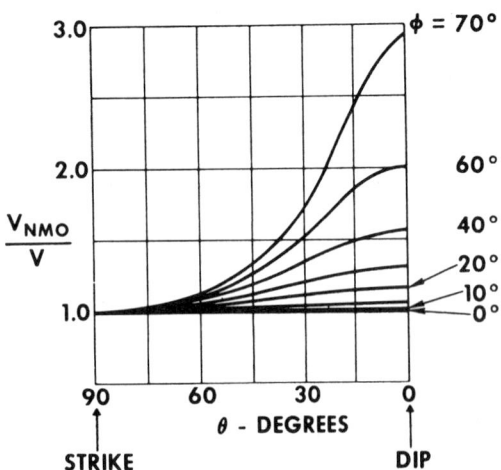

FIG. 3. Ratio of V_{NMO} to V as a function of θ for several dip angles.

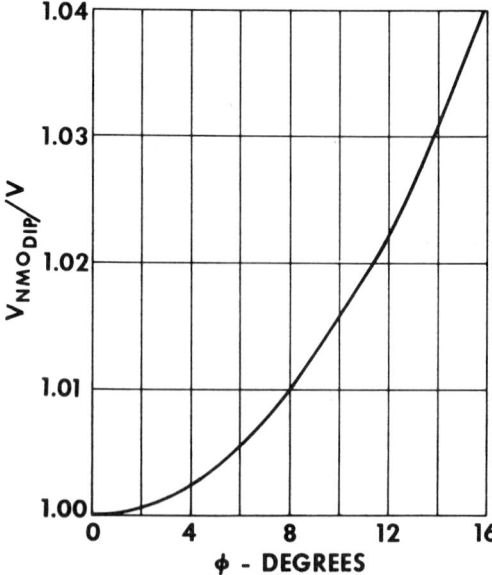

FIG. 4. Ratio of $V_{NMO_{dip}}$ to V for small angles of dip.

For the simple section of Figure 1, the conversion is easy. If we want depth measured along a line perpendicular to the reflector, we use the velocity V. On the other hand, if we want depth measured along a line perpendicular to the surface and if the profile is a dip line, we use V_{NMO}. Note that the line must be a dip line. An incidental result of the difference between V and V_{NMO} is the following: for a dip line, if V_{NMO} is used to compute the dip, the apparent dip from a record time section is the actual dip.

In practice, a geophysicist finds V_{NMO} by applying a velocity determination program to CDP-gathered data. To convert to depth, he will want V. If he is lucky enough to have cross-spreads, V and ϕ may be found by well-known procedures (Slotnick, 1959). If, as is almost always true in marine seismology, he has only single-ended spreads, V may still be found if the field data acquisition procedures involved split-spread shooting or if the data can be reassembled in split-spread form. The method is discussed at length by Slotnick (1959). We designate as t_+ and t_- the reflection times measured by geophones equally distant from but on opposite sides of the source. We define an average time τ by

$$\tau^2 = \frac{t_+^2 + t_-^2}{2} = \left(\frac{2d}{V}\right)^2 + \frac{X^2}{V^2}. \quad (7)$$

V is found from a plot of τ^2 against X^2. The $\tau-X$ hyperbola is symmetrical about the source but the $t-X$ hyperbola has its head displaced in the up-dip direction. As a practical method, the application of equation (7) is less attractive than it first appears, since the technique demands that a time-distance plot be available and, unlike modern velocity determination computer programs, does not operate directly on recorded data.

We indicated earlier that D does not intersect the reflector at the reflection point (Figure 1). The reflection point, which depends both on the spread length X and on D, is up-dip relative to the intersection point. If we call the distance between the two points Δ, we find (see appendix) that

$$\Delta = \frac{X^2}{4D} \sin\alpha \cos\alpha = \frac{X^2}{8D} \sin 2\alpha$$
$$= \frac{X^2}{4D} \sin\phi \cos\theta \sqrt{1 - \sin^2\phi \cos^2\theta}. \quad (8)$$

For a strike line, $\Delta = 0$. For a dip line,

$$\Delta_{dip} = \frac{X^2}{8D} \sin 2\phi. \quad (9)$$

If the dip is small, we may replace $\sin 2\phi$ by 2ϕ. To get some idea as to the magnitude of Δ, we consider a 10,000 ft source-geophone separation, a reflector at 10,000 ft, and a 30 degree dip. $\Delta \approx 1100$ ft. It is doubtful if the idea of a reflection point moving along the reflector should be taken too seriously, since wavefronts and not rays are reflected. The relation between the two is discussed by Trorey (1970).

Due to our overly simplified model, we've ignored many important effects. Most important, of course, are all effects due to refraction at interfaces between the surface and the reflector being considered. However, CDP data stack even when dips are appreciable. Hence, calculations of the type described by Taner and Koehler (1969) are likely to modify but not invalidate the conclusions drawn here.

MULTIPLE REFLECTIONS

The mathematics developed for the primary reflection carries over with very little change when multiples are being considered. Again image sources are used to get the time-distance relation

$$t_n^2 = \left[\frac{2D}{V} \frac{\sin(n+1)\phi}{\sin\phi}\right]^2$$
$$+ \left[\frac{1 - \cos^2\theta \sin^2(n+1)\phi}{V^2}\right] X^2. \quad (10)$$

In equation (10), n is the order of the multiple; for the primary reflection, $n=0$. We define a stacking velocity V_{NMO_n} as

$$V_{NMO_n}/V = [1 - \cos^2\theta \sin^2(n+1)\phi]^{-1/2}. \quad (11)$$

$V_{NMO_n}/V = 1$ for a strike line and the maximum deviation from unity occurs for a dip line. For a dip line,

$$V_{NMO_{n_{dip}}}/V = \sec(n+1)\phi. \quad (12)$$

Plots of this function (Figure 5) show that the ratio of velocities increases rapidly with multiple number, so that even for very small dips, the multiple reflection stacking velocity is appreciably different from V. The effect is even more pronounced when the data of Figure 5 are replotted as Figure 6. (The curves in Figure 6 are drawn to assist the eye; only the values at integer n's have physical meaning.) Figure 7 differs from

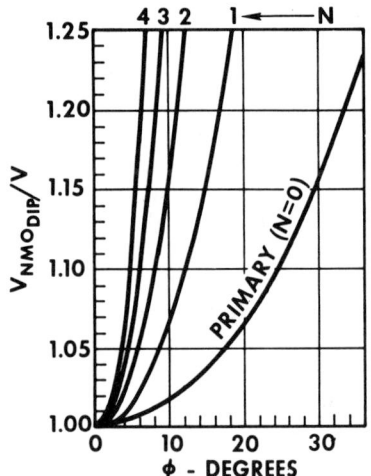

FIG. 5. Ratio $V_{\text{NMO}_{n\text{dip}}}/V$ as a function of dip angle for $n=0$ to $n=4$.

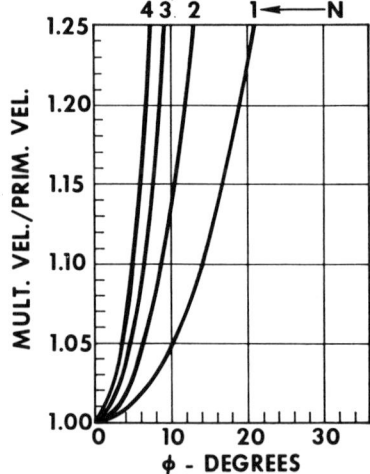

FIG. 7. Ratio $V_{\text{NMO}_{n\text{dip}}}/V_{\text{NMO}_{0\text{dip}}}$ as a function of dip angle for first four multiples.

Figure 5 in that the ratio $V_{\text{NMO}_{n\text{dip}}}/V_{\text{NMO}_{0\text{dip}}}$ is plotted against ϕ. It is this ratio, rather than the ratio of equation (12), that an interpreter is able to measure.

In contrast to their behavior for primary reflections, time intercepts for multiple reflections depend on the dip angle. The intercept

$$t_{0_n} = \frac{2D}{V}\frac{\sin(n+1)\phi}{\sin\phi} \quad (13)$$

from equation (10) is independent of the direction of the profile line relative to the direction of dip, since t_{0_n} corresponds to $X=0$. As the multiple number increases, the time separation of successive multiples decreases (Figure 8). The reader, distressed by a time intercept independent of dip for primary reflections but dip dependent for multiple reflections, should realize that the apparent anomaly results from our defining t_{0_n} in terms of D, a distance that, for the primary reflection, "pins" the reflector at a dip-independent distance from the spread midpoint.

Those who have had occasion to study light fringes in a wedge will recognize equation (13).

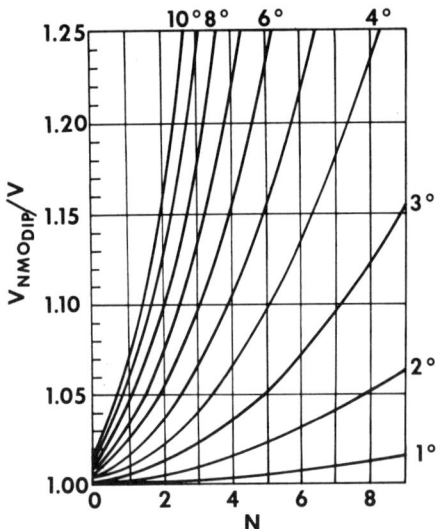

FIG. 6. Ratio $V_{\text{NMO}_{n\text{dip}}}/V$ for small dip angles, $n=0$ to $n=9$.

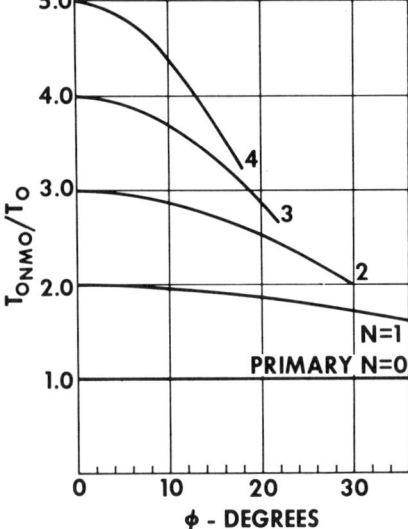

FIG. 8. Time intercepts as a function of dip for primary reflection and first four multiple reflections.

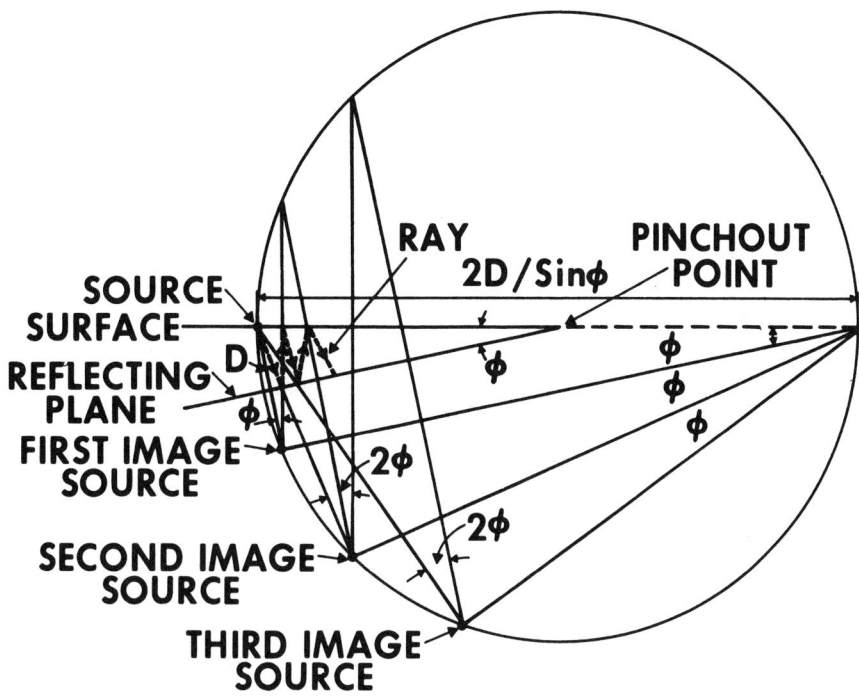

FIG. 9. The geometry involved in the concept of image sources or image geophones for multiple reflections.

Physicists (Born and Wolf, 1964) have long known that the image sources lie on a circle whose center is the edge of the wedge (the outcrop of the reflector) and whose radius passes through the point being examined (Figure 9). This construction and the fact that angles inscribed in a semicircle are right angles are sufficient to yield equation (13).

Corresponding to equation (7) for primary reflections, we can form for split spread data

$$\tau_n^2 = \frac{t_{n+}^2 + t_{n-}^2}{2} = \left(\frac{2d\,\sin(n+1)\phi}{V\sin\phi}\right)^2 + \frac{X^2}{V^2}. \quad (14)$$

The slope of the $X^2 - \tau_n^2$ curve gives V regardless of dip but the time intercept is identical to that found with CDP-ordered data.

All the reservations expressed in our discussion of primary reflections with regard to our oversimplified picture of the subsurface are equally valid for multiple reflections. There is one situation for which our model is a good approximation of reality. In marine exploration over a slope, the bottom in some cases is such a good reflector that the water layer is effectively decoupled from the underlying section; the water wedge can then be thought of as modelled by Figure 1. The velocity increase and time separation decrease with multiple number predicted by equations (12) and (13) are seen in the surface-to-bottom multiple reflection velocities and times.

REFERENCES

Born, M., and Wolf, E., 1964, Principles of optics: New York, Pergamon Press.
Brown, R. J. S., 1969, Normal-moveout and velocity relations for flat and dipping beds and for long offsets: Geophysics, v. 34, p. 180–195.
Cressman, K. S., 1968, How velocity layering and steep dip affect CDP: Geophysics, v. 33, p. 399–411.
Slotnick, M. M., 1959, Lessons in seismic computing: Tulsa, Society of Exploration Geophysicists.
Taner, M. T., and Koehler, F., 1969, Velocity spectra-digital computer derivation and applications of velocity functions: Geophysics, v. 34, p. 859–881.
Trorey, A. W., 1970, A simple theory for seismic diffractions: Geophysics, v. 35, p. 762–784.

APPENDIX

All of the relations in the body of the report can be derived by the use of geometry alone; however, since some of the readers may have the author's difficulty with seeing rays in three dimensions, I'll sketch the analytical geometry of the derivations.

We start with the plane

$$x \cos \alpha + y \cos \beta + z \cos \phi = d, \quad (1)$$

a source at $(0, 0, 0)$, and a geophone at $(X, 0, 0)$. We compute the position of an image geophone at the same distance below the plane along the perpendicular to the plane from $(X, 0, 0)$ as the geophone is above the plane. Since this type of computation reoccurs repeatedly, I'll give it in detail. We want the intersection point (X_1, Y_1, Z_1) which lies both on the plane and the normal to the plane:

$$(X_1 - X)/\cos \alpha = Y_1/\cos \beta = Z_1/\cos \phi$$

and

$$X_1 \cos \alpha + Y_1 \cos \beta + Z_1 \cos \phi = d.$$

Hence, $(X_1, Y_1, Z_1) = [(d - X\cos\alpha)\cos\alpha + X, (d - X\cos\alpha)\cos\beta, (d - X\cos\alpha)\cos\phi]$ and the image geophone is at

(X_2, Y_2, Z_2)
$= [2(d - X \cos \alpha) \cos \alpha + X, (d - X \cos \alpha) \cos \beta,$
$\qquad 2(d - X \cos \alpha) \cos \phi].$

The distance traveled by the reflection is the distance from the source at $(0, 0, 0)$ to the image geophone at (X_2, Y_2, Z_2) or

$$V^2 t^2 = X_2^2 + Y_2^2 + Z_2^2 = 4d^2 + X^2 - 4dX\cos\alpha. \quad (2)$$

$$D = d - X \cos \alpha / 2 \text{ or } d = D + X \cos \alpha / 2. \quad (1a)$$

Substituting equation (1a) into equation (2), we get

$$t^2 = 4D^2/V^2 + \frac{X^2 \sin^2 \alpha}{V^2} = \left(\frac{2D}{V}\right)^2 + \frac{X^2}{V^2_{\text{NMO}}}. \quad (4)$$

A little manipulation shows that $\cos \alpha = \sin \phi \cos \theta$. The expression for V_{NMO} becomes

$$V_{NMO} = V(1 - \sin^2 \phi \cos^2 \theta)^{-1/2}. \quad (5)$$

D is measured along a line perpendicular to the plane. To find the distance from center point of the spread, $(X/2, 0, 0)$, to the plane along a line perpendicular to the surface, we need only find the intersection of the line having the equation $x = X/2$, $y = 0$ with the plane. We call the intersection point $(X/2, 0, Z')$.

$$X/2 \cos \alpha + Z' \cos \phi = d$$

or

$$Z' = (d - X/2 \cos \alpha)/\cos \phi.$$

We let D' be the distance measured along the perpendicular to the surface.

$$D' = Z' = D/\cos \phi. \quad (2a)$$

If we define $t_0 = 2D/V$ and recall that $V_{\text{NMO}_{\text{dip}}} = V/\cos \phi$, we find that

$$t_0 = \frac{2D'}{V_{\text{NMO}_{\text{dip}}}}. \quad (3a)$$

To derive equation (8) for Δ, we must find the coordinates (X_0, Y_0, Z_0) of the point at which D intersects the plane and the coordinates (X_3, Y_3, Z_3) of the reflection point. (X_0, Y_0, Z_0) are found by the same procedure discussed in the second paragraph of the appendix.

$$(X_0, Y_0, Z_0) = (D \cos \alpha + X/2, D \cos \beta, D \cos \phi). \quad (4a)$$

(X_3, Y_3, Z_3) is the intersection point of the line connecting the source at $(0, 0, 0)$ and the image geophone at (X_2, Y_2, Z_2).

$$(X_3, Y_3, Z_3) = \left(D \cos \alpha + X/2 + \frac{X^2}{4D} \sin^2 \alpha \cos \alpha, D \cos \beta - \frac{X^2}{4D} \cos^2 \alpha \cos \beta, D \cos \phi - \frac{X^2}{4D} \cos^2 \alpha \cos \phi\right). \quad (5a)$$

$$\Delta = [(X_3 - X_0)^2 + (Y_3 - Y_0)^2 + (Z_3 - Z_0)^2]^{1/2}$$
$$= \frac{X^2}{4D} \sin \alpha \cos \alpha. \quad (8)$$

To compute the travel paths of multiple reflections, we extend the method used to find the travel path of the primary reflection. The image geophone for the first multiple is found by reflecting (X_2, Y_2, Z_2) in the surface and then reflecting this second image in the dipping plane. Higher-order multiples require us to repeat the process of reflecting first in the surface and then in the plane.

# A VELOCITY FUNCTION INCLUDING LITHOLOGIC VARIATION*

L. Y. FAUST†

ABSTRACT

Assuming velocity (V) a function of depth (Z), geologic time (T), and lithology (L) the resistivity log is an approach to the determination of L. Since general knowledge of water resistivity values (R_w) is lacking, the values of true resistivity (R_t) against $V/\alpha(ZT)^{1/6}$ were compared for 670,000 feet of section widely distributed geographically. Variations in R_w were presumably averaged out thereby, and the results indicate that statistically $L = [R_t]/T$ and $V = 1948\,(ZTL)^{1/6}$. This formula was applied to an additional 270,000 feet of section more localized geographically to observe its accuracy in predicting vertical travel time. If a correction map for R_w variations is applied the results are encouraging but less accurate than good velocity surveys.

Examination of an inconclusively small amount of data with more careful measurements of R_t suggests that accuracy comparable to direct measurement may be attainable. The cooperation of other investigators and of the electric-logging specialists is desired.

INTRODUCTION

This paper describes an investigation of longitudinal seismic velocity in sedimentary rocks as a function of depth, geologic time, and lithology. A tentative lithologic parameter is designed utilizing data available from conventional resistivity logs. The variables are analyzed in nearly one million feet of section represented by one hundred and fifty velocity surveys. The goal of the investigation is the prediction of velocity from well logs with accuracy equal to direct measurement. While this accuracy has not been obtained, the results are encouraging. The latter portion of the paper examines the nature of the errors of prediction of the derived velocity function and suggests that the desired end may be attainable. A cooperative effort is proposed for the extension of the investigation.

A LITHOLOGIC PARAMETER

A previous paper[1] assumed that velocity v could be expressed as:

$$V = f(Z, T, L) \tag{1}$$

---

* Presented before the Geophysical Society of Tulsa December 11, 1952 and before the Annual Meeting at Houston March 24, 1953. Manuscript received by the Editor January 8, 1953.
† Amerada Petroleum Corporation, Tulsa.

with Z the depth, T elapsed time since deposition, and L a lithologic variable. It was argued that by averaging many velocity measurements in essentially non-calcareous sections, the variable L was held constant and

$$V = \alpha(ZT)^{1/6}, \qquad L = L_1 \tag{2}$$

with L_1 representing "an average shale and sand section." The value of α was determined as 125.3 when Z was measured in feet, T in years, and V in feet per second.

In the previous paper evidence was presented that suggests a velocity formula including lithologic variation L would contain equation (2) as a member of a more generalized equation and that the variable L would appear either additively or as a coefficient of α. Since no generally accepted quantitative measure of lithologic variation L is available, a necessary first step is the formulation as a lithologic parameter of a quantity related to those lithologic variables which influence velocity. This quantity is found to be related to electric resistivity measurements. The following discussion will define first the factor in a velocity equation attributable to lithology. Next will be a description of the quantities influencing both this lithologic factor and resistivity. Then the steps necessary for the discovery of the desired lithologic parameter will be described in sequence.

The Lithologic Factor in a Velocity Equation

A Lithologic Factor K can be defined as:

$$K \equiv \frac{\Delta Z}{\Delta t} / \alpha(ZT)^{1/6} \tag{3}$$

where $\Delta Z/\Delta t$ is the actual measured velocity in the interval ΔZ and the denominator is the appropriate value for velocity from equation (2). Then $K=1$ for the lithologic conditions specified in equation (2) and in general K is that factor in velocity attributable to lithologic variation. It is therefore a quantity against which any tentative lithologic parameter can be tested. Previous to the definition of equation (3) an attempt was made to describe the effect of lithology on velocity as a term to be added to the right hand member of equation (2). The measurements to be described later were tested against that assumption with no correlative results.

The Resistivity Formation Factor

Archie has shown empirically that for permeable sections[2] the relation between true formation resistivity R_t and the resistivity R_w of the water impregnating the formation is given by

$$R_t/R_w = F = p^{-m} \tag{4}$$

where F is the "resistivity formation factor," p is proportional to porosity, and m is termed a "cementation factor." The number F derived from resistivity measurements is unaffected by the mineralogical constituents of the formation matrix whose water-free resistivity approaches infinity but is a function of the porosity and cementation of the formation.

That the same quantities affect the factor K in velocity is well known. Some evidence on these facts can be found in the writer's previously cited paper and Vogel[3] mentions the inverse relationship of porosity and velocity when other variables are held constant.

There is reason therefore to anticipate that the desired lithologic parameter should be related to F, the resistivity formation factor of equation (4). Unfortunately while the values of R_t can be determined with some accuracy from Resistivity Departure Curves, the corresponding values of R_w are generally unknown. Wyllie[4] has indicated a solution for R_w for permeable formations, but this method does not apply to all sections.

In the absence of specific knowledge of R_w, if a sufficiently large number of measurements of R_t is averaged over a wide areal distribution, the corresponding averaged value of R_w should either remain constant under differing values of other variables or should vary in functional relationship to such variables. As a first step in this investigation it will be assumed that under such averaging the value of R_w remains constant. For ease of measurement it will be assumed that this constant value of R_w is unity. A quantity can then be defined

$$[R_t] \equiv \frac{1}{N} \sum_{}^{N} R_t \bigg/ \left(\frac{1}{N} \sum_{}^{N} R_w = 1\right) \qquad (5)$$

where N is the number of measurements and $[R_t]$, the magnitude of the average value of R_t, is dimensionless. Actually $[R_t]$ will be in a relationship to the average corresponding F value determined inversely by the relationship of $1/N \sum^N R_w$ to the variables to be studied, namely velocity, depth, geologic time, and lithology. Thus $[R_t]$ as defined in equation (5) is assumed to measure F or a quantity analogous to F.

Measurements

Measurements of R_t are derived from the value of R_a, the apparent resistivity, using bit size for the hole diameter correction and the appropriate value of mud resistivity, R_m, corrected for temperature. As a practical expedient some simplifications seem desirable. The values of R_a are taken, usually measured by the lateral device, over intervals (ΔZ) of approximately one thousand feet. The average value of R_a is taken from a smooth curve so drawn that the areas of positive difference between this curve and the measured trace are equal to the areas of negative difference. The extent of the segment of the curve is limited essentially by a maximum range of five to one in R_a variations. When larger variations occur the values of R_a and corresponding R_t are read over intervals less than one thousand feet and an average value of R_t determined for the desired interval. The departure curves[5] used are those for beds of infinite thickness and no mud filtrate invasion.

While the inaccuracies introduced by these simplifications could not be tolerated by the petroleum engineer in the evaluation of thin oil bearing formations, simplified techniques still require a considerable amount of time and appear justifiable as a first approach to the problem.

Measurements of 670,000 feet of section corresponding to 96 velocity surveys have been taken from all available regions exclusive of Alberta and California. As wide lithologic variations as possible are included. The sections measured are, with minor exceptions, different from those reported previously. These data are then tabulated for value of geologic age T, depth interval ΔZ (usually one thousand feet), mean depth of burial Z, measured velocity $\Delta Z/\Delta t$, and the true formation resistivity R_t. Other quantities such as the appropriate velocity values from equation (2) and the value of K are derived and tabulated.

The Determination of a Lithologic Parameter

The obvious approach to the formulation of a tentative lithologic parameter involving $[R_t]$ is the study of the variation of $[R_t]$ when $K=1$. All data can be selected from the tabulated measurements where $K=1.00\pm0.10$. This specification is well within the limits of the data from which equation (2) was derived. The data corresponding to this restriction of K are then grouped by geologic time T. $[R_t]$ is determined from the data of each group. These results are plotted on logarithmic coordinates in Figure 1 and indicate that $[R_t]$ is nearly proportional to T. An attempt to split these data to investigate $[R_t]$ against Z for each value of T showed even poorer relationships but probably indicates that $[R_t]$ does not vary appreciably with depth. Although the relationship between $[R_t]$, Z, and T was not clearly determined, the assumption will be made and later verified that $[R_t]/T$ is independent of Z and T.

The quantity R_t/T, tabulated for all measurements, can be classified by value of T and subclassified for each T by the value of K in ranges of 0.10 from $K<0.90$ to $K>1.60$. The average K, the corresponding average $[R_t]/T$, and N, the number of measurements involved is summarized for each subclassification in Table I. Values involving less than ten measurements are considered unreliable and are not included. The data of Table I are plotted on logarithmic coordinates in Figure 2 and indicate a close approach to a linear relationship between the logarithms of K and of $[R_t]/T$. The values of $[R_t/]T$ show no tendency to deviate as a function of T. Figure 2 indicates that the required lithologic parameter L is defined by

$$L \equiv [R_t]/T. \tag{6}$$

A VELOCITY FUNCTION INCLUDING LITHOLOGY

The lithologic parameter L being determined, a velocity function including lithology can be formulated by comparison of L with K the lithologic factor. In

TABLE I. TABULATION OF DATA OF FIGURE 2

(Values of K, $[R_l]/T$ (reciprocal years) and number of measurements of thousand foot sections N arranged by values of Geologic Time T (years) and Lithologic Factor K.)

$10^{-6}T$		$< K < 0.90$	$0.90 \leq K < 1.00$	$1.00 \leq K < 1.10$	$1.10 \leq K < 1.20$	$1.20 \leq K < 1.30$	$1.30 \leq K < 1.40$	$1.40 \leq K < 1.50$	$1.50 \leq K < 1.60$	$K \geq 1.60$
26	K $10^6[R_l]/T$ N		0.96 0.056 15	1.02 0.083 12						
43	K $10^6[R_l]/T$ N	0.87 0.043 12	0.94 0.048 16							
70	K $10^6[R_l]/T$ N		0.95 0.046 21	1.04 0.070 27	1.14 0.18 24	1.23 0.18 12	1.37 0.32 14			
192	K $10^6[R_l]/T$ N		0.97 0.089 13	1.04 0.113 19	1.15 0.19 19	1.24 0.19 23	1.35 0.41 30	1.44 0.55 30	1.55 0.78 26	1.71 1.32 25
220	K $10^6[R_l]/R$ N		0.96 0.064 44	1.04 0.072 53	1.14 0.240 22	1.25 0.560 15	1.35 0.490 15			
245	K $10^6[R_l]/T$ N					1.28 0.375 11				

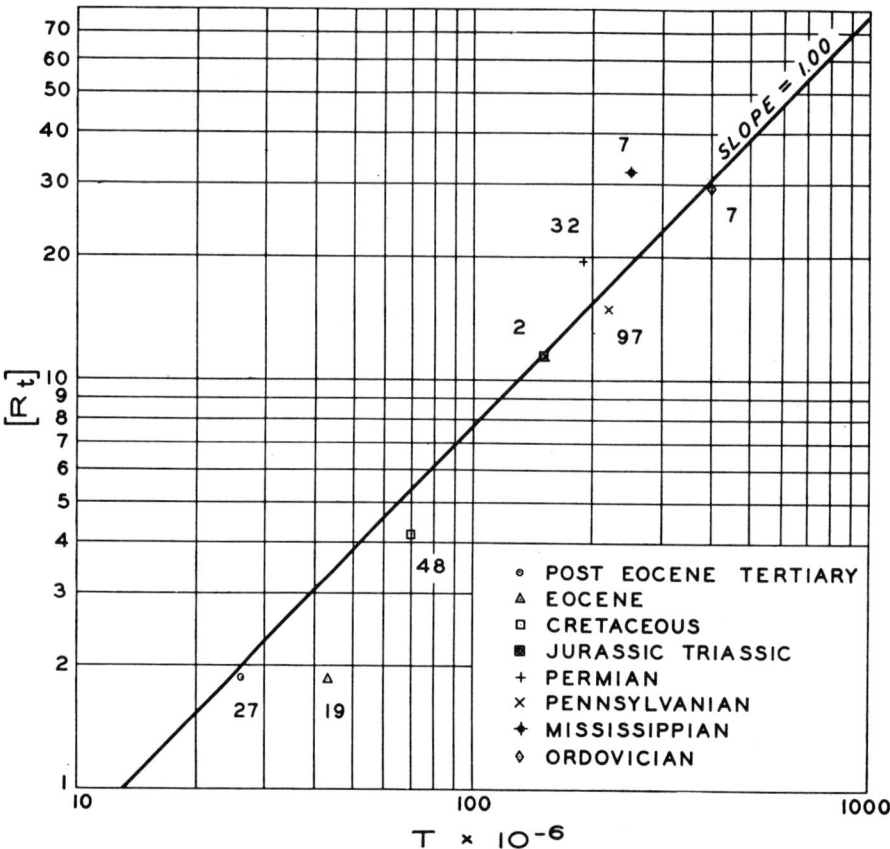

Fig. 1. $[R_t]$ as a function of geologic time T for an "average shale and sand section" ($K = 1.00 \pm 0.10$). The numbers above the plotted points show the number of thousand foot sections used in the determination. $[R_t]/T$ is nearly constant.

the following sub-sections will be shown first, the velocity formula and second, the predictability of this formula in individual wells. Due to the lack of knowledge of the individual values of R_w some error is to be expected. In the third part, an attempt will be made by analysis of additional data to show that the prediction errors are ascribable to variations of R_w and to suggest one method of correction.

The Velocity Formula

As a result of equation (6) all individual measurements of K can be grouped according to range of L regardless of age T. As a consequence one hundred and twenty measurements excluded from Table I could be included in these classifications. The average of L in each of five ranges together with the corresponding averages of K are plotted on logarithmic coordinates in Figure 3. These values of $L \times 10^6$ and K are respectively: (0.048, 0.92), (0.075, 1.03), (0.183, 1.15), (0.487, 1.35), (1.03, 1.61). The number of measurements in each range is shown above each plotted point. A relationship of the form

$$K = CL^{1/n} \quad (7)$$

describes these points when $n=6$. From equations (3) and (7) replacing $\Delta Z/\Delta t$ by V

$$V = \gamma(ZTL)^{1/6} \quad (8)$$

where

$$\gamma = 1948.$$

The type of averaging and classifying necessary to arrive at the results of Figure 3 should eliminate variations of R_w whether functional or random. This assumption is partly verified by the fact that Figure 2 confirms equation (7) although the method of grouping individual measurements was different. From equation (6) evidently

$$V = \gamma(Z'[R_t])^{1/6}. \quad (9)$$

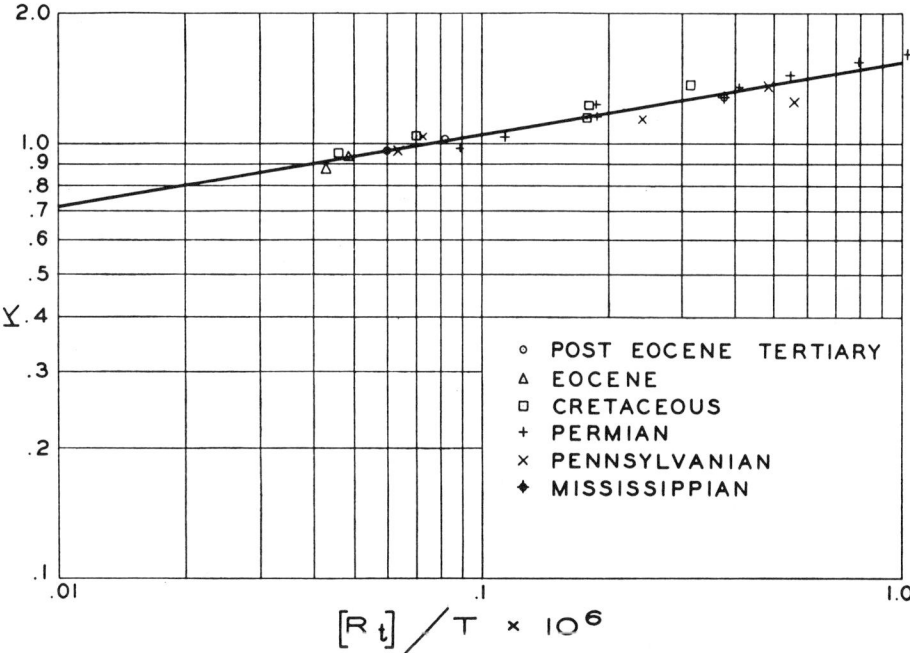

FIG. 2. The data of Table I plotted on logarithmic scale suggests that the lithologic factor K is proportional to $([R_t]/T)^{1/n}$ and that $[R_t]/T$ is independent of T.

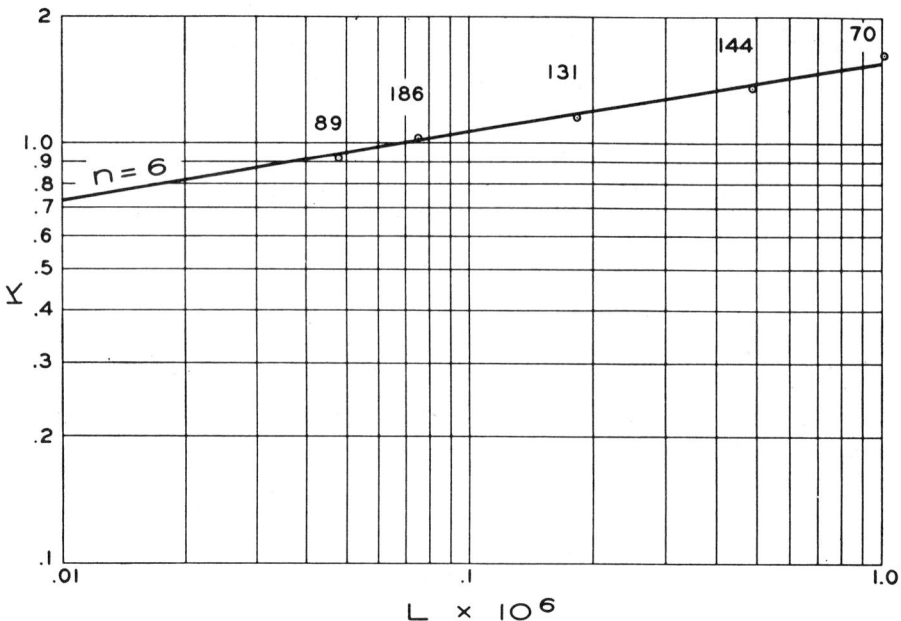

Fig. 3. Average values of the lithologic factor K versus the average value of the lithologic parameter L grouped by value of L show that $K = CL^{1/n}$ where $n = 6$. Numbers above the plotted points show the number of thousand foot sections averaged.

Equation (9) describes statistically a velocity function including lithology. This equation represents the limit of the investigation until R_w values can be determined. Given values of R_w, L in equation (8) can be redefined to make that equation one form of a specific velocity law. Meanwhile the general utility of equations (8) or (9) can be investigated by determining their predictability for individual velocity surveys.

A possible implication of equation (9) is that the effect of geologic time T is lithologic. Heiland[6] in discussing the effect of age on velocity stated "An increased age merely increases the probability that it has undergone a greater degree of dynamometamorphism."

Possibly more surprising than the relationship of T and L is the apparent independence of $[R_t]$ and Z. This is checked in Table II where the average ratio of $\gamma(ZTL)^{1/6}$ to $\Delta Z/\Delta t$ is shown for the indicated number of measurements according to range of depth Z. There is some tendency for the ratio to increase with depth. In view of the present inadequate knowledge of R_w variations, this could be explained as a variation of R_w with depth or as evidence of an incompletely averaged lateral variation of R_w.

In defense of this latter supposition it should be pointed out that when equation (8) is tested in individual wells in the following discussion a correction

TABLE II. RATIO OF PREDICTED TO MEASURED VELOCITIES AGAINST DEPTH

(Average of indicated ratio for N observations classified by depth Z (feet). The slight increase with depth is not substantiated by other evidence.)

$Z =$	500	1,500	2,500	3,500	4,500	5,500	6,500	7,500	8,500	9,500	10,500	11,500
$\gamma(ZTL)^{1/6}/\dfrac{\Delta z}{\Delta t}$	0.947	0.970	0.928	0.929	0.933	0.983	1.009	1.018	1.017	1.047	1.099	1.073
N	16	38	54	60	61	55	46	38	29	21	12	8

in R_t for depth Z increases the errors of prediction. $[R_t]$ is considered therefore to be independent of Z.

Possibly the work of Gassmann[7] may be helpful in explaining the independence of the Z term.

Predictability

The data illustrated in Figures 2 and 3 are the average of measurements having a wide range of values. While the deviations could have been included in Tables I and II, the predictability of equation (8) should be a more useful measure of the errors. The evidence indicates that equation (8) represents a velocity law, but other variables not considered or a wide range of R_w values might produce such scattering of predicted versus measured velocities as to make equation (8) useless. When equation (8) or (9) is used to predict velocity in individual wells, the assumption of equation (5) of the constancy of R_w is no longer valid.

By a summation for each well of the time increments predicted by use of equation (8), the equality may be written:

$$\sum_{Z_1}^{Z_2} \frac{\Delta Z}{\gamma(ZTL)^{1/6}} = (t_2 - t_1) + \epsilon \tag{10}$$

where Z_1 and Z_2 are respectively the shallow and deep limits of available measurements in a well, t_1 and t_2 the corresponding measured vertical times, and ϵ the prediction error.

A group of fifty velocity surveys from the ninety-six used in the determination of equation (8) are analyzed in this manner with the results plotted on linear coordinates in Figure 4. The ordinate of each point is the value of the left hand member of equation (10) and the abscissa is the corresponding (t_2-t_1) value. The deviation from the Line of Correlation is the error ϵ. While the scattering is not too bad, one measured time (t_2-t_1) at 0.588 seconds has an ϵ of -0.090 second. The greatest positive value of ϵ is 0.054 second and the mean value of $|\epsilon|$ is 0.026 second.

If ϵ is a function of ΔR_w, a total variation in this function of five to one is sufficient to reduce ϵ to zero for the extreme deviations. Available measurements[8] of R_w show variations of greater than one hundred to one. Since these latter are observed in permeable sections, the water resistivities measured are not necessarily of water laid down with the formation. On occasion the water present in a permeable formation may be meteroic. Nevertheless the range of five to one necessary to explain these ϵ's is conceivable.

If the prediction errors may be explained as the effect of variations in R_w, then

$$V = \gamma'(Z\Phi)^{1/6} \tag{11}$$

may be the final form of a law governing velocity where Φ is proportional to R_t and is a function of R_w. Φ is analogous to or equal to F.

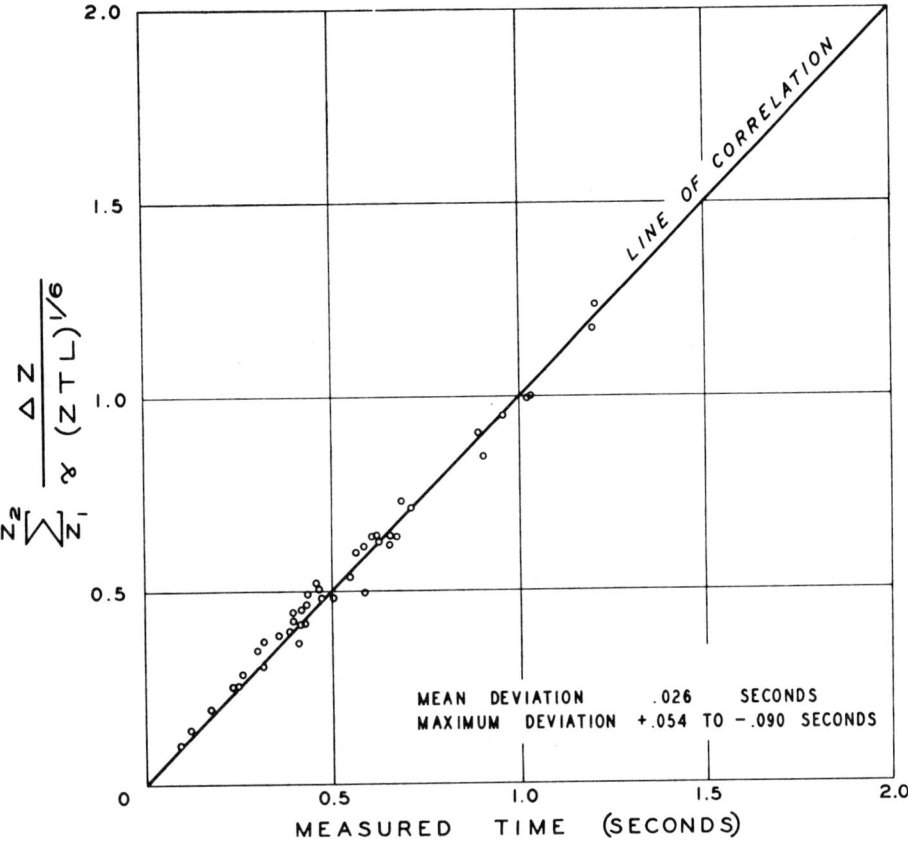

FIG. 4. Scattering of predicted one way travel times versus measured travel times for fifty wells taken over a wide geographical distribution. The deviations could be explained by a 5 to 1 range in the effect of R_w.

The Variation of Water Resistivity

The validity of the assumption upon which equation (11) is suggested may be verified in part by other indirect evidence. The values of ϵ shown in Figure (4) are from data taken over a wide geographical distribution with consequent variations in depositional environment. If sufficient data were available in a more localized area, the range of R_w variation should be more restricted. Such a condition exists in Alberta where forty-six velocity surveys not used in the development of equation (8) represent 270,000 feet of additional section. All of these surveys include measurements of some Paleozoic section as well as younger formations. If the range of ϵ's for this group of surveys were considerably less than those of Figure 4, the argument that $\epsilon = f(\Delta R_w)$ would be sustained in part.

The short time span over which the Alberta wells were drilled insures that

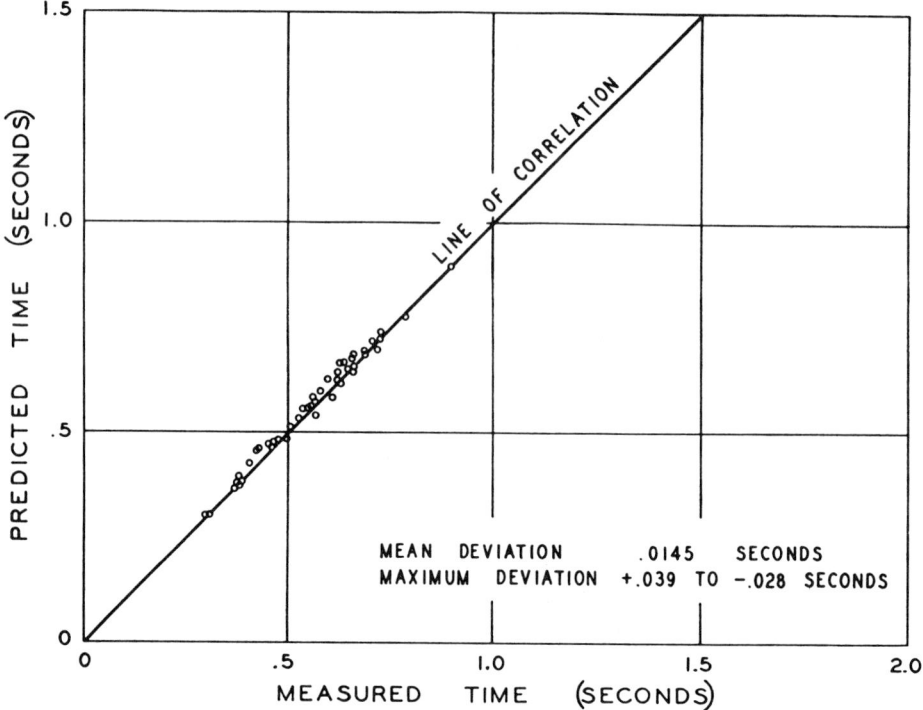

Fig. 5. A comparison similar to Figure 4 for forty-six Alberta wells not used in the derivation of equation (8). Variations of R_w should be less in this more restricted area and the deviations are smaller.

all the electric logs used in this study represent a single phase of development. The Alberta surveys show a deviation from equation (2), ($K=1$) from $+0.075$ seconds to -0.186 second with twelve of the surveys having deviations of a few milliseconds from that formula. The area is therefore useful also in evaluating the properties of L.

The results of the application of equation (9) to the forty-six Alberta surveys are plotted on linear coordinates in Figure 5 and show ϵ values ranging from $+0.039$ to -0.028 second with a mean $|\epsilon|$ of 0.0145 second. While tending to confirm the assumption that $\epsilon = f(\Delta R_w)$, the range of the ϵ's is unsatisfactorily large since it is assumed that good direct measurements have an accuracy of five milliseconds.

The distribution of the forty-six surveys in Alberta is shown in Figure 6. Although restricted to one general basin the dimensions of the area are approximately seven hundred by two hundred miles. Some variation of R_w may be expected over that extent. A rough measure of the assumed effect of R_w variation can be determined by indicating $\Delta\epsilon/\Delta Z$ for each well. The signed numbers in Figure 6 show this measure in milliseconds per thousand feet. Generalized contours of these quantities are shown. Although possibly fortuitous, the shape of the contours is similar to that of the epicontinental seas receding in this area through Mesozoic time. The easternmost well shown in Figure 6 and a well (not shown) off the south edge of the map were used for contour control only and are not included in the analysis. The data of Figure 5 are corrected by interpolation of the contours of Figure 6 except that all values within the $+5$ contour are taken as $+5$. The results are shown in Figure 7. The mean $|\epsilon|$ is 0.0079 second with a range of ϵ's from $+0.020$ to -0.018 second.

The wells indicated as A through F in Figure 6 are shown in cross section in Figure 8 where the uncorrected ϵ's are indicated by circles connected by dashed lines and the ϵ's corrected by Figure 6 are indicated by asterisks connected by solid lines. Three of the wells, namely A, C, and F, have corrected ϵ's of five milliseconds or less while the other three wells have the maximum variations of Figure 7. Evidently no generalized recontouring of Figure 6 would improve materially these errors, and therefore the errors shown in Figure 7 represent the limitation of the correction for R_w.

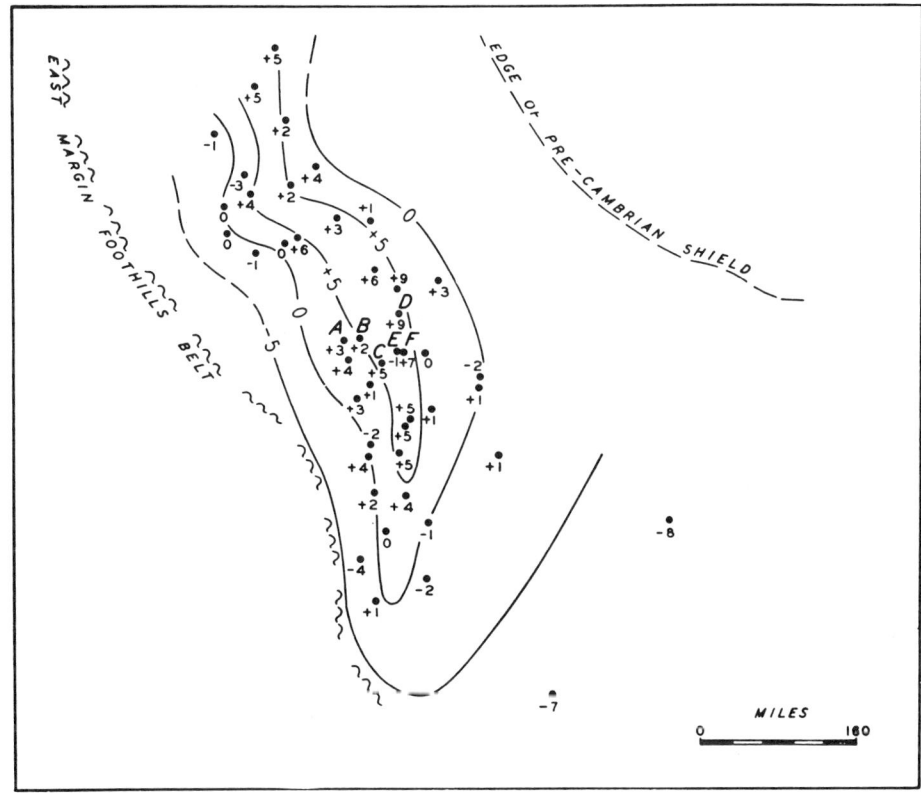

Fig. 6. Generalized contours of the deviations per thousand feet of the wells of Figure 5.

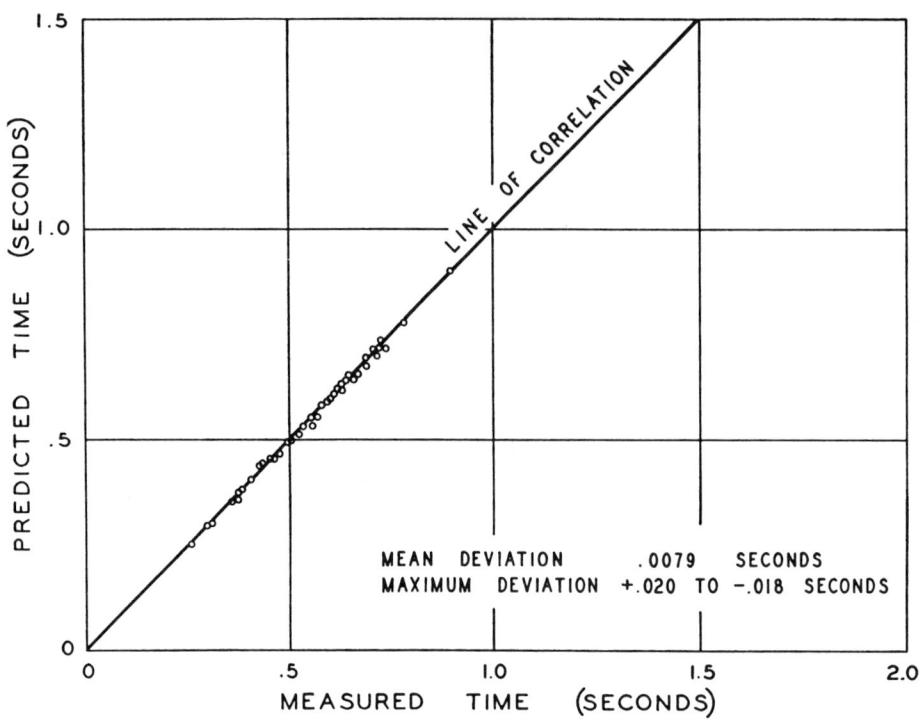

Fig. 7. A comparison for the same wells as shown in Figure 5 corrected by the contoured values of Figure 6.

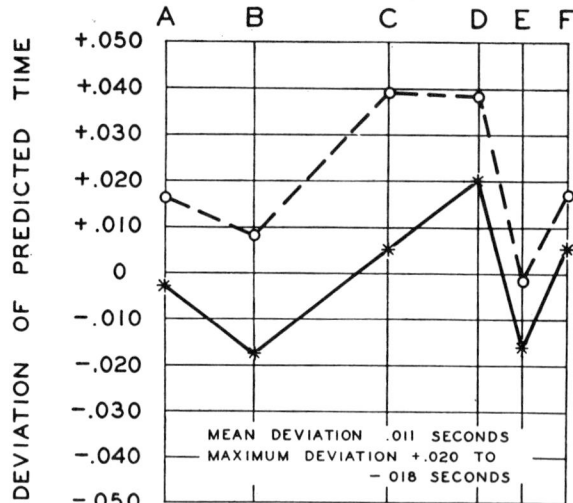

Fig. 8. A cross section of the wells indicated A-F of Figure 6 showing the uncorrected deviations (dashed lines) and the corrected deviations (solid lines). The variations between adjacent wells show that the limit of correction for R_w has been reached.

Since the method of determining R_t as discussed previously involved the introduction of some error, a more careful measurement should reduce the errors if equation (8) were exact when corrected for R_w but would have at least equal probability of resulting in larger ϵ's if the law were approximate only. The data of the six surveys of Figure 8 are re-evaluated by consideration of bed thickness, mud filtrate invasion, and the use of one hundred foot or smaller intervals of measurement. The revised data are shown in Figure 9. Although there is no improvement in the uncorrected values, the corrected ϵ's vary from $+0.008$ to -0.007 second with a mean $|\epsilon|$ of 0.005 second. Two other wells originally

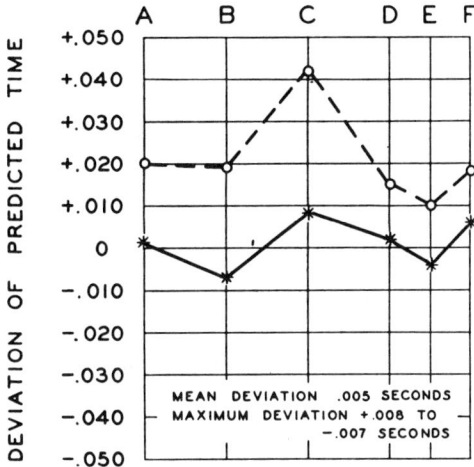

FIG. 9. The same wells as in Figure 8 showing the deviations when R_t is determined with greater accuracy. These corrected deviations are comparable in accuracy to direct measurement.

showing corrected ϵ values of -0.016 and -0.017 respectively have ϵ's of plus four and plus one milliseconds by the more accurate measurements.

It would be absurd to imply that the small errors in these eight wells are a real measure of the achievements of this study since the accumulation of more data will almost certainly disclose some large departures.

It is recognized that the arguments relating the prediction errors to variation of R_w do not prove conclusively that this is the correct explanation. Other possibilities have not been excluded. It is demonstrated however that the prediction errors are essentially systematic rather than random.

This paper has described the results of all but six surveys which have been used in testing equation (8). These six surveys represent a more recent effort by the writer to test the equation against the most anomalous velocity conditions. All six have checked satisfactorily. The most anomalous of this group is a deep survey in the Gulf Coast called to the writer's attention by O. B. Manes. This survey has a measured vertical time in excess of the predicted time from equation

(2) by nearly three-tenths of a second. Equation (8) predicts the measured time to twenty-one milliseconds (uncorrected).

Evaluation of Results

It is believed that equation (8) or its equivalent, equation (9), may be regarded as a statistical law governing velocity. While the improvement in prediction accuracy with each refinement is evidence favoring the conclusion that acceptable accuracy may be attainable, the results of the study of the Alberta velocities should not be generalized without further confirmation.

Since the measured vertical time over the greatest possible depth range represents usually the most accurate measurement of a conventional velocity survey, the predictions of equation (8) have been referred only to such measurements. It is arguable that in some shorter interval the predicted velocities could be the more accurate. K. S. Cressman under the general supervision of the writer has been successful apparently in using equation (8) in short intervals in a study of the composite nature of reflections. However resistivity measurements across small intervals of varying resistivity are inexact. While Summers and Broding reported a generally good correspondence[9] between five foot interval velocity logs and resistivity, Broding has stated in his discussion of the first presentation of this paper[10] that in some wells a total lack of correspondence has been observed over certain intervals.

The problem of the formulation of Φ in equation (11) requires consideration not only of the errors of the velocity formula but also of those errors involved in direct velocity measurement and especially of the errors in resistivity measurement.

The writer is not qualified to discuss the significance of this study from the viewpoint of the resistivity specialist. The work of Owen in emphasizing the importance of cementation[11] on resistivity may have an important bearing on this problem.

A CO-OPERATIVE INVESTIGATION

Two approaches to the practical use of these relationships seem possible. The preparation of correction maps with present techniques could be instituted immediately. The improvements in resistivity measurements and techniques for determining R_w variations await future development.

The Use of the Present Techniques

Much more investigation will be required before the general accuracy of these empirically derived formulae is ascertained. Independent work in this field is desirable. One suggested project is an attempt to duplicate Stulken's[12] velocity maps in the Bakersfield region by means of electric logs. It should he mentioned however that the re-evaluation of the data shown in Figure 9 required the equivalent of one work week. It may be preferable therefore to institute further investigation on a cooperative basis. The confidential nature of most velocity information might seem to bar such a venture. The writer has available a library of more than one thousand velocity surveys and would be willing to serve as a clearing house in the dissemination of non-confidential material. Correspondence could be confined to surveys available both to the writer and to the particular investigator. Maps of the areas of interest could be prepared similar to that shown in Figure 6. It is evident that such a correction map should be made within each general unconformity. These correction maps, being non-confidential, could be distributed to all collaborators. A first requirement would be an argeement on a standard method of measuring R_t determined preferably with the advice of specialists in electric logging.

Improvement of the Method

Improvements in the method depend primarily on the electric logging specialists and service companies. It would be helpful to show the zero deflection at both the start and end of a run. The deflection corresponding to one standard resistivity would be desirable. The example in Figure 4 of $\epsilon = -0.090$ is suspected of being an error in scale since the readings were abnormally high for all depths. Lacking good evidence on this point however the value was included. The clarification of the multiple scale traces would be useful. In Gulf Coast areas of low resistivity, a complete set of amplified curves would improve accuracy.

The newer logging devices may be expected to improve accuracy, especially in West Texas. However a measurement of R_w for the complete section is of paramount importance. Offers of aid in such a project have been received. Dr. Leendert de Witte has suggested[13] an extension of Wyllie's method for correcting for R_w to a first approximation. It is hoped that it will now be possible for the writer to secure the collaboration of an electric log specialist in comparing the results of this suggested correction with the data of Figure 6.

ACKNOWLEDGMENTS

Appreciation is expressed to Dr. B. B. Weatherby for permission to publish this paper. The writer is indebted to E. E. Finklea for information concerning resistivity measurements, to F. F. Campbell for his valuable aid in the writing of this paper and especially to K. M. Lawrence and Paul Lyons for their encouragement. The figures were prepared by the Amerada Drafting Department.

REFERENCES

(1) L. Y. Faust, "*Seismic Velocity as a Function of Depth and Geologic Time,*" Geophysics, XVI, 2 (1951), 192–206.
(2) G. E. Archie, "*The Electrical Resistivity Log as an Aid in Determining Some Reservoir Characteristics,*" Petroleum Development and Technology, 146 (1942), 54–62.
(3) C. B. Vogel, "*A Seismic Velocity Logging Method*" Geophysics, XVII, 3 (1952), 586–597.
(4) M. R. J. Wyllie, "*A Quantitative Analysis of the Electrochemical Component of the S P Curve,*" Petroleum Technology, 1, 1 (1949).

(5) Schlumberger Document Number 3.
(6) C. A. Heiland, "*Geophysical Exploration,*" Prentice-Hall (1940), 475.
(7) Fritz Gassman, "*Elastic Waves Through a Packing of Spheres,*" Geophysics, XVI, 4 (1951), 673–685.
(8) Water Resistivity Cards, A.I.M.E. Special Publication (1950).
(9) G. C. Summers and R. A. Broding, "*Continuous Velocity Logging,*" Geophysics, XVII, 3 (1952), 598–614.
(10) Presented before The Geophysical Society of Tulsa, December 11, 1952.
(11) J. E. Owen, "*The Resistivity of a Fuid-Filled Porous Body,*" Petroleum Transactions, A.I.M.E. 195 (1952), 169–174.
(12) E. J. Stulken "*Seismic Velocities in The Southeastern San Joaquin Valley of California,*" Geophysics, VI, 4 (1941), 327–355.
(13) Personal communication, December 29, 1952.

Seismic data indicate depth, magnitude of abnormal pressures

Gulf Coast studies suggest that overpressured formations can be anticipated in any basin area in which compaction has occurred

E. S. Pennebaker, Jr., Division Staff Engineer
Humble Oil & Refining Co., Corpus Christi, Texas

15-second summary

Well log interpretation improvements and drill cuttings analysis permit measurement of formation pressures after hole is drilled. Such information derived prior to drilling was long restricted to that estimated from nearby wells. This article shows how routine seismic field records are used to predict both the depths to abnormal-pressure formations and approximate pressure magnitudes.

ANALYSES of numerous well velocity surveys revealed a close correlation between interval velocity (interval travel time) and pressure—and that interval travel time varies exponentially and predictably with depth.

Departure from this normal trend signifies abnormal pressure or gross lithologic changes. A predictive method results from the fact that interval velocities can be calculated from seismic data using well-known geophysical techniques. Results compare favorably to those obtained by well velocity surveys.

This method also permits determination of pressures with enough accuracy so that the proper casing can be available at the drillsite, optimum drilling mud-weighting schedule and hole-size selection can be made and the proper rig can be chosen. The method has been used with good results over a wide area along the Texas coast, including continental shelf locations.

Methods of measuring pore pressures from well logs have greatly improved the accuracy of downhole pressure predictions in areas where there are sufficient wells for adequate control. In areas of poor geological control, deep wildcatting can be extremely hazardous and costly for lack of adequate pressure information.

Considerable discussion of abnormal-pressure technology has appeared in recent literature. Quantitative relations between various log parameters measured in drilling wells and pore pressures have been developed.[1,2] All are based on the relationship between shale compaction and abnormal pressure.

Seismic velocity—fluid pressure relations. Role of velocity in seismic work. The reflection seismograph measures time between the earth's surface and various subsurface reflecting horizons. If average velocity of seismic energy through the sedimentary column to a reflecting horizon is known, depth to the reflector can be determined. Thus, if enough velocity information is available, knowledge of subsurface structure may be obtained.

Techniques for determining velocities are well known in geophysics. The method most commonly used is the well velocity survey. A series of shots is detonated near the well, and travel times are recorded for energy to travel from the surface to a geophone placed at successive depths (usually 500 to 1,000-foot increments). Differences in arrival times at the various geophone

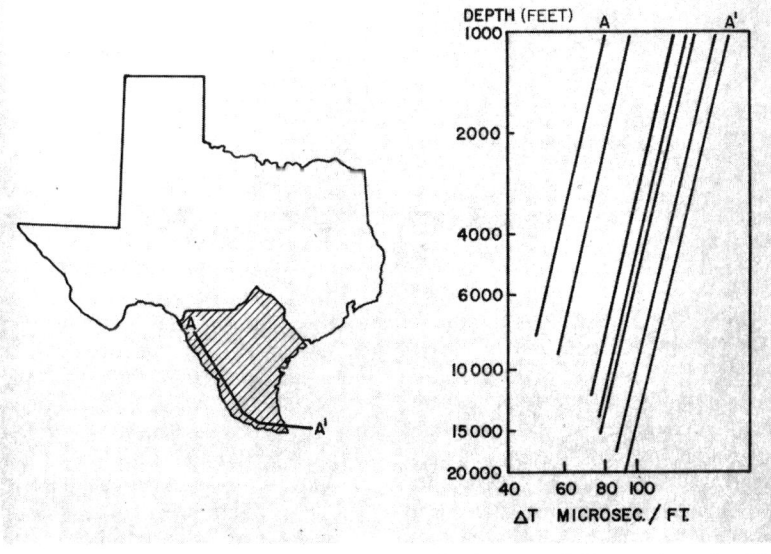

FIG. 1—Trend of interval travel time versus depth in south Texas along the Rio Grande River from Maverick County to Offshore from Cameron County.

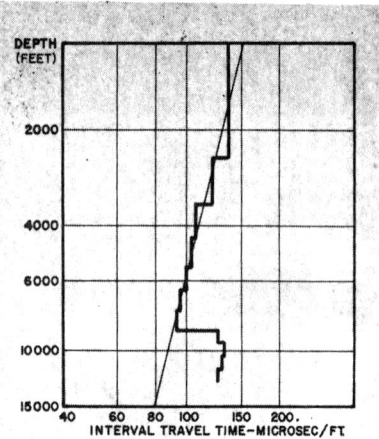

FIG. 2—Logarithmic plot of interval travel time, Well A, Kleberg County, Texas, showing abnormal pressures below 8,500 feet.

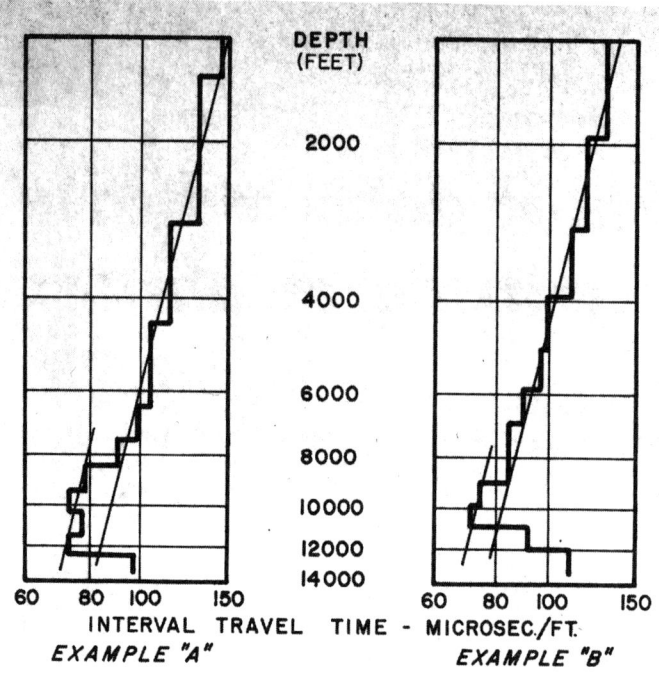

FIG. 4—Examples showing shifts in normal base line due to calcareous sands and shales.

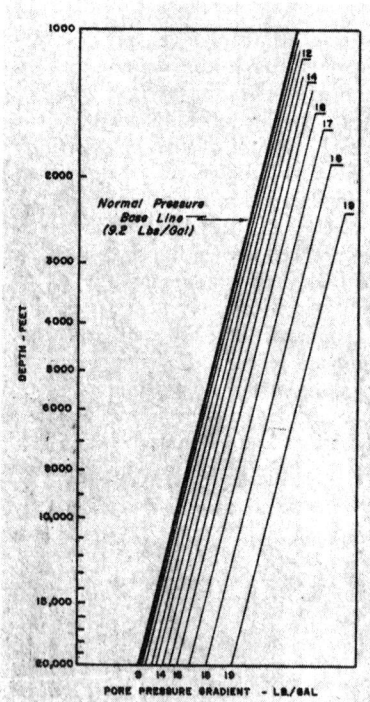

FIG. 3—Abnormal pressure as related to degree of departure from normal pressure base line.

locations can be used to develop an average interval travel time or velocity profile.

A second technique is a more indirect method which uses data from seismic field records to compute interval travel time profiles. This method is the basis for this predictive technique. Noteworthy improvements in accuracy of such measurements and resulting calculations have been attained with the advent of new field procedures, digital recording and machine data analysis.

Though geophysicists generally think in terms of velocities, this article will refer mostly to travel time (reciprocal of velocity), since that is the term used in sonic logging. The interval travel time profile is in effect a sonic log averaged over fairly long (500 to 1,000-foot) vertical intervals.

The effect of depth on velocity has long been recognized by geophysicists,[3,4,5,6,7] who have generally agreed that interval velocity increases exponentially with depth. A study of 350 wells in the Gulf Coast area in which well velocity surveys were run confirmed this relation. So in terms of interval travel time, we can expect a decrease with depth in a straight-line relation when plotted on log-log paper. Slope is ¼. All wells exhibit this same slope as long as only normal-pressure sand-shale sequences are encountered.

It was found that the wells with greater interval travel times were associated with wells located toward the basin's center. Those with lesser interval travel times were associated with wells toward the edge.

Fig. 1 shows the trend in the interval velocity vs. depth relation found in normal-pressure formations in South Texas from Maverick County to about 55 miles offshore Cameron County, along line A-A'. Interval travel times offshore from Cameron County are similar to those encountered in certain areas offshore from Louisiana. Slopes of the lines in Fig. 1 remain constant, but the position shifts to the right as the wells approach the Gulf. Intuitively we expect this shift from inland, with older sediments, to more recently deposited, less compacted sediments closer to the Gulf.

Effect of abnormal pressure. Geophysicists have long been aware of anomalies in the interval velocity-depth trend in certain areas. For instance in the well of Fig. 2, interpreters would have noted that the interval travel times decrease in a normal fashion to about 8,500 feet, then begin to increase. The beginning of these "velocity inversions" were observed to coincide with the tops of certain formations, such as marine Frio and Vicksburg and massive shale sections.

These inversions define zones of abnormal pressure. Velocities are abnormally low because the formations are uncompacted.

The degree of departure from the normal interval travel time-depth line is directly proportional to the abnormal pressure. Fig. 3 shows this relationship, which was developed from abnormal-pressure information obtained from bottom-hole pressure surveys, wire-line tests, and in some instances from induction and acoustic well logs. Of the 350 wells studied, 148 encountered abnormal pressures before reaching total depth.

A transparent overlay of Fig. 3 serves as a convenient means for measuring abnormal pressures from the interval travel time-depth plots. To measure pore pressures the overlay is placed over the plot (of Fig. 2 for instance) so that the depth reference lines coincide. Then the overlay is moved laterally until the normal pressure (9.2 ppg equivalent pore pressure) line coincides with the data in the shallow formations having normal pressure. Pore pressure gradients are then read from the superimposed lines of equal pore pressure. For instance, at 10,000 feet, 18.6 ppg equivalent pore pressure is noted.

Pore pressures so determined should be within 1.0 ppg of the exact pressure. However, better accuracy can often be achieved.

Abnormal amounts of limy or calcareous sands and shales decrease interval travel times. In many areas of South Texas and Louisiana a thick section of highly calcareous sand and shale overlies deep, abnormal-pressure zones. This calcareous material shows up on well velocity surveys as an anomaly of low interval travel times. Examples are shown in Fig. 4.

In Example A such a zone occurs between 8,000 feet and the top of the abnormal-pressure zone at 12,500 feet. Example B shows a thinner zone of calcareous material. To measure pore pressures in such a case the normal-pressure base line of the overlay must be shifted to the left to coincide with a new base line established immediately above the abnormal-pressure zone.

Interval travel times from seismic reflection data. Determination of subsurface velocities from seismograph field data is a well-known procedure in geophysical work. But prior to recent development of better field procedures and data recording and processing techniques, accuracy of these computed velocities was considered less than satisfactory, particularly when data quality is poor.

Several methods of computing interval velocities from seismic data exist. All are based on the same elementary reflection problem, as follows.

Elementary reflection problem. In Fig. 5, let SS represent the earth's surface. Assume the shot point O to be at the surface. When explosives at the shot point are detonated, acoustic energy is created in the form of compressional waves. This seismic energy moves equally in all directions. The vertically traveling energy strikes the subsurface plane RR and is reflected back to the surface SS along vertical path OPO. Energy from the shot also propagates along innumerable diagonal paths to the plane RR in the subsurface, (e.g., path OT) and is reflected back to the surface on path

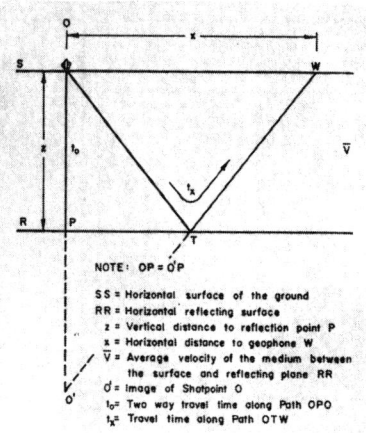

FIG. 5—Elementary reflection problem.

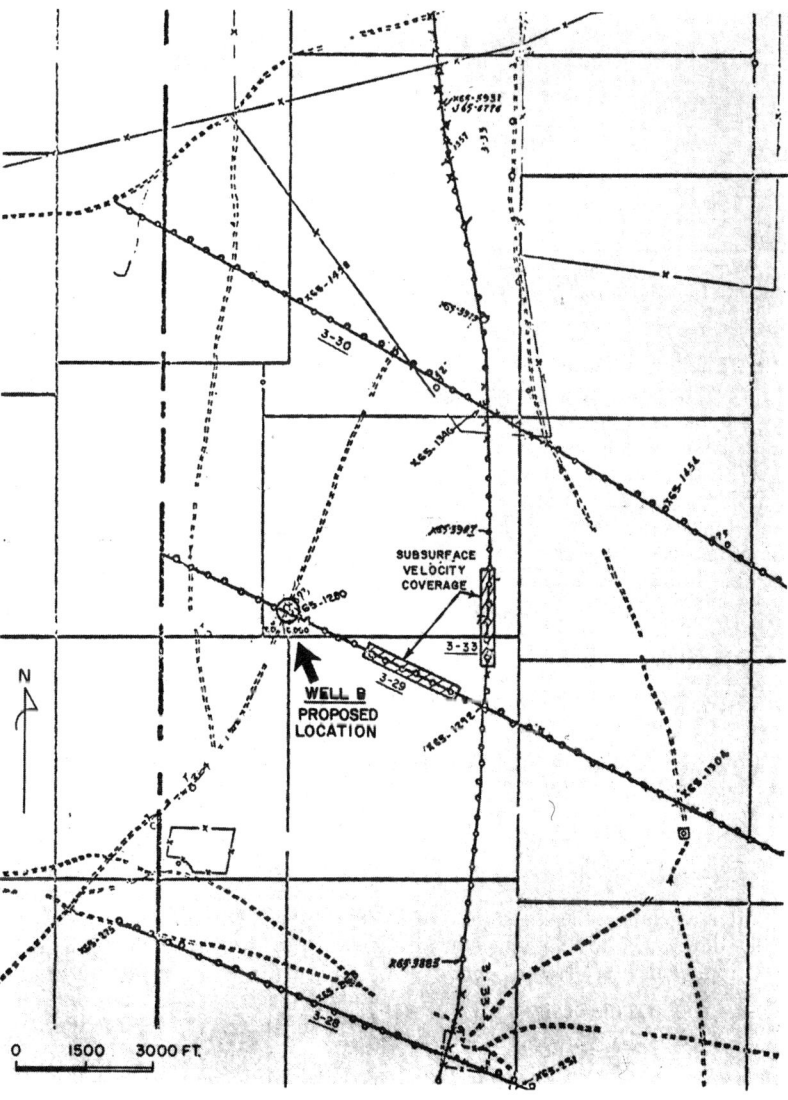

FIG. 6—Seismic surveys near Well B.

TABLE 1—Two-Way Reflection Time and Average Velocity Determinations
Computed from Seismic Line 3-33
Shot Points x65-3889 to x65-3905

Event No.	T_o Sec.*	Average Velocity, Ft./Sec.	Depth, Ft.
1	1.060	7,600	4,028
2	1.104	7,400	4,085
3	1.534	8,500	6,520
4	1.651	8,860	7,314
5	1.700	8,850	7,523
6	1.895	9,100	8,622
7	1.935	9,300	8,998
8	2.000	9,200	9,200
9	2.175	9,600	10,440
10	2.235	9,720	10,862
11	2.280	9,800	11,172
12	2.328	9,820	11,430
13	2.439	10,050	12,256
14	2.480	10,100	12,524
15	2.530	10,030	12,688
16	2.675	10,130	13,549
17	2.733	10,120	13,829
18	2.915	10,250	14,939
19	2.968	10,350	15,359
20	3.014	10,300	15,522
21	3.028	10,500	15,897
22	3.137	10,360	16,250
23	3.190	10,220	16,301
24	3.289	10,215	16,799
25	3.505	10,400	18,226
26	3.520	10,400	18,304
27	3.559	10,400	18,507
28	3.614	10,300	18,612

* T_o is the two-way reflection time from datum (sea level) to a point on a subsurface reflector directly beneath the shot.

TW. The time required for the energy to travel the two ray paths is recorded by geophones at points O and W, separated horizontally by distance X.

With this information, depth to the reflecting horizon can be calculated and the average velocity in the medium between the surface and the reflecting horizon as follows:

t_o = travel time along path OPO

t_x = travel time along path OTW

$\overline{V}$ = apparent average velocity from surface to reflecting horizon

From the relation that Distance = Average Velocity × time

$$OPO = \overline{V} \times t_o \quad (1)$$

$$OTW = \overline{V} \times t_x \quad (2)$$

After extending the line OP vertically downward to the image point O', from elementary laws of optics:

$$OTW = O'W$$
$$OPO = O'O$$

and

$$\overline{OTW}^2 = \overline{O'O}^2 + \overline{OW}^2 \quad (3)$$

Substituting Equation 1 and 2 in Equation 3

$$(\overline{V}t_x)^2 = (\overline{V}t_o)^2 + X^2$$

$$\overline{V}^2(t_x^2 - t_o^2) = X^2$$

$$\overline{V} = \sqrt{\frac{X^2}{t_x^2 - t_o^2}} \quad (4)$$

Depth to the reflecting bed may then be found by the relation:

$$Z = \overline{V} \times \frac{t_o}{2} \quad (5)$$

Velocity profiles from seismic data. By making similar time measurements to other subsurface reflecting horizons, a curve of $\overline{V}$ vs. various values of t_o can be developed. The final desired curve, interval travel time vs. depth for 1,000-foot increments of depth, can then be readily obtained.

In practice a geophone is not usually placed at the shot location, but there are ways to achieve the same result by extrapolating reflection hyperbolae to the zero x distance. Also, to minimize false reflections and enhance accuracy, data from more than one shot are usually combined for each velocity computation.

Sufficient data are available in records obtained during every routine seismograph survey to compute an interval velocity profile beneath the line centered at horizontal intervals of several hundred feet if desired.

▶ **Field example—Well B**

A routine survey was conducted consisting of a pattern of seismic lines running approximately northwest-southeast and north-south (Fig. 6). Shot points for the survey were spaced every 400 feet. For each shot, 24 geophone patterns were located 400 feet apart along the line, starting 400 feet from the shot point and extending out to 9,600 feet. Two of these lines, 3-29 and 3-33, helped define a deep-seated structure with its apex directly beneath shot point x65-1279 on line 3-29.

The structure was found to dip eastward toward line 3-33, a strike line running north and south along the flank. The contour of the mapped horizon beneath line 3-33 is 400 feet downdip from the high point. Shot point x65-1279 on the high point of the structure and on the dip line 3-29 was selected as the location for Well B. Information on the depth to abnormal pressure and maximum expected mud weight required to drill to 16,000 feet were desired.

Unfortunately, field records of the portion of the line directly over the proposed location were missing, so data were selected from seismic records about ¾-mile down the line as shown. Due to the steep dip along this portion of the line and the presence of diffractions from a nearby fault, good reflections above 11,000 feet at this location were limited. Hence the excellent data from strike line 3-33 are presented here. Conclusions substantiated the first survey but provided more detail.

Table 1 presents results of computations made from the seismic record representing subsurface coverage from shot points x65-3889 through x65-3905 on line 3-33. Each velocity shown in the table is the average velocity computed from datum to the depth of that particular reflecting event. The final interval travel time-depth profile obtained from these data is plotted for 1,000-foot vertical increments in Fig. 7.

Using the calibration curve overlay, Fig. 3, it is estimated from Fig. 7 that a formation fluid pressure gradient equivalent to 12.0 ppg fluid would be encountered at 11,100 feet in the well, 17.5 ppg at 13,000 feet and about 18.0 ppg at 16,000 feet. Seismic data thus suggested that a good depth for setting protective casing would be below 11,100 feet, and that a protective liner probably would be required before the mud density could be raised safely to the maximum 18.0-plus ppg required at total depth.

Shift of the profile to the left at about 9,000 feet was interpreted as a calcareous sand and shale interval immediately above the abnormal-pressure zone.

Comparisons of predicted and actual mud and casing programs. Drilling crews would be alerted that a calcareous sand and shale interval would be encountered at about 9,000 feet and that indications of abnormal pressure would be noted at about 10,600 feet. They would also be forewarned that at about 11,100 feet mud density of about 12.5 ppg would be required. Depth for protective casing would be below 11,100 feet. Also, since ultimate fluid pressure of 18.0 ppg is predicted, a protective liner would probably be needed before total depth was reached.

From actual fluid pressure measurements after the well was drilled, estimated top of the abnormal-pressure zone was at about 10,600 feet. Fluid pressure gradients reached 12.0 ppg between 10,525 and 11,306 feet. Casing was set at 10,875 feet. There were indications (e.g., partial loss of returns) that this setting was shallow. Protective liner was set at 11,640 feet and the well was then continued to 16,050 feet total depth with maximum 18.1-ppg mud.

Limitations. The degree of accuracy demonstrated may not be experienced in all areas. However, less accuracy may be adequate for planning purposes for which the predictive technique is designed. Accuracy of velocity computations depends on quality of the original seismic records and on careful data playback and analysis. In general the more recent surveys using long spreads and improved recording techniques are superior, particularly where information at extreme depths is required.

This method has been most successful along the Gulf Coast, where abnormal-pressure zones have been accurately predicted at depths as shallow as 6,500 and as deep as 12,000 feet. The technique is limited to sedimentary basins in which abnormal pressures result from compaction phenomena.

Other sources of information, e.g., paleontological data, log plots of nearby wells, regional geology and drilling experience in the area should be used for comparison during the planning phase. Density measurements of cuttings, well kicks, penetration rate, temperature measurements, hole instability, and other indications should be watched while approaching the pressure transition zone and used to pick the exact casing depth required for optimum control.

This article is taken from the paper, "Detection of Abnormal-Pressure Formations from Seismic Field Data," presented to the Southern District Meeting of API, San Antonio, March 6-8, 1960.

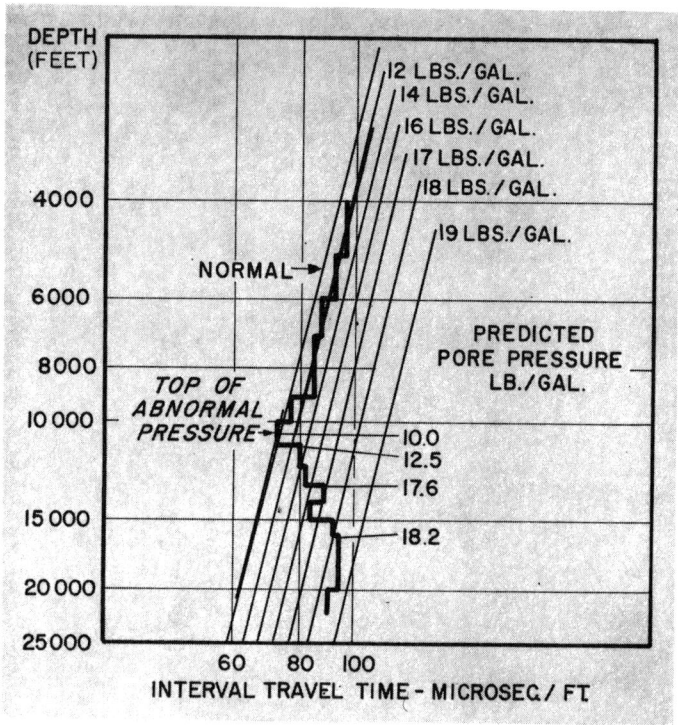

FIG. 7—Interval travel time profile computed from field seismic data at Well B location. Abnormal pressure was predicted below 10,000 feet.

About the author
E. S. PENNEBAKER, JR., *is division staff engineer in Humble Oil & Refining Co.'s Corpus Christi, South Texas Division office, Production Department, currently working in subsurface engineering, development and research applications. A native of Memphis, Tenn., he was raised in Dallas Texas, attended Tulane University and Antioch College, and was awarded the B.S. degree in petroleum engineering in 1939 by The University of Texas. Following U.S. Navy duty as a radar officer, he joined Humble in 1945. He has worked in hurricane wave force studies and offshore marine drilling equipment design and construction in Humble's early offshore drilling programs, and in various district and division engineering assignments. He is a member of SPE.*

LITERATURE CITED

[1] Hottmann, C. E. and Johnson, R. K., "Estimation of Formation Pressures from Log-Derived Shale Properties," *Journal of Petroleum Technology*, June 1965, pp. 717-722.
[2] Wallace, W. E., "Abnormal Subsurface Pressures Measured from Conductivity or Resistivity Logs," *Oil and Gas Journal*, Vol. 63, No. 27, July 5, 1965, pp. 102-106.
[3] Dix, C. H., "Seismic Velocities from Surface Measurements," *Geophysics*, Vol. 20, No. 1, January 1955, pp. 68-86.
[4] Faust, L. Y., "Seismic Velocity as a Function of Depth and Geologic Time," *Geophysics*, Vol. 15, No. 2, April 1950, pp. 192-206.
[5] Kaufman, H., "Velocity Functions in Seismic Prospecting," *Geophysics*, Vol. 18, 1953, pp. 289-297.
[6] West, S. S., "Dependence of Seismic Wave Velocity Upon Depth and Lithology," *Geophysics*, Vol. 15, 1950, pp. 653-662.
[7] Sarmiento, R., "Geological Factors Influencing Porosity Estimates from Velocity Logs," *The American Association of Petroleum Geologists Bulletin*, Vol. 45, No. 5, May 1961, pp. 633-644.

Synthetic sonic logs—a process for stratigraphic interpretation

R. O. Lindseth*

Modern seismic reflection data may be processed to approximate closely the reflection coefficient series of a sedimentary section. Inversion of the series will produce a low-cut filtered impedance log. Extension of the technique to include density correction and replacement of missing low-frequency components leads to generation of a synthetic sonic log having dimensions and characteristics similar to a conventional borehole sonic log.

A synthetic sonic log section provides several advantages over conventional seismic sections, particularly for stratigraphic exploration. Direct depth display largely eliminates the vertical scale distortion inherent in time sections and simplifies integration with other subsurface data. Resolution is generally improved.

The process is well adapted to automated generation of stratigraphic sections. Individual rock units are mapped by contours of constant transit-time, to show facies changes and depositional sequences in detail. The detailed measurements of velocity may be used to interpret rock type. Comparison of synthetic sonic log sections to the conventional seismic data from which they were derived, clearly demonstrates improved understanding of subsurface geology.

INTRODUCTION

Digital recording and processing have been responsible for substantial improvements in the quality of seismic reflection data, but many of these benefits are achieved only because of an earlier invention; specifically, summation ("stacking") of common-depth-point traces (Mayne, 1962) which can separate seismic signals from noise of the same frequency. When stacking is combined with digital processing capabilities, it can extend the range of useful seismic frequencies into regions of the spectrum that have not been generally utilized in the past. First, an improved method to utilize the effective resolution contained in modern broad-band seismic reflection data, and second, a new technique for interpretation will be described.

In outline, the basic processing operations are (1) derivation of an approximate reflection coefficient sequence for each seismic trace by digital processing, (2) inversion of the reflection coefficients to band-limited velocity coefficients (impedance logs) and, (3) use of (2), following a density correction, to modulate low-frequency velocity components, obtained from reflection velocity analyses, to produce synthetic sonic logs (Seislogs®).

The results can provide a reasonable approximation to a borehole sonic log.

BASIC PRINCIPLES

The inversion of seismic trace data to obtain time-domain impedance logs is not new. The basic procedure first described by Delas et al (1970) and published by Lindseth (1972), Lavergne (1975), and others, is essentially the reverse of that used to produce synthetic seismograms from borehole sonic logs. Comparison of the two procedures can provide a better understanding of the inversion process.

A reflection coefficient sequence RC_i can be computed from a borehole sonic log. If density information is available, a better approximation to the reflection coefficients can be obtained from

$$RC_i = (\rho_{i+1}V_{i+1} - \rho_i V_i)/(\rho_{i+1}V_{i+1} + \rho_i V_i), \quad (1)$$

®Registered trademark of Teknica, Ltd.

Manuscript received by the Editor August 8, 1977; revised manuscript received April 21, 1978.
*Teknica Resource Development Ltd., 412 339-6th Ave. S.W., Calgary, Alberta, Canada T2P0R8
0016-8033/79/0101-0003$03.00. © 1979 Society of Exploration Geophysicists. All rights reserved.

where ρ_i and V_i are, respectively, the density and velocity at time index i. In practice, density logs frequently are unavailable because they are not normally included in borehole logging programs or are run only over the zones of interest. Fortunately, the effects of density appear to be quite minor in many cases, as the computation of synthetic seismic traces from sonic logs alone has proven to be quite successful in the majority of sedimentary environments.

Gardner et al (1974) have shown that density is closely proportional to the one-quarter power of velocity, the relation varying slightly with rock type. It is also possible to derive a general empirical linear relationship between acoustic impedance ρV and velocity,

$$V = 0.308 \rho V + 3460, \qquad (2)$$

(Figure 1) and from this,

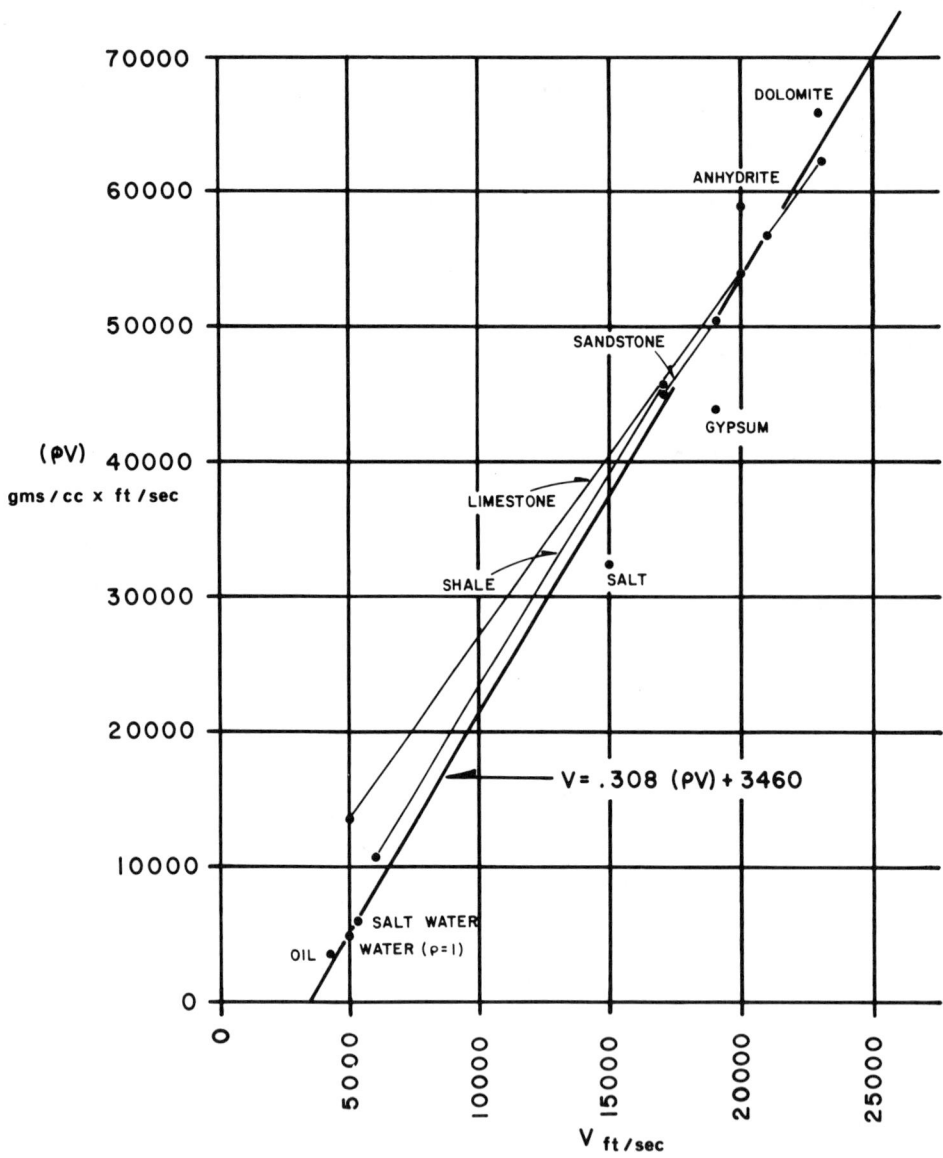

FIG. 1. Values of acoustic impedance (ρV) plotted against rock velocity (V) for common rock types follow an approximately linear function (from Gardner et al, 1974).

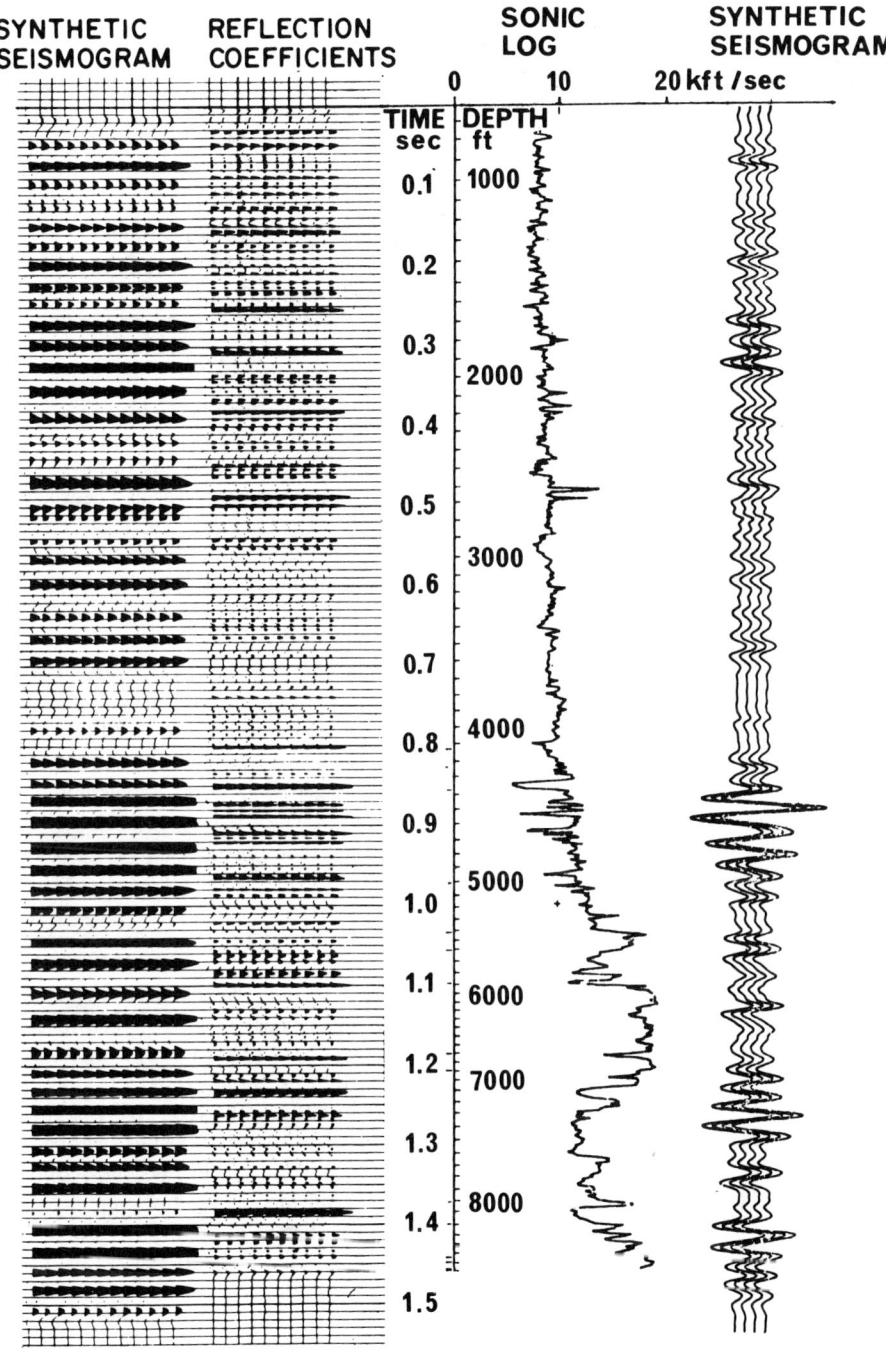

FIG. 2. A time-corrected ("integrated") sonic log and synthetic seismogram (right) derived from this log. The same synthetic seismogram and corresponding reflection coefficients are plotted (left) in variable area reverse polarity. The reflection coefficients are a direct transformation of the sonic log and, consequently, correlate closely to it. The poorer correlation of the synthetic seismogram illustrates the result of bandwidth reduction and phase distortion which occurs in the field recordings.

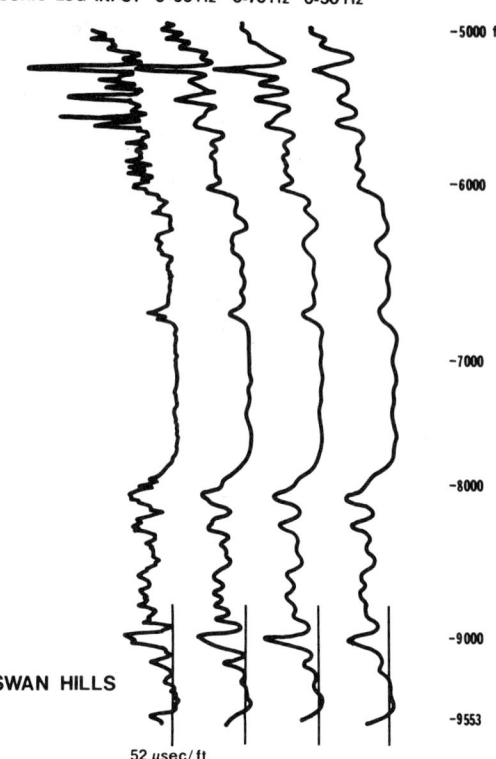

FIG. 3. Removing successively increasing amounts of high-frequency components from sonic logs correspondingly results in decreased resolution. A 50 Hz cut-off is often used on seismic sections.

$$\rho V = (V - 3460)/0.308, \quad (3)$$

where V is in ft/sec and ρ is in g/cm³. When density data is lacking, equation (3) may be substituted in equation (1) to give

$$RC_i = (V_{i+1} - V_i)/(V_{i+1} + V_i - 6920). \quad (4)$$

The inversion of seismic data is based upon rearranging the terms of (1) to give the impedance log series,

$$\rho_{i+1} V_{i+1} = \rho_i V_i (1 + RC_i)/(1 - RC_i). \quad (5)$$

The empirical relationship (2) then may be applied to derive a velocity series from the impedance log. The inversion requires the initial value of ρV to be known. If the correct initial value is not known a simple scalar correction factor and shift can be applied subsequently to all velocity values.

Frequency bandwidth

The above basic inversion procedure is used in the following development, but for this to be successful in practice, the field trace data must yield a close approximation of a reflection coefficient series when processed. Most likely this goal will be only partially achieved.

A major limitation is the limited frequency bandwidth of most seismic recordings. The effects of bandwidth reduction on seismic signals are illustrated in Figure 2, which compares a sonic log to its corresponding series of reflection coefficients and resulting synthetic seismogram.

The reflection coefficients are a sample-for-sample transformation of the sonic log samples and, as such, have the same resolution (and bandwidth). Accordingly, there is a direct correlation between the reflection coefficients and the sonic log.

Convolution of the reflection coefficients with a wavelet to simulate the filter and phase shift effects of the earth and recording system produces the synthetic seismogram. Whereas the reflection coefficients in this example have a bandwidth extending from 0 to 250 Hz, the synthetic seismic trace is band-limited to approximately 15 to 50 Hz.

Although the synthetic seismogram is noise-free, it is no longer possible to correlate it to the sonic log with precision and confidence. Widess (1973) demonstrated the time-shift and distortion of band-limited reflections from thin beds (closely spaced reflection coefficients). Comparison of the reflection coefficients of Figure 2 to the synthetic seismogram reveals an abundance of similar effects. Schoenberger and Levin (1976) described the phenomenon of very high amplitude response of band-limited reflections from critically spaced reflectors. This effect is very well illustrated by the seismic response to the shaly section between 3000 and 4000 ft on Figure 2. The broadband reflection coefficients define this nearly homogeneous section with high fidelity, but the seismogram contains several misleading high-amplitude reflections.

A useful procedure to illustrate the significance of these effects is suggested by the work of Branisa (1974). The frequency bandwidth of seismic data can be easily identified. The bandwidth of a sonic log can then be modified, by appropriate filters, to a similar condition to demonstrate the loss of geological information that occurs in the seismic reflection process. If such a filter, matched to the seismic data spectrum, removes significant geological information from sonic logs, it is obviously futile to expect such information to be present on corresponding seismograms.

Progressive high-cut filtering will illustrate the re-

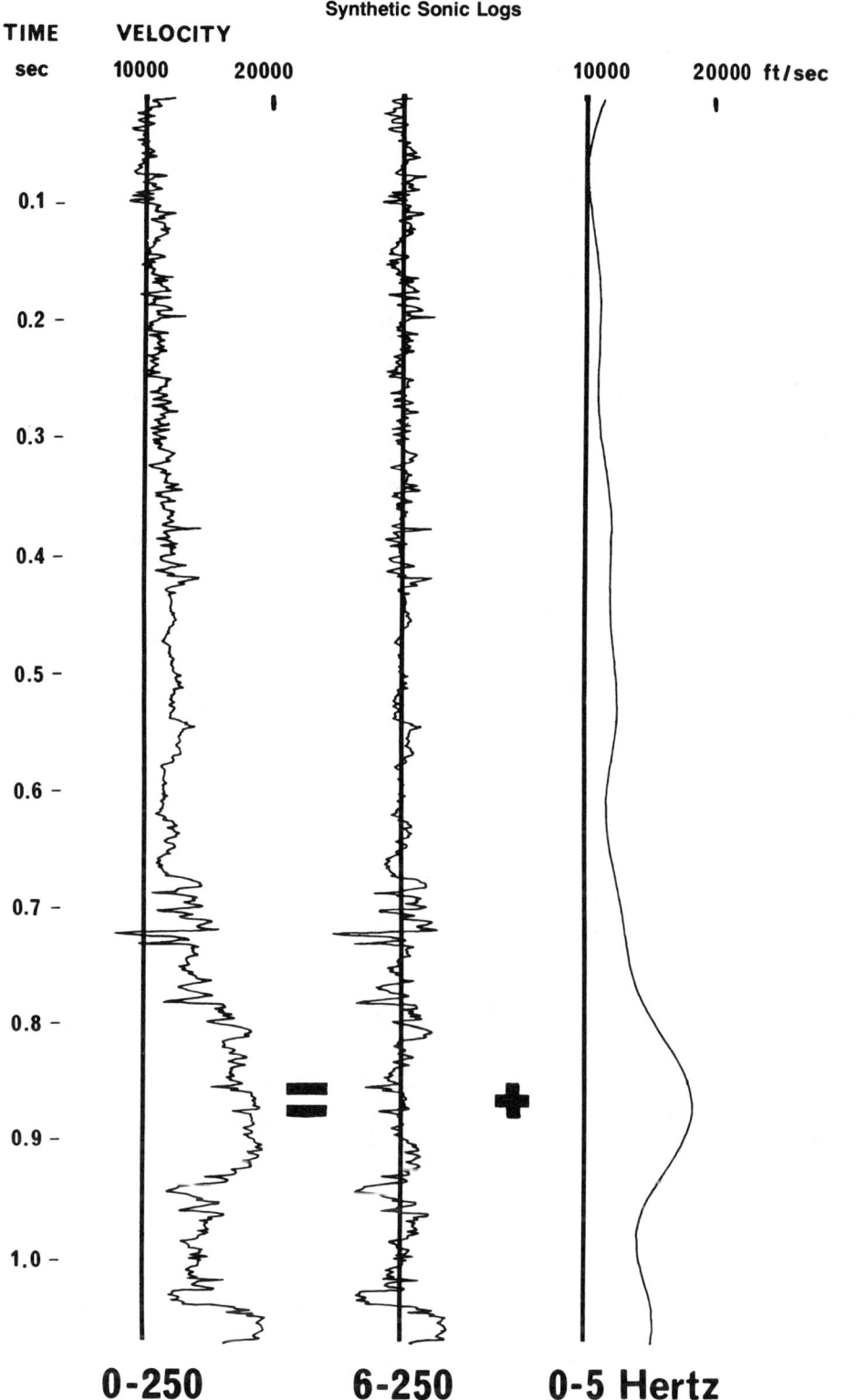

FIG. 4. A sonic log may be considered to be the sum of a gross velocity function (0–5 Hz) and a detailed velocity function (6–250 Hz).

lation of high frequency to resolution. A log, high-cut filtered at 50 Hz in Figure 3, shows the approximate resolution of the average seismic recording after inversion. Obviously, considerable detail is lost from a seismic section conventionally high-cut near 50 Hz.

A low-cut filter (6–250 Hz), applied to a sonic log in Figure 4, illustrates the effect of missing low-frequency components, which evidently contain much of the essential velocity information. The sonic log may be considered as a low-frequency function modulated by a higher-frequency function; i.e., as a sum of frequency components above 5 Hz added to those below 5 Hz. Without low-frequency components the synthetic sonic log values obtained from inverting seismic data do not closely approximate velocities measured on borehole sonic logs.

In many areas careful field operations, with conventional seismic equipment and procedures, can preserve valid high-frequency reflection spectral components approaching 100 Hz. Careful time-correction and stacking of common-depth-point traces can remove much of the noise, leaving essentially uncontaminated signals. These can be enhanced by deconvolution. The result is seismic data of greatly increased resolution.

While recovery of moderately high-frequency reflection components is primarily the result of careful field procedure and data processing, the recovery of low-frequency reflection components is another matter. On land, reflection components are often low-cut near 12 to 15 Hz, since the response of many field geophones is designed to decrease rapidly below about 15 Hz in order to attenuate low-frequency ground-roll. Deconvolution may extend reflection spectra to lower frequencies within the roll-off portion of the geophone response, but even in areas where geophones have response extending down to 10 or 8 Hz, seismic data will rarely contain spectral components below 5 Hz. Data processing procedures can be used to derive many of the missing low-frequency components, from reflection-derived velocity analyses, and also to enhance the high-frequency response of high-quality reflection records.

DATA PROCESSING

Gain recovery

Cumulative transmission losses in field-recorded reflection signals are commonly assumed to increase exponentially with reflection time and are compensated for by application of an exponential gain curve. This simple procedure may not always be valid. The rate of signal decay may be high where the signal traverses a thin-bedded laminated section, but lower through a nearly homogeneous section. Reflection normal moveout also causes substantial variation in the rate of decay from near to far-trace ranges.

In all cases the average amplitude of the signal will drop step-wise to a lower level immediately following the arrival of each reflection. The higher the reflection amplitude, the greater the subsequent drop. These sudden losses must be recognized and proper compensation applied if correct amplitudes are to be recovered. One method is to apply a short time-average operator to the trace and adjust the amplitude for the difference before and after each reflection burst.

Unfortunately, the result of any data-dependent scheme for compensating transmission losses is distorted by noise superimposed on reflections. One practical solution is to apply an arbitrary function for the early stages of processing. Later, following removal of some of the noise, the arbitrary correction function can be reversed and the data-dependent amplitude correction function can be applied.

Deconvolution

If there is a single key process for successful seismic data inversion [and for improved reflection resolution (Schoenberger, 1974)] it is a deconvolution process which yields the best possible approximation to a zero-phase spike signal representing each reflection coefficient. Such an approach requires a reliable measurement of source signal spectra. But this is not readily obtained from field seismic data, especially on land. It is only slightly less difficult to obtain from marine data (Stoffa et al, 1974), unless the source signal has been recorded separately. Thus, the well-known and much more tractable minimum-phase assumption (Rice, 1962; Robinson, 1957) has been the basis for the majority of existing seismic data deconvolution routines. Unfortunately, this approach does not provide adequate phase compensation for seismic data inversion and in some cases may whiten the amplitude spectrum excessively.

Several new deconvolution techniques, grouped under the broad classification of wavelet processing, provide results better suited to inversion. For marine data, when the proper wavelet source signature has been recorded, application of a matched filter and a two-sided inverse operator converts this wavelet into a close approximation of a zero-phase spike within the limitations imposed by the bandwidth.

Variations on the homomorphic technique (Oppenheim and Schafer, 1975) can provide good results if the phase spectrum can be determined reliably, which is not the usual case. Assumptions and approximations are necessary and, like most deconvolution operations, compromise and trade-offs in the selection of a technique are necessary. The reader is referred to Lines and Clayton (1977), and Lines and Ulrych (1977) for a more detailed discussion.

In some cases, it appears that the major effect on the amplitude and phase spectra of the seismic trace is attributable to the instrument response. A preliminary correction may be applied to the amplitude and, in particular, to the phase response characteristics of field instrument filters, as published by manufacturers.

Deconvolution of the data before normal moveout corrections should be followed by deconvolution after trace stacking. Invariably, the stacking process modifies the amplitude spectrum of the input traces. As a result, it is necessary to apply this second "trim" deconvolution after stacking to restore the amplitude spectrum and to make any required minor adjustments to the phase spectrum.

After trace stacking, conventional amplitude scaling and filtering procedures may be applied to produce a conventional seismic section. The broadband unscaled output from the stacking process, deconvolved, is the input to the inversion process. If stacking did not remove a satisfactory percentage of the noise, it may be helpful to apply a fan filter (Embree et al, 1963), provided the data will respond to that type of treatment.

The signal bandwidth should be from three to four octaves. Generally, the bandwidth will extend to 75 Hz or higher. Bandwidth of conventional marine seismic data, recorded at a 4 msec sample interval, is limited by the antialias filter to an upper limit near 62 Hz. The low-frequency end of the spectrum will lie somewhere between 5 and 15 Hz depending upon the response of the geophones and/or any low-cut filter applied in the field recording. Deconvolution will have extended the low-frequency end of the spectrum slightly, within the frequency response roll-off zone of the geophones and/or field filters, but the very low frequencies, which represent the gross velocity information, must be obtained from another source.

Reflection-derived velocity analysis for low frequencies

If it were possible to record reliable seismic data as low as 6 Hz, and occasionally it is, the inversion would still produce no better approximation to a sonic

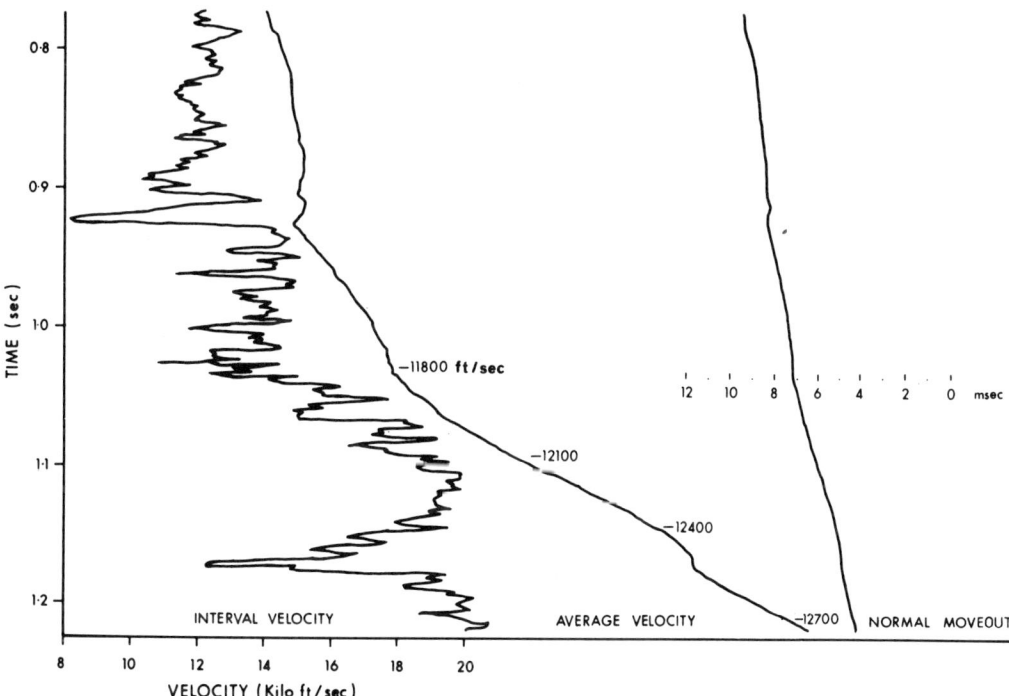

FIG. 5. The continuously changing interval velocities of the sonic log on the left yield the average-velocity and normal moveout curves shown. Conventional velocity analysis attempts to recover the interval velocities from the normal moveout, a difficult and imprecise process.

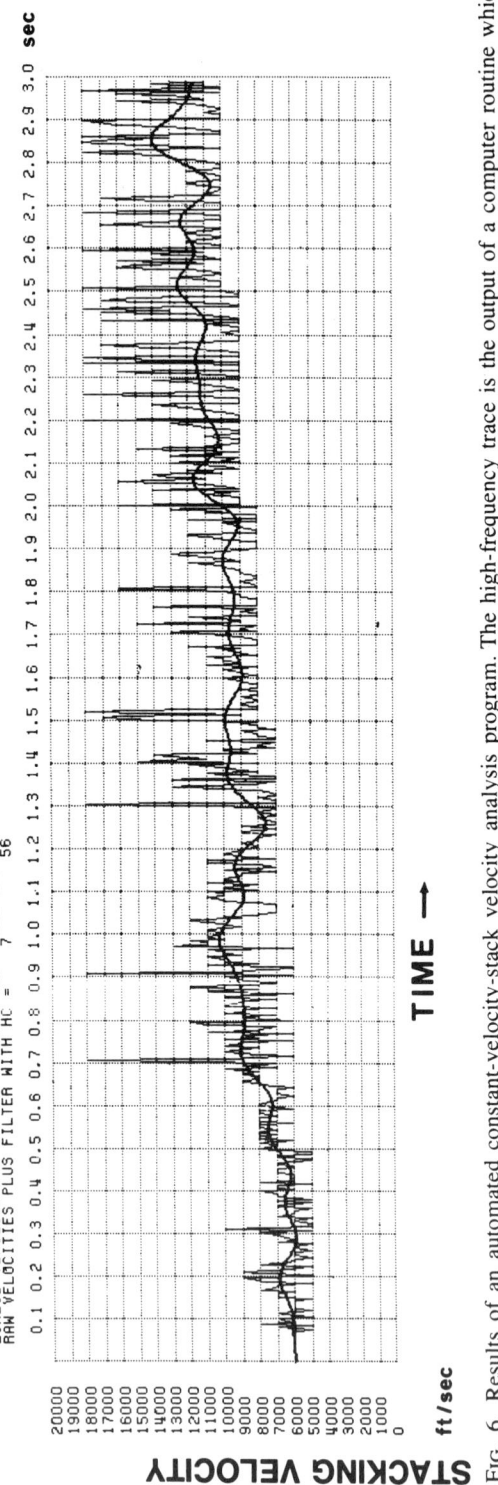

FIG. 6. Results of an automated constant-velocity-stack velocity analysis program. The high-frequency trace is the output of a computer routine which derives a stacking velocity at each time increment. Application of a 7 Hz low-pass filter yields a smooth line estimate of stacking velocity.

log than the 6 to 250 Hz log of Figure 4. The missing 0 to 5 Hz velocity components, which form a smooth continuous function of instantaneous velocity versus time, can be recovered from reflection-derived velocity analyses.

Lavergne and Willm (1977) introduced interval velocities derived from stacking velocities to compensate for the missing elements, but these require a preliminary interpretation of the major velocity intervals to be made and introduce serious discontinuities in the data. Interval velocities, derived from stacking velocity determinations based on reflection moveout, are commonly calculated for intervals between distinct seismic reflections from important mapping horizons. At best, strong reflections mark large velocity contrasts within the section; at worst, they represent nothing more than constructive interference of band-limited signals, while the major velocity boundaries may have rather poor reflections (Figure 2).

Thus, the calculated velocity intervals may not correspond to the proper geologic intervals. Another potential error in this approach is due to the fact that, at the usual depth of interest, very small changes in normal moveout result from large changes in the velocity. Figure 5 compares instantaneous (interval) velocities (the desired type of information) of a sonic log, to average velocity (the initial derivation), and to normal moveout (the source data). Small errors in measurement of normal moveout can represent a high percentage of the total moveout and, consequently, large velocity errors result.

Lindseth (1975) suggested a procedure by which normal-moveout analysis may be refined to obtain the velocities required to complete the synthetic sonic log. A computer is programmed to select a value of stacking velocity for each increment of reflection time on a conventional constant-velocity stack, i.e., successive summations of CDP trace sets using a series of constant velocities for normal moveout correction. The result of such an operation is the high-frequency velocity trace plotted in Figure 6. This trace is then low-pass filtered to a selected frequency to yield a smoothly varying stacking-velocity-versus-time function similar to the heavy line plotted in Figure 6.

If a single velocity-versus-time function is derived where data quality is poor or distorted by noise, diffractions, or multiple reflections, it probably will not be reliable. Hence, the process should be repeated at frequent locations along a line, randomly varying the series of constant velocities used for the stacking-velocity analyses. Using a fixed set of constant velocities for all analyses may bias the results

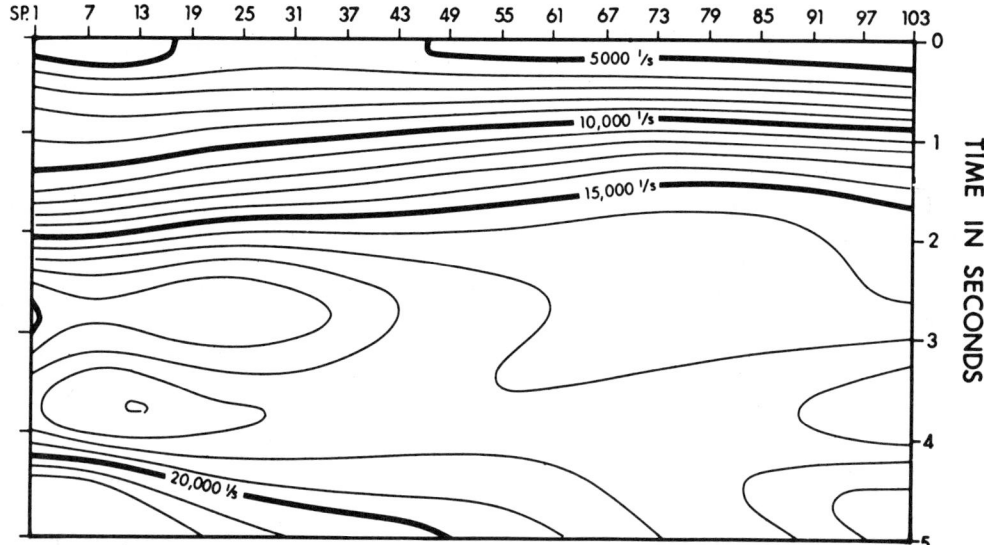

FIG. 7. The final estimate of an instantaneous-velocity surface obtained by (1) filtering a stacking-velocity analysis at each numbered shotpoint station, (2) converting each result to vertical average velocity by ray-tracing, (3) fitting a mathematical high-order surface to the vertical velocities, and (4) calculating interval velocities for each time sample.

toward these fixed values. Also, it should be remembered that on land, reflection-derived velocities are correct only when referred to the ground surface, not to some datum above or below the surface.

The next step is to convert the stacking velocities to vertical average velocities. If the sedimentary section is relatively flat, and has few major velocity changes, the Dix (1955) method for deriving approximate vertical velocities from stacking velocities has been shown by Woeber and Penhollow (1975) to provide good results.

In highly layered sequences, particularly for extremely long spreads (Al-Chalabi, 1974), or areas of more than moderate dip (Everett, 1974; Gangi and Yang, 1976), better results can be obtained by using ray-tracing or minimum-path procedures to derive the vertical-velocity-versus-time function. The result is vertical average velocity as a function of time at the location of the analysis.

The entire suite of vertical-average-velocity-versus-time functions is then smoothed by applying areal surface-fitting techniques (Street and Lindseth, 1974). Where a number of lines are available, the process can be extended to 3-D to obtain a continuous velocity function, $V(x,y,t)$, describing a hypersurface of vertical average velocity at any location and reflection time. A new function representing continuous instantaneous (interval) velocity versus time can be derived by calculating the interval velocity for each sample of the vertical-average-velocity-versus-time series along any line. The resulting instantaneous-velocity surface (Figure 7) represents an approximation of the low-frequency velocity components that are absent from the seismic field traces.

Evidently this procedure will not provide correct gross velocity information in all cases. Velocity surfaces in faulted areas, for example, may be smoothed across fault planes.

A velocity function derived in the manner described can be compared directly at a borehole tie point to a sonic log properly scaled and filtered to the same frequency range. In fact, if borehole logs are available in the area, filtered velocity functions derived from such should be included in the construction of the velocity surface or, where abundant, may constitute the main source of velocity information for calculation of low-frequency interval-velocity-versus-time surfaces.

Synthetic sonic log construction

At this stage, all material necessary to construct the synthetic velocity log has been gathered. Low-frequency data, in the form of a continuous instantaneous-velocity function, has been derived and low-pass filtered to some cutoff value, here termed the crossover frequency, approaching the low-frequency cutoff of the seismic data. Seismic trace data have been processed to the best possible approx-

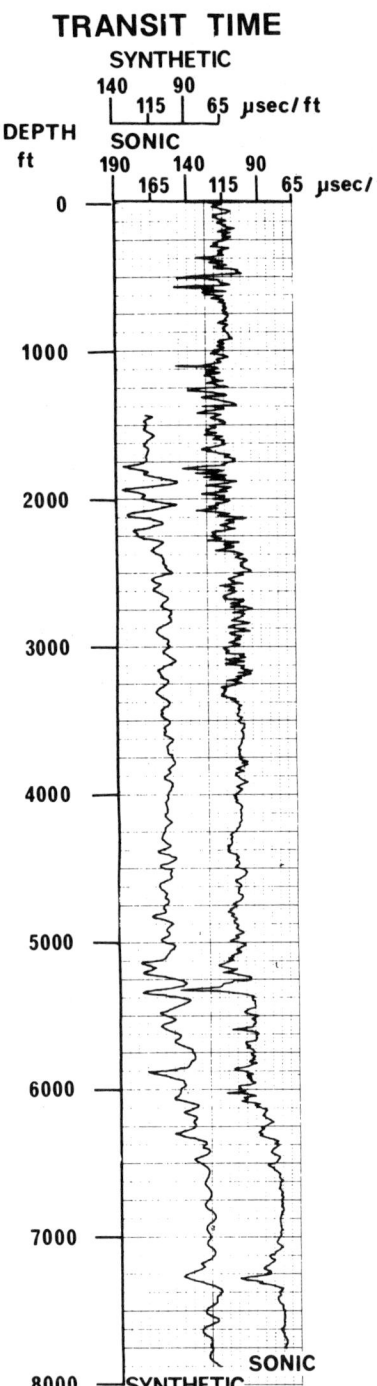

FIG. 8. Comparison of a nearby borehole sonic log to a synthetic sonic log derived from a good quality seismic trace. Some detail is missing on the synthetic log because the frequency bandwidth of the seismic data input is limited to about 100 Hz in the shallow beds and about 75 Hz at depth.

imation of reflection coefficients. These are next inverted to obtain the instantaneous-velocity components above the crossover frequency.

The approximate reflection coefficient-versus-time series must be scaled to an appropriate amplitude level. This can be done by setting the rms amplitude value of the data equal to that of the reflection coefficients derived from a nearby sonic log which have been low-cut filtered to the crossover frequency. Alternatively, the rms value can be estimated from experience, usually between 0.005 and 0.05. It will be found that within this range the factor is not critical. The main effect of a change is a minor variation in the shape of the final log. If the factor is too large, the results will be unstable.

Application of equation (5) to invert the approximate reflection coefficient series produces an approximate acoustic impedance series. These results, scaled to correct for density, and properly positioned in time, are then simply added to the low-frequency velocity function obtained from reflection and/or sonic log data, to produce a synthetic sonic log.

At this stage, the vertical scale is in time. This scale is converted to depth by multiplying the interval velocity of each sample by the time increment (one-way) to obtain the depth increment for that sample. The reciprocal of the velocity then yields the transit-time (μsec/ft) for each depth sample. Plotting these results to a suitable scale produces the synthetic sonic log.

A crucial test of the process is to overlay a sonic log from a borehole at a tie location, if such is available, on the synthetic sonic log. Under favorable conditions the two compare well as shown in Figure 8. It should be stressed that the synthetic logs are affected by all the normal seismic phenomena such as noise, diffractions, multiple reflections, reverberations, migration, and density variations that have not been adequately removed or corrected in the processing sequence, plus any distortion introduced by processing. For instance, some residual effects of density may remain, but a more precise correction would require the lithology of the section to be known first and the compensation applied as a variable depending upon rock type.

These imperfections should not exceed those contained in the corresponding conventional seismic section. In fact, in some cases the inversion procedure may be a useful aid to detecting or measuring the effects of such disturbances where a direct comparison to a borehole sonic log can be made. For example, features on the synthetic log which do not appear on the sonic log are usually due to either multiple reflections in the seismic input data or local density variations.

APPLICATIONS

It is much easier to identify geologic sequences on a sonic log than on a conventional seismic trace, because the velocities measured are usually directly related to rock type. Inversion of the seismic data to synthetic sonic logs provide a similar advantage. Presentation parameters of transit-time versus depth are the same as sonic log parameters and thus are familiar to geologists. Resolution generally is increased and lithologic units are identified distinctly with measurable thicknesses. Continuous depth sections of synthetic sonic logs are largely free of the vertical depth scale distortion on seismic time sections as well as of distortion resulting from lateral velocity changes. False "pull-up" and "pull-down" structures caused by local velocity variations are minimized.

The interpretation of synthetic log sections is similar to the geological interpretation of stratigraphic sections constructed from sonic logs, except the synthetic logs are spaced a few hundred feet apart rather than a few thousand feet, as is often the case with borehole logs. However, some seismic interpreters may be unfamiliar with the manner in which stratigraphic features are expressed and other interpreters, using the results for the first time, may not recognize the effects of distorting phenomena such as band-limited data, phase shifts, and noise on the sections. Several of these considerations are discussed by Beitzel et al (1977).

Correlation of the synthetic log to the sonic log gives a direct measure of the confidence that may be placed in the former. Much can also be learned from any differences which appear between the two. For example, experience with the phase adjustments required in the seismic data inversion technique to improve the correlation to sonic logs has led to a better understanding of the effects of that phenomenon. It is well known that a phase difference can cause imperfect ties between different sets of seismic data but it is also quite common to limit the definition of the problem to "normal" or "reverse" seismic trace polarity. Comparison of synthetic and sonic logs, however, suggests that the simple binary choice of normal and reversed polarity rarely describes phase distortions of seismic data. Phase differences, as indicated by synthetic and sonic log comparisons, are usually some amount other than 180 degrees and often vary with frequency. The fact that close agreement often can be obtained between syn-

FIG. 9. Poor correlation at tie points can be caused by variations in phase response of different instruments. A synthetic log (left) which does not match a reference synthetic log at the tie point (center) is phase-shifted 50° to yield a corrected log (right) which matches the reference more closely.

thetic and sonic logs merely by adjusting phase provides valuable insight into deconvolution processes and is an important factor in obtaining high-fidelity synthetic sections that adequately represent geology.

A simple illustration drawn from a real situation may be used to demonstrate the effect of phase distortion. Two lines were recorded with identical procedures, geophones, and filter settings, but with different instruments, both of which were checked for polarity. The appearance of the two seismic line cross-sections was quite similar at the intersection but the tie suffered small reflection-time differences which varied at different levels in the section. Identical processing of the seismic and inversion data revealed a phase difference of approximately 50 degrees at the level of interest (Figure 9, left). Application of a uniform phase shift to the tie log produced

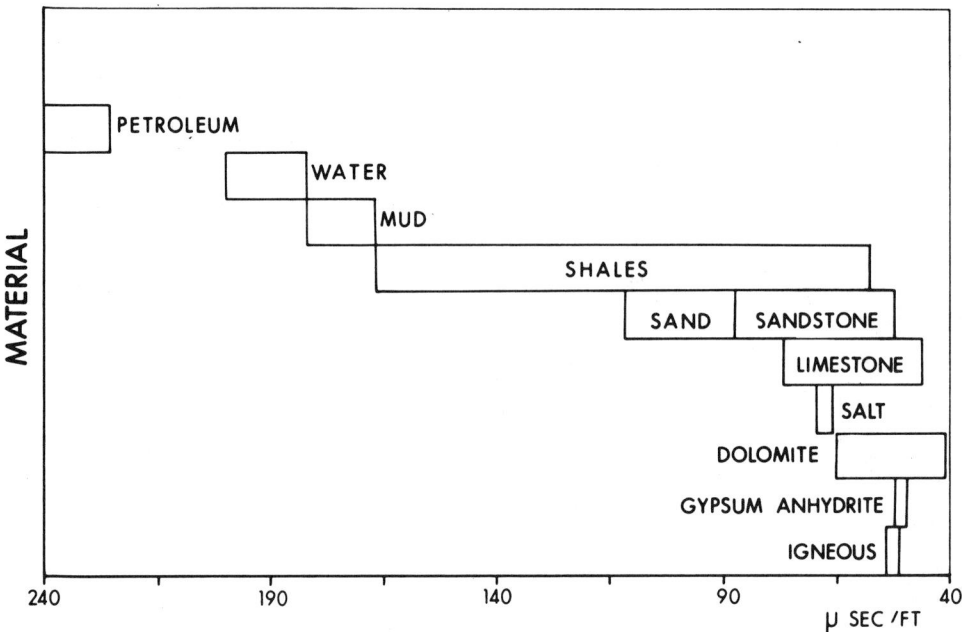

FIG. 10. Normal range of sonic transit-times for common subsurface rocks and fluids.

the greatly improved match illustrated (Figure 9, right).

The single most useful feature of the synthetic logs is the direct measurement of velocity within those lithologic units sufficiently thick to be resolved by the log. However, it should be remembered that interpretation of the synthetic log, like that of any log curve including the borehole sonic, will be ambiguous in some respects, since a change in response may be due to a number of different geological conditions. For this reason, conventional borehole logging programs usually record the gamma-ray log simultaneously with the sonic log to help distinguish between shale and porosity, to which the sonic log responds similarly and the gamma-ray log differently. Seismic data inversion produces only one curve type, an approximation to the sonic log, and the criteria for interpretation are velocity (expressed as its reciprocal, transit-time) and the log character.

Velocity and lithology are closely related (Gardner et al, 1974; Guyod and Shane, 1969). Data from logging handbooks (Dresser Atlas, 1971; Schlumberger, 1972) are compiled in the chart of Figure 10 to illustrate the range of velocities for each major rock type. The ranges encompass nearly all possible conditions in which each broad classification of rock may be found. In local areas, or through a limited thickness of section, the range for a given rock type is generally much narrower. Porosity in a rock formation lowers the velocity, particularly when the pore space is filled with gas. Mixtures of rock types produce intermediate velocities that can be similar to those of other types. Many sediments, particularly shales, respond to increased depth of burial and age by increased velocity (Faust, 1951). All of these factors compound the ambiguity in the response of velocity to lithology. Nevertheless, reasonably reliable velocities, combined with other knowledge of the sedimentary section, should narrow the range of possible lithologies so that, for example, a shale layer rarely would be mistaken for a carbonate layer.

The synthetic logs are very sensitive to changes in velocity, and variations of 2.5 μsec/ft can be detected without difficulty. Reliable as relative values may be, the accuracy of the absolute value is highly sensitive to continuity of the spectrum at the low-frequency end. The spectral bandwidth of seismic data is generally limited by geophone response, and possibly by field instrument filters, to frequencies above 12 to 15 Hz. Coincidentally, it may be found that the quality of the reflection-derived velocity functions is such that reliable low-frequency spectral components do not exist above 5 Hz. The synthetic logs obtained by combining inverted and reflection-derived velocity data would then have a gap in the spectrum from 6 to, say, 11 Hz. The absence of this low-frequency portion of the spectrum (a frequent occurrence in practice) will distort the synthetic log response in a fairly predictable manner, making some velocities too high, others too low. The effect is illustrated in Figure 11, a sonic log which has been filtered with a notch from 6 to 11 Hz. When this band of spectral components is missing from the log, some velocities, particularly those in the deeper portion of the section, appear to be alternately higher and lower than the correct values.

Fortunately, and most importantly, such deviations will tend to remain constant within a given area and changes in measured velocity can be related with reasonable confidence to changes in lithology.

Facies interpretation

The constancy and reliability of synthetic log interval velocities, even over thin beds, provide the basis for a new technique for seismic data interpretation based on the direct measurement of quantitative differences occurring in observed synthetic sonic log data.

Synthetic logs are plotted in standard sonic log coordinates of depth and transit-time (Figure 12). Transit-time is plotted horizontally on a scale of 25 μsec/ft/inch, referenced to a 90 μsec/ft baseline located at the shotpoint identifier. Transit-time decreases to the right to correspond to increasing velocity in that direction. Depth is plotted vertically, usually on a scale of 250 ft/inch. Horizontal excursions of the log across the vertical grid lines, which represent transit-time, make it possible to measure a change of 2.5 μsec/ft on the log at any depth. Each distinct lithologic unit in the section has a characteristic velocity. Ideally, the log curve will shift to the appropriate transit-time as each unit is traversed. Thus, points of equal velocity (transit-time) on a suite of log traces can be joined to obtain continuous isotransit-time contours parallel to the top and base of these lithologic units which are characterized by a change in velocity. Changes of velocity observed within such units will, in most cases, be related to changes in lithology and/or porosity.

Contour lines developed in this manner have been found to map individual lithologic units with remarkable fidelity in several exploration areas, outlining not only the physical shape of the unit but indicating lateral variations in lithology as well. Continuous mapping of synthetic log velocities can provide detailed information regarding lithologic and stratigraphic changes in the sedimentary section revealing facies changes in a manner not possible with conventional seismic presentations. Figure 12, a section of synthetic logs derived from seismic traces, will be used to illustrate the procedure. Most of the contour and grid lines that would normally appear have been removed for clarity. The transit-time scale at the top of the figure corresponds to the center log which has been drawn heavier for easy reference, as have two other logs toward the edges of the section. If the transit-time scale at the head of the center log is shifted one vertical grid line to the right, it will then correspond to the next log on the right, and so on. For example, shifting the scale twelve positions right and left will provide scales for the two other heavy logs. The vertical grid lines

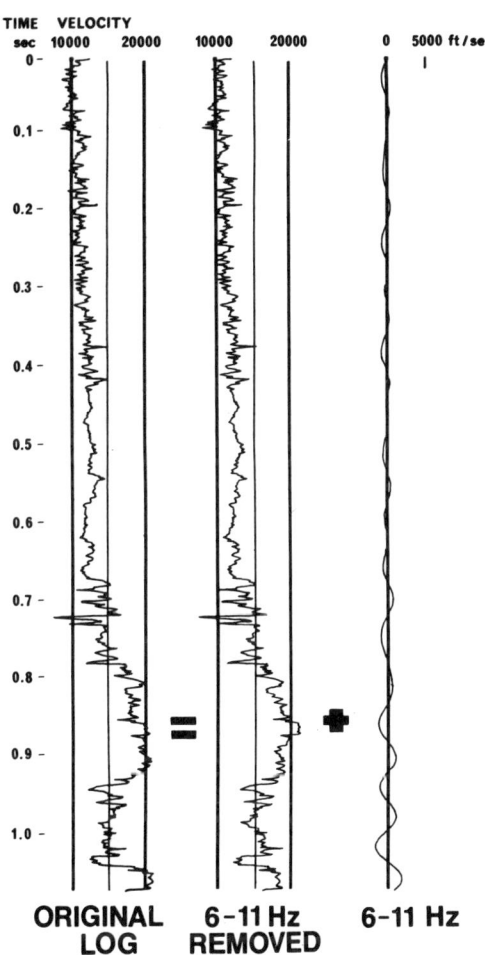

FIG. 11. The absence of a band of frequencies from 6 to 11 Hz causes log velocities to be distorted alternately higher and lower than those of the unfiltered log.

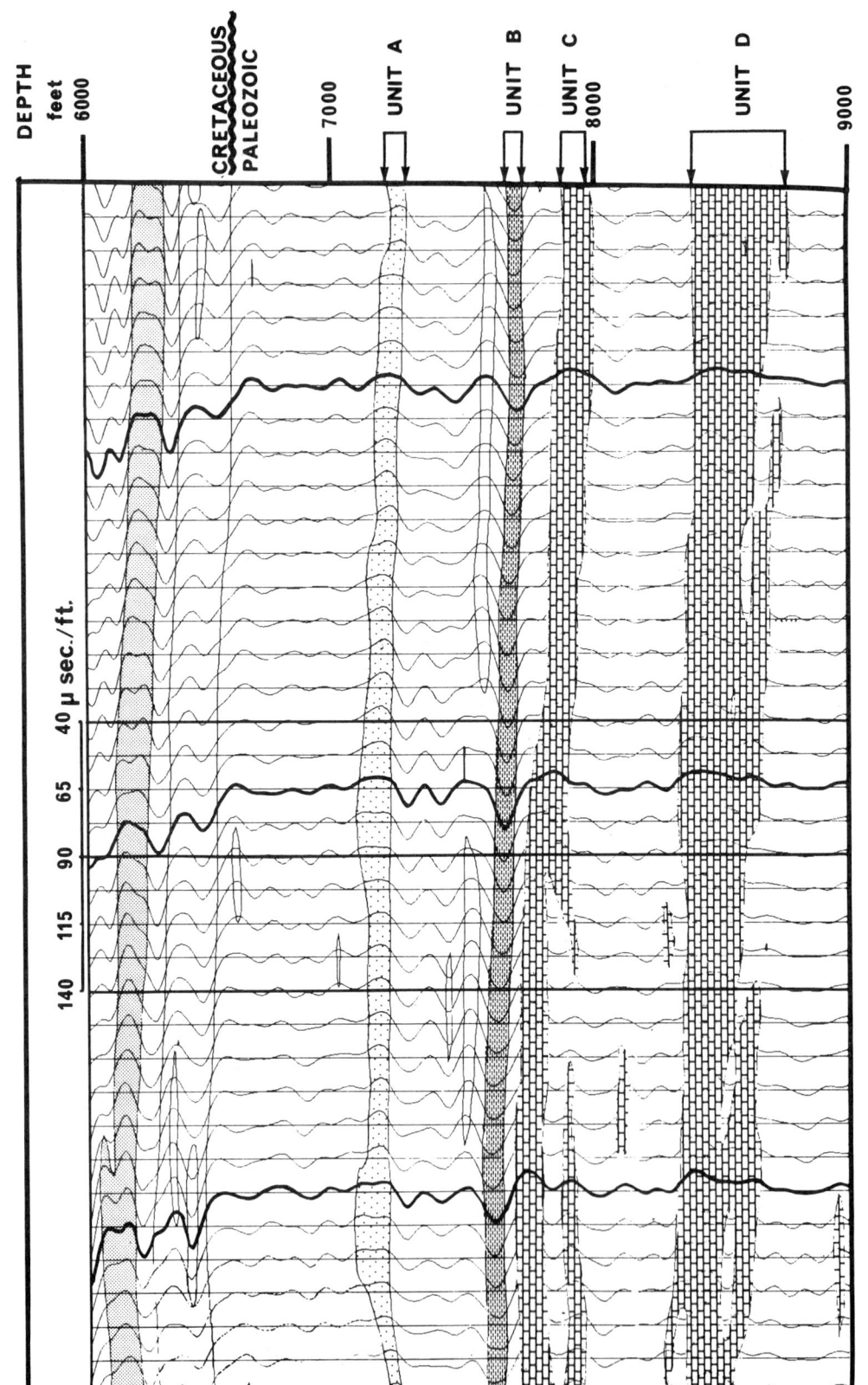

FIG. 12. Isotransit-time contours of synthetic sonic log values can define formation contacts, outline individual stratigraphic units, and reveal depositional patterns. The operation can be automated readily for machine contoured output as shown here.

serve both to mark the transit-time reference for the log excursions and to locate the geographic position of the shotpoint to which the log trace corresponds. Thus, horizontal excursions of the log represent variations in transit-time while a horizontal jump from one log to the next represents a shift in shotpoint location.

The section in Figure 12 corresponds to a Paleozoic carbonate sequence capped by lower Cretaceous sandstones interbedded with shale layers. The major age division in the upper part of the section is made clearly visible by an abrupt excursion on all logs to intersect the 65 μsec/ft line, which line they then more or less follow with increasing depth.

Immediately above the level of the major shift in transit-time a line has been drawn through a point on each log where the transit-time is 72.5 μsec/ft. The continuous line joining these points is the contact between the lower velocity Cretaceous section above, and the higher velocity Paleozoic section below. Thus, here a continuous transit-time velocity contour defines the top of a horizon (the unconformity) quite precisely.

Within the Paleozoic section, each log more or less coincides with the 65 μsec/ft vertical grid line for that particular log. About half-way down the section, near the top of unit A, the logs make a small excursion to the right of the grid line. A point at 62.5 μsec/ft has been marked on each log and a continuous line joining equivalent points at this level on all logs appears as a smooth contour. The line defines the top of stratigraphic unit A within the section. A similar line, having the same transit-time value as the contour above, is drawn at the base of the unit. The two isovelocity lines map the top and the base of a single, approximately homogeneous lithologic unit. Compare the rather flat line at the base of this unit with the undulating form of the line at the top. The upper contour displays the detail of a natural depositional surface of a stratigraphic unit which is less than 100 ft thick.

Deeper in the section a sharp low-velocity excursion, marking the top of unit B, occurs on each log. Another pair of isovelocity contour lines is used to outline this excursion. The value selected in this case is 72.5 μsec/ft. The resulting contour lines are very nearly parallel, with a slight convergence to the right, outlining a shale member which gradually thins to the right side of the section. This member is 75 ft thick or less which, on a conventional seismic section is no more than one-third of the amount of section ordinarily encompassed by the wavelength corresponding to one cycle of the usual seismic reflection wavelet.

A thick high-velocity unit C, a limestone, lies immediately below the well-defined thin shale, unit B. On the left side of the example, the 62.5 μsec/ft contour defines the top of high-velocity facies within unit C. The log response at this level is almost identical over some fourteen traces from the left side of the section, then changes rather abruptly. The velocity of the rock at the same depth is reduced, as demonstrated by the shift of the logs toward the 65 μsec/ft grid line, while the contour defining the top of the higher-velocity unit C falls deeper in the section. This suggests the facies immediately below unit B changes laterally over the right half of the section to increasing amounts of shale or to a more porous limestone unit. The reader may wish to sketch in a contour at the intersection of the logs with the 65 μsec/ft baseline above and below unit C to observe for himself the manner in which this procedure reveals significant information of the depositional sequence.

Finally, a thick unit D of high-velocity material is defined. The top of this unit is rather smooth, whereas the base of the unit appears to cycle in a regular pattern of progradation across the section.

The lower part of the section in Figure 12 is over 8000 ft deep, yet the contours outline units that appear as thin as 50 ft. The transit-time and thickness of these thin beds may not be defined correctly; however, changes in thickness and position in the section correspond reliably with those observed on well logs. Compare these features with similar features on the filtered borehole logs of Figure 3.

The above procedure is easily automated. The contours of the example in Figure 12 are machine-drawn, demonstrating the potential for mapping in detail those stratigraphic units for which a velocity contrast exists at the boundary between adjacent units.

Examples of interpretation

Whereas structural traps can, in most cases, be mapped with adequate precision by conventional seismic sections, the procedures outlined have been found to be of particular value in the location of stratigraphic traps, whether or not they are related to structure. The application will be illustrated by a number of specific examples.

Carbonate porosity.—A large area of the Western Canadian basin contains a massive lower Devonian reef bank termed the Swan Hills (see Figure 3). The thickness of the bank averages about 350 ft and, with the exception of the edges of the

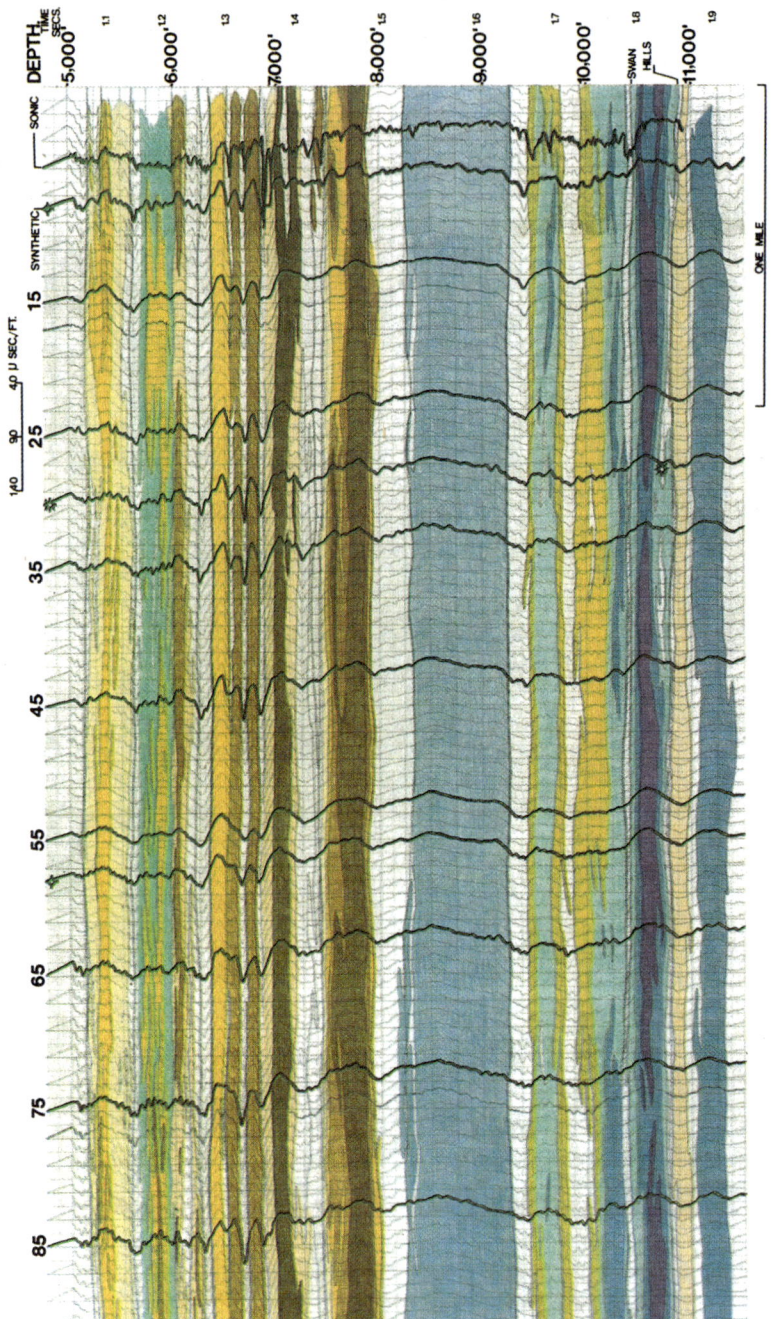

FIG. 13. Synthetic sonic log section. A borehole sonic log from a dry hole superimposed on the section at the right end near a tie point correlates well to the synthetic log, demonstrating the effectiveness of the inversion process. The line crosses gas production obtained from a 65-ft interval of porosity in Swan Hills carbonate at a depth of 10,600 ft near shotpoint 30. Porosity appears as low-velocity intervals (white) in otherwise high-velocity limestones (blue).

bank and channels that cut into it, the thickness is nearly uniform over wide areas. Most of the bank is composed of dense, tight carbonate, but occasionally porosity occurs which is frequently filled with gas or water and sometimes with oil. Porous zones are localized, variable in thickness, and may lie at almost any level within the bank. Rarely do they extend laterally over considerable distance and they have no apparent relationship to structure.

On a seismic section the time interval corresponding to the bank occurs between two reflections separated by a few tens of msec. In this area, "bright spots" (reflection amplitude increases associated with gas) have been used in recent years as an indicator with some degree of success. Unfortunately, other gas accumulations occur where bright spots do not occur and, conversely, bright spots are observed where no porosity exists.

A line of seismic data in a representative area was obtained using dynamite as a source, recording with 12-Hz geophones, binary gain instruments, and with no filter applied in the field other than the 125 Hz antialias filter corresponding to the 2-msec sampling interval.

Reprocessing of the data through the broadband wavelet processing sequence and subsequent inversion of the data yielded the synthetic log section shown in Figure 13. The section extends in depth from 5000 ft to approximately 11,500 ft. On the right-hand side is a sonic log from a nearby borehole, plotted to the same depth scale as the synthetic log traces. The correlation of the synthetic log to the sonic log is excellent in the upper part of the section, deteriorates slowly with depth, and is not particularly good below 10,000 ft, the level of interest. A small depth mis-tie is noted near 9500 ft below a massive carbonate layer, due likely to a slight undercorrection for density of the carbonate. However, this has been subsequently corrected by an overcorrection for density of a shale lying between 9500 and 10,500 ft. Hence, depths are reliable near the zone of interest at 10,500 to 11,000 ft.

Close comparison of the sonic log and synthetic log in the depth interval 9000 to 11,000 ft, Figure 14, reveals that the synthetic log displays the basic form but not the detail of the sonic log. When the sonic log is high-cut filtered, in this case at approximately 75 Hz, the correlation between the two is much improved. Notice, however, that the synthetic log trace appears to have a low-frequency oscillation, shifting excessively to the left at 10,000 ft and then too far to the right within the Swan Hills zone. This is the result of missing spectral components in the vicinity of 10 Hz. This deviation between the synthetic log and the sonic log is nearly the same for other comparisons in the area. Hence, synthetic log traces reliably exhibit changes in the section.

The Swan Hills carbonate bank zone has been contoured at increments of 2.5 μsec/ft in Figure 13, and each interval between contours has been color-coded. The bank has a fairly continuous core of high-velocity material over the length of the line. However, within the interval between shotpoints 25 and 45, the velocity near the base of the bank decreases as indicated by the shift of the traces to the left. Contours outline two low-velocity members (uncolored) which, although separate here, may possibly join outside of the line of section. A total shift of 7.5 μsec/ft occurs between the highest and lowest

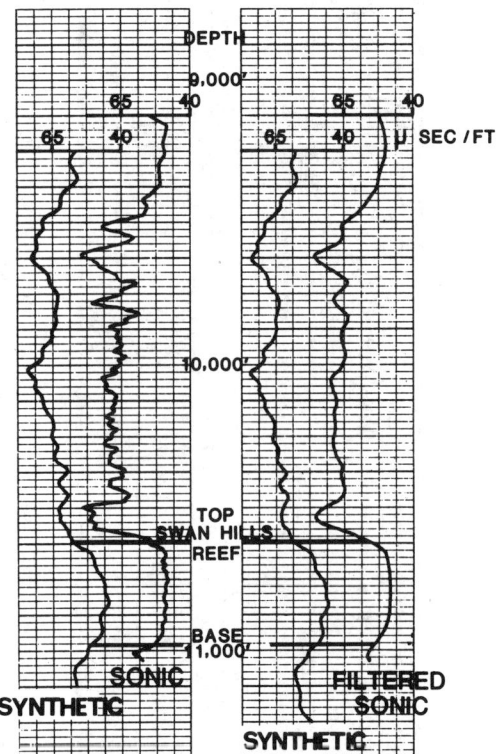

FIG. 14. Detail comparison between synthetic log and unfiltered borehole sonic (left) of Figure 13 in the depth interval 9000 to 11,000 ft. Comparison is poor because of narrow frequency bandwidth of synthetic log at this depth. Filtering sonic log (right) to approximate bandwidth of synthetic improves correlation. Other differences are due to the absence of frequencies in a notch at about 10 Hz on synthetic log.

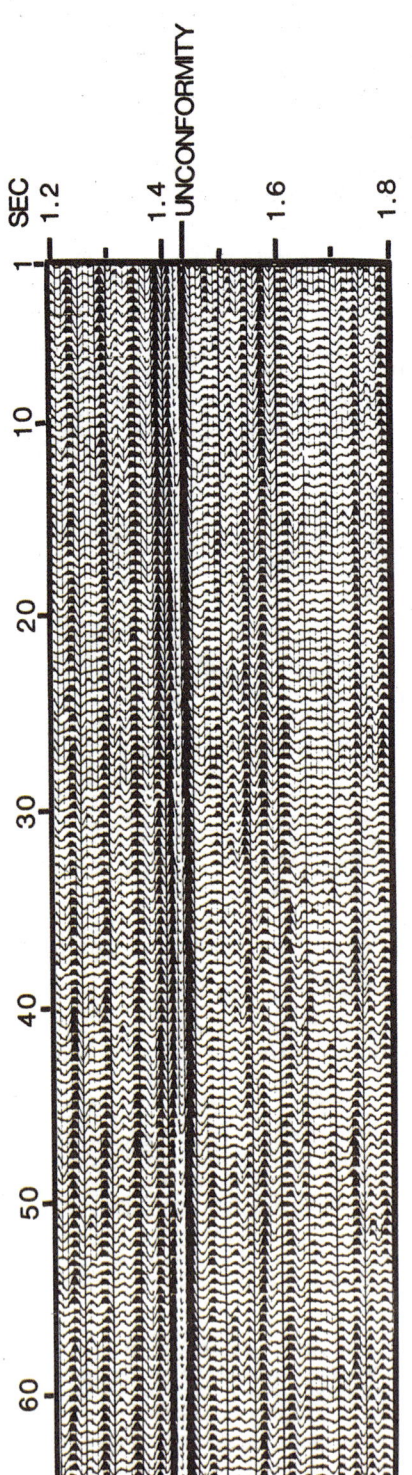

FIG. 15. An unconformity on the conventional seismic section lies at about 1.44 sec, but it is usually mapped on the heavy reflection indicated because the lower reflections do not have good continuity.

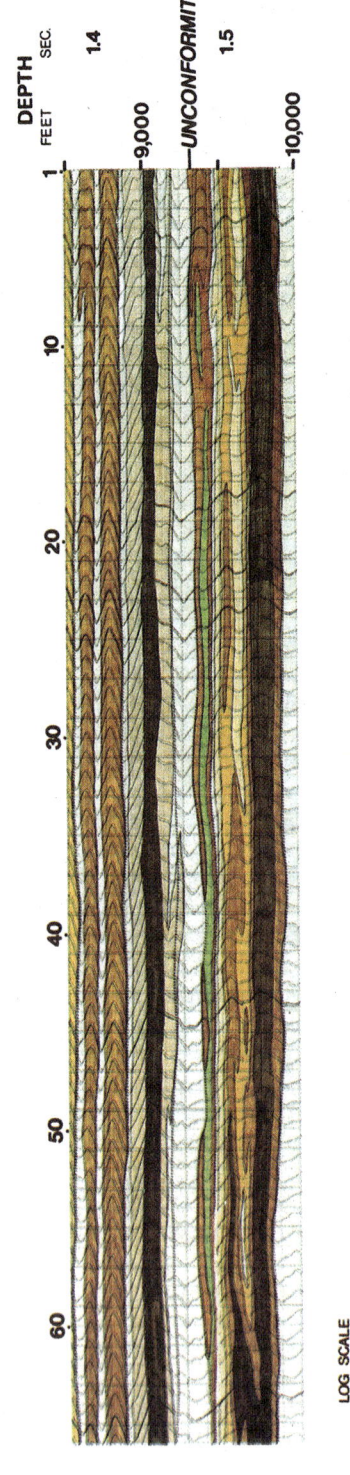

FIG. 16. Synthetic logs show that the unconformity lies deeper in the section than the level usually mapped on the conventional seismic section of Figure 15 and provide detail of the facies which subcrop at the unconformity.

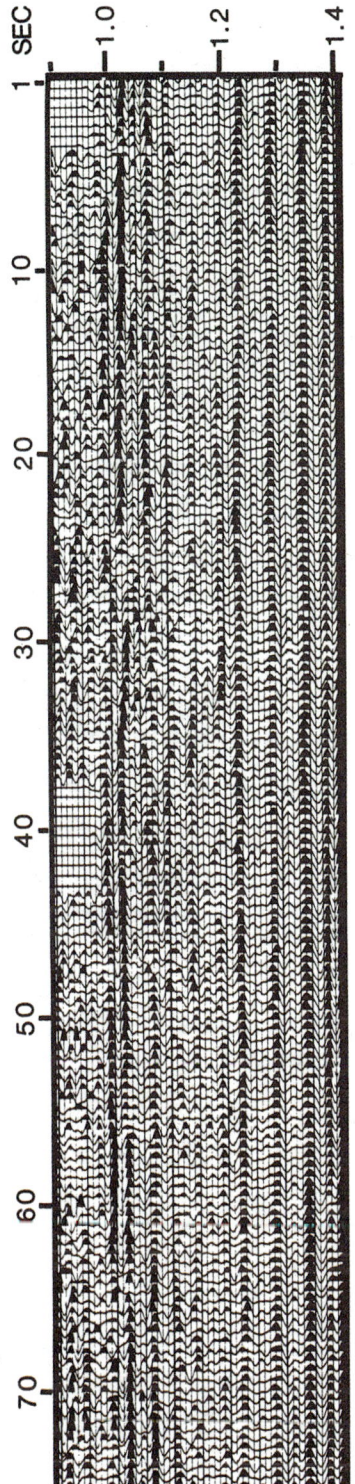

FIG. 17. A weak seismic reflection near 1.1 sec maps the top of the Cardium zone, a series of thin sands. The seismic data appear of rather poor quality but are typical for this zone.

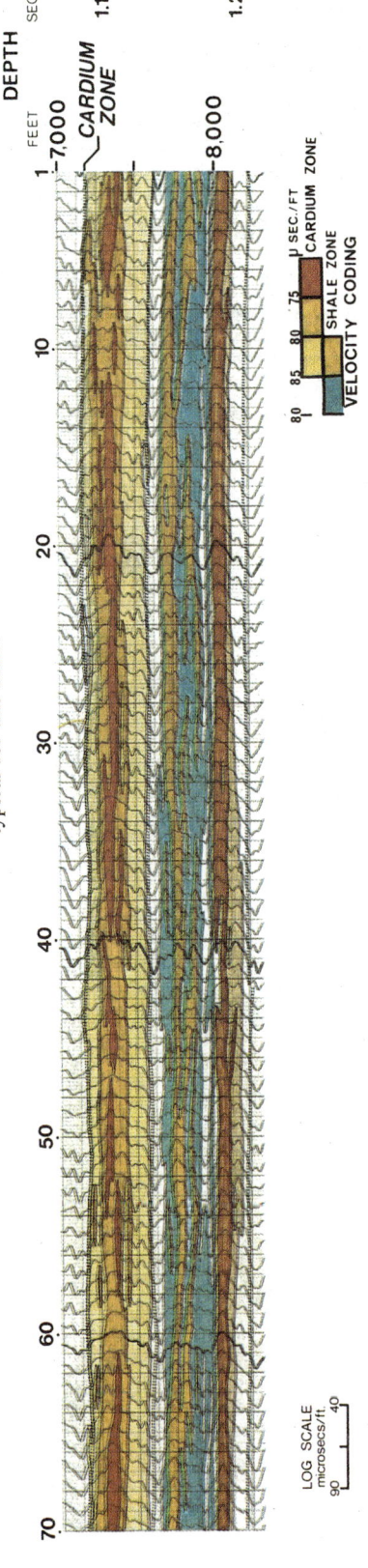

FIG. 18. Contouring synthetic logs corresponding to data of Figure 17 easily outlines the high-velocity prospective sandstones in the Cardium zone. The apparent poor quality of the seismic input data might suggest the lenticular pattern of sands is due to noise, but only a few msec deeper the group of sediments below 7500 ft contains several very thin continuous members which offlap from another continuous unit above; evidence that depositional environment, rather than noise, is the source of the pattern.

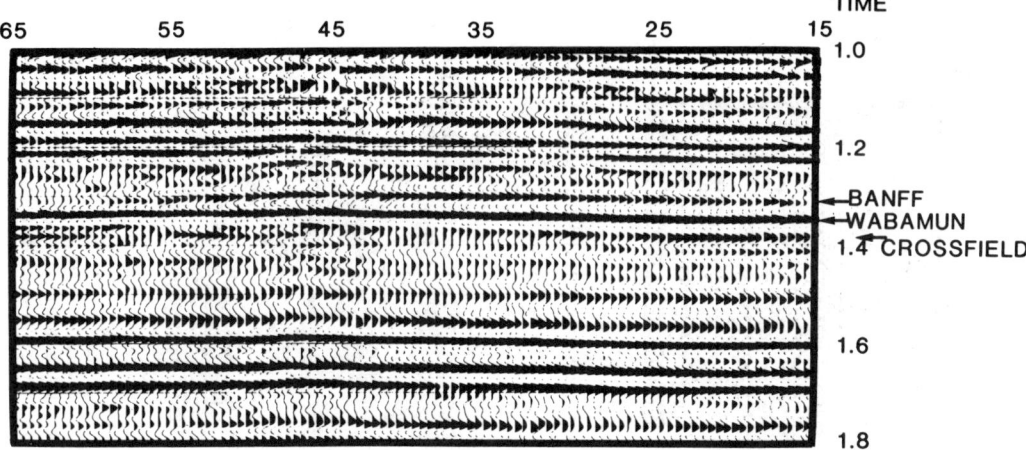

FIG. 19. A seismic section across a gas field producing from a local porous zone in the Crossfield carbonate. The extent of the field defined by previous drilling is from beyond the left side of the example to approximately shotpoint 55.

velocity material within the bank. The difference, in this area, is a reliable indicator of porosity in the bank. The well drilled at the position indicated uncovered excellent porosity and production, although the well was not located on the basis of this study.

Unconformity.—The second example, Figure 15, illustrates the problem of mapping an unconformity within a section when the bedding above and below is essentially parallel. In the area of the example, mapping of the unconformity is usually accomplished by following the strong reflection near 1.44 sec. However, this reflection is in fact from a member of the lower Cretaceous, which drapes over the Paleozoic unconformity, conforming to local structural changes at the unconformity only in a gross manner. The actual unconformity lies somewhat deeper, at a very weak reflection which cannot always be mapped on conventional seismic sections with confidence.

Inversion of the data is demonstrated in Figure 16, which is a synthetic log section over the zone of interest (8500–10,000 ft). Approximate reflection times have been marked on this section for correlation back to the seismic time section. The unconformity is represented by a contour line separating a zone of low-velocity material from a somewhat higher velocity zone below. An extensively eroded argillaceous limestone overlies a thin shale which subcrops at the unconformity. Between shotpoints 38 and 41, and again between shotpoints 48 and 50, erosion has removed the limestone, and beyond shotpoint 62 both the thin shale and the limestone have been completely eroded.

Thin sandstone beds.—The next example demonstrates the ability of the isotransit-time contours to map thin sandstone beds and to indicate the probable manner of deposition. The seismic section in Figure 17 has been processed by conventional techniques and illustrates the results which would normally be available for interpretation. The same data were reprocessed using broadband techniques and inverted to obtain the depth section illustrated in Figure 18. Two zones of deposition are illustrated. The upper is the Cardium zone, a group of sediments which may produce oil when they are sufficiently clean and well developed. Below this zone is a shale with minor sandstone developments.

The Cardium zone is bounded by distinct low-velocity shales at the top and base, as clearly illustrated on the synthetic logs. Within the Cardium zone the velocity increases as the sandstones become less contaminated by shale. The darkest color, representing the highest velocity, is clean sandstone that can be productive. In this area, however, the sandstones contain only brine.

At first glance, it might be concluded that the apparent discontinuous nature of the bedding within the Cardium zone might be merely the effect of noise or lack of resolution, particularly in view of the rather poor input data (Figure 17). Confidence that the pattern represents stratigraphy may be increased by examination of the deeper zone.

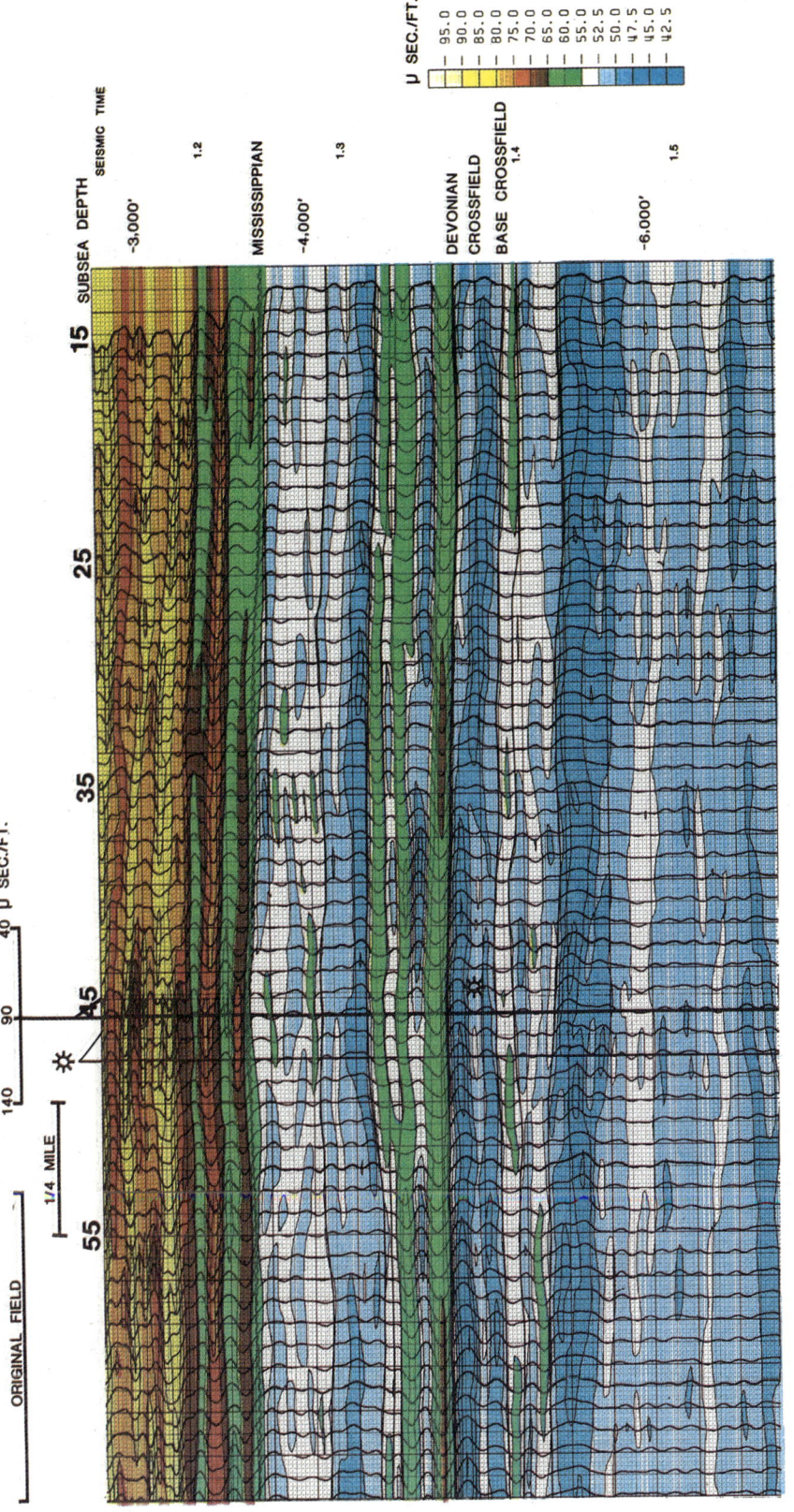

FIG. 20. Synthetic sonic logs produced by broadband processing and inversion of data in Figure 19. Porosity development is indicated by a change from high to low velocity in the Crossfield member near shotpoint 40, extending the definition of the producing area beyond the limit previously indicated by drilling to be near shotpoint 55.

The second zone, below the Cardium, is composed largely of shale, with a few poorly developed sandstones. Isovelocity contours in this part of the section exhibit a pattern quite different to that in the section lying only 50 msec above. Near the base of this zone, thin sandstone beds, varying from 25 to 50 ft in thickness, prograde from left to right across the section and are capped by a thicker continuous band of sediments of similar velocity. The continuity of the very thin sandstone members at this level likely would not be maintained if the section were contaminated by noise. Contrast the regular pattern of deposition here with the short, lenticular pattern of the Cardium sandstones above. The difference in depositional patterns, only 50 msec apart in time, tends to support the conclusion that the effects observed on the section are geologically meaningful.

Carbonate porosity.—The final example is from a study which correctly predicted a recent discovery in Alberta, Canada. The seismic section in Figure 19 crosses an existing gas field. Production is obtained from porosity in the Crossfield carbonate member of the Devonian Wabamun formation at about 5000 ft subsea elevation. The production extends from beyond the left end of the figure to about shotpoint 55. A reflection correlated to the Crossfield member occurs at a time of approximately 1.39 sec.

The seismic section was inverted to produce the synthetic log section in Figure 20. The Crossfield zone is at an elevation of approximately −5000 ft. The traces show very high velocity material, indicative of dense carbonates, between shotpoints 15 and 38. At shotpoint 39 a lower-velocity zone, corresponding to transit-time greater than 52.5 μsec/ft occurs in the Crossfield member and develops rapidly into a well-defined feature about 75 ft thick, extending to the previous limit of development near shotpoint 55. This feature has been left uncolored for clarity.

It is now evident, from subsequent results, that the constriction near shotpoint 55 may actually be non-porous, providing separation between the newly discovered feature to the right and existing production to the left. A test drilled near shotpoint 45 found a separate reservoir.

Of particular note is the imperfect correlation between the reservoir outlined on the synthetic log section and the character of the seismic section reflections. A change in the character of the corresponding seismic reflection is evident on Figure 19, but it does not define the reservoir. To demonstrate this further, the extreme left end of the seismic line

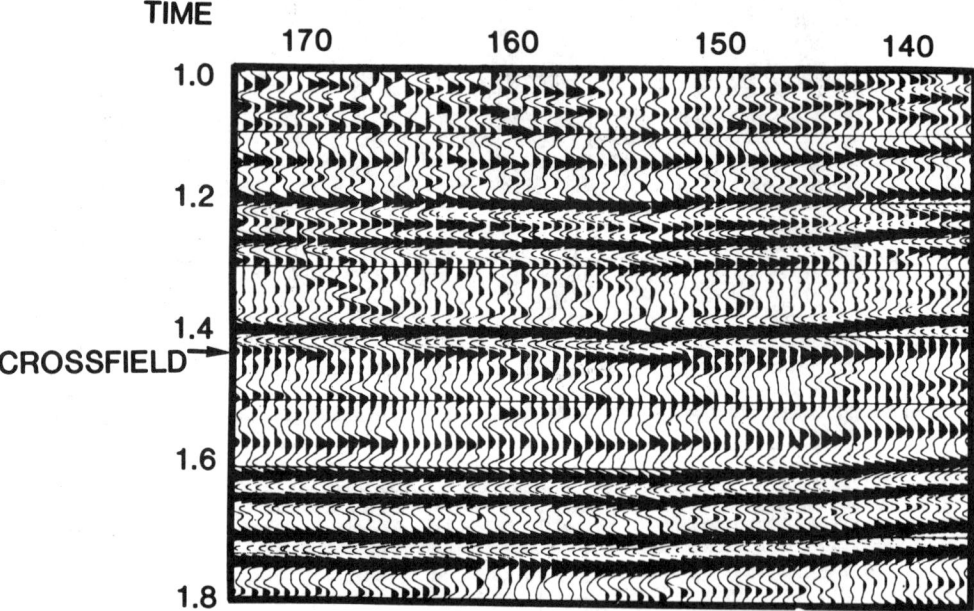

FIG. 21. The seismic section at the extreme end of the line containing the section in Figure 19 shows reflection character at the Crossfield level beyond the production similar to that associated with production.

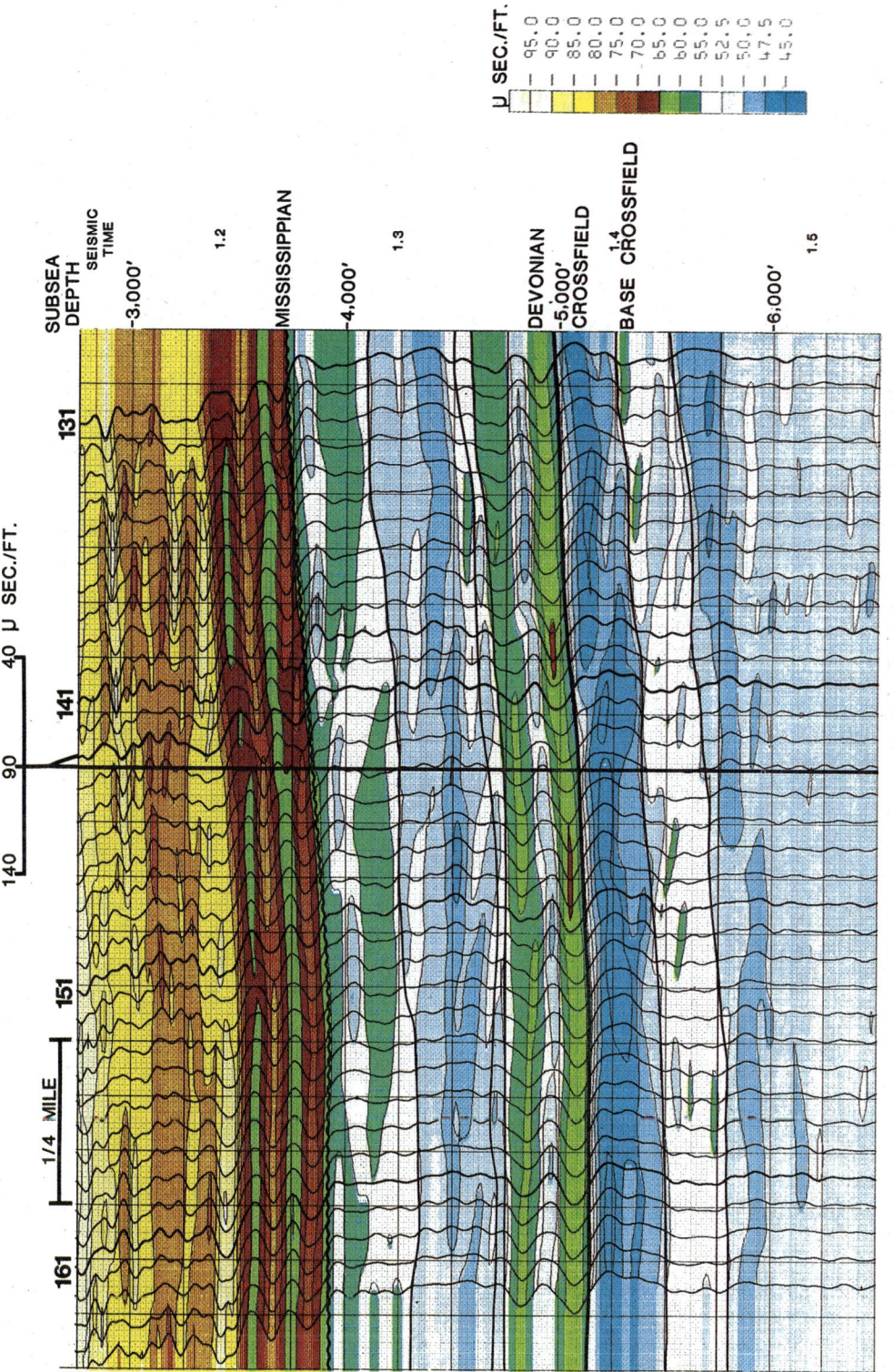

Fig. 22. Synthetic logs developed from the data of Figure 21 show no Crossfield porosity to be present, contrary to the conclusion that might be drawn from the conventional seismic presentation.

containing the section in Figure 19, has been reproduced in Figure 21. Although this portion of the line is beyond the gas field, the Crossfield reflection appears to have character similar to that associated with production. The corresponding synthetic log section of Figure 22, shows only limited velocity change at the level of interest, indicating insufficient porosity to provide a reservoir.

CONCLUSIONS

A new method of processing, displaying, and interpreting seismic data has been presented which increases the amount of geologic information available to the interpreter. Modern seismic field data are deconvolved by wavelet processing, inverted and combined with low-frequency velocity components derived from reflection analyses to produce synthetic sonic logs.

The synthetic logs provide consistent detailed velocity values which are then contoured to outline individual units. If the assumption is made that a given rock unit will maintain a uniform velocity until it undergoes some change in facies or porosity, a cross-section of individual velocity/lithology units can be developed, to provide an insight into the nature of the sedimentary section not generally obtainable from seismic data.

Finally, it should be stressed that the process is a transform of seismic data, and the results may contain the same errors, propagation phenomena, migration effects, or other aberrations in the seismic data.

ACKNOWLEDGMENTS

The many contributions of my colleague, A. V. Street, are acknowledged with deep appreciation. The author also wishes to thank the reviewers for their very effective suggestions and comments. Data examples are provided through the courtesy of Sigma Explorations Ltd.

REFERENCES

Al-Chalabi, M., 1974, An analysis of stacking, rms, average, and interval velocities over a horizontally layered ground: Geophys. Prosp., v. 22, p. 458–475.
Beitzel, J. E., Cone, R. M., Dees, J. L., and Thompson, D. E., 1977, Understanding seismically derived impedance logs: Presented at the 47th Meeting of SEG, September 23 in Calgary.
Branisa, F., 1974, Filtering of well-log curves: Geophysics, v. 39, p. 545–549.
Delas, C., Beuchamp, J. B., de Lombares, G., Fourmann, J. M., and Postic, A., 1970, An example of practical velocity determination from seismic traces: Presented at the 32nd Meeting of the EAEG, Edinburgh.
Dix, C. H., 1955, Seismic velocities from surface measurements: Geophysics, v. 20, p. 68–86.
Dresser Atlas, 1971, Log interpretation: Houston, Dresser Industries Inc.
Everett, J. E., 1974, Obtaining interval velocities from stacking velocities when dipping layers are included: Geophys. Prosp., v. 22, p. 122–142.
Embree, P., Burg, J., and Backus, M., 1963, Wideband velocity filtering—the pie-slice process: Geophysics, v. 28, p. 948–974.
Faust, L. Y., 1951, Seismic velocity as a function of depth and geologic time: Geophysics v. 16, p. 192–206.
Gangi, A. F., and Yang, S. J., 1976, Traveltime curves for reflections in dipping layers: Geophysics, v. 41, p. 425–440.
Gardner, G. H. F., Gardner, L. W., and Gregory, A. R., 1974, Formation velocity and density—the diagnostic basis for stratigraphic traps: Geophysics, v. 39, p. 770–780.
Guyod, H., and Shane, L. E., 1969, Geophysical well logging: Houston, Hubert Guyod.
Lavergne, M., 1975, Pseudo diagraphies de vitesse en offshore profond: Geophys. Prosp., v. 23, p. 695–711.
Lavergne, M., and Willm, C., 1977, Inversion of seismograms and pseudo velocity logs: Geophys. Prosp., v. 25, p. 231–250.
Lindseth, R. O., 1972, Approximation of acoustic logs from seismic traces: J. Canadian Well Logging Society, v. 5, p. 13–26.
——— 1975, Acquisition of velocity information in terms of sampling theory: Presented at the Midwestern SEG Meeting in Tulsa.
Lines, L. R., and Clayton, R. W., 1977, A new approach to Vibroseis deconvolution: Geophys. Prosp., v. 25, p. 417–433.
Lines, L. R., and Ulrych, T. J., 1977, The old and the new in seismic deconvolution and wavelet estimation: Geophys. Prosp., v. 25, p. 512–540.
Mayne, W. H., 1962, Common reflection point horizontal data stacking techniques: Geophysics, v. 27, p. 927–938.
Oppenheim, A. V., and Schafer, R. W., 1975, Digital signal processing: Englewood Cliffs, Prentice-Hall Inc.
Rice, R. B., 1962, Inverse convolution filters: Geophysics v. 27, p. 4–18.
Robinson, E. A., 1957, Predictive decomposition of seismic traces: Geophysics, v. 22, p. 767–778.
Schlumberger, 1972, Log interpretation; Vol. 1—Principles: New York, Schlumberger Limited.
Schoenberger, M., 1974, Resolution comparison of minimum-phase signals: Geophysics, v. 39, p. 826–833.
Schoenberger, M., and Levin, F. K., 1976, Reflected and transmitted filter functions for simple subsurface geometries: Geophysics, v. 41, p. 1305–1317.
Stoffa, P. L., Buhl, P., and Bryan, G. M., 1974, The application of homomorphic deconvolution to shallow-water marine seismology: Geophysics, v. 39, p. 401–426.
Street, A. V., and Lindseth, R. O., 1974, Continuous velocity mapping from seismic reflection data: Presented at the 36th EAEG Meeting, in Madrid.
Widess, M. B., 1973, How thin is a thin bed?: Geophysics, v. 38, p. 1176–1180.
Woeber, A. F., and Penhollow, J. O., 1975, Depth prediction from velocity profiles on the Texas Gulf Coast: Geophysics, v. 40, p. 388–398.

VELOCITY SPECTRA—DIGITAL COMPUTER DERIVATION AND APPLICATIONS OF VELOCITY FUNCTIONS†

M. TURHAN TANER* AND FULTON KOEHLER‡

Multifold ground coverage by seismic techniques such as the common reflection point method provides a multiplicity of wave travel path information which allows direct determination of root-mean-square velocities associated with such paths. Hyperbolic searches for semblance among appropriately gathered arrays of traces form the basis upon which velocities are estimated. Measured semblances are presented as a velocity spectral display. Interpretation of this information can give velocities with meaningful accuracy for primary as well as multiple events. In addition, the velocity data can help correctly label events. This paper outlines the fundamental principles for calculating velocity spectra displays. Examples are included which demonstrate the depth and detail of geological information which may be obtained from the interpretation of such displays.

MOTIVATIONS FOR AND HISTORICAL BACKGROUNDS OF VELOCITY ANALYSIS

Proper indentification of primary reflections amidst a background of noise and energy from various multiple reflections was a major problem when all seismic work was single coverage and is still a problem in the common reflection point multifold coverage method. Many analog and digital techniques have been developed to reduce this undesirable interference and have been utilized with varying degrees of success. Time-varying inverse filtering and common reflection point techniques are among the most effective procedures. See for instance Mayne (1962) and Courtier and Mendenhall (1967).

The objective of improved signal-to-noise ratio and attenuation of multiples utilizing the CRP method can be accomplished only by applying the proper time corrections to the seismic data, so that the primary reflection signals will be moved in-phase and stacked properly. It is first necessary, therefore, to determine corrections for all primary reflections for all the traces (with different shot-to-geophone distances) which will be stacked into a single trace.

This work develops and discusses the velocity spectral display as a powerful tool for identifying primary reflections and stacking velocities. Information relevant to lithology is also shown to be present by careful study of these displays. Underlying the method is the well-established principle of employing a correlation-like technique upon an assemblage of seismic traces which contain redundant information so that an improved estimate of some desired quantity may be made. Rieber (1936), Trorey (1961), Picon and Utzmann (1962), and Simpson (1967) are among the authors who have also recognized and used this idea. In fact, the velocity analysis of Schneider and Backus (1968), which was developed contemporaneously with our method, also made use of these fundamentals.

THE VELOCITY SPECTRA DISPLAY, CONSTRUCTION

In the past, time corrections have been estimated from previously measured nearby well velocities utilizing simple straight-ray or curved-ray methods, and resulting misalignments were corrected by applying residual normal moveout corrections. This procedure is somewhat unsatisfactory because it is time consuming and, since the data is handled more than once, it is costly. It becomes more unsatisfactory when there is no

---

† Presented at the 37th Annual International SEG Meeting, Oklahoma City, November 2, 1967. Manuscript received by the Editor December 17, 1968; revised manuscript received August 1, 1969.

* Seismic Computing Corp., Houston, Texas 77036.

‡ Seismic Computing Corp., Houston, Texas 77036, and University of Minnesota, Minneapolis, Minnesota 55455.

Copyright ©1970 by the Society of Exploration Geophysicists.

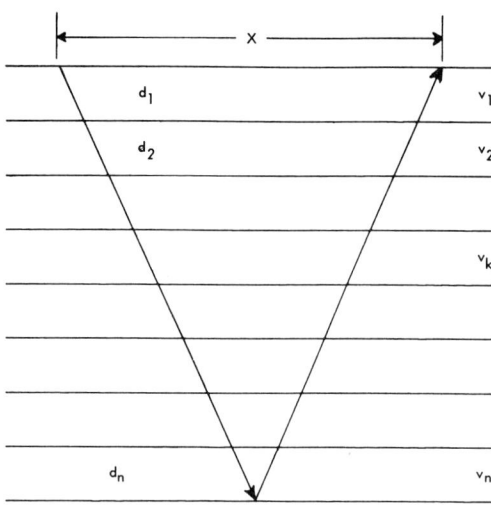

FIG. 1. Straight ray analysis in horizontally layered medium.

well data available. In any case, the time corrections from known data are generally generated from the simple case of horizontally layered media and effects of dipping beds are not considered.

Let us now consider a horizontally layered medium (Figure 1) where,

v_k = interval velocity at kth layer,
d_k = thickness of kth layer,
t_k = two-way traveltime within the kth layer, $t_k = 2d_k/v_k$,
X = horizontal distance between the energy source and detecting device.
$T_{0,n}$ = two-way vertical traveltime to the base of nth layer

$$T_{0,n} = \sum_{k=1}^{n} t_k = 2 \sum_{k=1}^{n} \frac{d_k}{v_k}, \quad (1)$$

$V_{a,n}$ = average velocity to the base of nth layer

$$V_{a,n} = \frac{\sum_{k=1}^{n} v_k t_k}{T_{0,n}} = \frac{2 \sum_{k=1}^{n} d_k}{T_{0,n}}, \quad (2)$$

$T_{x,n}$ = arrival time of a reflection from the base of nth layer at X distance.

Now, if we assume that the wavefront is traveling along the shortest (distance) path between the energy source, reflector, and geophone, total traveltime, which we will call arrival time, $T_{x,n}$ is given by the familiar relationship

$$T_{x,n}^2 = T_{0,n}^2 + \frac{X^2}{V_{a,n}^2}. \quad (3)$$

However, we know that the wavefront is traveling along the shortest time path, in accordance with Fermat's principle, hence the arrival time will be given by the infinite series of the form (Figure 2)

$$T_{x,n}^2 = c_1 + c_2 X^2 + c_3 X^4 + c_4 X^6 + \cdots \quad (4)$$

where the coefficients $c_1, c_2, \cdots$, depend on layer thicknesses $d_1, d_2, \cdots, d_n$, and the interval velocities $v_1, v_2, \cdots, v_n$. Derivation of equation (4) is presented in Appendix A, and it is in accordance with Snell's law. Studies have shown that, for the X distances generally encountered in practice, the first two terms of equation (4) give an accuracy of about 2 percent which is adequate for seismic exploration purposes. Therefore, we can neglect all the terms containing higher powers of X^2, and simplify the expression to,

$$T_{x,n}^2 = c_1 + c_2 X^2, \quad (5)$$

where

$$c_1 = \left(\sum_{k=1}^{n} t_k \right)^2 = T_{0,n}^2 \quad (6)$$

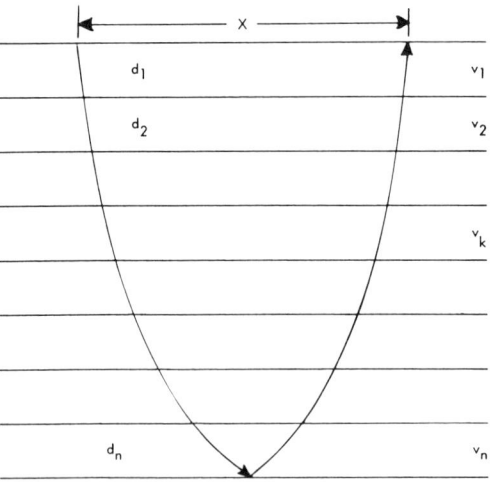

FIG. 2. Snell's law ray analysis in horizontally layered medium.

$$c_2 = \frac{\sum_{k=1}^{n} t_k}{\sum_{k=1}^{n} t_k v_k^2} \equiv \frac{1}{\overline{V}_n^2}, \quad \text{or,}$$

$$\overline{V}_n^2 = \frac{\sum_{k=1}^{n} v_k^2 t_k}{T_{0,n}}. \qquad (7)$$

Therefore, equation 5 becomes,

$$T_{x,n}^2 = T_{0,n}^2 + \frac{X^2}{\overline{V}_n^2} \qquad (8)$$

which is very similar to equation (3), with the notable difference that, instead of average velocity $V_{a,n}$, we are using $\overline{V}_n$, as defined by equation (7). This apparent velocity is the same one as given in a formula by Dix (1955), and represents the time-weighted, mean-square velocity. In order to differentiate it from the linear average velocity, we will call this the rms (root-mean-square) velocity. This expression suggests that, in the areas where the beds have gentle dips, the time corrections could be computed fairly accurately by

$$\Delta T_n = \sqrt{T_{0,n}^2 + \frac{X^2}{\overline{V}_n^2}} - T_{0,n}, \qquad (9)$$

where

$$\Delta T_n = T_{x,n} - T_{0,n}. \qquad (10)$$

When we consider the problem for the general case of arbitrarily dipping beds, it becomes practically impossible to obtain an expression similar to equation (4) that would give the explicit relation between time and distance. However, we could obtain this relation by studying mathematical models and ray tracing using Snell's law. Figure 3 shows an example of the model study where CDP traces for a horizontally layered case have been simulated. Observed and computed arrival times, distances, and rms and average velocities are tabulated for comparison. It should be noted that the average velocity is lower than the rms velocity in all cases.

Single coverage record simulation for dipping interfaces is shown in Figure 4. In this example, the shotpoint is assumed to be at the center of the spread. Time-distance relations are still nearly hyperbolic, however, the curves are translated in the updip direction. Figure 4 also contains computed raypaths for a few typical geophone locations. There is a unique and most important raypath among all of these, that is, the one for which shotpoint and recording geophone are in the same location, and the travel path is perpendicular to the reflector so that incident and reflected raypaths become the same. We will call this path the

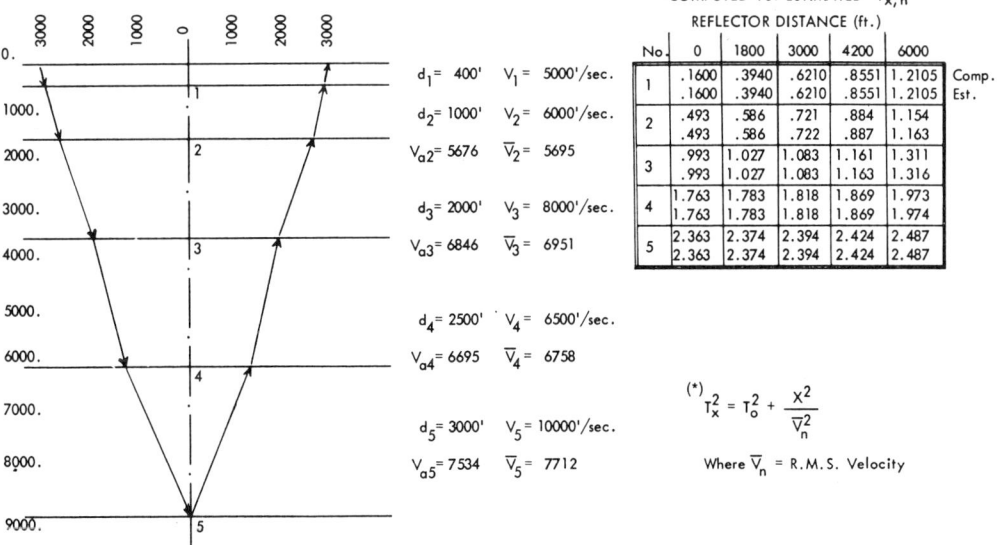

Fig. 3. Horizontally layered model study.

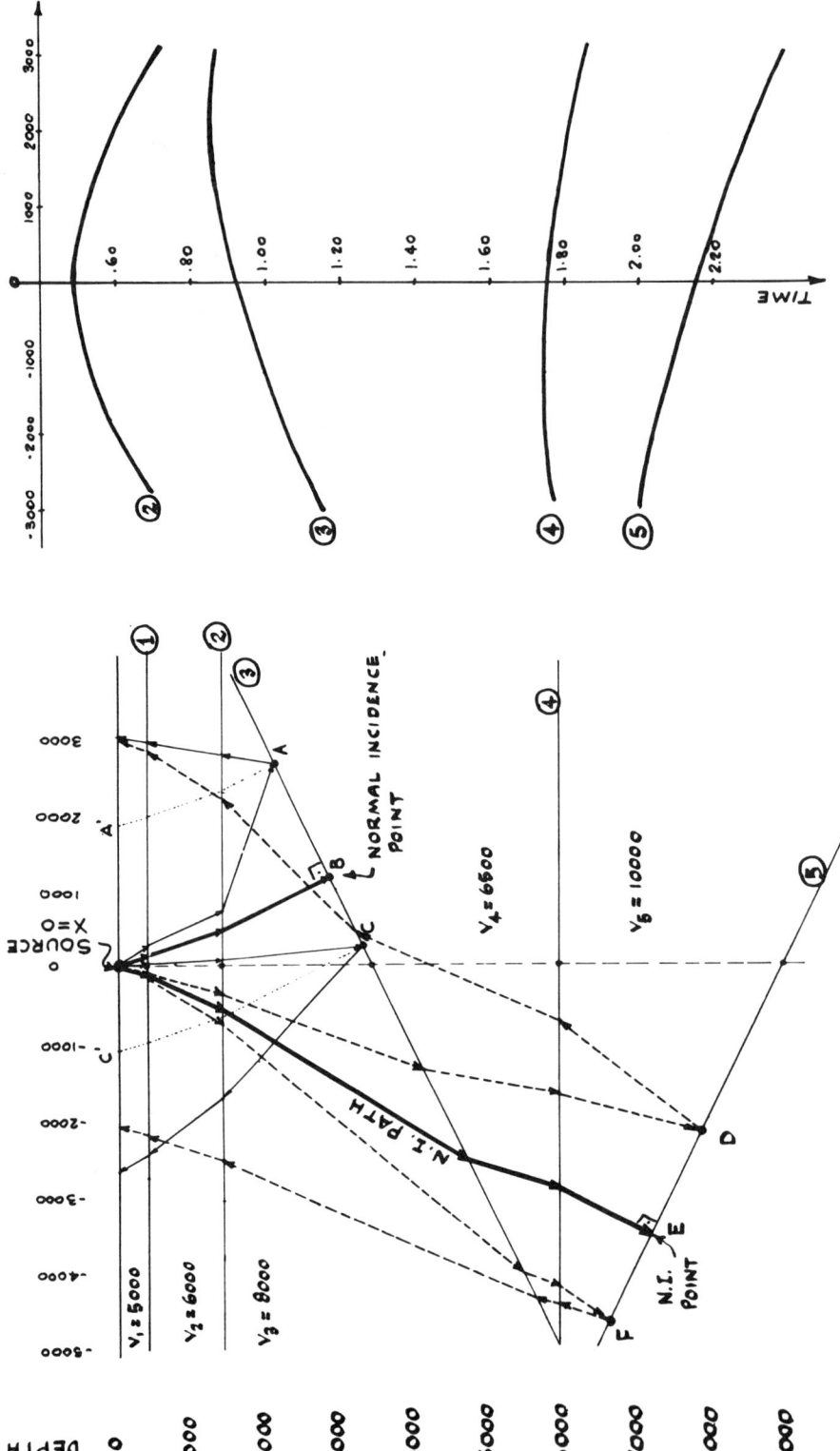

FIG. 4. Single coverage model with dipping reflectors—depth and CRP time section.

normal incidence path (NIP) and the time taken to travel this path (from shot to reflector to geophone) the normal incidence time (NIT).

The time correction (ΔT_n) in the case of single coverage records is the time difference between the total traveltime (from shot to a particular reflection point to the geophone) and normal incidence time for that particular reflection point. By applying a time correction to a given trace, we are generating a new trace whose shot and recording positions coincide at the ground position of the normal incidence path going through the reflection point.

Now let us look at the common-depth-point simulation as shown on Figure 5. These travel paths are generated by assuming shot and recording locations are kept symmetrical with respect to the ground position. In this example we are keeping all conditions the same as in the previous example with one exception. In this case the previous shot location now becomes the common ground point on which the stack is based.

Experiments using differing dips and interval velocities indicated that arrival time-distance relations remain nearly hyperbolic. This has been further verified by the closeness of least-squares fit hyperbolas to synthetically generated time-distance values. If we assume that the parameters of these hyperbolas are similar to those as shown in equations (5) and (8), we would then define the values of the apparent velocities observed from common reflection point traces. Investigation has also shown that, if we keep all the parameters the same, increasing dips produce increased apparent velocities or flatter time-distance curves. One interesting and important point is that the travel path between the shot and recording with respect to any reflecting bed is reversible. That is, if we exchange the shot and recording position, the travel path (although reversed) and traveltime stay the same. Therefore, the time-distance relation represents a symmetric curve and its apex is always at $X=0$. Furthermore, the arrival time at the apex (minimum arrival time) is equal to the normal incidence time at the common ground point. The time correction (ΔT_n) in the case of common depth point traces, is the time difference between the total traveltime for a given shot-geophone position and the normal incidence time at the common ground point.

By applying time corrections (NMO) to the traces obtained with shooting and recording symmetrically disposed about a ground point, we are actually simulating a new set of traces that theoretically were shot and recorded at the same place, the common ground point. This time correction is different from the time correction applied to the single coverage records, and the main confusion about time corrections for stacked data stems from misunderstanding this difference (see Nugent, 1967). Since the time-distance relations are still nearly hyperbolic, records should be stacked just as properly in the dipping bed cases as they would be stacked in horizontal bed areas, provided that the parameters of the hyperbola are determined properly. This could be done without the remedial procedure of residual moveout correction.

One of the generally practiced methods for determining corrections is the T^2 versus X^2 study of common reflection point traces. However, this normal-visual method is somewhat limited by the seismic interpreter's ability to recognize and analyze visible strong reflectors. If we display the input traces to a common reflection point gather according to their X distances (Figure 6), reflections coming from a common reflector should appear along a particular hyperbola. Also, each unique set of reflections would have a normal incidence time and an apparent rms velocity (apparent owing to the assumption of horizontal layering and neglect of dip).

This display which we are used to observing, shows the reflected energy as a function of two variables, time and distance. Hence, mathematically we are looking at a reflection energy surface over the time-distance domain. It is, therefore, possible to display this reflected energy in a more convenient two-variable domain, normal incidence time—rms velocity. This can be accomplished by transforming the reflected energy from one domain (energy as a function of time and distance) into another (energy as a function of normal incidence time and apparent rms velocity).

Details of the transformation can, of course, vary; however, if a process generically related to correlation with a suite of curves is used, the output may be referred to as a spectrum. We note that the familiar Fourier calculation is based on correlation with sinusoids. Clearly then the search for a velocity with a constant normal

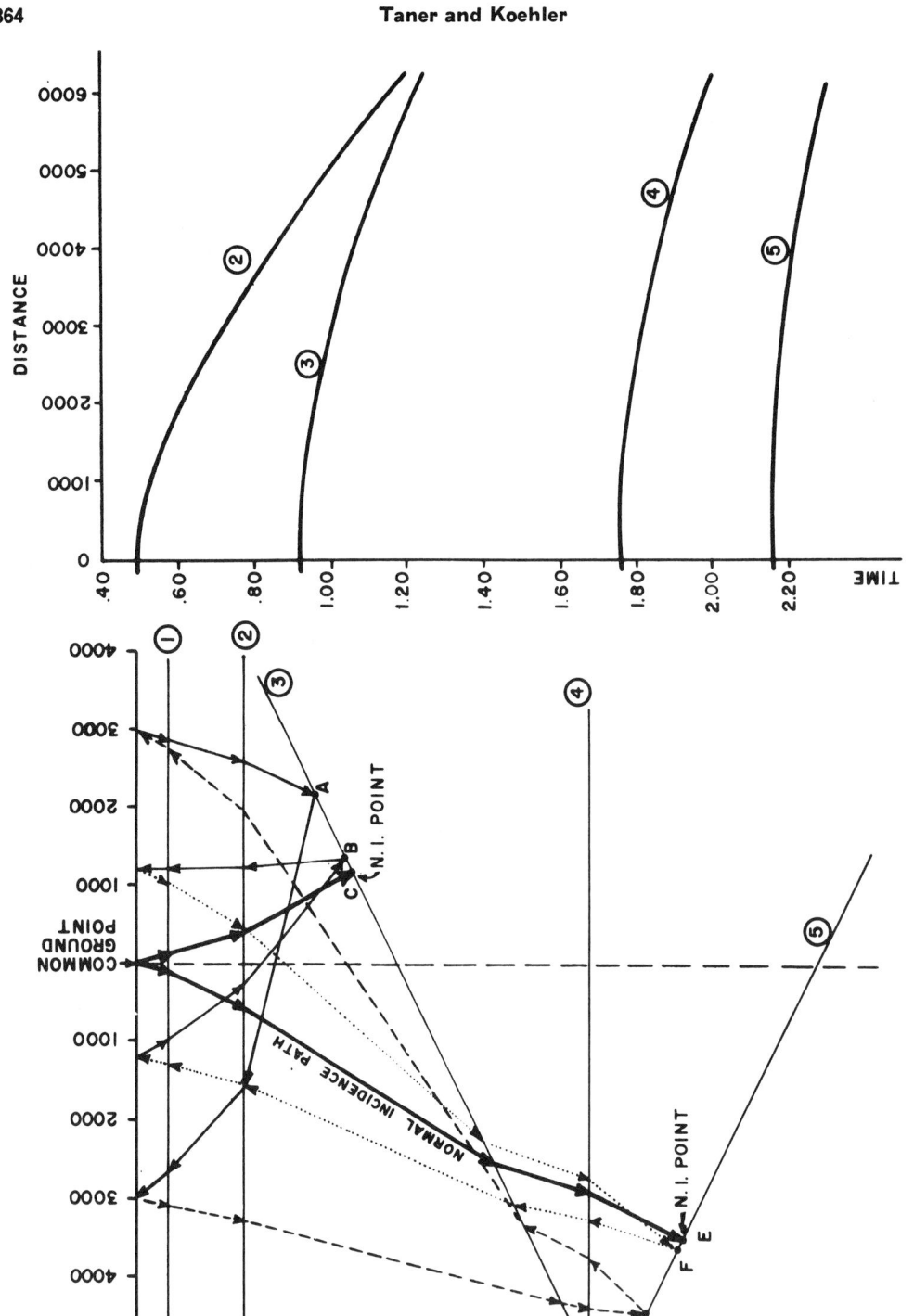

Fig. 5. Common reflection point model with dipping reflectors.

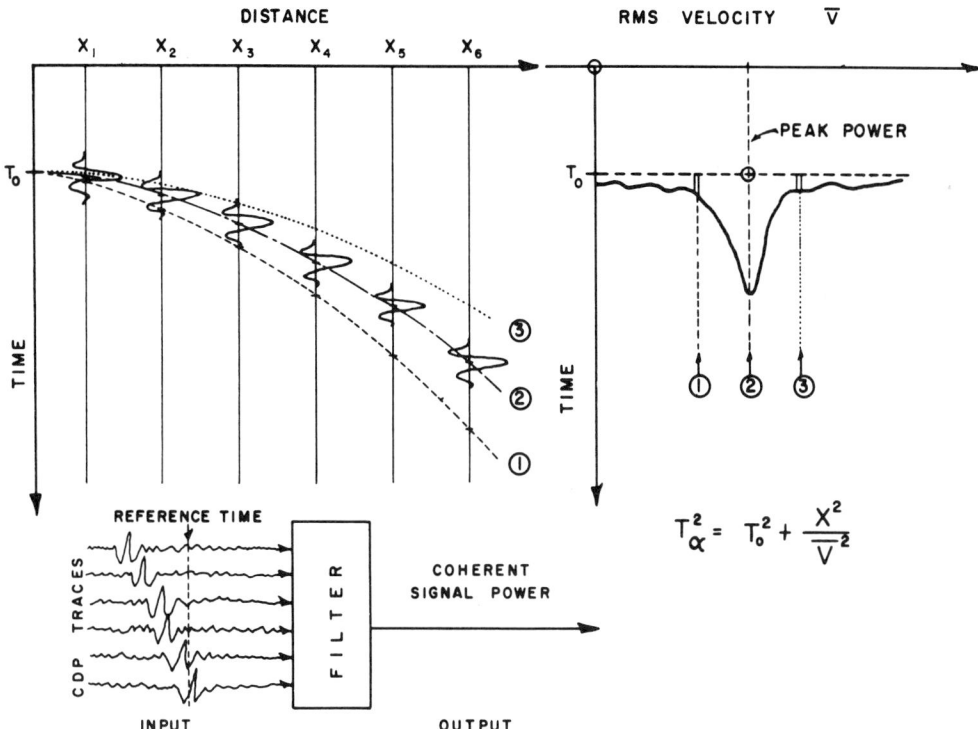

Fig. 6. Trace gather for a velocity search.

incidence time gives a velocity spectrum and the display summarizing such searches for all normal incidence times constitutes velocity spectra.

The transformation must serve to measure the power of reflections arriving according to various paths as given by equation (8). For this purpose, a specially designed multichannel filter was utilized. This filter, a semblance filter, is a normalized crosscorrelation function (see Appendix B) and measures the common signal power over the channels according to the specified lag pattern. A schematic diagram of this filter is shown on Figure 6. After the alignment of the input data with respect to a given hyperbolic delay pattern, a digital filter is computed for each trace that would selectively pass events common to all traces. Then, these are stacked together to give the best estimate of the input signal. The power of this estimate is then computed by summing the amplitude squares within a specified time gate around the reference time and this power is then displayed on the velocity spectra. Computation of power spectra for a given set of traces is done on a grid pattern. First, a reference time $T_{0,n}$ is chosen. The time is kept constant and $\overline{V}_n$ is then varied at regular intervals between a minimum and maximum $\overline{V}$ for a given area. Each set of $T_{0,n}$ and $\overline{V}_n$ will give a particular hyperbolic pattern. Each trace is displaced an amount corresponding to the X distance, passed through the semblance filter, and output power is displayed with respect to its corresponding $T_{0,n}$ and $\overline{V}_n$ values as shown on Figure 6. Peaks on the spectrum thus indicate the arrival of reflected energy at this particular $T_{0,n}$ time with an apparent velocity indicated by the corresponding $\overline{V}$ coordinate.

To generate a velocity spectrum then, we sweep the traces with various hyperbolas but always keep the apex at one point, $X=0$. After examining a predetermined range of velocities, a new $T_{0,n}$ is chosen at a time equal to half the time gate down the traces. Another sweep of predetermined velocities is made at this new $T_{0,n}$ and results in all reflections being analyzed. In other words, a complete spectrum (velocity sweep) is computed at constant intervals down the record. This procedure starts at a given time and is repeated down the record until the desired final analysis time is reached.

Figure 7 shows a typical velocity spectra display computed from 0.200 sec to 4.000 sec for velocities varying from 5000 ft/sec to 10,000 ft/sec at 100 ft/sec intervals. This is generated from 24 traces (shown on the right side of the figure) composed of 4 sets of 6 traces from 4 adjacent common ground points. Increasing power is shown as displacement in the direction of increasing time. Velocity spectra are interpreted in a manner similar to the interpretation of conventional seismic traces. In this case the velocity function is obtained by connecting the interpreted primary energy arrivals as indicated on the figure. These spectra also show some multiple energy around 3.2 to 3.8 sec with 7000 to 8000 ft/sec velocity. Since the spectra are computed for overlapping time gates and very close velocity intervals, the final display will contain the results of analysis of all the reflections, primaries, multiples, and other types of coherent energy.

THE VELOCITY SPECTRA DISPLAY, USES

Several uses of the velocity spectra display suggest themselves.

Velocity spectra displays help to determine the velocity function needed for optimum stacking. In order to determine this, the spectra are interpreted and primary reflections are identified. If the chosen velocity function goes through the primary reflections where they appear on the spectra, we are assured that they will be stacked properly and will appear on the final stacked section.

They may be used to check the final section against any procedural, interpretational, or computational error. For instance, if for some reason a reflection chosen as a good primary reflection on the velocity spectra display is not found on the stacked section and the 100 percent section shows improper time correction, this is a good indication that the spectra were interpreted improperly or some other error occurred during the processing. Instead of arbitrarily applying a residual move-out correction, spectra can be reinterpreted and all of the input data rechecked.

They aid in determining the effect of multiple interference. Since velocity spectra encompass all reflections, it is, in many cases, possible to determine the timing, order, apparent velocity, and relative power of multiples. These, in comparison with the primary reflections, will give us a good indication of the multiple content of the stacked section. Looking back at Figure 7, we can conclude that the section down to 2.8 sec will consist primarily of strong primary reflections, and the possible blurring effect around 3.4 to 3.8 sec is due to strong multiples.

The velocity spectra can be used to determine two-way normal incidence time and rms (apparent) velocity of major reflectors for later use in interpretation. We can estimate the interval velocities between major reflectors by the expression,

$$v_n^2 = \frac{\overline{V}_n^2 T_{0,n} - \overline{V}_{n-1}^2 T_{0,n-1}}{T_{0,n} - T_{0,n-1}}. \quad (11)$$

Extensive experience, such as Cook's (1967), has shown that in areas of gentle dips (less than 5 degrees), equation (11) gives interval velocities 2 percent to 3 percent higher than actually measured in well-velocity surveys and can be used in the initial interpretation stage. These values, normal incidence time and rms velocities, from different velocity spectral analyses are later used to compute more precise interval velocities, reflection depths, layer thickness, dip and other parameters for two or three dimensional migration.

They can be used to obtain stratigraphic and structural information. Experience over the past two years has shown that much stratigraphic information is also contained on the spectra. Changes of character of primaries on velocity spectra displays within a given area serve as a good indicator that changes have taken place in the stratigraphy. Figure 8 shows a velocity spectra display in which primary energy appears to stop at about 2.24 sec. Below this line we see a series of multiples that persist to about 4.0 sec where another strong reflection with about a 9200 ft/sec velocity appears. This disappearance of primary reflections between 2.24 and 4.0 sec is due to a large shale body which is known to exist in this particular area.

ILLUSTRATIONS OF THE USE OF VELOCITY SPECTRA

Figure 9 shows a stacked section of an experimental marine Vibroseis line from offshore Louisiana and further shows the kind of information given by velocity spectra. Figure 10 shows three velocity spectra displays computed along

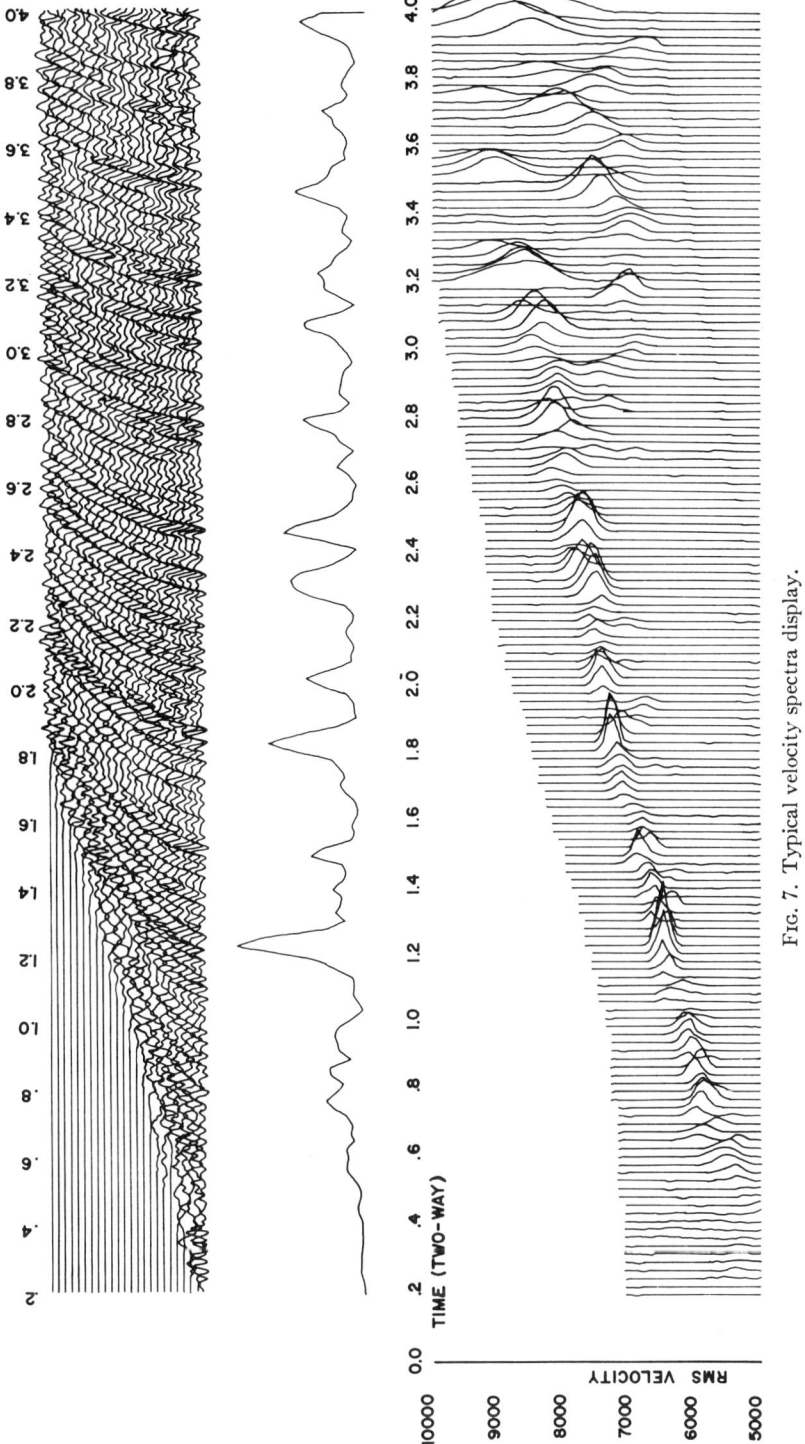

Fig. 7. Typical velocity spectra display.

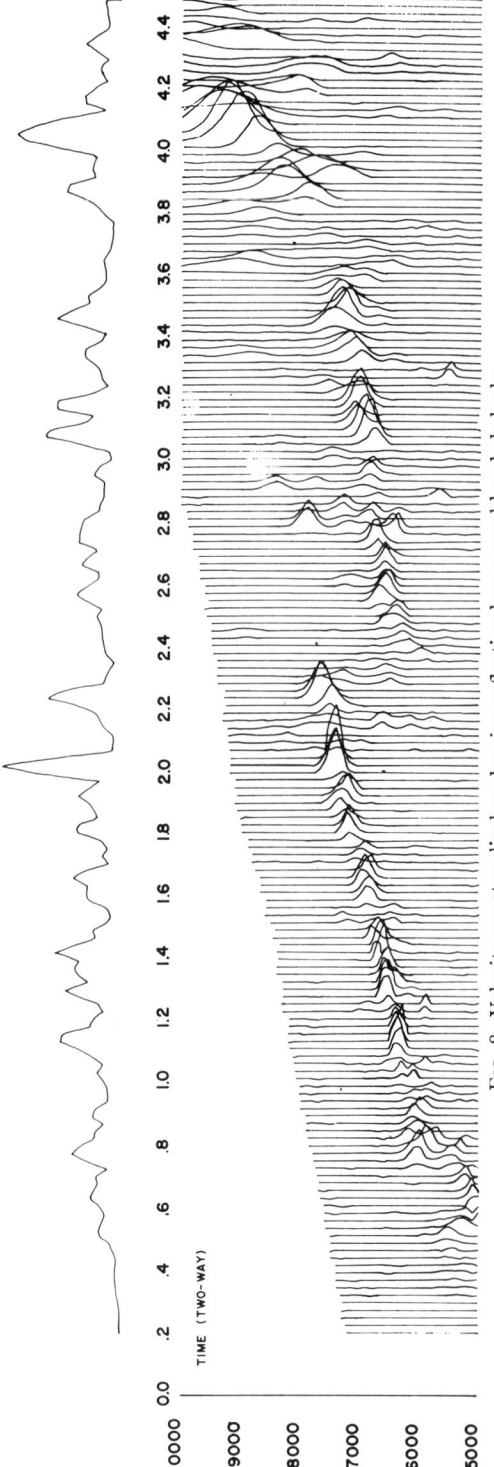

FIG. 8. Velocity spectra display showing reflection loss caused by shale body.

this line at three of the locations indicated on Figure 9. These spectra are plotted with the same time scale as the final sections so that geophysicists can lay the spectra on the section as an aid in interpretation. These spectra are also used in the computation of time-varying-multichannel filters that pass only the desired primary reflections and attenuate multiple reflections and random noise. Such a section is shown in Figure 11. Spectra in Figure 10 indicate that the section contains mostly primary energies down to 2.6 to 3.2 sec, and also shows a gentle velocity gradient. For example, the rms velocity at 1.4 sec on the west side of the section is 6400 ft/sec, on the east side it is 5900 ft/sec. A single velocity function would, therefore, result in under or over-corrected records.

In actual practice time corrections between spectra locations are interpolated from the two nearby spectra. This allows us to follow the velocity gradient closely without an abrupt change of velocity functions. Character correlation on the spectra also sometimes help in the structural interpretation. For example, the reflection at 3.16 sec on the spectra display at SP 6 has a character similar to that of the reflection at 3.04 sec on the spectra at SP 73. Comparing this with the section, we see that both are coming from the same interface. Moving over to the spectra at SP 233, we see that similar character appears at 2.60 sec.

Figure 12 shows a velocity spectra display computed on land data from South Texas. Initially, the reflection around 2.0 sec was thought to be the deepest primary in the area. The spectra show some additional primaries between 2.0 and 3.0 sec which were identified as sand layers in a later drilled well.

North Sea data generally contain very strong reflections, as well as strong surface and peg-leg multiples. Figure 13 shows various spectra from the North Sea. In some instances, where strong multiples are present, it is difficult to see or identify the primary energy below the strong reflector. After these multiples are identified, they are removed from the traces and new (multiples eliminated) spectra are computed. An example of this application is shown in Figure 14. Eliminations of reflections (multiple or primary) is accomplished with the same type multichannel filters used in spectra computation. However,

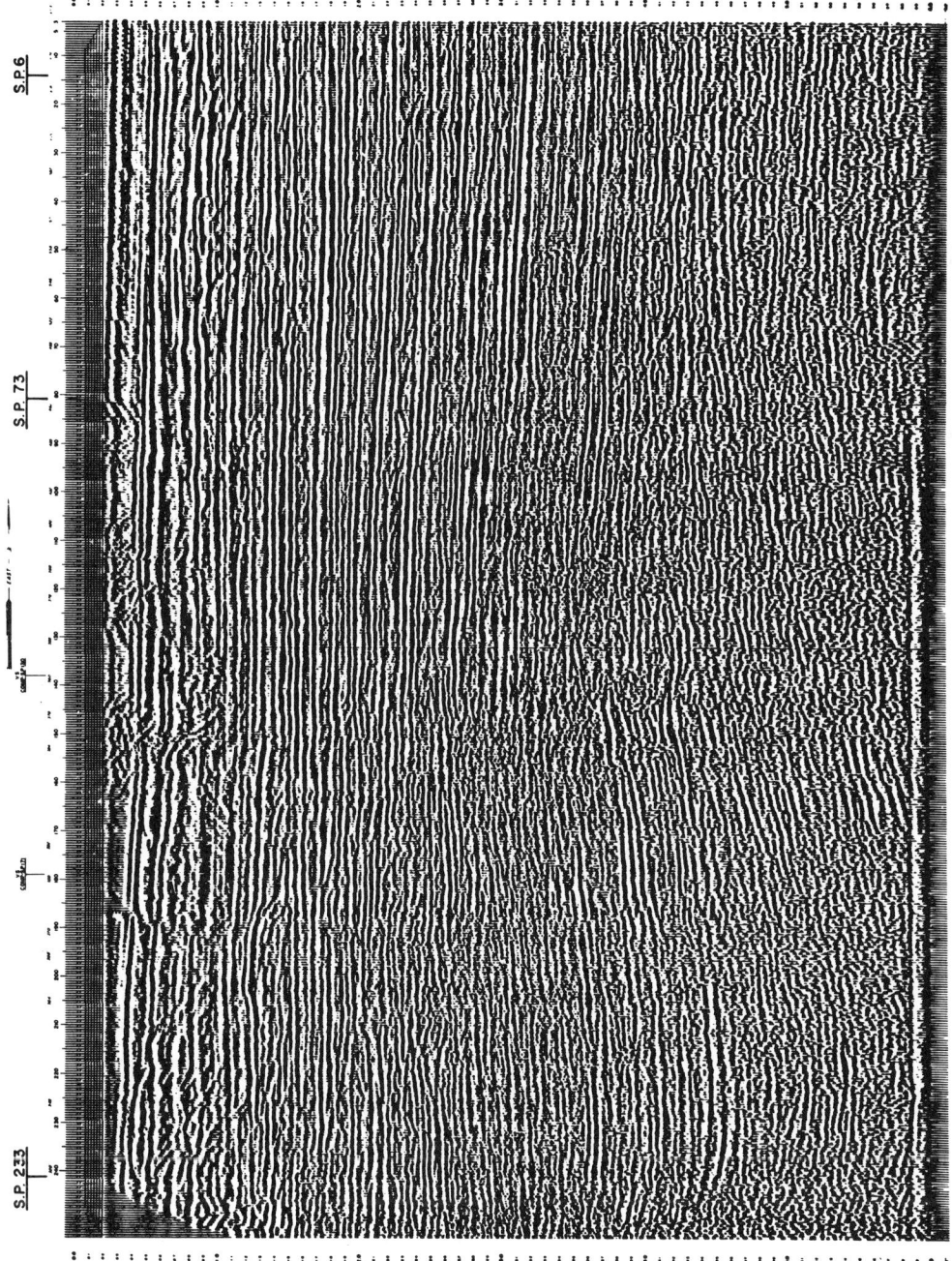

FIG. 9. Stacked section using primary velocities derived from spectral display (Offshore Louisiana).

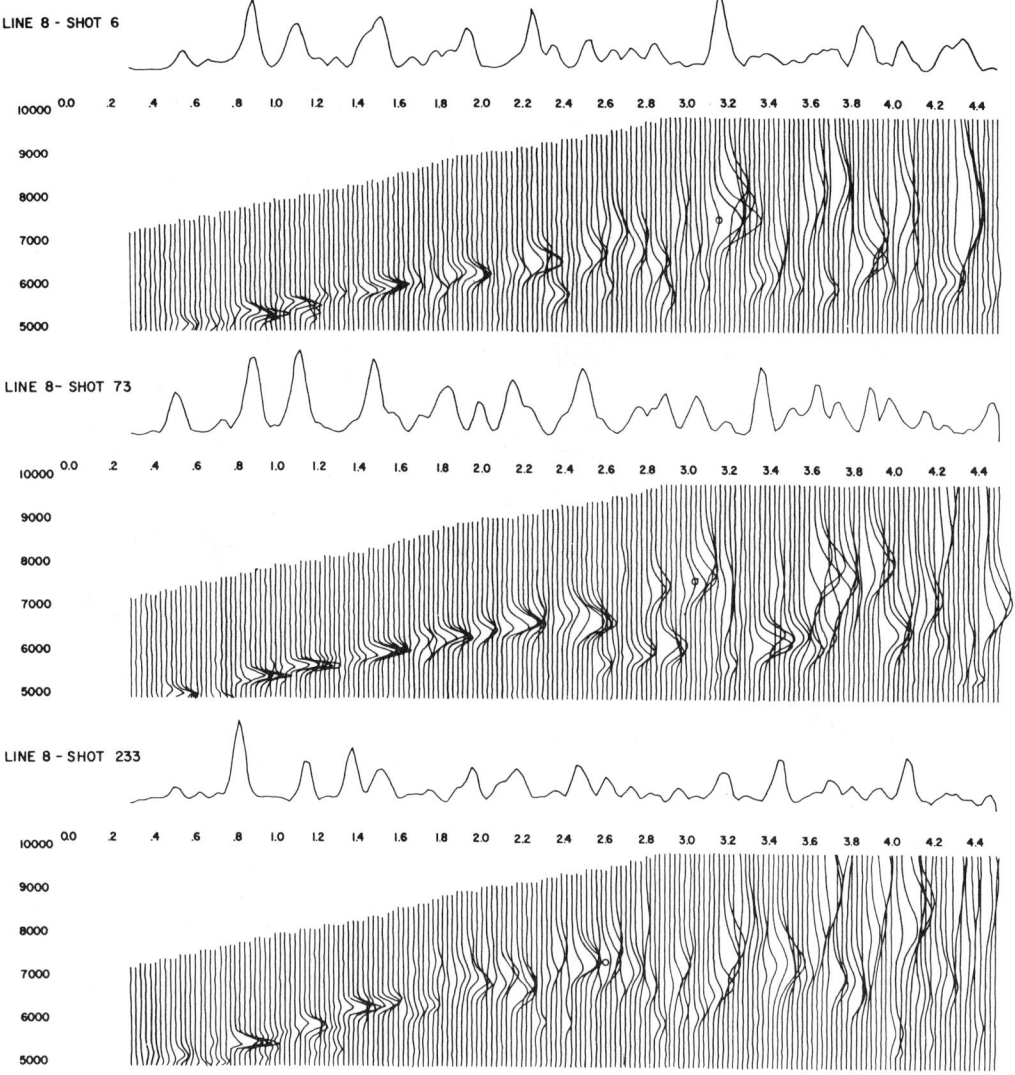

FIG. 10. Spectral displays illustrating continuity of reflections and character in spectral domain.

these filters are in this case modified to attenuate the semblant energy arriving at a given time and with a given velocity.

One of the most difficult interpretational problems is the interpretation of a velocity inversion. Figure 15 shows three spectra displays computed in an area with such an inversion, whose existence was later verified by drilling. The trends of the spectra warn us of the existence of a possible inversion. Interpretation of deeper reflections becomes more difficult owing to the existence of various multiples in the primary energy zone.

During the past two years, velocity spectra displays have been computed on data from all parts of the world and for all types of energy sources. Figure 16 shows a set of spectra computed from gas gun records. Figure 17 shows spectra computed from marine Vibroseis records in the deep sea.

VELOCITY SPECTRA FOR EARTH MODELS

Earlier in the paper it was shown that arrival times and apparent velocities from single or

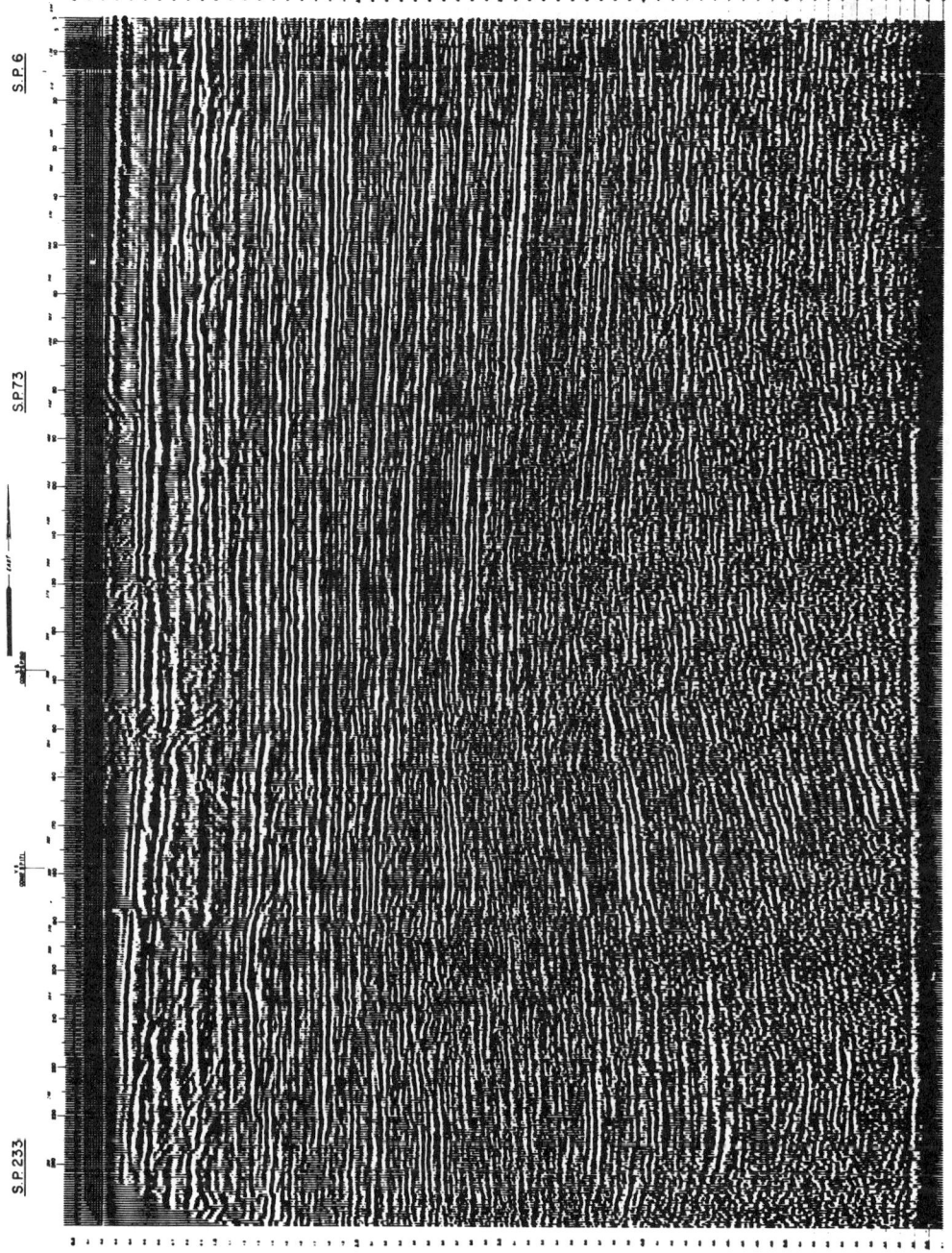

FIG. 11. Seis/Stack[1] section (Offshore Louisiana).

[1] Seismic Computing Corp. Trademark.

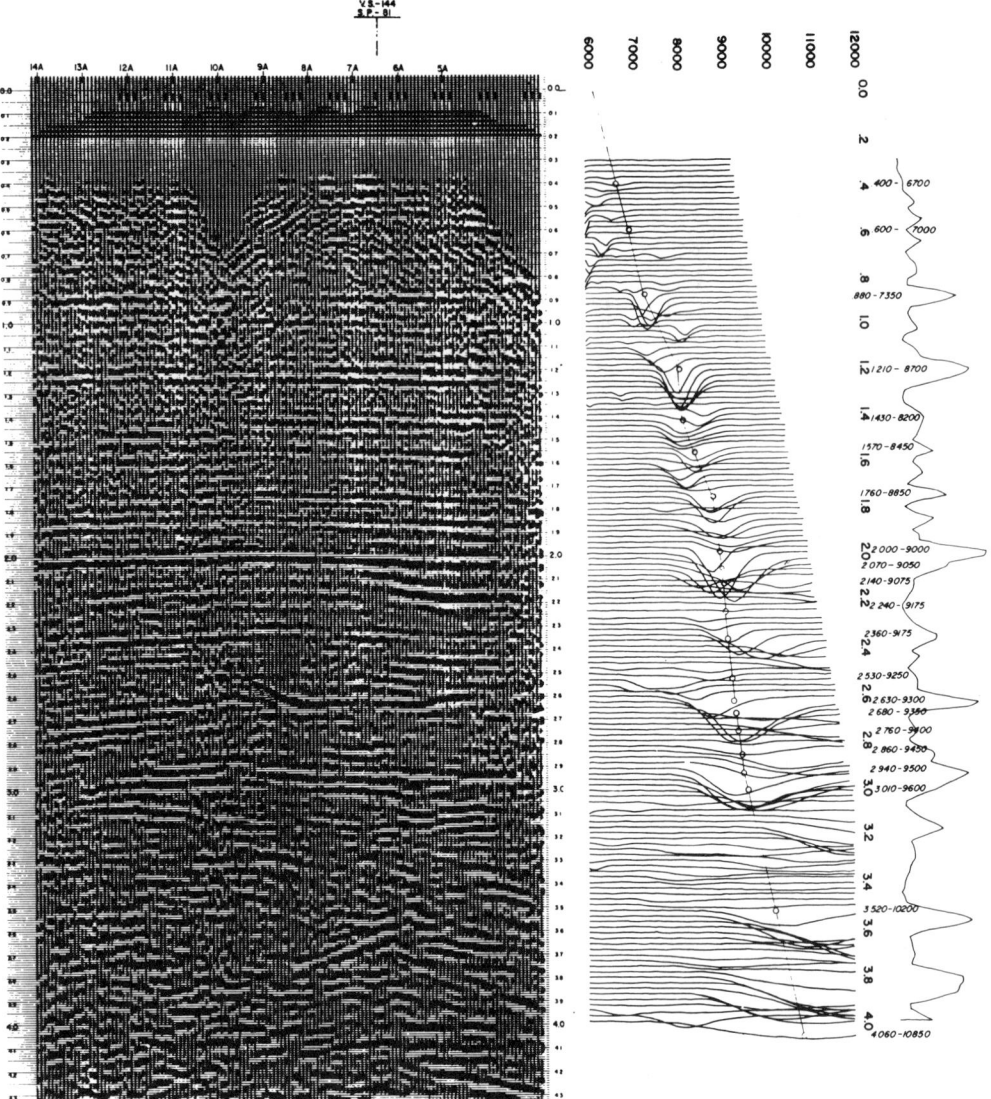

FIG. 12. Spectral display and section for land data (South Texas).

multiple coverage records could be computed by generating a simulated mathematical model from known interval velocities, dips, and layer thicknesses. Conversely, it is also possible to compute the interval velocities, dips, and layer thicknesses by generating a model for which the arrival times and apparent velocities are the same as of those observed from field records. This method is presently used in dip, depth, and interval velocity computation from velocity spectra and a sample output of such a computation is shown on Figure 18. This procedure is programmed so that the computer builds the model one layer at a time, and by iteration it minimizes the difference between the observed and computed values of arrival times and apparent rms velocities. The iteration procedure is terminated when the differences are less than a desired value. The accuracies of these computed dips, depths, and interval velocities are completely dependent on the accuracies of the observed arrival times and apparent velocities on the velocity spectra.

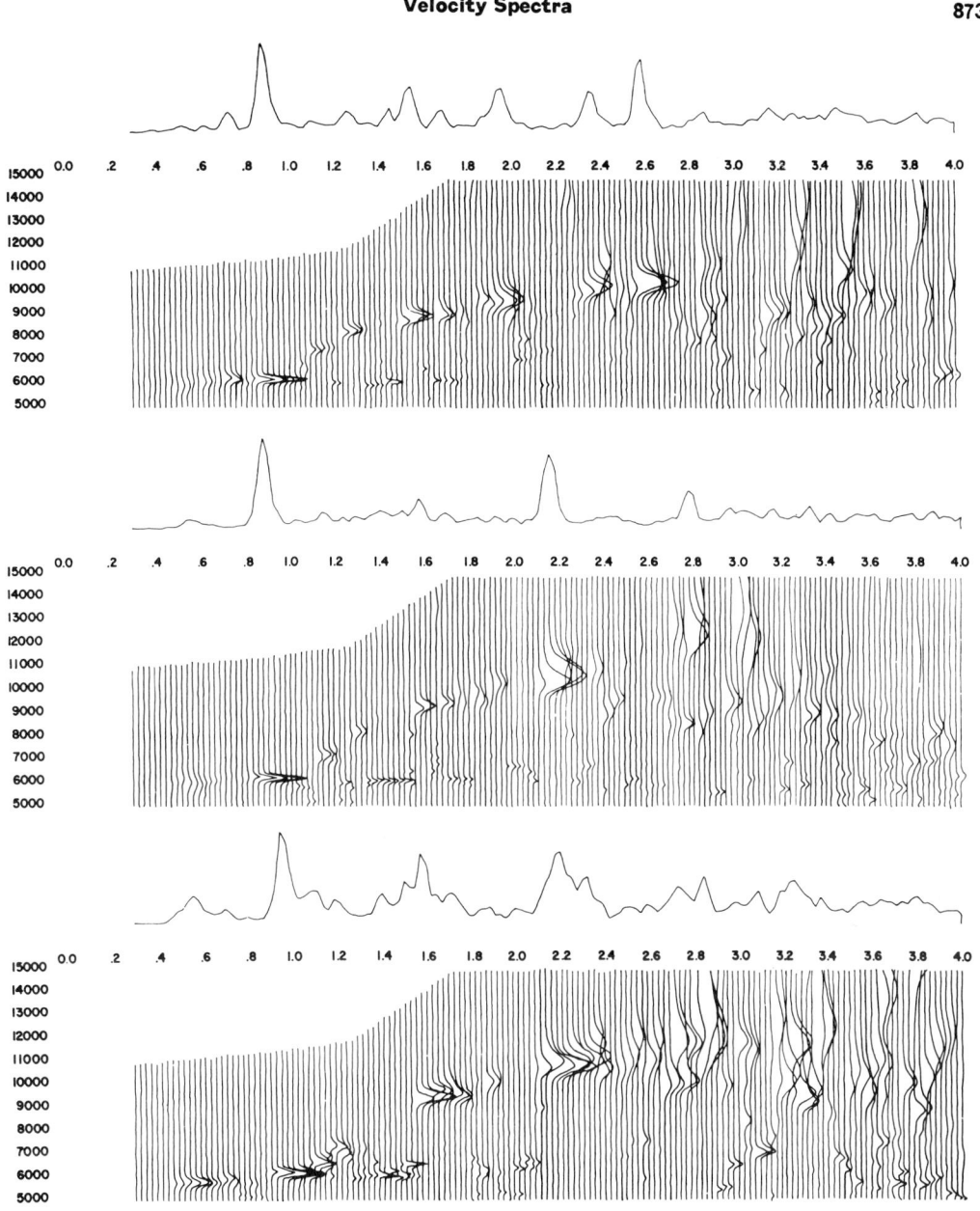

FIG. 13. North Sea velocity spectra displays.

SUMMARY

This paper outlines the fundamental principles for calculating velocity spectra displays. Information about coherent events is derived by means of hyperbolic searches of common reflection point trace gathers. A semblance criterion is used to measure this energy. Where maxima of coherent energy correspond to primary reflections, the display conveniently shows their relation to stacking velocities.

The interpretation of velocity spectra displays is similar to the interpretation of seismic sections in that both should be done by experienced geophysicists. Velocity spectra are an additional tool in the hands of the geophysicist, and they aid him in geophysical as well as geological interpretation

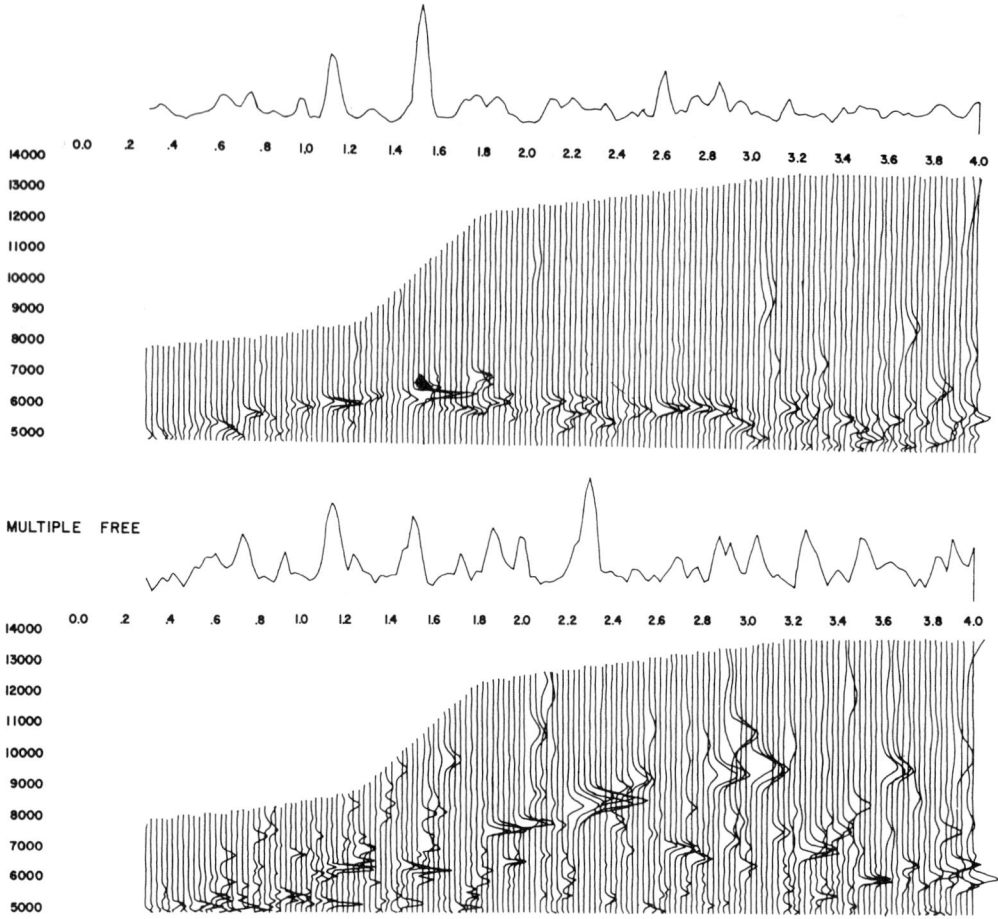

Fig. 14. Velocity spectra displays after multiple suppression. The spectra on the bottom are multiple free.

by allowing him to see his primary reflection events in terms of their velocities. In some cases it was seen that velocity spectra permit character correlations across faults that supplement and confirm those made on the sections. Since the clarity by which the reflections are identified depends on the differential arrival times, a long spread length with single end cable configuration is recommended. Based on the experience gained from velocity spectra, additional interpretational procedures are now being developed to aid the geophysicist in the stratigraphic and structural interpretation.

ACKNOWLEDGMENTS

We would like to take this opportunity to thank Louisiana Land and Exploration Company and Signal Oil and Gas Company for making their data available for presentation in this paper. We would also like to thank Dr. Norman S. Neidell for his comments and his help in the preparation of a final manuscript.

APPENDIX A

TIME-DISTANCE FORMULAS FOR A HORIZONTALLY LAYERED EARTH

Let a coordinate system be chosen with the x axis horizontal and the z axis vertically downward (Figure 1A). Using the method of raypaths, we can compute the time for a ray, starting from the origin, to traverse n horizontal layers, and after reflection return to the surface at the point $(x, 0)$. This is a problem whose solution is well known, being usually found by the use of Snell's law. An alternative method, based on Fermat's principle of least time, is given here.

Let the thicknesses of the n layers be d_1, d_2,

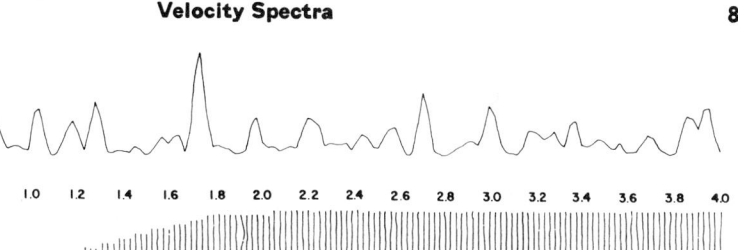
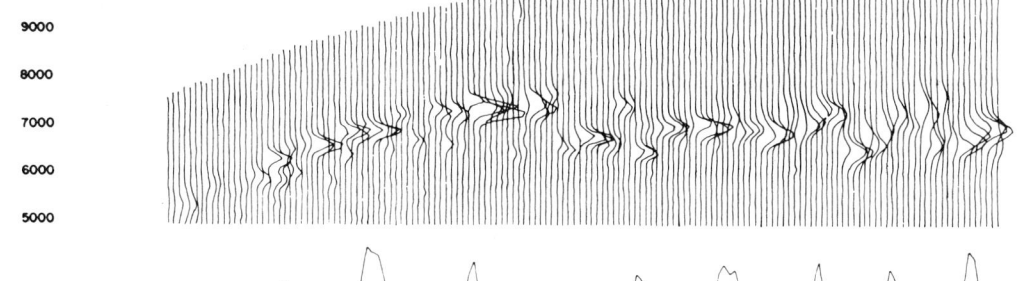
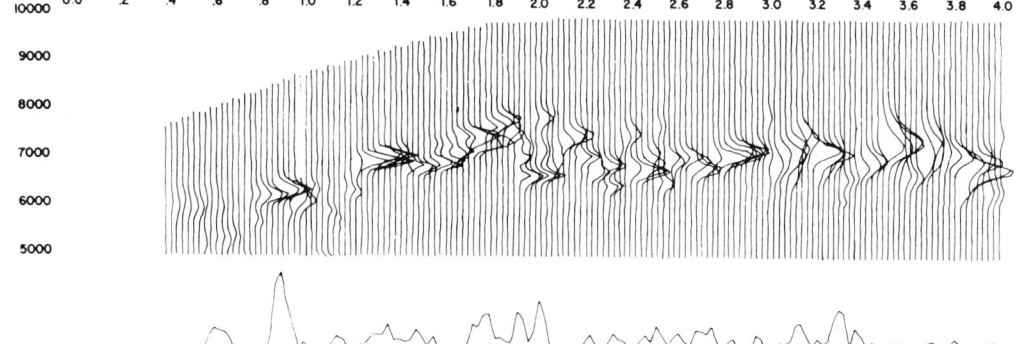
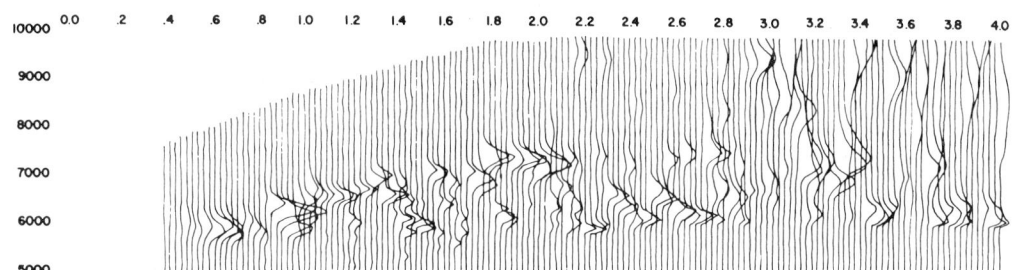

Fig. 15. Velocity spectra displays of velocity inversion.

$d_3, \cdots, d_n$ and the velocities $v_1, v_2, v_3, \cdots, v_n$. The segment of the downward traveling ray within the kth layer will be assumed to have length p_k, with horizontal components x_k and angle to the vertical θ_k as indicated in the figure. The total down-up traveltime T_x is determined, from Fermat's principle and from symmetry about the vertical line through the point of reflection, by the formulas:

$$T_x = 2 \sum_{k=1}^{n} \frac{p_k}{v_k} = \text{minimum}, \quad (A1)$$

$$p_k^2 = d_k^2 + x_k^2, \quad (A2)$$

$$2(x_1 + x_2 + \cdots + x_n) = x. \quad (A3)$$

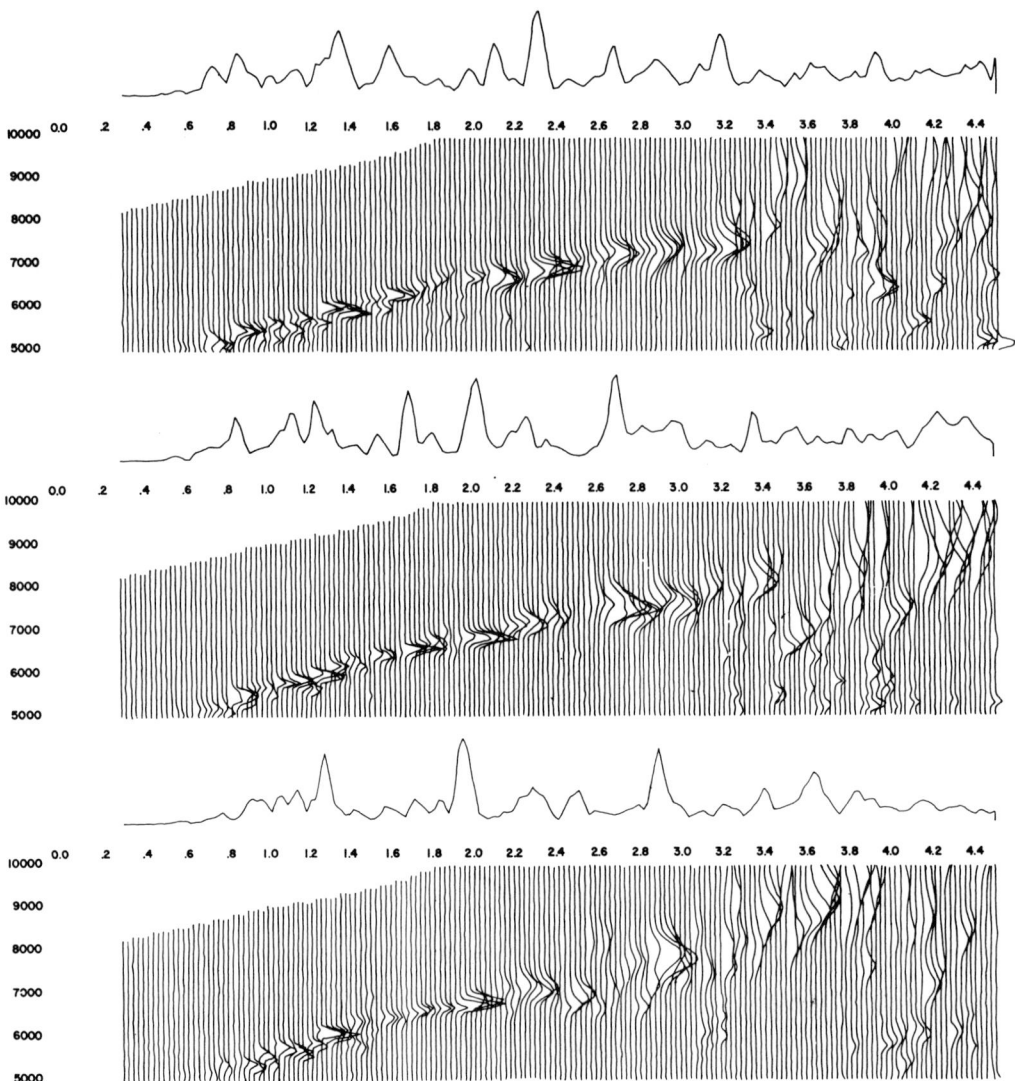

FIG. 16. Velocity spectra displays from gas gun data.

The minimum requirement (A1) under the constraint (A3) leads to the equations,

$$\frac{\delta T_x}{\delta x_k} = \frac{2x_k}{p_k v_k} = 2\lambda, \quad k = 1, 2, \cdots, n \quad (A4)$$

where λ is a Lagrange multipler. From (A4) and (A2) we get,

$$x_k^2 = \frac{\lambda^2 v_k^2 d_k^2}{1 - \lambda^2 v_k^2}, \quad (A5)$$

and this yields, by substitution in (A3) and (A1),

$$x = 2\lambda \sum_{k=1}^{n} \frac{v_k d_k}{\sqrt{1 - \lambda^2 v_k^2}} \quad (A6)$$

$$T_x = 2 \sum_{k=1}^{n} \frac{d_k/v_k}{\sqrt{1 - \lambda^2 v_k^2}}. \quad (A7)$$

Equations (A6) and (A7) are the parametric form of the time-distance relationship with λ as parameter as given in Slotnick (1959). As λ varies from 0 to $1/v$, where $v = \max(v_1, v_2, \cdots, v_n)$, x

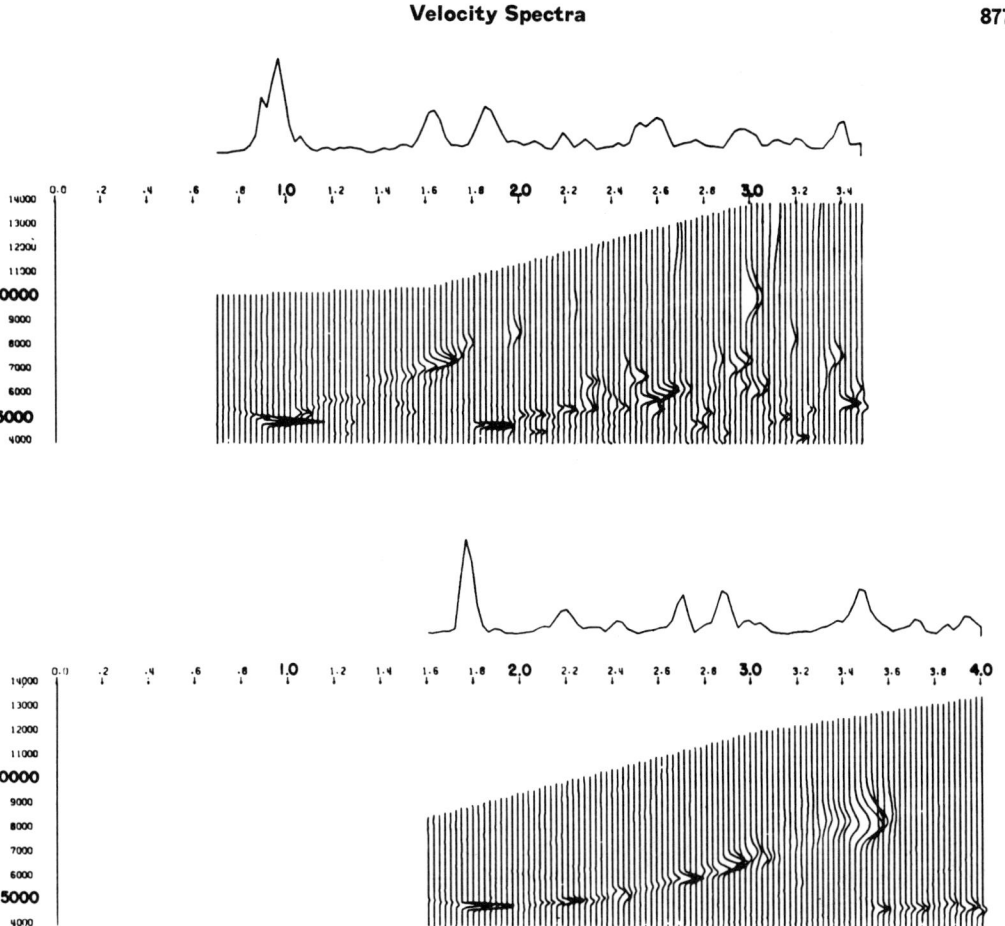

FIG. 17. Velocity spectra displays from marine Vibroseis in the deep sea.

TEST CASE FOR SIMULATOR WITH 6 LAYERS

SHOT POINT = 101

REFLECTOR NO.	RMS TIME	AVE VELO	INTERVAL VELO	DIP ANGLE	DIP SLOPE	VERT DEPTH	LAYER THICK.	HORIZ DISP.	REFLECTION DEPTH
			5000.				100.		
1	.040	5000.	5000.	.000	.00000	100.		0.	100.
			5999.				1499.		
2	.539	5940.	5925.	2.925	.05109	1599.		-81.	1595.
			7499.				1502.		
3	.925	6736.	6596.	11.368	.20106	3101.		-550.	2991.
			5998.				1000.		
4	1.273	6487.	6439.	.013	.00023	4101.		124.	4101.
			7998.				1000.		
5	1.495	6867.	6695.	-11.390	-.20145	5101.		982.	4903.
			10001.				1595.		
6	1.829	7462.	7267.	-5.649	-.09892	6696.		802.	6617.

FIG. 18. Sample output of interval velocity modeling program.

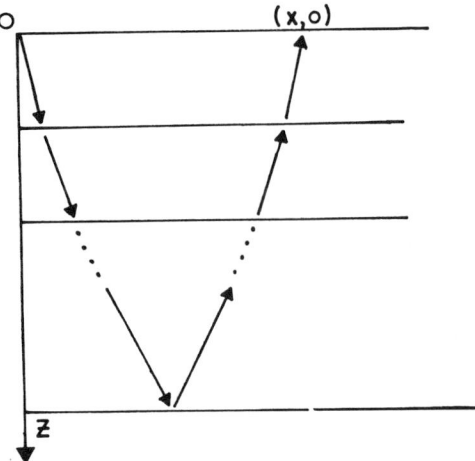

FIG. 1A. Coordinates of a Snell's law raypath in a horizontally layered medium.

will vary from 0 to ∞. The physical significance of the parameter λ can be seen from (A4) and Figure 2A; namely,

$$\lambda = \frac{\sin \theta_k}{v_k}, \quad k = 1, 2, 3, \cdots, n, \quad (A8)$$

which is Snell's law.

We now consider the problem of deriving an explicit time-distance formula from the parametric equations (A6) and (A7). In the special case where $v_1 = v_2 = \cdots = v_n = v$ these equations reduce to

$$x = \frac{2\lambda v d}{\sqrt{1 - \lambda^2 v^2}}, \quad T_u = \frac{2d}{\sqrt{1 - \lambda^2 v^2}},$$

where $d = d_1 + d_2 + \cdots + d_n$, and elimination of λ gives the formula

$$T_x^2 = \frac{4d^2}{v^2} + \frac{x^2}{v^2}. \quad (A9)$$

From general function-theoretical considerations (Copson, 1935), it can be seen that in the general case there will exist a generalization of the elementary formula (A9) of the form,

$$T_x^2 = c_1 + c_2 x^2 + c_3 x^4 + c_4 x^6 + \cdots \quad (A10)$$

where the expression on the right-hand side is an infinite series with coefficients c_n which depend on $d_1, d_2, d_3, \cdots, d_n; v_1, v_2, \cdots, v_n$.

The coefficients c_n in (A10) can be expressed by Lagrange's formula or by equivalent contour integrals in the complex plane. A shorter method of computation, however, can be developed using only algebraic operations with power series. The latter method is to express T_x^2, x^2, x^4, $x^6 \cdots$ as a power series in λ^2, derived from (A6) and (A7); substitute these series in (A10); and equate the coefficients of like powers of λ, thus getting a system of equations which can be solved one at a time for $c_1, c_2, c_3 \cdots$. The formulas are as follows: Let

$$q_1 = 1, \quad q_k = \frac{1 \cdot 3 \cdots (2k-3)}{2 \cdot 4 \cdots (2k-2)} \quad (A11)$$
$$(k = 2, 3, \cdots),$$

$$a_m = 2 \sum_{k=1}^{n} v_k^{2m-3} d_k \quad (m = 1, 2, 3, \cdots), \quad (A12)$$

$$b_m = q_m a_{m+1} \quad (m = 1, 2, 3, \cdots), \quad (A13)$$

$$\gamma_m = q_m a_m \quad (m = 1, 2, 3, \cdots). \quad (A14)$$

From (A6) and (A7) we get

$$x = \sum_{k=1}^{\infty} b_k \lambda^{2k-1}, \quad (A15)$$

$$T_x = \sum_{k=1}^{\infty} \gamma_k \lambda^{2k-2}. \quad (A16)$$

Squaring (A15) and (A16), we get

$$x^2 = \lambda^2 \sum_{k=1}^{\infty} B_{k1} \lambda^{2k-2}, \quad (A17)$$

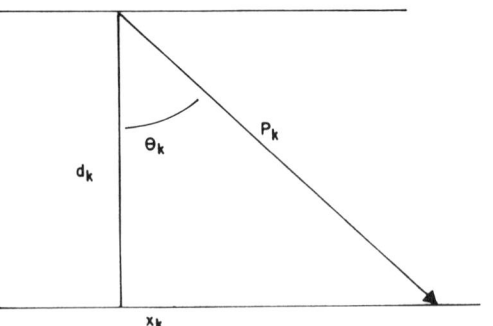

FIG. 2A. Decomposition of a slant path through a layer.

$$T_x^2 = \sum_{k=1}^{\infty} A_k \lambda^{2k-2}, \quad (A18)$$

$$B_{k1} = b_1 b_k + b_2 b_{k-1} + \cdots + b_k b_1,$$
$$(k = 1, 2, \cdots), \quad (A19)$$

$$A_k = \gamma_1 \gamma_k + \gamma_2 \gamma_{k-1} + \cdots + \gamma_k \gamma_1$$
$$(k = 1, 2, \cdots). \quad (A20)$$

From (A17) we can obtain series for x^4, x^6, x^8, $\cdots$.
Let

$$x^{2n} = \lambda^{2n} \sum_{k=1}^{\infty} B_{kn} \lambda^{2k-2} \quad (n=1, 2, \cdots). \quad (A21)$$

Then the coefficients B_{kn} are determined recursively by

$$B_{kn} = B_{k,n-1} B_{11} + B_{k-1,n-1} B_{21}$$
$$+ B_{1,n-1} B_{k1}$$
$$\text{for, } (n = 2, 3, 4, \cdots), \quad (A22)$$
$$(k = 1, 2, 3, \cdots).$$

$$c_3 = \frac{a_2^2 - a_1 a_3}{4 a_2^4} \quad (A28)$$

$$c_4 = (2a_1 a_3^2 - a_1 a_2 a_4 - a_2^2 a_3)/8 a_2^7 \quad (A29)$$

$$c_5 = \frac{(24 a_1 a_2 a_3 a_4 - 24 a_1 a_3^3 - 5 a_1 a_{2a}^2 + 9 a_2^2 a_3^2 - 4 a_2^3 a_4)}{64 a_2^{10}}. \quad (A30)$$

Substituting (A18) and (A21) in (A10) and equating coefficients gives

$$c_1 = A_1$$
$$c_m B_{1,m-1} + c_{m-1} B_{2,m-2} + \cdots$$
$$+ c_2 B_{m-1,1} = A_m \quad (m = 2, 3, 4, \cdots). \quad (A23)$$

The coefficients $c_1, c_2, c_3, \cdots$ can be computed in recursive fashion from (A23). Explicit formulas for c_1, c_2 are,

$$c_1 = a_1^2 = \left(2 \sum_{k=1}^{n} \frac{d_k}{v_k} \right)^2, \quad (A24)$$

$$c_2 = \frac{a_1}{a_2} = \sum_{k=1}^{n} \frac{d_k}{v_k} \Big/ \sum_{k=1}^{n} v_k k_k. \quad (A25)$$

Let

$$t_k = \frac{d_k}{v_k} = \text{one-way traveltime for a vertical ray to cross the } k\text{th layer.} \quad (A26)$$

It can be seen from (A24) that c_1 is the square of the two-way traveltime for a vertical ray. The reciprocal of c_2 can be written, from (A25) and (A26), as

$$\frac{1}{c_2} = \bar{v} = \frac{\sum_{k=1}^{n} v_k d_k}{\sum_{k=1}^{n} \frac{d_k}{v_k}} = \frac{\sum_{k=1}^{n} t_k v_k^2}{\sum_{k=1}^{n} t_k}. \quad (A27)$$

Hence, $1/c_2$ is the weighted average of the squares of the interval velocities, the weights being equal to the vertical traveltimes. The coefficient $1/v^2$ of the x^2 term in the elementary formula (A9) is thus replaced in the general formula (A10) by the reciprocal of a time-weighted mean square velocity. This same formula is given by Dix (1955).

Formulas for the next three coefficients in (A10) are,

There does not appear to be any way in general of determining the radius of convergence of the power series (A10). However, in actual cases the numerical results from (A10) and from the parametric formulas (A6) and (A7) can be compared to get some indication of the convergence of (A10). In many cases of practical interest that have been investigated by ourselves and our coworkers, the convergence of (A10) has been found to be quite rapid for all values of x of interest; in fact, so rapid that the first two terms of (A10) give sufficient accuracy for practical purposes in a great many cases.

It is of some interest to note that, when the two-term approximation in (A10) is used, the first neglected term, $c_3 x^4$, is always negative or zero. This follows from (A28) and (A12):

$$A_2^2 = 4\left(\sum_{k=1}^{n} v_k d_k\right)^2$$

$$= 4\left(\sum_{k=1}^{n} \sqrt{v_k^3 d_k}\,\sqrt{\frac{d_k}{v_k}}\right)^2$$

$$\leq 4\left(\sum_{k=1}^{n} v_k^3 d_k\right)\left(\sum_{k=1}^{n} \frac{d_k}{v_k}\right)$$

by the Schwartz inequality. Hence,

$$a_2^2 = a_1 a_3,$$
$$c_3 \leq 0.$$

The case $c_3=0$ occurs if, and only if, $v_1 = v_2 = \cdots = v_n$.

APPENDIX B

SEMBLANCE AND OTHER COHERENCY CRITERIA

Let us assume that a number of traces $f_i(t)$ ($i=1, 2, \cdots, m$) corresponding to different values of x, all contain a common signal $S(t)$ from the same reflector but arriving at different times. If the lag times are t_i and if the traces contain only the signal, the lag times can be found theoretically as the solution of a maximization problem.
Let

$$E(\tau_1, \tau_2, \cdots, \tau_m) = \Sigma_t [\Sigma_i f_i(t+\tau_i)]^2. \quad (B1)$$

Then the maximum of $E(\tau_1, \cdots, \tau_m)$ is attained for $\tau_i = t_i$. For actual traces containing noise in addition to the signal $S(t)$, the criterion of maximizing $E(\tau_1, \cdots, \tau_m)$ can still be used to give an estimate of the arrival times of the coherent signal.

Instead of the function defined in (B1) the sum of the crosscorrelations between pairs of traces with variable lag times can be used; i.e.,

$$C(\tau_1, \tau_2, \cdots, \tau_m) = \Sigma_t \Sigma_{i,j} f_i(t+\tau_i) f_j(t+\tau_j), \quad (B2)$$

the inner summation being taken over all distinct pairs of different indices (i, j). The E and C functions are connected by

$$2C = E - \Sigma_t \Sigma_i f_i^2(t). \quad (B3)$$

A normalized form of the E function called the semblance would be:

$$P(\tau_1, \tau_2, \cdots, \tau_m) = \frac{E}{m \Sigma_t \Sigma_i f_i^2(t)} \quad (B4)$$

which has a possible range from 0 to 1. Such a normalized function, the semblance coefficient, gives a measure of coherency which is, in a sense, independent of the joint power level of the traces. More precisely, the P function is unchanged when a set of traces $f_i(t)$ is replaced by a set $kf_i(t)$, for any value of the scale factor k. It is, however, sensitive to the variations within the set of traces, unlike normalized crosscorrelation.

The maximization of any of the functions E, C, P when m is equal to, say 6, presents a numerical problem of considerable magnitude and can be solved only by a long sequence of successive approximations. The problem is greatly simplified if it can be assumed that the lag times depend on only one or two parameters.

For applications in identifying velocities, the simplest assumption is based on the two-term approximation to (A10) with a selected value for c_1 of $T_{0,n}^2$ and with c_2 being $(1/\bar{V}^2)$ where,

$T_{0,n}=$ two-way normal incidence traveltime [see (1) and (A24)]

$\bar{V}=$ time-weighted mean square velocity [see (7) and (A27)]

In this case it is a relatively simple matter to compute the value of E (or, C, or P) as a function of $\bar{V}$ and to pick the value of $\bar{V}$ which gives the maximum.

If the x values in (A10) are so large that the two term approximation is too crude, or if account is to be taken of sloping interfaces in a single coverage trace gather, the lag times t_i must be assumed to depend on more than two parameters. With sloping interfaces the formula for τ^2, analogous to (A10), will contain odd powers of x as well as even ones.

REFERENCES

Cook, E. E., 1967, Geophysical reconnaissance in the Northwestern Caribbean: Presented at the 37th Annual International S.E.G. Meeting, October 31, 1968, Oklahoma City, Oklahoma.

Copson, C. T., 1935, Theory of functions of a complex variable: Oxford University Press, p. 121–125.

Courtier, W. H., and Mendenhall, H. L., 1967, Experiences with multiple coverage seismic methods: Geophysics, v. 32, p. 230–258.

Dix, C. H., 1955, Seismic velocities from surface measurements: Geophysics, v. 20, p. 68–86.
Mayne, W. H., 1962, Common reflection point horizontal stacking techniques: Geophysics, v. 27, p. 927–938.
———— 1967, Practical considerations in the use of common reflection point techniques: Geophysics, v. 32, p. 225–229.
Nugent, L. E., Jr., 1965, Critical analysis of stack seismic systems: Bull. Amer. Assoc. Petrol. Geol., v. 51, p. 915–956.
Picon, C., and Utzmann, R., 1962, La "Coupe sismique vectorielle" un pointé semi-automatique: Geophys. Prosp., v. 10, p. 497–516.
Rieber, F., 1936, A new reflection system with controlled directional sensitivity: Geophysics, v. 1, p. 97–106.
Sattlegger, J., 1965, A method of computing interval velocities from expanding spread data in the case of arbitrary long spreads and arbitrarily dipping interfaces: Geophys. Prosp., v. 13, p. 306–318.
Schneider, W. A., and Backus, M., 1968, Dynamic correlation analysis: Geophysics, v. 33, p. 105–126.
Simpson, J. K., 1967, Traveling signal-to-noise ratio and signal power estimates: Geophysics, v. 32, p. 485–493.
Slotnick, M. M., 1959, Lessons in seismic computing: Tulsa, SEG, p. 194.
Trorey, A. W., 1961, The information content of a Rieber Sonogram: Geophysics, v. 26, p. 761–764.

The Effect of Cracks on the Compressibility of Rock[1]

J. B. Walsh

Department of Geology and Geophysics
Massachusetts Institute of Technology, Cambridge
and Woods Hole Oceanographic Institution
Woods Hole, Massachusetts

Abstract. Compressibility of porous material is greater than that of solid material of the same composition, and the difference is shown to be equal to rate of change of porosity with pressure, for any pore shape or concentration. Expressions for compressibility are given for two special cases for material of low pore concentration: for spherical pores and for narrow cracks. Comparison of the two cases shows that a crack increases compressibility nearly as much as a spherical pore of the same diameter as the length of the crack, although porosity in these two cases differs enormously. For material in which all porosity occurs as narrow cracks, it is shown that porosity can, in certain cases, be determined quite precisely from compressibility measurements.

Introduction

Measurement of the deformation of rocks under hydrostatic pressure shows that compressibility—the fractional volume decrease per unit of pressure—often is not a constant, as would be expected for an elastic material, but varies with pressure and with the type of test. For example, *Adams and Williamson* [1923] and *Zisman* [1933] found that compressibility of samples jacketed with material impervious to the pressurizing medium usually decreases markedly with increasing pressure, whereas compressibility of unjacketed rocks is nearly constant[2]. A similar effect was reported by *Birch* [1960], who found from studies of compressional wave velocity in rocks that, when the pressure surrounding jacketed rocks is raised, velocity increases more rapidly than for unjacketed rocks. Also, variations in velocity with direction of propagation, which often are present at low confining pressure, are generally found to disappear or diminish greatly when the confining pressure is raised.

These and other departures from the behavior of conventional materials, such as polycrystalline metals and glass, are usually explained by saying that 'solid rock' is not solid at all but contains many small fissures, created presumably upon removal of the high pressure and temperature present at formation. These flaws in the rock structure decrease the apparent moduli until reapplication of pressure to jacketed samples closes the flaws and makes the rock appear stiffer. When the rock sample is not jacketed, the fluid is free to enter most of the pores, the cracks have little tendency to close, and measurements reflect the deformation of the rock grains, which are nearly perfectly elastic. The presence of flaws and their closing under pressure, as first suggested by *Adams and Williamson* [1923] and now generally accepted, thus seems to offer a qualitative explanation for the large initial change in the elastic moduli with pressure.

In the analysis presented in this paper, we have evaluated theoretically the effect of cracks on deformation of rock by making use of a mathematical model in which rock is portrayed as an elastic isotropic material containing randomly oriented narrow cracks. At first glance, the use of an isotropic material as a model would not seem acceptable because of the notoriously anisotropic behavior of rock at low pressures. However, the anisotropic behavior appears to be due more to the effect of cracks themselves than to anisotropy in the crystalline framework of the rock. *Birch* [1961]

---

[1] Contribution 1553 from the Woods Hole Oceanographic Institution.

[2] Actually, compressibility of single crystals and other nonporous solids changes slightly with pressure, by a few per cent in 10 kb, but we ignore this effect here.

found in the measurement of sonic velocity in mutually perpendicular directions that raising the confining pressure on jacketed samples of even highly 'anisotropic' material such as slate renders them nearly isotropic to the passage of sound waves. To be sure, one property—compressive strength—has been shown by *Donath* [1961] for a slate to depend on orientation of the sample for any confining pressure. However, *Walsh and Brace* [1964] show that this behavior can be explained by assuming that the crystalline framework of the rock is elastic and isotropic but that the distribution of cracks is anisotropic; i.e., cracks parallel to the plane of anisotropy are longer than those in other orientations. Thus, anisotropy is generally caused by anisotropic crack distribution, and the assumption in our analysis that the material between cracks is isotropic and elastic appears justified.

ANALYSIS

General. An expression for the effective compressibility β_{eff} of a porous elastic material can be derived in terms of the compressibility β of the solid material and the rate of change of porosity η with external pressure p. Consider a body whose external boundaries enclose a volume V_0 containing cavities of total volume V_c. *Chree* [1892] shows, using the reciprocal theorem, that the decrease in volume of the solid material, $\Delta(V_0 - V_c)$, due to an increase in pressure Δp is given by $\beta \Delta p V_0$. In differential form,

$$d(V_0 - V_c) = \beta V_0 \, dp \quad (1)$$

Differentiating and rearranging gives

$$(1/V_0)(dV_0/dp) = \beta + (1/V_0)(dV_c/dp) \quad (2)$$

The first term in (2) is the definition of effective compressibility, β_{eff}. The last term, for strains small enough so that V_0 may be considered constant, is the rate of decrease of porosity with external pressure. Using the convention that an increase in porosity is positive, we have

$$\beta_{eff} = \beta - d\eta/dp \quad (3)$$

Equation 3, based only on the requirement that the crystalline framework of the rock be linearly elastic, is independent of the shape or total volume of the pores. Experimental verification of (3) is not currently possible because no experiments are reported in which both the compressibility and the porosity have been measured as a function of pressure. It is interesting to note that β_{eff} should depend on the rate of change of porosity rather than on porosity directly. For some pore shapes (e.g., spherical pores), it can be shown that $d\eta/dp$ is a function of η, and in such cases elastic moduli and porosity are uniquely related. In general, however, η and $d\eta/dp$ are unrelated, and no correlation between the elastic properties and porosities of a suite of samples of the same rock should be expected

Spherical pores. The simple case of a body containing spherical cavities serves to illustrate (3). Consider the body to be separated into irregularly shaped regions, each containing one pore and each having the same porosity. Assume that the distance between pores is great enough so that the stress at the boundary of each region is very nearly the same as that at the boundary of the body. For the purpose of finding the change in the volume v_c of a small cavity due to the application of pressure p at the boundary of the region, the regime may be approximated by a sphere with volume v equal to that of the region. The tangential stress $\sigma_{\theta\theta}$ at the boundary of the cavity is found from Lamé's solution for thick-walled spheres [see *Timoshenko and Goodier*, 1951, p. 59] and can be expressed in the form

$$\sigma_{\theta\theta} = -(3p/2)(1 - v_c/v)^{-1} \quad (4)$$

The tangential strain $e_{\theta\theta} (1 - \nu)(\sigma_{\theta\theta}/E)$, and the volumetric strain $\Delta v_c/v_c$ is $3e_{\theta\theta}$. Thus, the fractional change in the volume of the cavity for a differential change in pressure is

$$\frac{dv_c}{v_c} = -\frac{9}{2}\frac{dp}{E}\frac{1-\nu}{1-v_c/v}$$

and the rate of change of porosity with pressure is

$$\frac{d\eta}{dp} = \frac{dv_c}{v_0 \, dp} = -\frac{9(1-\nu)}{2E}\frac{\eta}{1-\eta} \quad (5)$$

Combining (5), (3), and the identity $\beta = 3(1-2\nu)/E$ results in

$$\beta_{eff} = \beta\left(1 + \frac{3}{2}\frac{1-\nu}{1-2\nu}\frac{\eta}{1-\eta}\right) \quad (6)$$

Equation 6 is identical with results obtained by *MacKenzie* [1950], *Eshelby* [1957], and *Hashin* [1959], each using different methods but relying on elastic theory for infinitesimal strains. It is strictly valid only for small porosity, although *Hashin* [1959] has considered finite concentration of cavities.

Duckworth et al. [1950], *Coble and Kingery* [1956], and *Hasselman and Fulrath* [1964], in tests on samples of porcelain, alumina, and glass, in which the porosity occurred as nearly spherical cavities, found excellent agreement between predicted and experimental values of elastic moduli. On the other hand, experimental values of moduli when no precautions are made to ensure spherical pores in the samples (e.g., *Knudsen* [1962] for alumina) were found to be lower than those predicted by the sperical-pore theory. Similarly, sonic velocity measurements on a suite of sandstones of varying porosities by *Shimozuru et al.* [1957] indicated a definite decrease in modulus with increasing porosity, although the scatter was considerable. The measured sound velocity for a suite of low-porosity granites by *Ide* [1936], however, bore virtually no relation to the porosities of the samples. The explanation lies in the likelihood that the sandstone and granite samples contained cracks which influenced the moduli without affecting the porosity significantly. Indeed, the porosity of the granite samples was probably due almost entirely to narrow cracks, and no correlation between elastic behavior and porosity should be expected.

Narrow cracks. The effect of a random distribution of narrow cracks upon the compressibility of an elastic body is treated as in the spherical pore case, by first dividing the body into regions, each of which contains one crack. One difficulty not present in the spherical-pore case immediately arises—that of deciding the boundary conditions to be applied to each region. If the stress at the region boundary is assumed to be equal to the applied hydrostatic pressure, the deformation of the boundary, although having a symmetry corresponding to that of the crack, is not uniform. Thus, continuity of displacements cannot be maintained at the boundary of two regions containing cracks of different orientations. Similarly, if the strain at the boundary of a region is assumed to be uniform and equal to that of the body, the stress on the region boundary is not uniform, and equilibrium between adjacent regions is, in general, not satisfied. This same problem has occurred in computing the elastic constants for polycrystalline aggregates from the moduli of the components. *Voigt* [1928] suggested averaging the elastic moduli, which is equivalent in the present problem to assuming a uniform strain on the boundary of each region, whereas *Reuss* [1929] averaged the compliances, which corresponds to the assumption of uniform stress on the region boundary. *Hill* [1952] showed that the Voigt and Reuss methods are upper and lower bounds on the moduli; the separation of the bounds depends on the degree of anistropy of the components. The Voigt-Reuss bounds are considered adequate for the present analysis, since the anistropy of individual regions, which depends on the ratio of crack length to region size, is small for the assumed low concentration of cracks.

The shape of the crack must be specified in order to calculate its effect upon the compressibility of a region. We shall investigate three possibilities—the penny-shaped crack, the elliptical crack in plane stress, and the elliptical crack in plane strain. The results will be found to be quite similar.

The compressibility of a region is found somewhat more conveniently by using, instead of (3), equation 2 rearranged as follows:

$$\beta_{eff} = (\beta V_0 p\, dp + p\, dV_c)/V_0 p\, dp$$

or, for a single region,

$$\beta_{eff} = (\beta v_0 p\, dp + p\, dv_c)/v_0 p\, dp \qquad (7)$$

The first term in the numerator on the right-hand side of (7) is the strain energy associated with the region if there is no crack, and the second term is the increase in strain energy dw_c due to the crack. This second term is of interest in the Griffith fracture theory and therefore has been calculated for several cases—the penny-shaped crack by *Sack* [1946] and the elliptical crack in plane stress and plane strain by *Griffith* [1920]. For very narrow cracks of half-length and half-depth c, with uniform pressure p on the region boundary, the expressions are

$$dw_c = \beta \frac{16(1-\nu^2)c^3}{9(1-2\nu)} p\, dp \quad \text{(penny-shaped)}$$

$$dw_c = \beta \frac{4\pi(1-\nu^2)c^3}{3(1-2\nu)} p\, dp \quad \text{(plane strain)}$$

$$dw_c = \beta \frac{4\pi c^3}{3(1-2\nu)} p\, dp \quad \text{(plane stress)} \quad (8)$$

The total strain energy dW_c due to all cracks is found by summing the values dw_c in (8) over the total number of regions N. For the penny-shaped crack, for example,

$$dW_c = \beta \sum_N \frac{16(1-\nu^2)c^3}{9(1-2\nu)} p\, dp \quad (9)$$

Similarly, the expression for β_{eff} is found from (9) and (7), where $dW_c = p\, dV_c$:

$$\beta_{eff} = \beta\left[1 + \frac{16(1-\nu^2)}{9(1-2\nu)V_0} \sum_N c^3\right] \quad (10)$$

If we define an average crack length $\bar{c}$, where $N\bar{c}^3 = \Sigma_N c^3$ and an average region volume $\bar{v}$, where $N\bar{v} = V_0$, the effective compressibility for the Reuss assumption is found to be

$$\beta_{eff} = \beta\left(1 + \frac{16}{9}\frac{1-\nu^2}{1-2\nu}\frac{\bar{c}^3}{\bar{v}}\right) \quad \text{(penny-shaped)}$$

$$\beta_{eff} = \beta\left(1 + \frac{4\pi}{3}\frac{(1-\nu^2)}{1-2\nu}\frac{\bar{c}^3}{\bar{v}}\right) \quad \text{(plane strain)}$$

$$\beta_{eff} = \beta\left(1 + \frac{4\pi}{3(1-2\nu)}\frac{\bar{c}^3}{\bar{v}}\right) \quad \text{(plane stress)} \quad (11)$$

The expressions for β_{eff} derived from the Voigt assumption require a somewhat longer calculation, since expressions for the strain energy of a cracked solid with a uniform-strain boundary condition are not available in the literature and must be derived. From an analysis in the appendix, the effective compressibility using the Voigt model approaches that of the Reuss model for very small crack concentrations. The relationship between the Voigt and Reuss values of the effective compressibility (equation A-10) is

$$(\beta_{eff})_V = (\beta_{eff})_R - \beta \sum_N \frac{2A^2(v/V)}{9(1-2\nu)(1+\nu)} \quad (12)$$

where A, defined by equation A-1 for various crack types, is a parameter proportional to c^3/v. Equation 12 shows that the Voigt value for compressibility is less than the Reuss value and that the difference between the two decreases with decreasing concentration as predicted by *Hill* [1952]. However, since the crack concentration has been assumed to be small, the summation term is of second order and, to the order of approximation of this analysis, the Voigt and Reuss values should be assumed equal.

Crack closure. The compressibility of a region after the crack has closed is the same as that of uncracked material, and it remains essentially constant with further changes in pressure. Thus the rock gradually becomes stiffer as cracks close under increased confining pressure. For rocks where the entire porosity is caused by cracks, the compressibility eventually decreases to a nearly constant value equal to that for an uncracked sample, or, equivalently, that of unjacketed samples where the fluid can freely enter the pores. For rocks like sandstone, where cracks are responsible for a relatively small proportion of the total porosity, the compressibility of the sample again decreases to a nearly constant value. However, the final compressibility is not that of a 'solid' sandstone but that of a porous sandstone, since porosity other than that caused by rather narrow cracks is virtually impossible to remove even with very high confining pressures. This can be illustrated by finding the pressure required to close a penny-shaped crack with major axis $2\bar{c}$ and a ratio of minor to major axis α. The volume of the ellipsoidal cavity is $(4/3)\pi\bar{c}^3\alpha$, so the rate of change of porosity with pressure, where the region volume is $\bar{v}$, is

$$d\eta/dp = (4\bar{c}^3/3\bar{v})\, d\alpha/dp \quad (13)$$

The rate of change of porosity can also be found by equating (3) and (11) for the penny-shaped crack:

$$d\eta/dp = -16\beta(1-\nu^2)\bar{c}^3/9(1-2\nu)\bar{v} \quad (14)$$

Equations 13 and 14 are equated, and the resulting equation is integrated to give the pressure p_c required to reduce α to zero.

$$p_c = \pi E\alpha/4(1-\nu^2) \quad (15)$$

Deriving similar expressions for the elliptical crack in plane stress and plane strain gives similar results: to an order-of-magnitude approximation, the pressure required to close a crack is $E\alpha$. Now consider a porous sandstone composed of quartz grains, for which Young's modulus is about 1000 kb. The pressure required to close cavities with aspect ratios of one-tenth is 100 kb, a pressure beyond the capabilities of most apparatus. Some cavities are likely to be more nearly spherical and to require even a higher pressure. Thus, the only cavities which close under pressure must be very narrow cracks. The experiments discussed by *Birch* [1961], for example, where application of several kilobars served to close the cracks, indicate that the cracks had an aspect ratio of the order of 1 to 1000.

The closing of cracks under hydrostatic pressure provides a useful method for determining the porosity of rock for which all porosity is in the form of cracks. For such rocks, the porosity, which is less than 1% or so and difficult to measure, can be found from the plot of volmetric strain versus pressure taken during a compressibility test. First, consider (2) integrated to give an expression for the change in porosity $\Delta \eta$.

$$-\Delta V_c / V_0 = \Delta \eta = \Delta V_0 / V_0 - \beta p \quad (16)$$

The total change in porosity when all cracks have closed is equal to the initial porosity η_0. If we assume that all cracks are closed at a pressure p_1 corresponding to a volumetric strain $(\Delta V/V)_1$, (16) becomes

$$\eta_0 = (\Delta V_0 / V_0)_1 - \beta p_1 \quad (17)$$

Reference to Figure 1, a plot of pressure versus volumetric strain, shows that the porosity η_0 in (17) is simply the intercept on the volumetric strain axis of the straight-line portion of the curve extrapolated back to zero pressure. The change in porosity $\Delta \eta$ due to application of pressure p_2 less than that required to close all the cracks can be shown by using (16) to be the intercept on the volumetric strain axis of a line dropped at an angle equal to the compressibility. It is necessary, of course, that the slope of the straight-line portion (see footnote 2) of the volumetric strain-pressure curve correspond to the compressibility of a sample with zero porosity for this graphical method to be valid. This can be checked by verifying that the slope of the straight-line portion is equal to the slope of the plot for an unjacketed sample where fluid is free to enter all pores.

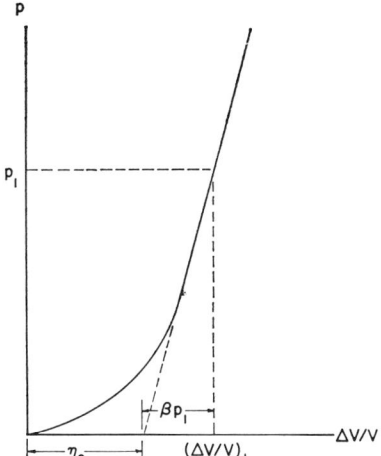

Fig. 1. Graphical determination of porosity (see text).

Brace [1965] compared porosity, determined graphically as described above, with that measured directly for granites from Westerly, R. I., and Stone Mountain, Ga. The graphical method, applied to three mutually perpendicular samples cored from a single large block of uniform material, gave the quantity called 'crack porosity,' η_c, with an uncertainty of about 0.0005 (Table 1). Total porosity, η, of the two granite specimens was measured by standard methods [*Dreyer*, 1958], again with an uncertainty of about 0.001 (Table 1). For the Stone Mountain granite, η_c was equal to η to within the errors of the measurements. For Westerly granite, however, η_c was much smaller than η. By comparison of intrinsic compressibility, calculated from mineral content, with observed com-

TABLE 1. Porosity of Two Granites*

Sample	From Compressibility Data			Measured
	η_c ±0.0005	η_p ±0.003	η	η ±0.001
Granite Stone Mt., Ga.	0.0035	0	0.0035	0.003
Granite Westerly, R. I.	0.0020	0.012	0.014	0.011

* After *Brace* [1965].

pressibility at pressures of 5 to 9 kb, the reason for this discrepancy was clear. Evidently, a rather large amount of porosity persisted at these high pressures in Westerly, whereas Stone Mountain granite had none. An estimate of this 'pore porosity,' η_p, from (6), shown in Table 1, appeared to account for the difference noted above at least to within the uncertainty in measurement of η_p (0.003).

Discussion

In the preceding analysis, an equation was derived which shows that the difference between the compressibilities of a porous body and a solid body of the same composition is equal to the rate of change of porosity with externally applied pressure for the porous body. This relationship is independent of the shape or concentration of the pores and relies only on the assumption of linear elasticity and small strains. Expressions for the compressibility of porous bodies are found by assuming a pore shape and concentration for which the rate of change of porosity can be determined theoretically. Two cases were considered—low concentrations of spherical pores and narrow cracks.

A comparison of spherical pores and narrow cracks illustrates the effect of pore geometry upon compressibility. If we assume that all spherical pores have the same diameter $\bar{c}$, the compressibility (6) for low porosity is

$$\beta_{\text{eff}} = \beta\left(1 + 2\pi \frac{1-\nu}{1-2\nu} \frac{\bar{c}^3}{\bar{v}}\right) \quad (18)$$

Equation 18 is identical with the expressions derived for narrow cracks (11), with the exception of the magnitude of the coefficient of the last term. Evaluating the coefficient for various values of Poisson's ratio shows that spherical holes are some two to three times more effective than cracks of the same radius in increasing compressibility. Thus we find that the effect upon compressibility of a certain concentration of narrow cracks equals the effect of the same number of spherical pores with a diameter roughly two-thirds that of the cracks. The porosity of the two samples differs enormously, of course. Also, the analysis shows that compressibility is independent of α, the aspect ratio of a crack, so that a crack which is about to close is as effective as a fully open crack of the same length.

The compressibility is strongly affected by the presence of a few long cracks. As shown in (11), in which crack length appears to the third power, one crack ten units long has the same effect as 1000 cracks one unit long. The strong influence of the longer cracks accounts for the apparent anisotropy at low pressure of seemingly isotropic rocks because of the improbability of a small collection of long cracks having a random character.

An interpretation of some interesting measurements of linear compressibility (linear strain per unit pressure) as a function of pressure by *Brace* [1965] summarizes the effects of cracks and crack closure derived above. Linear compressibilities measured parallel ($||$) and perpendicular ($\perp$) to foliation of a gneiss sample were found to have the unusual pressure dependence shown in Figure 2, in which $\beta_\perp$ is first larger, then smaller, and finally again larger than $\beta_{||}$ as the pressure is increased. The rock contained about 10% mica in the form of long flakes with a strong preferred orientation; the remainder of the rock was quartz and feldspar with virtually random distribution of crystallographic axes. If we assume that grain boundaries form the sites of cracks, here is a rock with a few highly oriented long cracks among a large number of randomly oriented shorter cracks. In Figure 2, at the lowest pressure, $\beta_\perp$ is higher because the cracks perpendicular to that direction are longer. At the highest pressures all cracks have presumably closed, but $\beta_\perp$ is still higher than $\beta_{||}$ because mica is more compliant normal than parallel to (001), the plane of anisotropy. Between the highest and lowest pressures, the two curves cross. Here we see the effect of crack shape upon closure. The longer cracks tend to close at a lower pressure than the shorter, causing $\beta_\perp$ to fall to its intrinsic value at a lower pressure than $\beta_{||}$.

Appendix

Consider a region of volume v with uniform strain e on the boundary produced by stress R, S, and T, where R and T are in the plane of the crack and S is normal to the crack. The strain energy of the region is considered to be composed of two parts—the amount dw_s if there is no crack plus an amount dw_c attributed to the

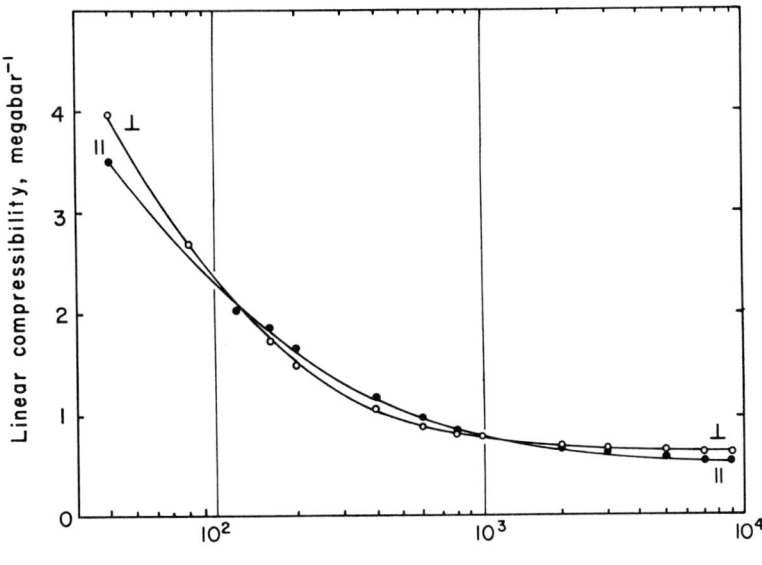

Fig. 2. Linear compressibility of Torrington gneiss parallel (∥) and perpendicular (⊥) to foliation (after *Brace* [1964]).

presence of the crack. *Sack* [1946], for the penny-shaped crack, and *Griffith* [1920], for the crack in plane stress and plane strain, show that the strain energy, dw_c, associated with the crack is independent of the values of R and T and is given by

$$dw_c = AvS\, dS/E$$

where

$$A = 16(1 - \nu^2)c^3/3v \quad \text{penny-shaped}$$
$$A = 4\pi(1 - \nu^2)c^3/v \quad \text{plane strain}$$
$$A = 4\pi c^3/v \quad \text{plane stress} \quad \text{(A-1)}$$

The strain energy dw_s for solid region of volume v is

$$dw_s = \{[R - \nu(S + T)]\, dR$$
$$+ [S - \nu(R + T)]\, dS$$
$$+ [T - \nu(S + R)]\, dT\}v/E \quad \text{(A-2)}$$

The total strain energy for a region is the sum of (A-1) and (A-2). The average strain in each principal direction is given by

$$v^{-1}\, \partial w/\partial R \qquad v^{-1}\, \partial w/\partial S \qquad v^{-1}\, \partial w/\partial T$$

or

$$Ee = R - \nu(S + T)$$
$$Ee = S(1 + A) - \nu(R + T)$$
$$Ee = T - \nu(R + S) \quad \text{(A-3)}$$

The stresses can be found in terms of the boundary strain e by solving (A-3).

$$R = T = \frac{3e}{\beta} \frac{1 + A/(1 + \nu)}{1 + A(1 - \nu)/(1 + \nu)(1 - 2\nu)}$$
$$S = \frac{3e}{\beta}\left[1 + \frac{A(1 - \nu)}{(1 + \nu)(1 - 2\nu)}\right]^{-1} \quad \text{(A-4)}$$

The strain energy for a region is found by substituting (A-4) in the sum of (A-2) and (A-1).

$$dw = \frac{9E\,de}{\beta}$$
$$\cdot \frac{1 + \dfrac{A(5 - 7\nu)}{3(1 + \nu)(1 - 2\nu)} + \dfrac{2A^2(1 - \nu)}{3(1 + \nu)^2(1 - 2\nu)}}{[1 + A(1 - \nu)/(1 + \nu)(1 - 2\nu)]^2} v \quad \text{(A-5)}$$

The total strain energy is found by summing (A-5) over all N regions.

$$dW = \frac{9e\,de}{\beta}$$
$$\cdot \sum_N \frac{1 + \dfrac{A(5 - 7\nu)}{3(1 + \nu)(1 - 2\nu)} + \dfrac{2A^2(1 - \nu)}{3(1 + \nu)^2(1 - 2\nu)}}{[1 + A(1 - \nu)/(1 + \nu)(1 - 2\nu)]^2} \quad \text{(A-6)}$$

The total strain energy is work done by the pressure on the total volume V, or

$$dW = \beta_{eff}(Vp\,dp) \quad (A\text{-}7)$$

Combining (A-6) and (A-7), where $e = \beta_{eff}p/3$, gives an expression for β_{eff}.

$$\frac{1}{\beta_{eff}} = \frac{1}{\beta} \cdot \sum_N \frac{1+\dfrac{A(5-7\nu)}{3(1+\nu)(1-2\nu)}+\dfrac{2A^2(1-\nu)}{3(1+\nu)^2(1-2\nu)}}{[1+A(1-\nu)/(1+\nu)(1-2\nu)]^2}\frac{v}{V} \quad (A\text{-}8)$$

An approximate expression for β_{eff}, correct to the order of approximation of the analysis, can be found by expanding (A-8).

$$\frac{1}{\beta_{eff}} = \frac{1}{\beta}\left[\sum_N \frac{v}{V}\left(1 - \frac{A}{3(1-2\nu)} + \frac{A^2}{3(1+\nu)(1-2\nu)^2} + \cdots\right)\right]$$

or

$$\beta_{eff} = \beta\left(1 + \sum_N \frac{A(v/V)}{3(1-2\nu)} - \sum_N \frac{2A^2(v/V)}{9(1-2\nu)(1+\nu)} = \cdots\right) \quad (A\text{-}9)$$

It can be shown that the first two terms on the right-hand side of (A-9) give the effective compressibility for the Reuss assumption. Thus, the Voigt and Reuss values of the compressibility are related by

$$(\beta_{eff})_V = (\beta_{eff})_R - \beta\sum_N \frac{2A^2(v/V)}{9(1-2\nu)(1+\nu)} \quad (A\text{-}10)$$

Acknowledgments. W. F. Brace has had a strong influence on this research and was especially helpful in preparing the final manuscript.

The research reported in this document has been sponsored by the Air Force Cambridge Research Laboratories, Office of Aerospace Research, United States Air Force, Bedford, Mass., under contract AF 19(628)-3298.

References

Adams, L. H., and E. D. Williamson, The compressibility of minerals and rocks at high pressures, *J. Franklin Inst., 195*, 475–529, 1923.

Birch, Francis, The velocity of compressional waves in rocks to 10 kilobars, 1, *J. Geophys. Res., 65*, 1083–1102, 1960.

Birch, Francis, The velocity of compressional waves in rocks to 10 kilobars, 2, *J. Geophys. Res., 66*, 2199–2224, 1961.

Brace, W. F., Some new measurements of linear compressibility of rocks, *J. Geophys. Res., 70* (2), 1965.

Chree, C., Changes in the dimensions of elastic solids due to given systems of forces, *Trans. Cambridge Phil. Soc., 15*, 313–337, 1892.

Coble, R. L., and W. D. Kingery, Effect of porosity on physical properties of sintered alumina, *J. Am. Ceram. Soc., 39*, 377–385, 1956.

Donath, F. A., Experimental study of shear failure in anistropic rocks, *Bull. Geol. Soc. Am., 72*, 985–990, 1961.

Dreyer, W., Uber die Genauigkeit der Porositatsbestimmung aus Absolutund Raumdichte, *Tonind. Ztg. Keram. Rundschau, 82*, 72–79, 1958.

Duckworth, W. H., et al., Mechanical properties of ceramic bodies, *Rept. R-209*, Rand Corporation, 1950.

Eshelby, J. D., The determination of the elastic field of an ellipsoidal inclusion, and related problems, *Proc. Roy. Soc. London, A, 241*, 376–396, 1957.

Griffith, A. A., The phenomena of rupture and flow in solids, *Phil. Trans. Roy. Soc. London, A, 221*, 163–197, 1920.

Hashin, Z., The moduli of an elastic solid, containing spherical particles of another elastic material, in *Nonhomogeneity in Elasticity and Plasticity*, edited by W. Olszak, pp. 463–478, Pergamon Press, New York, 1959.

Hasselman, D. P. H., and R. M. Fulrath, Effect of small fraction of spherical porosity on elastic modulus of glass, *J. Am. Ceram. Soc., 47*, 52–53, 1964.

Hill, R., The elastic behavior of a crystalline aggregate, *Proc. Phys. Soc. London, A, 65*, 349–354, 1952.

Ide, J. M., Comparison of statically and dynamically determined Young's modulus of rocks, *Proc. Natl. Acad. Sci. U. S., 22*(2), 81–92, 1936.

Knudsen, F. P., Effect on porosity of Young's modulus of alumina, *J. Am. Ceram. Soc., 45*, 94–1962.

Mackenzie, J. K., The elastic constants of a solid containing spherical holes, *Proc. Phys. Soc. London, B, 63*, 2–11, 1950.

Reuss, A., Berechnung der Fleissgrenze von Mischkristallen auf Grund der Plastizitats bedingung fur Einkrisalle, *Z. Angew. Math. Mech., 9*, 49, 58, 1929.

Sack, R. A., Extension of Griffith's theory of rupture to three dimensions, *Proc. Phys. Soc. London, A, 58*, 729–736, 1946.

Shimozuru, Daisuke, On the elasticity of rocks up to 2000 bars—The interpretation of the velocity relation, *Mem. Fac. Sci., Kyusyu Univ., B, 2*(3), 98–103, 1957.

Timoshenko, S., and J. N. Goodier, *Theory of Elasticity*, McGraw-Hill Book Company, New York, 1951.

Voigt, W., *Lehrbuch der Kristallphysik*, B. G. Terebner, Leipzig, 1928.

Walsh, J. B., and W. F. Brace, A fracture criterion for brittle anisotropic rock, *J. Geophys. Res., 69*, 3449–3456, 1964.

Zisman, W. A., Compressibility and anistropy of rocks at and near the earth's surface, *Proc. Natl. Acad. Sci. U. S., 19*, 666–679, 1933.

(Manuscript received October 29, 1964.)

MIGRATION

Reprinted by permission of Elsevier Applied Science
Publications from *Developments in Geophysical Exploration Methods—2*, ed. A. A. Fitch (1981), p. 151-230.

From
Developments in Geophysical Exploration Methods—2

edited by A. A. Fitch

Chapter 6

MIGRATION

P. Hood
Geophysics Research Branch, British Petroleum Co. Ltd, London EC2Y 9BU, UK

SUMMARY

Seismic migration is one of the most rapidly changing fields in data processing. During the last twelve years three major methods and a host of minor methods have appeared on the scene, each with its own range of applicability. In this article we examine in detail the three mainstream migration methods, i.e. the diffraction stack, F–K migration, and finite difference migration; we scrutinise their strengths, weaknesses and relative merits in terms of practical migration problems. We also look at some of the new techniques which have been discussed in the literature and which are, potentially, the migration methods of the future—these include hybrid finite difference/Fourier methods, direct velocity inversion techniques, and stack enhancement by partial migration.

NOTATION

c	Velocity of seismic wave propagation
c_H	Horizontal velocity
$\bar{c}$	Frame velocity
CMP	Common midpoint, i.e. point half way between the shot and geophone
D	Down-going wave
f	Temporal frequency

h	Offset coordinate, i.e. $2h = x_g - x_s$
k_h	Wavenumber in the h direction
k_x	Wavenumber in the x direction
k_z	Wavenumber in the z direction
n	Unit normal vector to a surface
P	Pressure amplitude (assuming that this can be obtained from the particle velocity (geophone) or is recorded (hydrophone))
$P^n_{j,k}$	Discrete representation of the pressure at coordinate $(j\Delta x, k\Delta z, n\Delta t)$.
$\tilde{P}$	Fourier transform over time or pseudodepth, i.e. $$\int P \exp(-i\omega t)\, dt \quad \text{or} \quad \int P \exp(-ik_d d)\, dd$$
$\hat{P}$	Fourier transform spatially over x or z, i.e. $$\int P \exp(ik_x x)\, dx \quad \text{or} \quad \int P \exp(ik_z z)\, dz$$
$\tilde{P}'$	Time retarded version of $\tilde{P}$, i.e. $\tilde{P}' = \tilde{P}\exp(-i\omega z/\bar{c})$
p	Snell's law parameter $= \sin\theta/c$
$\tilde{Q}$	Input pressures convolved with a shaping operator
R	Vector with magnitude $[(z-z_0)^2 + (x-x_0)^2 + (y-y_0)^2]^{1/2}$
$R(x,z)$	Earth reflectivity map
T	Time shift in shifting equation
T_0	Apex time on a hyperboloid surface in 3D migration
t	Time coordinate
$t_{\max}$	Maximum recorded time on a section
$\bar{t}$	Time of travel for a wave from a point scatterer to the surface geophone
t'	Time coordinate of downward continued data
U	Up-going wave
x	Horizontal coordinate along dip line
y	Horizontal coordinate along strike line
z	Depth coordinate measured downwards from the surface of the earth
(x_0, z_0)	Coordinates of a buried scatterer
(x_g, z_g)	Coordinates of a geophone
(x_s, z_s)	Coordinates of a shotpoint
α_r	Angle of dip of a plane layer on the migrated depth or earth reflectivity section
α_t	Angle of dip of a plane layer on a time section
ε	Perturbation parameter used in Cohen and Bleistein's theory

θ	Parameter used in a skewed finite difference formula
	0 Forward difference in z
	0·5 Crank–Nicholson central difference in z
	1·0 Backward difference in z
ρ	Density
τ	Two-way time coordinate measuring vertical travel time
ω	Angular velocity in rad s^{-1}.

1. INTRODUCTION

Seismic migration is one of the last of the processes to be applied in the data processing sequence. Its purpose, briefly stated, is to transform a seismic wave field recorded at the earth's surface (time section) to an earth reflectivity map (depth section). Up to the late 1960s this was achieved by manual methods on a few picked horizons using ray tracing and timing calculations. Then around 1970 the first of the 'diffraction stack' migration methods became commercially available. Again, this was based on ray tracing concepts and the scalar diffraction theory of Huygens and Fresnel,[1] but in this case the method could be applied to complete common midpoint (CMP) sections. In the 1970s several major developments took place. One was the use of wave rather than ray theory. The key figure in this movement was Professor Jon Claerbout at Stanford University, who currently runs a major project called the 'Stanford Exploration Project'. This is financed by the oil industry and it aims to look into new exploration techniques. Another development was the understanding that the diffraction stack method could be improved by referring to Kirchhoff integral theory rather than the ray theory which approximates to it. This led in turn to a better application of processing parameters in the method.

Over the past decade three main processing techniques have emerged: these are known as diffraction stack (or Kirchhoff) migration, finite difference (or wave equation) migration and F–K (or wavenumber) migration. The use of these epithets is however a little confusing, since people tend to refer to all three of the methods by the single term 'wave equation migration'. This is because all the methods are based on solutions to the scalar wave equation. There have been some further developments recently which also look promising: one of these is the direct inversion of the surface wavefield to obtain the velocities and structure. This represents a further move away from ray theory to a complete wave theoretical description of migration process. Until recently migration of data followed either from a velocity field which had been derived from stacking velocities, or occasionally from velocities based on ray tracing studies of a two-dimensional earth model, and relating the predicted with the observed wavefields. These latest inversion methods offer the interesting possibility of circumventing the iterative loop of migration from model-derived velocities. So far we have mentioned the 'wave equation' without being at all specific. In practice the following equation is the most generally used:

$$\nabla^2 P = (1/c^2)(\partial^2 P/\partial t^2) \qquad (1.1)$$

where $P(x, y, z, t)$ = pressure amplitude at coordinates (x, y, z) and time t; $c(x, y, z)$ = propagation speed of the acoustic wave. This equation describes the spatial and temporal evolution of the pressure field (but not the displacement or the particle velocities). Equation (1.1) is known as the scalar wave equation. It is assumed that, although the velocity can vary, the density of the medium is a constant which does not enter the calculation. This assumption is a reasonable one; modelling of well borehole data shows that it is the sonic velocities which largely determine the shape of the synthetic seismogram; the densities normally reinforce rather than alter the picture. There is of course a further problem in that the densities are not generally available. In the case where the density is known, the wave equation is modified by the presence of an extra density term as follows:

$$\nabla^2 P - (1/c^2)(\partial^2 P/\partial t^2) - \nabla \log \rho \nabla P = 0 \qquad (1.2)$$

This equation is rarely used in geophysical exploration; however, one example of its potential use is in the direct inversion of one-dimensional velocity and density profiles.[2]

Further complexity may be added into the migration picture if the earth's elastic constants are known. Such an ideal state of affairs has not been given serious attention until recently.[3] With increasing emphasis being placed nowadays on recording shear wave data, then perhaps some progress along this path may take place in the future. As it is, eqn. (1.1), which is valid for fluid media, can be used to model the usual diffraction and refraction (Snell's law bending) effects of either shear or compressional waves separately, but will not of course predict mode conversions between the two types of wave, nor the correct variation of reflection coefficient with angle of incidence, for which purpose it is necessary to solve the elastic wave equation. Currently, the most important limitations remaining on correct migration of seismic data via eqn. (1.1) lie in the imprecise knowledge of the velocity c, and the band limitation and noise corruption of the recorded data. The errors in velocity of course lead to errors in the migration of data; where dips are steep, such errors can be significant both in terms of migration and CMP mis-stacking. The band limitation of the data affects the ultimate quality and resolution of the migrated output, or, in a complete inversion scheme, manifests itself in a certain non-uniqueness of the inverted solution. Noise corruption, if large, tends to sabotage efforts to obtain the velocities accurately, and after migration will obscure signal underneath 'noise smiles'. To gain as complete a picture as possible from seismic data, modern methods offer no panacea for poor field acquisition; indeed, best results will always be obtained from wide-bandwidth, noise-free data.

In this article we start off in Section 2 with a review of some of the fundamental concepts of migration; in Sections 3–5 the mainstream migration techniques will then be covered. In Section 6 some of the most recent developments will be discussed. Then in Section 7 we give an overview of the various migration methods available and make specific recommendations in the choice of processing methods. No great effort has been made towards originality of material; however, some of the sources are obscure and this article will hopefully give these the attention they deserve. Some of the material is new, and particular thanks must go to certain contracting companies for permission to use their as yet unpublished diagrams. It would not be appropriate to single out individual companies here, and so in the final section full acknowledgement is given to each company.

2. FUNDAMENTAL CONCEPTS

There are a number of concepts and assumptions made in migration theory which are fundamental to a clear understanding of present day practice; these are discussed in this section. It is assumed that the reader already has a fair knowledge of migration, and so little time will be spent in discussing what migration is. It was mentioned earlier that migration was a mapping from surface recorded acoustic data to an earth reflectivity section. This process is sometimes referred to as 'depth migration'. However, the reason for the title is not the nature of the final section, but the fact that the migration process has tracked the wavefield in depth taking full account of reflection curvature and of refraction and diffraction effects. More often than not, the resulting reflectivity section is convolved with a wavelet, since imprecise knowledge of the original recorded wavelet prevents perfect deconvolution. Another common presentation of migrated data is in terms of a time section. In this case the earth reflectivity section can be converted to a time section using a suitable velocity field for the conversion, or, in the case of a 'time migration', then time coordinates are the most natural output coordinates of the migration. In 'time migration' diffraction effects are considered, but not those refraction effects which are due to lateral changes in velocity.[4,64]

2.1. The Earth Model

Until comparatively recently, seismic data was shot and recorded along single lines. These lines tended to follow, as far as possible, either the dip or the strike of the structure. There is less reluctance nowadays to shoot areal surveys where the underlying structure is truly three-dimensional. Nevertheless, for the majority of cases it is sufficient to consider the earth structure as locally two-dimensional. Mathematically this simplification is not required; the main trouble always arises in moving from one to two dimensions. But, having derived the algebra in two dimensions, extension to three or more dimensions is trivial. Furthermore the numerical techniques developed in two dimensions can just as easily be extended to three. However, the number of calculations involved in a correct 3D migration means that there is considerable advantage to be obtained in splitting the migration down into a series of 2D migrations. For this reason most of the discussion in this article will centre around a 2D earth model in which the earth does not vary in the direction normal to the survey line (y axis).

Another point which is well worth bearing in mind is that, if a 2D survey has been shot at an angle to the dip line, or if the structure is plunging along the y axis, it is still not necessary to use a full 3D migration procedure. As French[5] has pointed out, 2D migration in such cases is quite sufficient, provided that the velocities have been adjusted by a simple scaling factor.

There are some other assumptions which are frequently made regarding the earth model. The most useful of these goes under the name of the Born approximation. Essentially this approximation means that propagation is in a single direction from source to receiver and back again, i.e. there are no multiple reflections. Multiple reflections can theoretically be removed by solution of the wave equation during migration; however, there are no commercially available programs to achieve this. Another approximation made relates to the source. This is often assumed to be two-dimensional (line source) rather than three-dimensional (point source) in nature. Where this is so, the times on the migrated section will be correct although the amplitudes will be in error. Recorded data can however be 'corrected' to simulate cylindrical rather than spherical divergence.

2.2. Ball Bearing Model

The ball bearing model[6] is a useful model in that it offers a simple pictorial analogue to the diffraction stack process of migration. In this model a reflecting horizon is assumed to consist of a number of ball bearings spaced extremely close together (Fig. 1). The main response on the time section due to a single ball bearing embedded in a homogeneous medium lies along a 'diffraction' hyperboloid. The apex of the hyperboloid lies at the two-way time and position of the migrated 'time section'. If several ball bearings are placed side by side in a plane the reflected waveforms interfere constructively at short distances and destructively at greater distances. The result of this is, in the limit, that a plane can be considered as the sum of closely spaced point scatterers. This concept indicates that migration may be achieved by summing amplitudes on the time sections over individual hyperboloids and placing the resultant sum at the respective apexes. In essence this is the basis of the diffraction stack method of migration.

2.3. Up-going and Down-going Waves

The idea of separation of the seismic wavefield into up- and down-going components is a useful one, as we shall see later. Down-going waves refer to those waves which either emanate directly from the shot or are generated by multiple reflections, and which consequently propagate in the downwards direction into the earth (Fig. 2). Up-going waves refer to those waves which have been generated from the upward reflection of the down-going waves by changes in acoustic impedance, and which travel towards the surface of the earth where they are recorded. In both cases the significant contribution to the energy spectrum comes from waves with small angular deviation from the vertical (z axis) direction. This is because most horizons have small dip, and significant energy normally comes from specular rather than diffuse reflection; furthermore, geophone or shot pattern response will effectively eliminate waves with a large horizontal wavenumber from the time section. These, it must be emphasised, are generalities; steep fault planes or salt domes with dips up to or exceeding 90° can be encountered. Even so, the concept of up-going and down-going waves is useful, although in this case the categorisation into up-going or down-going waves depends on the overall trend of wave movement. Some confusion appears to reign when considering reciprocity and it is perhaps appropriate to deal with this point here. With some qualifications,[65] this important principle says that if a

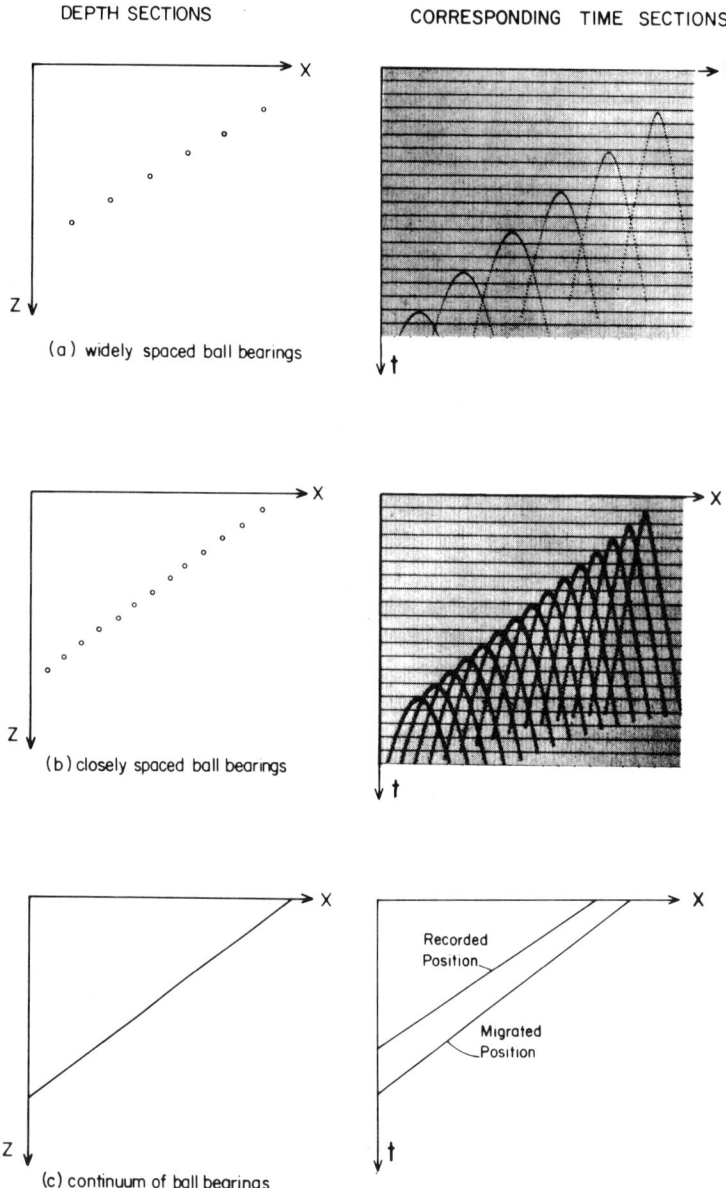

FIG. 1. Reflection as a sum of diffractions—ball bearing model.[6]

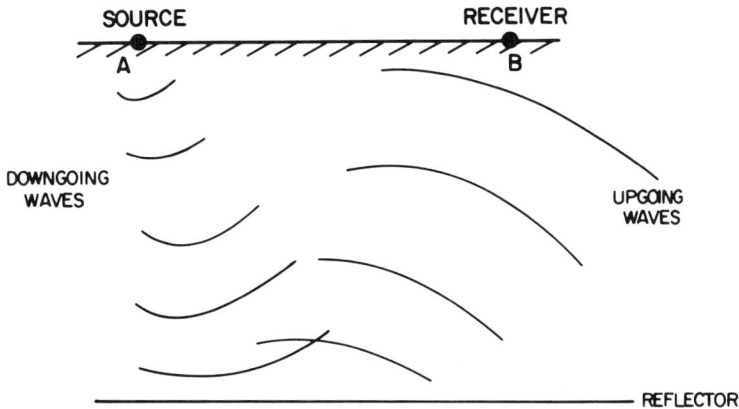

FIG. 2. Separation of wavefield into up- and down-going wavefields.

wave is generated at A and recorded at B (Fig. 2), the same recording would be received had the wave been initiated at B and recorded at A. Note that, in a practical application of this principle, source and receiver directivity effects and differences in ground coupling are neglected. Reciprocity is used in some methods of migration effectively to extend the notional area of surface coverage in the sense that it is possible to determine the seismogram, which would be recorded at the shot points, from data actually recorded at the geophones. In fact this trick increases neither the fold of cover nor the line extent. The confusion lies in the treatment of this rearranged shot point recorded data as down-going wave energy. It is of course notionally recorded up-going wave energy and must be treated as such.

2.4. Downward Continuation and Datumming

For a wavefield recorded on the earth's surface, application of Huygen's principle or solution of the wave equation will permit the reconstruction of the seismogram at a different datum level in the earth. This process of moving data from one level to the next is known as downward continuation. Mathematically, it is the derivation of the wavefield $P(x, z + \Delta z, t')$ from the known wavefield $P(x, z, t)$. There is a slightly more general process described by Berryhill,[7] known as datumming. In this case the surfaces on which the data are recorded may be irregular, i.e. $z = z(x)$, and similarly the datum to which the data is to be downward continued may also be irregular.

2.5. Imaging Conditions

If we pursue the idea of up and down-going waves further, this leads to the

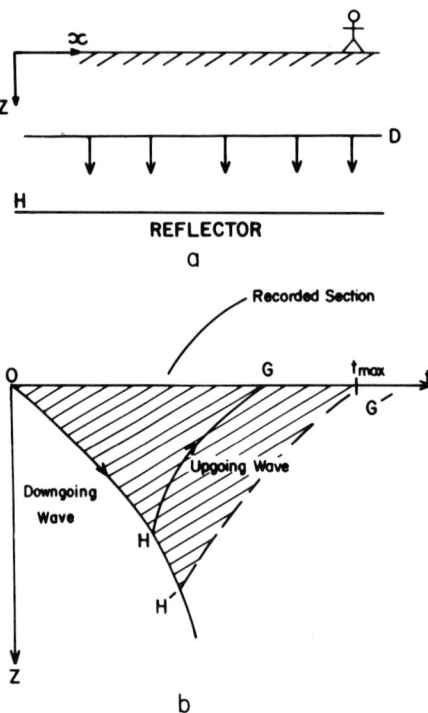

FIG. 3. Derivation of imaging conditions: (a) earth model; (b) depth–time trajectories.

idea of depth–time trajectories and eventually to Claerbout's imaging principle.[8] For this purpose we consider a vertically travelling down-going plane wave D and a single horizontal reflector (Fig. 3(a)). It will be assumed that the earth is laterally homogeneous. After striking the reflector the down-going plane wave creates an up-going plane wave U and continues its motion downwards. The resulting depth–time trajectory is shown in Fig. 3(b). The x coordinate is not displayed since velocities are laterally invariant. If t_{max} represents the last time recorded on the section, then H' represents the deepest level about which we can deduce information, and clearly that part of (z, t) space lying to the right of the curve G'H' will be of no interest. Similarly the line OHH' defines the left most extremum of interest, since no acoustic energy can travel faster than the medium velocity. Migration is thus concerned exclusively with the 'shaded' region bounded by OHH'G'G.

The process of downward continuation may be used to extrapolate the up-going wavefield recorded at $z = 0$ back into the earth (and earlier in time) until it meets the down-going wave trajectory OHH'. Similarly, if the shot waveform is known, downward continuation of the down-going wavefield on the path OHH' may be achieved. At H the down-going wavefield is time- and space-coincident with the up-going wavefield. The wavefield is said to image the reflector at this point and the division of $U(x,z,t,)/D(x,z,t)$ will yield an estimate of the reflection coefficient $R(x,z)$. If the shot waveform is unknown, and an impulsive down-going wavefield of unit amplitude is downward continued, this ratio will yield instead the reflection coefficient convolved with the shot waveform.

We are now in a position to set forth Claerbout's imaging principle. This states that 'reflectors exist at points in the ground where the first arrival of the down-going wave is time coincident with an up-going wave'. This imaging principle really applies to the situation envisaged above where both shot and received wavefields are downward continued. Both wavefields are required if multiply reflected waves are to be removed during migration; this is almost exclusively relevant in terms of plane wave theory. In more usual applications, the imaging condition is not used as it stands, since only up-going waves are downward continued. The imaging condition used depends entirely on the model.

One of the most important models used in migration today is the exploding reflector model.

2.6. The Exploding Reflector Model

The idea of the exploding reflector model was first introduced by Loewenthal *et al.*,[9] and for this reason it is sometimes called the Loewenthal model. Instead of the sources being on the surface, each reflector is considered to be composed of a series of point sources (Fig. 4(a)); the magnitude of each source equals the value of the reflection coefficient at that point. All the sources are set off at zero time, and eventually their emanations are received at the surface. There are two variations on this theme: one is the original Loewenthal model in which a zero offset section is considered. In ray theoretical terms this corresponds to a single ray being traced up to the surface from each source point, with a starting angle normal to the reflecting surface. If the travel times are doubled (or the velocities are halved), then the result is similar to a zero offset section. Migration consists of the inversion of this forward model. Downward continuation proceeds and the reflectors are imaged at time $t = 0$ in the exploding reflector time coordinates, since this is the time of initiation of the shots.

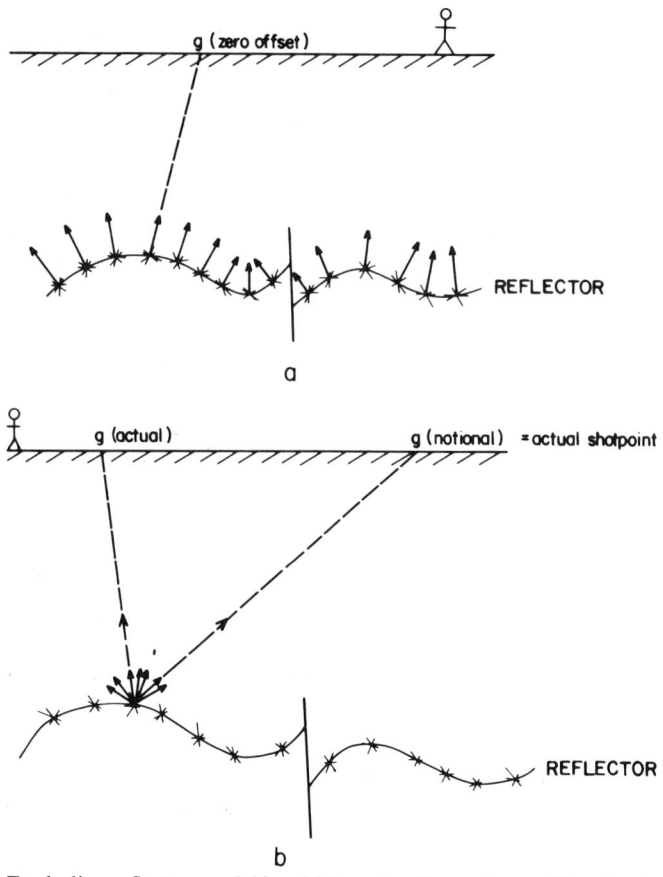

FIG. 4. Exploding reflector model in which each reflector is considered to be made up of a number of point sources: (a) zero offset case—a single ray path is considered, starting at right angles to the reflector; (b) Non-zero offset—twin families of ray paths are considered, the first set travel up to the actual geophone locations and the other set travel up to the shot point locations.

The other variation of the exploding reflector model arises in the application to migration before stack; in ray theory this model would correspond to tracing twin families of rays from each source point on the reflector up to the surface (Fig. 4(b)). One set of rays travels up to the geophones, and the other travels up to the 'notional' geophones located at the surface position of the actual shots (using reciprocity); account can therefore be taken of variations in earth velocity on both up- and down-going ray paths, and the multibranching focusing effects (Fig. 5) which

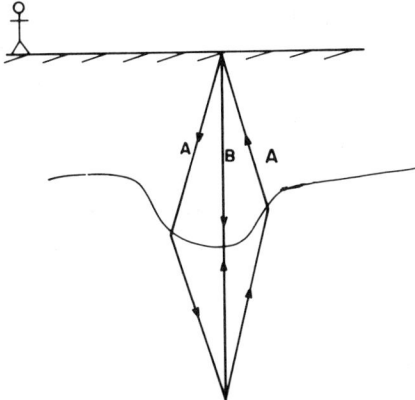

FIG. 5. In the zero offset case ray path B is the only path considered, whereas in non-zero offset case ray paths of types A and B are considered.

are ignored in the original Loewenthal model. Imaging is therefore conceptually more difficult to grasp. The total travel time t for two typical rays may be expressed as

$$t = t_s + t_g \qquad (2.1)$$

where:

t_s = travel time from the exploding reflector to the notional geophone (actual shot) position

t_g = travel time from the exploding reflector to the actual geophone position.

If the wavefields, received on both shot and geophone gathers, are downward continued using the wave equation, the image again occurs at a time $t = 0$ since this was the time of the reflector explosion. From eqn. (2.1), it may be noted that this implies that $t_s = t_g = 0$, since t_s and t_g are both greater than or equal to zero. The mechanics of this type of migration will be discussed later on.

It must be mentioned that there are yet other possibilities in the before-stack exploding reflector model: for instance line or plane wave rather than point sources might be used. The general recipe for imaging in every case is obtained from travel time considerations and spatial coincidence of the up-going waves with the exploding reflector. Finally, it is clear that since downward continuation and imaging in the exploding reflector model concerns itself purely with up-going waves, no account can be taken of multiple reflections, since to do so would require treatment, in addition, of down-going waves.

3. FINITE DIFFERENCE MIGRATION

3.1. Introduction

Finite difference or 'wave-equation' migration was introduced in the early 1970s by Claerbout in a remarkable series of papers.[10,11,8,12] The technique uses the downward continuation process in a numerical solution of the scalar wave eqn. (1.1). This process is based on the finite difference method. These early concepts have been extensively developed, and in this section we will be looking at the latest techniques for the migration of both stacked and unstacked data. There is a need to consider migration before stack since velocity inhomogeneities, if overlooked, tend to cause misstacking just as much as incorrect migration.

Finite difference methods have several strengths and some weaknesses. In early applications, so many approximations were made to the wave equation that the final equation behaved very badly when applied in a heterogeneous medium. Latterly, the approximations which caused the trouble have been identified,[13] and the tendency now is to make very few approximations to the wave equation. With these better approximations, it is possible to migrate data correctly in regions with quite severe lateral velocity variations; indeed in this situation the technique currently performs better than any other method in the same price range, provided that the dips are not excessive (e.g. $>50°$). At very steep dip, finite difference methods do not perform well, and there are problems with both attenuation and dispersion of steeply dipping waves. On the migrated section these appear as a weakened main event and several nearby 'ghosts'.

3.2. One-way Wave Equations

It is quite standard practice to develop specialised wave equations which propagate energy within a small angle about a given axis, usually upwards or downwards. The reasons for doing so are largely mathematical and arise from the less stringent nature of the boundary conditions demanded by a unidirectional wave equation (Dirichlet BC) relative to those for the full wave equation (Cauchy BC). The price to be paid for developing unidirectional equations is that the transmitted refracted wave at an acoustic interface, although positionally correct, will not be diminished in energy with respect to the incident wave and so will incur an error in amplitude (unless unidirectional waves travelling in the opposite direction are correctly coupled). The full wave equation has been used in modelling by Alford et al.[14] and in migration by Deregowski;[15] in the latter case special initial conditions were used and only those parts of the total solution which were of interest were considered.

Historically, one-way wave equations have been derived in a number of ways. Splitting matrices are one method used in underwater acoustics,[16] whilst in seismic work three methods have been used. In the first method, due to Claerbout and Johnson,[12] the unidirectional equation is derived from the wave equation by transforming to a coordinate frame moving at the speed of sound in a given direction. Certain terms can be dropped from this equation which are 'small' for waves travelling in this direction but 'large' for waves in the opposite direction. The effect of dropping the 'large' terms is to annihilate waves travelling in the opposite direction, while effectively leaving the waves propagating in the given direction undamaged. Another approach, also due to Claerbout,[11] starts by taking a Fourier transform over the time coordinate of eqn. (1.1). This is then

$$\partial^2 \tilde{P}/\partial x^2 + \partial^2 \tilde{P}/\partial z^2 = -\omega^2/c^2 \qquad (3.1)$$

where

$$\tilde{P} = \int_{-\infty}^{\infty} P \exp(-i\omega t)\, dt \qquad (3.2)$$

Claerbout derives a one-way wave equation governing propagation in the direction of the z axis by taking the square root of eqn. (3.1):

$$\partial \tilde{P}/\partial z = i\left(\frac{\omega^2}{c^2} + \frac{\partial^2}{\partial x^2}\right)^{1/2} \tilde{P} \qquad (3.3)$$

This equation is an exact one-way wave equation, and in the jargon is termed a 90° equation, since it propagates off-axis energy exactly in all directions right out to the full 90° limit. Unfortunately equation (3.3) is only valid if c is a constant; for a space-variant velocity such a decomposition is not correct. Certain approximations have to be made to the square root term in eqn. (3.3) in order to obtain a solution. In the time domain these correspond to similar approximations which are made to the $\partial^2 P/\partial z^2$ or P_{zz} term. One obvious way in which eqn. (3.3) might be tackled is to approximate the square root by means of the binomial expansion. A less obvious way has been proposed in which the square root is approximated by means of the continued fraction expansion:

$$S = (1 + X^2)^{1/2} = 1 + \cfrac{X^2}{2 + \cfrac{X^2}{2 + X^2 \ldots}} \qquad (3.4)$$

This expression is developed, for example, by Lapidus[17] and Muir[18] in the Stanford Exploration Project (see also the work by Clayton and Engquist[19]). Truncation of the expansion gives rise to approximate one-way wave equations. For example,

$$S^{(3)} = 1 + \frac{X^2}{(2 + X^2)/2} = 1 + \frac{2X^2}{4 + X^2} \quad (3.5)$$

generates an equation which is called a 45° approximation (see Appendix 1). The angles describing approximations are those beyond which the effective velocity of wave propagation is in error by more than 1%—assuming no discretisation error. The discrete step sizes in x, z and t introduce further errors which can reduce this approximation angle significantly.

It is not an uncommon practice to substitute more general coefficients into these approximate equations. For instance, eqn. (3.5) might be replaced by

$$S^{(3)'} = 1 + \alpha X^2/(\beta + \gamma X^2) \quad (3.5)'$$

where α, β and γ are selected so that the desired wave propagation properties are obtained. For example, effort might be directed towards an increase in accuracy in the 45–60° range of dips,[20,21] with a slight but insignificant loss of accuracy at lesser angles.

Buchanan[22] has introduced yet a third way of producing one-way wave equations for seismic work based on Dirac spinor theory. Unfortunately the method involves expressing the boundary conditions in terms of a summation, which makes application of his method somewhat difficult. Nevertheless it represents a novel approach to the problem.

3.3. Finite Difference Approximations

In the finite difference method, differential equations such as eqn. (1.1) are solved by approximation of the partial derivatives by means of difference equations. Thus, for example, if we have a regular gridwork of values of the pressure $P(j\Delta x, k\Delta z, n\Delta t) \equiv P_{j,k}^n$, where $(\Delta x, \Delta z, \Delta t)$ represent the sample spacing in the (x, z, t) directions respectively, we can express derivatives like $\partial P/\partial z$ as follows:

$$\partial P/\partial z \approx (P_{j,k+1}^n - P_{j,k}^n)/\Delta z \quad (3.6)$$

Similarly

$$\partial^2 P/\partial x^2 \approx (P_{j+1,k}^n - 2P_{j,k}^n + P_{j-1,k}^n)/\Delta x^2 \quad (3.7)$$

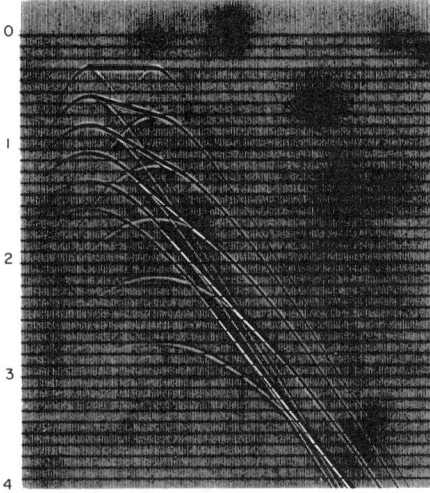

FIG. 6. Synthetic zero offset time section consisting of a number of dipping planes with dips in the range 0–50° in 10° intervals. The earth velocity is constant at 10^4 ft s^{-1}.

Substitution of expressions like eqns. (3.6) and (3.7) into the governing differential equation will yield a set of equations for the unknown values of pressure at a new grid position in terms of the initially recorded (or previously calculated) pressures. The finite difference method therefore provides a convenient technique for downward continuation of the surface wavefield.

The approximations made in replacing partial derivatives by their corresponding difference operations, as in eqns. (3.6) and (3.7), lead to errors in the numerical solution. These errors normally increase with frequency and wavenumber, so that steeply dipping beds will incur a relatively large migration error. This gives rise to a dispersion effect where low frequencies are separated from the high-frequency components, and each bed will be broken up into a main low-frequency event and its associated high-frequency ghosts (see Figs. 6 and 7(d)). The steeper the dip, the wider the separation of the ghosts from the true position; whilst at zero dip there is normally zero separation.

The user may have two parameters at his disposal which he can manipulate to reduce this dispersion. The first of these is the step size Δz (more usually $\Delta \tau \equiv 2\Delta z/c$) used in the downward continuation. This size is normally set at about $\Delta \tau - 24$–40 ms; if this parameter is chosen larger than 40 ms there may be some loss in angular accuracy of the one-way wave

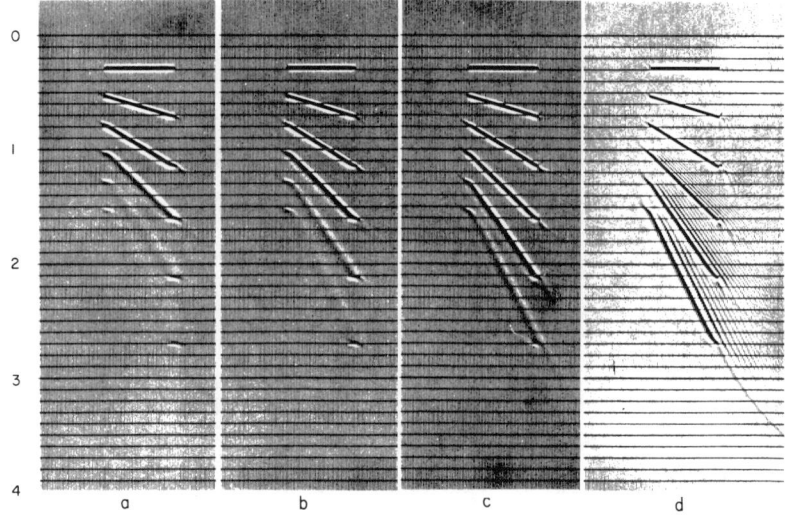

FIG. 7. Migrated version of Fig. 6 by the finite difference method using a 45° equation, with various values of θ: (a) 1·0; (b) 0·75; (c) 0·6; (d) 0·52.

equation. The second parameter which can be used is sometimes called a θ parameter.[21] This parameter effectively allows a bias away from the centred finite difference approximation normally applied (see glossary). θ values in the range 0·5–0·52 are common. Values of θ less than 0·5 lead generally to unstable formulations, whilst with values of θ greater than 0·5 the process acts as a dip filter and can attenuate dipping events rather severely (see Fig. 7). Note that $\theta = 0.5$ corresponds to an unbiased or central difference formulation.

A quite novel approach to dispersion errors has been discussed by Whittlesey and Quay[23] and also by Stolt.[24] In this method the finite difference representation of an approximate wave equation is cast in terms of generalised parameters. A minimisation or least-squares procedure is then set up whose aim, quite simply, is to make the finite difference approximation as close as possible to the exact wave equation at certain selected propagation angles and frequencies. This procedure thus takes two sources of error into account: the error produced by solving an approximate rather than a correct wave equation, and the error caused by the finite difference representation of the partial derivatives. Unlike more conventional approaches to finite difference migration this approach can handle steep dips relatively well, although this is at the price of slight errors at zero dip. Probably most of the better programs use this type of

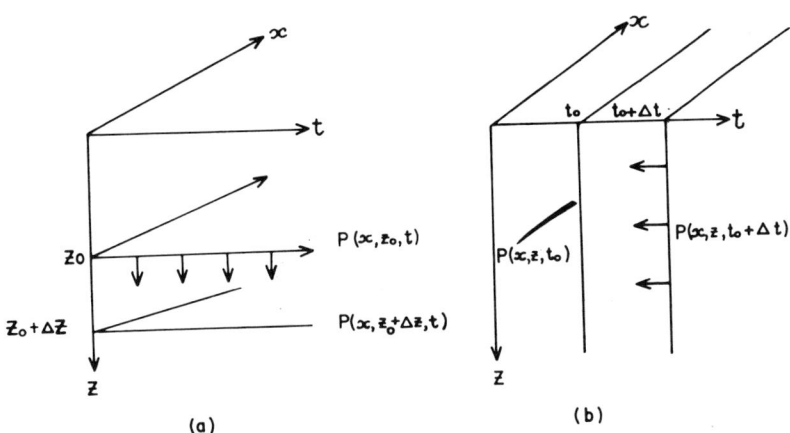

FIG. 8. Migration may be achieved by propagation of energy: (a) in depth—known as downward continuation; or (b) temporally—known as wave tracking. The initial plane $P(x, z, t_{max})$ is set to zero. As wave tracking proceeds surface recorded data $P(x, 0, t)$ get fed in as boundary values.

algorithm, which should not take significantly more computer time than normal finite difference methods.

3.4. Migration of Common Midpoint Stacked Sections

There are two ways in which downward propagation of the surface wavefield may be carried out. In the first the wavefield is propagated from the surface (defined as the plane $z = 0$) to the next surface (lying on the plane $z = \Delta z$) and so on: a process which is called downward continuation. The second method considers wavefields at a given time instant $t_0 + \Delta t$ and propagates this to the next time instant t_0, in a process which may be termed wave tracking (see Fig. 8). Each plane in this method represents a 'snapshot' of the wave at a given instant in time. The two procedures are, of course, physically equivalent if the propagation velocity is assumed to be constant.

The procedure for migration of zero offset data by downward continuation is rather simpler than that for wave tracking. In essence the velocities are halved (see Section 2.6) and, after the finite difference approximations have been inserted, eqns. (A1.11) and (A1.10) are alternately applied to lower the effective recording plane. After each step Δz some of the data will have moved to zero time; these data are then fully migrated. To avoid an excessive number of depth steps Δz is chosen somewhat larger than $c\Delta t/2$, where Δt is the time sampling interval. Consequently an interpolation procedure is required to obtain the data on the $t = 0$ plane (see Fig. 9). This

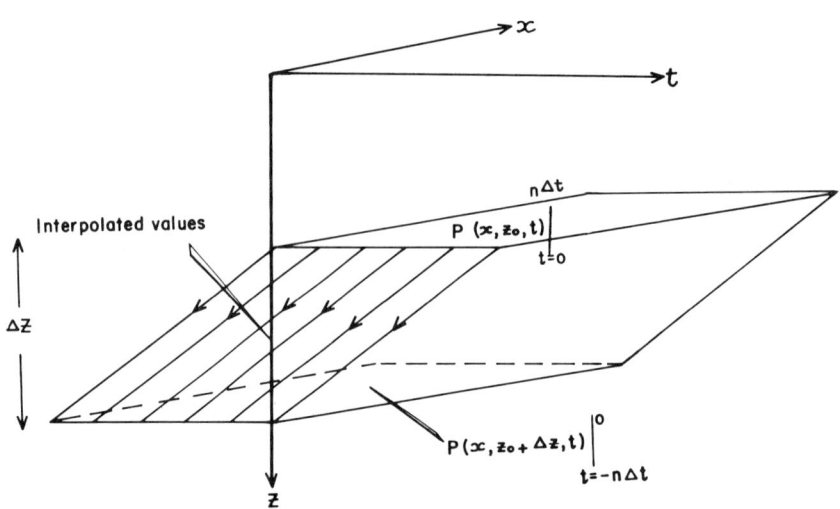

FIG. 9. Interpolation is required in downward continuation, due to the finite depth steps used, to obtain data at intermediate points on the z axis. Data which are propagated to times less than zero are slightly over-migrated; after the interpolation procedure, samples at negative times are not considered further.

can be achieved by linear velocity interpolation assuming a constant velocity within the interval Δz. Because velocity varies laterally it is desirable that Δz should be small.

The output migrated section will lie on the $(x, z, 0)$ plane. Sometimes interpreters prefer a time to a depth section so that data do appear compressed at early and stretched at late parts of the section. This may be achieved by dividing the depth axis by a function $\bar{c}(z)$—an arbitrary function of the depth coordinate only. Ideally this should be chosen to be representative of the mean in the x direction of the average velocity so that times on the migrated and unmigrated sections will be close.

Another way of looking at migration is in terms of rejuvenation of constant time slices. Data at a fixed time is propagated to earlier times by means of the wave equation. The operations involved are the exact counterpart to those used in downward continuation; however, here it is the $\partial^2 P/\partial t^2$ or P_{tt} term which causes difficulty rather than the P_{zz} term. A quite radical approach to this problem has been proposed by Deregowski,[15] in which the P_{tt} term is retained and so the full wave equation is solved. The starting point for the derivation is the acoustic wave eqn. (1.1), in which the velocity is halved, i.e.

$$\nabla^2 P = (4/c^2)(\partial^2 P/\partial t^2) \qquad (3.8)$$

Define a coordinate transformation as follows:

$$x' = x$$

$$t' = t$$

$$\tau = t + \int_0^z \frac{2}{\bar{c}} \, dz \qquad (3.9)$$

where $\bar{c}$ is a frame velocity which is independent of x. This coordinate transformation defines a new coordinate τ in place of depth, which moves with a 'frame' velocity $\bar{c}/2$. In other words, if an observer is moving in the direction of the negative z axis at frame velocity $\bar{c}/2$ then the time τ is stationary. When the frame velocity is close or equal to the acoustic wave velocity some interesting changes occur in the transformed wave equation. Substituting eqn. (3.9) into (3.8) and dropping the primes gives

$$\frac{4}{c^2}\frac{\partial^2 P}{\partial t^2} = \frac{\partial^2 P}{\partial x^2} + 4\left(\frac{1}{\bar{c}^2} - \frac{1}{c^2}\right)\frac{\partial^2 P}{\partial \tau^2} - \frac{8}{c^2}\frac{\partial^2 P}{\partial t \, \partial \tau} + 2\frac{\partial}{\partial z}\left(\frac{1}{\bar{c}}\right)\frac{\partial P}{\partial \tau} \qquad (3.10)$$

In making the substitution it was assumed that $\bar{c}$ was independent of x. Even when $\bar{c}$ is a function of x, Deregowski justifies the derivation of eqn. (2.10) provided that the depth steps $\Delta \tau$ are small. In the case where $\bar{c}$ is a constant, the last term in eqn. (3.10) disappears and the equation may be split into 'diffracting' and shifting or 'refracting' parts respectively as follows:

$$\partial^2 P / \partial t \, \partial \tau = -\tfrac{1}{2}(\partial^2 P / \partial t^2) + c^2/8(\partial^2 P / \partial x^2) \qquad (3.11a)$$

$$\partial P / \partial t = \tfrac{1}{2}[(c^2/\bar{c}^2) - 1] \partial P / \partial \tau \qquad (3.11b)$$

To solve eqn. (3.11a) by conventional methods would involve dropping or approximating the P_{tt} term to obtain a one-way 15° or 45° equation. After substituting in the finite difference approximations, eqn. (3.11a) could be solved in the usual manner. Similarly, eqn. (3.11b) could be integrated to yield a solution which is the direct counterpart to eqn. (A1.8). Migration by means of wave tracking could therefore be carried out in a manner precisely analogous to downward continuation.

Deregowski approached the problem rather differently, and it is outside the scope of the article to discuss this method fully. However, he introduced some useful concepts which can be summarised here. First of all he discovered that the P_{tt} term could be retained in eqn. (3.11a) without approximation. Leaving this term in the finite difference equation means that, for a solution, two or more wave tracking planes are required to initiate the time marching scheme. For the first step only of the process, the

P_{tt} term was dropped, and so only one wave tracking plane was required as initial data. On the second and subsequent steps of the wave tracking procedure, there were therefore at least two planes of wave tracked data available so that the P_{tt} term could be retained without approximation. Since wave tracking starts at the maximum recorded time on the section and continues through to zero one-way time, any down-going waves generated by the algorithm are automatically dropped from the calculation as it proceeds, because these waves evolve by moving to later rather than earlier times. Furthermore, down-going wave energy is rapidly dispersed by the finite difference calculations in a retrogressive coordinate frame. The other interesting feature of Deregowski's work was that the finite difference grid sampling size was relatively coarse in the t direction (i.e. greater than the recording sampling interval Δt) and fine in the τ direction. This arrangement meant that, at each step of the wave tracking procedure, several new samples of recorded data (rather than a single sample) were interpolated onto the top of the wave tracking frame. After the data had been tracked to the time $t = 0$, the migrated wavefield could be obtained without interpolation since the z step chosen was effectively:

$$\Delta z = (c/2)\Delta t$$

Deregowski's work thus gainsays much current thinking on grid sampling size in the t direction, and on the retention of all the terms in the wave equation.

3.5. Absorbing Boundary Conditions

The surface recorded data is generated from reflection points which may not lie within the domain bounded by the extreme shot point and geophone point positions. Downward continuation in such a case will cause data to image outside the extremities of the computational domain. Unless the computational domain is extended to include these imaged positions, then data may be artificially reflected from the side boundaries as downward continuation or wave tracking proceeds, and may destroy useful images within the domain. As implied by the previous sentence, extra zero filled traces can be added to the computational domain and will generally prove sufficient to tackle the problem. It is possible, however, to develop specialised equations at the domain boundary so that most of the outgoing energy is absorbed by the boundary rather than reflected, as Clayton and Engquist[19] have shown.

Between the extremes of a simple-minded padding out of the domain with dead traces and the use of a modified boundary equation lies a method discussed by Deregowski.[25] His method consists of predicting the extreme boundary values at time $t - K\Delta t$ from the gradient and values of the wave surface in two previous wave tracking frames, i.e. $P(x, z, t)$ and $P(x, z, t + K\Delta t)$. These predicted values are then used as Dirichlet conditions on the new wavefield. His method is moderately successful and can be used to reduce the number of extra traces required by the computation.

3.6. Finite Difference Migration before Stack—An Introduction

Migration before stack is necessary when dips are sufficiently steep or velocities are so rapidly varying that the assumptions of a flat layered earth model patently fail. In this case the common midpoint stack will be a poor approximation to a zero offset section. There are a number of definite levels on which this problem can be tackled. The first level may be termed a partial pre-stack migration. The aim here is to map data recorded at a finite offset to zero offset. All the resulting zero offset sections may then be summed together to obtain a corrected CMP stack, before a conventional post-stack migration is applied. This method will be covered in Section 6 under the heading of stack enhancement. The next level on which this problem may be approached is in terms of migration of constant offset sections. A finite offset section can be pictured as being created by an exploding reflector model based on an elliptic wavefront.[26] This elliptic wavefront has a constant interfocal distance equal to $2h$ (the offset) and a semimajor axis of $\frac{1}{2}vt$. The semimajor axis therefore expands at the same rate as the radius of the circular wavefronts used in the zero offset exploding reflector model. All the offset sections are migrated together using a generalised wave equation—commonly called the double square root equation (Section 4.4). To migrate the offsets independently requires a diffraction stack (Section 5.4) rather than a downward continuation procedure, unless the latter approach is limited in accuracy to 15° type equations. If higher-order terms are included in the finite difference equations, these imply a coupling between the offsets,[27] with the result that downward continuation of each offset cannot be achieved independently from the others. The 15° equations are developed in Claerbout[28] and appear as eqn. (11-3-19) in the latter publication. It is understood that some contractors have obtained steeper angle equations than this, under the assumption that the velocities are weakly laterally variable, and that the double square root equation can be approximated by a single square root equation—the one-way wave equation at zero offset with a special velocity. In this case, normal move-out corrections are applied to the non-zero offset sections, and a migration velocity is developed which would collapse diffraction pseudo-hyperbolae to a 'focus' at each offset.[29] Since in fact the response at a finite offset to a point scatterer is somewhat flattened as compared with a hyperbola,[30] whereas in this finite difference procedure a true hyperbola is implied, then the foci will be diffuse. Each constant offset section is migrated, therefore, with its own migration velocity using a steep dip wave equation. The process is, by virtue of the approximations made, a 'time' migration rather than a 'depth' migration. We would not expect that this method would yield any significant improvement over pre-stack partial migration followed by a 'depth' migration and will not discuss it further.

Slant stack migration is a rather different approach to the problem; it is included in the category of pre-stack migration since, although some partial stacking is involved in the formation of the slant stacks, the main stacking occurs after migration. This method is discussed in some detail in Section 3.7. The main drawback of the method is that it is restricted to sections where the velocity is a function of depth only.

The final approach which will be mentioned is migration of shot and geophone gathers in a downward continuation process. This technique represents one of the best methods available for migration in a variable velocity medium. Not only can velocity variation within a spread length be included but the differences in velocity on both up- and down-going wave paths can be considered. It is, unfortunately, also one of the most expensive methods, which is a strong deterrent against general use. A discussion of the method appears in Section 3.8.

3.7. Slant Stack Migration

The idea of slant stacking was first introduced by Claerbout in Stanford Exploration Project reports as early as 1974; it was made public in a paper by Schultz and Claerbout.[31] The idea is to simulate plane wave rather than point sources. This is achieved by summing the output, at a single geophone, produced by a series of closely spaced equi-amplitude point sources. It is assumed that the sources are regularly spaced and extend to infinity. In essence this is rather like the ball bearing model for reflectors discussed earlier (Section 2.2). Obviously the assumptions made are not realised, and various truncation and aliasing effects arise in practice since the shot points are spaced some distance apart, and are of uneven amplitude, and the spread length is finite.

Using the methods outlined by Schultz and Claerbout we might sum all the traces in a common geophone gather without any time delay between traces. This is called a vertical stack and approximates to the trace which would be obtained from a plane wave travelling vertically downwards into

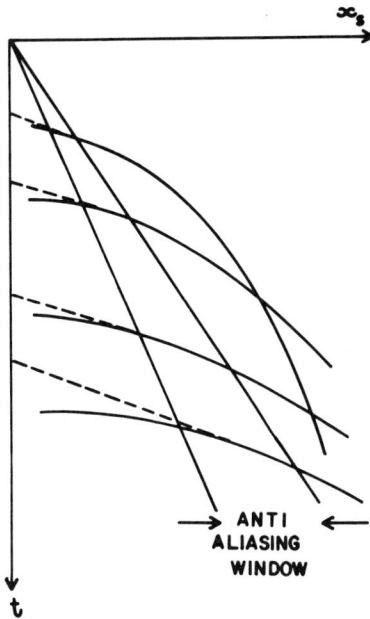

Fig. 10. A common geophone gather showing summation lines for a typical slant stack. Data within the anti-aliasing window is summed together and placed at the shot–geophone zero offset trace at the time given by the intersection of the dashed line on the time axis.

the earth. If the traces from each shot are given an arbitrary delay then any shape of wavefront can be simulated. In particular, if the time delay is linear with respect to the shot distance coordinates then a plane wave travelling at an angle to the vertical is simulated (Fig. 10). The traces derived in this way from each common geophone gather may be displayed side by side and the section so obtained is called a slant stack. In this display the time coordinate is shifted so that arrival times from a horizontal bed appear at the same time on each trace. This time transformation is

$$t = t' + x_g p \tag{3.12}$$

where:

t = time measured from the instant a plane wave at angle θ to the horizontal arrives at the origin (Fig. 11);
t' = time coordinate in the slanted frame;
x_g = horizontal coordinate of each common geophone location;
p = effective horizontal slowness ($1/c_H$) of plane wave.

The time coordinate t' is in fact the natural output coordinate from the stack over each gather, and the parameter p is more commonly known as the Snell's law parameter. We can form a number of stacks in which the incident wavefront has differing arrival angles. These may be characterised by the Snell's law parameter p, since this is related to the wave angle by the expression

$$p = \sin\theta/c \qquad (3.13)$$

Velocity analysis can be carried out using p gathers, as Schultz and Claerbout point out, but this is outside the scope of this article.

To migrate slant stacked data using downward continuation and imaging, it is necessary to construct first of all the imaging condition. In Fig. 11 the travel time t for a plane wave to reach the point scatterer is given by

$$t = \int_0^{z_0} \frac{\cos\theta(z)}{c(z)}\,dz + x_0 p \qquad (3.14)$$

Defining a further time coordinate $\bar{t}$ which measures time from the instant the plane wave strikes the scatterer gives a total travel time

$$t = \int_0^{z_0} \frac{\cos\theta(z)}{c(z)}\,dz + x_0 p + \bar{t} \qquad (3.15)$$

When we migrate, the imaging time is given by $\bar{t} = 0$, i.e. by eqn. (3.14). Transforming eqn. (3.15) to the slant frame time at the geophone (using eqn. (3.12)) gives

$$t' = \int_0^{z_0} \frac{\cos\theta(z)}{c(z)}\,dz + (x_0 - x_g)p + \bar{t} \qquad (3.16)$$

After downward continuation of the slant stacked data to the time $\bar{t} = 0$, the geophone will be spatially coincident with the scatterer, i.e. $x_0 = x_g$ in eqn. (3.16). Thus the imaging conditions in the slant frame time coordinates are given by

$$x_0 = x_g$$
$$t' = \int_0^{z_0} \frac{\cos\theta(z)}{c(z)}\,dz \qquad (3.17)$$

and by eqn. (3.13):

$$t' = \int_0^{z_0} (1 - p^2 c^2(z))^{1/2}/c(z)\,dz \qquad (3.18)$$

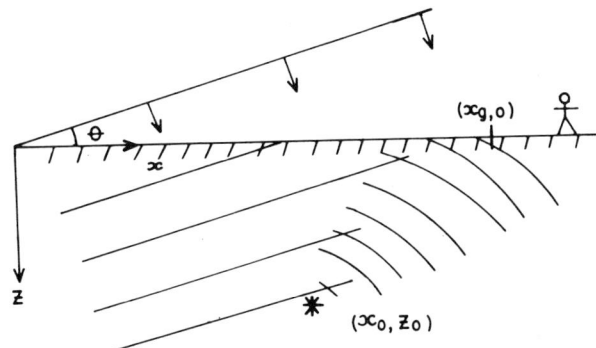

FIG. 11. Geometry of wavefront propagating into the earth, shown in real time. The slant stack section transforms real time to a new coordinate in which horizontal beds will all appear at the same time on each trace. Also shown is a point scatterer at (x_0, z_0) which reflects energy back to the surface diffusely.

The imaging condition in eqn. (3.17) can be calculated by numerical integration. After this the transformation from the image at the scatterer to a surface related coordinate is achieved by a simple skewing of the data, given by

$$x = x_0 - \int_0^{z_0} \tan \theta(z) \, dz$$
$$= x_0 - \int_0^{z_0} \frac{pc(z) \, dz}{(1 - p^2 c^2(z))^{1/2}} \quad (3.19)$$

Migrated sections with different p values can be superimposed after this transformation has been done.

Although the imaging conditions and post-imaging transformation are more complicated than usual in the slant frame, no special methodology is required in the downward continuation. Standard techniques as outlined in Appendix 1 may be used, with a transformation to slant frame coordinates as appropriate.

The relations developed in this section are applicable only when the earth model is depth stratified, i.e. $c = c(z)$. Although the slant stack procedure still remains valid in a medium with lateral velocity variation, the imaging conditions have to be determined by a ray tracing procedure, or by downward continuing a unit amplitude plane wave impulse, and imaging at the point of time and space coincidence of this down-going plane wave and the downward continued up-going wave. The latter method has been extended to include removal of multiple reflections by Estevez[32] for the case of a constant velocity medium during the migration. Note that this method should not be confused with multiple removal by deconvolution techniques, which exploit the constant step-out timing relationships on slant stacked data.

More generally, it is considered that slant stack migration is a good but expensive method in a depth stratified medium, and is a potential wave equation technique for multiple removal in this case. Once the medium has significant lateral velocity variations the method has nothing to recommend it.

3.8. Migration of Shot and Geophone Gathers

We now come to one of the best techniques in existence for migration of data in a medium with strong lateral velocity variations. The procedure, which is described by Schultz and Sherwood,[33] consists of alternately downward-continuing shot and geophone gathers back into the earth. The use of geophone gathers requires that reciprocity of shot and receiver positions be valid. After each Δz step downwards, the data is regathered into shot and geophone gathers and the process continues (Fig. 12) until the data is imaged on the plane ($x_s = x_g, z, t = 0$)—see Section 2.6. This is the zero offset plane and data is simply collected here without any stacking being necessary.

Since the processes involved correspond to an actual rather than a conceptual physical experiment, downward continuation is possible using the unmodified wave eqn. (1.1). Thus we solve alternately

$$\nabla^2 P_g - \frac{1}{c^2}\frac{\partial^2 P_g}{\partial t^2} = 0 \qquad (3.20)$$

and

$$\nabla^2 P_s - \frac{1}{c^2}\frac{\partial^2 P_s}{\partial t^2} = 0 \qquad (3.21)$$

where P_g and P_s are the pressure amplitudes recorded on common shot and geophone gathers respectively. Unidirectional wave equations may be developed to solve this pair of equations along the lines discussed in Appendix 1.

The only drawback of this method (apart from cost) is that data may be undersampled in the horizontal direction if shots are widely spaced, with consequent aliasing problems. The main advantage is its ability to handle velocity variations on both paths, from scattering point to shot and to receiver.

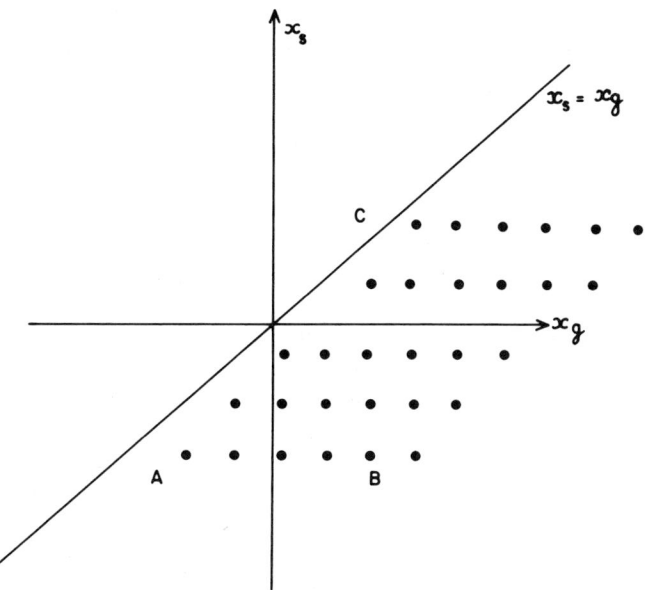

FIG. 12. (x_s, x_g) plane showing arrangement of geophone points. The shots are set off on the line $x_s = x_g$ and the recorded positions are indicated by dots. Common shot gathers lie on lines paralleled to AB, while common geophone gathers lie on lines parallel to BC.

4. F–K MIGRATION

4.1. Introduction

Up to the early 1970s migration was conventionally performed in the 'time domain' or in (x, z, t) space, but there were of course no conceptual reasons why the same operations could not have been achieved in wavenumber (spatial frequency) or (k_x, k_z, ω) space. Indeed, as early as 1972, Maginness[34] applied Fourier reconstruction methods to ultrasonic imaging. However, the method proposed by Maginness, although it could be applied to a depth stratified media, was costly computationally since it involved forward and inverse Fourier transforms at each step in the propagation, since the interest was in reconstruction of the total wave field at a remote plane. This differs from the seismic imaging objective, which is concerned with only the wavefield at zero time on the remote planes (see Section 2.6). Interest in Fourier methods was awakened as a result of an excellent paper by Stolt,[24] similar work was published in the context of holographic reconstruction at around the same time by Booer et al.[35] Stolt's method consisted of a double forward Fourier transform, a modification of the phase and amplitude of each Fourier component, followed by an inverse Fourier transform. Although Stolt's method was extremely fast computationally, its natural application was to constant velocity media, where it was and still remains the best method available for migration. In extension to media in which the velocity was a slowly varying function of position, Stolt had to make several assumptions; these led to a significant loss in migration accuracy and made the general application of his method doubtful.

There have been further developments of Fourier techniques by Gazdag[36,37,38] which permit accurate migration in a depth stratified media and fairly accurate migration in a heterogeneous media. Some of Gazdag's proposals are discussed in Section 4.3, and his hybrid finite difference/Fourier methods are discussed in Section 6. Another area of interest is in migration before stack, and Phinney and Frazer[39] have produced an article on this subject which is discussed in Section 4.4.

In general, because of their economy and very low computationally generated noise, Fourier methods are attractive as a preliminary migration tool. Application of these methods in heterogeneous media is not yet free from problems, because they naturally work best in a constant velocity medium. If, as is anticipated, these problems can be overcome, then Fourier methods will almost certainly supersede finite difference techniques as the major migration tool in the future.

4.2. Geometrical Interpretation of Fourier Migration Methods

We start this development by looking at ray theory. Suppose that we consider a single ray with wavenumber $\mathbf{k}$, where $|\mathbf{k}| = 2\pi/\lambda$, this vector has components along the x and z axes of k_x and k_z respectively (see Fig. 13). This ray may be considered typical of the many rays which can be constructed from a plane wave front in the direction AB. In seismic terms, if we have an earth reflectivity series then the double Fourier transform of this series would yield the spectral decomposition into its planar components, and correspondingly this can be pictured in terms of the rays normal to each plane. If we look at the recorded time section, a double Fourier transform will again decompose this into its spectral planar components (and hence 'rays' normal to each plane). One of the earliest migration techniques—known as a swinging arm technique—exploited the relation which exists between dip on the time section and dip on the migrated depth section or earth reflectivity series. By picturing the Fourier transformation as nothing

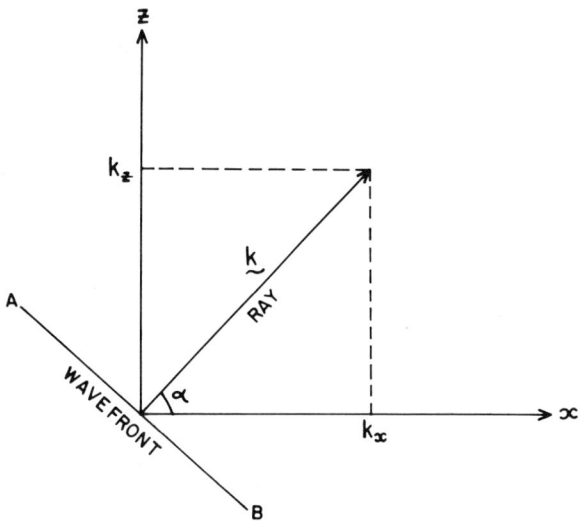

FIG. 13. Ray vector **k** is composed of two components k_x and k_z.

more than a decomposition into the normals to each plane on the section, it can be seen that the swinging arm technique can be applied in the Fourier domain.[40] In Fig. 14 we show a semi-infinite plane in a constant velocity medium, dipping at an angle α_r on the reflectivity section and α_t on the time section. Assuming an acoustic wave speed c, point P maps into point P' on the zero offset time section, at a time t given by

$$t = z_0/\cos\alpha_r$$

By trigonometry $OS = O'S' = x_0 + z_0 \tan\alpha_r$, and

$$c(\tan\alpha_t) = c \cdot \frac{S'P'}{O'S'} = \frac{z_0}{\cos\alpha_r(x_0 + z_0\tan\alpha_r)} = \sin\alpha_r \quad (4.1)$$

Equation (4.1) relates the dips on the reflectivity and time sections. In the Fourier transform domain, OP has spectral components k_x, k_z related by

$$\tan\alpha_r = k_x/k_z \quad (4.2)$$

$$\sin\alpha_r = k_x/(k_x^2 + k_z^2)^{1/2} \quad (4.3)$$

The normal to O'P' has spectral components k'_x, ω which are related by

$$c\tan\alpha_t = ck'_x/\omega \quad (4.4)$$

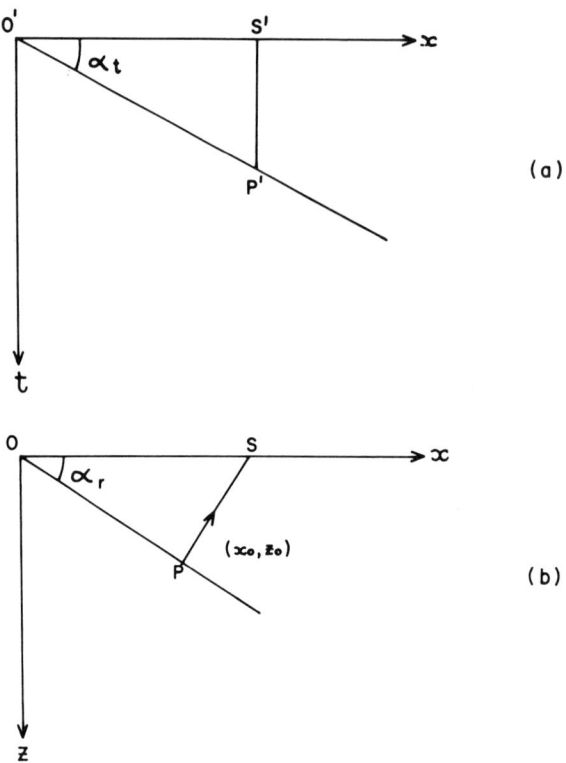

FIG. 14. (a) A dipping plane O'P' on the time section is generated by (b) a dipping plane OP on the depth section.

Thus the migration mapping maps the Fourier components by virtue of eqns. (4.1), (4.3) and (4.4) as follows:

$$\tilde{\tilde{P}}(k_x', 0, \omega) \to \tilde{\tilde{P}}(k_x, ((\omega/c)^2 - k_x^2)^{1/2}, 0) \qquad (4.5)$$

where

$$\tilde{\tilde{P}}(k_x, z, \omega) = \iint P(x, z, t) \exp\left[i(k_x x - \omega t)\right] dx\, dt$$

$$\tilde{\tilde{P}}(k_x, k_z, t) = \iint P(x, z, t) \exp\left[i(k_z z - \omega t)\right] dz\, dt$$

This mapping, due to Booer et al.[35] is illustrated in **Fig. 15**. The surface recorded data is double Fourier transformed over x and t and lies on the plane ABCD. Data is projected out to the dispersion curve of the medium, which is a cone defined by

$$\beta(k_x, k_z, \omega) = (k_x^2 + k_z^2)^{1/2} - (\omega/c) = 0 \qquad (4.6)$$

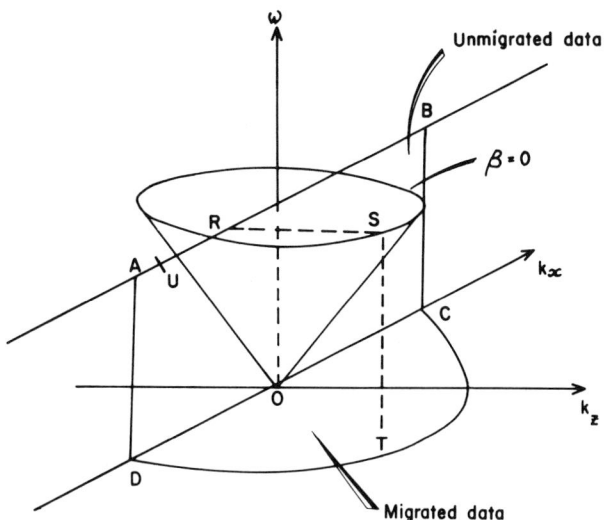

FIG. 15. Unmigrated data lying in the plane ABCD is mapped onto the surface of a cone $\beta = 0$ and then onto the migrated plane defined by the region DTCO. A point R maps onto S and then to T. A point U which lies outside the cone is not mapped by this process—it corresponds to non-real data for which $|k_x| > \omega/c$. A similar mapping occurs for negative values of ω.

and then a perpendicular is dropped onto the k_x, k_z plane. The imaged data is the inverse double Fourier transform of data in the k_x and k_z plane. The mapping in a dispersive but constant velocity media is analogous, but in this case a distorted conical surface β is used.

4.3. Stolt's Theory in 2D

The geometrical illustration of F–K migration describes in essence all that is involved in the migration process. It is simply a mapping from the $(k_x, 0, \omega)$ plane to the $(k_x, k_z, 0)$ plane (see Fig. 15). However, by developing the method from a wave theoretical rather than a ray theoretical basis, a $\cos \alpha_r$ directivity factor appears in the mapping (which is identical to that obtained in Kirchhoff integral theory). This development is demonstrated in Appendix 2; we will only concern ourselves with the final relation (A2.10):

$$P(x, d, d) = \frac{1}{4\pi^2} \iint \hat{\tilde{P}}[k_x, 0, (k_d'^2 + k_x^2)^{1/2}]$$
$$\times \frac{k_d'}{(k_d'^2 + k_x^2)^{1/2}} \exp\left[i(k_x x + k_d' d)\right] dk_x\, dk_d' \qquad (1.7)$$

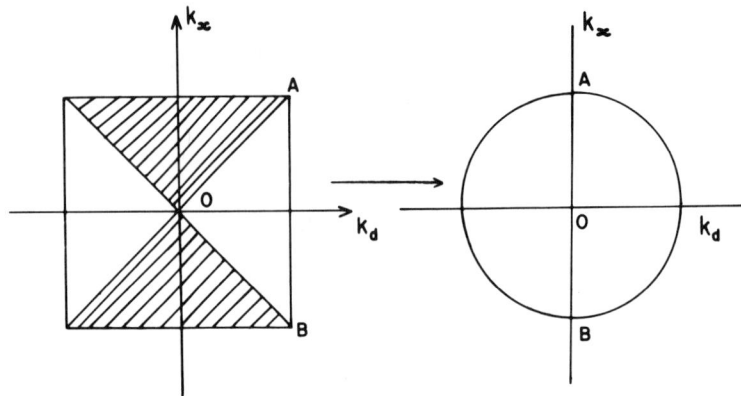

FIG. 16. The Stolt mapping of data from a square to a circular region in k_x, k_d space. Data in the shaded area is not mapped since it corresponds to values for which $|k_x| > |k_d|$.

The coordinate d represents a distance type of coordinate–Stolt transforms from a time section to a 'depth' section by multiplication of the times by the constant section velocity. Migration is a mapping from the original Fourier transformed data as follows.

$$\hat{\tilde{P}}(k_x, 0, k_d) \to \hat{\tilde{P}}(k_x, 0, (k_d^2 - k_x^2)^{1/2}) \tag{4.8}$$

and is represented by a mapping from a square region in (k_x, k_d) space to a circular region (see Fig. 16) in the same space. This is identical in effect to Booer's construction. There is also a directivity term in eqn. (4.7) which is the multiplying factor

$$\cos \alpha_r = \frac{k_d'}{(k_d'^2 + k_x^2)^{1/2}} \tag{4.9}$$

In considering a variable velocity medium Stolt converts the time section to a 'depth' section using a pseudo-velocity function. A simple 'depth' conversion using average velocities tends to distort the flanks of diffraction hyperbolae, relative to the shoulders, and Stolt has devised a special velocity which attempts to correct for this effect. The change in coordinates from time to depth uses the relation

$$d = \left(2 \int_0^t c_{\text{RMS}}^2 t \, dt\right)^{1/2} \tag{4.10}$$

where d represents a depth coordinate. The purpose in transforming to a depth coordinate is that the effects of variable velocity are considerably reduced in these coordinates. Migration can therefore be carried out with a constant velocity.

Stolt obtains an equation in which all the effects of velocity are lumped together in a single parameter W. Stolt suggests that, in most cases, a constant W may be used in migration. Trials by this author have borne this out to a certain extent, with $W = 0.5$ being quite a useful value. However, the main conclusion reached is that with uncertainties in effect of the pseudo-depth conversion followed by the use of a W factor Stolt's method is unattractive as a migration tool except as a preliminary migration in a detailed migration study, or where the velocities can be treated as constant, e.g. in modelling sea bottom multiples.

Having painted a sombre picture of Stolt's method, it nevertheless remains popular, and so perhaps some final caveats on practical processing procedures are in order. First of all, since the method is based in practice on fast Fourier transform algorithms, the number of traces to be processed is padded out with zero traces to a power of 2. There is a requirement in any case for some zero traces to be appended to avoid data imaging outside the computational domain (Section 3.5), but one should beware of 'just missing' a power of 2 and thereby doubling computation time. Another problem arises if the velocities are not thoroughly smoothed across the section to be processed. Undesirable distortions of the pseudo-depth section are liable to occur with disastrous effects on the final migrated section. Lastly, experimentation with synthetic sections may lead to the conclusion that a single W parameter is insufficient. It seems feasible to merge the results from several migrations, each with its own W parameter. This is analogous to the suggestion of Chun et al.[41] for multivelocity migration, in which the migrations with differing constant velocities are combined.

4.4. Gazdag's Phase Shift Method

Gazdag[36] has presented an alternative Fourier based method which is exact in a depth stratified medium. In the case of a constant velocity medium his method reduces to that of Stolt, whilst in a laterally heterogeneous medium certain approximations are required which render his method rather less useful. Unfortunately, with Gazdag's formalism it is not possible to use a double inverse Fourier transformation after the migration mapping; consequently his method is somewhat slow in comparison with Stolt's, but nevertheless it should be comparable in speed to finite difference computations.

Gazdag starts from the scalar wave equation and, after a change of variables in which the depth z is replaced by a two-way vertical time τ, he obtains (eqn. (5.3)) a result which in our notation is

$$P(x, \tau, t = \tau) = \iint dk_x\, d\omega\, \hat{\tilde{P}}(k_x, 0, \omega)$$
$$\times \exp\left(ik_x x + i\omega \int_0^t \left(1 - \frac{c^2 k_x^2}{4\omega^2}\right)^{1/2} d\tau\right) \quad (4.11)$$

This result, which is reminiscent of the earlier one by Stolt, implies that data can be migrated by applying a phase shift to each Fourier component, followed by a summation over ω and a fast inverse Fourier transform over k_x.

In the case of a weak lateral velocity variation, Gazdag approximates the wave equation by a 15° equation, and obtains an integro-differential expression for the migrated result. Bearing in mind how poorly a 15° equation behaves on steeply dipping beds we cannot believe that this is the way to approach the problem. Rather a better method, in our view, is to propagate through each layer using eqn. (4.11), assuming local homogeneity within the layer. A phase mask is then applied to this result which takes care, to a large extent, of deviations from homogeneity. This approach, due to Estes and Fain,[42] has been applied in underwater acoustics and certainly warrants examination by the oil industry.

Gazdag has produced some other interesting techniques which are based on hybrid finite difference/Fourier methods; we discuss these in Section 6.

4.5. Migration Before Stack

Stolt[24] has discussed F–K migration before stack for data arranged in CMP and offset coordinates. His algorithm is developed for migration in a constant velocity medium. In Stolt's development, migration of all the offsets simultaneously is implicitly required, and so we prefer the development by Phinney and Frazer[39] which embraces the particular cases of migration of vertically and slant stacked data, as well as migration before stack of monochromatic sections. Again, unfortunately, the restriction is to a constant velocity; but small-scale fluctuations about the constant value are permitted. However, in this case, migration of any two-dimensional subset of the original three-dimensional spectrum of recorded data (viz. offset, CMP, time) can be used to determine the earth reflectivity series, which in turn is a function of only two dimensions (viz. CMP, depth). The correlation of different estimates of the reflectivity series permits noise suppression or velocity analysis to be achieved.

Phinney and Frazer's method is very similar in final effect to Stolt's method, but it does differ in two important respects. First of all, the spectrum of the source and receiver response is implicitly included in a premultiplying term. Secondly, the equations for dealing with stacked data are different in that Phinney and Frazer's method permits migration of data stacked with a constant move-out (i.e. slant stack) as opposed to normal move-out stacked data required by Stolt's method. For unstacked data both methods result in what is termed a double square root equation, so called because of the two square roots appearing under the integral. As an example, Stolt's equation for migration is

$$P(x, z = ct/z, h, t = 0) = \frac{1}{(2\pi)^{3/2}} \int d\omega \int dk_x \int dk_h P^*(k_x, z = 0, k_h, \omega)$$
$$\times \exp\left[-i[k_x x - (q_s + q_0)ct/2]\right] \quad (4.12)$$

where

$$x = (x_g + x_s)/2$$
$$q_s = [(\omega/c)^2 - \tfrac{1}{4}(k_x - k_h)^2]^{1/2}$$
$$q_0 = [(\omega/c)^2 - \tfrac{1}{4}(k_x + k_h)^2]^{1/2}$$
$$P^*(k_x, z, k_h, \omega) = \frac{1}{(2\pi)^{3/2}} \int\int\int dt\, dx\, dh P(x, z, h, t) \exp\left[i(k_x x + k_h h - \omega t)\right]$$

Neither of these two approaches to the problem is suited to real migration problems. A great deal of effort needs to be expended before migration prior to stack using Fourier techniques becomes a practical proposition in a heterogeneous earth.

5. DIFFRACTION STACK MIGRATION

5.1. Introduction

Diffraction stack migration is a well tried and tested technique and was, in the early 1970s, the most popular migration method. It has gone somewhat out of favour in the last few years, but nevertheless it is still widely used. The developments in this area have been steady rather than innovative, and have largely stemmed from a clearer theoretical understanding of the relation between the diffraction stack process and the Kirchhoff integral solution to the wave equation which it approximates (see for example Larner and Hatton[43]). This has led to a better use of weighting and directivity factors, and to a filter for the correction of phase shifts.[44]

Another step forward has been the application of the 'datumming' techniques of Berryhill,[7] which uses a mixture of downward continuation and diffraction stack processing. Datumming techniques are used to project data recorded on the surface of an arbitrarily segmented earth model down to different datum levels which may be irregular in shape; in each segment the velocity is approximately constant. This is related to, but distinct from, the concept of recursive Kirchhoff migration, in which the segment velocities can be variable but the datum layers are parallel—as discussed by Berkhout and Palthe.[45]

Finally, the 'Hubral' correction must be mentioned.[4] This correction deals with the case where migration has collapsed energy to the apexes of the diffraction hyperbolae. This energy, being misplaced due to laterally varying refraction effects, is repositioned by tracing 'image' rays down from the earth's surface. This process may be regarded as a first step towards a complete ray tracing procedure to determine the summation trajectories for the diffraction stack. Although the 'Hubral' correction can be applied to any migration scheme which ignores laterally varying refraction effects, in practice the better finite difference algorithms now in use include these effects, and the F-K migration Stolt algorithm copes with these refractions to a limited degree, so that the Hubral correction is only really useful in terms of the diffraction stack process, when velocities vary laterally. Note that refraction effects due to vertical variation in velocity do not displace the apexes of diffraction hyperbolae and so the Hubral correction is not required in this case.

Despite its poor reputation, diffraction stack migration still offers, in our view, quite an acceptable migration—particularly in areas of steep dip—provided that all the corrections and filters are properly applied. The main criticisms which its detractors make regarding the process are that it:

(1) appears to produce a great amount of migration noise from horizontal beds;
(2) loses high frequencies from the data;
(3) introduces a large amount of noise when data is spatially undersampled;
(4) produces a larger amount of 'smile' patterns from noise on the section than other methods.

While there is some truth in all of these allegations, the effects can be overcome in part by a judicious choice of processing parameters. Indeed it is our belief that it is the misapplication of weighting, directivity factors, filters and mutes which has bought the Kirchhoff integral method into its present disrepute.

In this section we shall be looking at these developments in our theoretical understanding of the migration process, and their relevance to the choice of processing parameters. Also we will examine 3D migration from the Kirchhoff viewpoint, and discuss when it is valid to split this into a series of 2D migrations in alternate directions. Although the question of splitting is relevant to all other migration techniques, it is perhaps easiest to comprehend the assumptions made in a diffraction stack process than any other.

5.2. Development of the Diffraction Stack via the Kirchhoff Integral

The development of the diffraction stack method was initially based on ray tracing concepts and the scalar diffraction theory of Huygens and Fresnel. Later on, it was discovered that the diffraction stack process could be related to the Kirchhoff integral solution to the wave equation,[5,46] and this provided the basis for weighting factors in the summation.[47] Finally, to complete the circle of interrelationships, it was shown that the F–K Stolt[24] algorithm expresses precisely the same operation in the frequency domain as the diffraction stack process,[48,49] and Berkhout and Palthe[45] have related finite difference algorithms to the Kirchhoff integral.

Schneider[46] has derived the Kirchhoff integral in terms of the free surface Green's function for the wave equation. He obtained the following form of the 3D Kirchhoff integral formula:

$$P(\mathbf{r}, t) = \frac{1}{2\pi} \int\int dA \frac{\cos \alpha_r}{|\mathbf{R}|c} \left[\frac{\partial P}{\partial t}(\mathbf{r}_0, t_0) + \frac{c}{|\mathbf{R}|} P(r_0, t_0) \right]_{t_0 = t + (|\mathbf{R}|/c)} \tag{5.1}$$

This relates the wavefield $P(\mathbf{r}_0, t_0)$ observed on the plane $z = 0$ to its value at a point $P(\mathbf{r}, t)$ in the earth's subsurface (see Fig. 17) at an earlier time. In the seismic application, the second term in the square brackets is normally ignored since it is small, but, as Schneider points out, its inclusion is trivially achieved. The operations implied by eqn. (5.1) are simply weighting, scaling and phase shifting of data on a hyperboloid. The term $\cos \alpha_r$ represents a directivity term which falls off from its value of unity at the apex of the hyperboloid to a lesser value on the flanks. Various other directivity terms have been discussed by Kuhn.[50] In the conventional application, $\cos \alpha_r$

directivity works well, provided that the data is not near the spatial aliasing limit (i.e. it is valid for gentle dips). Under the limiting conditions, Kuhn recommends instead a $\cos^3 \alpha_r$ directivity term coupled with a 'beam steering' approach to migration. In this case diffraction stack migration requires precalculated dip angles, ray path distances, migration velocities weights and mute patterns. It is clear that such complicated calculations make Kuhn's suggestion somewhat unattractive, although if the data is at

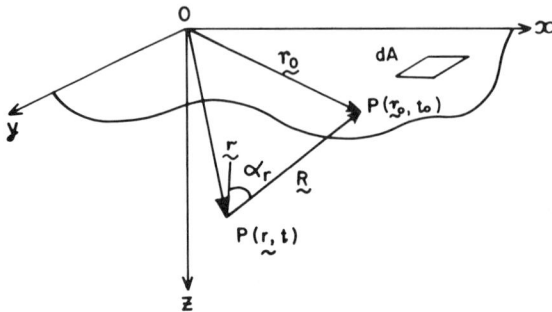

FIG. 17. Geometry for the 3D Kirchhoff integral solution.

the spatial aliasing limit then this price may have to be borne. The more usual directivity term $\cos \alpha_r$ is identical to that obtained in Stolt's F–K migration, and the angle α_r represents the dip angle of the earth reflector which is to be imaged.

The other features of interest in eqn. (5.1) are the factor $1/|\mathbf{R}|c$ which represents a true amplitude scaling factor, and the differentiation of the pressure with respect to time. Differentiation, when examined in the frequency domain, represents nothing more than a $\pi/2$ phase shifting operation together with a linear high-frequency boost—this is the 'Newman' filter. In practice the pressure amplitude rather than its derivative is summed over the hyperboloid; if the derivative of pressure is summed then no Newman filter is required, otherwise a Newman filter can be applied before migration.

For two-dimensional structures the area integration in eqn. (5.1) can be reduced to a line integral,[51] by integration over one of the variables (y for example). This reduction is somewhat tedious and results in the following integral:

$$P(x, z, t) \approx \int_{-\infty}^{\infty} \frac{\cos \alpha_r}{(R_2 c)^{1/2}} \left[\frac{\partial^{1/2} P}{\partial t^{1/2}} (x_0, 0, t_0) \right]_{t_0 = t + (R_2/c)} dx \quad (5.2)$$

where

$$R_2 = [(x_0 - x)^2 + z^2]^{1/2}$$

In eqn. (5.2) we have dropped the last term in the square brackets of eqn. (5.1). The square root differentiation in eqn. (5.2) is not defined, except in the frequency domain, where it represents a non linear high-frequency boost followed by a $\pi/4$ phase shifting operation. This is the two-dimensional version of the Newman filter. The other factors appearing in eqn. (5.2) are the two-dimensional counterparts to those in eqn. (5.1). If the 3D migration is split into alternating direction 2D migrations, then the 2D Newman filter will be applied twice and produce the same effect as the 3D filter; however, the amplitude weighting factor will be in error since in eqn. (5.2) it was derived for a line rather than a point source.

In the practical implementation of eqns. (5.1) and (5.2), the integration is replaced by summation and the infinite limits on the integrals by finite limits. The replacement of integration by summation results in discretisation errors (due to the discrete nature of sampling on the ground), whilst the termination of the summation after a finite number of terms manifests itself in terms of an error which may be called 'truncation' error. Provided that the ground sampling interval Δx and the fold of the diffraction stack are chosen with due regard to reflector dips, neither of these effects will be significant. When this is not the case, the application of high cut filters before migration, and trace weighting in the migration, will overcome the worst of the troubles,[52] but only at the expense of some loss in definition. Indeed, if sampling is near the spatial aliasing limit, then the beam steering method proposed by Kuhn[50] appears to offer the best solution.

It is easy to see how discretisation noise arises with the diffraction stack method. Consider a plane horizontal reflector (Fig. 18(a)). Application of the diffraction stack method will move data in the direction indicated by the arrows, for a central output trace. This output trace is shown in Fig. 18(b) and contains the main broadened pulse, preceded by a long tail. This effect is aggravated further when the beds are dipping. To remove these effects a time and offset varying top cut filter should be applied, before and during the migration. This explains why the finite difference method, which needs no pre-migration filtering, performs more satisfactorily for gently dipping beds. However, on the steeper dipping beds, dispersion effects produced by some numerical algorithms do tend to reverse the balance in favour of diffraction stack migration.

It was mentioned previously that the early termination of the infinite limits on the integrals of eqns. (5.1) and (5.2) caused some undesirable

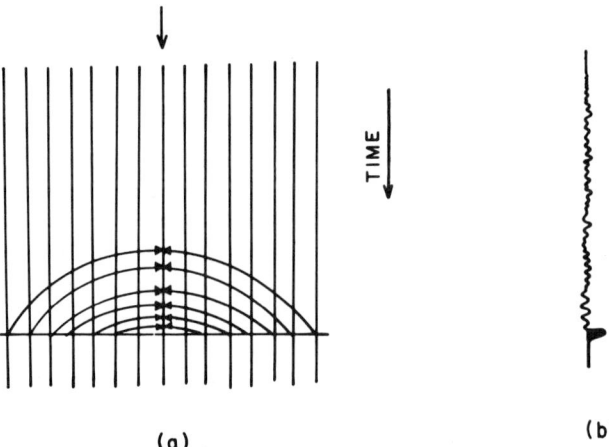

FIG. 18. (a) Energy data from a plane horizontal reflector are moved, in the diffraction stack process, in the direction given by the curved arrows to a central output trace. (b) This gives rise to a broadened output pulse preceded by a long tail.

truncation errors. Nevertheless the onus remains on the user to reduce the number of terms retained in the discrete representation of the integrals in order to achieve the following:

(1) Rapid processing—the fewer the terms the faster the computation runs.
(2) Limitation of noise smiles produced from isolated bursts of noise on the unmigrated section.
(3) Accurate representation of dipping beds—the number of terms included in the summation must be sufficient to cope with the maximum dip on the time section, but no more, since this tends to introduce noise as above, and also the construction of offset-dependent filters becomes more complicated at the larger offsets.

Having truncated the number of terms in the diffraction summation, there is a penalty to be paid in that a number of 'ghosts' will appear before each reflector on the migrated output. Although similar in appearance to the 'discretisation noise', Safar[53] has shown that their cause is indeed due to truncation of the Kirchhoff integral rather than its approximation by a discrete summation. These effects may be reduced by tapering the $\cos \alpha_r$ directivity factor at the end of the swings. Another penalty is that, with fewer terms than are required to achieve migration, i.e. short apertures and large dip angles, the migrated output will appear in the wrong place—

'under-migrated'. Since it may not be obvious what the maximum dips involved on the time section are, one may well migrate with an aperture which is smaller than that which is required.

5.3. Variable Velocity Migration and Datumming

What we have said so far applies strictly to a constant velocity medium, since Kirchhoff theory is only valid in this case. To apply the diffraction stack process in a practical situation means that some approximations are required. In a depth stratified medium ray tracing and timing would indicate that we could still use the previous theory, but with the constant velocity c replaced by the RMS velocity occurring at the apex of the diffraction hyperbola. In the case of a medium with a weak lateral velocity variation, the diffraction stack migration should be followed by a Hubral correction.[4] The Kirchhoff integral is employed to deal with the diffraction effects, and the Hubral correction with the refraction effects or ray bending. The ray bending is such that the minimum travel time path for a point diffractor does not emerge at the surface directly above it, but at a point displaced to the side with higher velocity. The correction made by Hubral entails tracing down 'image rays', i.e. rays emerging at an angle of 90° to the earth's surface, back into the earth and accumulating travel times and corresponding lateral displacements. These corrections are then applied to the migrated output of the diffraction stack process.

When the earth has strong lateral velocity variations this type of procedure breaks down, and one might then apply a downward continuation procedure in conjunction with the diffraction stack process. This is called recursive Kirchhoff migration by Berkhout and Palthe,[4,5] who suggest the method. Another technique, which is particularly useful when the earth can be pictured as a series of arbitrarily shaped segments each with its own constant velocity, has been proposed by Berryhill.[7] Instead of projecting the seismogram downwards in regular depth slices, these slices are irregular and follow the interfaces between the constant velocity segments. Given that the pressures at the top of each segment are recorded, or have been previously computed, the method computes the values at the bottom of the segment—the new datum level—by summation and weighting over the traces at the top of each segment in a method based on the Kirchhoff integral. Thus Berryhill derives an expression for calculation of the output traces $P(x, z(x), t)$ as follows:

$$P(x, z(x), t) = (1/\pi) \sum_i \Delta x_i \cos \theta_i (t_i/|\mathbf{r}_i|) Q(x_i, z_i(x), t - t_i) \quad \cdot \quad (5.3)$$

where:

P = output trace

Q = input trace at location $(x_i, z_i(x))$ delayed by the travel time t_i and convolved with a 5–10 sample length shaping operator—the Newman filter

t_i = time for pressure wave to travel in a straight line between input and output locations

θ_i = angle between the normal to the input horizon and the line joining input and output locations

Δx_i = trace separation between input traces at the ith location.

The geometrical quantities occurring in eqn. (5.3) are shown in Fig. 19. This expression is naturally enough reminiscent of the diffraction stack process; with weighting, directivity, amplitude and trace filtering all included. Note that this method can be applied in a variable velocity segment provided that the travel times t_i can be accurately computed. Berryhill claims that the naive approximation in which curved ray paths (such as the dashed path in Fig. 19) are replaced by straight ray paths gives adequate results for a practical procedure. In this case the travel time t_i is computed by integrating the ratio $dr_i/c(x, z)$ over the straight line joining input and output locations. Although Berryhill's approximation will lead to some error, there is no reason, provided that the segments are thin, to dispute Berryhill's claims.

It must be borne in mind that the output from this datumming technique is not the migrated output *per se*, but the time section which would be recorded at each datum level. Zero time on the output level refers to the time at which shots placed on the actual datum level are set off and recording starts. The method is therefore particularly useful when there is irregular sea bottom topography, which tends to cause all manner of complications

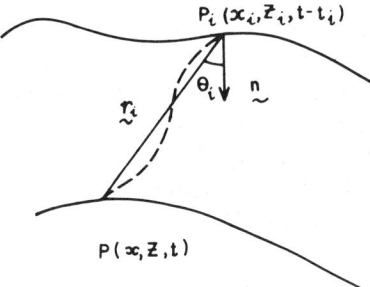

FIG. 19. Geometry used in Berryhill's datumming method.

on the recorded section. Berryhill's method will remove these effects and, after datumming to a horizontal reference level, conventional migration processing is then possible.

5.4. Migration before Stack of Constant Offset Sections

Although constant offset migration via the diffraction stack process is possibly one of the most widely available of the pre-stack migration processes, there appears to be virtually no discussion of the method in the literature. For this reason more space will be devoted to this technique than its importance would warrant, to redress this surprising imbalance. It was noted, in discussing the finite difference method (Section 3.6), that the wave equation could not be separated out over the offset direction unless a low-accuracy wave equation was used, or, alternatively, a steep-angle wave equation could be used provided that distorted diffraction hyperbolae on a finite offset section were approximated by true hyperbolae with the aid of 'pseudo-migration' velocities. Again in discussing F–K migration methods it was concluded that migration before stack was only relevant in the practically uninteresting case of a constant velocity medium. It is clear that migration of constant offset section is most readily performed via diffraction stack methods.

The theoretical basis for the migration procedure is again the wave equation, but for simplicity ray tracing and timing calculations are used here, although these do not give directivity factors (which wave theory predicts). Suppose that we consider a constant offset section where the half offset value is h (Fig. 20). If the medium is homogeneous and has a constant velocity c, then ray path considerations tell us that the two-way travel time to a point scatterer located at $(0, z_0)$ is

$$t = \frac{1}{c}(SO + OG) = \frac{1}{c}\{[(x - h)^2 + z_0^2]^{1/2} + [(x + h)^2 + z_0^2]^{1/2}\} \quad (5.4)$$

A sample occurring at time t on the constant offset section will be repositioned at a time t_0 on the output trace, where $t_0 = 2z_0/c$. This operation includes migration and NMO corrections. Residual NMO corrections can be applied in a refined velocity analysis. However, it is sometimes convenient to apply migration without correcting for NMO. In this case the samples are repositioned to t' where

$$t'^2 = t_0^2 + (x^2/c^2) \quad (5.5)$$

A conventional velocity analysis may be applied and the data stacked in the usual fashion.

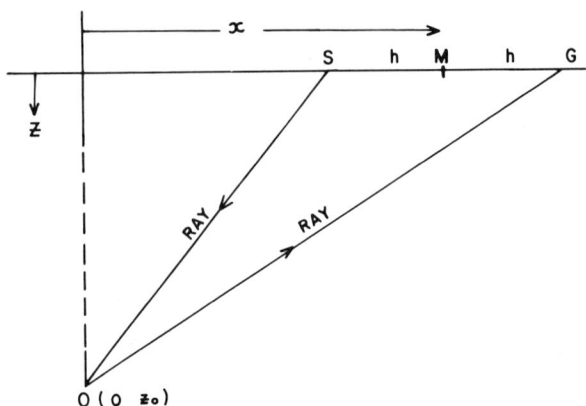

FIG. 20. A point scatterer at O is illuminated by a sound wave emanating from S. Scattered energy is received at a geophone G. The common midpoint is located at M.

Ray tracing calculations can be used to determine the number of terms to be included in the migration scans. As in the zero offset, this depends on the dip angle α_r of the dipping horizon. This may be related to the (one-way) time dip angle α_t on the zero offset section (compare with eqn. (4.1) by

$$\alpha_r = \sin^{-1}(c \tan \alpha_t) \tag{5.6}$$

In Appendix 3 it is shown that the minimum half-scanwidth x_m needed to migrate correctly a bed with dip angle α_r is given by (see Fig. A3.1)

$$x_m = (ct_0/2)[\tan(\alpha_r - \beta)] + h \tag{5.7}$$

where

$$\beta = \tan^{-1}\left\{\frac{-ct_0 + [(ct_0)^2 + 4h^2(1 - \cos^2 2\alpha_r)]^{1/2}}{2h(1 - \cos 2\alpha_r)}\right\} \tag{5.8}$$

To gain some insight into the distortions produced by diffraction stack migration, it is instructive to examine an article by Gardner et al.[6] It is shown here that an initial wavefield with a wavelength of λ, given by

$$P = \cos 2\pi ct/\lambda \tag{5.9}$$

will be distorted after migration into a new wavefield given by

$$P' \approx \frac{(z\lambda)^{1/2}}{2}\cos\frac{2\pi}{\lambda}\left(2z + \frac{h^2}{z} + \frac{\lambda}{8}\right) \tag{5.10}$$

The distortion can be removed by multiplication by a nonlinear frequency boost and amplitude scaling term:

$$\frac{2}{\sqrt{z\lambda}} \equiv \frac{f^{1/2} 2\sqrt{2}}{(c^2 t_0)^{1/2}}$$

and applying a 45° phase shift (compare with the term $2\pi\lambda/8\lambda$ in eqn. 5.10). So it appears that the ordinary 2D Newman filter is appropriate to migration before stack. Equation (5.10) further predicts a change in wavelength after migration to λ', where

$$\lambda' = \lambda[1 + (h^2/2z^2)] \qquad (5.11)$$

In all migration schemes the frequency after migration is lower than before migration. This is easiest to see in F–K migration, which represents a mapping from high to low frequencies. Another way of illustrating this is in terms of a dipping bed. Migration essentially preserves the thickness of individual beds as it rotates them to steeper angles. The display of both the migrated and unmigrated bed takes vertical slices (traces) through the bed, which appear of course thicker as the dip of the bed becomes steeper. The only bed whose frequency content is unchanged by migration is a bed of zero dip. The theory of Gardner *et al.* is derived for zero dip and eqn. (5.11) predicts that, at finite offset, there will be a lowering of the frequency after migration even at zero dip—NMO stretch. Equation (5.11) thus provides the basis for deriving corrective filters or a mute region in the migration to avoid excessive pulse stretching. Suppose that each pulse is permitted a maximum stretch of 25%, then by eqn. (5.11)

$$h^2/2z^2 \leq \tfrac{1}{4}$$

or since

$$z = ct_0/2$$
$$t_0 \geq 2h/c \qquad (5.12)$$

At a given offset h, the equality sign in eqn. (5.12) determines the minimum time on the output trace from which we would expect to receive contributions from the input section.

All of the relations developed in this section extend in the obvious way to a depth stratified medium. The velocity c generalises to the RMS velocity c_{RMS} at the output time t_0. The only exception to this rule is eqn. (5.12), where, since the velocity on the right-hand side of the expression is itself a function of t_0, an iterative procedure is required to define t_0. Here and elsewhere c_{RMS} is appropriate only for angles of dip less than about 60°.[46] For steeper angles of dip, just as in CMP stacking, the inclusion of fourth-order terms in defining the summation trajectory is appropriate. In the case of lateral velocity variation, there are only heuristic procedures available for choosing a migration velocity. The mean RMS velocity between input and output trace positions at a time t_0 is one possibility. Alternatively, ray tracing may be used to determine the summation trajectory.

Finally, it must be remarked that it is common practice, in the interests of economy, for groups of constant offset sections to be lumped together before migration into substacks. There are no fixed rules which dictate at what level of earth complexity migration of a stacked CMP section will fail, since the definition of the onset of such failure is a matter of subjective appraisal. Similarly it is uncertain what number of substacks should be used in a migration before stack procedure. Newman[54] has, however, produced guidelines for determining the number of offsets to be stacked in a substack on the basis of subsurface coverage. Near-offset sections have a greater density of subsurface areal coverage than far offset sections and consequently the nearer offsets are stacked together whilst the far offsets are migrated singly or in small groups. At present constant offset migration has been demonstrated to offer advantages (generally in the early part of the section) over post-stack migration only when the CMP stacking process has failed in some respect; usually this occurs in regions of complex and steeply dipping structure.

5.5. 3D Migration—Splitting Techniques

Diffraction stack migration in three dimensions is a generalisation of 2D migration. The diffraction hyperbola from a single point scatterer (Fig. 21)

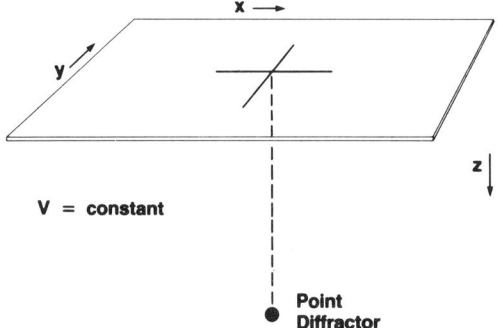

FIG. 21. Point diffractor in a constant velocity medium.

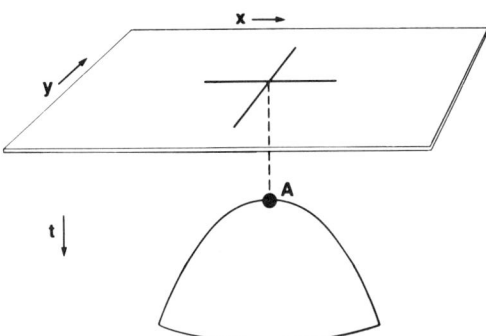

FIG. 22. Hyperboloidal diffraction pattern.

thus becomes a hyperboloid (Fig. 22), and migration consists of the collection of amplitudes over all points of the hyperboloid and placing at the apex A (Fig. 23). As was discussed in Section 5.2, weighting directivity and frequency filtering are required.

A considerable economy can be achieved in 3D migration by splitting it down into a series of 2D migrations. In the Kirchhoff method, for example, these savings are typically of the order of a factor of 100. The penalty to be paid is a slight reduction in migration accuracy, as Gibson et al.[55] have demonstrated. In a splitting method, amplitudes are collected in two stages (Fig. 24); they are moved first along the path to $(x, 0, t_0)$ and finally to $(0, 0, T_0)$. Provided that the velocity is constant, no errors are introduced. For a depth stratified medium it is possible to identify the source of error by

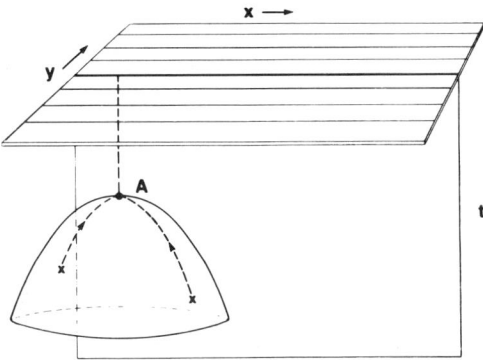

FIG. 23. Heuristically, 3D migration involves placing the sum of all amplitudes on the hyperboloid at its apex. Allowance for geometric spreading, directivity and frequency-dependent effects are required as well.

200   P. HOOD

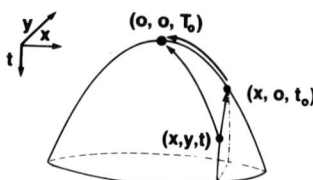

FIG. 24. The amplitude at any input point (x, y, t) can be brought to the intermediate point $(x, 0, t_0)$ at the apex of the hyperbola with x constant, prior to doing a second stage summing along the hyperbola on the plane $y = 0$.

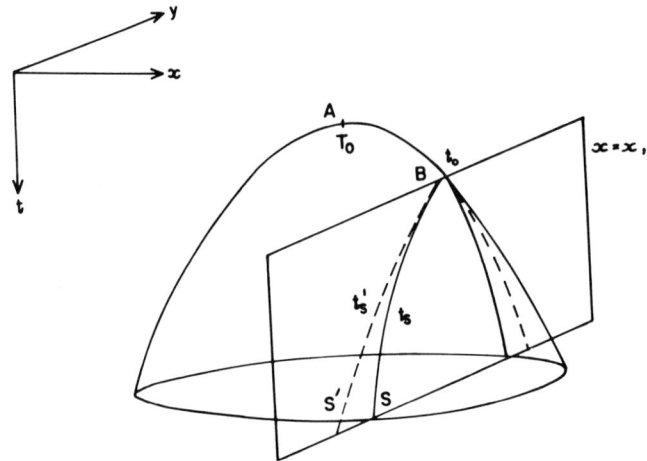

FIG. 25. In a splitting method amplitudes are summed along the curve S'; a diffraction hyperbola whose velocity depends on the apex time t_0. In the full 3D migration, amplitudes are summed along the curve S lying on the hyperboloid surface; this curve is also a hyperbola but the velocity in question comes from the apex of hyperboloid T_0 and not t_0.

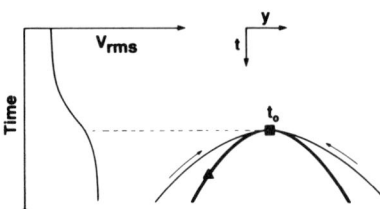

FIG. 26. We wish to perform two-step migration to move the amplitude at ▲ to ■ and finally to ● on Fig. 23. The problem when velocities are varying with time, however, is that the correct diffraction surface is characterised by the velocity at the apex position T_0, but during the first step of the migration the process does not know of the apex time T_0 and moves amplitudes instead along the thin lines.

examining the trajectory over which amplitudes are summed. This is defined by

$$t^2 = T_0^2 + (4x^2/c_{RMS}^2(T_0)) + (4y^2/c_{RMS}^2(T_0)) \quad (5.13)$$

A plane $x = x_1$ which intersects a typical hyperboloidal surface is shown in Fig. 25. The intersection of this plane with the hyperboloid on the curve S is itself a hyperbola, with a minimum time on the apex of t_0. The equation defining S is (from eqn. (5.13))

$$t_s^2 = T_0^2 + (4x_1^2/c_{RMS}^2(T_0)) + (4y^2/c_{RMS}^2(T_0)) \quad (5.14)$$

Now suppose A, B lie in the plane $y = 0$; the travel times t_0 and T_0 are then related by

$$t_0^2 = T_0^2 + (4x_1^2/c^2(T_0)) \quad (5.15)$$

Substituting in eqn. (5.14) gives

$$t_s^2 = t_0^2 + (4y^2/c_{RMS}^2(T_0)) \quad (5.16)$$

Equation (5.16) is the correct summation curve for 3D migration. If we perform a splitting method then the summation curve used will be along a different hyperbola S', where the travel time t_s' is defined by

$$t_s'^2 = t_0^2 + (4y^2/c_{RMS}^2(t_0)) \quad (5.17)$$

The difference in eqns. (5.16) and (5.17) lies only in the time at which the RMS velocity is defined. In the usual case of velocities increasing with depth, the summation will be along the thin curve in Fig. 26, rather than the correct bold curve. The resulting migration error depends on dip and on variation in magnitude of velocity, since S' varies with both of these quantities from the true curve S. The error is zero when either x or y lie on the strike direction and is largest when either of these axes lie at 45° from strike.

It is generally true that migration with the incorrect velocity will result in positional (Δx) and temporal (Δt) errors in placement of a dipping reflector (see Fig. 27). The reverse is also a useful concept. We can interpret errors in x or t as an equivalent error in velocity. Gibson et al.[55] have produced a very useful study on these equivalent velocity errors and their results are reproduced here. Note that there is a slight change of notation convention; in their figures V is used for velocity rather than the letter c used in this text, and x, y coordinates may not necessarily follow dip and strike directions. Taking as a model a single dipping fault plane (Fig. 28), they simulated migration of this plane using several different (but realistic)

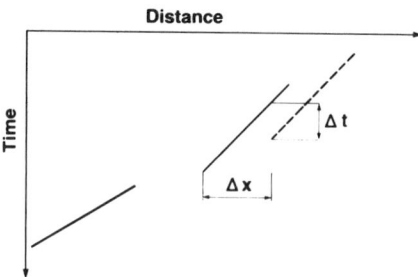

FIG. 27. Migration moves the dipping plane on the left of the figure to the right and places it at an earlier time. In general, migration with the incorrect velocity causes a dipping reflection to migrate to the wrong place (characterised by the wrong lateral position) and/or the wrong time (indicated by the dashed event).

velocity functions, and converted known timing errors on a given output trace into the equivalent velocity error. The four velocity models which were used in this study are shown in Fig. 29, which may be compared with two typical velocity functions for the North Sea and the Gulf of Mexico (Fig. 30). The percentage velocity error is displayed as a function of reflection time for a steeply dipping reflector at the worst (i.e. 45°) azimuth direction (Fig. 31). The largest relative error ($\sim 1\%$) occurs for the lowest velocity function in Fig. 29, and gets progressively smaller for the higher-velocity models. In Fig. 32 the dip dependence of the errors in the splitting method (labelled 'fast') is compared with that of the full 3D approach (labelled 'full'). Errors in the 'full' approach are solely attributable to approximation of a distorted hyperboloidal surface by a hyperboloid (i.e. 4th order and higher terms in offset are neglected in the calculations of the surface). Even

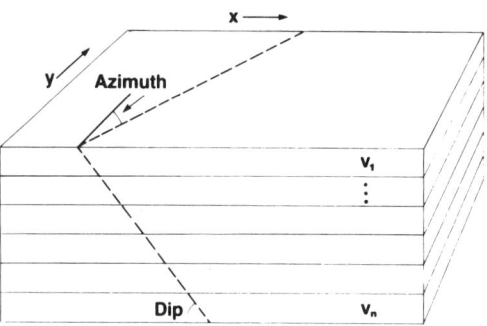

FIG. 28. Gibson et al.[55] simulated migration of a dipping fault plane using different velocity functions. The notation used in the text is defined in this figure.

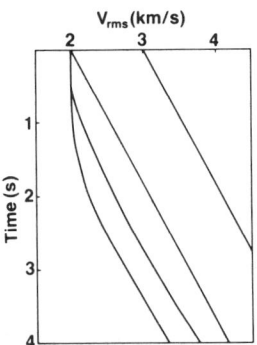

FIG. 29. r.m.s. velocity functions for the four velocity models studied.

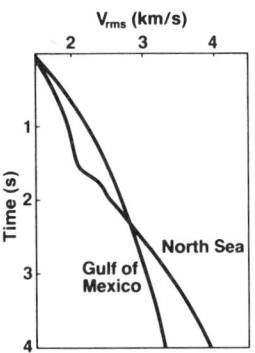

FIG. 30. Typical velocity functions for the North Sea and the Gulf of Mexico.

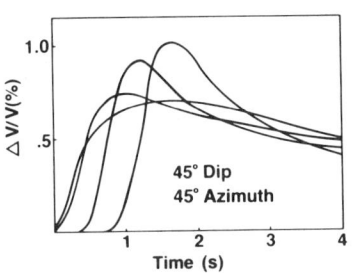

FIG. 31. Percentage equivalent velocity error as a function of reflection time for a steeply dipping reflector at the worst azimuth orientation. The worst error ($\sim 1\%$) occurs for the slowest velocity function in Fig. 29. The errors are progressively smaller for the higher-velocity models.

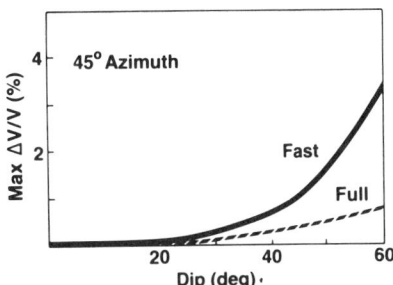

FIG. 32. Maximum percentage error (the maximum point in Fig. 31) as a function of dip. Even for quite steep dip, the maximum error is modest in comparison with uncertainties in migration velocity typically encountered in practice. The curve labelled 'Fast' pertains to the splitting method. The curve labelled 'Full' pertains to the full 3D migration approach. Errors in the full approach are attributable solely to the approximation to the true diffraction surface by a hyperboloid.

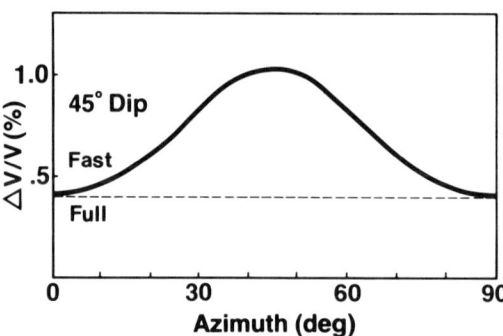

FIG. 33. The error in the splitting method is a function of the orientation of the dip direction relative to the x, y processing directions. The error in the full 3D method is independent of azimuth.

for quite steep dip the errors are reasonably small. Finally, Fig. 33 shows that the error in the splitting method for a fixed dip angle is a function of the azimuthal angle of the x, y processing directions. The error in the full 3D method is of course independent of azimuth.

This study, by Gibson et al.,[55] is particularly reassuring and in our view lends support to the idea of splitting 3D into a series of 2D migrations. The maximum errors involved, when expressed as equivalent velocity errors, are modest in comparison with other uncertainties in migration velocity which can occur in practice.

6. MISCELLANEOUS TECHNIQUES

6.1. Introduction

The previous sections have concentrated on three of the most important methods for migration of seismic data. In this section we bring together some recent developments of theoretical interest, which will no doubt feature on a more practical level in the future. The first of these methods to be examined uses a hybrid finite difference/Fourier method—a technique which is already in use commercially on a small scale. This applies finite difference methodology to the downward continuation of the Fourier transformed surface data. Most of the published articles have concentrated on data which have been Fourier transformed in time t[11,28,37,45] rather than spatially over x.[24] The reason for this preference is that velocity is normally space variant, and so a Fourier transform over the space variable is not necessarily a good idea. On the other hand, hybrid methods which work with a transform over t can be applied in a heterogeneous medium relatively simply.

Another area, which has recently caused some interest, has been pioneered by Cohen and Bleistein,[56] and is called velocity inversion theory. By this is meant the determination, from the surface recorded seismic data, of the earth velocities; this results, at the same time, in a migration. Their method is based on a perturbation expansion to determine a small parameter ε which measures the departure from a constant reference velocity; it also assumes a constant earth density. In the case of one-dimensional problems there is no limit placed on the magnitude of ε,[57] but for two-dimensional or higher problems ε is constrained to be 'small'—and a 20% variation from the reference velocity is quoted by Cohen and Bleistein[56] as being a reasonable bounding limit. Apart from these restrictions on velocity, there is a more serious practical limitation in the number of computer operations involved with the method. The determination of ε requires evaluation of a fivefold integral, and in comparison with other migration methods such as Stolt's (with its twofold integration), the method is uneconomical.

At present, Cohen and Bleistein's technique is no more than a theoretical curiosity, but it is possible that future developments could make this a truly important procedure. It is considered feasible for example that a perturbation expansion about a spatially varying velocity could be made.[58] Approximations which reduce the number of integrations are also being examined; these could lead to better computation times.

Another area for possible future exploitation is the field of underwater acoustics theory. Already some geophysicists have reported the use of what are termed split step techniques[59] in connection with migration. Another practical approach to underwater acoustics has been put forward by Estes and Fain,[42] which again could have geophysical application. Their technique consists of a two-part propagation: propagation through a homogeneous interval followed by a correction due to the fact that the medium in the interval is heterogeneous.

Stack enhancement techniques form a subject which is closely allied to migration. Rather than performing a complete migration before stack these methods transform, using the wave equation, finite offset sections to zero offset sections. It must be noted that the output from this procedure is a set of unmigrated zero offset time sections, even though a 'partial' migration is involved in the formation of this set. Because of their economy in relation to full migration before stack, it is expected that stack enhancement by partial migration will become more popular in the future.

6.2. Hybrid Methods

Claerbout[11] first introduced the idea of hybrid methods in conjunction with seismic migration, using the space–frequency domain variables (x and ω) as opposed to space–time coordinates (x, t). The use of these coordinates holds some advantages. For a start, finite difference approximations to the time derivatives become multiplications by ω; and a time shift over a non-integral number of samples, required for example by eqn. (A1.10), becomes a phase shift in the frequency domain. Both these operations are performed much more accurately in the frequency domain than in the time domain. Another advantage of these coordinates is that only frequencies of seismic interest need be considered; this results in some economy.

In common with his other approaches, Claerbout[11] approximated the full scalar wave equation before attempting any solution. Later developments by Kjartansson[60] and Gazdag[38] used similar approximations to the wave equation. The difference between the various approaches lies in the way each one approximates the 'square root' term in the wave equation (see Appendix 1); all result in roughly similar equations. Kjartansson's method, for example, results in a pair of equations: a 'diffracting' part (see eqn. (A1.6))

$$\left(\frac{2\omega}{c} + \frac{c}{2\omega}\frac{\partial^2}{\partial x^2}\right)\frac{\partial \tilde{P}'}{\partial z} = i\frac{\partial^2 \tilde{P}'}{\partial x^2} \quad (6.1)$$

and a phase shifting part (from eqns. (A1.4) and (A1.2)):

$$\tilde{P}(x, z + \Delta z, \omega) = \exp i(\Delta z \omega/c)\tilde{P}'(x, z, \omega) \quad (6.2)$$

where

$$\tilde{P}'(x, z, \omega) = \tilde{P}(x, z, \omega)\exp(-i\omega z/\bar{c})$$

In this scheme the wavefield is advanced to greater depths via a solution of eqn. (6.1) followed by eqn. (6.2) (see Fig. 34). After each step the migrated output on the plane $t = 0$ is found by inverse Fourier transformation. Since the term $\exp(i\omega t)$ in this operation is unity at zero time, then inverse Fourier transformation reduces to a simple summation.

The operations in Kjartansson's method are summarised below:

(1) Fourier transform the surface data from time to frequency

$$P(x, z = 0, t) \to \tilde{P}(x, z = 0, \omega)$$

(2) Downward continue using: (a) eqn. (6.1)—a diffraction equation which propagates energy from level z to level $z + \Delta z$ (using a finite

MIGRATION

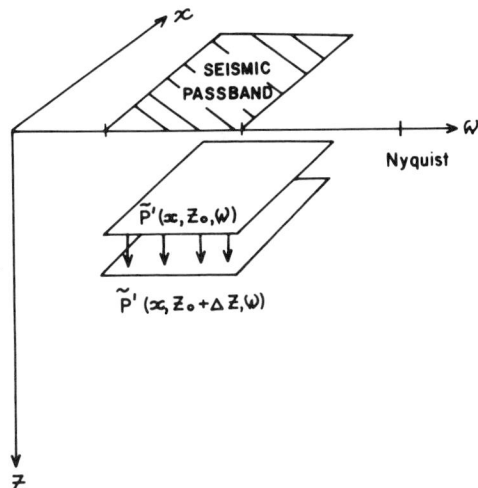

FIG. 34. Data are projected from the plane at level z_0 to $z_0 + \Delta z$; each frequency component is treated separately.

difference method); (b) eqn. (6.2) which takes account of propagation differences in a variable velocity medium.

(3) Synthesise data by summing over all the frequencies:

$$P(x, z + \Delta z, t = 0) = \sum_\omega \tilde{P}(x, z + \Delta z, \omega)$$

(4) Subtract the 'd.c. term' $P(x, z = z + \Delta z, t = 0)$
(5) Go to 2.

As Kjartansson has noted, operation (4) subtracts the wavefield from each frequency; this is to avoid wrap-around. In downward continuation of time domain data, once data has moved through zero time to negative times it is no longer considered. Similar effects occur in the frequency domain, except that data wraps around as it is downward continued. The subtraction of the term $P(x, z = z + \Delta z, 0)$ cures this wrap-around effect by removing the 'd.c. term' after each step.

Gazdag[38] starts from a binomial expansion of the square root term and obtains

$$\frac{\partial \tilde{P}}{\partial z} - \frac{i\omega}{c}\tilde{P} + \frac{c}{2\omega}\frac{\partial^2 \tilde{P}}{\partial x^2} - \frac{c^3}{8\omega^3}\frac{\partial^4 \tilde{P}}{\partial x^4} \qquad (6.3)$$

Instead of advancing the wavefield using a conventional finite difference approach, Gazdag downward continues data using a truncated Taylor expansion:

$$\tilde{P}(x, z + \Delta z, \omega) = \tilde{P}(x, z, \omega) + \Delta z \frac{\partial \tilde{P}}{\partial z}(x, z, \omega)$$

$$+ \frac{\Delta z^2}{2} \frac{\partial^2 \tilde{P}}{\partial z^2}(x, z, \omega) + \frac{\Delta z^3}{6} \frac{\partial^3 P}{\partial z^3}(x, z, \omega) \qquad (6.4)$$

The z derivatives in eqn. (6.4) are evaluated from repeated differentiation of eqn. (6.3) with respect to z at the level z. This results in quite an accurate representation of the wavefield at the new level $z = z + \Delta z$. The surprising feature of Gazdag's method is that the evaluation of x derivatives in eqn. (6.3) is not done by an accurate finite differencing scheme but via a further Fourier transform over x and using the relation

$$\frac{\partial^2 \tilde{P}}{\partial x^2} = -\sum_{k_x} k_x^2 \hat{P}(k_x, z, \omega) \exp(ik_x x) \qquad (6.5)$$

Although Gazdag's approach avoids the problems of finite difference dispersion errors, the use of eqn. (6.5) to evaluate the x derivatives must make this method uneconomical.

6.3. Velocity Inversion Procedures

As applied to seismic data, velocity inversion procedures are a means of obtaining the subsurface velocities (and hence migrating the data) directly from surface recorded measurements. Migration is often mistakenly referred to as an inverse problem. It is nevertheless a forward problem; the surface recorded wavefield defines the initial conditions on the acoustic wave equation, and the wave equation is downward continued with a prescribed velocity field. Mathematically this problem is rather trivial, and in practice the real difficulties lie in obtaining efficient and accurate numerical solutions. Velocity inversion, on the other hand, demands that the velocity used in the acoustic wave equation is derived directly from the data itself.

In theory the one-dimensional problem can be completely solved for arbitrary velocity variation, but constant density; however, in two or more dimensions there are restrictions on the permitted variation in velocity. Cohen and Bleistein[57] laid the foundation for current approaches to the problem. In one dimension they transform the wave equation to the Schrödinger equation and solve the inverse problem via the

Gelfand–Levitan integral equation. The propagation speed is then derived in terms of the potential for the Schrödinger equation. For two or more dimensions these authors express the unknown velocity in terms of a constant reference value and a small (up to 20%) perturbation from it.[56,57,61] An integral equation is derived for the velocity perturbation which involves (for the 2D seismic case) a very expensive five-fold integration of the observed data. Cohen and Bleistein consider that practical restrictions such as noise, discretisation error and finite bandwidth, etc., are of more concern than the theoretical limitation on velocity perturbations. We cannot entirely agree, and whilst we believe that their method has limited practical application in terms of sensitivity to noise and amplitude errors, both the costs and the permitted velocity variation are severe limitations on their approach. If, as Kennett believes to be the case, a perturbation expansion about a velocity field defined by conventional velocity analysis could be made, then such a technique would be extremely powerful. The cost could be reduced if some of the integrations involved were approximated; just how this can be done is not clear.

In one dimension, Raz[2] has derived a useful extension of the above theory to obtain both velocity and density information from field data. The method permits arbitrary variation of both velocities and densities from the reference values. Another article by the same author[62] considers the question of multiple reflections in a 1D velocity inversion scheme, but in this case the density information is not obtained.

6.4. Stack Enhancements

When the subsurface reflectors are steeply dipping or have high curvature, it is known that conventional CMP stacking will not be satisfactory. For instance, the stacking velocity appropriate for a dipping bed will be inappropriate for a horizontal bed and *vice versa*, so that it will not be possible to stack crossing events such as fault planes or diffractions on a time section optimally. Migration of the CMP stack will not therefore be adequate. There are a number of measures which can be taken to avoid these undesirable effects. The simplest approach has been called a 'broad dip band stack' by Western Geophysical Corp. One stacking velocity function is used to stack up the flat events, and a second or third velocity function is used to enhance various dipping effects. The final stack is the sum of these intermediate stacks. In Fig. 35 we show a conventional stack of a growth fault area, and in Fig. 36 a 'broad dip band stack'. The corresponding migrations using Gazdag's phase shift method (Section 4.3) are shown in Figs. 37 and 38. Improvements are subtle and particularly

FIG. 35. CDP stack of a line over the Brazos Ridge, offshore Texas.

relate to the greater clarity of the fault plane. This may be compared with the much more costly migration before stack using Kirchhoff summation (Section 5.4) shown in Fig. 39, where the large growth fault stands out very clearly.

An advance on the broad dip and stack method is a partial migration scheme which transforms non-zero offset time recorded data to zero offset data. If we examine the response on a hypothetical zero offset and far offset section to a buried point scatterer (Fig. 40), it is seen that the NMO process moves the apex of the diffraction hyperbola, but not the tails, to the correct

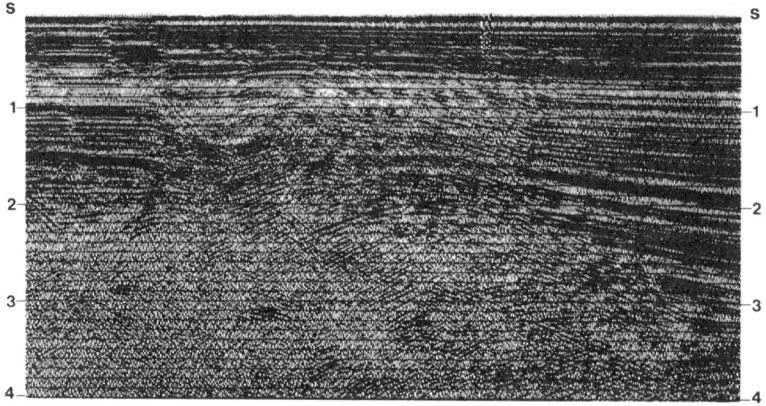

FIG. 36. Broad–dip–band stack of the data used to create the CDP stack in Fig. 35.

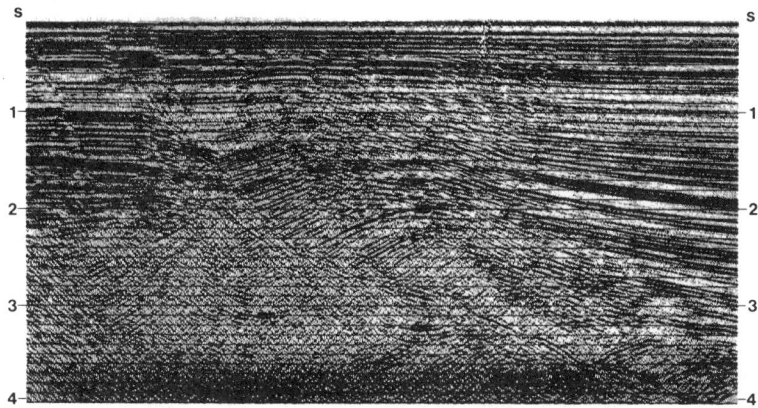

FIG. 37. Migration of the CDP stacked data in Fig. 35 (frequency–wavenumber method).

zero offset position. To obtain an optimum stack requires a partial migration of data (either before or after application of NMO). Digicon Geophysical Corp. have developed a procedure which goes some way along these lines;[63] they have termed the process 'DEVILISH'. The main restriction of this procedure is that it assumes that lateral velocity variations can be ignored in the partial migration procedure. This may not be important since there is some evidence that pre-stack partial migration is remarkably insensitive to lateral velocity changes. In Fig. 41 various stacks of a salt dome region are shown. The stacks before 'DEVILISH' are quite

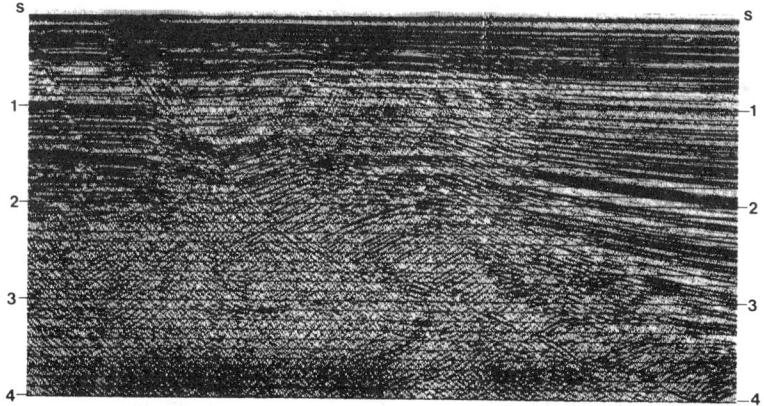

FIG. 38. Migration of the Broad–dip–band stack in Fig. 36 (frequency–wavenumber method).

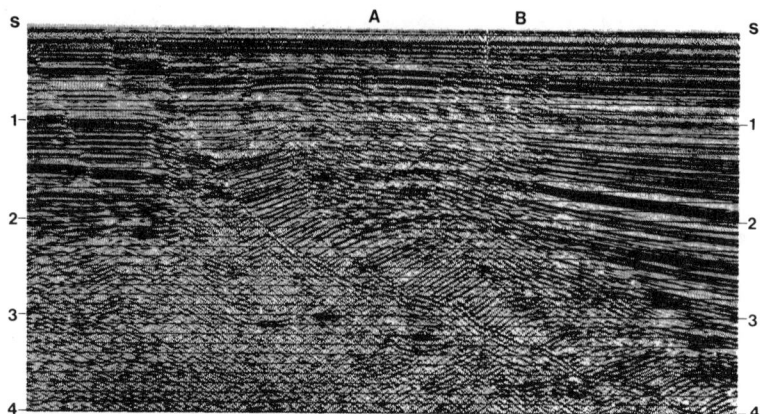

FIG. 39. Migration before stack (Kirchhoff summation). Note the reflections from the major fault at 2·8 s beneath location A. The zone of weak amplitudes in Fig. 37 now shows beneath B a sequence of beds dipping into the fault at 3·0 s. Also, reflections now appear from the steeply dipping adjustment faults in the shallow section.

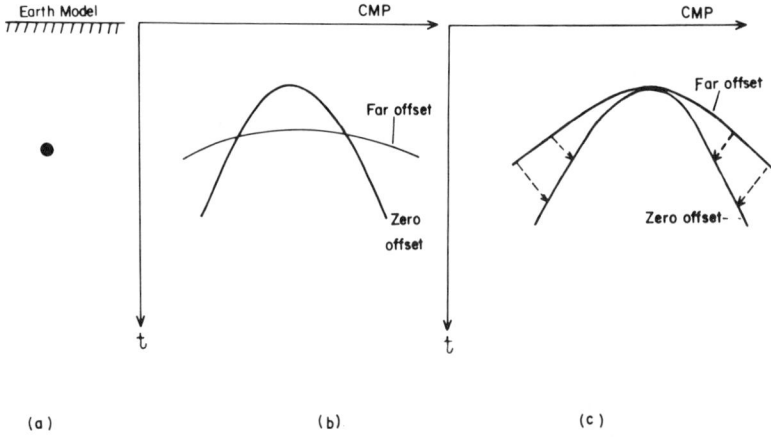

FIG. 40. (a) A buried point scatterer in a homogeneous medium gives rise to (b) the corresponding time section. Zero offset and far offset sections have been superimposed. (c) After NMO the peaks of the diffraction response will coincide but the tails will not. To obtain a good zero offset stack, far offset data must be mapped in the direction shown by the arrows.

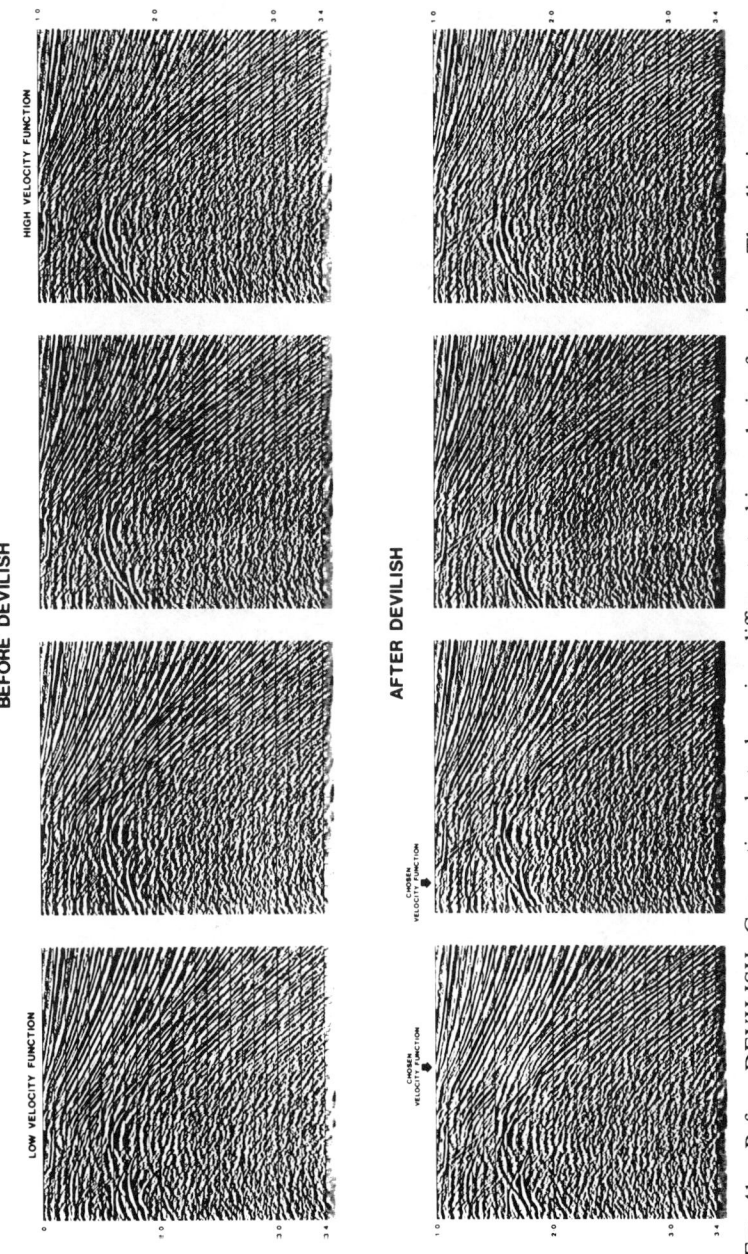

FIG. 41. Before DEVILISH. Conventional stacks using different stacking velocity functions. The dipping events appear stronger at a higher stacking velocity than the more gentle dipping events. After DEVILISH. Stacks after DEVILISH are less sensitive to velocity, and dipping events are more continuous than with conventional stacking.

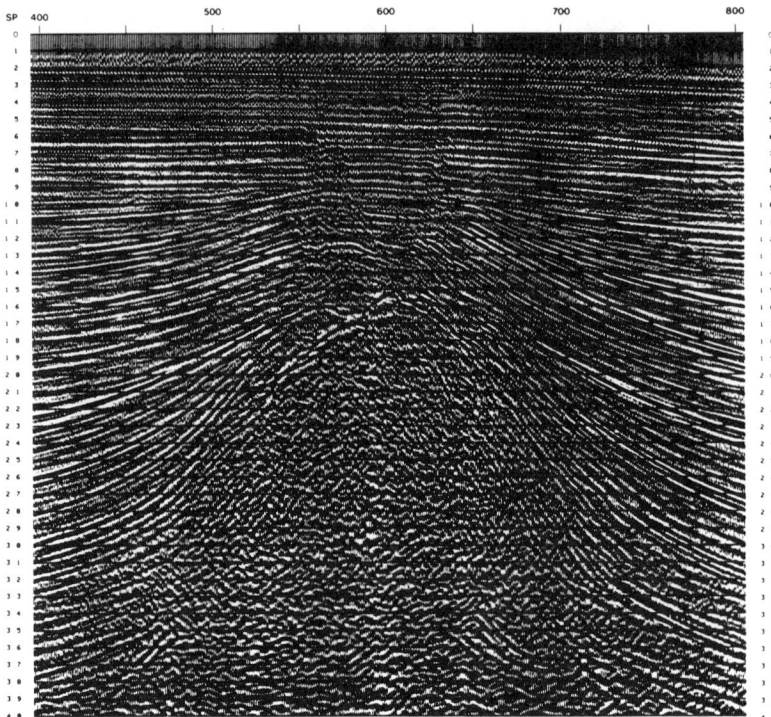

FIG. 42. Migration of conventionally produced stack. The left flank of the salt dome at about 2·0–2·2 s is broken up.

sensitive to the stacking velocity, with some events being stacked out by too low or too high a velocity. Analogous results after 'DEVILISH' are less sensitive to the chosen velocity function. In Figs. 42 and 43 the migrated stacks are displayed. The continuity of the left flank of the salt dome at about 2·0 s is noticeably improved by the 'DEVILISH' procedure.

Yilmaz has completed an important study of pre-stack partial migration in which he has developed finite difference procedures which will handle lateral variations in velocity.[30] After transforming constant offset sections to zero offset sections, the equations developed imply a lateral shift along the section to handle variations in velocity: a splitting which is reminiscent of the 'diffraction' and 'shifting' equations in depth migration by finite differences (Section 3.4).

Stack enhancements will no doubt become more widely used in the future. Judson *et al.* report the cost of their procedure 'DEVILISH' plus stack at only twice the cost of conventional NMO plus stack. The

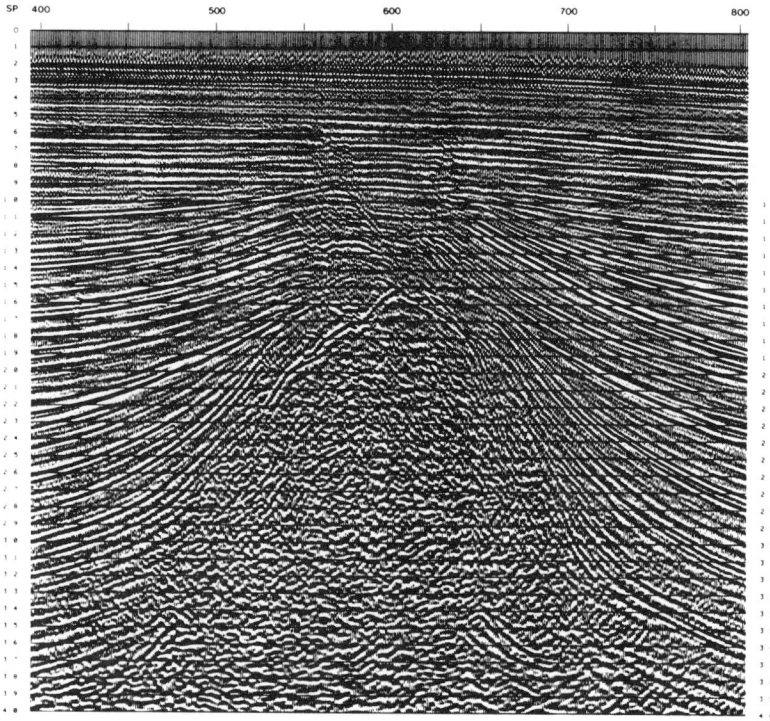

FIG. 43. Migration of a stack after DEVILISH. Note the continuity of the salt dome at 2·0–2·2 s in comparison with Fig. 42. Greater clarity of the dipping events is also obtained.

importance of this pre-stack partial migration is that the resulting zero offset section has a better signal to noise ratio than a conventional stack. Post-stack migration should then stand an improved chance of resolving detail on the section. Stack enhancements of course only go so far in improving the zero offset section. When severe focusing effects are present, or the velocity varies significantly within a spread length, migration before stack by downward continuation of shot and receiver gathers will almost certainly be required (Section 3.8).

7. OVERVIEW OF THE VARIOUS MIGRATION TECHNIQUES

In this chapter we have covered most of the migration techniques which are currently available. Faced with such a plethora of methods, the

TABLE 1

RECOMMENDED MIGRATION METHOD FOR VARIOUS EARTH VELOCITY FUNCTIONS

Velocity variation	Methods recommended	Refer to Section
1. Constant	F–K	4·3
2. Depth stratified	Gazdag's phase shift method	4·4
	Kirchhoff time migration	5·2
3. Depth varying and slow lateral changes (significant changes over several spread lengths)		
(a) Gentle dip	Kjartansson's method	6·2
	Finite difference depth migration	3·4
	Kirchhoff + Hubral correction	5·2–5·3
(b) Steep dips	Stack Enhancements followed by:	6·4
	Kjartansson's method	6·2
	Finite difference depth migration	3·4
	or	
	Kirchhoff migration before stack + Hubral correction	5·4
4. Rapidly varying lateral velocity (significant changes within a spread length)	Depth Migration before stack of shot and geophone gathers using:	3·8
	Kjartansson's method	6·2
	Finite difference method	3·4

geophysicist may well not recall the particular strengths or weaknesses of one method as opposed to another. We address this problem here, and attempt to lay down some guidelines for choosing the appropriate migration algorithm. In general, the progression from a simple time migration to a depth migration before stack should be undertaken in stages. It is quite possible that interpretation can be done on the basis of a cheap migration such as F–K migration; in this case there is no point in progressing to a better method. Even if a complete interpretation is not possible, a better idea of the velocity and earth structure should be obtainable after migration, and these can be used in a more detailed migration study.

The single most important factor which controls the type of migration required is the earth velocity. In Table 1 we recommend, in order of merit, specific migration methods which will give acceptable results for the various

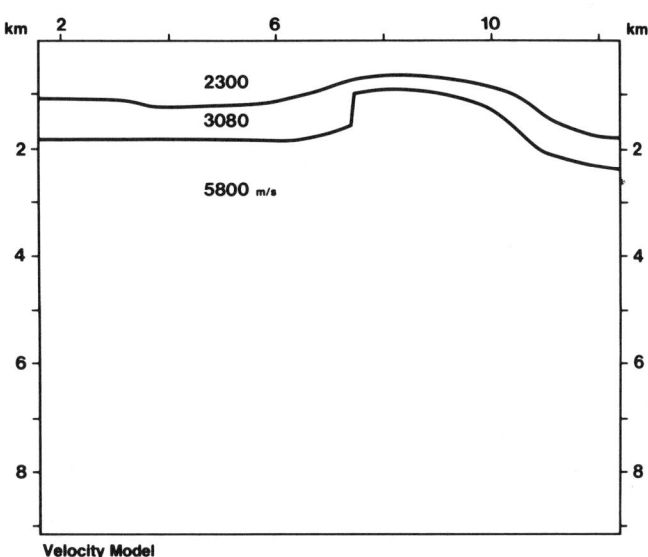

FIG. 44. Velocity model for migration of the stack in Fig. 45. Note the large velocity contrast across the second interface.

earth models listed. When using this table it must be noted that, although Kirchhoff time migration followed by a Hubral correction is a possibility when velocities are slowly varying laterally, there can be problems when interfaces are curved. In this case depth migration or Kjartansson's method is required. In Fig. 44 we show a model which was based on the time section shown in Fig. 45. It might be imagined that lateral velocity variations are slight. In Fig. 46 a time migration is shown, and it behaves quite satisfactorily up to about 1·5 s, but, due to refraction effects, fails in the lower half of the section. The image ray plot (shown in Fig. 47) has severe focusing effects at about 6 km along the line, and so the two-step time migration plus Hubral correction breaks down. A comparison may be made between the conventional time migrated section converted to depth (Fig. 48) and the depth migrated section (Fig. 49). Down to 2 km in depth the results are similar, but the horizon at 8 km is quite severely broken up by the time migration. The 'smile' appearing above this reflector on the depth migrated section is thought to be an artefact whose cause may be traced back to the original CMP stack. The deep reflectors lack the diffraction tails associated with the termination of the beds. These have been destroyed by stacking; the migration of the truncated reflections produces this artificial

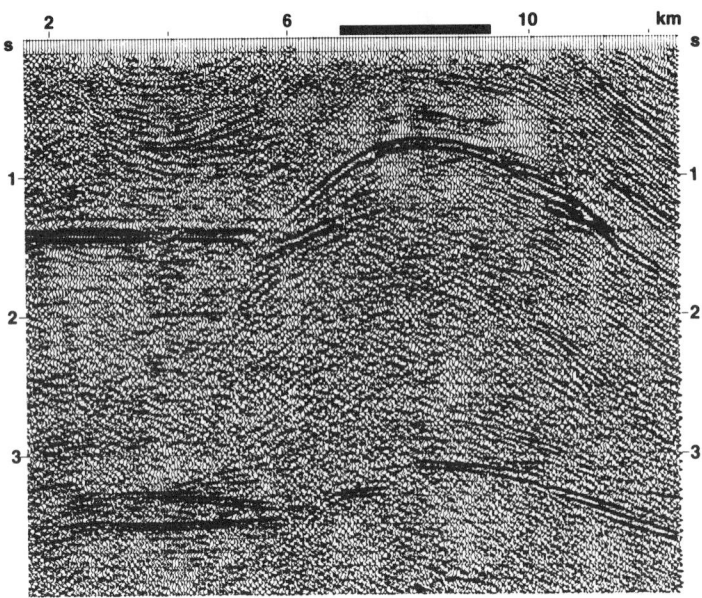

FIG. 45. Stack section from Guatemala. It is likely that an anomalous overburden causes the apparent anomaly at depth.

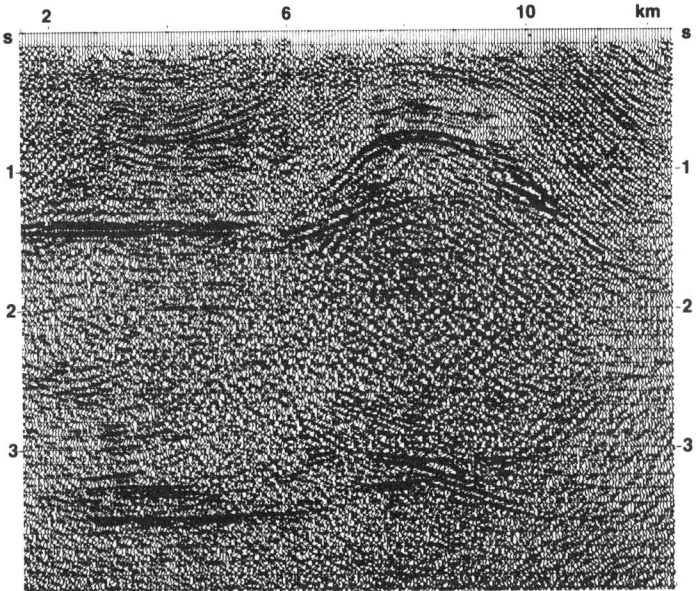

FIG. 46. Conventional time migration (finite difference algorithm). Appears overmigrated at depth.

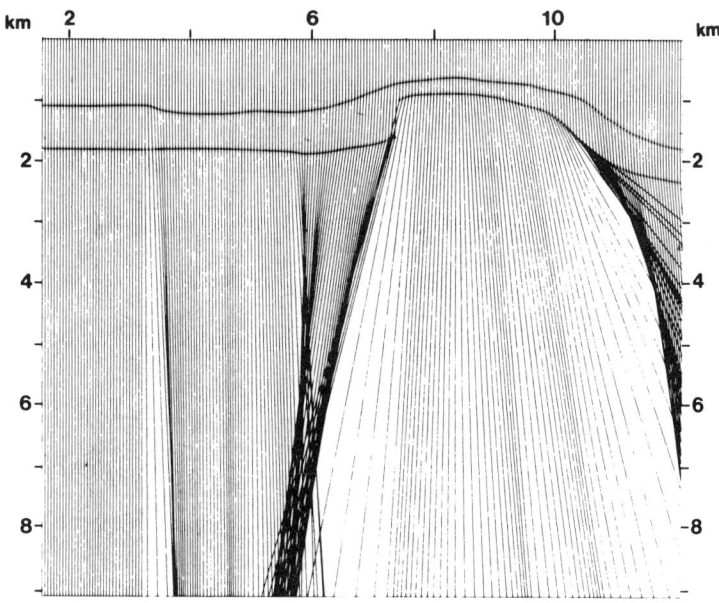

FIG. 47. Image ray paths on a depth plot. The diverging image rays below lateral position 7 to 9 km. will be unable to unscramble the complex deep reflection in Fig. 46.

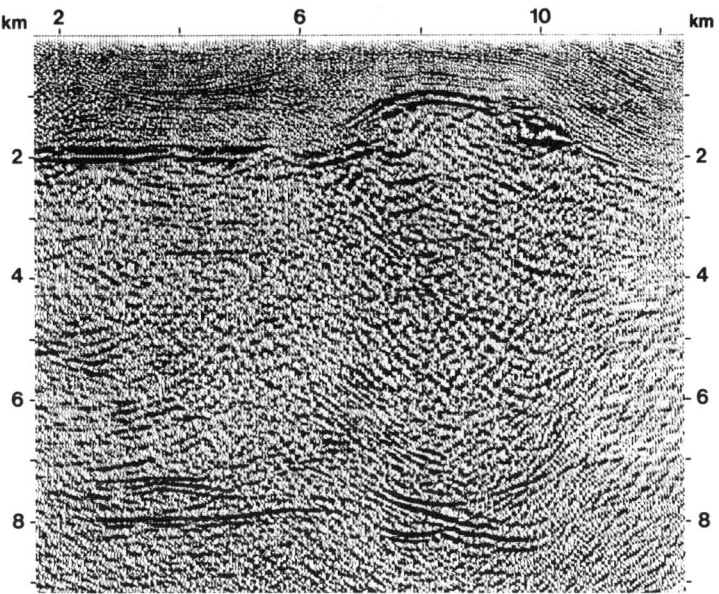

FIG. 48. Conventional depth section (vertical stretching of the time migrated section).

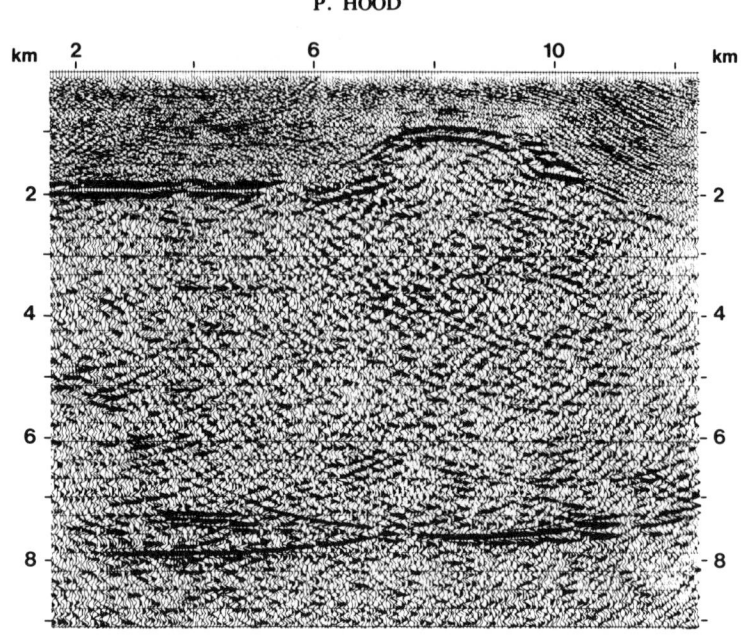

FIG. 49. Depth migration. The deep reflection is well imaged. (The upward smile event rising from the deep reflection is thought to be an artefact related to shortcomings of the CMP stack in the vicinity of the anomaly.)

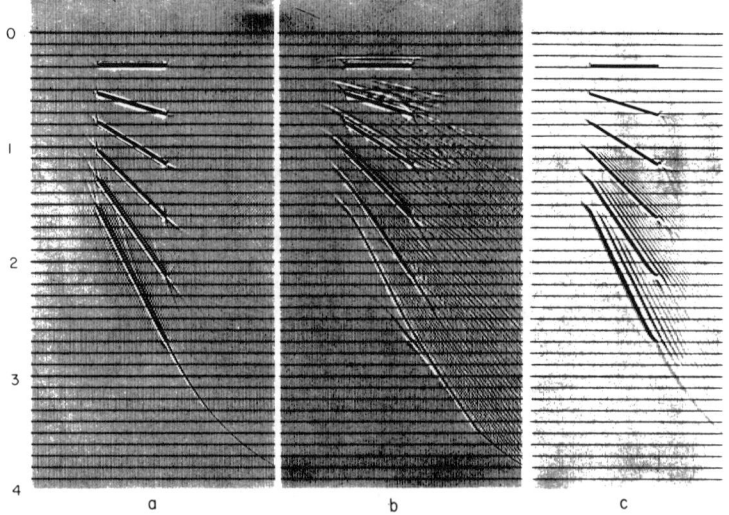

FIG. 50. Comparison between migrated synthetic seismogram of Fig. 6: (a) F-K migration; (b) diffraction stack migration; (c) finite difference migration.

TABLE 2
STRENGTHS AND WEAKNESSES OF THE MAIN MIGRATION METHODS

Migration method	Strengths	Weaknesses
Finite difference method	1. Can, though does not usually, handle rapid lateral variations in velocity 2. Quite economical 3. Low computational noise from gently dipping events	1. Ghosting and grid dispersion errors from steep dip events 2. Regular grid of data points is required 3. Effect of controlling parameters is not apparent to the user, and may not be known except to the original programmer
F–K methods	1. Few control parameters 2. Can, though does not usually, handle depth stratification exactly 3. Very low migration noise from steeply or gently dipping events 4. Economical	1. To avoid aliasing effects sometimes requires double length transforms 2. Lateral refraction effects are not correctly handled
Diffraction stack methods	1. Can handle steep dips 2. Equally spaced grid of data is not required 3. Effect of control parameters is readily understood	1. Liable to produce 'break-up' noise and smiles 2. Cannot handle lateral velocity variations without a ray tracing procedure 3. An expensive method

smile. This example serves to illustrate that caution should be exercised when using this table.

It should be noted that the remarks, here and elsewhere, concerning the breakdown of the Hubral correction, apply only to the automatic application of the image ray traced corrections to the migrated section. These failures are manifested in an incorrect amplitude prediction and in the deformation of the wavelet shape when the migration has failed to bring diffractions to a focus. Another widespread use of the Hubral method is to define a velocity/depth model for depth migration. In this application the

Hubral procedure is used in time to depth conversion of certain picked horizons on a time-migrated section, and irrespective of horizon curvature or image ray focusing the method succeeds.

With this caveat in mind, it can be seen that we would recommend only a few migration methods. Of these it is seen that a finite difference depth migration or Kjartansson's method can be applied in most realistic situations, whether pre- or post-stack. There are certain cosmetic differences between the main migration methods, which we summarise in Table 2, and these should be self-explanatory. This table can be read in conjunction with a demonstration of the output obtained from migrating the synthetic seismogram in Fig. 6 by each of the methods (Fig. 50). Unfortunately, it was not possible to include Kjartansson's method in this comparison, but it is expected to be a cross between the F–K migration and the finite difference migration in appearance and effect.

8. ACKNOWLEDGEMENTS

The author would like to thank the Chairman and Directors of The British Petroleum Company Limited for permission to publish this chapter, and to his colleagues in the Company for their assistance and advice. The author also acknowledges his debt to the management of Seismograph Service (England) Limited for much of his early grounding in migration theory. Particular gratitude is expressed to the Western Geophysical Company of America, and to Dr. K. Larner for his assistance in providing Figs. 21–4, 26–33, 35–9, 44–9 along with the captions, and for his advice on the subject of splitting 3D migration into a series of 2D migrations. Dr. L. Hatton of Merlin Geophysical Company Limited is acknowledged for his discussion on finite difference depth migration. Dr. J. W. C. Sherwood is thanked for his discussion on finite difference migration and the DEVILISH procedure, and Digicon Inc. for providing Figs. 41–3. Finally, the editor of *Geophysical Prospecting* is acknowledged for permission to reproduce Figs. 6, 7 and 50, which came from an earlier publication by the author.

REFERENCES

1. HAGEDOORN, J. G., A process of seismic reflection interpretation, *Geophys. Prospecting*, **2**, pp. 85–127, 1954.
2. RAZ, S., Direct reconstruction of velocity and density profiles from scattered field data, *Geophysics*, submitted for publication, 1980.

3. MARFURT, K. J., Elastic wave equation migration—inversion, PhD Thesis, Columbia University, 1978.
4. HUBRAL, P., Time migration—some ray theoretical aspects, *Geophys. Prospecting*, **25**, pp. 738–45, 1977.
5. FRENCH, W. S., Computer migration of oblique seismic reflection profiles, *Geophysics*, **40**, pp. 961–80, 1975.
6. GARDNER, G. H. F., FRENCH, W. S. and MATZUK, T., Elements of migration and velocity analysis, *Geophysics*, **39**, pp. 811–25, 1974.
7. BERRYHILL, J. R., Wave equation datumming, *Geophysics*, **44**, pp. 1329–44, 1979.
8. CLAERBOUT, J. F., Toward a unified theory of reflector mapping, *Geophysics*, **36**, pp. 467–81, 1971.
9. LOEWENTHAL, D., LU, L., ROBERSON, R. and SHERWOOD, J. W. C., The wave equation applied to migration, *Geophys. Prospecting*, **24**, pp. 380–99, 1976.
10. CLAERBOUT, J. F., Course grid calculations of waves in inhomogeneous media with application to delineation of complicated seismic structure, *Geophysics*, **35**, pp. 407–18, 1970.
11. CLAERBOUT, J. F., Numerical holography, *Acoustical holography*, Vol. 3, ed. A. F. Metherell, pp. 273–83, Plenum Press, New York, 1970.
12. CLAERBOUT, J. F. and JOHNSON, A. G., Extrapolation of time dependent waveforms along their path of propagation, *Geophys. J., Roy. Astron. Soc.*, **26**, pp. 285–93, 1971.
13. HATTON, L., LARNER, K. and GIBSON, B., Migration of seismic data from inhomogeneous media, *presented at 41st Mtg of EAEG, Hamburg*, 1979.
14. ALFORD, R. M., KELLY, K. R. and BOORE, D. M., Accuracy of finite difference modelling of the acoustic wave equation, *Geophysics*, **39**, pp. 834–42, 1974.
15. DEREGOWSKI, S. M., A finite difference method for CDP stacked section migration, *presented at 40th Mtg of EAEG, Dublin*, 1978.
16. MCDANIEL, S. T., Parabolic approximations for underwater sound propagation, *J. Acoust. Soc. Am.*, **58**, pp. 1178–85, 1975.
17. LAPIDUS, L., *Digital computation for chemical engineers*, McGraw-Hill, New York, 1962.
18. MUIR, F., *Stanford Exploration Project*, Vol. 8, p. 54, 1976.
19. CLAYTON, R. and ENGQUIST, B., Absorbing boundary conditions for acoustic and elastic wave equations, *Bull. Seis. Soc. Am.*, **67**, pp. 1529–40, 1977.
20. MARSCHALL, R., Derivative of two-sided recursive filters with seismic applications, *presented at 48th Ann. Mtg of SEG, San Francisco*, 1978.
21. HOOD, P., Finite difference and wavenumber migration, *Geophys. Prospecting*, **26**, pp. 773–89, 1978.
22. BUCHANAN, D. J., An exact solvable one way wave equation, *presented at the 48th Ann. Mtg of the SEG, San Francisco*, 1978.
23. WHITTLESEY, J. R. B. and QUAY, R. G., Wave equation migration operators using 2-D Z-Transform theory, *presented at 47th Ann. Mtg of SEG, Calgary*, 1977.
24. STOLT, R. H., Migration by Fourier transform, *Geophysics*, **43**, pp. 23–48, 1978.
25. DEREGOWSKI, S. M., *Report on the finite difference method*, BP Company Limited (in preparation), 1979.
26. DEREGOWSKI, S. M., Private Communication, BP Company Limited, 1980.

27. DOHERTY, S. M., *Structure independent seismic velocity estimation*, PhD Thesis, Geophysics Department, Stanford University, Ca., 1975.
28. CLAERBOUT, J. F., *Fundamentals of geophysical data processing*, McGraw-Hill, New York, 1976.
29. SHERWOOD, J. W. C., Private communication, Digicon Inc., 1980.
30. YILMAZ, O., *Pre-stack partial migration*, PhD Thesis, Department of Geophysics, Stanford University, Ca., 1979.
31. SCHULTZ, P. S. and CLAERBOUT, J. F., Velocity estimation and downward continuation by wavefront synthesis, *Geophysics*, **43**, pp. 691–714, 1978.
32. ESTEVEZ, R., *Wide angle diffracted multiple reflections*, PhD Thesis, Geophysics Department, Stanford University, Ca., 1977.
33. SCHULTZ, P. S. and SHERWOOD, J. W. C., Depth migration before stack, *Geophysics*, **45**, pp. 376–93, 1980.
34. MAGINNESS, M. G., The reconstruction of elastic wavefields from measurements over a transducer array, *J. Sound and Vibration*, **20** (No. 2), pp. 219–40, 1972.
35. BOOER, A. K., CHAMBERS, J. and MASON, I. M., Numerical holographic reconstruction by a projective transform, *Electron. Lett.*, **13**, pp. 569–70, 1977.
36. GAZDAG, J., Wave equation migration with the phase-shift method, *Geophysics*, **43**, pp. 1342–51, 1978.
37. GAZDAG, J., Extrapolation of seismic waveforms by Fourier methods, *IBM J. Res. Dev.*, **22**, pp. 481–6, 1978.
38. GAZDAG, J., Wave equation migration with the accurate space derivative method, *Geophys. Prospecting*, **28**, pp. 60–70, 1980.
39. PHINNEY, R. A. and FRAZER, L. N., On the theory of imaging by Fourier transform, *presented at 48th Ann. Mtg of SEG, San Francisco*, 1978.
40. CHUN, J. H. and JACEWITZ, C. A., Fundamentals of frequency domain migration, *presented at 48th Ann. Mtg of SEG, San Francisco*, 1978.
41. CHUN, J. H. and JACEWITZ, C. A., A fast multi-velocity function frequency domain migration, *presented at 48th Ann. Mtg of SEG, San Francisco*, 1978.
42. ESTES, L. E. and FAIN, G., Numerical technique for computing the wide angle acoustic field in an ocean with range-dependent velocity profiles, *J. Acoust. Soc. Am.*, **62**, pp. 38–43, 1977.
43. LARNER, K. and HATTON, L., *Wave equation migration: two approaches*, Offshore Technology Conference, paper OTC-2568, Houston, 1976.
44. NEWMAN, P., Amplitude and phase properties of a digital process, *presented at 37th Mtg of EAEG, Bergen, Norway*, 1975.
45. BERKHOUT, A. J., and PALTHE, D. W. VAN W., Migration in terms of spatial deconvolution. *Geophys. Prospecting*, **27**, pp. 261–91, 1979.
46. SCHNEIDER, W. S., Integral formulation for migration in two and three dimensions, *Geophysics*, **43**, pp. 49–76, 1978.
47. KUHN, M. J. and ALHILALI, K. A., Weighting factors in the construction and reconstruction of acoustical wavefields, *Geophysics*, **42**, pp. 1183–98, 1977.
48. BOLONDI, G., ROCCA, F. and SAVELLI, S., A frequency domain approach to two dimensional migration, *Geophys. Prospecting*, **26**, pp. 750–72, 1978.
49. GARIBOTTO, G., 2-D recursive filters for the solution of two-dimensional wave equations, *IEEE Trans. on Acoust. Speech and Signal Processing*, ASSP-27, pp. 367–73, 1979.

50. KUHN, M. J., Acoustical imaging of source receiver coincident profiles, *Geophys. Prospecting*, **27**, pp. 62–77, 1979.
51. DEVEY, M. G., *Derivation of the migration integral*, Technical Note TN451, BP Company Ltd,.Exploration and Production Department, 1979.
52. HOSKEN, J. W. J., *Improvements in the practice of 2D diffraction stack migration, Report No.* EPR/R1247 BP Company Limited, Exploration and Production Department, 1979.
53. SAFAR, M., Private communication, The British Petroleum Company Ltd, 1980.
54. NEWMAN, P., Geometrical aspects of migration before stack, *presented at 40th Mtg of EAEG, Dublin*, 1978.
55. GIBSON, B., LARNER, K. L., SOLANKI, J. J. and NG, A. T. Y., Efficient 3D migration in 2 steps, *presented at 41st Mtg of EAEG, Hamburg*, 1979.
56. COHEN, J. K. and BLEISTEIN, N., Velocity inversion procedure for acoustic waves, *Geophysics*, **44**, pp. 1077–87, 1979.
57. COHEN, J. K. and BLEISTEIN, N., An inverse method for determining small variations in propagation speed, *Soc. Ind. Appl. Math., J. Appl. Math.*, **32**, pp. 784–99, 1977.
58. KENNETT, B. L. N., Private communication, Department of Geodesy and Geophysics, University of Cambridge, 1980.
59. TAPPERT, F. D. and HARDIN, R. H., *A synopsis of the AESD workshop on acoustic propagation modelling by non-ray tracing techniques*, AD-773 741, AESD Tech. Note TN-73-05, 1973.
60. KJARTANSSON, E., The effect of Q on bright spots, *presented at 48th Ann. Mtg of SEG, San Francisco*, 1978.
61. GRAY, S. H., BLEISTEIN, N. and COHEN, J. K., *Direct inversion for strongly depth dependent velocity profile*, Report MS-R-7902, Department of Mathematics, University of Denver, Denver, Colorado, 1978.
62. RAZ, S., An approximate propagation speed inversion over a prescribed slab, *Acoustic imaging*, Vol. 9, Plenum Press, New York, in press, 1980.
63. JUDSON, D. R., SCHULTZ, P. S. and SHERWOOD, J. W. C., Equalising the stacking velocities via DEVILISH, *presented at 48th Ann. Mtg of SEG, San Francisco*, 1978.
64. JUDSON, D. R., LIN, J., SCHULTZ, P. S. and SHERWOOD, J. W. C., Depth migration after stack, *Geophysics*, **45**, pp. 361–75.
65. RAYLEIGH, J. W. S. (1877). *The theory of sound*, Sections 107–11, Dover Publications, London, 1945.

APPENDIX 1: DERIVATION OF A 45° WAVE EQUATION

We use the square root approximation $S^{(3)}$ defined in eqn. (3.5) with the following substitution:

$$X = \frac{c}{\omega} \frac{\partial}{\partial x}$$

Thus eqn. (3.5) becomes

$$S^{(3)} = 1 + \frac{2[(c/\omega)\partial/\partial x]^2}{4 + [(c/\omega)\partial/\partial x]^2} \tag{A1.1}$$

Substituting eqn. (A1.1) into eqn. (3.3) gives

$$\frac{\partial \tilde{P}}{\partial z} = \frac{i\omega}{c}\left[1 + \frac{2[(c/\omega)\partial/\partial x]^2}{4 + [(c/\omega)\partial/\partial x]^2}\right]\tilde{P}$$

Now transform to a moving coordinate frame by means of the substitution

$$\tilde{P} = \tilde{P}' \exp(i\omega z/\bar{c}) \tag{A1.2}$$

where $\bar{c}$ is a constant velocity. Then

$$\frac{\partial \tilde{P}'}{\partial z} = i\left(\frac{\omega}{c} - \frac{\omega}{\bar{c}}\right)\tilde{P}' + \left(\frac{i(2c/\omega)\partial^2/\partial x^2}{4 + (c/\omega)^2 \partial^2/\partial x^2}\right)\tilde{P}' \tag{A1.3}$$

This equation may be split (with slight error) into two equations:

$$\frac{\partial \tilde{P}'}{\partial z} = i\left(\frac{\omega}{c} - \frac{\omega}{\bar{c}}\right)\tilde{P}' \tag{A1.4}$$

and

$$\frac{\partial \tilde{P}'}{\partial z} = \left(\frac{i(2c/\omega)\partial^2/\partial x^2}{4 + (c/\omega)^2 \partial^2/\partial x^2}\right)\tilde{P}' \tag{A1.5}$$

Rearranging eqn. (A1.5),

$$\left[-(i\omega)^2 + \frac{c^2}{4}\frac{\partial^2}{\partial x^2}\right]\frac{\partial \tilde{P}'}{\partial z} = i\omega \frac{c}{2}\frac{\partial^2 \tilde{P}'}{\partial x^2} \tag{A1.6}$$

Equation (A1.4) can be solved directly to give

$$\tilde{P}'(x, z + \Delta z, \omega) = \tilde{P}'(x, z, \omega) \exp\left[i\omega \int_z^{z+\Delta z}\left(\frac{1}{c} - \frac{1}{\bar{c}}\right)dz'\right] \tag{A1.7}$$

Transforming back into the time domain, eqn. (A1.7) becomes

$$P'(x, z + \Delta z, t) = P'(x, z, t + T) \tag{A1.8a}$$

where

$$T = \int_z^{z+\Delta z}\left(\frac{1}{c} - \frac{1}{\bar{c}}\right)dz' \tag{A1.8b}$$

and eqn. (A1.6) becomes

$$\frac{\partial^3 P'}{\partial z \partial t^2} - \frac{c^2}{4}\frac{\partial^3 P'}{\partial x^2 \partial z} + \frac{c}{2}\frac{\partial^3 P'}{\partial x^2 \partial t} = 0 \tag{A1.9}$$

Equations (A1.8) and (A1.9) can be used to downward continue data. Equation (A1.9) is applied first to move the data through a distance Δz, followed by eqn. (A1.8) which represents a correction for the departure over the interval Δz of the velocities from a constant reference velocity $\bar{c}$. To avoid error, this constant should approximate to the local velocity $c(x, z)$ averaged over x. The reason for the transformation (A1.2) should now be clear—it is used to eliminate large data shifts within each depth step.

These equations can be cast into a form appropriate to the zero offset exploding reflector model by dividing every occurrence of velocity in eqns. (A1.2), (A1.8) and (A1.9) by 2. The relevant equations in this case are

$$P'(x, z + \Delta z, t) = P'(x, z, t + T) \tag{A1.10a}$$

where

$$T = 2 \int_{z}^{z+\Delta z} \left(\frac{1}{c} - \frac{1}{\bar{c}}\right) dz' \tag{A1.10b}$$

and

$$\frac{\partial^3 P'}{\partial z \, \partial t^2} - \frac{c^2}{16} \frac{\partial^3 P'}{\partial x^2 \, \partial z} + \frac{c}{4} \frac{\partial^3 P'}{\partial x^2 \, \partial t} = 0 \tag{A1.11}$$

Again eqns. (A1.11) and (A1.10) are solved alternatively in each depth step; eqn. (A1.10) deals with all the refraction effects and eqn. (A1.11) deals with diffraction effects.

APPENDIX 2: DERIVATION OF F–K MIGRATION THEORY IN TWO DIMENSIONS

The development of the equations in this Appendix will be restricted to migration of two-dimensional CMP stacks; an exploding reflector model is appropriate in this case, hence the following wave equation is used:

$$\frac{\partial^2 P}{\partial x^2} + \frac{\partial^2 P}{\partial z^2} = \frac{4}{c^2} \frac{\partial^2 P}{\partial t^2} \tag{A2.1}$$

with initial data of $P(x, 0, t)$ recorded on the plane $z = 0$. We define a new coordinate system

$$d' = \frac{ct}{2} + z$$

$$x' = x$$

$$z' = z \tag{A2.2}$$

and the wavefield $P'(x', z', d') = P(x, z, t)$. In these new coordinates the wave equation becomes

$$\frac{\partial^2 P}{\partial x'^2} + 2\frac{\partial^2 P'}{\partial d' \partial z'} + \frac{\partial^2 P'}{\partial z'^2} = 0 \qquad (A2.3)$$

with initial data $P'(x', 0, d')$. The imaged section will be obtained at time $t = 0$ and by (A2.2) this will be at the point $d' = z$, so $P'(x', d', d')$ defines the migrated image. Let $\hat{\hat{P}}'$ be the double Fourier transform of the wavefield (primes will be dropped hereafter):

$$P(x, z, d) = \frac{1}{4\pi^2} \iint \hat{\hat{P}}(k_x, z, k_d) \exp[-i(k_x x - k_d d)] \, dk_x \, dk_d \quad (A2.4)$$

Substituting eqn. (A2.4) into eqn. (A2.3) and treating each component separately, we obtain

$$-k_x^2 \hat{\hat{P}} + \frac{d^2 \hat{\hat{P}}}{dz^2} + 2ik_d \frac{d\hat{\hat{P}}}{dz} = 0 \qquad (A2.5)$$

This ordinary differential equation has the solution

$$\hat{\hat{P}}(k_x, z, k_d) = \hat{\hat{P}}(k_x, 0, k_d) \exp\{-i[k_d - (k_d^2 - k_x^2)^{1/2}]z\} \quad (A2.6)$$

Substitution of eqn. (A2.6) into eqn. (A2.4) and taking $z = d$ gives

$$P(x, d, d) = \frac{1}{4\pi^2} \iint \hat{\hat{P}}(k_x, 0, k_d) \exp\{i[k_x x + (k_d^2 - k_x^2)^{1/2} d]\} \, dk_x \, dk_d$$
$$(A2.7)$$

The limits on this integral are infinite and may be split into three regions:

$$I = I_1 + I_2 + I_3 \equiv \int_{-\infty}^{\infty} dk_x \left[\int_{k_d = |k_x|}^{\infty} dk_d + \int_{k_d = -\infty}^{-|k_x|} dk_d + \int_{k_d = -|k_x|}^{|k_x|} dk_d \right]$$
$$(A2.8)$$

I_3 vanishes for all but very small depths and can be ignored. I_1 and I_2 are handled by a change of variables from k_d to k_d', where

$$k_d = \text{sgn}(k_d)(k_d'^2 + k_x^2)^{1/2} \qquad \text{for } |k_d'| > 0 \qquad (A2.9)$$

and in this coordinate system eqn. (A2.7) becomes

$$P(x, d, d) = \frac{1}{4\pi^2} \iint \hat{\hat{P}}(k_x, 0, \sqrt{k_d'^2 + k_x^2})$$
$$\times \frac{k_d'}{(k_d'^2 + k_x^2)^{1/2}} \exp(ik_x x + ik_d' d) \, dk_x \, dk_d' \quad (A2.10)$$

The mapping defined by eqn. (A2.9) is only unambiguous for $|k'_d| > 0$; for $k'_d = 0$ it is convenient in eqn. (A2.10) to take $P(k_x, 0, k_x)$ as zero.

In practical application of the method a modified directivity term is sometimes used in eqn. (A2.10) to reduce the effect of noise smiles. The modifications might typically lead to a directivity term

$$\frac{k'_d}{\sqrt{\gamma k_x^2 + k_d'^2}} \quad \text{where} \begin{cases} \gamma > 1 \text{ attenuates dipping beds and noise smiles} \\ \gamma < 1 \text{ boosts dipping beds} \end{cases} \quad \text{(A2.11)}$$

For a variable velocity medium, Stolt[24] obtains a modified equation in which velocity dependence resides in a single parameter W where $0 \leq W \leq 1$. Pursuing the analysis yields the following relation for the migrated wavefield:

$$P(x, d, d) = \frac{1}{4\pi^2} \int dk_x \int dk'_d A(k_x, k'_d) \exp[i(k_x x + k'_d d)] \quad \text{(A.212)}$$

where

$$A(k_x, k'_d) = \frac{1}{2(W-2)} \left\{ (W-1) - \frac{k'_d}{[k_d'^2 + (2-W)k_x^2]^{1/2}} \right\}$$

$$\times \hat{P}\left\{ k_x, \frac{k'_d(W-1) - [k_d'^2 + (2-W)k_x^2]^{1/2}}{2(W-2)} \right\}$$

Migration in this case involves, as before, a simple shifting and scaling in the Fourier transformed domain.

APPENDIX 3: DERIVATION OF THE SCANWIDTH IN A CONSTANT OFFSET DIFFRACTION STACK

Consider the construction shown in Fig. (A3.1). A ray leaves the shot point S and, after striking the reflector at O, is specularly reflected and received back at the surface at G. In the constant offset framework the CMP position is at M. The distance x_m given by AM on the figure is the required minimum scan size for the diffraction stack operation. By geometry

$$x_m = z_0 \tan(\alpha_r - \beta) + h \quad \text{(A3.1)}$$

and

$$OS/\sin[90° - (\alpha_r + \beta)] = 2h/\sin 2\beta \quad \text{(A3.2)}$$

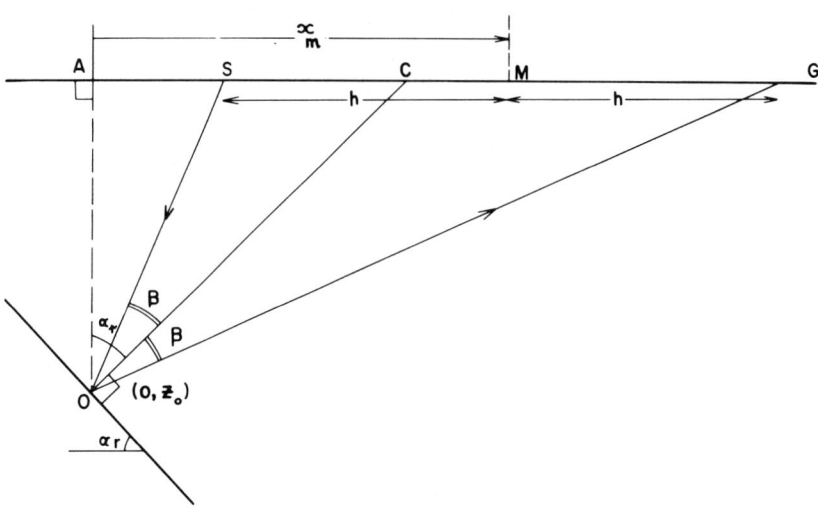

FIG. A3.1. Geometry for dipping plane.

Now

$$\mathrm{OS} = z_0/\cos(\alpha_r - \beta) \qquad (A3.3)$$

Substituting eqn. (A3.3) into eqn. (A3.2) gives

$$z/\cos(\alpha_r - \beta)\cos(\alpha_r + \beta) = 2h/\sin 2\beta \qquad (A3.4)$$

Making the further substitution $b = \tan \beta$ into eqn. (A3.4) and solving the resulting quadratic equation yields

$$b = \frac{-ct_0 + [(ct_0)^2 + 4h^2(1 - \cos^2 2\alpha_r)]^{1/2}}{2h(1 - \cos 2\alpha_r)} \qquad (A3.5)$$

GEOPHYSICS

TWO-DIMENSIONAL AND THREE-DIMENSIONAL MIGRATION OF MODEL-EXPERIMENT REFLECTION PROFILES

WILLIAM S. FRENCH[*]

A reflection profile represents an unfocused picture of the subsurface. In areas of rapid structural change, this unfocused picture may not reveal directly the true geometry of subsurface structures. Computer processing techniques, collectively called migration, have been used by many companies to focus 2-D reflection data. A description of the migration process can be given which allows immediate generalization to three-dimensions with arbitrary source and receiver positions.

Reflection profiles digitally recorded in the laboratory over known acoustically semitransparent structural models establish the effectiveness of migration. Processed reflection data over 3-D models demonstrate that 3-D migration eliminates many of the lateral correlation ambiguities caused by "sideswipes" and "blind structures."

Structure maps developed from the results of 3-D migration of reflection data give a true and precise picture of 3-D models. When the same data are processed using 2-D migration, the mapped structures are distorted.

In structurally complex areas it is desirable to collect 3-D reflection data. Single profiles cannot, and conventional grids may not, reveal adequate cross-dip information.

INTRODUCTION

In areas of rapid structural change, unprocessed seismic reflection profiles may not reveal directly the true geometry of subsurface structures. In some cases, it is possible to identify an isolated structure on the basis of its characteristic reflection pattern (e.g., a fault or syncline). In general, however, it may be necessary to employ some technique to "focus" the data from complex areas.

Two calculational procedures to focus seismic reflection data have been discussed in the literature: downward continuation of moveout corrected seismograms (Claerbout and Doherty, 1972) and digital migration (Schneider, 1971). In essence, the methods utilize a surface-recorded reflection profile to calculate the hypothetical profile that would be obtained if the geophones could be placed in a plane deep in the earth. Their usefulness results from the fact that the closer the profile is to a structure, the more the reflection patterns take on the geometry of the structure. Both techniques are founded on the mathematics of wave propagation and should provide identical results if properly executed. The first technique is a numerical solution to the differential equations of wave motion. The second technique is based upon numerical solution of integral equations describing the wave motion (Maginness, 1972). The integral equations admit a construct which proves to be of great value to the intuition,

---

Presented at the 43rd Annual International SEG Meeting October 25, 1973, Mexico City. Manuscript received by the editor December 6, 1973.

[*] Gulf Research & Development Co., Pittsburgh, Penn. 15230.

© 1974 Society of Exploration Geophysicists. All rights reserved.

and such an approach will be adopted in this paper; the term migration will be used as a synonym for the construct.

In this paper the assumptions inherent to migration will be stated, the technique will be extended to handle the case of 3-D objects with arbitrary source-receiver arrays, and the results of both 2-D and 3-D migration of seismic model data will be presented.

ASSUMPTIONS OF DIGITAL MIGRATION

The typical seismic record is the result of recording at several locations the reflected waves generated by an impulsive source. A profile consists of a collection of such records for several source locations. For convenience of discussion we will deal with the case of marine profiles. Thus the detectors are hydrophones and the profile consists of the pressure-time records $P(s_j, r_j^i, t)$, where

$s_j = (x_j, y_j, z_j)$ is the location of the jth shotpoint,

$r_j^i = (x_j^i, y_j^i, z_j^i)$ is the location of the ith hydrophone recording the jth shot, and

t is a time variable which is reset to zero at the instant of each successive shot.

At this point, no particular shooting geometry has been assumed (i.e., the shotpoints and receiver locations do not necessarily lie in the same line or even in the same plane). The task is to interpret $P(s_j, r_j^i, t)$ in terms of subsurface structural geometry. Migration will be used to accomplish this end.

Computer migration schemes represent a search for scattering centers (i.e., diffraction or reflection points). The process involves assigning to each subsurface point a number which is a measure of the probability that scattered energy emanated from that point. The number is determined by summing the recorded data for all shotpoints and receiver locations at times where energy from that subsurface point could arrive. A number with large absolute value (positive or negative depending upon the wavelet shape) indicates that scattered (reflected or diffracted) energy probably did come from that particular subsurface point. In other words, the coherence of the recorded data along a traveltime versus distance surface appropriate to the source-receiver geometry and subsurface point determines the assigned probability that a reflector or diffractor exists at that subsurface point.

The assumptions of the migration method used in this paper are as follows:

(a) Shear waves can be ignored.
(b) Each subsurface point represents a possible scattering center.
(c) Reflecting surfaces can be considered a continuum of scattering centers.
(d) The pulse shape for the scattered waves is the same for all directions. The pulse is short enough so that the delay in arrival time of later portions can be neglected.
(e) Traveltime versus distance surfaces can be calculated to sufficient accuracy by averaging horizontal velocity variations. That is, a vertical path is considered the least-time travel path from surface to subsurface point, and moveout times for other surface points can be calculated on the basis of surface distance and a root-mean-square velocity-depth function (see Taner and Koehler, 1969).
(f) A coherent signal summed along the appropriate time-distance surface leads to an average with large absolute value. Any noise summed along the time-distance sursurface will lead to a small average value due to the equal probability of positive and negative numbers in the noise field.

It is known that some of the above assumptions are not strictly true [e.g., the shape of a pulse scattered by a small sphere depends upon the scattering angle (Morse and Ingard, 1968)]. However, they are consistent with the usual assumptions of seismic data processing.

GENERAL EQUATIONS FOR MIGRATION

The above description of migration can be stated mathematically. We assume that conditions (a) through (f) hold and that we have available the pressure-time data $P(s_j, r_j^i, t)$. Let the time required for a pulse to travel the dashed line path of Figure 1 [from the source at s_j to the subsurface point $r = (x, y, z)$ and subsequently to a receiver at r_j^i] be given by $t_j^i(r)$. As before, we establish a measure of the probability that a scattering center exists at r by forming the sum

$$M(r) = \sum_j \sum_i P[s_j, r_j^i, t_j^i(r)]. \qquad (1)$$

Equation (1) represents 3-D migration for arbitrary source and receiver arrays. If the source and receiver lie in the xy-plane, then [according to condition (e)]:

$$\overset{i}{t_j}(r) = \left\{ T_{1/2}^2(r) + \frac{(x_j - x)^2 + (y_j - y)^2}{V^2(z)} \right\}^{1/2} + \left\{ T_{1/2}^2(r) + \frac{(x_j^i - x)^2 + (y_j^i - y)^2}{V^2(z)} \right\}^{1/2}, \quad (2)$$

where $T_{1/2}(r)$ is the one-way vertical traveltime to r and $V(z)$ is the rms velocity function as described by Taner and Koehler (1969).

If we consider the special case where there is but one receiver for each source and both source and receiver are at the same location, then equation (2) becomes

$$\overset{1}{t_j}(r) = t_j(r) \quad (2a)$$

$$= \left\{ T_0^2(r) + \frac{4[(x_j - x)^2 + (y_j - y)^2]}{V^2(z)} \right\}^{1/2},$$

where T_0 is the two-way vertical traveltime to r.

Equation (1) involves an implicit time-to-depth conversion since the rms velocity is given as a function of depth. In practice, we know the rms velocity more accurately as a function of time. Therefore, in the following applications of equation (1) we replace z by T_0 and plot migrated time sections rather than depth section. Equations (1) and (2a) have been used to focus data collected with both 2-D and 3-D models.

Two-D models have thickness variations only in the direction of the profile—cross-sections perpendicular to the profile line are of constant

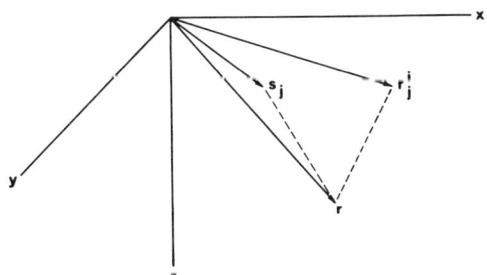

FIG. 1. Coordinate system for migration equation.

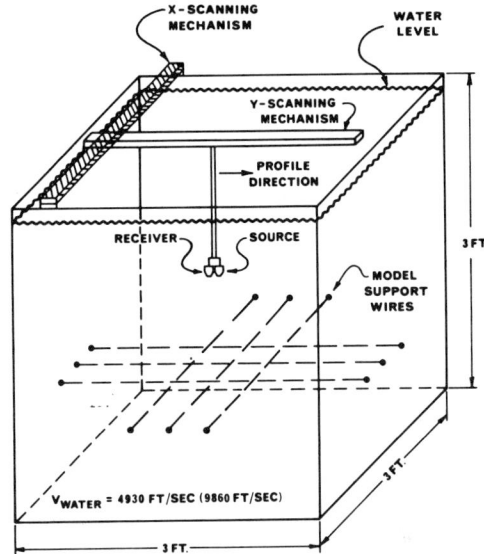

FIG. 2. Water tank arrangement.

thickness. For such models, only a single profile of data is required. Three-D models have thickness variations both in-line with and perpendicular to the profile line; these models require an areal coverage of reflection data.

A 3-D model was studied, and in order to set a standard for the quality of results desired from the 3-D processing, an experiment was run on a 2-D model whose cross-section displayed the same structures and dips found in certain cross-sections of the 3-D model. Results of these experiments are given in the following sections along with others for an additional 2-D experiment which displays a pertinent phenomenon.

EXPERIMENTAL ARRANGEMENT

Ultrasonic pulse-echo electronics with digital recording was used to obtain the reflection data from a model supported in a water tank by fine wires (Figure 2). Two-D or 3-D single coverage reflection data were recorded at some predetermined height above the model. The 2-D data consisted of the usual single coverage profile while the 3-D data consisted of numerous parallel 2-D profiles resulting in a square grid of source-receiver locations. The model dimensions are such that the data represent simulated field data. The scaling is indicated by the following equivalent dimensions and velocities:

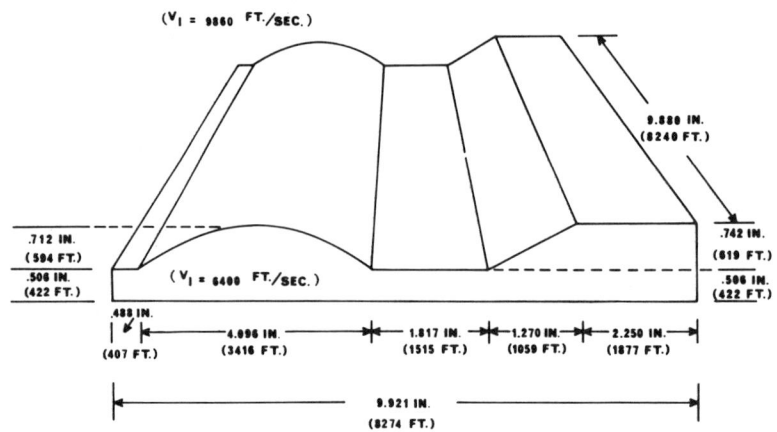

FIG. 3. Ridge and fault model with equivalent field dimensions. The equivalent field velocities are shown.

	Model	Field
Time	.2 μsec	1 msec
Length	6.336 inches	1 mile
Velocity	v	$2v$

Hereafter, equivalent field values will be given in parentheses. In all cases the source-receiver separation was .827 inches (675 ft). The data were processed, however, as if the source and receiver were coincident. The actual sound speed in the material of which the models were constructed was less than that of water. Thus, the models represent structures with a velocity less than that of the overburden.

TWO-DIMENSIONAL RESULTS

The 2-D model consisted of a rounded ridge and an angular step (or normal fault) whose dimensions and corresponding field values are given in Figure 3. A single coverage reflection profile taken 4.11 inches (3430 ft) above the top of the step is shown in Figure 4. This profile was focused by 2-D migration according to equations (1) and (2a). The migration results are shown in Figure 5. The simple process described by equations (1) and (2a) generates some background noise seen in the figure. This migration was carried out using a constant field-equivalent rms velocity for all depths [i.e., $V(z) = 9860$ ft/sec in equation (2a)]. As a result, the top surface of the model is properly in focus but the velocity-anomaly generated structures of the bottom surface are slightly out of focus. Furthermore, due to the fact that the same rms velocity-depth function was used for all traces (i.e., horizontal velocity variations were ignored), the migrated profiles generated are what we have called "true vertical time sections" rather than "true structural geometry sections." Thus, velocity-anomaly generated pseudostructures are not eliminated by the process. They are, in fact, brought into approximate focus.

A clear example of this pseudostructural focus-

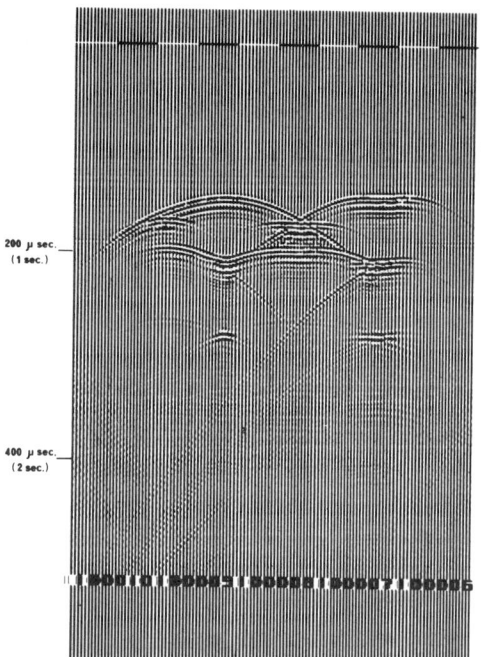

FIG. 4. Profile over ridge-and-fault model. Shotpoint separation = .133 inch (111 ft).

cross-over pattern obtained from a real buried focus.

Further processing is necessary in order to eliminate velocity-anomaly generated pseudostructures (i.e., in order to flatten the reflection from the bottom of the model in this case). The pseudostructures are caused by interval velocity variations and can be corrected through time-depth conversion. The methodology is known and straightforward, but accurate velocities are required for the conversion. It is not the purpose of this paper to discuss time-depth conversion problems. We are concerned here with the proper 3-D focusing of time structures.

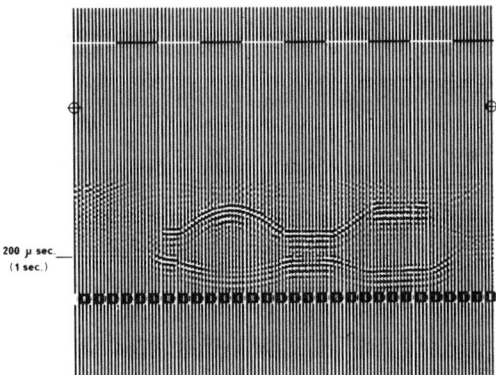

FIG. 5. 2-D migrated time section of data from ridge-and-fault model (Figure 4).

Returning to Figure 5, we see that the migrated results give a true picture of the upper surface of the model. This was verified by comparison with measurements taken directly from the model. (From here on we will be concerned only with the upper surface of the models as the pseudostructures from the lower surfaces have been improperly focused.)

ing is presented in Figures 6, 7, and 8. Figure 6 gives dimensions for the 2-D model of a low-velocity layer with a high curvature ridge. Note that the bottom of the model is flat. A reflection profile taken at a height of 8.9 inches (7440 ft) above the center of this model is shown in Figure 7, and the migrated result using the same $V(z)$ for all traces is shown in Figure 8. A velocity-anomaly generated pseudosyncline in the reflection from the base of the model is brought into focus by the migration just as is the true ridge on the surface of the model. There is a three-to-one exaggeration of the vertical scale in these profiles which explains the stubby appearance of the migrated result when compared to the model shown in Figure 6. An incidental result (apparent in Figure 7) is the fact that the velocity-anomaly generated syncline produces the same reflection

Figure 5 represents a 48 trace-scan migration (24 to each side), i.e., data from 48 traces were summed to give each migrated trace. A field equivalent 30–60 hz, 24 db/octave band-pass filter was applied to the results (the equivalent bandwidth of the somewhat "ringing" source used was 30–60 hz). The figure indicates the quality of focusing obtainable when equations (1) and (2a) are applied to simple 2-D model data. This 2-D experiment was carried out as a control for the 3-D experiment (to be described in the next section), and the processing parameters used for the 2-D migration (trace-scan distance, velocity,

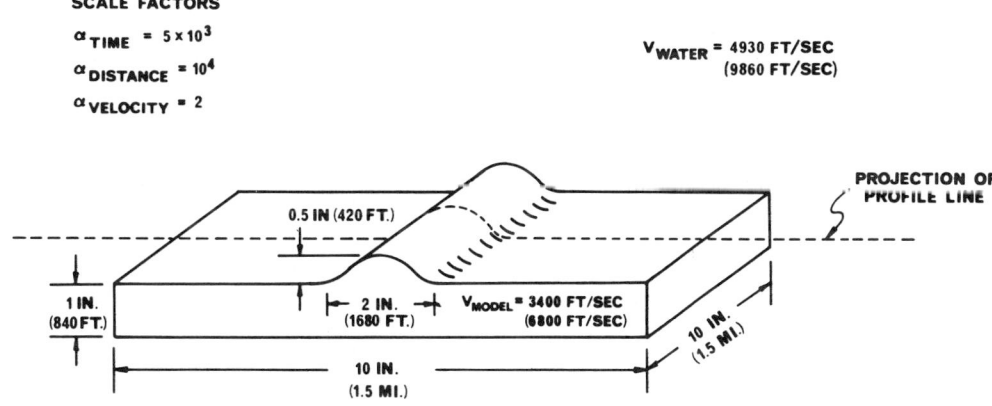

FIG. 6. 2-D high curvature ridge model.

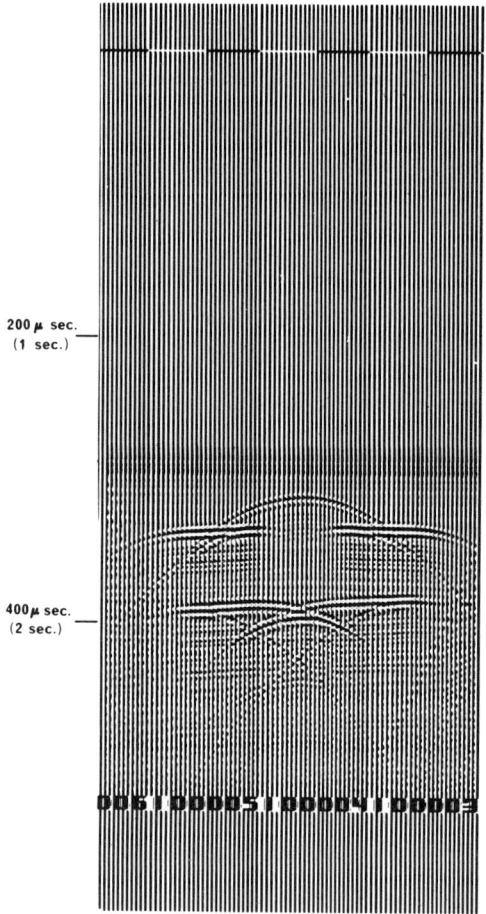

FIG. 7. Profile over model shown in Figure 6. Shotpoint separation = .2 inch (167 ft).

locations for the first of 96 profiles shot over the model. Numbers along the top of Figure 10 represent the locations of the 96 profiles. The dashed lines which are numbered at the bottom of Figure 10 show the locations of the cross-sections which were calculated for the purpose of constructing a structure map. Our choice of these thirteen 3-D migrated cross-sections was quite arbitrary. The total number of 3-D migrations performed on field data will depend only upon the number of true cross-sections required to delineate the structures. Figure 11 is a photograph of the model showing relief along the cross-sections of interest.

Results for cross-sections 3, 7, and 11 are shown in Figures 12 to 14. Each figure represents a different cross-section, and the display labeled A is the 2-D raw data profile over that cross-section. The display labeled B is a conventional 2-D migration of the raw data. Display C of each figure is the 3-D migration result for that cross-section. The single raw data profile shown is used in the 2-D migration. The 3-D migration, on the other hand, uses data from a number of profiles on either side of the reconstruction plane of interest. We chose to collapse the data from 48 consecutive profiles (24 on each side) into each of the 3-D migrated cross-sections shown in Figures 12–14.

shotpoint distances, final filter parameters) were also used for the 3-D migration.

THREE-DIMENSIONAL RESULTS

The model used for the 3-D experiment is shown in Figure 9 (see Figure 11 for a perspective view). The purposes of the experiment were to study 3-D complications in reflection profiles and to provide data from a known structure for testing 3-D migration as defined by equations (1) and (2a). Reflection data were recorded over the model and processed to produce 13 different 3-D migrated cross-sections of the model.

Figure 10 represents a plan view of the data collection scheme with the model shown schematically in solid lines. The numbers on the left of Figure 10 represent the successive source-receiver

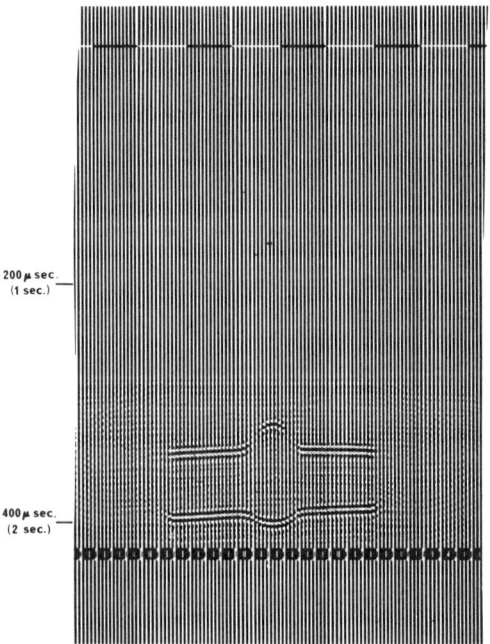

FIG. 8. Migration of profile shown in Figure 7.

Two- and Three-Dimensional Migration

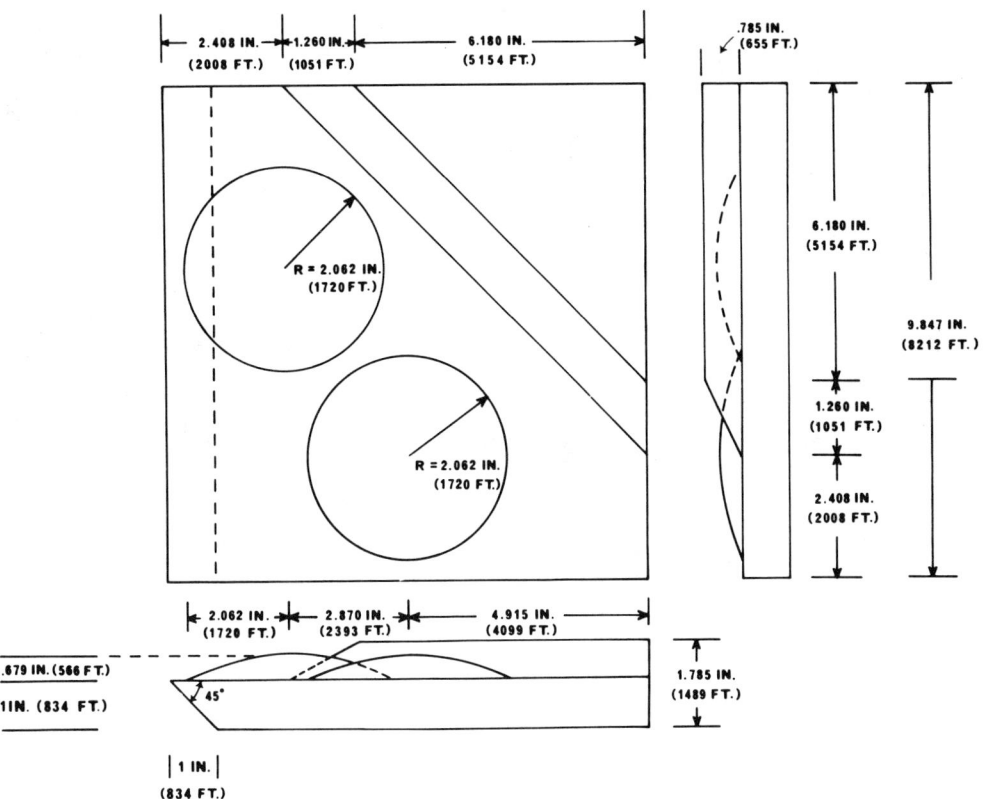

FIG. 9. 3-D model with equivalent field dimensions. See Figure 11 for perspective view.

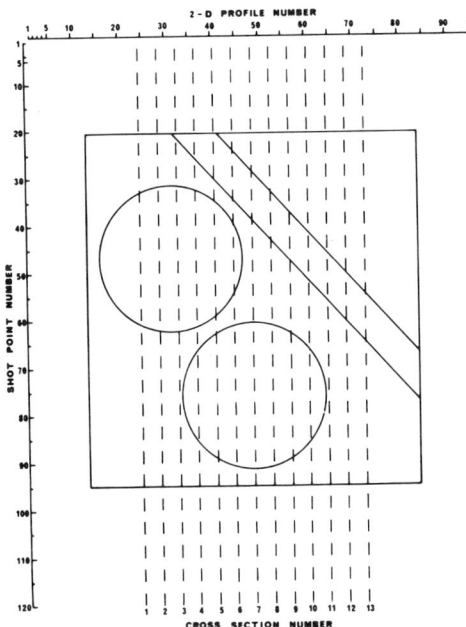

FIG. 10. 3-D data grid.

The quality of the results indicates that less data probably could have been used without serious deterioration of the results. The 96 parallel raw data profiles were taken .133 inches (111 ft) apart while the 3-D migrated profiles were reconstructed at intervals of .532 inches (444 ft).

It is most instructive to examine each of Figures 12 to 14, comparing the raw data, 2-D migrated data, and 3-D migrated data with the true cross-section (Figure 11). Here, the central cross-section (no. 7 in Figure 11) and the corresponding data of Figure 13 are singled out for specific discussion.

Several statements can be made concerning the results shown in Figure 13.

(1) The raw data are extremely difficult to interpret due to diffraction and sideswipe events. Of course, when one starts with a knowledge of the model geometry, it is easy to account for all the arrivals in the raw data.

(2) True structures in the profile plane can be

FIG. 11. Photograph of 3-D model.

absent in the raw data due to a component of dip perpendicular to the profile plane (blind structures). For example, no reflected energy from the portion of the fault which lies in the plane of the profile is recorded in the raw data profile.

(3) Conventional 2-D migration of the raw data profile does not eliminate sideswipe events or enhance blind structures. For example, the central hump in Figure 13b is a sideswipe event and has not been removed by conventional migration. This sideswipe causes an ambiguity in lateral correlation. (One could even go so far in this example as to assume the horizontal feature under the hump to represent a gas-fluid contact under an anticlinal trap.) Notice also that arrivals from the correct fault plane position are absent. Correlation across this blind zone is difficult.

(4) Two-D migration produced an increase in the background noise. This is probably a result of using the simplified integration scheme represented by the sum in equation (1).

(5) Three-D migration eliminated sideswipes and brought out blind structures. The resultant profile, shown in Figure 13c, is devoid of any interpretational ambiguities caused by these phenomena.

(6) The background noise created by the 3-D migration is similar to that of the 2-D migration and can probably be reduced by use of more sophisticated numerical techniques.

Similar statements could be made concerning the cross-sections displayed in Figures 12 and 14.

STRUCTURE MAPS FROM 2-D AND 3-D RESULTS

The upper surface of the model was mapped by first using a more conventional 2-D shooting and migration plan and then using the 3-D migration results. These maps are shown, respectively, in Figures 15 and 16. Contour values used to construct the maps were taken from every fourth shot point as this interval is equal to the distance between the processed profiles. The location of layer terminations and the fault-plane boundaries were read (to the nearest shotpoint) from the profiles.

Several statements can be made concerning the mapping of these experimental data:

(1) Structure could not be determined from the raw data profiles; migration was necessary.

(2) The structural map constructed from the 3-D migrated profiles resulted in a true and precise picture of the upper surface of the model (compare Figure 16 with Figure 10).

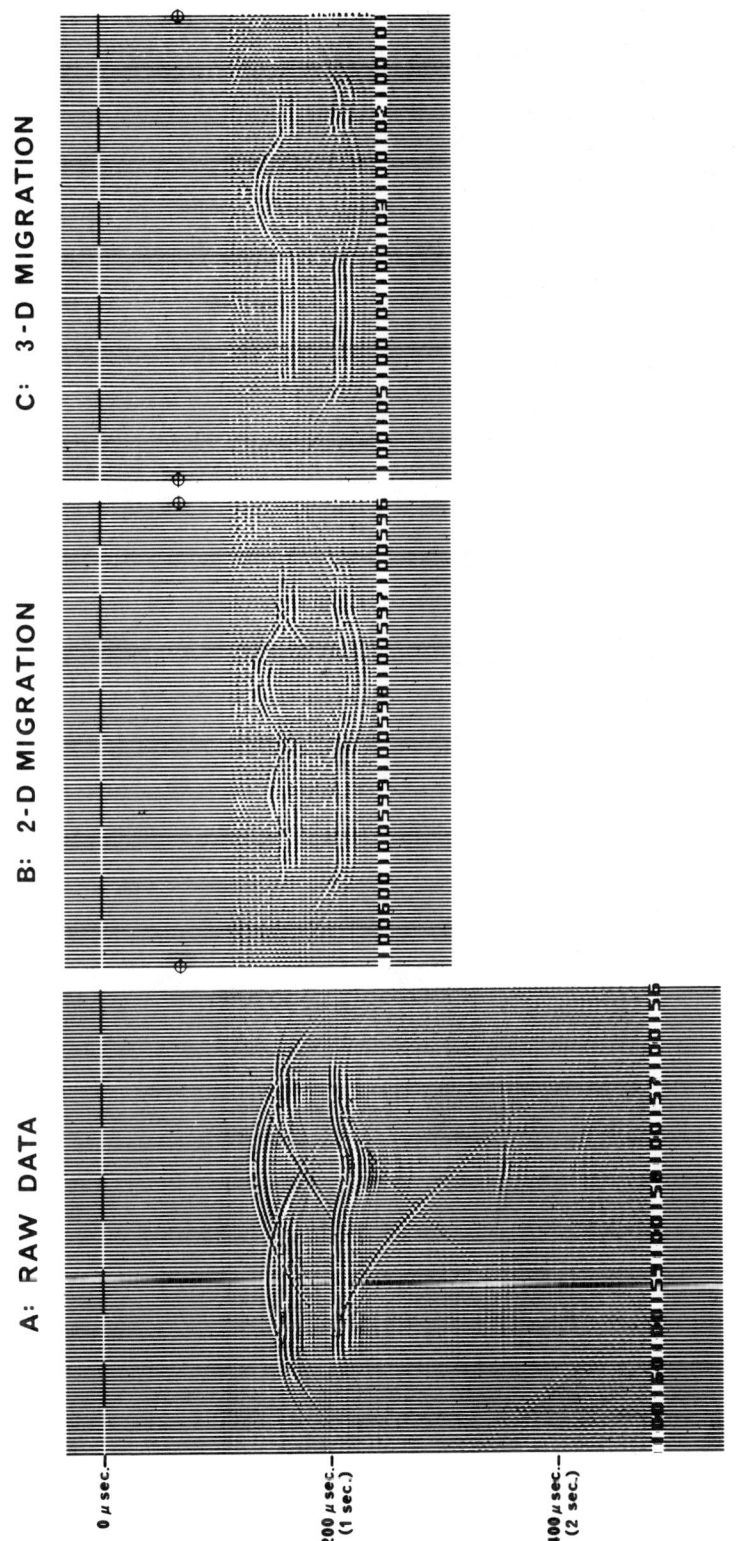

FIG. 12. Results for cross-section no. 3.

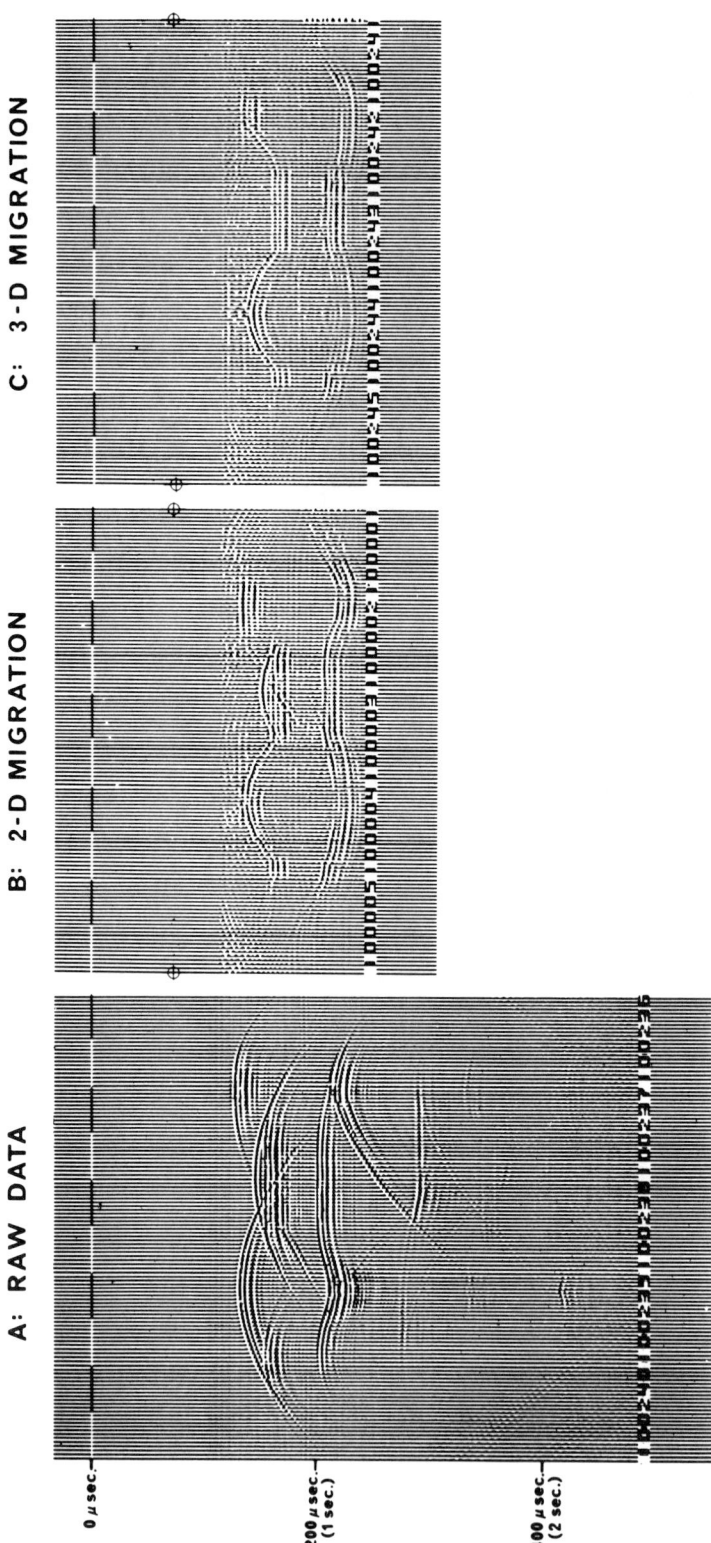

FIG. 13. Results for cross-section no. 7.

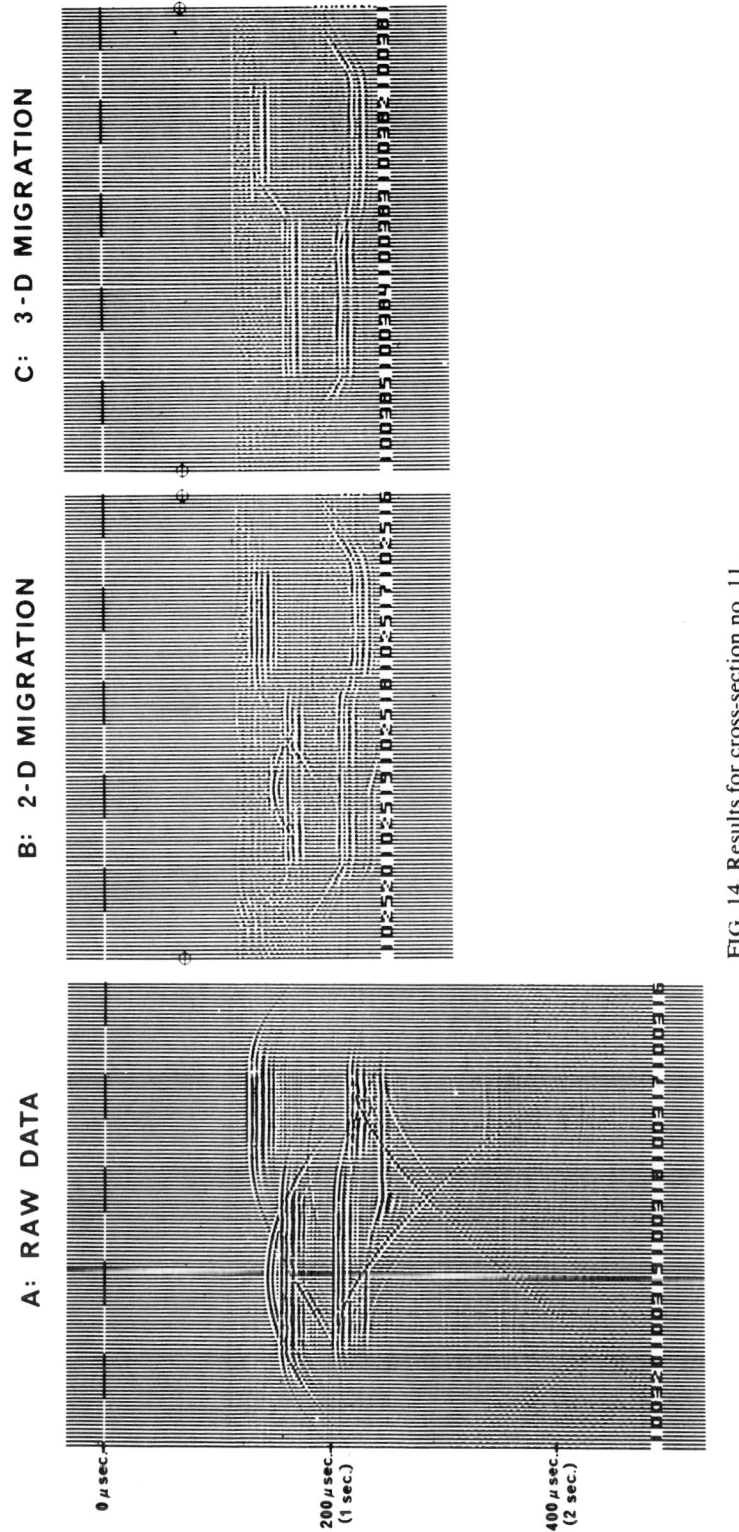

FIG. 14. Results for cross-section no. 11.

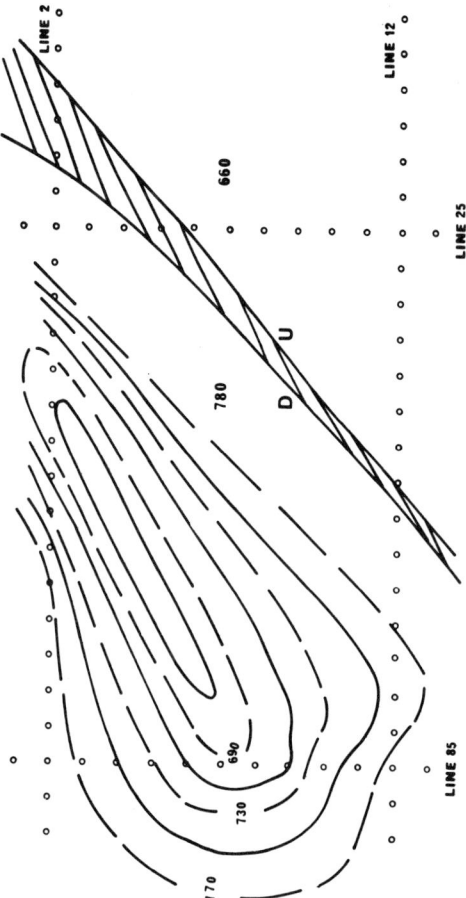

FIG. 15. Structure map constructed from migration of conventional 2-D profiles (contours in msec).

would have been inadequate. The advantages of 3-D processing will probably be even more evident in layered cases where greater correlation ambiguities exist. The 3-D process will also allow us to image features which are some distance removed from our data-gathering area.

CONCLUSIONS

The following conclusions can be drawn from the results and discussion presented above:

(1) The migration method of processing seismic-reflection data can be extended to handle 3-D exploration problems with arbitrary source and receiver arrays. A mathematical description is given by equations (1) and (2) above.

(2) When an average velocity-depth function is used over an entire section, false structures caused by lateral velocity variations in the overburden are not eliminated but, rather, are brought into focus by migration.

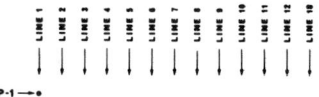

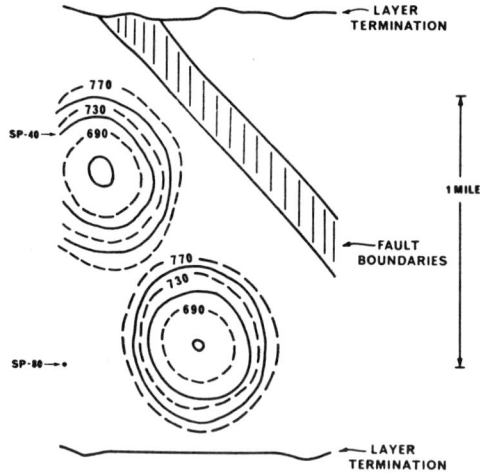

FIG. 16. Structure map constructed from 3-D migrated profiles (contours in msec).

(3) The structural map made from the 2-D migrated profiles is distorted in shape (compare Figure 15 with Figure 10).

With dense enough 2-D coverage and 2-D migration, structural distortions in the 2-D data can be corrected through careful interpretation. As the interpreter develops his map, he can detect which events are sidewipes on individual 2-D profiles and apply appropriate areal displacements. In other words, this experiment does not demonstrate that 3-D migration is essential for delineating the features of this model, although it clearly shows that at least 2-D migration is indispensable. Furthermore, the gathering of dense areal data is essential to an accurate resolution of the model; conventional 2-D recording

Subsequent trace-by-trace, time-to-depth conversion will eliminate these pseudo-structures but a rather detailed knowledge of the section velocity is required.

(3) When single coverage reflection profiles over 2-D models are processed by conventional migration, an exact reproduction of a true vertical-time section of the model results. The background noise is, however, increased. It has been shown that these noise problems can be eliminated by better numerical techniques (Gardner et al, 1973).

(4) Single coverage profiles recorded over 3-D models are almost totally uninterpretable unless a migration process is applied.

(5) When single coverage reflection profiles over 3-D models are processed by conventional 2-D migration, correlation ambiguities may arise due to sideswipes and blind structures.

(6) Structure maps constructed from a conventional survey with profiles independently processed by 2-D migration show distortions of the structures. However, the distortions can be reduced by increased coverage and careful interpretive techniques.

(7) Simultaneous 3-D migration of a sequence of parallel reflection profiles eliminates the correlation ambiguities caused by sideswipes and blind structures.

(8) Structure maps constructed from the 3-D results give a true and precise picture of the model.

REFERENCES

Claerbout, Jon F., and Doherty, Stephen M., 1972, Downward continuation of moveout corrected seismograms: Geophysics, v. 37, p. 741–768.

Gardner, G. H. F., French, W. S., and Matzuk, T., 1973, Elements of migration and velocity analysis: Presented at the 43rd Annual International SEG Meeting, October 25, 1973, Mexico City.

Maginness, M. G., 1972, The reconstruction of elastic wave fields from measurements over a transducer array: J. of Sound and Vibration, v. 20, no. 2, p. 219–240.

Morse, P. M., and Ingard, K. U., 1968, Theoretical acoustics: New York, McGraw-Hill Book Co., Inc.

Schneider, William A., 1971, Developments in seismic data processing and analysis (1968–1970): Geophysics, v. 36, p. 1043–1073.

Taner, M. Turhan, and Koehler, Fulton, 1969, Velocity spectra—digital computer derivation and applications of velocity functions: Geophysics, v. 34, p. 859–881.

A SIMPLE THEORY FOR SEISMIC DIFFRACTIONS[†]

A. W. TROREY[*]

Existing diffraction theory is often cast in such a way as to preclude a ready qualitative understanding of diffraction phenomena. This difficulty can be overcome by making simplifying but realistic approximations which permit the diffractive response of an arbitrary subsurface with point-source excitation to be obtained in a simple closed form. The main approximations are that the subsurface behaves as an acoustic medium, that its average velocity is constant, and that its reflectivity is low. An objective of this paper is to provide the field interpreter with a practical understanding of diffraction behavior.

INTRODUCTION

Much diffraction theory is formulated in such a way as to make an understanding of the detailed behavior of diffractions difficult. This difficulty can be overcome by making simplifying but realistic approximations concerning the seismic behavior of the subsurface and developing a simple theory based, in essence, on Huygen's principle. This approach permits solutions to be obtained in a closed form, these solutions being of such a nature that they provide considerable insight into the behavior of diffractions. The present paper is divided into three main parts. In the first part, we develop simple mathematical expressions for the response of a seismic subsurface. In the second part, we examine the behavior of the diffractions predicted by these expressions. In the third part, examples of diffractions for some simple structures are given.

MATHEMATICAL DEVELOPMENT

In order to achieve a straightforward mathematical development, 6 main approximations are made. To start with, we make

Approximation 1: The subsurface behaves as an acoustic medium

This means that all shear waves will be ignored in our solution. In a real earth, this approximation is reasonable provided angles of incidence are not too large (the usual case for *reflection* seismology, but not for *refraction* seismology).

Approximation 1 permits us to represent the seismic wave field by means of a scalar potential and to use the scalar wave equation

$$\nabla^2 \varphi = (p/v)^2 \varphi, \quad (1)$$

where

φ is the Laplace transform of the scalar potential,

p is the Laplace transform variable,

v is the velocity in the medium.

Equation (1) is valid in a source-free region. So φ may be thought of, for example, as pressure or as a scalar such that its gradient is displacement. In this paper, we shall examine only the behavior of φ itself and so may think of φ as being a pressure field.

For a φ satisfying equation (1), Helmholtz's equation (see, for example, Longhurst, 1957) applies:

$$4\pi\varphi_p = \int_S e^{-pr/v} \left[\frac{1}{r} \frac{\partial \varphi_s}{\partial n} - \varphi_s \frac{\partial \left(\frac{1}{r}\right)}{\partial n} + \frac{p\varphi_s}{rv} \frac{\partial r}{\partial n} \right] dS, \quad (2)$$

[†] Manuscript received by the Editor April 20, 1970.
[*] Chevron Oil Field Research Company, La Habra, California 90631.
Copyright © 1970 by the Society of Exploration Geophysicists.

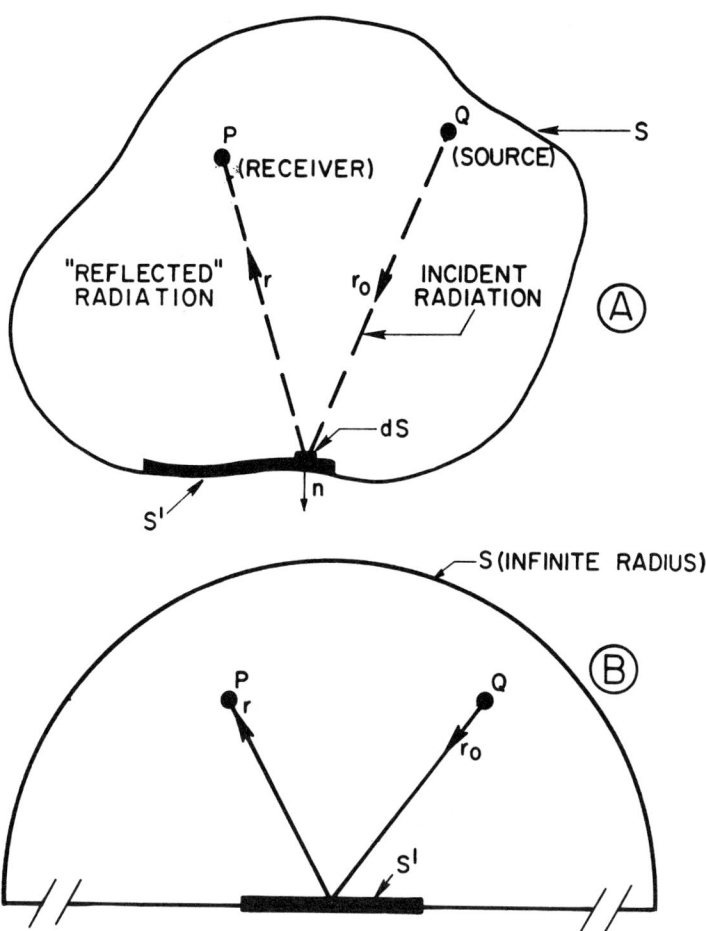

FIG. 1. Figure 1A shows the parameters needed for solution of Helmholtz's equation. P is a point totally enclosed by the surface S, r is the distance from P to an element dS of S, and n is the outward normal of S at dS. In solving equation (2), we will assume that P is a point receiver and Q is a point source (Q may be inside or outside S). The seismic reflector/diffractor will occupy the portion S' of S. From the characteristics of Q, ϕ_s will be estimated over S'. Figure 1B shows the particular case of S being a hemisphere of infinite radius with S' being a plane included in the base.

where

φ_p is the potential at a point P,
S is a surface enclosing P,
φ_s is the potential on S.
r is the distance from P to an element of S,
n is the outward normal of S.

These items are schematically illustrated in Figure 1A.

If a reflecting/diffracting interface (hereafter referred to simply as a "reflector") occupies some part S' of our surface S (see Figure 1A), equation (2) would be soluble if we could estimate φ_s on S' and assume it to be zero on the rest of S where the reflector is nonexistent. Since, as indicated in Figure 1, φ and φ_s are to be representations of only the reflected or reradiated field and are not to include the incident field, this procedure is reasonable provided the surface S is shaped in such a way that the reflected field from S' is zero or small at all other points on S. A surface S shaped so as to satisfy this condition is shown in Figure 1B.[1] Clearly any energy reflected from S' will be zero on the spherical part of S, since it is at infinite distance from S'. Owing to the fact that S' is infinitely thin, we would intuitively expect

---

[1] In effect, we will be ignoring multiple reflections.

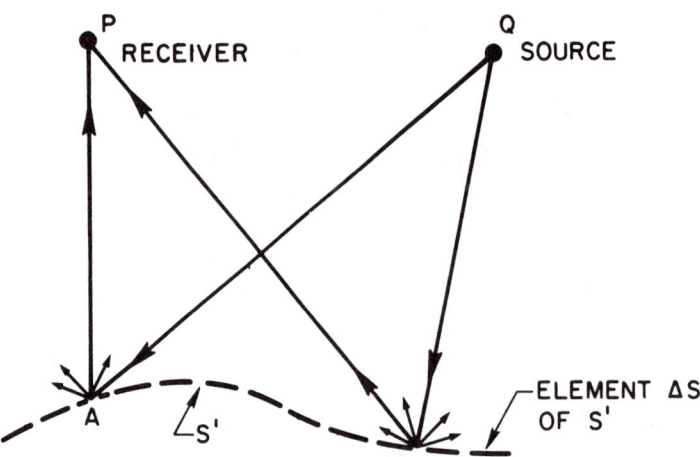

Fig. 2. Illustrating that a reflector S' can be approximated by an ensemble of elementary diffractors. Consequently Snell's law (as usually applied) does not hold, since the path QAP does not have equal angles of incidence and reflection.

energy reflected from S' to any other points in the base of S to be small for finite frequencies and for reasonable angles of incidence. Moreover, this conclusion is born out by existing work in diffraction theory (Biot and Tolstoy, 1957). Consequently, if φ_s for the reflected wave can be estimated over a *plane* surface S', we may assume φ_s to be zero elsewhere on S and integrate equation (2) only over S'. This leads us to

Approximation 2: An actual reflector can be approximated by an ensemble of plane surfaces. The total response will be the sum of the responses from the individual plane surfaces.

Provided the actual and approximating surfaces differ in location by amounts reasonably small compared to a wavelength, it is clear that primary energy (reflections plus diffractions) will be well approximated but that multiple reflections between the plane elements (multiple reflections, multiple diffractions, reflection-diffractions) will be ignored. In principle they could, of course, be incorporated; however, for the purposes of this paper, they are an unnecessary complication. Since surfaces of low reflectivity will have but a small amount of multiple energy associated with them, we thus make

Approximation 3: The reflection coefficient of the reflector is small.

The last ingredient we require is an estimate of φ_s on S'. In essence, we use a simple extension of St. Venant's Principle (Longhurst, loc cit) resulting in

Approximation 4: The reflected field on S' is the same as that which would be present if S' were infinite in extent

This approximation is reasonable for a reflector of low reflectivity. It permits us to assume that φ_s is produced by the *image* with respect to S' of Q reduced in strength by the reflection coefficient R (assuming R to be constant over S' and independent of the angle of incidence).

It is desirable to pause momentarily to get a physical picture of the situation so far. Equation (2) could be approximately evaluated by a summation rather than an integration over small elements ΔS of S', where ΔS is small compared to a wavelength. Thus S' may be thought of as being an ensemble of diffractors, each producing diffracted rays in all directions when struck by an incident ray from the source Q (see Figure 2). Those rays which diffract towards the receiver P are summed to obtain the response at P. Thus *each* point on a reflector sends energy towards *all* receivers. As we will show later, for a plane reflector of infinite extent, this approach yields the well-known simple mirror reflection (equal angles of incidence and reflection) in which we often think of all the energy as being reflected at the reflecting point. This viewpoint is not correct: all points on a reflector produce the reflection. For truncated or nonplanar surfaces, diffractions

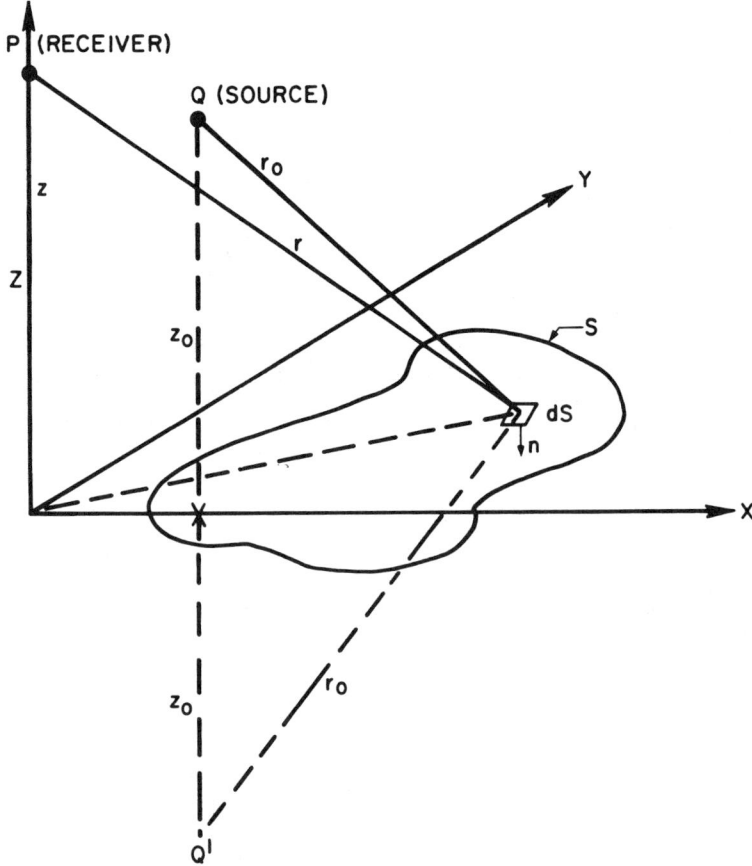

Fig. 3. Geometry for finite plane reflector S lying in XY plane. P is distance r from an element dS and distance z from the XY plane. Q is distance r_o from dS and z_o from the XY plane. Q' is the image point of Q with respect to XY plane. The outward normal to dS is n.

always occur. We may, in fact, argue that the term reflection is a misnomer in that all reflections are in reality an ensemble of diffractions.

In passing, it is worthwhile noting that, since equation (2) is an integral over a surface, if the surface has no area, it has no response. This means that a line or a point cannot produce a diffraction. Any diffractor must have dimensions comparable to or greater than a wavelength to have any significant response. Thus, for example, a cone produces a diffraction apparently emanating from the tip of the cone. This diffraction, however, is actually produced by the body of the cone. The tip itself does not produce the diffraction.

With this background out of the way, we can now solve equation (2) for the finite plane surface S' (hereafter referred to simply as S). Figure 3 shows the geometry. To find φ_s (assumed to be the reflected field at S if S were infinite), we make the fifth approximation.

Approximation 5: The velocity from Q to dS is constant.[2]

Then φ_s is the field due to Q', the image of Q. The solution of equation (1) for the point source Q' in any region not including Q' is $e^{-pr/v}/r$. Thus, on S where $r = r_0$, we have

$$\varphi_s = Rf(p)e^{-pr_0/v}/r_0, \qquad (3)$$

[2] Equation (2) could be evaluated numerically on a computer for nonconstant v. By assuming v constant, we can evaluate equation (2) analytically and thus gain an understanding of the solution not possible with a numerical evaluation.

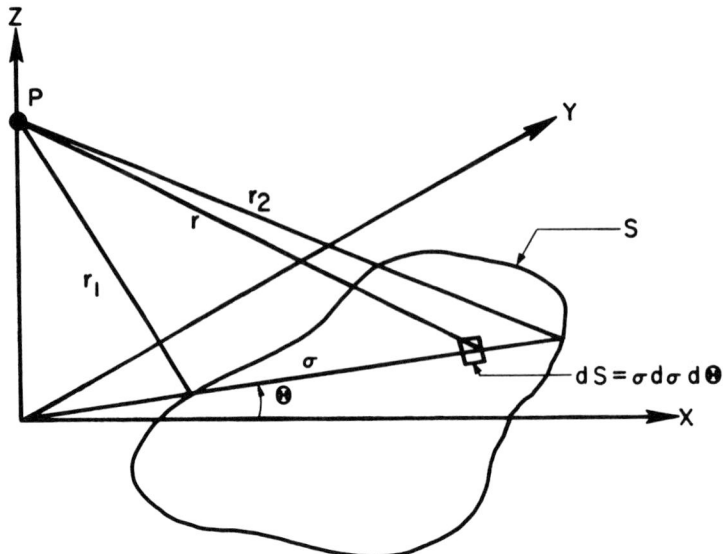

FIG. 4. Geometry for evaluating equation (4). The coordinate origin is the intersection of the normal to S passing through P (the Z axis) and the XY plane. θ and σ are the cylindrical coordinates of dS in the XY plane; r is the distance from P to dS; and r_1 and r_2 are the distances from P to the edges of S at the angle θ.

where

R is the reflection coefficient of S (assumed constant over S),

$f(p)$ is the Laplace transform of the source waveform associated with Q.

If we substitute equation (3) into equation (2) and then let r_o approach r, we then get our sixth approximation:

Approximation 6: The source and receiver are at the same location.

From Appendix I, equation (A-13), we find

$$2\pi\varphi_p = zRf(p) \int_S e^{-2pr/v} \left[\frac{1}{r^4} + \frac{p}{vr^3} \right] dS. \quad (4)$$

By assuming P and Q coincident, we are making but a minor approximation, since normal-moveout is usually removed from seismic data. Normal-moveout removal is a process which (except for regions of excessive Doppler shift, multiple reflections, and source/receiver waveform effects) approximates the trace we would have recorded had our source and receiver been coincident at the midpoint between the actual source and receiver location. Consequently, Approximation 6 is but a minor approximation for primary energy with records subjected to proper dynamic corrections.

Unlike many equations arising in wave propagation, equation (4) can be evaluated exactly without having to assume that P is a large number of wavelengths away from S. By using the coordinate system of Figure 4, we reduce equation (4) to an integral about the boundary of S. As shown in Appendix II, this integral is

$$4\pi\varphi_p = zRf(p) \int_\theta \left[\frac{e^{-2pr_1/v}}{r_1^2} - \frac{e^{-2pr_2/v}}{r_2^2} \right] d\theta, \quad (5)$$

where θ, r_1, and r_2 are defined in Figure 4. If we define ξ as the distance from P to A (Figure 5), for the boundary of S not enclosing the origin (Figure 5A), equation (5) becomes

$$4\pi\varphi_{pA} = zRf(p) \int_\theta \frac{e^{-2p\xi/v}}{\xi^2} d\theta, \quad (6)$$

where the integration is such that the point A in Figure 5 traverses the boundary of S completely in a clockwise direction. The subscript pA will denote a source/receiver location P such that the boundary of S does not enclose the origin.[3]

---

[3] The origin is defined by the intersection of S (or S extended) with the normal to S (or S extended) passing through P.

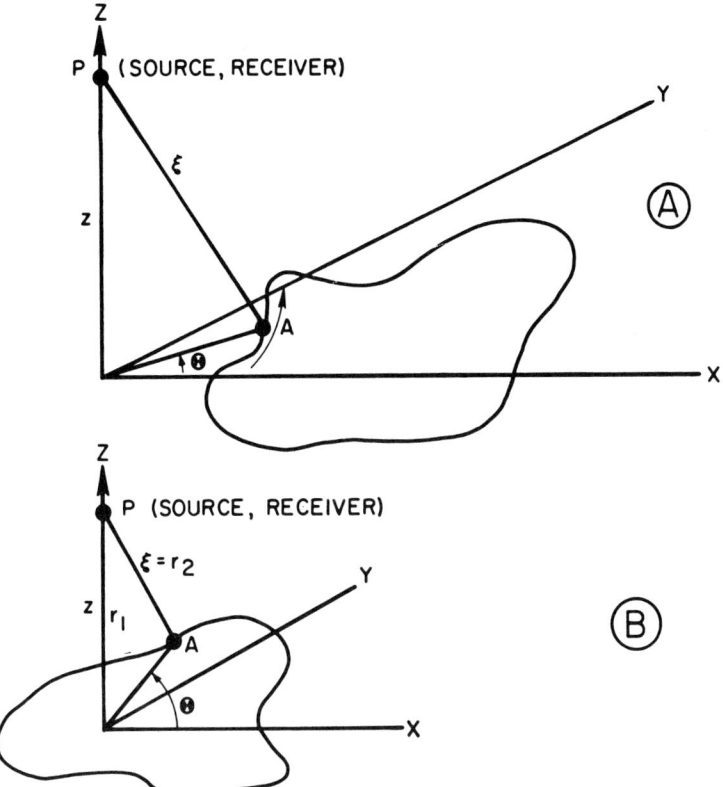

Fig. 5. Geometry for arbitrary shaped plane surface. The distance ξ is from P to a point A on the boundary of S. Full integration of equation (6) or (7) results when point A traverses the entire boundary clockwise. Figure 5A is the case when S does not enclose the origin (P not above S); Figure 5B is for the case of P above S.

If P is located such that the boundary of S does enclose the origin as in Figure 5B, referring to Figure 4, we see that $r_1 = Z =$ constant, $r_2 = \xi$, so that

$$4\pi\varphi_{pB} = \frac{2\pi R f(p) e^{-2pz/v}}{z} - z R f(p) \int_\theta \frac{e^{-2p\xi/v}}{\xi^2} d\theta. \quad (7)$$

Comparison of equations (7) and (3) (with r_o set to $2z$) shows that the first term in (7) is the simple reflection response (or image response) that would result if S were an infinite plane. For such a plane, ξ is infinite so that the second term is zero. Thus φ_{pB} reduces to the correct answer (as it must) for an infinite plane. Moreover, this observation permits us to identify the integrals in equations (6) and (7) as *diffracted arrivals*. Thus, *for P not above* S (φ_{pA}), *we have diffractions and no reflec-* *tions. For* P *above* S (φ_{pB}), *we have an infinite plane reflection plus a diffraction.*

While a number of general principles can be derived from equations (6) and (7) as they stand, it is perhaps more meaningful to examine these principles with respect to the particular case of rectilinear surfaces S. This examination is performed in the next section of this paper.

BEHAVIOR OF DIFFRACTIONS IN A TWO-DIMENSIONAL SUBSURFACE

The bulk of exploration seismology today involves shooting along a line. The data are frequently interpreted on the basis that there is no variation in the subsurface perpendicular to the line. Even though the above results are applicable to more general cases, in this section we will for the purpose of illustration assume that the dip is zero perpendicular to the "line of profiles," the latter consisting of a sequence of traces, each of which may be thought of (after dynamic

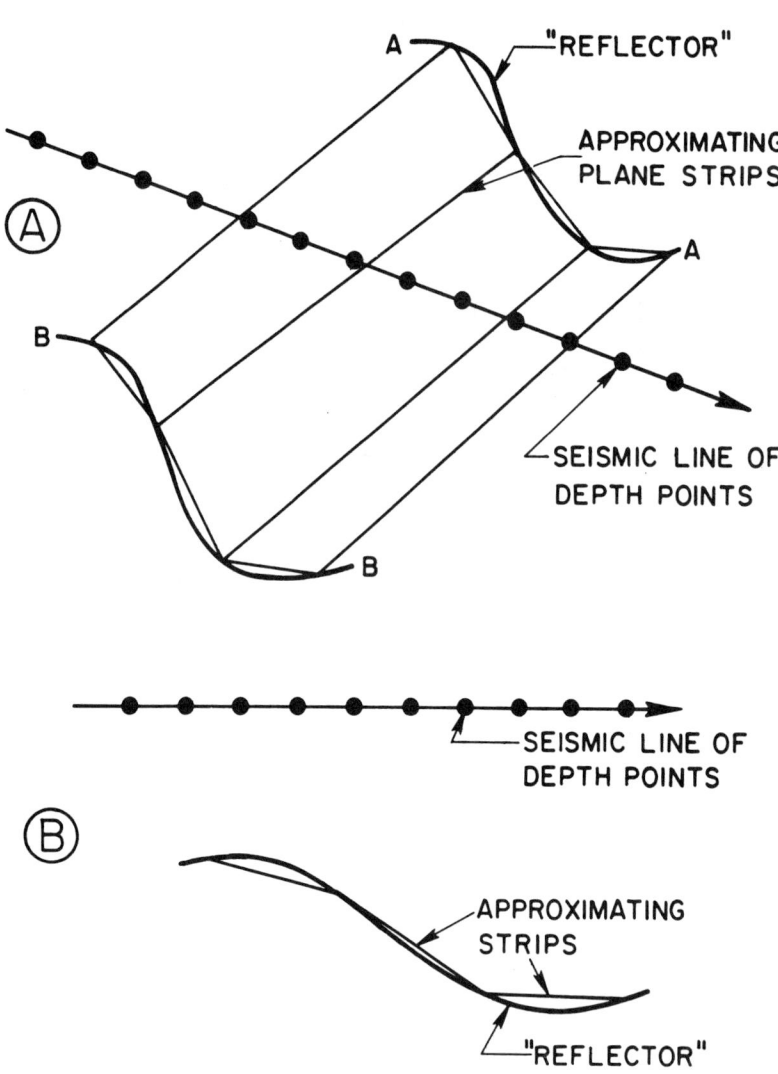

Fig. 6. Two dimensional subsurface. Figure 6A shows how a "reflecting" surface (with no dip normal to the seismic line) may be represented by plane strips. Figure 6B is an in-line vertical cross-section of Figure 6A.

corrections) as being recorded with a single shot into a single geophone at each depth point (at each point midway between an actual shot and receiver).

Under such assumptions, Figure 6 shows how an arbitrarily shaped two-dimensional reflector may be approximated by a sequence of plane strips. To compute the response for such a system, we need to evaluate equations (6) and (7) for each strip and sum the results. The response for other reflecting horizons would be found in the same way and simply added to that of the first.[4]

In order to evaluate equation (6) for a plane strip in which the edges AA and BB (Figure 6) are assumed to be at infinity and so have no contribution, we note that from Figure 7, ξ clearly can be expressed as a function of θ, so that in principle equation (6) can be integrated. The first conclusions we wish to draw, however, do not

[4] Since equations (6) and (7) assume v constant, we would in practice use the average velocity to each reflector.

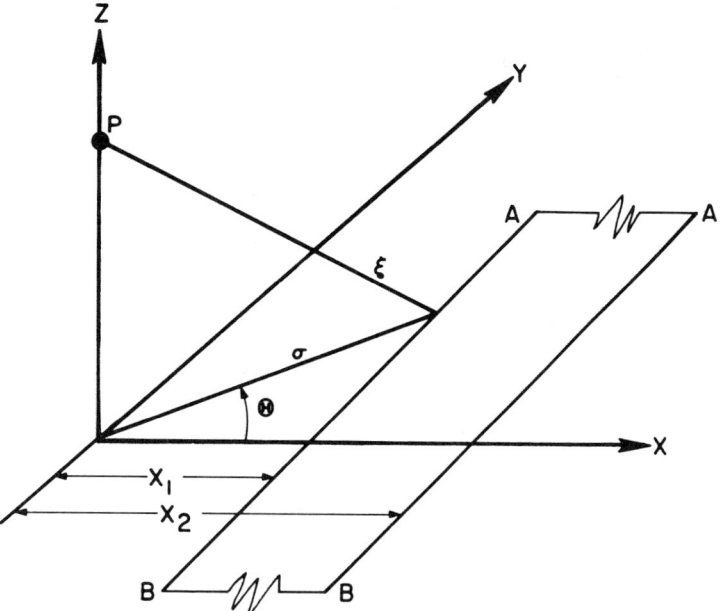

FIG. 7. Geometry for plane strip not enclosing origin. Strip is parallel to Y axis with edges distant x_1 and x_2 from the normal through P. The edges AA and BB are at infinity. If the strip has in-line dip, the line of depth points is not parallel to the X axis.

require that we actually perform the integration. Moreover, recall that φ_{pA} is the Lapace transform of our answer. As will be shown later in Appendix III, the time domain solution can be obtained directly from equation (6) without performing an integration.

Let D_2 be the contribution of equation (6) resulting from integrating from $-\pi/2$ to $+\pi/2$ along the left or x_1 edge in Figure 7 and let D_1 be the result from integrating from $-\pi/2$ to $+\pi/2$ along the right or x_2 edge. Then

$$\varphi_{pA} = D_2 - D_1. \qquad (8)$$

Clearly if $x_1 = x_2$, $D_1 = D_2$ and the response will be zero as it must be for a surface of zero area. If we let x_1 be negative so that P is above S, and if we let V denote the reflected term, clearly

$$\varphi_{pB} = V - D_2 - D_1. \qquad (9)$$

Comparing equations (8) and (9), we see that D_2 changes sign as P crosses the edge of the strip.

We can illustrate these last results by momentarily assuming S to have zero dip and imagining P to be moving parallel to S so that we go from the condition of equation (8) to that of equation (9). The interesting result is depicted in Figure 8. At each edge of the strip (at points A), *the diffraction must undergo a phase change of 180 degrees*. Physically this must be true, for if it were not, the response of the surface when it becomes of zero width would not be zero. The importance of this conclusion is that many discrete seismic diffractions we observe on dynamically corrected seismograms should behave in this manner. Moreover, the argument above is true if the reflector segment has dip.

Another physical argument that can lead to the same conclusion is that φ_p must be a slowly varying function of position. We shall instead use this argument to check our solution. Consider the behavior near one edge (region B in Figure 8). Just to the left of A we have D_2. Just to the right we have $(V - D_2)$. In order that φ_p be continuous, we must then have (close to the edge) $D_2 = V - D_2$ or

$$D_2 = V/2. \qquad (10)$$

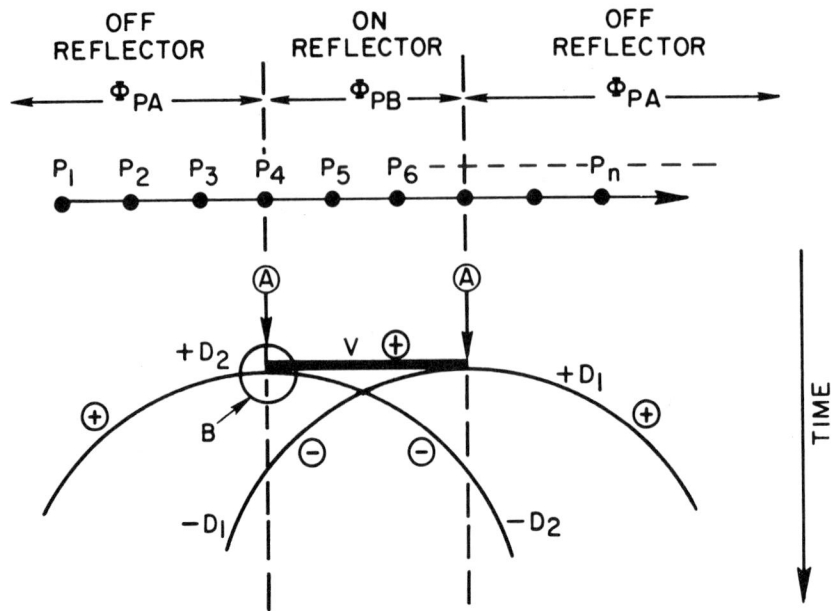

Fig. 8. Illustrating that a diffraction must undergo a 180 degree phase change on either side of a diffracting edge. P_1, P_2, etc., are successive source/receiver locations along a line. The diffractor is a plane strip with edges at A-A. D_1 and D_2 are diffractions; V is the reflection. The circled signs show the phases of the various parts of the response.

Therefore, *at the diffracting edge, the diffracted waveform is identical to the reflected waveform but is half the amplitude of the reflected waveform.* That our solution satisfies condition (10) can be readily seen from the definition of D_1, D_2, and V. We merely choose the x_1 edge for D_1 to be directly under P to verify this result from equations (6) and (7).

In practice (with a constant reflection coefficient and a normal seismic bandwidth) this effect will cause the reflection to appear as though its amplitude is decreasing as the diffraction edge is reached, since, at the diffraction edge, the amplitude of the total response is half the amplitude of the reflection response alone. This conclusion is also valid for a dipping reflector.

Further examination of the equations we have obtained can lead to more detailed conclusions. The most important conclusions of general applicability are the 180 degree phase change, the 50 percent amplitude reduction at the diffracting edge, and the fact that, near the diffracting edge, the diffraction waveform is the same as the reflection waveform. In the next section we will show a number of time-domain solutions for a few simple subsurfaces of interest.

EXAMPLES OF BANDWIDTH LIMITED DIFFRACTIONS

The results obtained so far in this paper have been in the form of Laplace transforms of the subsurface response. In practice, we are usually interested in bandwidth limited solutions in the time domain. As mentioned earlier, however, the transforms we have obtained can be directly inverted into the time domain in the form of closed solutions by the simple method described in Appendix III. For a nonimpulsive source, solutions of the form of equation (A-310) in Appendix III must be convolved with the source waveform, $g(t)$, as described in connection with equation (A-311). Since it is convenient to perform such convolutions on a digital computer, all examples in this section have been calculated on a computer, the computer in essence being used to evaluate equation (A-311) for two arbitrary $g(t)$ functions.

For simplicity of illustration, the examples are limited to geologic models which can be represented by large plane segments. The 4 models used are shown and described in Figures 9, 10, 11, and 12. For simplicity, we will designate these as "models 9, 10, 11, 12." The responses of these

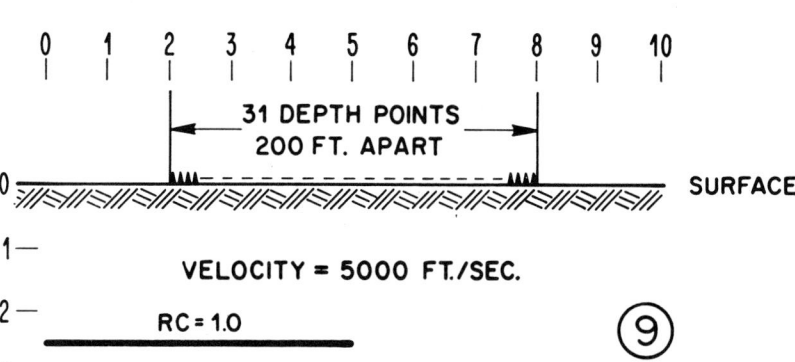

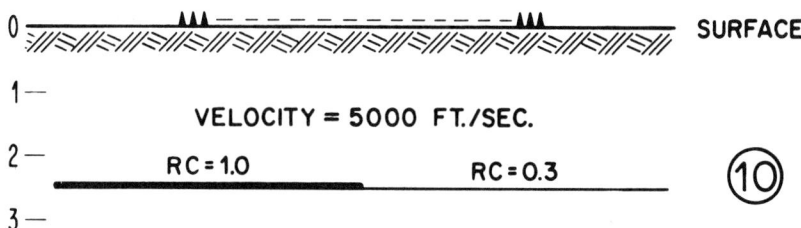

Figs. 9 and 10. Geometry for first two models. All models are "recorded" with an array consisting of 31 depth points spaced 200 ft apart as shown. The array is positioned so as to emphasize the response due to that part of the subsurface at a horizontal position of 5000 ft. Horizontal and vertical scales are in thousands of feet. All models have zero dip normal to the seismic line. Figure 9 shows a single reflector truncated at 5000 ft with a normalized reflection coefficient (RC) of unity. Figure 10 is the same, except that the reflector is extended to the right with a lower RC = 0.3.

models for a fairly wide-band source function are shown in Figures 13, 14, 15, and 16. In *all* these figures, the responses have for convenience been corrected for simple inverse spreading by multiplying by the quantity vt.

Figure 13, the response of model 9, clearly shows the 180 degree phase reversal and half-amplitude reduction predicted earlier. It is of interest to note that the rate at which the diffraction amplitude falls off is much greater than that of simple inverse spreading (this is because the diffracted energy does not actually emanate from the diffracting edge). We also note that there are no major changes in the shape of the diffracted waveform (this observation will be clearer when Figure 16 is examined).

Figure 14, the response of model 10, behaves as we would expect and merely illustrates that diffractions are produced by a change in reflection coefficient. Invoking the principles of Figure 8, it is clear that two diffractions, identical except as to amplitude and a 180 degree phase change, are destructively interfering so that the combined diffraction amplitude is fairly low. If the second segment of model 10 had an RC of 1.0 instead of 0.3, the two diffractions would exactly cancel leaving only the reflection.

Figure 15, the response of model 11, shows that the reflections are discontinuous. These reflections are joined, however, in a continuous manner (see the traces between 3600 and 5000 ft) by a multiple time branch consisting of the constructive interference of the two diffractions from the two plane segments.[5] Owing to this constructive interference, the diffraction amplitude joining the reflections is fairly constant.

[5] The discussion in connection with Figure 8 shows that the diffraction interference must be constructive.

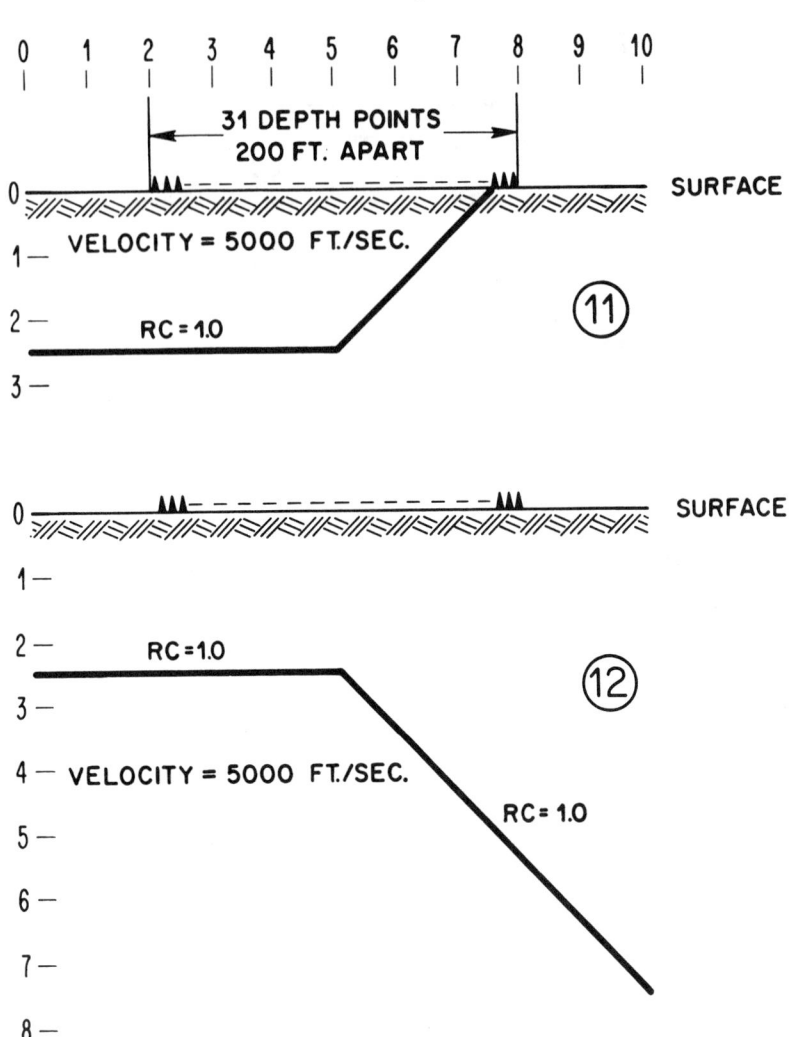

Figs. 11 and 12. Geometry for last two models. Up to 5000 ft, both models are identical to that of Figure 10. In Figure 11, the reflector is then bent upwards; in Figure 12, it is bent downwards.

The response of model 12 is shown in Figure 16. Here again, the two diffractions reinforce in the region between the two reflections. Moreover, there is little change in waveform with the result that the unwary interpreter may be led astray, particularly if he does not check his interpretation against the "curves of maximum convexity" described by Hagedoorn (1954).

In Figure 17, 18, 19, and 20 we have the same responses as Figures 13, 14, 15, and 16, this time, however, viewed with a rather narrow band source function. The general tendency is that the diffractions themselves are not so directly obvious and that consequently it would be possible to attribute their effects to other causes. In Figure 18, for example (for model 10), the diffractions are not immediately obvious. Note, however, that their presence causes a minimum to occur in the reflection amplitude at 5400 ft, a position which is to the right of the diffracting point at 5000 ft. (This corresponds to the loss of the trough at 5400 ft in Figure 14.) In Figure 19, the diffraction

(*Text continued on page 781*)

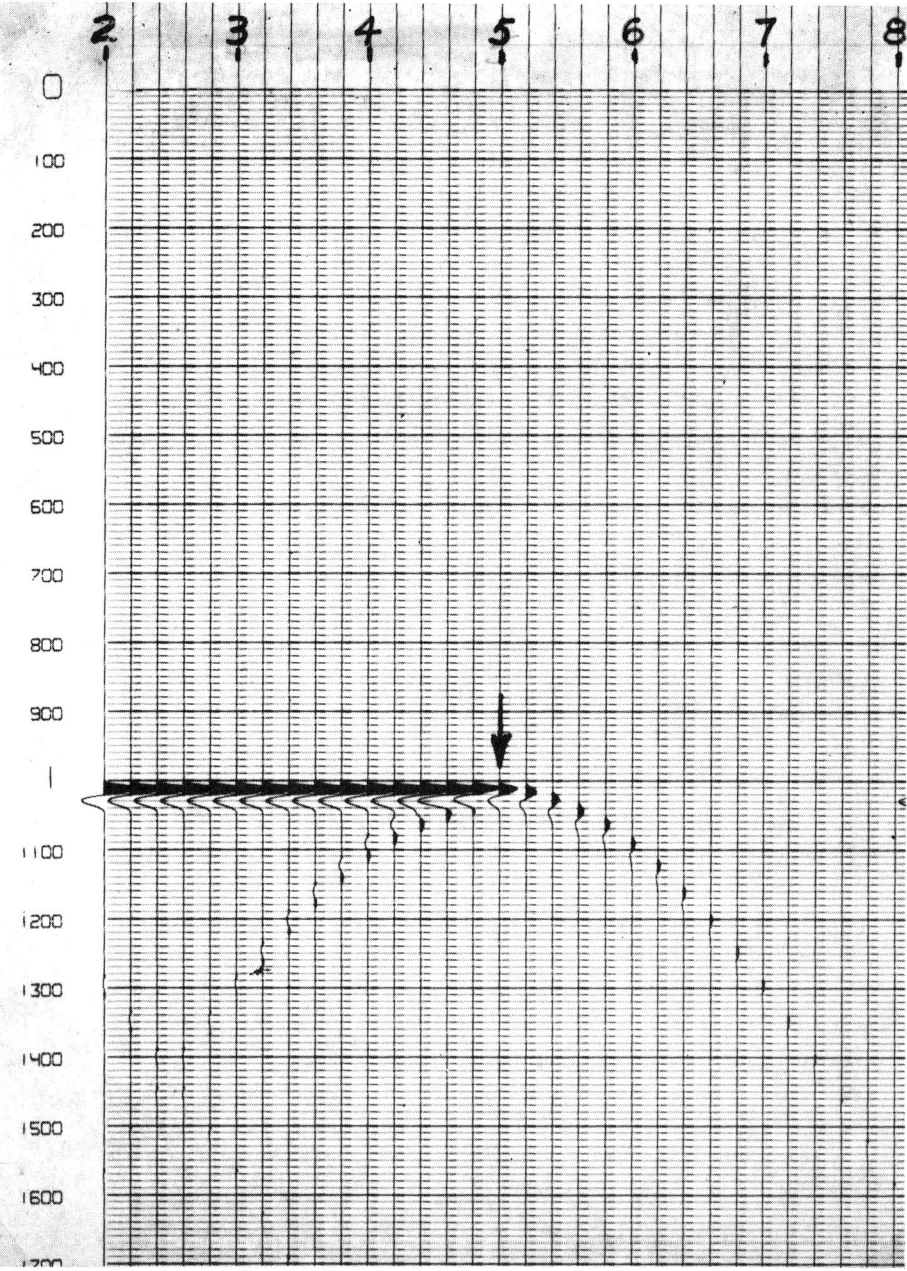

Fig. 13. Wide-band seismic response of model 9. The arrow indicates the lateral termination of the reflection. Note the 50 percent amplitude reduction at this point and the 180 degree phase change of the diffraction about this point. Note also that the diffraction waveform does not suffer any drastic changes but that its rate of fall off is considerably greater than that of simple inverse spreading (Figure 13 has already been corrected for inverse spreading).

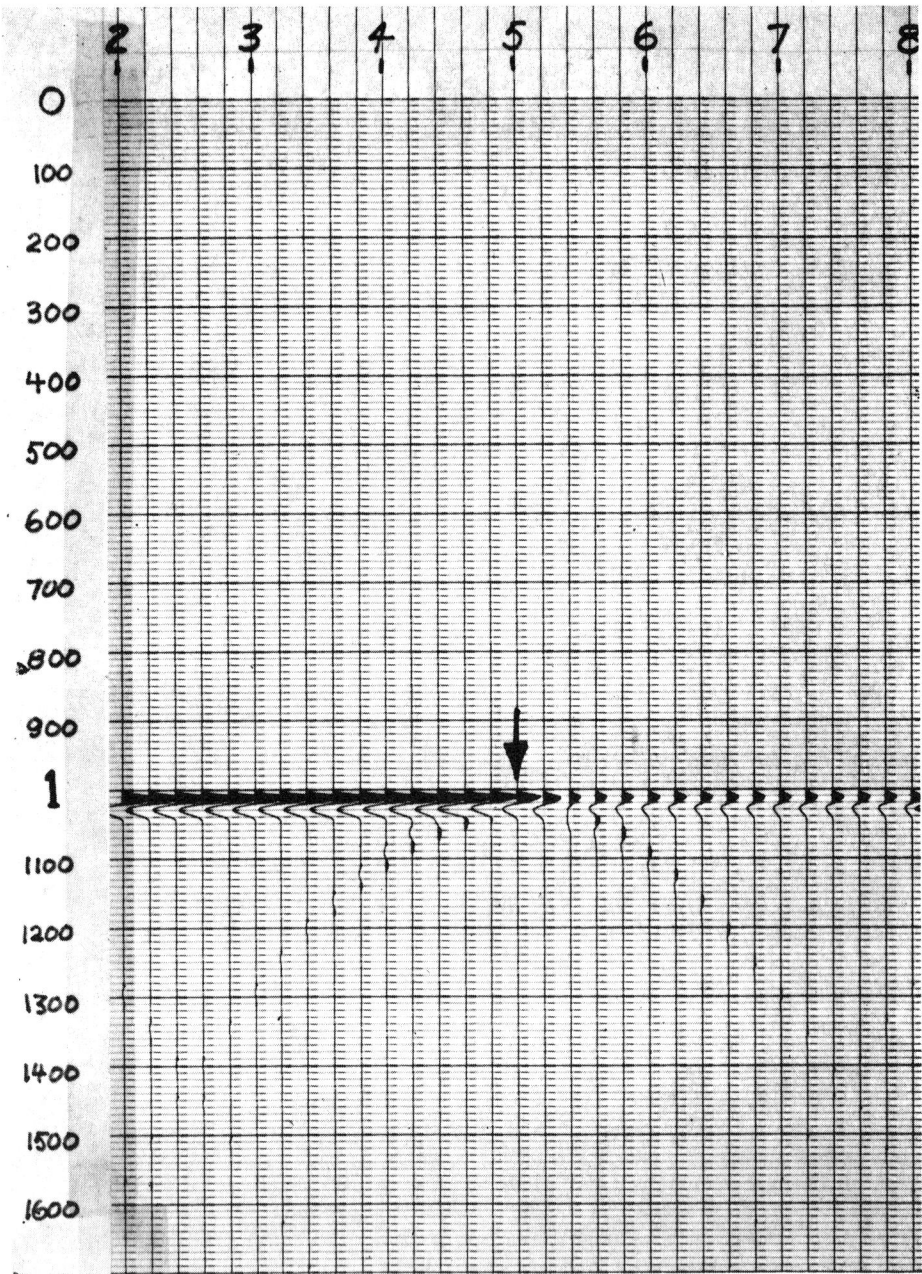

FIG. 14. Wide-band seismic response of model 10. The arrow indicates the point at which the reflection coefficient changes. The diffraction is of fairly low amplitude because it consists of the destructive interference of two diffractions (see text). The response has been corrected for inverse spreading.

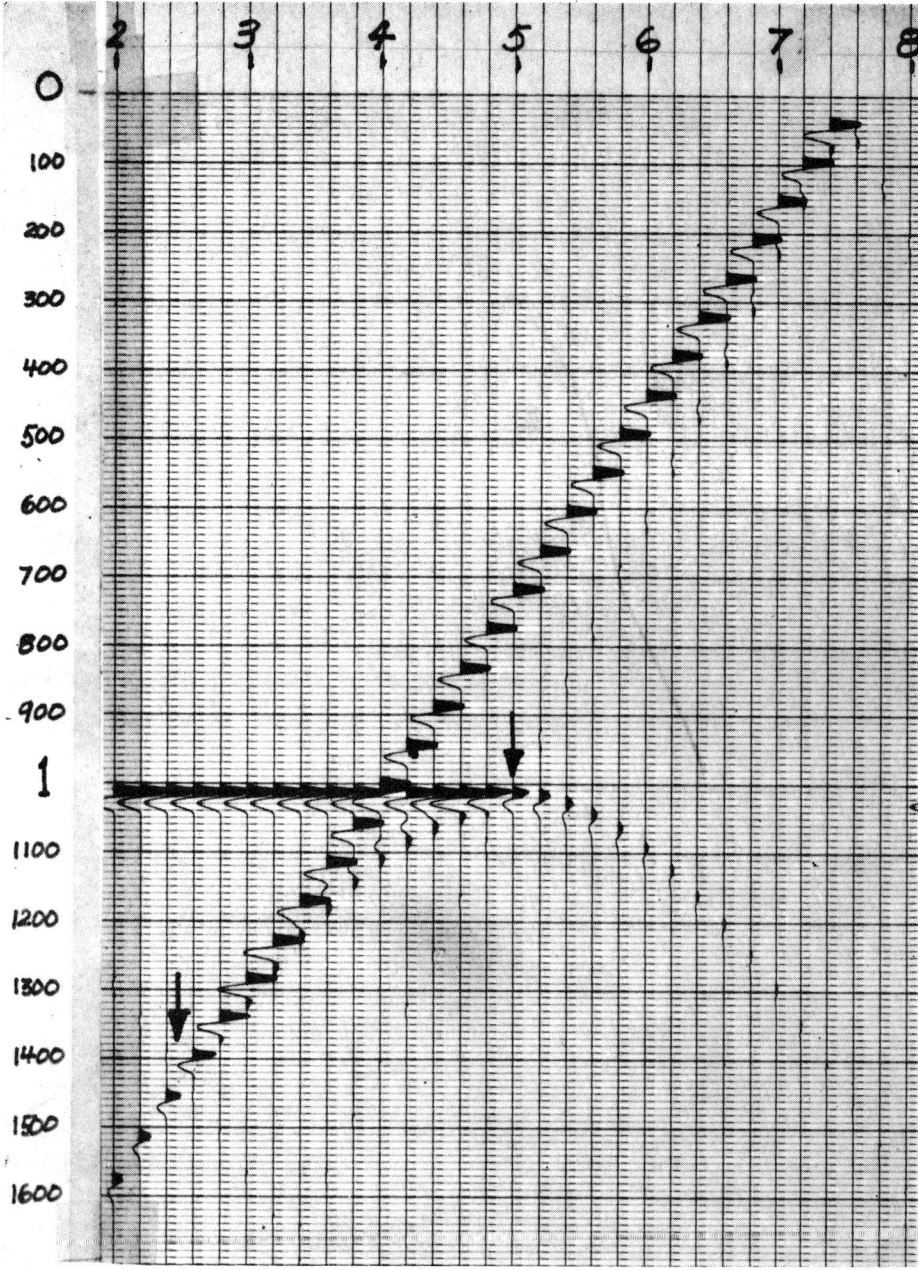

Fig. 15. Wide-band seismic response of model 11. The arrows mark the lateral positions at which the two reflections terminate. That part of the diffraction joining the two reflections is of fairly high amplitude because it consists of the constructive interference of two diffractions (see text). The response has been corrected for inverse spreading.

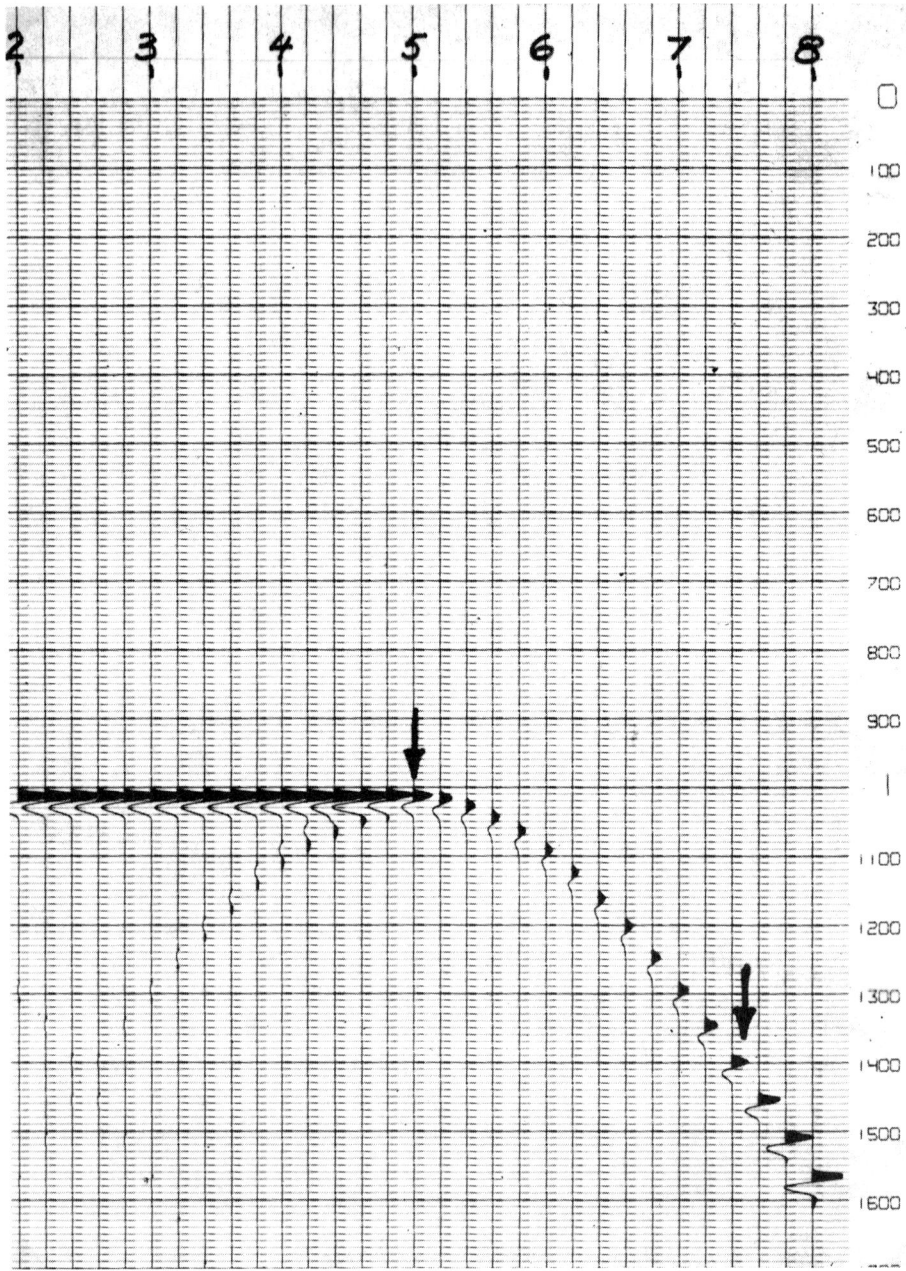

FIG. 16. Wide-band seismic response of model 12. The arrows mark the lateral positions at which the two reflections terminate. That part of the diffraction joining the reflections consists of the constructive interference of two diffractions and so is fairly large in amplitude. The response has been corrected for inverse spreading.

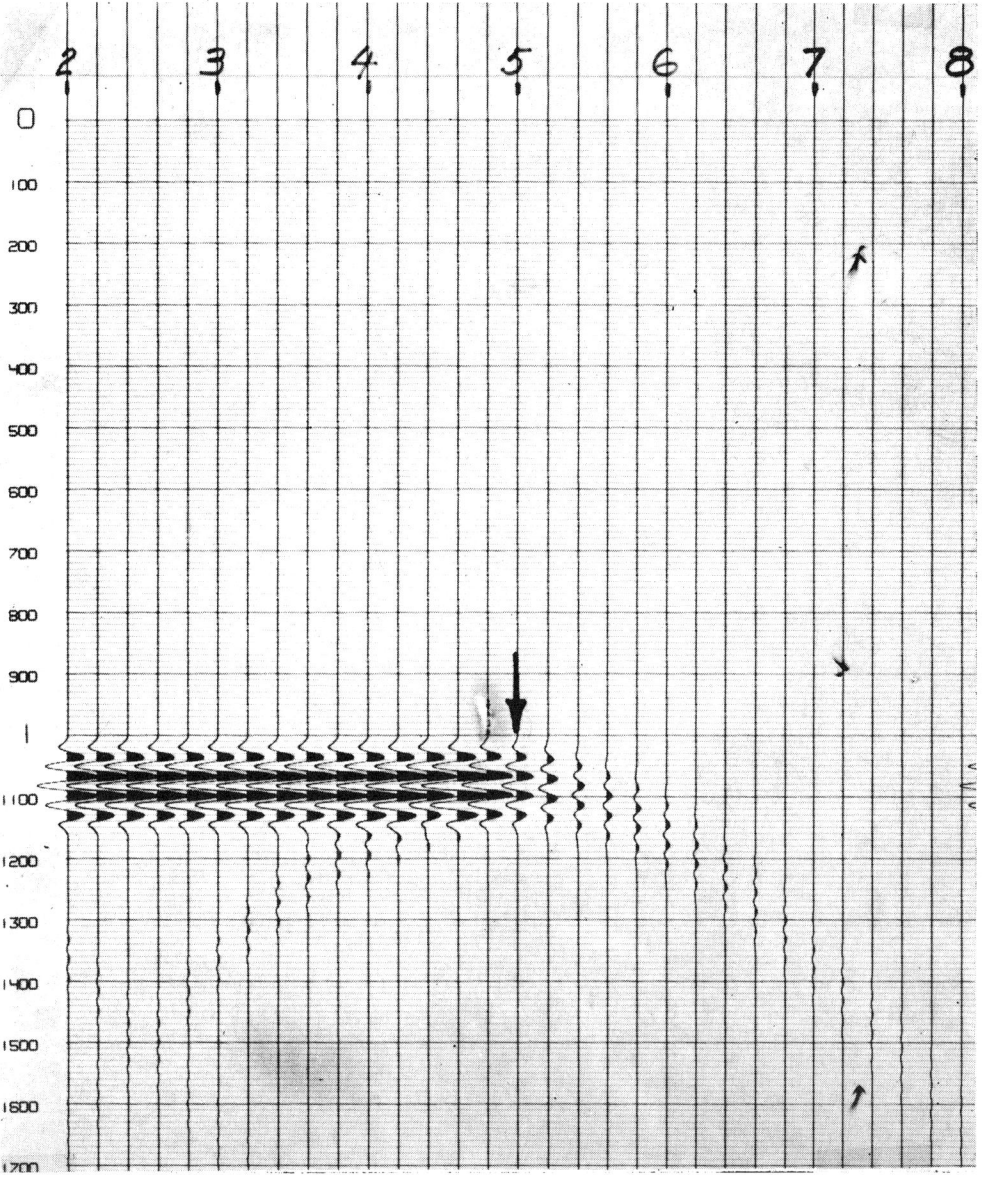

Fig. 17. Narrow-band seismic response of model 9. Compare to Figure 13. Although the isolated diffraction itself is not so obvious, its effects are still pronounced in that we still have a 50 percent amplitude reduction at the reflection termination (marked by the arrow).

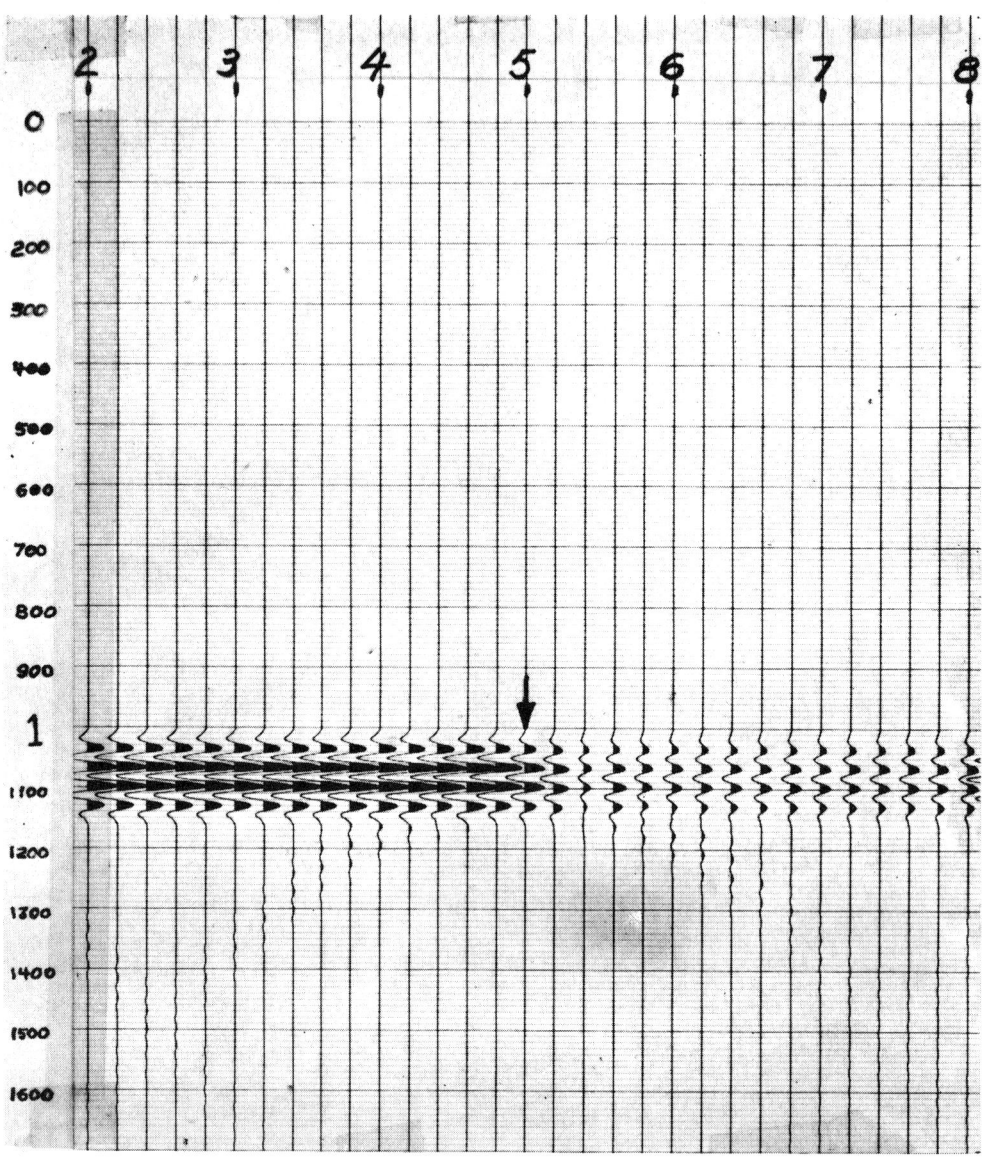

FIG. 18. Narrow-band seismic response of model 10. Compare to Figure 14. Note the reduction in trace amplitude two traces (400 ft) to the right of the reflection coefficient discontinuity (this latter point is marked by the arrow).

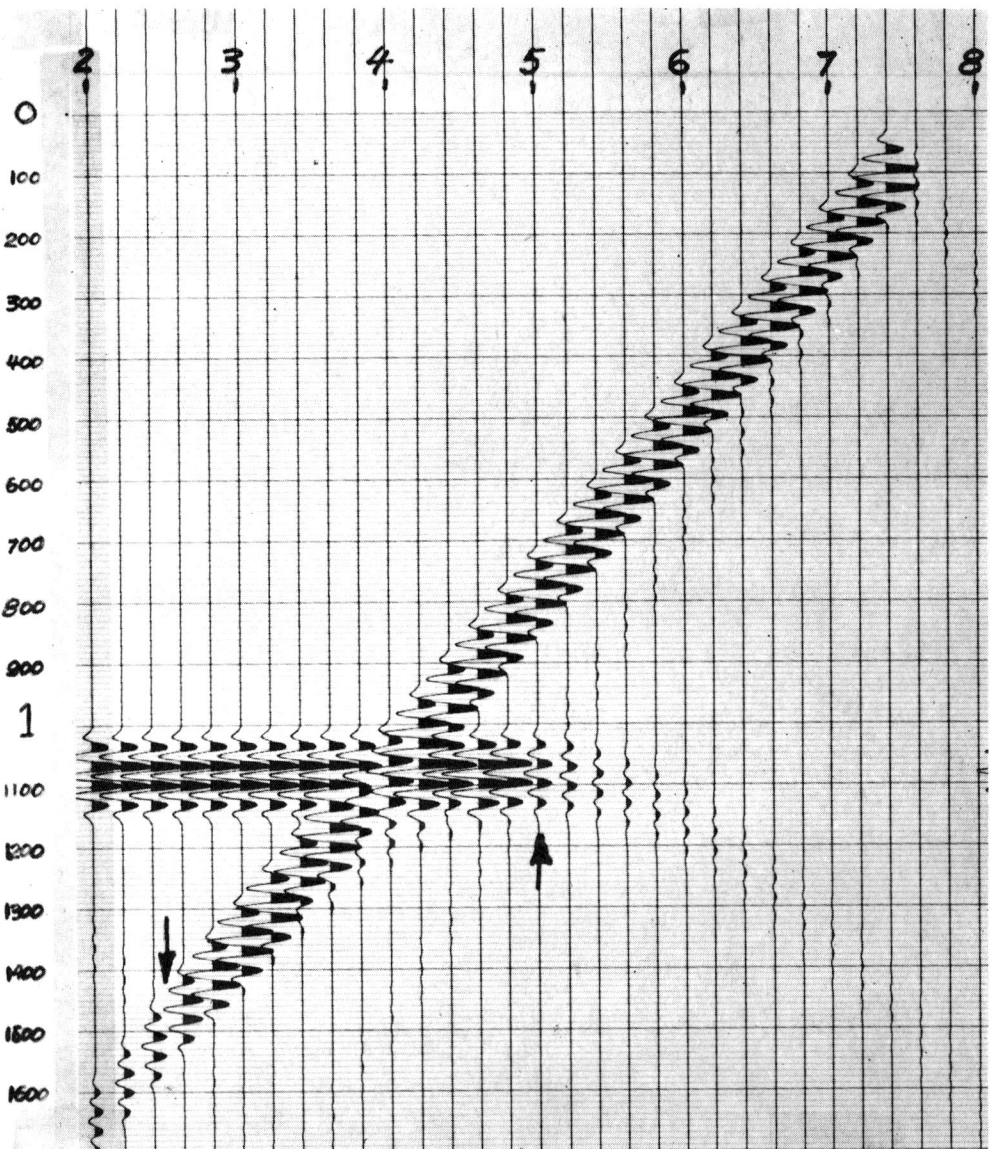

FIG. 19. Narrow-band seismic response of model 11. Compare to Figure 15. Note the complicated interference pattern involving reflections and diffractions. The arrows are as described in Figure 15.

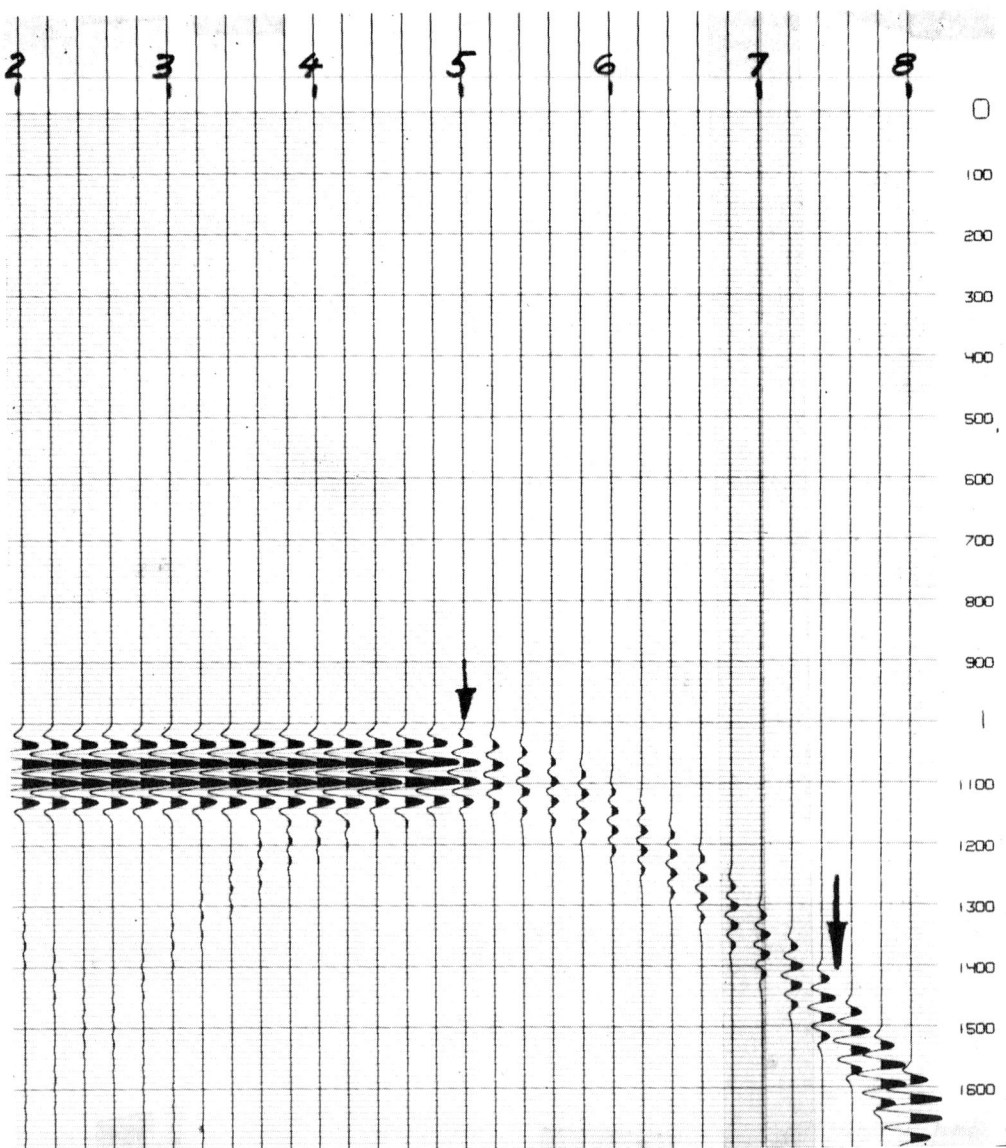

FIG. 20. Narrow-band seismic response of model 12. Compare to Figure 16. Observe that the predominant feature is a smooth, continuous variation from one reflection to another (the terminations of the reflections are shown by arrows).

mainly causes complex amplitude and phase changes, thus making a meaningful interpretation difficult. The salient features of Figure 20 remain unchanged.

CONCLUSIONS

A method has been devised for obtaining the point source diffraction and reflection response for an arbitrary acoustic subsurface having low reflectivity and constant velocity. Since the solution is in a closed form, the effect of changing various parameters in our model can be readily determined and studied.

We find that for a single truncated plane reflector, the phase of the diffraction suffers a reversal of 180 degrees on either side of the reflecting edge and that the total response at this edge is half the amplitude of the reflected part.

We have observed that any subsurface reflector other than an infinite plane of constant reflection coefficient will produce diffractions, the significance of the diffractions becoming greater the more the surface departs from being an infinite plane. We have also seen that diffractions are not produced from a diffracting edge: the entire reflector, not just its edge, produces the diffraction. This causes the amplitude of a diffraction from a single edge to decay more rapidly than would be predicted by simple inverse spreading.

Finally we note that the seismic response of a laterally discontinuous subsurface is laterally continuous—diffractions behave in such a way as to make the total response continuous. Thus sharp edges, abrupt truncations, faults, etc., cannot be determined by seeking abrupt changes on a seismogram. In looking for such features, an interpreter must bear in mind the behavior of the ever-present, but often obscure, diffractions.

ACKNOWLEDGMENT

The author would like to thank Mr. R. E. Doe of the Chevron Oil Field Research Company, Houston Laboratory, for kindly providing computer solutions for the examples shown in this paper.

REFERENCES

Biot, M. A., and Tolstoy, I., 1957, Formulation of wave propagation in infinite media by normal coordinates with an application to diffraction: J. Acoust. Soc. Am., v. 29, p. 381.
Hagedoorn, J. G., 1954, A process of seismic reflection interpretation: Geophys. Prosp., v. 2, p. 85–127.
Howes, E. T., et al, 1953, Seismic model study: J. Acoust. Soc. of Am., v. 25, p. 915–921.
Krey, Theodor, 1952, The significance of diffraction in the investigation of faults: Geophysics, v. 17, p. 843–858.
Kunz, B. F. J., 1960. Diffraction problems in fault interpretation: Geophys. Prosp., v. 8, p. 381–388.
Ledoux, Y., 1957, Quelques exemples de diffractions en sismique-refraction et leur application a la determination des vitesses verticales: Geophys. Prosp., v. 5, p. 392–406.
Longhurst, R. S., 1957, Geometrical and physical optics: London, Longman Green and Co.

APPENDIX I—INCORPORATION OF φ_s INTO HELMHOLTZ'S EQUATION

Referring to Figure 3, we have

$$\frac{\partial r}{\partial n} = \frac{z}{r}, \quad \frac{\partial r_0}{\partial n} = -\frac{z_0}{r},$$

$$\frac{\partial \left(\frac{1}{r}\right)}{\partial n} = -\frac{z}{r^3}. \tag{A-11}$$

The second expression has a negative sign because n is the downward normal and r_0 (with reference to Q') is positive upwards. We also have from equation (3)

$$\frac{\partial \varphi_s}{\partial n} = \frac{\partial \varphi_s}{\partial r_0} \frac{\partial r_0}{\partial n}$$

$$= \frac{z_0 R f(p) e^{-pr_0/v}}{r_0} \left[\frac{1}{r_0^2} + \frac{p}{r_0 v}\right] \tag{A-12}$$

If we put equations (A-11) and (A-12) into equation (2) and let P and Q become coincident ($z_0 = z$, $r_0 = r$), we have

$$2\pi\varphi_p = zRf(p) \int_S e^{-2pr/v} \left[\frac{1}{r^4} + \frac{p}{vr^3}\right] dS. \tag{A-13}$$

APPENDIX II—REDUCTION OF HELMHOLTZ'S EQUATION TO LINE INTEGRAL

From Figure 4, $r^2 = z^2 + \sigma^2$, $rdr = \sigma d\sigma$, so that

$$dS = \sigma d\sigma d\theta = rdrd\theta.$$

Thus, equation (4) becomes

$$2\pi\varphi_p = zRf(p) \int_\theta \int_{r_1}^{r_2} e^{-2pr/v} \cdot \left[\frac{1}{r^4} + \frac{p}{vr^3}\right] rdrd\theta. \tag{A-21}$$

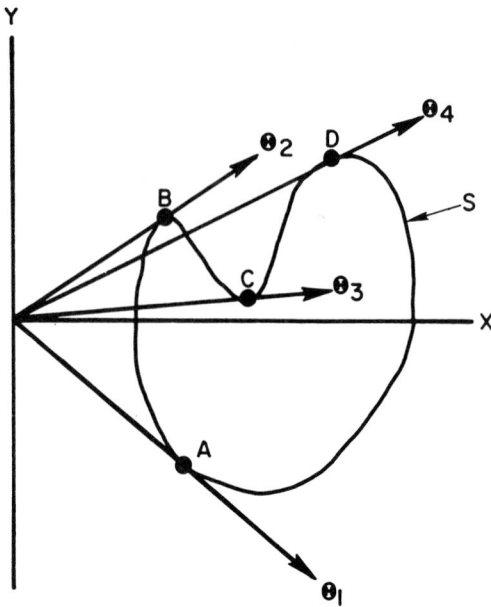

Fig. A-31. In evaluating equation (A-31), we must note that ξ may not be a single-valued function of θ. The equation can be broken into single-valued regions by performing the integrals θ_1 to θ_2, θ_2 to θ_3, θ_3 to θ_4, and θ_4 to θ_1. The results will involve the times τ_A, τ_B, τ_C, and τ_D which are the two-way traveltimes from P to A, B, C, and D, respectively.

Integration by parts yields

$$\int_{r_1}^{r_2} \frac{e^{-2pr/v}}{r^3} dr = \frac{-e^{-2pr/v}}{2r^2}\bigg|_{r_1}^{r_2} - \frac{p}{v}\int_{r_1}^{r_2} \frac{e^{-2pr/v}}{r^2} dr. \quad (A\text{-}22)$$

Substituting (A-22) into (A-21) results in

$$4\pi\varphi_p = zRf(p)\int_\theta \left(\frac{e^{-2pr_1/v}}{r_1^2} - \frac{e^{-2pr_2/v}}{r_2^2}\right) d\theta. \quad (A\text{-}23)$$

APPENDIX III—TIME DOMAIN SOLUTION

From equations (6) and (7) in the main text, we must evaluate integrals

$$\overline{L}(p) = \int_\theta \frac{e^{-2p\xi/v}}{\xi^2} d\theta. \quad (A\text{-}31)$$

If $\overline{L}(p)$ could be inverted to $L(t)$, $f(p)$ in equations (6) and (7) could be incorporated by simple time-domain convolution.

Normally ξ is not a single-valued function of θ. This difficulty can be overcome by the method described in Figure A-31. Thus, we must solve

$$\overline{L}(p) = \int_{\theta_1}^{\theta_2} \frac{e^{-2p\xi/v}}{\xi^2} d\theta, \quad (A\text{-}32)$$

where θ_1, and θ_2 are the bounds of any single-valued region. Let us make the change of variable $t = 2\xi/v$. Then

$$\begin{aligned}\overline{L}(p) &= \frac{4}{v^2}\int_{\theta_1}^{\theta_2} \frac{e^{-pt}}{t^2} d\theta \\ &= \frac{4}{v^2}\int_{\tau_1}^{\tau_2} \frac{e^{-pt}}{t^2} \frac{d\theta}{dt} dt,\end{aligned} \quad (A\text{-}33)$$

provided that $d\theta/dt$ is not infinite everywhere (if it is infinite, ξ is constant and equation (A-32) can be evaluated directly). The quantities τ_1 and τ_2 are simply the two-way traveltimes from the points on the boundary of S corresponding to θ_1 and θ_2 as described in Figure A-31. Within a region τ_1, τ_2 as defined in Figure A-31, t may not be a single-valued function of θ. If this is the case, a further subdivision of each τ_1, τ_2 region will be required so that within each subdivision t is a single-valued function of θ. The necessary subdividing of the boundary can always be performed.

Now by definition of the Laplace transform,

$$\overline{L}(p) \equiv \int_{\tau_1}^{\tau_2} e^{-pt} L(t) dt, \quad (A\text{-}34)$$

where $L(t)$ is the inverse of $\overline{L}(p)$ and is zero for t outside the range $\tau_1 < t < \tau_2$. Consequently, from (A-33) we have immediately that

$$L(t) = \frac{4}{(vt)^2} \frac{d\theta}{dt} \quad \tau_1 < t < \tau_2. \quad (A\text{-}35)$$

Thus, to find $L(t)$, we merely subdivide the S boundary into single-valued regions, τ_1 and τ_2 being the two-way times from P to the edges of these regions. We then write ξ as a function of θ, replace ξ by $vt/2$, and calculate $d\theta/dt$. Since $L(t)$ is an impulse response, it will typically have a singularity at $t = \tau_1$ and $t = \tau_2$. This singularity will be well behaved and will disappear when convolved with a band-limited excitation function.

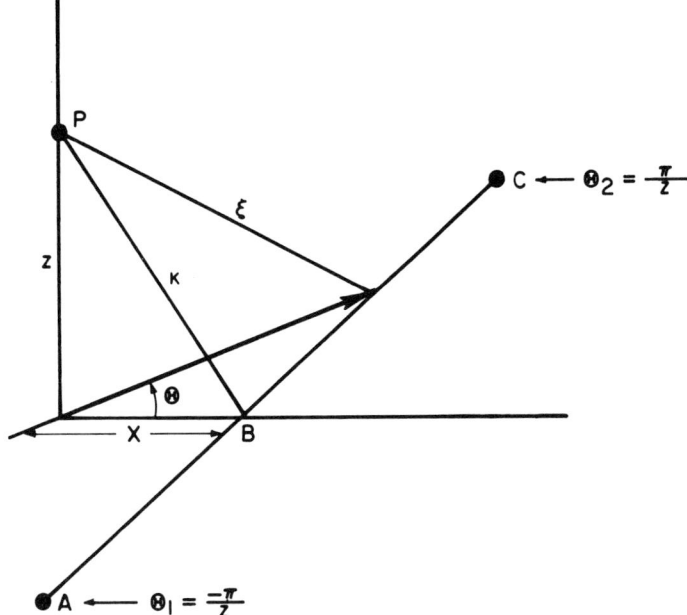

FIG. A-32. Geometry for time-domain solution of diffracting edge. The edge must be divided into the two subregions AB, BC in order to define single-valued subregions.

We will illustrate the method described above for one example—the straight diffracting edge described in the text. The geometry is shown in Figure A-32. The distance K is independent of θ. From Figure A-32, we have

$$\xi^2 = (x \tan \theta)^2 + K^2. \quad (A\text{-}36)$$

Setting $t = 2\xi/v$, $\tau_x = 2x/v$, and $\tau = 2k/v$, we have

$$t^2 = \tau_x^2 \tan^2 \theta + \tau^2 \quad (A\text{-}37)$$

Thus,

$$\frac{d\theta}{dt} = \frac{t \cos^2 \theta}{\tau_x^2 \tan \theta}. \quad (A\text{-}38)$$

By use of equation (A-37), we can express $\cos^2 \theta$ and $\tan \theta$ in terms of t, τ_x, and τ to obtain

$$\frac{d\theta}{dt} = \frac{t\tau_x}{(t^2 + \tau_x^2 - \tau^2)\sqrt{t^2 - \tau^2}}. \quad (A\text{-}39)$$

Note that τ_x is the two-way time from the origin to point B (Figure A-32) and τ is the two-way time from P to B. The solution is singular at $t = \tau$ as one would expect.

Before using equation (A-39) to obtain our complete solution, we must consider the single-valued regions on the edge ABC in Figure A-32. From equation (A-32) the range of θ is $-\pi/2$ (at A) to $+\pi/2$ (at C): A and C are at infinity. Equation (A-37) shows that t is not a single-valued function of θ over the entire range ABC. Over the range AB or the range BC, it is single valued. Thus, we integrate from B to C so that $\tau_1 = \tau$ and $\tau_2 = \infty$. Moreover, it is clear from the symmetry of this problem that the integral from A to B is the same as that from B to C, so that we merely need twice the integral from B to C. Thus, we have, from equation (A-35) that

$$L(t) = 8\tau_x/[v^2 t(t^2 + \tau_x^2 - \tau^2)\sqrt{t^2 - \tau^2}],$$
$$\tau > t, \quad (A\text{-}310)$$
$$= 0, \quad t < \tau.$$

Consequently, from equations (6) and (8) in the main text we have

$$D_2(t) = \left[\frac{2Rz\tau_x}{\pi v^2 t(t^2 + \tau_x^2 - \tau^2)\sqrt{t^2 - \tau^2}}\right] \quad (A\text{-}311)$$
$$* g(t),$$

where $g(t)$ is the inverse of the excitation function $f(p)$ and where $*$ denotes convolution.

From the main text, expression (A-310) should become an impulse as τ_x approaches zero. Defining an impulse as a function of infinite value at the singularity, finite area over the singularity, and zero elsewhere, we can, by noting that $\tau^2 = \tau_x^2 + (2z/v)^2$, indeed show that expression (A-310) reduces to an impulse by taking the limit as τ_x approaches zero, or that expression (A-311) reduces to

$$Rg(t - 2z/v)/4z,$$

namely half the reflection waveform, as must be the case from our earlier discussion in the text.

The significance of expression (A-310) is that the shape of the diffraction waveform differs more and more from that of the reflection waveform as we move away from the diffraction edge on a time section (as we move away from B in Figure 8).

Owing to the convolution in equation (A-311), realistic solutions involving bandlimited $g(t)$ are best evaluated on a computer. Examples of computer solutions for some simple structures are given in the main text.

Migration of seismic data from inhomogeneous media

Les Hatton*, Ken Larner‡, and Bruce S. Gibson‡

ABSTRACT

Because conventional time-migration algorithms are founded on the implicit assumption of locally lateral homogeneity, they leave events mispositioned when overburden velocity varies laterally. The ray-theoretical depth migration procedure of Hubral often can provide adequate first-order corrections for such position errors. Complex geologic structure, however, can so severely distort wavefronts that resulting time-migrated sections may be barely interpretable and thus not readily correctable. A more accurate, wave-theoretical approach to depth migration then becomes essential to image the subsurface properly. This approach, which transforms an unmigrated time section directly into migrated depth, more completely honors the wave equation for a medium in which variations in interval velocity and details of structural shape govern wave propagation. Where geologic structure is complicated, however, we usually lack an accurate velocity model. It is important, therefore, to understand the sensitivity of depth migration to velocity errors and, in particular, to assess whether it is justified to go to the added effort of doing depth migration.

We show a synthetic data example in which the wave-theoretical approach to depth migration properly images deep reflections that are poorly resolved and left distorted by either time migration or ray-theoretical depth migration. These imaging results are, moreover, surprisingly insensitive to errors introduced into the velocity model. Application to one field data example demonstrates the superior treatment of amplitude and waveform by wave-theoretical depth migration. In a second data example, deep reflections are so influenced by anomalous overburden structure that the only valid alternative to performing wave-theoretical depth migration is simply to convert the unmigrated data to depth.

When the overburden is laterally variable, conventional time migration of unstacked data can be as destructive to steeply dipping reflections as is CDP stacking prior to migration. A schematic example illustrates that when migration of unstacked data is judged necessary, it should normally be performed as a depth migration.

INTRODUCTION

In many areas of interest, seismic data are collected over geologic structures that have substantial lateral velocity variation. Very often, lateral inhomogeneity is directly related to steeply dipping beds and, in processing the seismic data, migration will almost always be performed.

The importance of velocity estimates to the success of migration has been well documented (e.g., use of a migration velocity that is too low results in the incomplete collapse of diffractions and the insufficient movement of dipping reflections). When the medium is inhomogeneous, the proper specification of velocity is of even greater importance. In that case, the geophysicist must make a commitment to a detailed model of overburden velocities in order to migrate properly. Furthermore, details of the migration algorithm itself must be carefully considered because a sophisticated migration algorithm is required to honor the detailed velocity information.

Hubral (1977) showed that the Kirchhoff summation migration of data from laterally inhomogeneous media fails to position reflected events properly. Larner et al (1981, this issue) show further that migration by *any* conventional technique cannot properly position subsurface features when the overburden has substantial lateral variation in velocity. Errors in position result from approximations made to the scalar wave equation, the foundation of all migration techniques commonly used today (finite-difference, Kirchhoff summation, or frequency domain). Approximations can be identified in each technique to explain why each approach misplaces reflected events in about the same way.

Even though they provide for gross lateral variation in velocity, conventional migration techniques fail to position reflections properly because all techniques include the implicit assumption that, locally, the velocity in the medium does not change in the horizontal direction. Following Hubral, we shall refer to migration algorithms that assume such local homogeneity as *time-migration* algorithms. A common characteristic of time-migration techniques is that their direct output is a seismic section in time.

Hubral's observation was that time-migration algorithms position reflected events at locations that have a simple relationship to their true locations. He identified those locations as the surface terminations of *image raypaths*. Exploiting Hubral's work, Larner

---

Presented at the 48th Annual International SEG Meeting October 31, 1978, in San Francisco. Manuscript received by the Editor November 29, 1979; revised manuscript received September 4, 1980.
*Formerly Western Geophysical Co., Houston; presently Merlin Geophysical Co., Morris House, Commercial Way, Woking, Surrey, England.
‡Western Geophysical Co., P.O. Box 2469, Houston, TX 77001.
0016-8033/81/0501—751$03.00. © 1981 Society of Exploration Geophysicists. All rights reserved.

et al (1981) present a ray-theoretical solution to the mispositioning problem.

In the ray-theoretical approach, the first step is a standard time migration. Next, velocities are specified for intervals between prominent horizons on the time-migrated section. Such a velocity model provides sufficient information to trace image raypaths through the medium. (Image raypaths leave the surface vertically and obey Snell's law at layer interfaces; velocity contrasts across dipping horizons thus cause these raypaths to be deflected laterally.) In the second step, the time-migrated section is converted to depth by stretching traces along the image raypaths. This process is called *depth migration* because the result is a depth section with properly positioned reflections.

An important aspect of ray-theoretical depth migration is that one need not adopt a final, detailed interval-velocity model until time migration has clarified the subsurface structure. Nevertheless, any velocity model derived for field data will be imperfect because typically it is constructed either from isolated well logs or from seismic velocity analyses that are themselves dependent upon a model. Larner et al (1981) use a Monte Carlo numerical experiment to demonstrate that where the subsurface is complex and a reasonably accurate velocity model can be constructed, ray-theoretical depth migration is a process of primary importance, i.e., corrections in lateral position can be significant and are subject to a relatively small probable error.

That two-step approach is, however, an artificial division of the full migration process. Although the ray-theoretical technique has been successful for field cases of moderate complexity, we shall exhibit a simple but severely inhomogeneous velocity model for which time migration cannot image the subsurface correctly, and thus ray-theoretical depth migration is inappropriate.

We shall then examine the scalar wave equation to reveal the source of positioning and imaging errors in time migration, and we suggest a cure. The proper treatment of the wave equation yields an algorithm that properly images an unmigrated time section and maps it directly into depth in one step. This technique we call *wave-theoretical depth migration* because of its direct development from the scalar wave equation. Superior treatment of both synthetic and field data by this wave-theoretical approach demonstrates that the technique honors the essentials of the scalar wave equation for inhomogeneous media. As in the ray-theoretical approach, however, we must ask again whether the added effort and cost of this more sophisticated approach are justified in the presence of imperfect velocity information. Applications of the technique with intentionally incorrect velocity models demonstrate a fortunate insensitivity to velocity uncertainty.

However, this extended migration algorithm does not properly account for all the effects of velocity inhomogeneity. In complex media, for example, raypath distortion can be so severe that conventional CDP stacking destroys much reflection information. Subsequent migration then cannot reconstruct an image of the subsurface. The conventional solution to such a problem is to migrate the unstacked data. We shall, however, demonstrate schematically that if the data are merely *time* migrated and then stacked, reflections can be damaged as much as in conventional CDP stacking. The proper solution is to *depth* migrate the unstacked, variable-offset data.

We observe that where the subsurface is inhomogeneous and structurally complex, it is usually complex in all three dimensions. The migration of data from inhomogeneous media is thus a problem of three-dimensional (3-D) migration. While we shall not address 3-D depth migration here, we see it as important and inevitable future work.

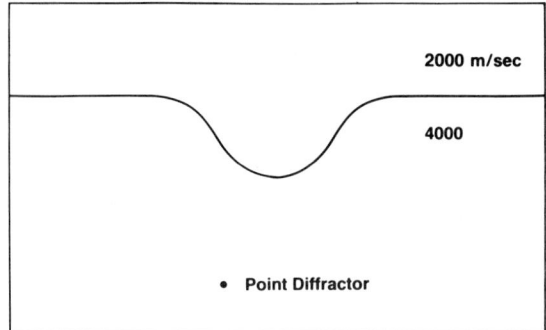

FIG. 1a. Depth model depicting a subsurface scatterer beneath an anomaly in the overburden.

SHORTCOMINGS IN RAY-THEORETICAL DEPTH MIGRATION

The success of ray-theoretical depth migration is predicated on the ability of time migration to form an adequate image of subsurface reflectors. When the medium is homogeneous, time migration is accurate both in positioning features and in imaging them. In addition, Larner et al (1981) indicate that for certain types of lateral heterogeneity, time migration mispositions reflected energy but does not severely degrade the quality of the image. We now show an example of inhomogeneity in which the issue of mispositioning is subordinate; the principal failure is the construction of the image.

Figure 1a shows a point diffractor in a two-layer medium. The interface between the two layers has a smooth depression with steep flanks. By ray tracing from the point diffractor, we obtain the traveltimes of its diffraction pattern at the surface (Figure 1b). The complex pattern in Figure 1b would be approximately that recorded by a coincident source-receiver survey over this model. For this example, the diffraction pattern is split into two hyperbola-like features. Perhaps worse yet, these features exhibit such large curvatures that their proper migration would seem to require

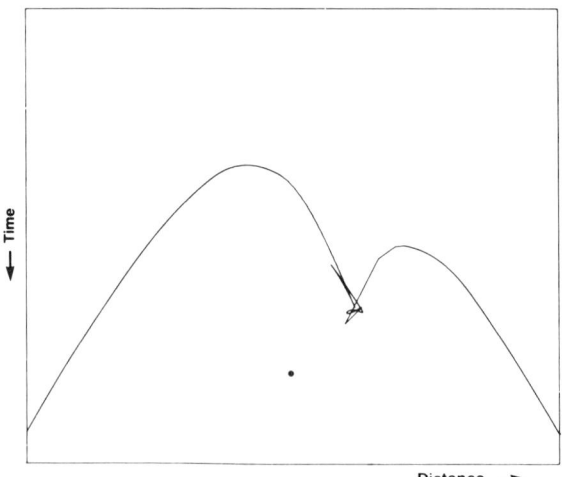

FIG. 1b. Unmigrated time section displaying the diffraction pattern, the curve of two-way traveltime between source-receiver points at the earth's surface, and the scatterer in Figure 1a. The dot marks the correct lateral position of the point scatterer.

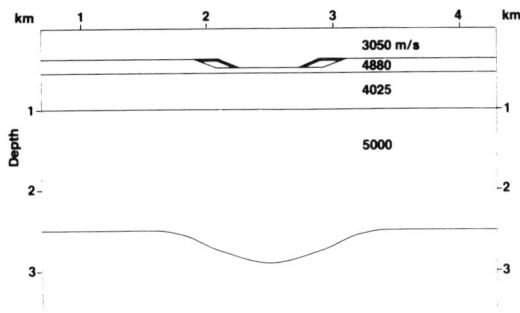

FIG. 2a. Depth model. The heavy lines on the upper interface indicate a perturbation of the original model used in the velocity sensitivity tests. The reflection coefficient of the interface between the 4880 and 4025-m/sec layers is zero. Reflection coefficients at all other interfaces are identical to one another.

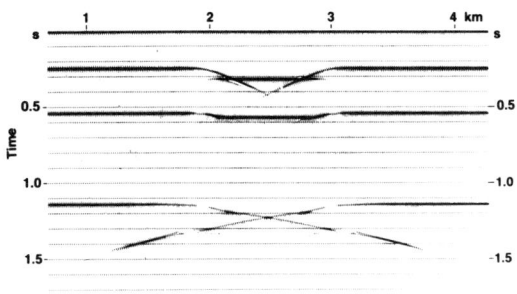

FIG. 2b. Normal-incidence time section for the depth model in Figure 2a. This zero-offset section was generated by the diffraction response technique of Trorey (1970) as modified by Larson and Hilterman (1976). Note the phantom diffractions on the second and third horizons.

velocities that are ridiculously low relative to the average velocity for the medium.

Imagine time migrating the intricate pattern of Figure 1b by the Kirchhoff technique of summation along hyperbolas. (The results obtained by any other method would be quite similar.) The best we might hope for is a partial collapse of energy near the two peaks of the pattern. In fact, the result will be more confused than that because of the complex of events between the peaks; the image after time migration will not remotely resemble that of a point scatterer. Subsequent depth conversion (whether conventional or image ray) will not remedy this situation. Clearly, what is required is a proper incorporation of the detailed velocity model into the initial migration step.

WAVE-THEORETICAL DEPTH MIGRATION

To understand the failure of time migration (and, consequently, ray-theoretical depth migration), we must make a brief examination of the scalar wave equation. The scalar wave equation in two dimensions can be written

$$\frac{\partial^2 P}{\partial x^2} + \frac{\partial^2 P}{\partial z^2} = \frac{4}{v^2(x, z)} \frac{\partial^2 P}{\partial t^2}, \quad (1)$$

where $P(x, z, t)$ is the observed pressure wave field, $v(x, z)$ is the velocity of the medium, and t is the *two-way* reflection time appropriate for zero-offset data (hence, the factor 4 on the right side of the equation). In order to make the migration computation tractable, Claerbout (1976, p. 234) proposed a family of coordinate transformations. The key element in any of Claerbout's transformations is that the frame of reference for the migration algorithm moves along with the wave field being migrated. Consider the transformation:

$$x' = x,$$
$$z' = z,$$

and

$$t' = t + 2 \int_0^z \frac{d\sigma}{\bar{v}(x, \sigma)}.$$

The motion of the reference frame is specified by a parameter called the frame velocity $\bar{v}$.

If we choose the frame velocity to be a constant, the scalar wave equation can be rewritten exactly (see Appendices A and B) as

$$P_{t'z'} + \frac{\bar{v}}{4} P_{z'z'} = A(\bar{v}) P_{x'x'} + B(v, \bar{v}) P_{t't'}, \quad (2)$$

where P and v are now functions of the new primed coordinates, and $P_{t'z'}$ denotes $\partial^2 P / \partial t' \partial z'$, etc. Here A and B are used to denote more complicated expressions that involve the parameters indicated [see equation (B–1), Appendix B]. If the frame velocity is not constant, other terms involving its derivatives must be added to the right side of equation (2).

In implementing migration by means of equation (2), we are free to choose the frame velocity as we wish, providing we choose it to be constant. The usual practice in time migration is to bend the rules and choose the frame velocity to be the velocity of the medium, *in spite of the spatial variance* of that velocity. This choice simplifies equation (2) by forcing term B on the right side to be identically zero, leaving only the so-called diffraction term A. Time migration, then, is performed with the equation

$$P_{t'z'} + \frac{v}{4} P_{z'z'} = A(v) P_{x'x'}. \quad (3)$$

Another common approximation is to set $P_{z'z'} = 0$ (c.f., Appendix B), yielding

$$P_{t'z'} = A(v) P_{x'x'}. \quad (4)$$

Since v is a function of x and z, equation (4) *apparently* allows the use of a laterally variable velocity, and one would think that it properly describes the propagation of waves in an inhomogeneous medium. In reality, this equation propagates waves locally as though the medium were horizontally layered. In bending the assumptions, we have, among other things, discarded the gradients of $\bar{v}$ which embody Snell's law for a laterally inhomogeneous medium (see Appendix A). In wave-theoretical depth migration, we assume a constant $\bar{v}$ and honor Snell's law by including the so-called shifting term B of wave equation (2). When we do so, the term $P_{z'z'}$ cannot be so casually dropped (see Appendix B).

Our studies with synthetic data indicate that *so long as the lateral velocity variation is not too severe*, neglect of the gradients of $\bar{v}$ is not highly detrimental to the *imaging* ability of time migration. As will be seen in the next example, however, for severe inhomogeneity such neglect is patently inappropriate.

Although our discussion of the wave equation has been along the lines used by Claerbout in developing his finite-difference migration techniques, the same (and some even more restrictive)

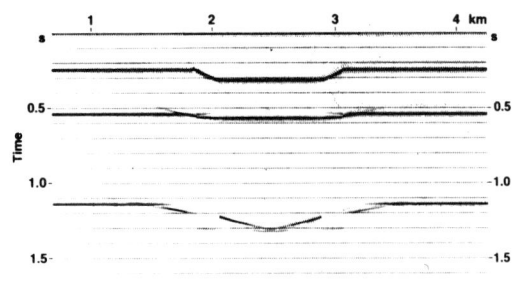

FIG. 2c. Time migration of the data in Figure 2b by the Kirchhoff summation technique.

approximations are made when migration is performed in the frequency-wavenumber domain (Stolt, 1978; Gazdag, 1978). In addition, Larner and Hatton (1976) showed that approximations made in conventional Kirchhoff summation migration are equivalent to those discussed here. Thus, our remarks on the positioning and imaging behavior of time migration apply regardless of the particular implementation.

Clearly, resolution of imaging and mispositioning problems requires incorporation of the shifting term in the migration algorithm. Consequently, $\bar{v}$ can be assumed constant and the conditions on its derivatives satisfied. Because the shifting term embodies the proper refraction of energy in laterally inhomogeneous media (Snell's law), this more complete form of the wave equation will solve both imaging and lateral positioning problems. For ease in implementation, we chose to perform our migrations by the finite-difference method.

The input to this extended technique is an unmigrated (stacked) time section. The output, obtained directly in one step, is a *fully migrated* depth section. We refer to this method as wave-theoretical depth migration to distinguish it from its ray-theoretical counterpart. [Judson et al (1980) and Gazdag (1980) present additional perspectives on depth migration.]

As in the ray-theoretical depth migration process, specification of migration velocity is again of central importance. Wave-theoretical depth migration requires that interval velocity be specified as a function of depth and horizontal position. Because the velocity function is not available in advance, we start with a reasonably close estimate of velocity as a function of those coordinates. The depth migration process will then provide information for improving the estimated structure. The relative insensitivity of the technique to imperfect velocity modeling usually allows the solution for structure to converge adequately after only one iteration.

SYNTHETIC DATA EXAMPLE

Consider the depth model in Figure 2a. The layers have constant velocities, and interfaces are basically horizontal except for a syncline at depth. The near-surface anomaly of this model constitutes the significant inhomogeneity whose disruptive effects will be evident. (If the top layer were less thick relative to the depths of deeper interfaces, time distortions caused by the near-surface anomaly could be corrected by conventional static time adjustments.) Dips in the near-surface exceed 30 degrees, and those in the syncline exceed 40 degrees. Figure 2b shows a synthetic, zero-offset section for this model, generated by a wave-theoretical, diffraction response technique.

The most important features in the unmigrated section are direct results of the near-surface inhomogeneity. Both the flat-lying event (just below 0.5 sec) and the buried focus from the syncline exhibit phantom diffractions and gaps that are the result of distortion of the wavefront during its transit through the anomalous near-surface.

Figure 2c displays the Kirchhoff time migration of the data in Figure 2b based on the known velocities in the medium. Note that the shallowest interface is correctly migrated (the medium is homogeneous above it), but the time migration has failed to treat the phantom diffractions on deeper horizons. Residual diffraction effects remain in the second reflection, and large gaps remain evident in the syncline. Note also the double events at the base of the syncline and at its edge. No form of depth conversion that involves stretching of traces—either vertically or along image raypaths—can convert these double events into the simple wavelet that should mark the interface.

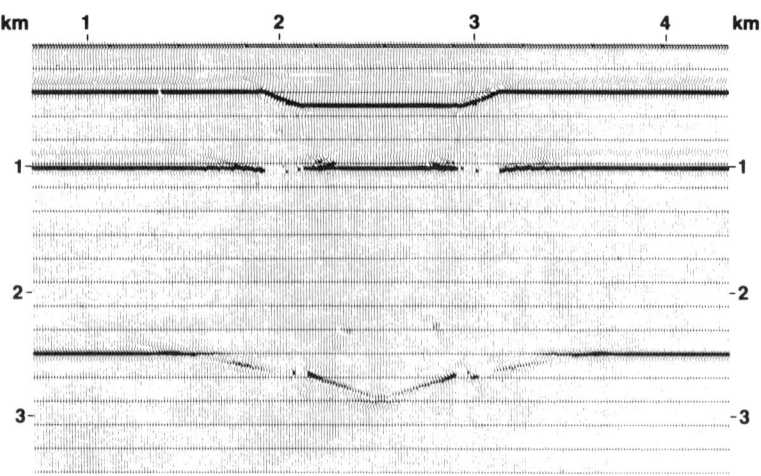

FIG. 3a. Ray-theoretical depth migration of the section in Figure 2b. Note that this is a depth section.

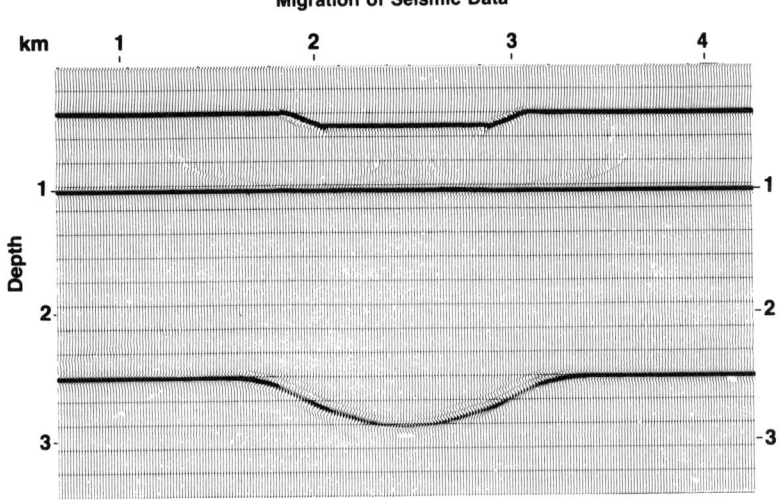

FIG. 3b. Wave-theoretical depth migration of the data in Figure 2b. Note that this is a depth section.

The double events are not the only problems confronting the image ray conversion from time to depth, as the ray-theoretical depth migration in Figure 3a vividly demonstrates. The strong velocity contrast and appreciable dip at the upper interface dictate large deflections for the image rays. With so pronounced an inhomogeneity, image-ray theory breaks down just as the imaging of time migration did. The subsequent redistribution of amplitudes into depth along the image rays actually creates more problems than it solves; large gaps are opened in the second reflection, and some diffracted energy is left essentially untreated.

Figure 3b is a depth section displaying the result of wave-theoretical depth migration of the data in Figure 2b. The process has yielded a section in which the positions and shapes of imaged interfaces compare well with their known positions (compare with the depth model in Figure 2a). Moreover, it has both properly imaged the syncline and correctly collapsed phantom diffractions.

In this example, the only velocity inhomogeneity of concern is that in the near-surface. A useful way to visualize migration is as a downward continuation process: the technique proceeds step by step to predict the seismic sections that would have been recorded had the sources and receivers of the survey been buried deeper and deeper. In finite-difference migration, the observation plane is in effect progressively lowered to the reflectors. At any particular stage of downward continuation, the seismic data above the observation plane are fully migrated, while the data below represent the (unmigrated) time section that would have been recorded at that depth of burial.

In Figure 3c, the data have been downward continued to a depth of just over 750 m (the dashed line). The resulting section is a hybrid; above the dashed line it is a (fully migrated) depth section of the near-surface, while below it is the time section that would have been recorded 750 m below the surface.

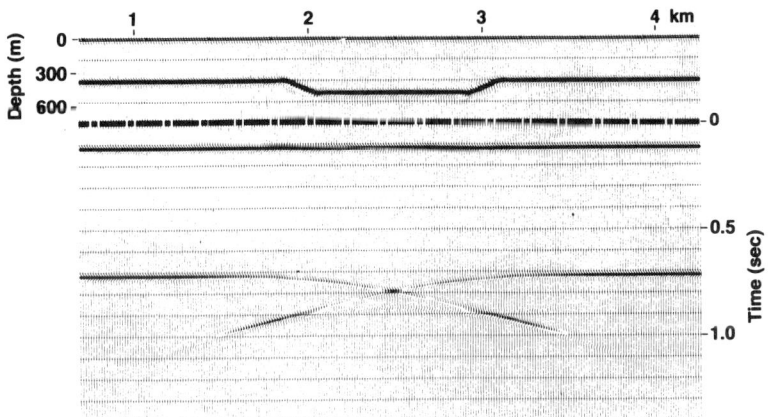

FIG. 3c. Partial downward continuation. The wave-theoretical depth migration was stopped with the source-receiver level at approximately 750-m depth (dashed line). Above the dashed line, the section is fully migrated in depth; the dashed line also corresponds to time zero on the time section (below the line) that would have been recorded had sources and receivers been buried at 750 m.

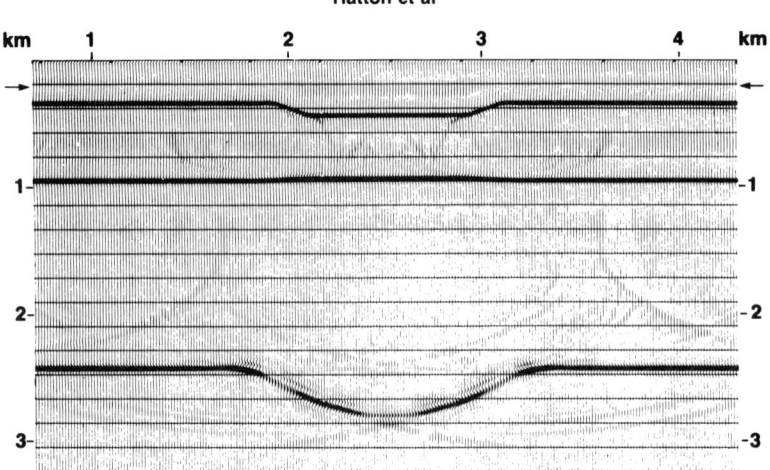

FIG. 4a. Wave-theoretical depth migration with the velocity of the uppermost layer specified as 2745 m/sec (10 percent lower than the known velocity for the model).

Note that the time section shows none of the distortion attributable to the near-surface anomaly. The middle reflector (just below the dashed line) is flat, and the phantom diffractions have healed completely. The buried focus from the syncline is also free of phantom diffractions. This time section could now be successfully treated with a conventional time-migration algorithm.

The advantage of such partial migration is clear. Often, we might expect to have a detailed velocity model for the near-surface from high-resolution seismic work, refraction shooting, or well surveys, whereas the velocity information for the deep section is less precisely known. Partial migration can remove the distorting influence of an inhomogeneous near-surface zone from deep data without the need for a full depth migration. In a sense, partial depth migration is thus a wave-theoretical extension of approaches to statics corrections for near-surface time anomalies. After performing partial depth migration to correct for near-surface variations, one can finish with time migration (based on whatever velocity information is available for the deeper section) and a final direct conversion to depth.

Again, the essential issue is the sensitivity of migration quality to inaccuracies in velocity. To study sensitivity, we performed wave-theoretical depth migration on the data of Figure 2b using velocity information that was known to be incorrect. The first two tests were based on the use of wrong velocities for the uppermost layer (correct velocity = 3050 m/sec); specifically, we performed depth migrations with that velocity 10 percent too low and then 10 percent too high.

Only the results of the 10 percent decrease are shown (Figure 4a). Surprisingly, the images of both the syncline and the flat reflector are still coherent and continuous across the section. Although the section exhibits some faint spurious diffraction effects, the syncline shape, wavelet character, and even amplitude treatment are quite good—far superior to the results of either time migration or ray-theoretical depth migration. The resulting reflectors are all

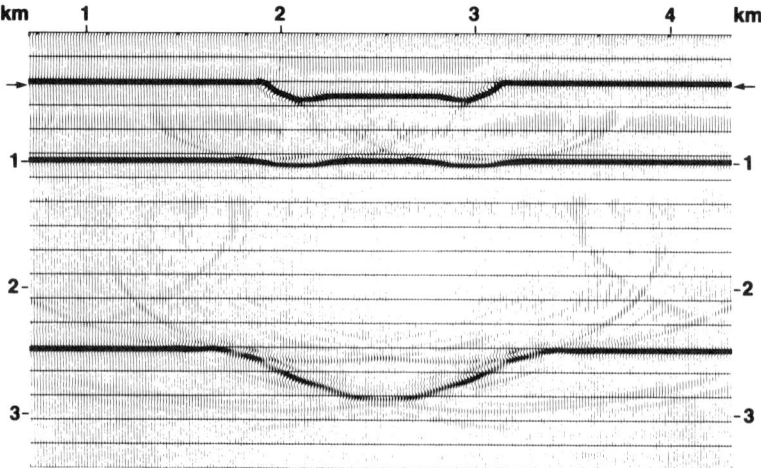

FIG. 4b. Wave-theoretical depth migration with the velocity model perturbed as shown by heavy lines in Figure 2a.

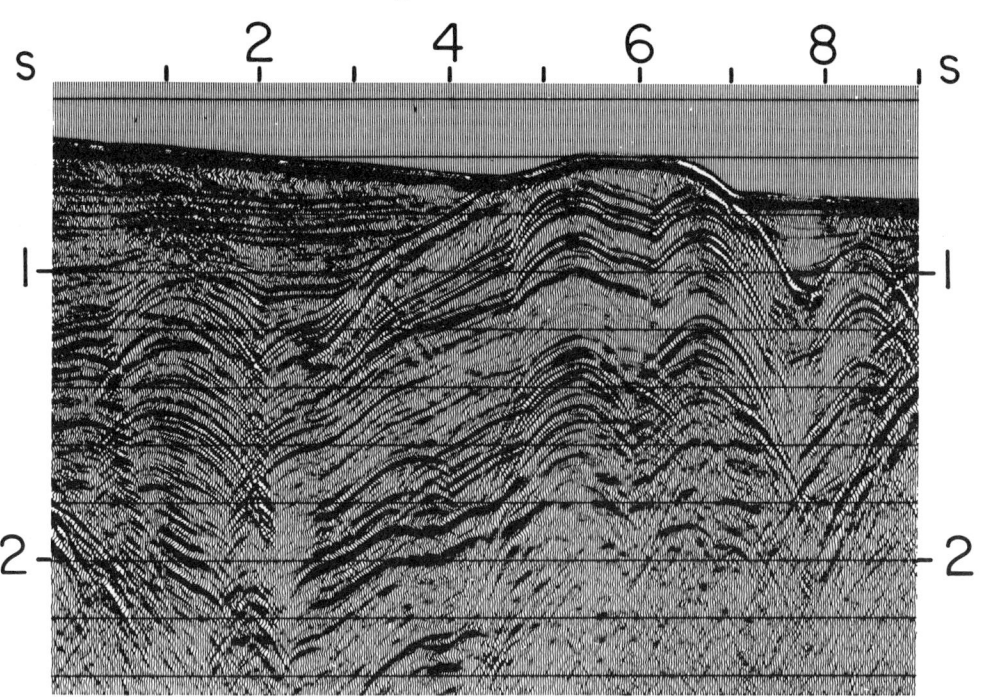

FIG. 5a. CDP stack section of data recorded in the Santa Barbara channel. Steepest dips are approximately 25 degrees.

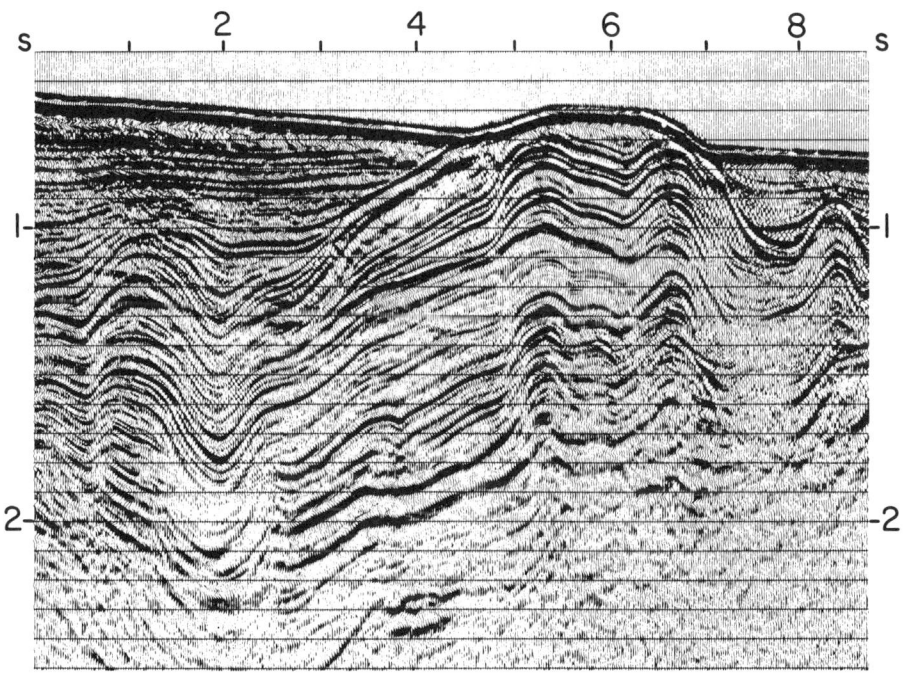

FIG. 5b. Finite-difference (time) migration of the Santa Barbara channel data.

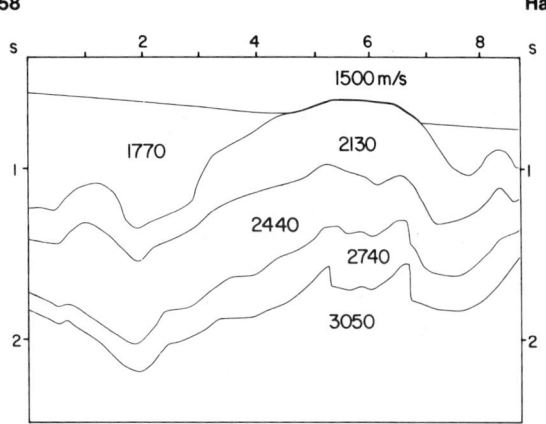

FIG. 5c. Interval velocity model used for performing depth migration on the data shown in Figure 5a.

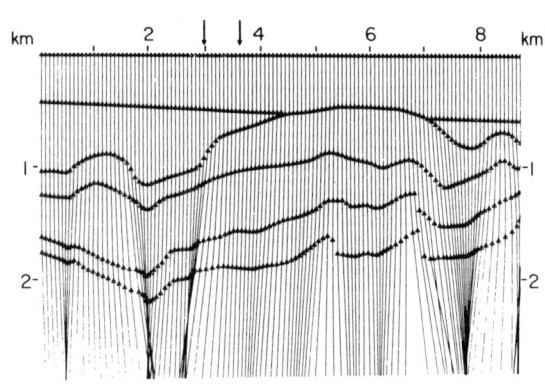

FIG. 5d. Image raypaths (plotted in depth) obtained by ray tracing a velocity model interpreted from the time-migrated section in Figure 5b.

slightly shallower than they should be. These depth errors are simply the familiar errors in converting from time to depth with the wrong velocity. Results of the 10 percent increase (not shown) were similar in character; the error in depth was naturally in the opposite direction.

Other types of errors can also contaminate the velocity model; in particular, interpreted shapes of layer interfaces could be inaccurate. To examine the effects of such an error, we perturbed the shape of the first interface as shown by the heavy lines in Figure 2a. Here, interpretation of the first interface was made from the unmigrated time section, certainly a crude thing to do for this model. In practice, the migrated time section yields a much more accurate estimate of the shallow structure. In this experiment, correct velocities were used for all layers.

Results of the wave-theoretical depth migration are shown in Figure 4b. The gross error in layer shape has caused considerable distortion in the deeper section and again introduced some spurious low-amplitude diffractions. Nevertheless, even for this severe error the reflectors are imaged rather coherently across the section. One can interpret the deeper section here better than on any of the sections that involved time migration as an intermediate step.

The results of this small sensitivity study parallel those of the image-ray uncertainty study of Larner et al (1981). Depth migration is less sensitive to the specification of velocity values than to interpreted shapes of layer interfaces. We consider the proper interpretation of layer shape as the keystone in approximating horizontal velocity gradients for a layered model. Horizontal gradients constitute the central issue in laterally inhomogeneous media.

FIELD DATA EXAMPLE: OFFSHORE CALIFORNIA

Figure 5a is a CDP stack of data from the Santa Barbara Channel, offshore California. The (finite-difference) time migration in Figure 5b shows the expected changes; crossing diffraction patterns are collapsed to synclines, and anticlines are constricted. In this profile, dips do not exceed about 25 degrees. We observe that the imaging has been reasonably successful in the sense that the time-migrated section presents a more plausible picture of the subsurface than does the original stack. Plausibility, however, is no guarantor of truth.

The unconformity surface between the gently dipping Quaternary section and the prominently folded Tertiary section could

FIG. 5e. A conventional depth section for the time-migrated data recorded in the Santa Barbara channel. Note the amplitude weakness in the synclinal features at lateral positions 2 and 7.5 km.

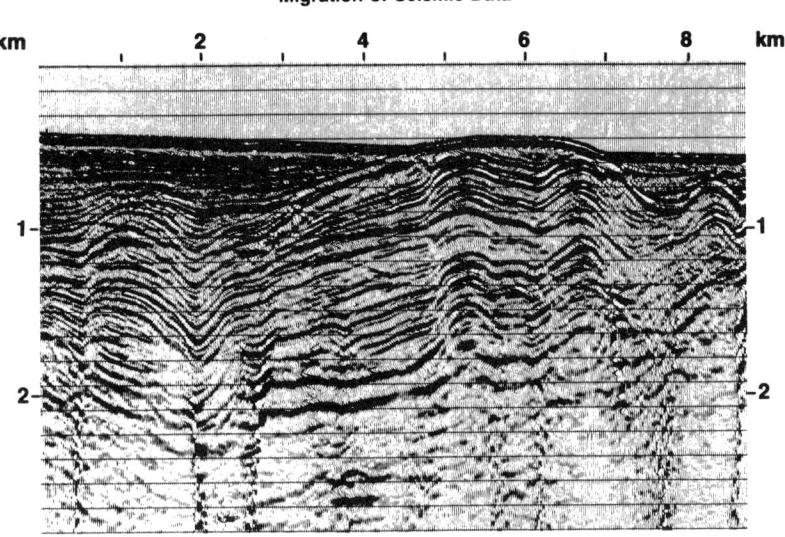

FIG. 5f. Depth section generated by stretching time-migrated data (Figure 5b) into depth along the image raypaths of Figure 5d. These steps constitute ray-theoretical depth migration.

provide the circumstances that lead to significant deflection of image rays from the vertical. If that is the case, we expect that (although plausible) the time-migrated section in Figure 5b is distorted. Using independently known velocities for the area and a digitized interpretation of several major horizons (including the unconformity, Figure 5c), we computed the paths of the image rays plotted in Figure 5d. This plot is in depth, and the triangular symbols denote intersections of the image raypaths with the (now fully migrated) horizons. In the model, velocities increased from about 1770 m/sec in the Quaternary section to about 2100 m/sec across the unconformity and then to about 2750 m/sec in the deep layer at 2 sec.

In the image-ray plot for this section (Figure 5d), rays tend to converge downward into synclines and to diverge beneath anticlines. Another model of this line (not shown here) differed from this one only in that velocities increased more rapidly with depth, attaining values 50 percent higher for the layer immediately above basement. For that quite different velocity model, the pattern of convergence and divergence of image rays was much like that shown in Figure 5d. As noted earlier, computed depths are highly sensitive to the model velocities, but the lateral position errors requiring correction by depth migration are more directly related to the interpreted horizon shapes. For this reason, it is important to understand whether the velocities of a particular section are

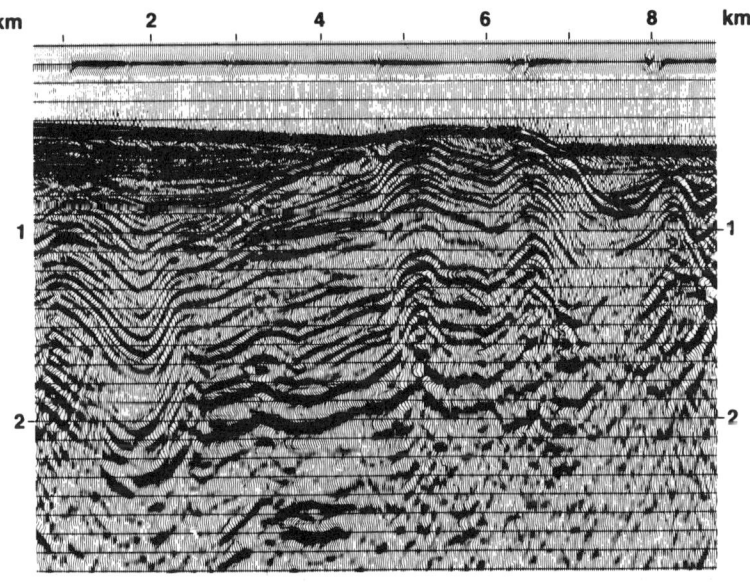

FIG. 5g. Wave-theoretical depth migrated section.

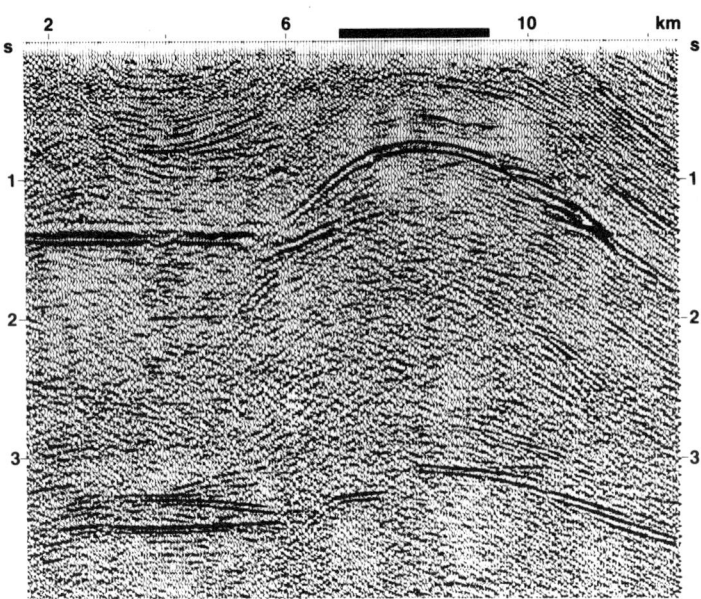

FIG. 6a. CDP-stacked section of data recorded in Central America. The prominent reflector at about 1.4 sec on the left side is the top of a thick, high-velocity carbonate/anhydrite sequence. The horizontal bar denotes the length of the spread.

controlled primarily by depth or by lithology, i.e., it is important to know whether or not velocity gradients trend normal to interpreted horizons.

Figure 5e is a conventional depth section obtained by stretching the traces in Figure 5b vertically from time to depth. Note the generally weaker amplitudes below about 1.5 km in depth at lateral positions of 2 km and 7.5 km. These zones coincide with the greatest degree of folding in the section.

Conversion to depth along image rays will tend to boost amplitudes in places where the rays converge, thereby compensating approximately for amplitude weakness. The image-ray depth conversion of the traces in Figure 5b is shown in Figure 5f. Inadequacies in amplitude treatment attributable to ray theory are evident, but the result is encouraging; amplitudes tend to be restored over those zones where they were originally weak.

In spite of substantial uncertainty in the available velocity information, the wave-theoretical depth migration shown in Figure 5g is superior to either the conventional or the image-ray depth section. In particular, note the generally improved continuity and more uniform treatment of amplitude throughout the syncline at

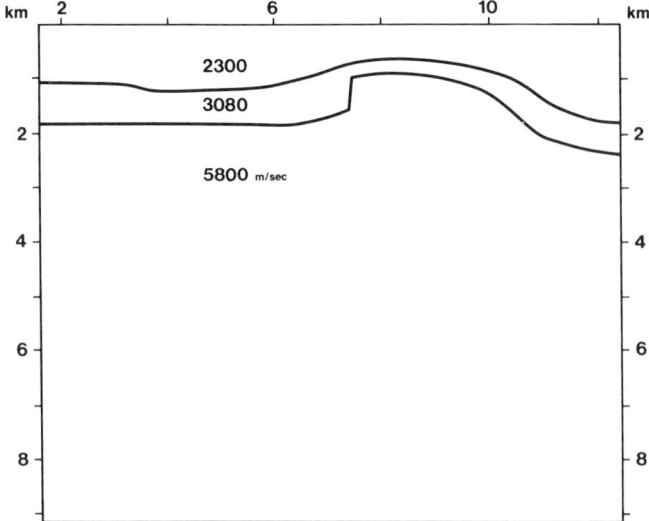

FIG. 6b. Interval velocity model used in processing the data in Figure 6a.

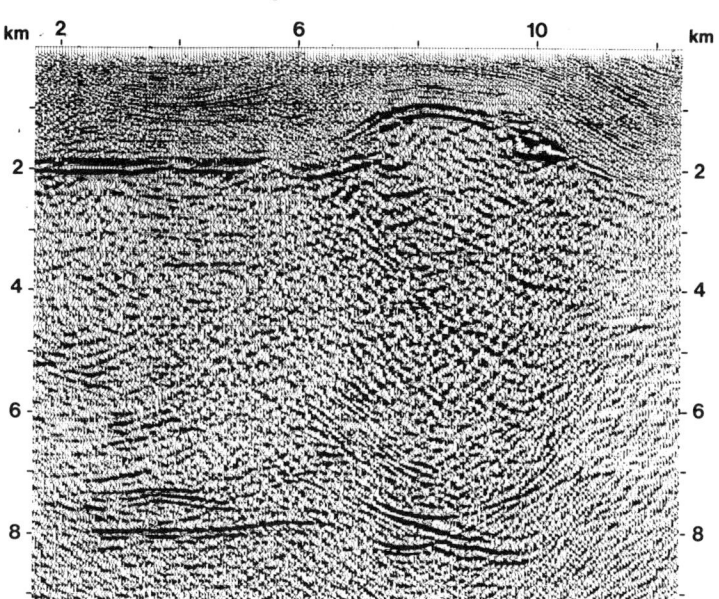

FIG. 6c. Depth section produced by time migration of the section in Figure 6a and simple conversion to depth. Both the time migration and conversion to depth used the velocity model in Figure 6c.

2 km. Again, the imaging of this depth migration process is relatively insensitive to inaccuracies in the velocities used for the computation.

Reviewing the sequence of sections from Figures 5a through 5g, we see an interesting progression near the lateral position of 2 km —from the implausible crossing of events before migration to a broad syncline in the conventional depth section and then back toward a tighter syncline in the depth-migrated sections. Note that differences exist in the velocities used to generate the two forms of depth-migrated data, and these differences account in part for the different shapes obtained for the syncline at 2 km. The basic conclusions, however, are unaffected: (1) realistic models for velocity lead to convergence of image rays beneath the synclines and resultant tightening of the synclines relative to the shapes on

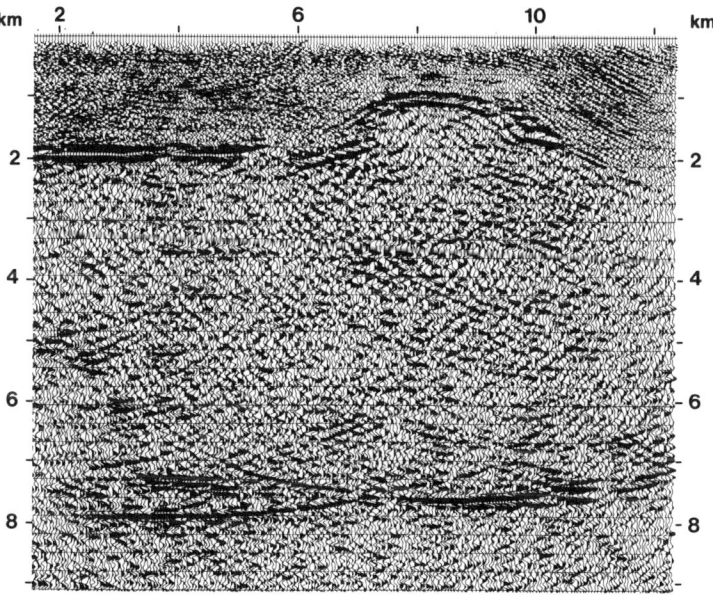

FIG. 6d. Wave-theoretical depth migration of the section in Figure 6a based on the velocity model in Figure 6b. Note the simplified image of the deep reflector.

the conventional depth section, and (2) from comparison of results obtained using different velocity models, we can put realistic bounds on the range of possible structural interpretations.

FIELD DATA EXAMPLE: CENTRAL AMERICA

Figure 6a shows an unmigrated stacked section of data collected in Central America. Regional tectonic forces have produced substantial folding and faulting in the upper part of the section. Above the prominent horizon (at 1.4 sec on the left side), the geologic section is comprised of Tertiary clastics with velocities in the neighborhood of 2500 to 3000 m/sec. The strong reflection marks the boundary between the Tertiary section above and a substantially higher-velocity Cretaceous carbonate/anhydrite sequence below. Velocities in the Cretaceous section are approximately 6000 m/sec.

This two-to-one velocity contrast, combined with the rollover and steeply dipping fault, result in greatly distorted deeper reflections. Specifically, regional information suggests that the deep reflector appearing at about 3.1 to 3.6 sec is essentially flat and continuous; the complexity it exhibits in this stacked section results from the distorting influence of the velocity variations in the overburden.

As noted earlier, the most critical step in imaging such a distorted reflector is the specification of a velocity model for the overburden. For the data here, the key issues are (1) the shape of the Tertiary-Cretaceous boundary, and (2) the interval velocities above and below. Well log information indicates that the Tertiary section is relatively homogeneous; hence, a preliminary time migration provides a good image of the Tertiary-Cretaceous horizon. Interpretation of the migrated data and well-log information suggests that the Tertiary section is acceptably described by a two-layer model and that the Cretaceous section is approximately homogeneous. The selected model (Figure 6b) has a slight, vertical velocity gradient in the uppermost layer and constant velocities in the other two.

For this section, the velocity model was developed with relative ease because the interval velocities were assumed reasonably well known and varied little within each layer. In more complex regions, interval velocities may have to be estimated from observed moveout reflection times by use of techniques such as those of Larner and Rooney (1972) or Hubral (1976). Such estimates, however, are often crude and may require careful interpretation (namely, editing and smoothing). The uncertainties in interval velocity estimates are related directly to the quality of surface-moveout analyses; such surface-velocity analyses will often be ambiguous when conducted over complex structures. For this reason, the use of independent velocity information, such as that from well logs, is always desirable and often essential. As indicated in the synthetic-data studies, depth migration is fairly insensitive to errors in the velocity model. We therefore expect that a preliminary time migration can clarify layer structure and thus be used to improve the velocity model.

Figure 6c is a conventional depth section obtained by vertically stretching a (finite-difference) time-migration result for the data in Figure 6a. Both the time migration and depth conversion used the velocity model of Figure 6b. As expected, the data down to the Cretaceous horizon are generally clarified; the deep reflection, however, is still quite complex, actually more confused than it was on the unmigrated data. A better image would have been gained by simply depth converting the original unmigrated section.

In contrast, the wave-theoretical depth migration (Figure 6d) produces a much simplified image of the deep reflector. The image here is essentially continuous and flat across the extent of

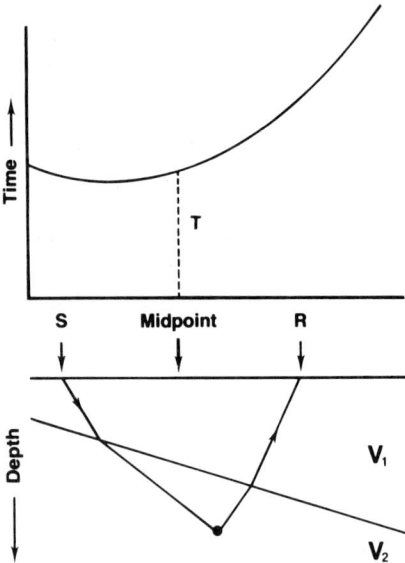

FIG. 7a. Depth model of a point diffractor beneath a plane dipping interface. A diffraction pattern for some nonzero source-receiver offset is also shown. Traveltime from source to scatterer to receiver is plotted at the source-receiver midpoint.

the section. The slight pull-up of that reflection near the center of the section is likely attributable to some imperfection in the model for the overburden and could be corrected by an adjustment to the velocity or shape of the upper layers. Most importantly, though, wave-theoretical depth migration has effectively compensated for the severe lensing caused by the shallow velocity structure.

While the deep reflection in Figure 6d is now approximately flat and continuous, a bowl-shaped artifact has appeared just above it. The presence of this pattern can be traced to the stacked section (Figure 6a). In that figure note that segments of the deep reflection lack diffraction tails that must exist in a recorded wave field. The tails likely have been destroyed in CDP stacking because complex propagation paths across the folded and faulted

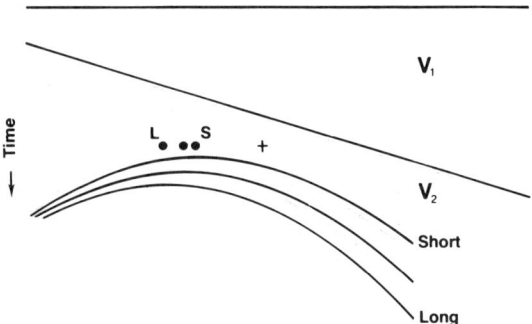

FIG. 7b. Diffraction patterns for the scatterer of Figure 7a as they would be seen on three common-offset time sections. L and S denote long and short offsets, respectively. The cross indicates the correct lateral position of the scatterer.

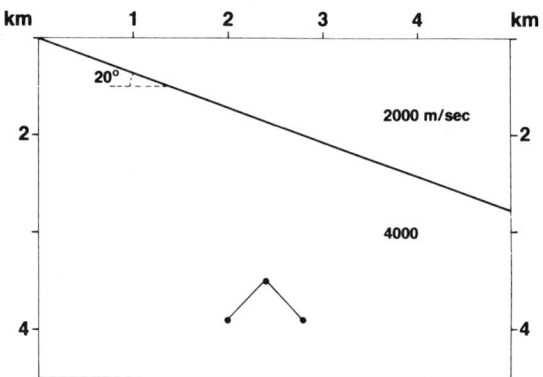

FIG. 8a. Small triangular feature (stylized reef or trap) beneath a dipping interface.

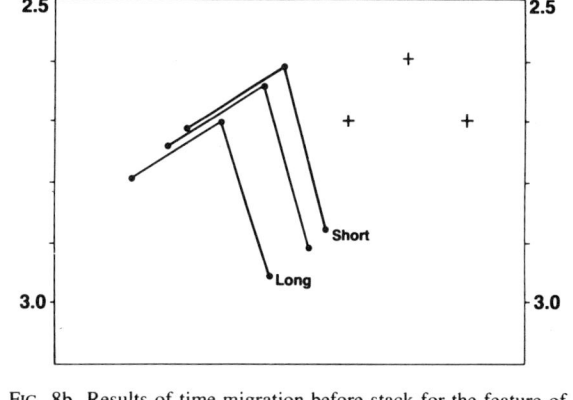

FIG. 8b. Results of time migration before stack for the feature of Figure 8a as seen on three different common-offset sections. Crosses mark the correct lateral positions of the three edges of the small feature.

structure have severely disrupted normal-moveout (NMO) relationships. The bowl-shaped pattern in Figure 6d then is a direct result of the migration of those truncated reflections. This artifact reminds us that a CDP-stacked section is not equivalent to a zero-offset section.

DEPTH MIGRATION BEFORE STACK

Since we are addressing various aspects of the migration of data from inhomogeneous media, it is fair to ask whether a poor migration for a complex section might be attributable to prior degradation by the CDP stacking process. Certainly, raypaths and traveltimes are likely to be complex and not well described by the horizontally layered model that underlies the stacking process.

Where a CDP stack is suspect, the conventional thinking has been that the data ought to be migrated before being stacked. Migration of unstacked data involves considerably more computation than migration applied after stack; hence, a close look at the assumed benefits of this more extensive effort is in order.

Consider first the simple case of a point diffractor beneath a dipping interface, as depicted in Figure 7a. Regroup traces from a conventional CDP survey so that they appear as though we had conducted a number of common-offset surveys. In each common-offset section, we would observe the diffraction pattern of traveltimes (from source to diffractor and back to receiver) plotted as a function of the point midway between the source and receiver. Figure 7b shows schematically three such diffraction patterns for small, medium, and large offsets. As in the normal-incidence case, the diffraction patterns are asymmetric with their apexes displaced laterally from the correct diffractor position.

Whereas for zero offset the position of the minimum time corresponds to the image-ray location, no such simple relationship exists for the minimum-time points on common-offset diffraction patterns. For the dipping interface model in Figure 7a, the apexes are displaced progressively farther updip as offset increases.

Now, suppose we performed conventional time migration on each offset section. Then, despite a failure to collapse the diffraction patterns completely, the process would condense amplitudes to a *different minimum point* for each offset section. With this lateral spread of minima, the time-migrated offset sections cannot be stacked properly.

Common-offset migration is not the only way to accomplish migration of unstacked data. An alternative approach is to downward continue, over each depth step, first the receivers associated with each shot and then the shots associated with each receiver (Clayton, 1978). As long as this process is performed without consideration of refraction through the overburden, it cannot resolve the lateral mispositioning problem any better than common-offset migration can.

We can illustrate this problem differently with a schematic model of a small structural feature beneath a dipping interface. In the depth section of Figure 8a, a small "trap" (the triangularly shaped feature) is embedded in an otherwise gently dipping section. Figure 8b shows the time-migrated positions (in close-up) of the trap as it would be seen on sections for three different offsets. The left dipping segments nearly coincide, but these features are not yet NMO-corrected. The processing steps that would follow the migration are velocity analysis and then stacking of the data. What velocities are required to correct the left and right segments of this feature for normal moveout? Along the left segment, the differential time from short offset to long is so small that the required stacking velocity might be as high as 6000 m/sec or more. In contrast, on the right segment, the required stacking velocity could be as low as 1400 m/sec or less. Between the two, the normal moveout may not be even approximately hyperbolic.

How could one hope to stack this small feature properly? Recall, our trap is a subtle structure in an otherwise gently dipping medium. In practice we would smooth variations in the stacking velocity over a distance of a kilometer or more. We would have little choice other than to correct for normal moveout with a velocity of around 3200 m/sec, not at all close to the velocities required to stack the time-migrated limbs of our small feature.

As this example illustrates, time migration before stack can be as damaging to reflections, particularly those from subtle features, as conventional CDP stacking. Thus, where the subsurface is sufficiently complex to warrant depth migration, the most appropriate solution often may be depth migration of the unstacked data. Conventional time migration of unstacked data, for all the added computational effort, may yield a result no better than migration after stack.

Schultz and Sherwood (1980) showed how depth migration of unstacked traces can be performed by the downward continuation of sources and receivers. Alternatively, Yilmaz and Claerbout (1980) discussed how depth migration can be accomplished for common-offset sections. In their technique, each offset section is partially migrated to convert it (approximately) to an unmigrated normal-incidence time section. The resulting sections are then stacked, and the stack can be depth migrated by the method discussed here. During each downward continuation step of the prestack migration process, the data are shifted laterally by an amount dependent upon the offset, reflection time, and degree of lateral heterogeneity. The lateral shifts compensate for the offset-dependent displacement of reflected and diffracted events.

Schultz and Sherwood (1980) demonstrated the application of depth migration to another common problem—the distortion and misstacking of deep reflection data by irregular, water-bottom topography.

CONCLUSIONS

Although virtually every seismic section ought to be migrated, not all need to be depth migrated. Where the overburden is known to be laterally inhomogeneous, however, depth migration can be a refinement of primary importance. Given a reasonably accurate estimate of velocity, the relatively efficient process of image-ray tracing (and velocity modeling) can be applied to the time-migrated data to determine whether the more complete, more accurate, and more costly wave-theoretical depth migration is warranted.

Information about detailed velocity structure will be imperfect, sometimes grossly inaccurate. Fortunately, like time migration, both ray-theoretical and wave-theoretical depth migration are reasonably insensitive to errors in the velocity model; nevertheless, the quality of depth migration is ultimately limited by the accuracy of velocity information available. We have demonstrated an algorithm that migrates data correctly *if* we know the velocity model. The primary task must be to improve our ability to estimate velocity.

Geophysicists are now squarely confronting seismic problems attributable to complex geologic structures and are finding that the issue of velocity is fundamental. We caution against applying too much effort to any single problem at the expense of others. At present, we artificially categorize depth migration, migration before stack, and 3-D migration as three *separate* tools for migrating data from inhomogeneous media. The problems they address are, however, deeply interconnected—they cannot be considered apart from one another. Thus, we must pursue a task that could not have been considered seriously only a short time ago—implementing 3-D depth migration before stack. Executed correctly, it will be based on the elastic wave equation.

ACKNOWLEDGMENTS

We gratefully acknowledge Ron Chambers for his expert computer programming, and Carl Savit for reviewing the manuscript. Stew Levin made significant contributions to the analysis shown in Appendix B. Our thanks go also to Brenda Edwards and Rhonda Boone for their creative drafting of the figures and Evelyn Fulford, Grace Bonaventura, and Dolores Meeks for their patient attention to the typing of the manuscript.

REFERENCES

Claerbout, J. F., 1976, Fundamentals of geophysical data processing: New York, McGraw-Hill Book Co., Inc., 274 p.
Clayton, R., 1978, Common midpoint migration: Stanford Expl. Proj. rep. no. 14, Stanford, p. 21–36.
Gazdag, J., 1978, Wave equation migration with the phase shift method: Geophysics, v. 43, p. 1342–1351.
——— 1980, Wave equation migration with the accurate space derivative method: Geophys. Prosp., v. 28, p. 60–70.
Hubral, P., 1976, Interval velocities from surface measurements in the three-dimensional, plane layer case: Geophysics v. 41, p. 233–242.
——— 1977, Time migration—Some ray-theoretical aspects: Geophys. Prosp., v. 25, p. 738–745.
Judson, D., Lin, J., Schultz, P., and Sherwood, J., 1980, Depth migration after stack: Geophysics, v. 45, p. 361–375.
Larner, K., and Hatton, L., 1976, Wave equation migration—Two approaches: Presented at 8th Annual Offshore Technology Conference, May, in Houston.
Larner, K., Hatton, L., and Gibson, B., 1981, Depth migration of imaged time sections: Geophysics, v. 46, this issue, p. 724–738.
Larner, K., and Rooney, M., 1972, Interval velocity computation for plane dipping multilayered media: Presented at the 42nd Annual International SEG Meeting, November 30, in Anaheim.
Larson, D., and Hilterman, F., 1976, Diffractions—Their generation and interpretation use: Presented at 29th Annual Midwestern SEG Meeting, March, in Dallas.
Mitchell, A. R., 1969, Computational methods in partial differential equations: New York, J. Wiley and Sons.
Schultz, P., and Sherwood, J., 1980, Depth migration before stack: Geophysics, v. 45, p. 376–393.
Stolt, R. H., 1978, Migration by Fourier transform: Geophysics, v. 43, p. 23–48.
Trorey, A., 1970, A simple theory for seismic diffractions: Geophysics, v. 35, p. 762–784.
Yilmaz, O., and Claerbout, J., 1980, Prestack partial migration: Geophysics, v. 45, p. 1753–1779.

APPENDIX A
IMAGE RAYS IN FINITE-DIFFERENCE ALGORITHMS

Larner et al (1981) show that both Kirchhoff summation and finite-difference time-migration algorithms fail to position events correctly in the presence of laterally varying velocity and indeed fail in similar ways. Hubral's (1977) elegant analysis introduces the concept of the image ray and its relationship to Kirchhoff summation. It is of interest, therefore, to identify image rays in the formalism of finite-difference algorithms.

An appropriate coordinate frame is useful when trying to identify a particular phenomenon. We start with the scalar wave equation

$$\frac{\partial^2 P}{\partial x^2} + \frac{\partial^2 P}{\partial z^2} = \frac{4}{v^2(x, z)} \frac{\partial^2 P}{\partial t^2}, \qquad (A-1)$$

where $P = P(x, z, t)$ is the disturbance as a function of x and z (the horizontal and vertical coordinates, respectively) and of two-way time t.

In the spirit of the coordinate transformations introduced by Claerbout (1976), we introduce the transformation:

$$x' = x,$$
$$z' = z,$$

and

$$t' = t + 2 \int_0^z \frac{d\sigma}{\bar{v}(x, \sigma)}. \qquad (A-2)$$

The rationale behind this choice is its intimate connection with ray tracing. Equation (A-1) can then be written with suffix notation for partial derivatives as

$$P'_{z't'} + \frac{\bar{v}'}{4} P'_{z'z'} = -\frac{v'}{4} P'_{x'x'} + \bar{v}' a P'_{x't'}$$
$$+ v' \left(\frac{1}{v'^2} - \frac{1}{\bar{v}'^2} - b\right) P'_{t't'} + \frac{\bar{v}'}{2} c P'_{t'}, \qquad (A-3)$$

where

$$P'(x', z', t') = P(x, z, t),$$
$$\bar{v}'(x', z') = \bar{v}(x, z),$$
$$v'(x', z') = v(x, z)$$

$$a(x', z') \equiv \int_0^z \frac{\bar{v}_x}{\bar{v}^2} d\sigma, \quad (A-4)$$

$$b(x', z') \equiv a^2(x', z'),$$

and

$$c(x', z') \equiv \frac{\bar{v}_z}{2\bar{v}^2} + \int_0^z \left[\frac{\bar{v}_{xx}}{\bar{v}^2} - \frac{2\bar{v}_x^2}{\bar{v}^3} \right] d\sigma.$$

Note that equation (A–3) is a more general form of equation (2) in the text (i.e., $\bar{v}$ is not assumed constant). Now neglect the term involving $P'_{z'z'}$, a common approximation in finite-difference migration. Equation (A–3) is therefore only approximate, but it will suffice to illustrate the derivation of image rays.

At this stage we will also set the frame velocity $\bar{v}'$ equal to the medium velocity v'.

A popular technique for solving equations like (A–3) is that of Marchuk splitting (Mitchell, 1969). In this method the initial equation is split into separate equations which are solved in parallel. Equation (A–3) can be written as

$$P'_{z't'} = -\frac{v'}{4} P'_{x'x'}, \quad (A-5)$$

and

$$P'_{z't'} = v' a P'_{x't'} - v' b P'_{t't'} + \frac{v'c}{2} P'_{t'}. \quad (A-6)$$

It can be shown that equation (A–5) describes the process of diffraction. As we shall see, the ability to refract at dipping interfaces is contained solely in equation (A–6).

The wave equation alluded to in conventional time migration is (A–5) or an equation with similar properties. Equation (A–6) or its equivalent is essential if propagation in a laterally heterogeneous medium is to be treated correctly.

Let us analyze equation (A–6): Further splitting reveals that the first term on the right is the image ray term, the second term on the right is a time-shifting term, and the third term on the right is a residual-amplitude term. We shall discuss the image ray term first.

With one t'-integration, the first term on the right of equation (A–6) behaves in split form as

$$\left(v' a \frac{\partial}{\partial x'} - \frac{\partial}{\partial z'} \right) P' = 0. \quad (A-7)$$

In the nontrivial case $v' a \neq 0$, equation (A–7) can be written

$$\frac{DP'}{Dx'} = 0, \quad (A-8)$$

where

$$\frac{D}{Dx'} = \frac{\partial}{\partial x'} + \frac{dz'}{dx'} \frac{\partial}{\partial z'}$$

is the Lagrangean derivative used frequently in describing the dynamics of fluids, and

$$\frac{dz'}{dx'} = -\frac{1}{v'a}. \quad (A-9)$$

Equations (A–8) and (A–9) indicate that P' should be kept constant along the trajectory given by (A–9). This is exactly the behavior we would expect along a raypath in the time-shifted frame described by equation (A–2). Let us further analyze the path described by equation (A–9).

Using equation (A–4), we write (A–9) as

$$\frac{dx'}{dz'} = -v' \int_{z'_0}^{z'} \frac{\partial v'/\partial x'}{v'^2} d\sigma, \quad (A-10)$$

and assume (x'_0, z'_0) is the starting point of the trajectory of interest. We suspect immediately that equation (A–10) is closely analogous to an image ray by observing that for $z'_0 = 0$, and within an underlying homogeneous layer (where $\partial v'/\partial x' = 0$), $dx'/dz' = 0$ and remains so until the depth is reached at which $\partial v'/\partial x' = 0$. Now, $dx'/dz' = 0$ implies that the corresponding trajectory departs $z'_0 (=0)$ normal to that surface. Furthermore, dx'/dz' is always constant across a homogeneous layer.

Consider the refraction of such a ray across the dipping interface in the model of Figure A–1. Initially, suppose that a transition zone centered at the dipping interface separates the homogeneous layers above and below. Also, let the velocity vary linearly across the transition zone (from V_1, the velocity in the upper layer, to V_2, the velocity in the lower layer). That is,

$$v'(x, z) = \begin{cases} V_1 & \text{for } z < z_b - \delta \\ V_0 + \lambda_1 x + \lambda_2 z & \text{for } |z - z_b| \leq \delta, \\ V_2 & \text{for } z > z_b + \delta \end{cases}$$

where $z_b(x)$ is the depth of the center line of the transition zone, δ is the vertical thickness of the zone, and V_0, λ_1, and λ_2 are constants. This model requires that

$$\lambda_2/\lambda_1 = -\tan \gamma,$$

where γ is the dip of the transition zone.

According to equation (A–10), the derivative dx'/dz' for the image ray trajectory at and beneath the transition zone is

$$\frac{dx'}{dz'} = -V_2 \int_{z_b - \delta}^{z_b + \delta} \frac{\lambda_1 d\sigma}{(V_0 + \lambda_1 x + \lambda_2 z)^2},$$

$$= (\lambda_1/\lambda_2) V_2 \left(\frac{1}{V_2} - \frac{1}{V_1} \right),$$

$$= (V_2 - V_1) \tan \gamma / V_1. \quad (A-11)$$

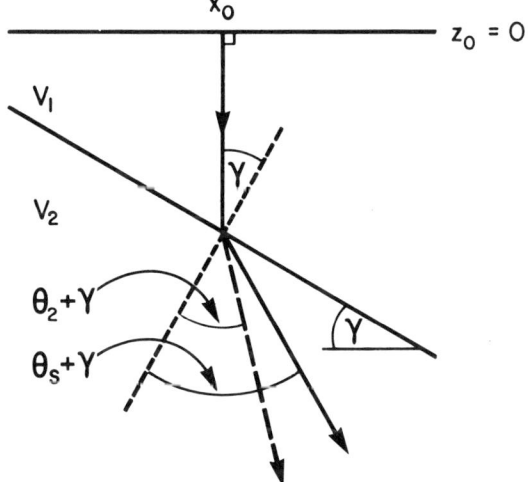

FIG. A–1. The raypaths resulting from a single dipping layer with velocity contrast as shown. $\theta_s + \gamma$ is the Snell's law angle, $\theta_2 + \gamma$ is the refraction angle associated with equation (A–10), and γ is the dip.

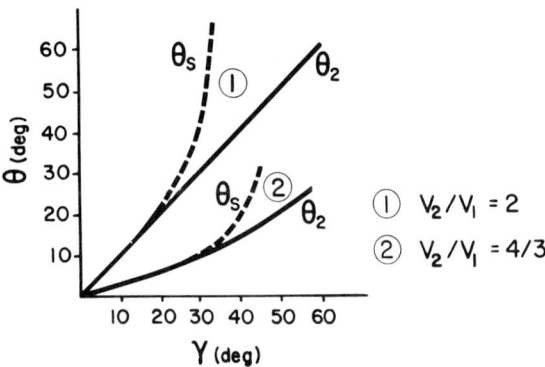

FIG. A–2. A graph of θ_2 and θ_s as functions of dip for two values of the ratio V_2/V_1.

Note that beneath the transition layer, the slope of the image raypath is constant (because V_2 is constant) and independent of the thickness of the transition layer. Letting δ go to zero gives the two-layer model shown in Figure (A–1). The slope $dx'/dz' = \tan \theta_2$, where θ_2 is the angle that the image raypath makes with the vertical. Some manipulation of (A–11) then gives

$$\frac{V_2}{V_1} = \frac{\sin(\theta_2 + \gamma)}{\sin \gamma \cos \theta_2}. \quad (A-12)$$

We may compare this result with Snell's law for the same model, which can be written

$$\frac{V_2}{V_1} = \frac{\sin(\theta_s + \gamma)}{\sin \gamma}. \quad (A-13)$$

Equations (A–12) and (A–13) then give the following relation between the Snell angle, $\theta_s + \gamma$, and the refraction angle of equation (A–10), $\theta_2 + \gamma$:

$$\sin(\theta_s + \gamma) = \frac{\sin(\theta_2 + \gamma)}{\cos \theta_2}.$$

Figure A–2 compares θ_2 and θ_s as a function of the dip γ for two choices of V_2/V_1. We see immediately that θ_2 is always less than θ_s; for small dip, however, θ_2 and θ_s are very close. That they differ at all is, in large part, a legacy of the paraxial approximation $P'_{z'z'} = 0$. For practical usage, however, the difference is slight. Should greater accuracy be required, it can be achieved by upgrading the quality of the paraxial approximation made to the wave equation.

In Appendix B, we approach the problem from a slightly different viewpoint. Instead of setting $\bar{v}' = v'$, we make $\bar{v}'$ constant (hence, not equal to v'). While that choice offers certain computational advantages, it requires special care in the treatment of the $P'_{z'z'}$ term.

For completeness, we discuss briefly the remaining two terms in equation (A–6). These can be written in split form as:

$$P'_{z't'} = -v' b P'_{t't'},$$

and

$$P'_{z't'} = \frac{v' c}{2} P'_{t'}.$$

One t'-integration of each yields

$$P'_{z'} = -v' b P'_{t'}, \quad (A-14)$$

and

$$P'_{z'} = \frac{v' c}{2} P'. \quad (A-15)$$

Equation (A–14) is equivalent in form to equation (A–7). The trajectories are now in the (z', t') plane and hence determine residual shifting in t'-time. The amount of shifting is different from that due to the term B of equation (2) in the text; it is termed residual because it is closely connected with the relatively small differences between integrating along z in equation (A–2) and integrating along the image raypath. Note that in the absence of lateral velocity variation, $b = 0$. Residual shifting most likely plays a role in the differences found between θ_2 and θ_s above.

Equation (A–15) has an obvious exponential solution that corresponds to small changes of amplitude with depth, providing of course that v' is not too heterogeneous [c.f., equation (A–4)].

APPENDIX B
THE PARAXIAL APPROXIMATION: $P_{z'z'} = 0$

In deriving equation (4) of the text, a key step was specifying $P_{z'z'} = 0$, a choice often called the paraxial approximation. This approximation is of considerable importance since P_{zz} cannot be obtained from surface seismic data [i.e., the wave field $P(x, z = 0, t)$]. Indeed, migration is not well posed with that term explicitly included in the wave equation. To make the migration problem tractable, it is necessary to reduce the wave equation from second order in z to first order. In so doing, one obtains a one-way wave equation that propagates waves in only the positive or negative z-direction but not both.

The neglect of $P_{z'z'}$ is justified on the grounds that upward traveling waves, as seen in the moving reference frame (x', z', t'), change very slowly. This approximate invariance is, in fact, true only for waves traveling near the vertical (the paraxial approximation is also called the 15-degree approximation). This restriction to small angles from vertical limits the quality of time migration attainable with equation (4).

Is there some better approximation than that $P_{z'z'} = 0$? For time migration, so-called 45- and 55-degree approximations have been proposed (see, e.g., Claerbout, 1976, p. 202) to extend the range of acceptable accuracy. The important point here, however, is that for *some* range of angles, accurate time migration is achievable even though the $P_{z'z'}$ term is *completely* neglected. For depth migration, on the other hand, we shall show that proper approximation of $P_{z'z'}$ is absolutely necessary if we are to obtain correct results, even for waves traveling vertically upward.

Let us return to the full, scalar wave equation in the moving coordinate frame [equation (A–3) in Appendix A]. Instead of forcing the frame velocity to equal the medium velocity as we did in Appendix A, let us now fix the frame velocity at some constant value. Then, all the terms in equation (A–3) involving derivatives of $\bar{v}$ vanish, leaving

$$P_{z't'} + \frac{\bar{v}}{4} P_{z'z'} = -\frac{\bar{v}}{4} P_{x'x'} + \bar{v}\left(\frac{1}{v^2} - \frac{1}{\bar{v}^2}\right) P_{t't'}. \quad (B-1)$$

We have dropped the primes on P, v, and $\bar{v}$ for clarity but retain them in the independent coordinates to avoid confusion.

If the $P_{z'z'}$ term could be ignored, the resulting equation could be treated conveniently by splitting it into two more simple equations (Marchuk splitting; Mitchell, 1969),

$$P_{z't'} = -\frac{\bar{v}}{4} P_{x'x'}, \quad (B-2)$$

and

$$P_{z't'} = \bar{v}\left(\frac{1}{v^2} - \frac{1}{\bar{v}^2}\right) P_{t't'}. \quad (B-3)$$

In time migration, $\bar{v} = v$; consequently, equation (B-3) vanishes leaving equation (B-2) to perform the required diffraction of waves. For depth migration, in which $\bar{v}$ does not everywhere equal v, the two equations would be solved alternately, once for each step of integration in z'. Equation (B-2) performs diffraction and equation (B-3), a shift in time, thereby allowing refraction to occur. But here difficulties arise. Equation (B-2) (a 15-degree equation) diffracts waves as though the medium velocity were the frame velocity $\bar{v}$, instead of the actual velocity v. Moreover, as we shall show, the time shifts specified by equation (B-3) are incorrect for waves propagating at any angle. The source of these difficulties is the dropping of the $P_{z'z'}$ term.

To see this, we first perform a t'-integration of equation (B-3), yielding an equation equivalent in form to equation (A-8):

$$\frac{DP}{Dz'} = P_{z'} - \bar{v}\left(\frac{1}{v^2} - \frac{1}{\bar{v}^2}\right) P_{t'} = 0. \quad (B-4)$$

Equation (B-4) indicates that P is kept constant along the trajectory given by

$$\frac{dt'}{dz'} = -\bar{v}\left(\frac{1}{v^2} - \frac{1}{\bar{v}^2}\right). \quad (B-5)$$

Stated differently, for each integration step $\Delta z'$, equation (B-3) shifts the time t' by an amount

$$\Delta t' = -\bar{v}\left(\frac{1}{v^2} - \frac{1}{\bar{v}^2}\right) \Delta z'. \quad (B-6)$$

To see that the time shift indicated by equation (B-5) is erroneous, consider migration of a horizontal reflection occurring on a surface-recorded seismic trace at time

$$t = t_0 = 2 \sum_{i=1}^{N} \frac{\Delta z}{v_i}.$$

Here, we have assumed the earth consists of N horizontal layers (down to the reflector of interest), each of thickness Δz and having depth-varying velocity v_i. After one integration step (i.e., one downward continuation step), the time of the reflection in the moving coordinate system governed by equation (A-2) should be

$$t' = \left(t_0 - \frac{2\Delta z}{v_1}\right) + \frac{2\Delta z'}{\bar{v}}$$
$$= t_0 + \Delta t'_1, \quad (B-7)$$

where, using the identity $\Delta z = \Delta z'$, we have

$$\Delta t'_1 = 2\left(\frac{1}{\bar{v}} - \frac{1}{v_1}\right) \Delta z'.$$

In general, the time shift at the ith integration step will be given by

$$\Delta t'_i = 2\left(\frac{1}{\bar{v}} - \frac{1}{v_i}\right) \Delta z'. \quad (B-8)$$

At any step, let us downward continue the wave field using equations (B-2) and (B-3) alternately. For horizontal reflections, the diffraction equation (B-2) has no effect on timing, i.e., the choice of $\bar{v}$ is immaterial, and equation (B-2) need not even be applied. In other words, the shifting equation (B-3) must bear the entire burden of migrating horizontal reflections. On comparing equations (B-6) and (B-8), however, we see that equation (B-3) will not provide the correct shift; hence, the migrated reflection will be incorrectly positioned in depth.

Equation (B-1), however, is *exact*. The discrepancy lies in the simple neglect of $P_{z'z'}$. How then should $P_{z'z'}$ be approximated? Of the several ways, the following is particularly illuminating. For convenience, Fourier-transform time in equation (B-1) to obtain the monochromatic wave equation:

$$i\omega P_{z'} + \frac{\bar{v}}{4} P_{z'z'} = -\frac{\bar{v}}{4} P_{x'x'} - \omega^2 \bar{v}\left(\frac{1}{v^2} - \frac{1}{\bar{v}^2}\right) P. \quad (B-9)$$

Now make the one-way wave assumption,

$$P_{z'} = \left(\sum_{k=0}^{\infty} \alpha_k \frac{\partial^k}{\partial x'^k}\right) P, \quad (B-10)$$

where the α_k are constants. (More generally, the α_k can be considered spatially variable but this line will not be pursued here.) Equation (B-10) can be used to express $P_{z'z'}$ in terms of P. Making that substitution into equation (B-9) replaces the explicit presence of $P_{z'z'}$. This approach is analogous to that employed in the development of higher-order approximations (45-degree, etc.) in time migration.

Equating coefficients of $\partial^j/\partial x^j$ results in the one-way equation

$$P_{z'} = 2i\omega\left(\frac{1}{v} - \frac{1}{\bar{v}}\right) P + \left(\left[\frac{-4\omega^2}{v^2} - \frac{\partial^2}{\partial x'^2}\right]^{1/2} - \frac{2i\omega}{v}\right) P. \quad (B-11)$$

We can now apply Marchuk splitting to equation (B-11) to obtain separate equations for diffraction and shifting. Doing so and transforming back into the time domain, we get

$$P_{z't'} + \frac{v}{4} P_{z'z'} = -\frac{v}{4} P_{x'x'} \quad \text{(diffraction)}, \quad (B-12)$$

and

$$P_{z'} = 2\left(\frac{1}{v} - \frac{1}{\bar{v}}\right) P_{t'} \quad \text{(shifting)}. \quad (B-13)$$

From a comparison of these equations with equations (B-2) and (B-3), we draw several conclusions. The diffraction equation, now expressed as a two-way wave equation, contains the $P_{z'z'}$ term explicitly. Note, however, it differs from equation (B-2) in that it contains the actual medium velocity v rather than the frame velocity $\bar{v}$. This result is more satisfying; equation (B-2) will not, in general, diffract properly since $\bar{v}$ can be chosen arbitrarily. The shifting equation (B-13) agrees with the simple argument for a horizontal reflector that led to equation (B-8). The influence of $P_{z'z'}$ is embedded implicitly in this equation.

Equations (B-12) and (B-13) could also have been derived by taking a correct one-way wave equation first in a fixed coordinate frame and then transforming to the moving frame.

To conclude, wave-theoretical depth migration must be implemented with a form of equations (B-12) and (B-13) in which $P_{z'z'}$ has been correctly approximated rather than with equations (B-2) and (B-3) in which the simplest assumption is made, viz., $P_{z'z'} = 0$. Although that simple assumption previously proved adequate for wave-theoretical *time* migration, more care must be exercised in developing *depth* migration.

TIME MIGRATION—SOME RAY THEORETICAL ASPECTS *

BY

P. HUBRAL**

Abstract

HUBRAL, P., 1977. Time Migration—Some Ray Theoretical Aspects, Geophysical Prospecting 25, 738-745.

Using an elementary theory of migration one can consider a reflecting horizon as a continuum of scattering centres for seismic waves. Reflections arising at interfaces can thus be looked upon as the sum of energy scattered by interface points. The energy from one point is distributed among signals upon its reflection time surface. This surface is usually well approximated by a hyperboloid in the vicinity of its apex. Migration aims at focusing the scattered energy of each depth point into an image point upon the reflection time surface. To ensure a complete migration the image must be vertical above the depth point. This is difficult to achieve for subsurface interfaces which fall below laterally inhomogeneous velocity media. Migration is hence frequently performed for these interfaces as well by the Kirchhoff summation method which systematically sums signals into the apex of the approximation hyperboloid even though the Kirchhoff integral is in this case not strictly valid. For a multilayered subsurface isovelocity layer model with interfaces of a generally curved nature this can only provide a complete migration for the uppermost interface. Still there are various advantages gained by having a process which sums signals consistently into the minimum of the reflection time surface. The position of the time surface minimum is the place where a ray from the depth point emerges vertically to the surface. The Kirchhoff migration, if applied to media with laterally inhomogeneous velocity, must necessarily be followed by a further time-to-depth migration if the true depth structure is to be recovered. Primary normal reflections and their respective migrated reflections have a complementary relationship to each other. Normal reflections relate to rays normal to the reflector and migrated reflections relate to rays normal to the free surface. Ray modeling is performed to indicate a new approach for simulating seismic reflections. Commonly occuring situations are investigated from which lessons can be learned which are of immediate value for those concerned with interpreting time migrated reflections. The concept of the 'image ray' is introduced.

Introduction

Migration is usually most useful in areas of complex geology if performed with 3-D migration schemes on 3-D recorded seismic data. The process aims at focusing unstacked or stacked seismic reflections into data more suitable for interpretation. These data are still presented as a function of two-way time.

---

* Received August 1976.
** Bundesanstalt für Geowissenschaften und Rohstoffe, 3 Hannover-Buchholz, Stilleweg 2.

Scattered energy is usually contracted and weak segmented reflections often appear in a continuous and geologically more reasonable form. Faults show up frequently more distinct. To account for all observed phenomena the theory of migration (Hagedoorn 1954, Claerbout and Doherty 1972, French 1974) must certainly involve more than can be reasoned with ray-theoretical considerations alone. Still, ray theory is entirely sufficient to investigate the relationship between travel times of unmigrated and migrated reflections as done in this work. The conclusions at which we have arrived are in fact quite simple. They have, to our knowledge, not been stated in this form previously elsewhere.

Exploration seismologists base their interpretational skill and judgement largely on the study of primary reflections for selected key horizons. These are normally available to them in either a stacked or time-migrated form. In the presence of structure both types of reflections may considerably differ from each other. Migrated reflections often provide a more realistic picture of the geology which may not that simply be inferred from stacked reflections. Migrated reflections need, however, by no means always provide a more truthful picture of subsurface reflectors. This results from extending the Kirchhoff summation process which sums the scattered signals into the apex to media with lateral inhomogeneous velocity. All migration schemes are related to the wave equation (Larner and Hatton 1976). As they transform one wave field into another they have been extensively studied from various wave theoretical points of view. Little attention has, however, been given to the relationship which the travel times of migrated primary reflections have to their actual reflectors at depth. Rays lack physical existence and seem to be most inappropriate for a theory concerned with diffraction phenomena. Still, they contribute in clarifying this particular aspect.

Theory

If the geology is complex it is most unlikely that it can be approximated by a 2-D model. It is for this reason that our theoretical considerations are initially based on 3-D models though subsequent examples have for reason of simplicity a two-dimensional nature. Fig. 1 shows a 3-D isovelocity layer model with interfaces of a generally curved nature. Three interfaces are shown from an arbitrary number that may be permitted. All media are isotropic. It is this earth model for which the theory discussed here is valid. Wavefronts are thus always perpendicular to rays. Some source-receiver pairs are placed upon the free surface. The position of the jth pair is indicated by R_j. One scattering point D has been selected on the third interface. Rays from various source-receiver pairs have been drawn to D. Along each path a wave travels to D and part of the energy is scattered back to the receiver. Rays obey Snell's

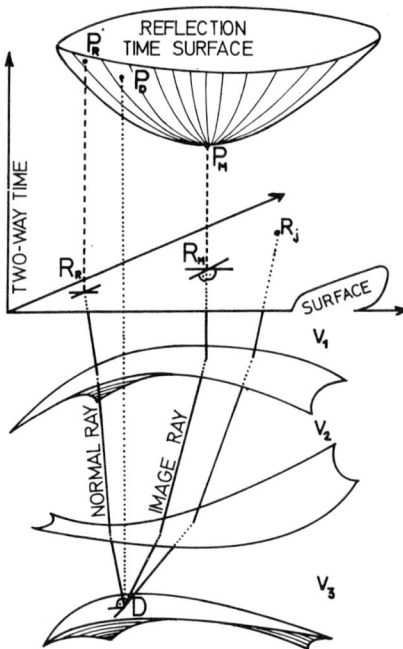

Fig. 1. 3-D subsurface isovelocity layer model featuring an interface scatterer D and its reflection time surface.

law at interfaces. Plotting the two-way time along a ray as a function of the source-receiver position provides the reflection time surface for D (fig. 1). Each subsurface scatterer usually results in a different reflection time surface. Three points upon it have special significance: P_R relates to a free surface point R_R where a ray normal to the interface in D emerges. The two-way time along this normal incidence ray is the primary reflection time to the interface in D recorded with a source-receiver pair in R_R. This particular time is usually well approximated by the stacked primary reflection time of a CDP gather with common mid point in R_R. P_D is a point within the time surface vertical above D. If the scattered energy of the coherent signals within a reflection time surface is summed into a signal in P_D then the migration is complete in the sense that a subsequent time-to-depth conversion is achieved by only correcting the migrated travel times. No further 'depth migration' is involved. A search for the image P_D within the reflection time surface requires, however, considerable decision making as its position with respect to the time surface minimum may differ for different depth points. It is thus for practical reasons that the 'image' is always chosen to coincide with a signal in P_M, the apex of the reflection time surface. The reflected signal recorded in R_R thus migrates to P_M rather than P_D. Practically, the image falls in fact into the apex of the approximation hyperboloid to the time surface. Both points are, however, very close to each other. It therefore appears reasonable to define *point image migration* as a process which establishes the depth point image *per definition* in the minimum of its reflection time surface even for models for which the Kirchhoff integral is not strictly valid. In this way one honours more the actual procedure commonly applied rather than the desired result which can not be obtained by this particular migration scheme. From this practical definition one can immediately conclude that a migration of the type considered is by no means complete for reflectors falling below interfaces which are not horizontal. Point image migration has hence to be followed by an additional time-to-depth migration if the true depth structure is to be recovered. This additional migration is necessary as P_M is usually not vertical above D (fig. 1). In spite of the apparent drawbacks this practical definition proposed here has also some interesting ray theoretical aspects. One can conclude that the position of P_M coincides with the position of the ray which emerges from D vertical to the free surface. This can be proven as follows:

The ensemble of rays of fig. 1 is necessarily identical with the one related to a wave originating in D. As wavefronts are perpendicular to rays the travel time of a wave from D to reach the surface is therefore half the time attributed to the reflection time surface. Both time surface minima have a horizontal tangent. Their position must therefore coincide with the position of the vertically emerging ray from D. *Migrated primary reflection times for a given depth model are consequently obtained by tracing rays vertically down from the free surface to the desired reflector at depth while plotting the two-way times at the respective surface positions of the rays.* Rays which emerge vertically at the free surface are subsequently referred to as *image rays*. They are naturally vertical only in a medium with constant—or only vertically inhomogeneous—velocity which may lie above (but not below) one with more complex velocity distribution. Below a curved interface they are refracted in a similar way as normal incidence rays. We have called them image rays as they connect a depth point with the surface position of its image. The signal positions and two-way times related to image rays are not affected by the migration process proposed here. Signals for all other rays (fig. 1) are migrated. *Unlike normal incidence rays image rays conform with each other irrespective of the interface to which they belong.* Both kinds of rays are complementary in the sense that normal rays are normal to the selected interface while image rays are normal to the free surface. The two-way primary reflection time for a shot-receiver pair in the free surface pertains to a ray normal to the selected interface. This time is transformed by the migration process into the time by which a wave reflected at the free surface is recorded by a shot-receiver pair within the normal incidence point of the normal ray. Based upon these simple considerations some commonly occurring 2-D models have been studied to demonstrate the importance of image rays as a means of interpreting and simulating migrated reflections. Image rays relate to migrated reflections in a similar way as normal incidence rays relate to stacked primary reflections (Taner, Cook, and Neidell 1970). They contribute to understanding the theory and limitations of the time migration process from a ray theoretical point of view. When one computes interval velocities from migration velocities they play the same fundamental role which normal incidence rays play in connection with computing interval velocities from stacking velocities. The complementary relationship between normal rays and image rays holds whenever the velocity distribution is isotropic and inhomogeneous.

Examples

Migration in its usually accepted sense poses no problem for the chosen earth model if an arbitrarily curved subsurface reflector falls below a system of plane horizontal isovelocity layers. Image rays are then strictly vertical down to the uppermost curved interface and a performed migration is complete. Quite often, however, curved interfaces fall below other curved velocity interfaces and image rays deviate from strictly vertical rays. The more refracted they are the less complete is in fact the migration. Though usually most effective if the structure is complex, migration may also be most incomplete in such cases. Errors introduced by predicting the true depths by only scaling the time-migrated reflections to depth can then be quite severe. As the following examples show it is quite likely that for migrated reflections puzzling situations may arise which are similar to those of stacked primary reflections of nonlinear subsurface reflectors (Taner et al. 1970). There is, however, one positive aspect which reduces the diversity of migrated primary travel time functions and deserves particular mentioning: It is that they remain invariant with the depth of the structure. This does not apply to stacked primary reflections which may be quite sensitive to the depth of a nonlinear reflector. It is well known that a trough near the surface may look very different on a stacked section from a trough at some depth. After section migration it will, however, look the same. Fig. 2 shows some plane horizontal interfaces below typical geological structures. Fig. 2b shows the normal incidence rays to the plane interfaces and fig. 2a shows the respective image rays. The two-way travel times along both types of rays are plotted in fig. 2c. The time-migrated primary reflections are dotted. Normal rays to the first interface are not shown. The structure of the uppermost interface changes the linearity of all other reflectors in a way which gets worse with the depth of the reflectors. The amount of travel time distortion for the migrated reflections becomes in

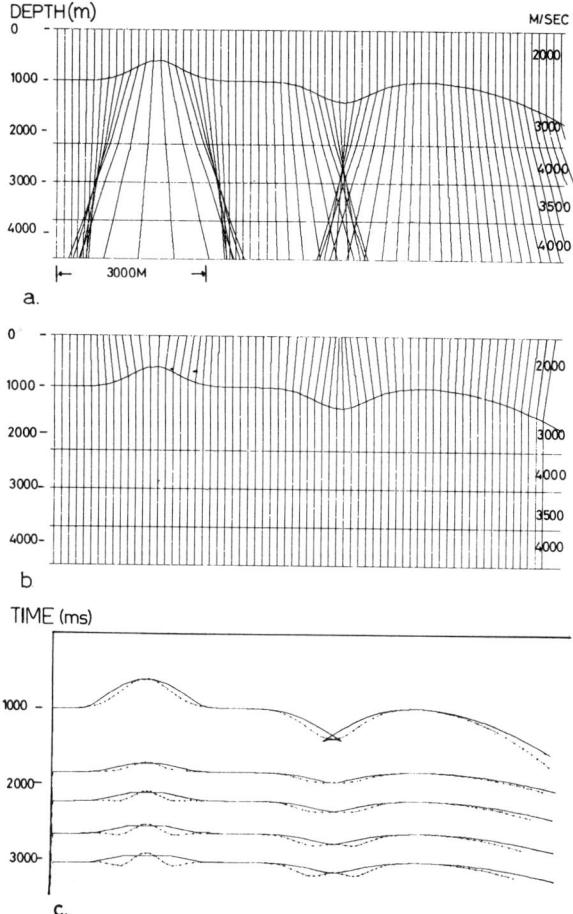

Fig. 2. a. Image rays to interfaces; b. Normal incidence rays to plane interfaces; c. Normal and time-migrated (dotted) reflections.

fact larger than that for stacked reflections. It is only the primary reflections for the first interface which are truely migrated in the conventional sense. All other depth reflectors are 'time-migrated' if one accepts the definition of 'migration' proposed in this work. To recover the linearity of all interfaces deeper than the uppermost one a further time-to-depth migration has to be applied in order to correct the non vertical image rays.

Fig. 3 shows a North German salt diapir where normal rays and image rays have been traced down to the Permian formation. The migrated primary reflections relate to *almost vertical* image rays. They thus provide a better approximation to the true depth model. As can be judged from the non vertical image rays there will, however, be some lateral distortion left in the structure

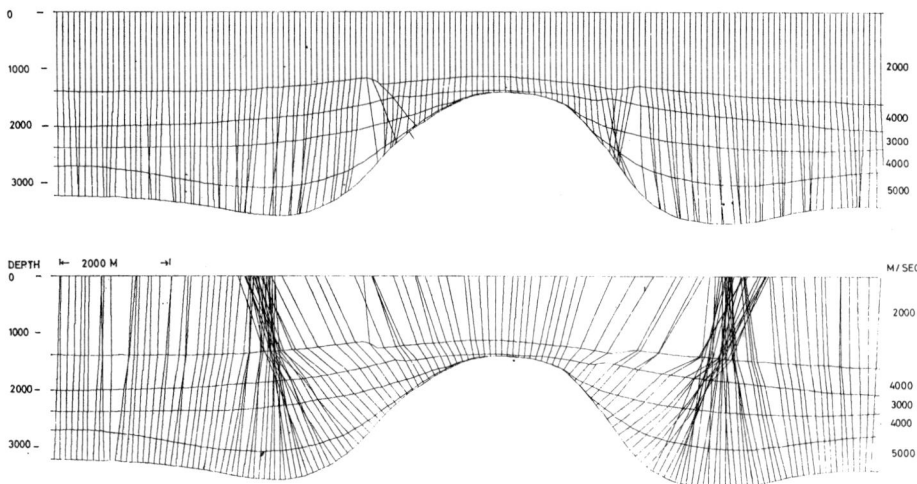

Fig. 3. a. North German salt dome model with normal incidence rays to Permian formation; b. North German salt dome model with image rays to Permian formation.

after time migration. The normal rays exaggerate largely the width of the salt dome on the stacked section. As revealed by the image rays the actual width of the dome will appear slightly decreased on the time migrated section. The deviation of image rays from the true vertical direction will often affect the interpretation of primary migrated reflections in a negative way. Caution must hence be exercised particularly if the reflectors used for interpretation fall below curved or dipping interfaces near the surface. The systematic summing of signals into the apex of reflection time surfaces may in fact occasionally create migration problems which are more severe than those existent on the original stacked section. An image ray describes the subsurface locations from which information is gathered that is displayed in a point image migrated trace. As image rays may cross each other one depth point may in fact be imaged into two different points on the section. This should disappoint only those readers who believe that the Kirchhoff summation can be applied to vertically and laterally inhomogeneous velocity media without having to pay any penalties. It cannot disappoint readers who accept the definition of point image migration given in this work—a definition valid for subsurface models more general than those upon which the Kirchhoff integral is based.

Conclusions

If signals scattered at subsurface points are summed into the minimum of their respective reflection time surfaces, the migration in its usual sense is only complete for depth points falling below plane horizontal isovelocity layers. Scattered signals for depth points below curved interfaces provide then no longer a complete migration when summed into the reflection time surface minima. A new ray theoretical definition is hence proposed for point image migration which extends the Kirchhoff summation method to all reflectors of the 3-D layered inhomogeneous earth model used in this work. As shown in this paper this process can be considered as pushing source-receiver pairs into the subsurface along image rays. It thus differs from the finite difference wave equation method which is generally viewed as a process which aims at pushing source receiver pairs vertically down into the subsurface. This aim is to our knowledge not as yet achieved by any time migration scheme. Point image migration as defined above is hence incomplete in its conventional sense. It offers, however, a strict complementary logic to the process of stacking. It provides consistent surface measurements which relate to subsurface reflectors in a unique way and which can be used for a further time-to-depth migration if the true depth-structure is to be recovered. The introduction of image rays contributes in our opinion much to clarifying the theory of migration from a ray theoretical point of view. Image rays offer themselves for interpretive seismic modeling. By accepting their existence one immediately realizes that time migrated sections—after having been scaled to depth—must not necessarily overlay depth sections which have been obtained by conventional time-to-depth migration methods which consider Snell's law and use known interval velocities and travel times of stacked primary reflections. Though a point image migration is not complete it resolves generally much better the reflecting horizons due to the contraction of disturbing diffraction patterns. Using primary time migrated reflection times and available interval velocities and performing a time-to-depth migration with image rays should result in a clearer picture of the true reflecting horizons as can generally be obtained from conventional time to depth migration methods that use stacked primary reflections and normal rays.

References

Claerbout, J. F. and Doherty, S. M., 1972, Downward continuation of moveout corrected seismograms, Geophysics 37, 741-768.

French, W. S., 1974, Two-dimensional and three-dimensional migration of model-experiment reflection profiles, Geophysics 39, 256-277.

Hagedoorn, J. G., 1954, A process of seismic reflection interpretation, Geophysical Prosp. 2, 85-127.

Larner, K. and Hatton, L., 1976, Wave equation migration: Two approaches, Paper presented at the 8th Annual Offshore Technology Conference, Houston, Texas, 1976.

Taner, M. T., Cook, E. E. and Neidell, N. S., 1970, Limitations of the reflection seismic method; lessons from computer simulation, Geophysics 35, 551-573.

THE WAVE EQUATION APPLIED TO MIGRATION *

BY

D. LOEWENTHAL, L. LU, R. ROBERSON, and J. SHERWOOD **

Abstract

LOEWENTHAL, D., LU, L., ROBERSON, R., and SHERWOOD, J., 1976, The Wave Equation Applied to Migration, Geophysical Prospecting 24, 380-399.

Claerbout's method has been implemented for the migration of stacked seismic data. A simplified description of the method is given together with an account of some of the practical programming problems and the types of inaccuracy encountered. Routine production results are considered to be comparable or superior to the results derived from alternative migration techniques. Particular advantages are 1) the possibility of using a detailed velocity model for the migration and 2) the preservation of the amplitude and character of the seismic events on the migrated time section.

Introduction

In recent years Jon Claerbout and his students at Stanford University have published many articles on numerical studies of approximations to the wave equation. Although their work has been strongly oriented towards applications in exploration seismology it seems that the geophysical industry has not been overly enthusiastic to experiment with or implement these techniques in routine data processing. With the conviction that maximizing the extraction of geophysical information from seismic data is ultimately dependent on the use of the wave equation, we felt that it was desirable to initiate an appropriate development program.

As an initial step we have implemented Claerbout's method for the migration of stacked seismic data. A simplified summary of the method will be given here, together with examples of typical results.

Assumptions and Objective

The data consist of a CDP stack section which we will define by its amplitude $A(x, t)$ as a function of the two-way travel time t and the position x along a straight surface line. Provided that reasonable stacking velocities and amplitude control have been used, the primary reflections on this section approximate

---

* Paper read at the 36th meeting of the European Association of Exploration Geophysicists, Madrid, Spain, June 1974.
** Digicon, Inc., 3701 Kirby Drive, Houston, Texas 77098.

what would be generated by a point source—receiver combination yielding independently recorded seimic traces at some constant increment along the x axis. For simplicity we will not discuss the effect of the migration on multiple reflections or on reflections from locations outside the plane of the section. Finally, an inherent assumption is that the basic seismic pulse is a simple impulse and that its amplitude is invariant as it is transmitted through the earth layers (see Foster 1975, for a pertinent discussion of this subject). One should note that this latter assumption and the neglect of multiples are compatible with an earth which is only weakly inhomogeneous.

Having made all these assumptions, our objective now is to modify the CDP section into a form more representative of a cross-section through the earth. This parallels the effort of research seismologists to estimate the detailed acoustic velocity and density variations in the subsurface. This is a difficult problem in statistical estimation and is beyond the scope of the present discussion. We will lower our objectives here and state that we wish to utilize our gross velocity estimates in order to derive from the CDP stack section $A(x, t)$ a function $B(x, z)$ where B is related to the variation of the earth's acoustic impedance with horizontal position x and depth z. In fact, because of the relative inaccuracy of our velocity information, we will content ourselves with deriving a function $B(x, \tau)$, where the depth variable z has been replaced by a less sensitive parameter τ representing the *vertical* two-way travel time.

In the following sections we will show that, for an earth with the assumed properties, a relatively simple transformation of $A(x, t)$ will yield a reasonable approximation to the desired migrated section $B(x, \tau)$.

METHOD

The CDP stack section may be considered as a wave field measured at the surface of the earth. Given the approximate velocity variations within the earth we are going to downward continue this wavefield into the subsurface and elucidate the source of the reflected and diffracted seismic events. The detailed procedure for this draws heavily on Claerbout's elegant and economic method for propagating a wave field using finite difference approximations to the wave equation (Claerbout and Johnson 1971, Claerbout and Doherty 1972). To achieve these favorable economics it is necessary for the significant components of the wavefield to be traveling within some small angular distribution around a given direction, which in our case is vertically upwards. At first sight, this requirement implies that Claerbout's technique is inapplicable to a CDP stack section, which is produced by the upward reflection and diffraction of originally downward traveling waves. Fortunately, this is not the case, since simplistic considerations permit us to visualize the CDP stack section as a superposition of upward traveling waves.

Let us quantize the subsurface so that it consists of small diffracting points, each with an appropriate scattering strength. Normally, of course, the strongest scatterers, or diffractors, will be distributed along strong reflecting interfaces.

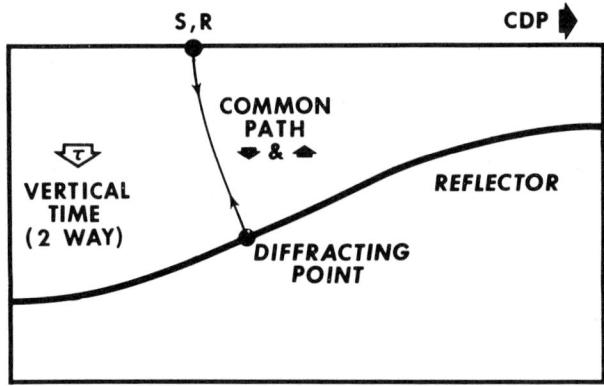

Fig. 1. The earth cross section.

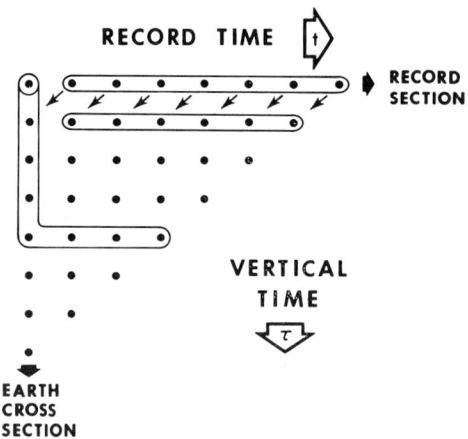

Fig. 2. Schematic diagram of the wave equation migration process.

Now consider a single point source and receiver combination, the impulsive source being initiated at time zero (see figure 1). The downward ray path and travel time $t/2$ to a diffracting point is identical with the return upward path and travel time $t/2$ to the receiver. The amplitude of the return will be proportional to the strength of the diffracting element. It is clear that an equivalent recording will result from a source initiated at time zero at the diffractor, with the strength of the diffractor, provided that the velocity $c(x, \tau)$ in the medium is halved. The CDP stack section can then be considered as due to upcoming waves from the totality of quantized diffractor sources, each initiated at time zero with an appropriate strength, and with the instantaneous velocity throughout the medium being $c(x, \tau)/2$.

A discussion of this equivalent mechanism for generating a CDP stack section will clarify the steps in the inverse procedure of migration. In figure 2, the horizontal axis is the observed two-way travel time t, the vertical axis is the *vertical* two-way travel time τ, and the CDP location x is perpendicular to the plane of the figure. The top left hand point is the origin $t = \tau = 0$. Thus, the vertical left hand column represents an earth cross-section composed of the strengths of the hypothetical diffractor sources initiated at time $t = 0$.

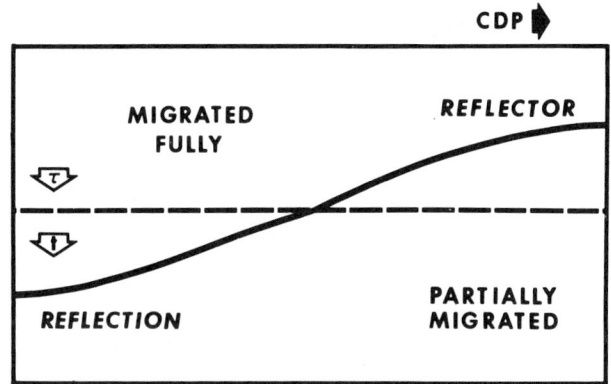

Fig. 3. An intermediate stage.

The top horizontal row corresponds to the CDP record section observed at the earth's surface, $\tau = 0$. In mathematical notation we can define the complete wave field as $A(x, t, \tau)$. Given that the wave-field is basically upward traveling, and given a particular initial source distribution $A(x, 0, \tau)$, it is possible to utilize Claerbout's algorithm to determine the complete wave field $A(x, t, \tau)$ using incremental steps Δt in the observational time direction. The surface observations $A(x, t, 0)$ correspond to a conventional seismic cross-section. Alternatively, if this seismic cross-section $A(x, t, 0)$ is given, it is possible to extrapolate the wave field downwards using incremental steps $\Delta \tau$. The subset of derived values $A(x, 0, \tau)$ corresponds to the diffractor source distribution and provides the desired migrated section result.

The actual order in which we choose to execute the computations is depicted by the L shaped outline in figure 2. This illustrates the data stored in the computer system at an intermediate stage of the processing. It consists of the completely migrated results down to a "depth" τ, followed by the remnant partially migrated data which still requires propagation through the deeper layers of the subsurface. This intermediate stage is exhibited with more clarity in figure 3. The upper part of the data field has been fully migrated, original reflection data having been transformed into the equivalent reflector positions. The lower part of the data is only partially migrated and is the seismic wave field that would be observed if the point source and detector combinations were positioned along the dashed line rather that at the surface.

Model Study Results

We will first examine the results of the wave equation migration process using simple synthetic data. Figure 4 shows an earth model, consisting of

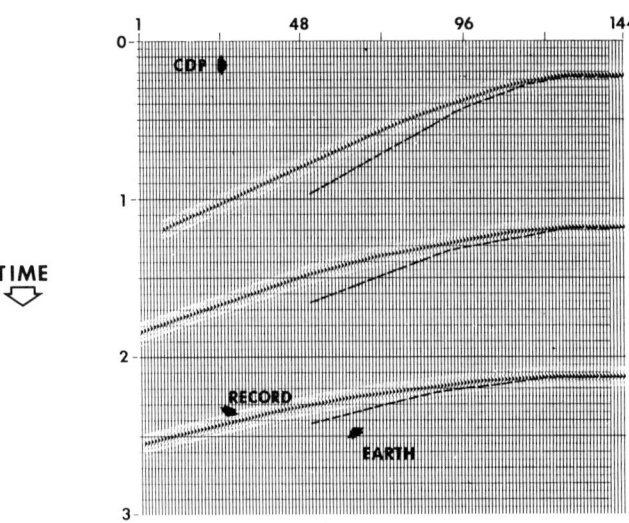

Fig. 4. Earth model.

three continuous reflectors $A(x, 0, \tau)$, superimposed on which is the corresponding seismic record section. The latter was derived from the earth cross-section by means of hand calculations using ray theory. The computer inputs were formed by inserting impulses at appropriate points (x, τ) and (x, t) and then convolving these impulses with a simple zero phase filter wavelet (8 Hz low cut, 30 Hz high cut) in order to avoid any drastic aliasing problems during the migration procedure. The model parameters were a constant instantaneous velocity of 3048 m/s, a horizontal or trace interval of $\Delta x = 24.4$ m, and a time dimension sampling rate of 4 ms. The maximum "dip" in the earth model is 12.5 ms/trace corresponding to a dip of about 39°. This leads to a reflection event on the record section having a dip of 10.0 ms/trace.

For the sake of economy, it is desirable to speed up the migration procedure by stepping down through earth layers which have individual two way travel times $\Delta\tau$ considerably greater than 4 ms. Figure 5 shows the result of migrating the synthetic record using a layer thickness $\Delta\tau$ of 200 ms after first resampling the data to an 8 ms increment. The comparison of the migrated result with the original earth model illustrates the inadequacy of these parameters. Distortion is particularly apparent for the higher frequencies at large dips. In this region the finite difference approximation to the wave equation propagates the wave disturbance at too slow a velocity, resulting in under-migration of the data. The discontinuities on the migrated record section are caused

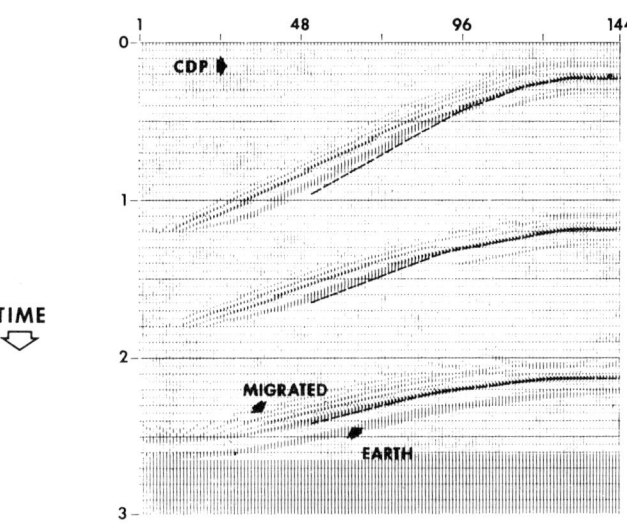

Fig. 5. Migration of the synthetic record section using $\Delta t = 8$ ms and $\Delta\tau = 200$ ms.

by use of the overly large layer thickness of 200 ms. It seems reasonable to believe that a finer sampling would produce more acceptable results. A quantitative investigation of the migration errors as a function of these sampling increments is appropriate at this point. For the convenience of the reader we will delegate detailed mathematics to the appendix and will discuss only the concept underlying the error analysis in the main text.

MIGRATION ERROR ANALYSIS

For simplicity we consider a uniform medium in which a plane wave with frequency f is propagating upwards with acoustic wave velocity c at an angle θ to the vertical. This plane wave corresponds to an exact solution of the wave equation for our idealized homogeneous, nondispersive medium. Now let us take this wave field as observed at the surface $z = 0$ and use Claerbout's finite difference equation as it would appear at the bottom of a layer of thickness

$\Delta z = c\Delta\tau/2$. The analysis in the appendix shows that this extrapolated wave field is consistent with a plane wave whose velocity and angle deviate from c and θ. The migration error for the frequency f can then be expressed as the relative decrease in velocity ($\Delta c/c$) and the decrease in angle $\Delta\theta$. This velocity

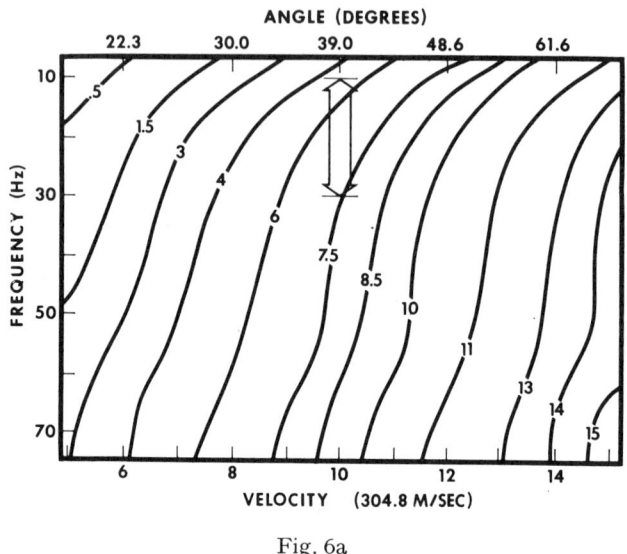

Fig. 6a

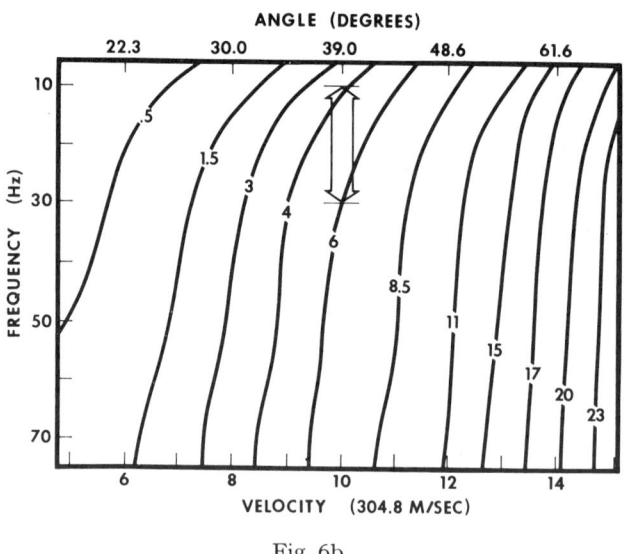

Fig. 6b

Fig. 6. Migration error estimates at dip = 10 ms/trace using $\Delta t = 8$ ms and $\Delta\tau = 200$ ms. (a) velocity error (%), (b) angular error (degrees).

dispersion $\Delta c/c$ is plotted in figure 6a for the highest synthetic record section dip of 10 ms/trace, and the parameters $\Delta x = 24.4$ m, $\Delta t = 8$ ms, and $\Delta \tau = 200$ ms. The vertical axis is frequency in Hz and the horizontal axis is acoustic velocity in units of 304.8 m/s. Since the dip is constant for this display and

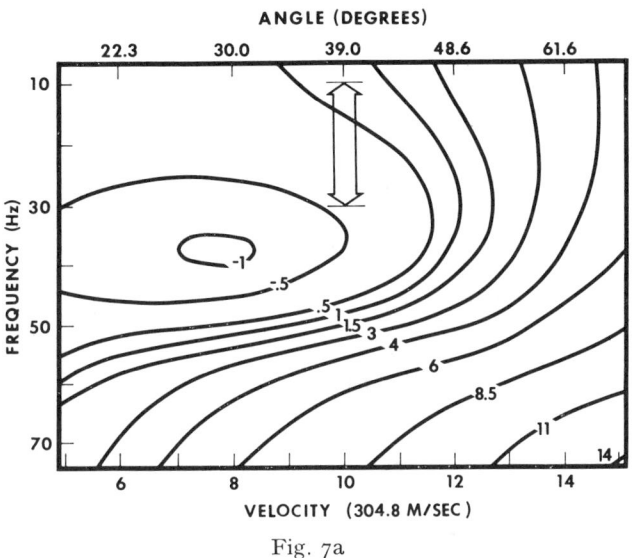

Fig. 7a

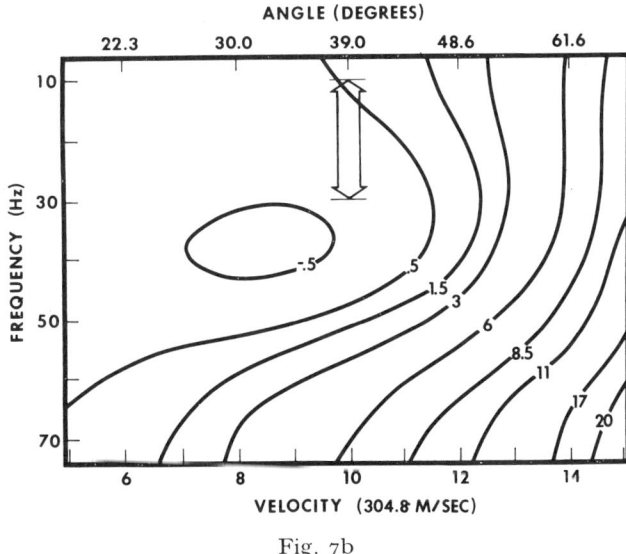

Fig. 7b

Fig. 7. Migration error estimates at dip = 10 ms/trace using $\Delta t = 4$ ms and $\Delta \tau = 20$ ms. (a) velocity error (%), (b) angular error (degrees).

is expressible as $(2\Delta x \sin \theta/c)$, we can also calibrate the horizontal axis linearly in $\sin \theta$. It is clear that the velocity dispersion increases with frequency and with acoustic velocity. The region indicated by the arrow shows the maximum velocity errors pertaining to the synthetic migration results in figure 5. It is emphasized that the 7-13% velocity dispersion refers to the steepest dip on the section in figure 4. The dispersion is less for shallower dips, becoming precisely zero for zero dip.

In figure 6b we show the corresponding plot of the angular error. The region indicated by the arrow again shows the maximum angular error in the previous model study. The steeply dipping event is propagated via the finite difference equations at an angle of 6 to 10 degrees less than the true angle of 39 degrees.

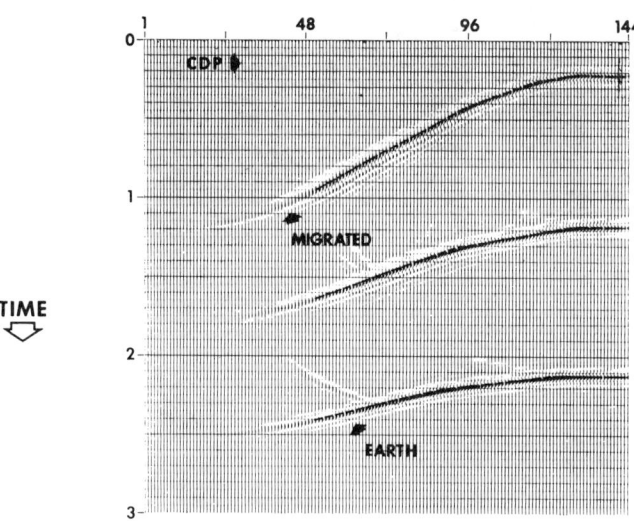

Fig. 8. Migration of the synthetic record section using $\Delta t = 4$ ms and $\Delta \tau = 20$ ms.

Errors of the above magnitude are totally unsatisfactory for quality processing and mainly result from using overly large grid increments Δx, Δt, and $\Delta \tau$. It is not generally considered convenient to diminish Δx in data processing. However, it is usually simple to change Δt and $\Delta \tau$, although at the cost of increased execution time on the computer. Figures 7a and 7b show the velocity dispersion and angular errors that result from changing Δt to 4 ms and $\Delta \tau$ to 20 ms. The errors for our steepest synthetic reflection are now diminished to a velocity dispersion of about 1%, with an angular error of 1 degree. It is believed that such errors are quite satisfactory for normal seismic processing. The result of using these improved values for the parameters Δt and $\Delta \tau$ are shown in figure 8. It is clear that most of the data have been satisfactorily migrated. For shallow dip the wavelet character is essentially unchanged by this wave equation method, a fact which is of considerable convenience for an interpreter wishing to correlate between events on the CDP record section and the migration display. For steep dips, especially at later times, some change in wavelet shape is apparent. Part of this event stretch is anticipated and correct. For example: in a uniform medium simple geometrical considerations show that a reflection event becomes stretched by a factor of ($1/\cos \theta$) when migrated, θ being the true dip of the layering. Thus part of the character change for the steepest reflector in figure 8 is valid. The remainder is due to errors arising partly from Claerbout's approximations to the wave equation and partly from the finite sampling increments Δt, Δx, and $\Delta \tau$.

Program Coding Considerations

The migration algorithm originally disclosed by Claerbout and Johnson (1971) and Claerbout and Doherty (1972) requires that the CDP stack data be multiplexed in time. In our original computer implementation it was convenient to manipulate only 140 seismic traces in this manner, mainly because of the limited random access storage available. Hence, we chose to partition the original seismic data into sections of this size and migrate each section independently, allowing the data to migrate outwards about 72 traces on each side of the section. Each resultant section would then be demultiplexed independently, prior to merging the now overlapping sections back together. At this point the migrated seismic data was in conventional trace sequential form and was ready for display or any additional processing.

This particular implementation of the process was satisfactory for obtaining initial experience and yielded good quality results for probably 75 per cent of our production workload. The limitation in the remaining cases was the inability of the data to migrate more than 72 traces away from the edge of a section.

For small values of Δx, say 25 m or less, the deeper high dip events which should have migrated far outside a section would instead reflect back into the section (a minor example of this is the weak spurious event with opposite dip which occurs around 2.2 seconds in the central part of figure 8). Decimation of the data in the x direction can obviously diminish this problem but in practice this proved to result in an inferior quality for the final migrated section.

In view of these initial results we transferred the migration algorithm to a computer with substantially more random access storage available. This enabled us to segment the data into sections of about 1520 traces, and the migrated results are allowed to expand 256 traces on each side. With this implementation we have not observed any significant problems. During production usage of it we have observed legitimate seismic reflections, at times of 4.0 to 5.0 seconds, migrate over horizontal distances of up to 7500 meters. For the common horizontal sampling increment of 25 meters, this corresponds to a migration over 300 seismic traces.

Migration Examples

Example 1

Figure 9a is a 36-fold conventional stack of our offshore Louisiana data recorded using a 3600 meter streamer. The data have been deconvolved before stack and filtered and equalized after stack. The sampling interval is 4 ms, the trace interval is 50 m. The horizontal scale is about 32 traces to a mile and the vertical exaggeration is roughly 3 to 1. The rms velocity in this area is generally 1830 m/s at 1 second, 2740 m/s at 4 seconds, and 3050 m/s at 6 seconds. In this section, there are two major depositional faults, with a shale sheath under the left fault.

Figure 9b is the migrated time section. It seems to clarify the original stack. First of all, the faults are better defined and the diffractions are clearly migrated back to the fault interface. A more simplified picture is obtained. For example, in the region indicated by a box on both sections, we notice that considerable energy has been moved up-dip. We believe that this fault plane is now closer

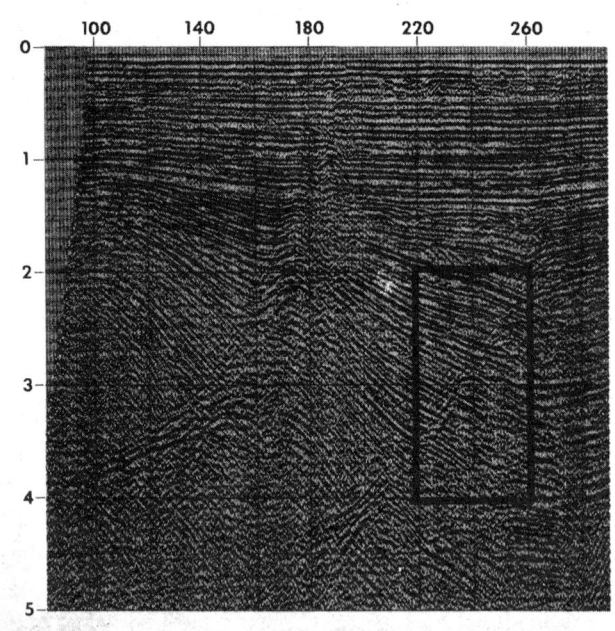

Fig. 9a. 36 fold stack.

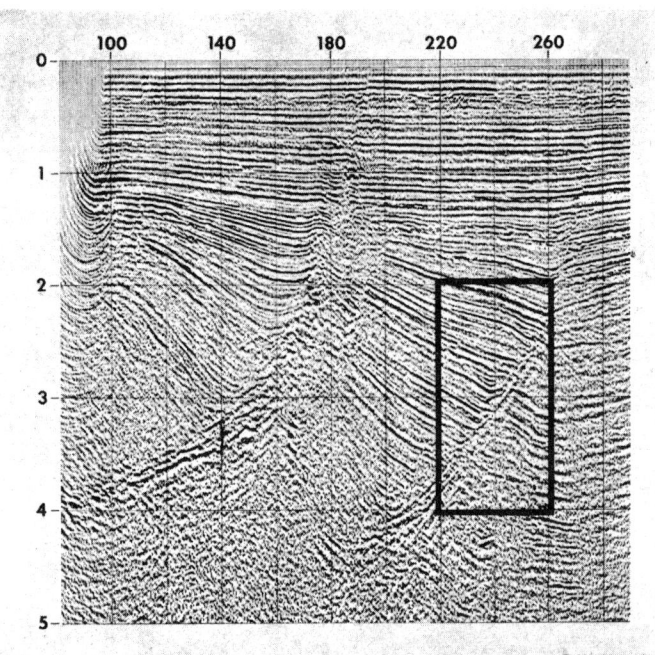

Fig. 9b. Wave equation migration.

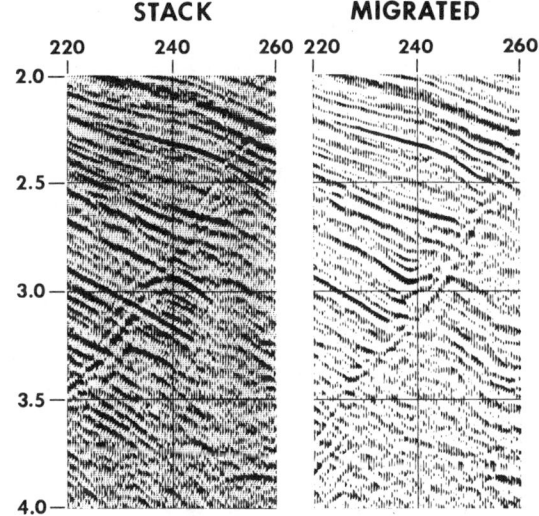

Fig. 9c. Delineation of fault.

to its actual location. The detailed comparison is shown in figure 9c. The strong energy in the taper region caused high amplitude artificial parabolas. The boundary effects always exist on both sides and at the bottom of a section migrated by the wave equation migration process. The migration errors in this example are less than 2% in velocity and 2° in angle for dips less than or equal to 10 ms/trace. A 4 ms sample rate and 100 ms layer thickness were used for the migration.

Example 2

In figure 10a and 10b, there are two interesting examples (indicated in boxes A and B) in addition to the delineation of faults as described in example 1. The first example (enlarged in figure 10c) shows focusing diffractions by migration. The energy in the diffractor region is apparently increased due to focusing. The second example (enlarged in figure 10d) shows delineation of a syncline by migration. The original section indicated very complicated events crossing one another. It is not obvious that this is a syncline or buried focus. The migrated section clearly shows the structure. Some of the steeply dipping events (20 ms/trace) in the original section might be dispersed during migration.

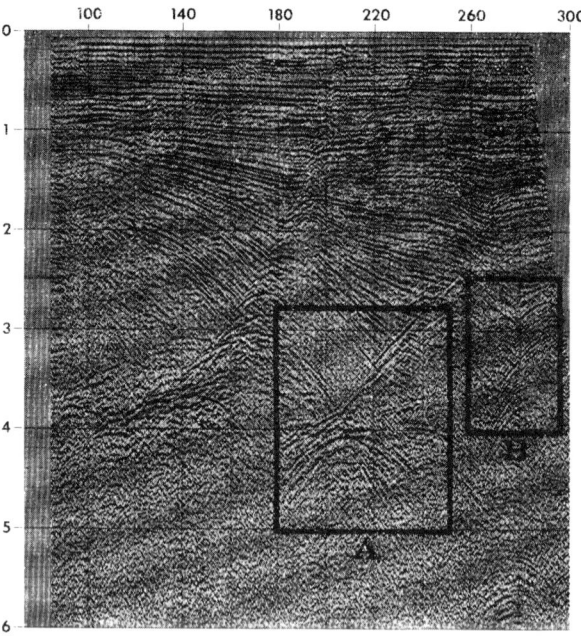

Fig. 10a. Stack.

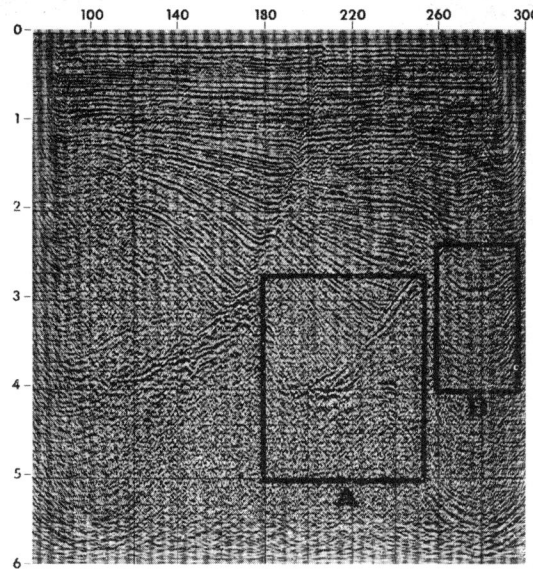

Fig. 10b. Wave equation migration.

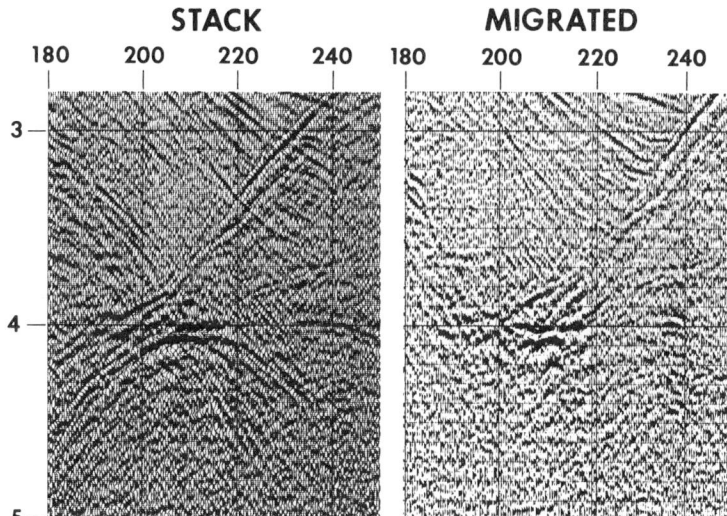

Fig. 10c. Focusing diffraction.

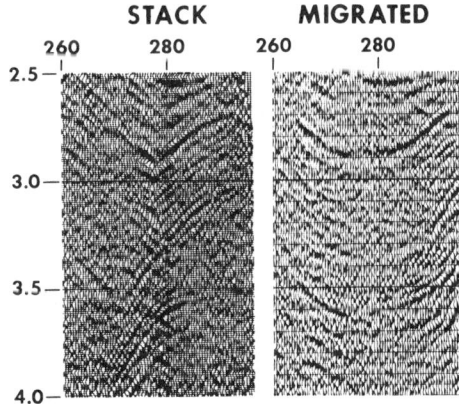

Fig. 10d. Delineation of syncline.

Example 3

Figure 11a shows a typical Gulf coast section with good delineation of a growth fault on the migrated section (figure 11b). The strong diffraction pattern occurring at shot point 70 around 4.2 seconds on the stack section is probably due to out-of-profile returns from the fault plane.

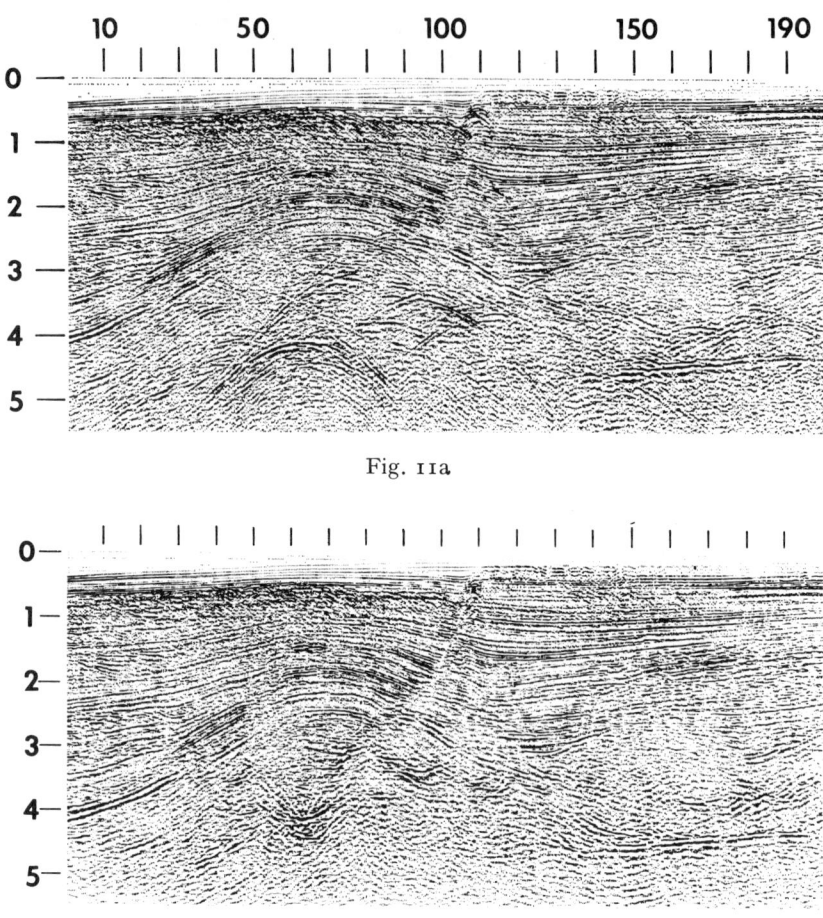

Fig. 11a

Fig. 11b

Fig. 11. Louisiana offshore data. a. 36 fold stack, b. Wave equation migration.

Example 4

Figure 12 is an interesting example of a Gulf coast graben, bounded by two growth faults. The migration better delineates not only faults but also the structure. First of all, it is easy to see that the anticline is more localized by curvature changes after migration, showing sharper structural relief. Apex

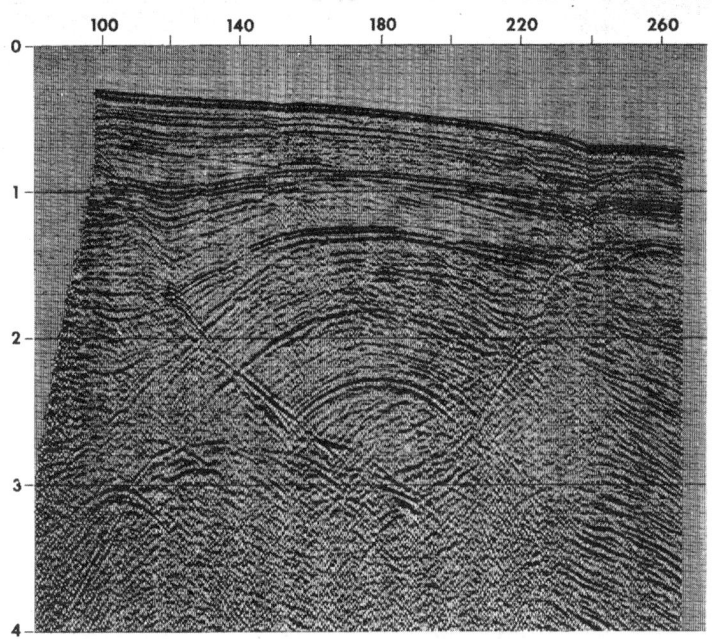

Fig. 12a

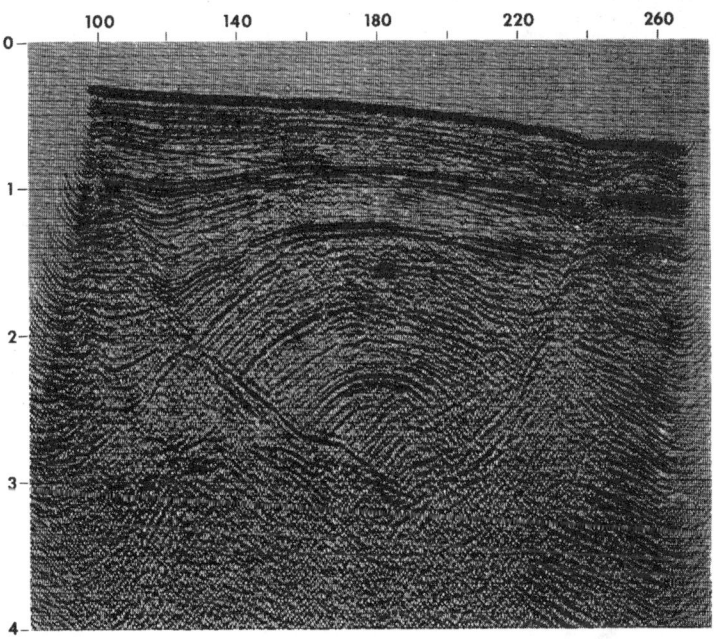

Fig. 12b

Fig. 12a, b. Comprised of stacked (a) and migrated data (b).

amplitudes on the migrated section are increased due to the curvature focusing. Some minor faults also become more obvious by focusing the diffractions. The boundary effects are also noticeable on both sides of this section.

Example 5

Figure 13 demonstrates the improved delineation of a salt dome flank and the preservation of wavelet character of the migrated events.

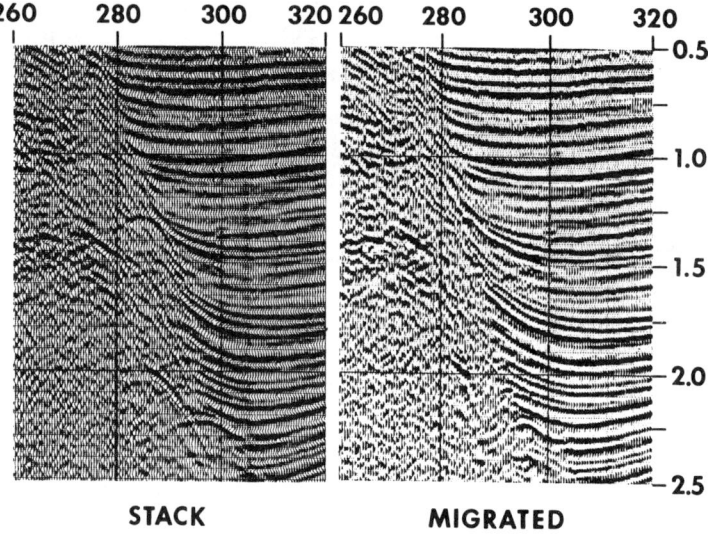

Fig. 13. Preservation of character.

Example 6

In figure 14, the migrated bright spot shows detailed event character which is nearly identical with that of the unmigrated data. This is the result which is

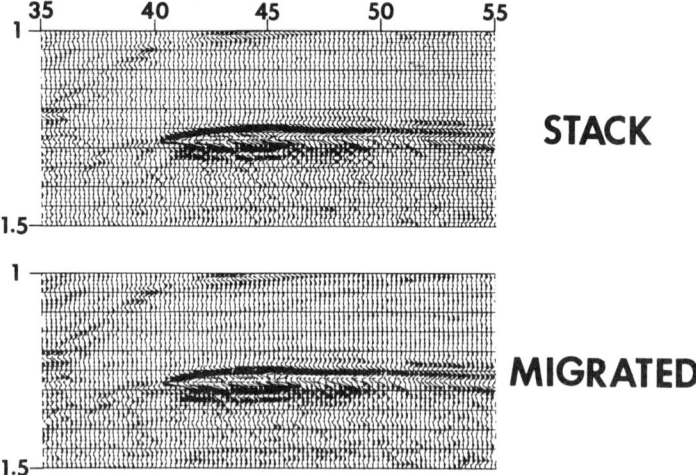

Fig. 14. Preservation of character.

expected because the migration process should not affect any flat event in a weakly inhomogeneous medium. Focusing effect due to weak diffractions is almost unnoticable in this example. Along with the amplitude preservation, we notice that the frequency content has been preserved as well. Due to the character preservation, it is now possible to model with migrated true amplitude sections.

CONCLUSION

Results have been shown of applying the wave equation form of processing to conventional exploration data. Resultant data quality has been very good and gives one an incentive for extending the use of this type of algorithm in production data processing. One obvious route to pursue is the application to the migration of unstacked seismic data, which could well lead to more accurate velocity analysis in addition to an improved migrated section (Sattlegger 1973). We believe that wave equation methods will play a significant role in the continual development of high quality, high resolution seismic surveys.

APPENDIX

In this appendix we briefly derive the migration error analysis, results of which were quoted in the main text.

Claerbout and Johnson (1971), in their equation (23), give the transfer characteristic across a layer of thickness Δz as

$$\exp(i\Phi) = \{1 - [aT(1+Z)/(1-Z)]\}/\{1 + [aT(1+Z)/(1-Z)]\}, \quad (1)$$

where
$$a = c\Delta t \Delta z / (8\Delta x^2). \quad (2)$$

This transfer characteristic applies to their simple finite difference approximation to the wave equation. Since there is no energy loss the transmission involves only a phase shift Φ at a frequency f. $(T/\Delta x^2)$ represents some finite difference approximation to the differential operator $(-\partial^2/\partial x^2)$. In our particular computer algorithm we have used a recursive operator for T, namely,

$$T = D/(1 - D/6), \quad (3)$$

where D is the double difference operator corresponding to the three point convolver $(-1, 2, -1)$.

The parameter Z denotes the conventional Z transform time delaying operator and is equivalent to

$$Z = \exp(-2\pi i f \Delta t). \quad (4)$$

We convert from the thickness of the layer Δz to the two way travel time across the layer using

$$\Delta z = c\Delta\tau/2. \quad (5)$$

We recall from the section on METHOD that the migration algorithm requires the replacement of the true velocity c in the medium by $(c/2)$. With these changes

$$a = c^2 \Delta t \Delta \tau / (32 \Delta x^2). \tag{6}$$

Now let us consider a plane wave traveling upward at some angle θ to the vertical. In the continuous medium this wave is expressible as:

$$A = \exp\left[2\pi i\, f\left(t - \frac{2x}{c}\sin\theta + \frac{2z}{c}\cos\theta\right)\right]. \tag{7}$$

The phase shift due to crossing a layer with thickness Δz is with (5)

$$\Gamma = 2\pi f \Delta \tau \cos\theta. \tag{8}$$

Claerbout and Johnson (1971) derive their finite difference equations using a time frame referenced to a vertically traveling wave ($\theta = 0$). Changing to a fixed time frame will cause the phase shift Φ, from equation (1), to be modified to

$$\psi = \Phi + 2\pi f \Delta \tau. \tag{9}$$

The phase shifts Γ and ψ in (8) and (9) will not normally be equal and their difference provides the error we wish to estimate. However, we prefer to investigate this error numerically from a somewhat different viewpoint.

In propagating the wave field on the finite difference grid the (x, t) dependence remains constant. This implies that the disturbance can still be visualized as a plane wave, but that its velocity and angle of propagation are distorted due to the error in the phase change across the layer. Let the apparent velocity be s and the angle be γ. Then the apparent plane wave becomes

$$B = \exp\left[2\pi i f \left(t - \frac{2x}{s}\sin\gamma + \frac{2z}{s}\cos\gamma\right)\right], \tag{10}$$

where, because of the invariance of the horizontal wave number, we have

$$\sin\theta/c = \sin\gamma/s. \tag{11}$$

The phase shift across the layer is $(4\pi f \Delta z \cos\gamma/s)$ which, when substituted into (9), yields

$$4\pi f \Delta z \cos\gamma/s = \Phi + 2\pi f \Delta \tau. \tag{12}$$

Substituting in equation (5) and rearranging gives

$$\frac{c}{s}\cos\gamma = 1 + \frac{\Phi}{2\pi f \Delta \tau}. \tag{13}$$

Equations (11) and (13) can be solved for the ratio of apparent velocity to true velocity:

$$(s/c) = \left[\sin^2\gamma + \left(1 + \frac{\Phi}{2\pi f \Delta \tau}\right)^2\right]^{-1/2} \quad (14)$$

Alternatively, (11) and (13) can be solved for γ, the apparent angle of propagation:

$$\tan\gamma = \sin\theta \bigg/ \left(1 + \frac{\Phi}{2\pi f \Delta \tau}\right). \quad (15)$$

Equations (14) and (15) were used in deriving figures 6a, 6b, 7a, and 7b. The phase is evaluated from equations (1), (3), (4), and (6), the quantity D in equation (3) being evaluated as

$$D = 2\left[1 - \cos\left(2\pi \frac{2f}{c} \sin\theta\, \Delta x\right)\right]. \quad (16)$$

Acknowledgement

We wish to thank Digicon Inc. for permission to publish this paper.

References

CLAERBOUT, J. F., 1970, Coarse grid calculations of waves in inhomogeneous media with application to delineation of complicated seismic structure, Geophysics 35, 407-418.

CLAERBOUT, J. F., 1971, Numerical holography, in Acoustical Holography, v.3, New York Plenum Press, p. 273-283.

Claerbout, J. F., 1971a, Toward a unified theory of reflector mapping, Geophysics 36, 467-481.

CLAERBOUT, J. F. and DOHERTY S. M., 1972, Downward continuation of moveout corrected seismographs, Geophysics 37, 741-768.

CLAERBOUT, J. F. and JOHNSON, A. G., 1971, Extrapolation of time dependent waveforms along their path of propagation, Geophys. J. R. Astr. Soc. 26, 285-293.

FOSTER, M., 1975, Transmission effects in the continuous one-dimensional seismic model, Geophys. J. R. Astr. Soc. 42, 519-527.

SATTLEGGER, J. W., 1973, Migration velocity determination in two and three dimensions, Presentation at Annual SEG Meeting in Mexico City.

WAVE-FRONT CHARTS AND THREE DIMENSIONAL MIGRATIONS*

ALBERT W. MUSGRAVE†

A computer is required to calculate the complex wave-front charts which are needed in many areas. On a medium size computer wave-front charts can be constructed using up to 40 layers. Each layer can be a constant velocity or can start with any velocity and have an increase in velocity with vertical time. These wave-front charts may be automatically plotted for use in migration in a vertical plane. At the same time that the wave-front chart is being obtained, a list may be made which shows the depth and offset for each reflection time and stepout value. This migration list may be used to migrate values from time maps in three dimensions. Before migrating, these time maps should have contours of all time values even though overlapping occurs as on buried foci of sharp synclines. Thus, it is a simple matter to make a migrated depth map from any time map regardless of the crookedness and discontinuity of the profiles or the lack of cross-line control.

INTRODUCTION

The geophysicist refers to the leading surface of the seismic disturbance as the wave front. The wave front is a spheroidal surface expanding from the shotpoint with increasing time. In isotropic media, raypaths are imaginary lines constructed perpendicular to the wave front at all times. The wave front is distorted from a perfect sphere by velocity changes in the earth through which it propagates.

Wave-front charts are made by the calculation of wave-front intersections, at a series of time increments, with a chosen set of raypaths in a vertical plane through the origin. In order to perform this calculation, assumptions about the subsurface velocity configuration must be mathematically describable. Before electronic computers were available, velocity configurations had to be limited to those which could be represented by simple mathematical functions.

VELOCITY ASSUMPTIONS

Velocity may vary in both the vertical and horizontal directions. Wave-front charts are usable only in areas where velocity does not vary horizontally or varies in particular ways as described in the author's doctorate thesis (Musgrave, 1952).

Horizontal Variations

Figure 1 shows the two basic isovelocity configurations as taken from this thesis. The diagram on the right shows an area where isovelocity surfaces are parallel to or equidistant from the reflector. Velocity has a horizontal gradient as a result of vertical changes, so wave-front charts and migration lists are difficult to use. Raypaths are straight lines in this case and the velocity changes near the origin. The special case of having small variations with depth will allow the use of fairly simple charts and lists. The diagram on the left shows an area where isovelocity surfaces are parallel to the datum plane. If the datum plane is horizontal, velocity has no horizontal gradient and a wave-front chart will resolve the two dimensional dip profiles. A migration list will resolve the dip in three dimensions.

Vertical Variations

Vertical velocity distribution in the earth is determined by velocity surveys. This determination can be made more accurately from continuous velocity logs. Velocity variations in the vertical direction are much more severe than in the horizontal direction. Several continuous functions have been used to approximate various vertical configurations. Electronic computer programs can use discontinuous mathematical functions which represent the vertical distribution of velocity with any amount of detail required. Figure 2 shows a time versus velocity plot for three velocity functions. Average velocity is shown by solid lines and instantaneous velocity is shown by

---

* Presented at the 29th Annual Meeting, SEG, Los Angeles, November 11, 1959. Manuscript received by the Editor April 18, 1961.
† Mobil Oil Company, Dallas, Texas.

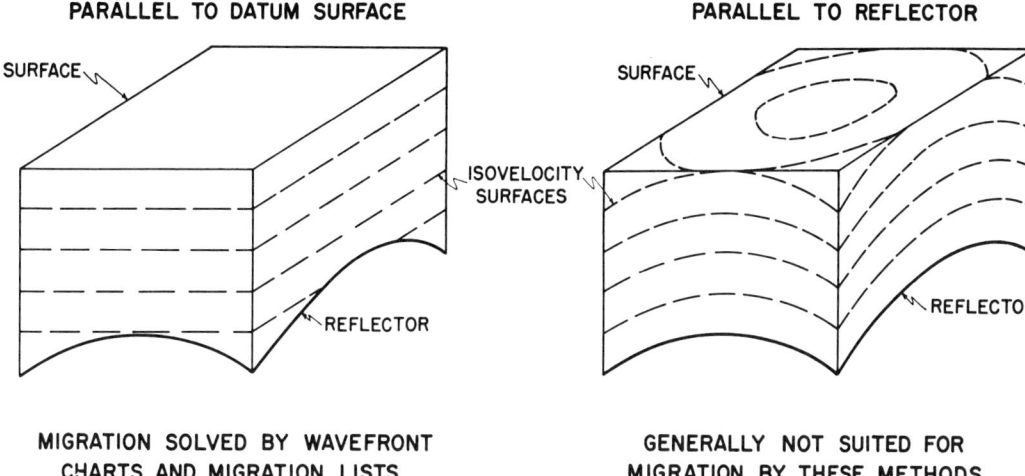

Fig. 1. Two general classifications of isovelocity configurations.

dashed lines. The curved lines represent a linear increase with depth function that is the basis of the wave-front chart shown in Figure 3. The straight lines represent a linear increase with time function (based on the same shallow velocity control as used for the linear with depth function) that is the basis of the chart in Figure 4. Finally, the lines originating at 6,800 ft/sec represent a two-layer discontinuous function that is the basis for the chart shown in Figure 5. The two layers have different accelerations or rates of velocity change.

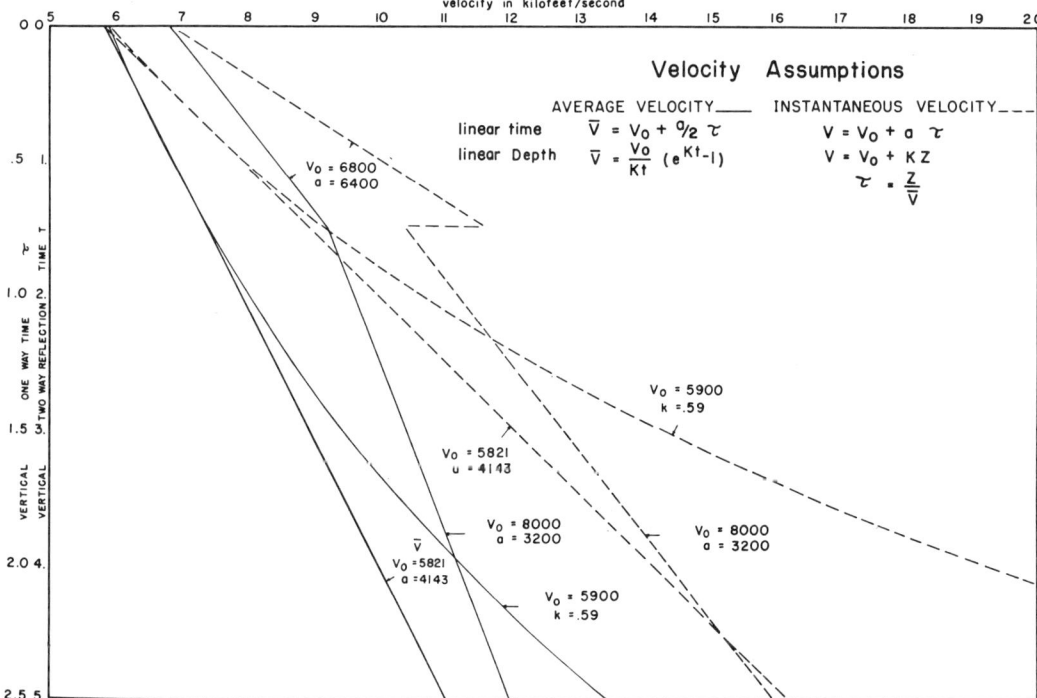

Fig. 2. Time vs velocity plot for three velocity configurations.

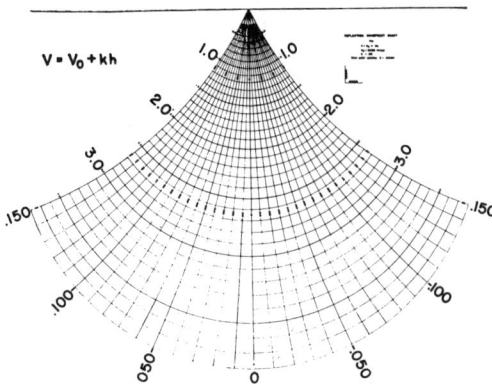

FIG. 3. A reflection wave-front chart made on the basis of a linear increase of velocity with depth.

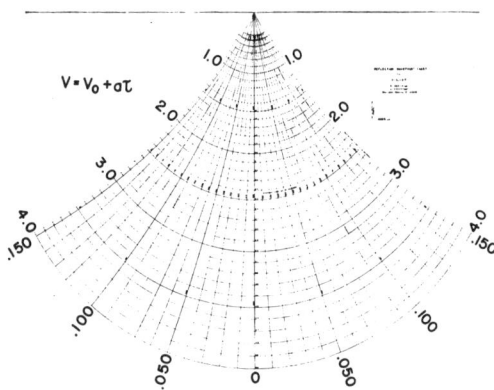

FIG. 4. A reflection wave-front chart made on the basis of a linear increase of velocity with vertical time.

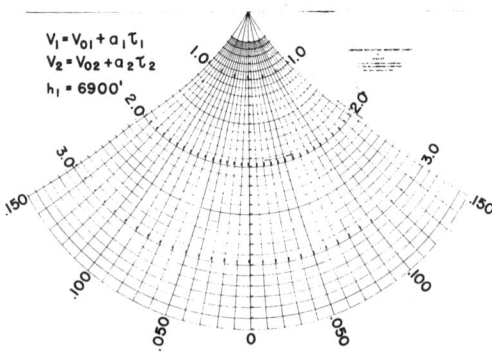

FIG. 5. A wave-front chart made using two layers with a different increase of velocity in each layer.

LINEAR INCREASE OF VELOCITY WITH TIME ASSUMPTION

Although the linear-with-depth function is more popular because of its simplicity, the linear-with-time function has several advantages as the basis of wave-front charts. Some of these advantages are: (1) a better empirical fit to actual velocity data is possible, (2) multilayer charts can be constructed, and (3) the constants are more easily determined than for a linear-with-depth function.

First Layer

Figure 6 shows a diagram of a raypath and the derivation of the equations used in calculating wave-front charts. The integral equations for time and distance are derived from Snell's law and trigonometric relations. Substitution of the linear increase of velocity with time assumption into these general equations gives the parametric equations for time, offset, and depth. Notice that the time along any raypath is directly proportional to the dip α, that the parameter θ equals 2α, and that the equations for offset and depth are those of a right cycloid whose axes are shifted by amounts x_0 and h_0.[1] A right cycloid is defined as a curve generated by a point on a circle rolling along a straight line as shown in Figure 7. θ is the angle of revolution of the circle and R is its radius. The circle generates the raypath by rolling along a line on the bottom side of a plane situated parallel to the datum plane and a distance above it which is coincident with zero on the velocity scale of the layer. The times and coordinates of depth and offset starting from zero for each wave-front-raypath intersection must be calculated. The values must be calculated all the way down to the depth of occurrence of the maximum velocity which is denoted by subscripts whose first character is M. The portion above the point of occurrence of minimum or beginning velocity is subtracted. The subtracted quantities are denoted by symbols having subscripts whose first character is zero. The second character of the subscripts refers to the layer being used. The character s denotes length of the subsurface coverage.

[1] In the oral presentation of the paper Dr. F. A. Van Melle brought to the author's attention that this curve is the same as the curve of the quickest descent of a particle under the influence of gravity and is called the brachistochrone.

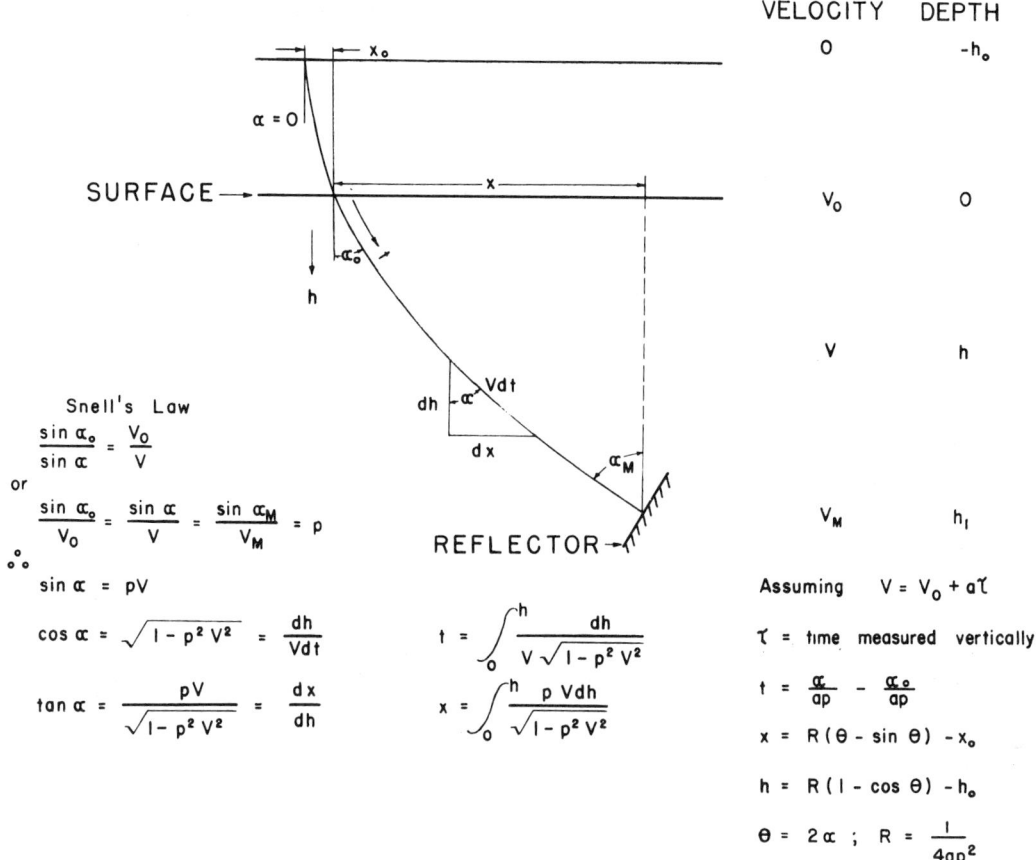

Fig. 6. The general relationships of time, offset, and depth for any continuous velocity function and specifically those for the assumption that velocity increases linearly with time.

Second Layer

Figure 8 shows the raypath entering the second layer, designated by subscripts whose second character is 2. This layer may have any velocity at its top and any rate of velocity increase, or acceleration, including zero. A change in these velocity properties from those of the first layer will change the size of the circle generating this second portion of the raypath. The height of the plane along which it rolls will also be changed. The plane would be repositioned at the height of zero on the velocity scale of the second layer. The raypath is made to maintain continuity across the interface by adding its second layer times and coordinates to the maximum corresponding values found on the raypath in the first layer.

COMPUTER PROGRAM

A continuation of this process is used in our IBM 650 program to carry out the tedious calculations required when velocity distributions become complex. Up to 40 layers can be handled by the program. Data input to the program for control is (1) a chart number, (2) the number of layers, (3) whether a refraction (1 way time) or reflection (2 way time) chart is to be calculated, (4) whether either or both list and plot are required, (5) whether special raypaths need to be calculated, and (6) information about plot symbols.

Data input to the program for calculation include (1) the initial and final raypath, (2) the raypath increment, (3) the minimum and maximum depth, (4) the maximum offset, (5) the increment of travel time, (6) the initial and final velocities, (7) the acceleration of each layer, (8) the depth of each layer, and (9) the scale factor for plotting if a plot is required.

The program calculates the coordinates for depth and offset of the intersection of each ray-

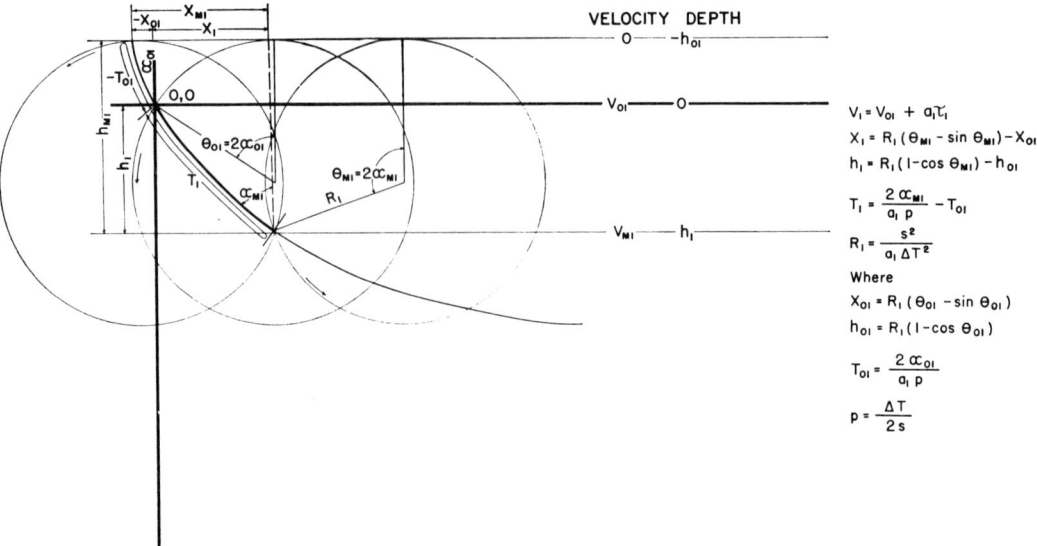

Fig. 7. Diagram and mathematical relations used to calculate the cycloidal raypaths through the first layer.

path with each increment of travel time within the limits of the maximum values. The output can be both a plot and a list card for each intersection calculated. Special plot symbols are assigned to intersections falling on designated raypaths and wave fronts, usually those having values that are multiples of five. In addition to the incremented raypaths, special raypaths may be calculated which are incident on each interface at very nearly the critical angle for the final velocity of each layer. These are needed to control the shape of the wave front near the inter-

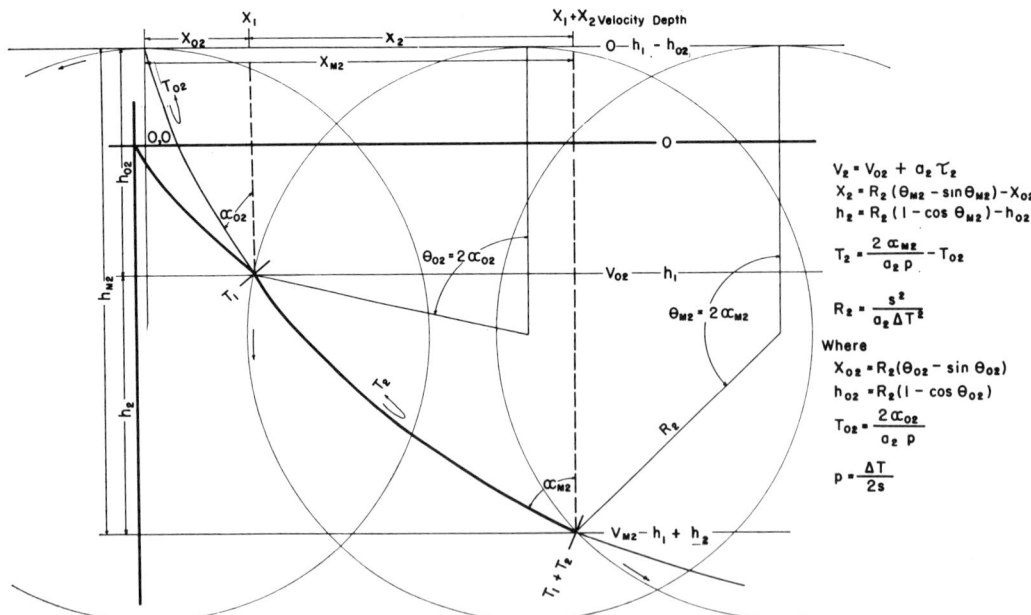

Fig. 8. Diagram and mathematical relations used to calculate the cycloidal raypath on through the second layer.

face. The raypaths used here are those defined by the ΔT across a chosen unit of length of subsurface coverage.

EXAMPLE OF COMPLEX WAVE-FRONT CHART

Figure 9 shows a plot of time versus velocity in a fairly complex configuration. This configuration was used to calculate a wave-front chart, a portion of which is shown in Figure 10. This chart was plotted by listing the IBM 650 output plot cards on an off-line IBM 407 tabulating machine. The platen drive of the 407 is modified for 10 lines to the inch, so plotting accuracy is plus or minus 1/20th of an inch. Accuracy can be improved by plotting to a large scale and photographically reducing the plots to the desired scale. The final wave-front chart is shown in Figure 11. The final plotting accuracy is plus or minus 1/40th of an inch or less than one millimeter.

Values of depth and offset from the list cards can be read directly into a Benson-Lehner Model S electroplotter. A plot prepared in this way is shown in Figure 12.

Both plots of this chart were calculated using the same velocity function but different raypath increments or lengths of subsurface coverage. This illustrates the utility of the program in adapting the charts to changes in length of the geophone cable used by the field crew.

Even though raypaths and wave fronts are not constructed with continuous lines, these charts are quite usable for plotting dips from reflection seismic records. The only manual drafting needed is the addition of identification and labels for the raypaths and the wave fronts.

MIGRATION BY WAVE-FRONT CHARTS

Reflection wave-front charts are used to plot depth cross-sections from the time and step out, or ΔT, taken from seismic records along a line of profile. The depth section can be made in two ways from split continuous records.

First Method

When using the long trace-to-long trace method, the cross-section paper is placed over the chart so that the shotpoint coincides with the

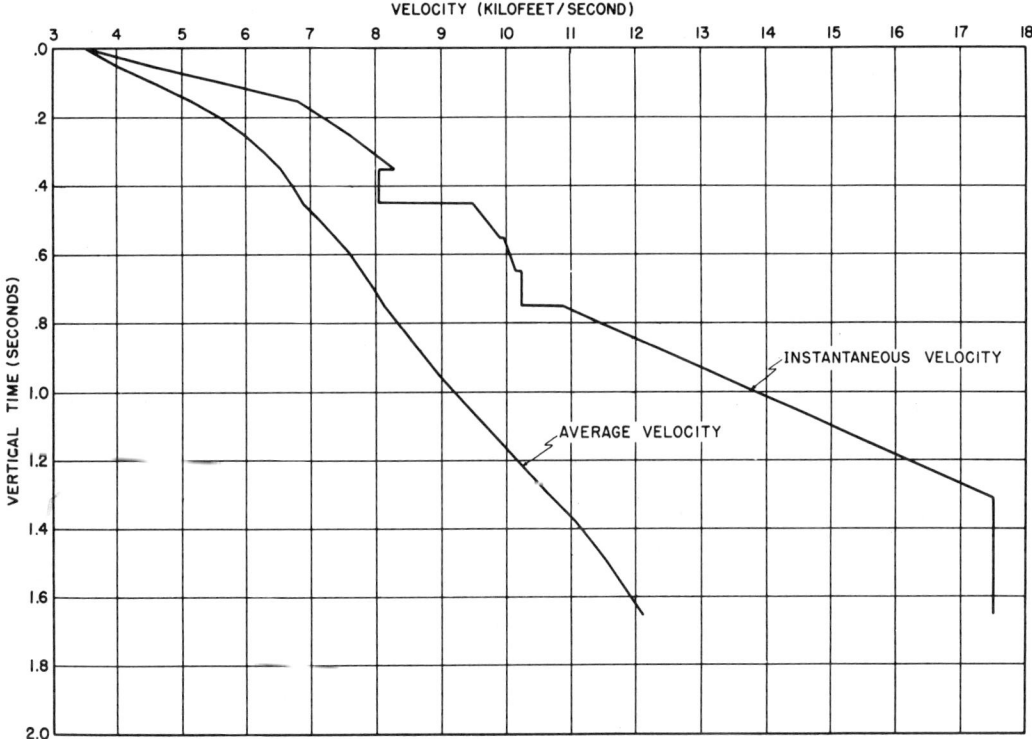

FIG. 9. Velocity configurations for a complex wave-front chart.

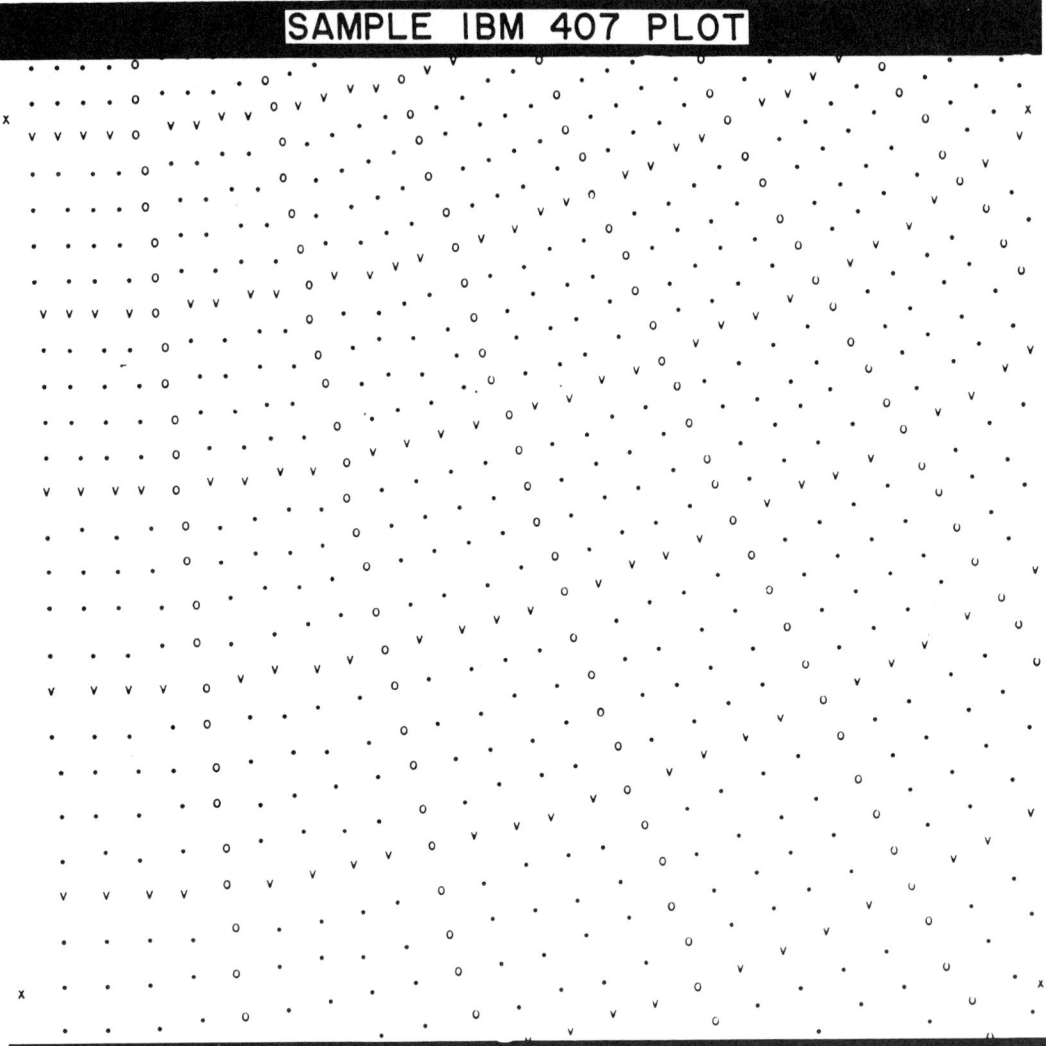

FIG. 10. A detailed portion of the wave-front chart shown in Figure 11 as plotted on the IBM 407.

origin of the chart. The travel time is taken as the corrected reflection time of the center traces. The ΔT is taken as the difference in corrected reflection times of the end traces. These two values locate the mid-point of the reflecting segment on the chart. The reflection segment is drawn tangent to the wave front with a length equal to one-half the distance between end traces. If irregular spread lengths are encountered, the end traces must be corrected for normal moveout and the ΔT normalized to the chart. This normalization is done by multiplying the moveout corrected ΔT by the ratio of the length of the spread to the unit of length chosen for the chart.

Second Method

When using the center-to-center method, the cross-section paper is placed over the chart so that its origin is midway between two shotpoints. Travel time is taken as the average of the corrected center trace times on the records taken at these shotpoints. The difference between the center trace times is the ΔT. These two values locate the reflecting segment on the chart. The segment is drawn as before with a length equal to the shotpoint spacing. If this spacing is not equal to that for which the chart was calculated, the ΔT must be normalized as described before, but no correction for normal moveout is required.

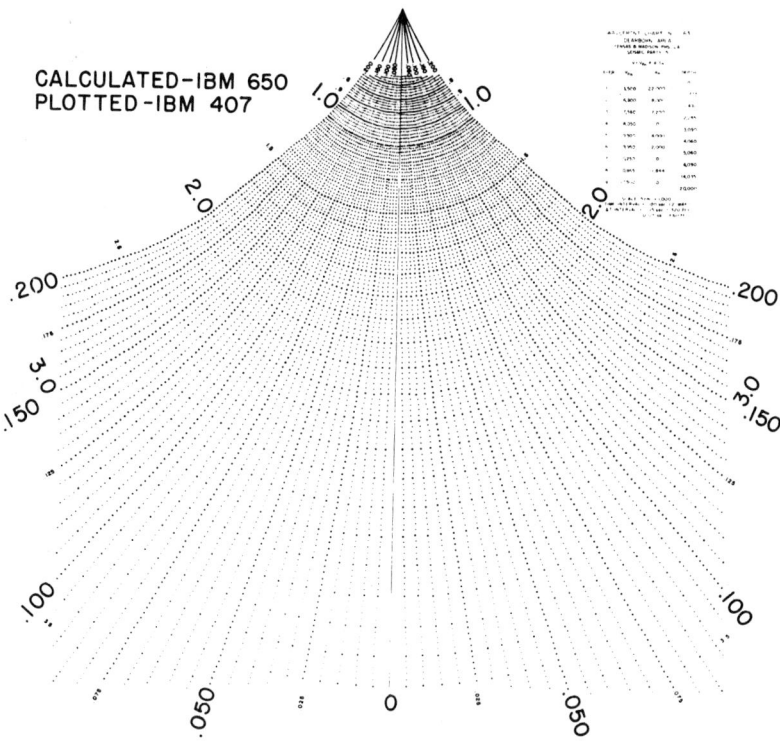

Fig. 11. A complete wave-front chart plotted by the IBM 407 for the velocity configuration shown in Figure 9 and for a subsurface coverage equal to 1,320 ft.

The advantage of the long trace-to-long trace method is that the ΔT is taken from a single record and is not affected by hole lag, variations in time break and uphole time, filter corrections, and ghost effect.

The advantage of the center-to-center method is that no normal moveout correction is necessary and that the datum correction can be calculated more easily.

EXAMPLE OF WAVE-FRONT CHART MIGRATION

Figure 13 shows a section where reflection times are plotted vertically below the shotpoints where they are observed. A record section would serve equally well. Any length of reflection segment may be migrated by normalizing the ΔT. These time data are migrated by a wave-front chart into the depth section shown in Figure 14. With some knowledge of the geology of the area and dips of shallower reflections, the depth section might be interpreted as shown in Figure 15. All of the time data turns out to be reflected from one horizon, the data having three buried foci and a point of diffraction. A buried focus of a reflection results from a syncline whose radius of curvature is smaller than the radius of curvature of the wave front.

MIGRATION LISTS

Three dimensional migration may be accomplished by using a printed list from the list cards output by the IBM 650 program. The information punched in the list cards is chart number, raypath, interface, travel time, depth, and offset. These cards can be sorted and listed as shown in Figure 16. The sort for this list was in order of raypath for each wave front or travel time. An alternate list is made by sorting in order of travel time for each raypath. Pages of this type of listing are put together to form the composite list shown in Figure 17. This list is most useful in resolving dip in three dimensions.

It is necessary to have an identifiable reflection. Before the list can be used, a time map must be made of this reflection. All possible reflection times must be picked and put on the map even

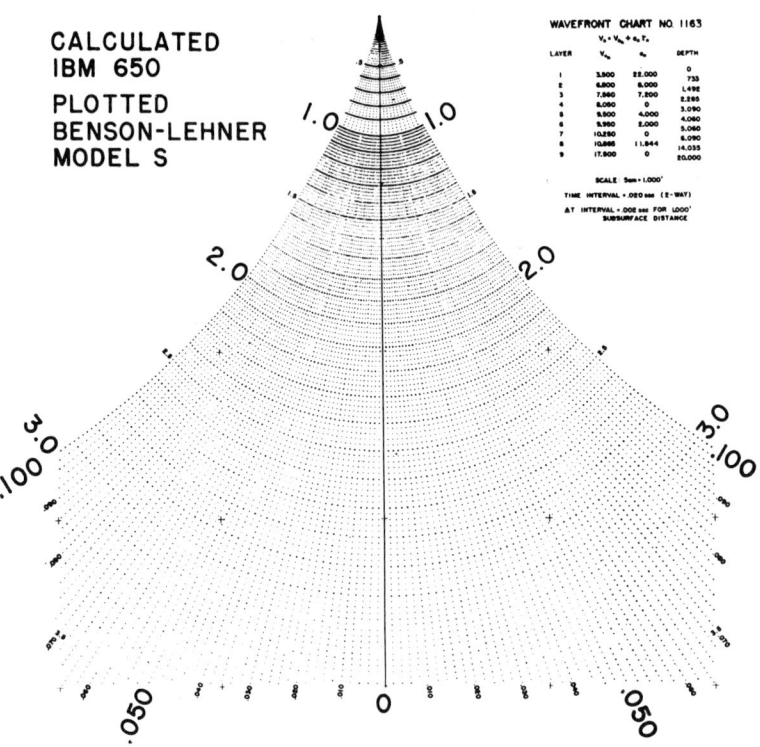

FIG. 12. A wave-front chart plotted by the Benson-Lehner Model S electroplotter using the velocity configuration shown in Figure 9 and a subsurface coverage equal to 1,000 ft.

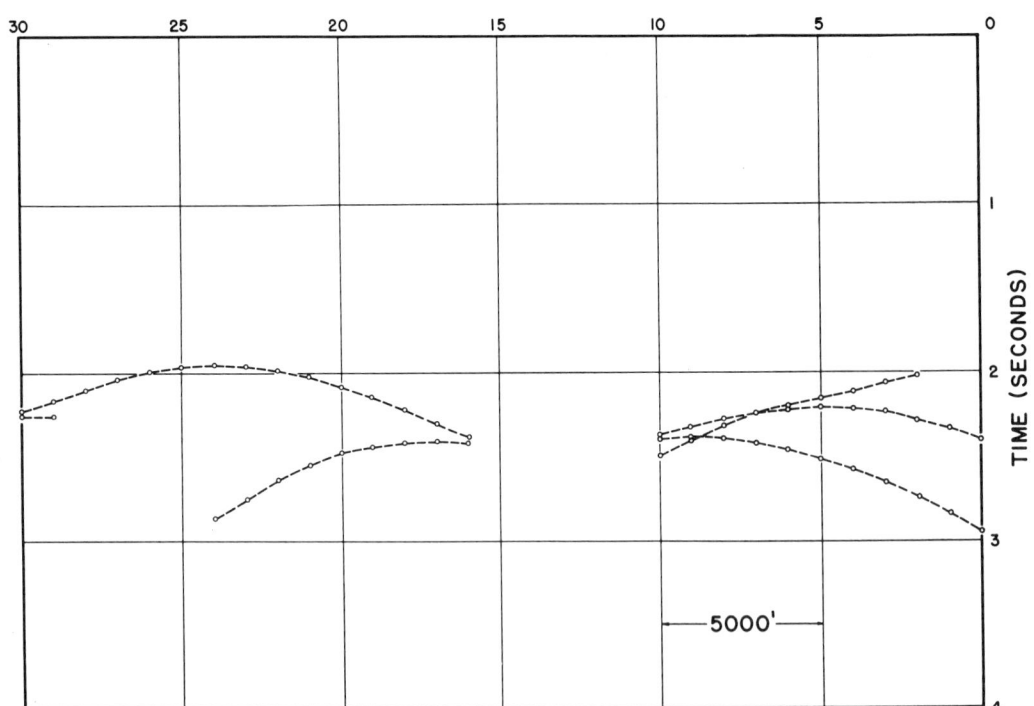

FIG. 13. A time cross-section made from reflection records in a complex area. A record section may be used just as well or better for this purpose.

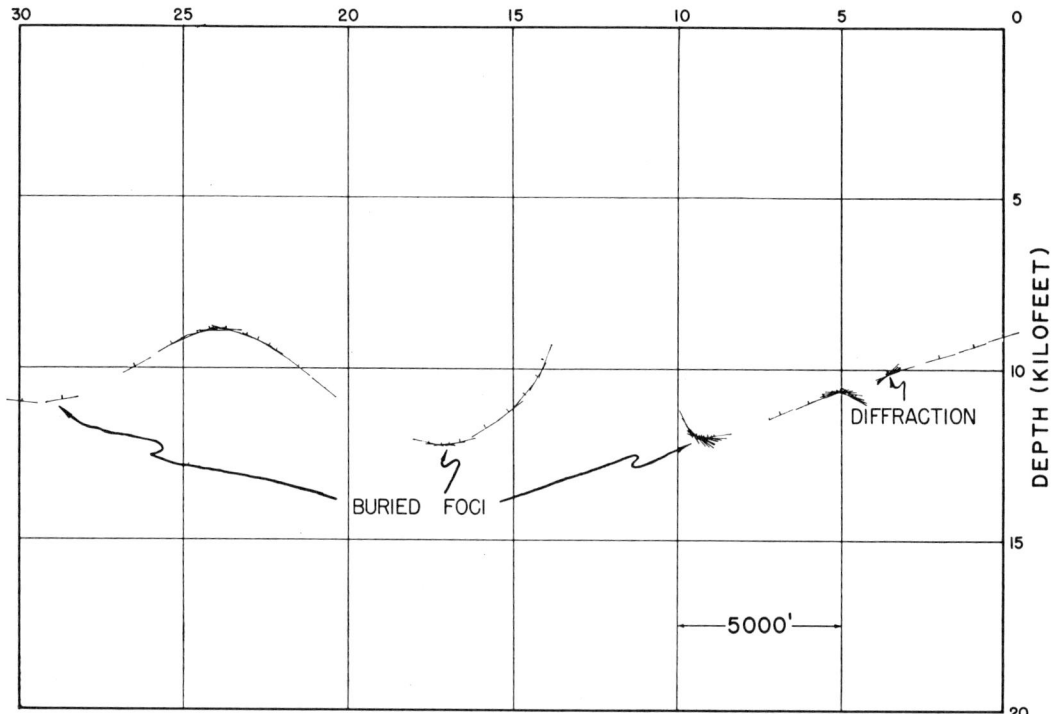

Fig. 14. A migrated depth section from the times shown in Figure 13.

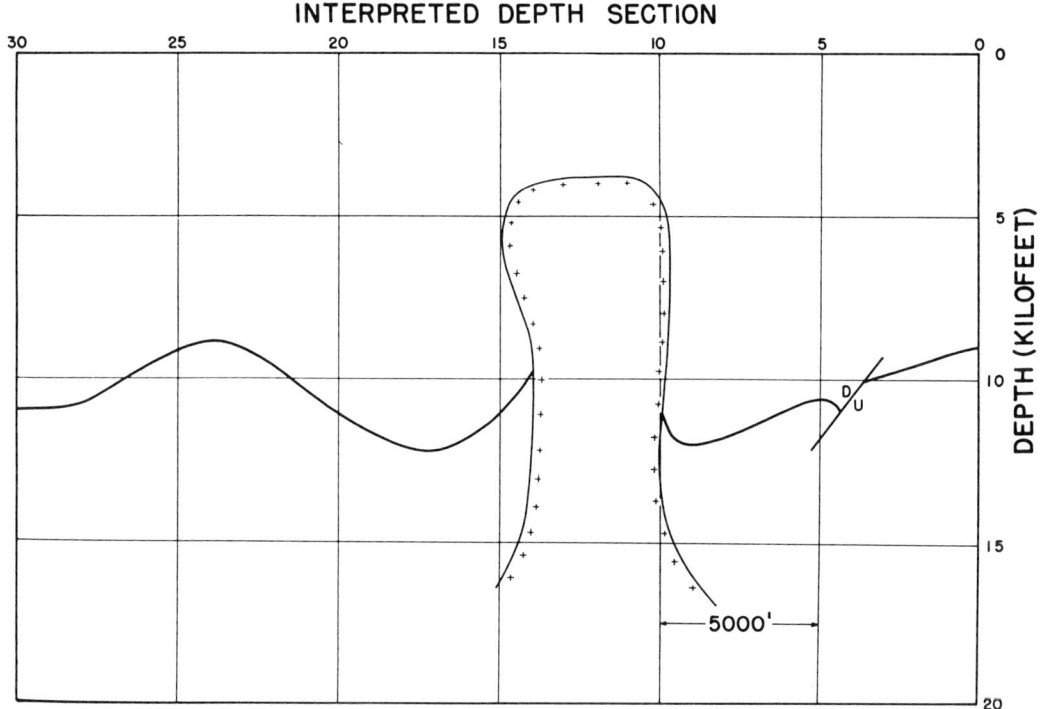

Fig. 15. An interpretation of the migrated section shown in Figure 14.

CHART NO.	RAYPATH IN SECS. PER M FT. OF HORZ. DIST.	INTER FACE	TRAVEL TIME SECONDS	DEPTH FEET	OFFSET FEET
1166	.002000	14	2.8600	11,321	93
1166	.004000	14	2.8600	11,321	187
1166	.006000	14	2.8600	11,320	280
1166	.008000	14	2.8600	11,312	372
1166	.010000	14	2.8600	11,309	466
1166	.012000	14	2.8600	11,304	558
1166	.014000	14	2.8600	11,296	651
1166	.016000	14	2.8600	11,289	744
1166	.018000	14	2.8600	11,280	837
1166	.020000	14	2.8600	11,270	929
1166	.022000	14	2.8600	11,259	1,022
1166	.024000	14	2.8600	11,248	1,117
1166	.026000	14	2.8600	11,236	1,208
1166	.028000	14	2.8600	11,220	1,299
1166	.030000	14	2.8600	11,206	1,389
1166	.032000	14	2.8600	11,189	1,480
1166	.034000	14	2.8600	11,173	1,573
1166	.036000	14	2.8600	11,156	1,665
1166	.038000	14	2.8600	11,137	1,756
1166	.040000	14	2.8600	11,119	1,844
1166	.042000	14	2.8600	11,097	1,937
1166	.044000	14	2.8600	11,073	2,026
1166	.046000	14	2.8600	11,055	2,115
1166	.048000	14	2.8600	11,027	2,207
1166	.050000	14	2.8600	11,004	2,295

FIG. 16. A representative portion of the IBM calculated migration list, sorted in order of increasing raypath or stepout for each time, showing depth and offset for each time and stepout.

though the reflection overlaps in synclines and appears several times on the same record. These overlaps may not be meaningful until they are migrated.

EXAMPLES OF THREE-DIMENSIONAL MIGRATION

Figure 18 shows a time map of an anticline made with values from records taken at all of the shotpoints indicated on the grid. The migrated depth map is made by reading travel times at the shotpoints and interpolating the stepout across the distance for which the list is calculated, measured normal to the time contours. One such time and offset is indicated on the map. These values serve as entry to the migration list and determine the values of depth and offset from Figure 17. Although straight traverse lines have been used for this synthetic example, the migration system is of even more benefit if cultural and/or topographic features require shooting crooked lines.

Figure 19 shows the completed depth map after all points have been migrated. Notice how much the depth points converge on the anticline in the areas of steep dip.

It is interesting to see what happens with a syncline of this same shape. Figure 20 shows a cross-section across the east-west axis of a shallow and a deep syncline. The synclines are narrower in time than in depth. The buried focus effect of the deep syncline causes some time overlap. Figure 21 shows a cross-section across the north-south axis of these same synclines. The steeper dips in this direction caused considerable time overlap which shows the buried focus effect more prominently. Note that the overlap increases with the sharpness of the syncline and with greater depth.

Figure 22 shows the time map of the shallow syncline. The steeper north-south dip causes more squeezing of the contours than the more gentle east-west dip. The small amount of time overlap, because of the buried focus, is shown in the center of the map.

Figure 23 shows a depth map of the syncline

STEPOUT (ΔT) →	0		.020		.040		.060		.080		.100	
TRAVEL TIME SECONDS	DEPTH FEET	OFFSET FEET	DEPTH FEET	OFFSET FEET	DEPTH FEET	OFFSET FEET	DEPTH FEET	OFFSET FEET	DEPTH FEET	OFFSET FEET	DEPTH FEET	OFFSET FEET
1.8800	8,368		8,316	793	8,162	1,568	7,909	2,311	7,570	3,014	7,145	3,656
1.9000	8,500		8,447	810	8,287	1,602	8,027	2,360	7,678	3,076	7,241	3,728
1.9200	8,633		8,578	828	8,413	1,636	8,145	2,409	7,786	3,138	7,337	3,802
1.9400	8,767		8,710	846	8,541	1,671	8,265	2,460	7,895	3,202	7,434	3,876
1.9600	8,902		8,844	864	8,669	1,706	8,385	2,511	8,005	3,267	7,531	3,952
1.9800	9,038		8,979	882	8,799	1,743	8,506	2,563	8,115	3,333	7,628	4,028
2.0000	9,176		9,114	901	8,929	1,779	8,628	2,616	8,226	3,400	7,726	4,106
2.0200	9,315		9,251	920	9,061	1,817	8,751	2,670	8,337	3,468	7,823	4,185
2.0400	9,455		9,389	940	9,193	1,855	8,874	2,725	8,449	3,537	7,921	4,265
2.0600	9,596		9,529	960	9,327	1,893	8,999	2,781	8,561	3,607	8,020	4,346
2.0800	9,739		9,669	980	9,461	1,933	9,124	2,837	8,674	3,678	8,118	4,429
2.1000	9,882		9,811	1,000	9,597	1,972	9,250	2,895	8,787	3,750	8,217	4,513
2.1200	10,027		9,953	1,021	9,733	2,013	9,376	2,953	8,901	3,824	8,315	4,597
2.1400	10,173		10,097	1,042	9,871	2,054	9,504	3,012	9,015	3,898	8,414	4,683
2.1600	10,320		10,242	1,064	10,009	2,096	9,632	3,072	9,130	3,973	8,514	4,770
2.1800	10,468		10,388	1,085	10,149	2,139	9,761	3,133	9,245	4,050	8,613	4,859
2.2000	10,618		10,535	1,108	10,289	2,182	9,890	3,195	9,361	4,128	8,712	4,948
2.2200	10,768		10,684	1,130	10,430	2,226	10,020	3,258	9,477	4,206	8,812	5,039
2.2400	10,920		10,833	1,153	10,573	2,270	10,152	3,322	9,593	4,286	8,912	5,131
2.2600	11,073		10,984	1,176	10,716	2,315	10,284	3,387	9,710	4,367	9,011	5,224
2.2800	11,228		11,135	1,200	10,860	2,361	10,416	3,452	9,828	4,449	9,111	5,318
2.3000	11,383		11,288	1,224	11,006	2,408	10,550	3,519	9,946	4,533	9,211	5,413
2.3200	11,540		11,442	1,248	11,152	2,455	10,684	3,587	10,064	4,617	9,311	5,510
2.3400	11,697		11,597	1,273	11,299	2,503	10,818	3,655	10,182	4,702	9,411	5,608
2.3600	11,856		11,754	1,298	11,447	2,551	10,954	3,725	10,301	4,789	9,511	5,707
2.3800	12,016		11,911	1,323	11,597	2,601	11,090	3,795	10,421	4,877	9,611	5,807
2.4000	12,178		12,070	1,349	11,747	2,651	11,227	3,866	10,540	4,965	9,711	5,908
2.4200	12,340		12,229	1,375	11,898	2,702	11,364	3,939	10,660	5,055	9,811	6,011
2.4400	12,504		12,390	1,402	12,050	2,753	11,502	4,012	10,781	5,146	9,911	6,114
2.4600	12,669		12,552	1,428	12,202	2,805	11,641	4,087	10,901	5,239	10,011	6,219
2.4800	12,835		12,715	1,456	12,356	2,858	11,781	4,162	11,022	5,332	10,111	6,325
2.5000	13,002		12,879	1,483	12,511	2,912	11,921	4,238	11,143	5,427	10,210	6,433
2.5200	13,171		13,044	1,512	12,667	2,966	12,061	4,316	11,265	5,522	10,310	6,541
2.5400	13,340		13,210	1,540	12,823	3,021	12,203	4,394	11,387	5,619	10,410	6,651
2.5600	13,511		13,378	1,569	12,981	3,077	12,345	4,473	11,509	5,717	10,509	6,761
2.5800	13,683		13,546	1,598	13,139	3,134	12,487	4,554	11,631	5,816	10,608	6,873
2.6000	13,856		13,716	1,628	13,299	3,191	12,630	4,635	11,754	5,917	10,707	6,986
2.6200	14,031		13,887	1,658	13,459	3,250	12,774	4,718	11,876	6,018	10,806	7,102

FIG. 17. A part of a migration list sorted in order of time for each stepout.

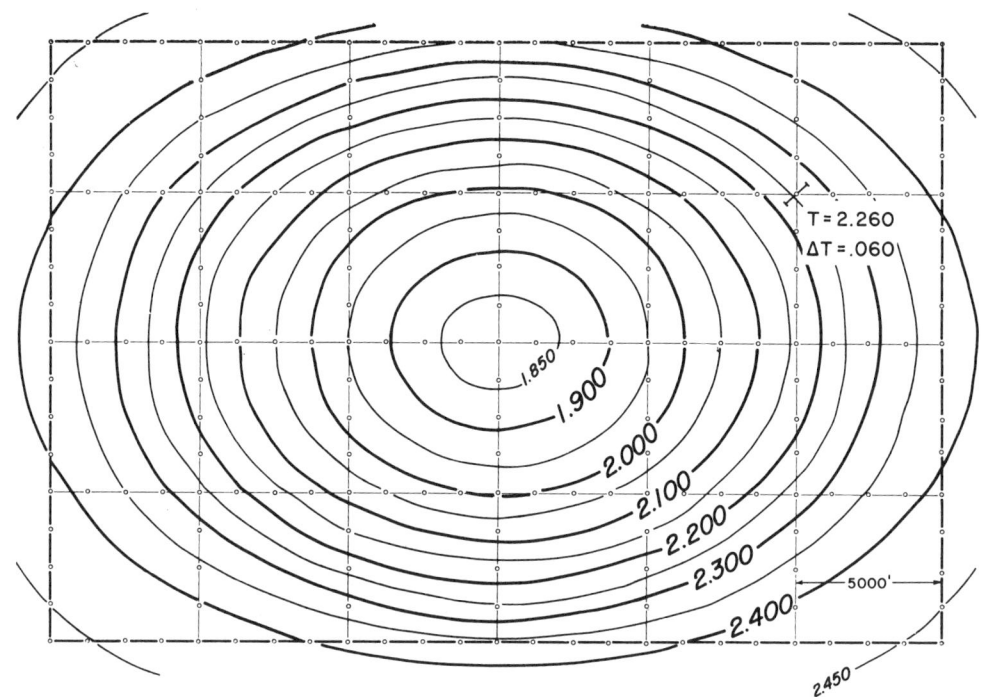

FIG. 18. A time map of an anticline showing the shotpoint grid with 5,000 ft between lines and 1,250 ft between shotpoints. An example time (T) and stepout (ΔT) are shown for which depth and offset information is obtained from the list shown in Figure 17.

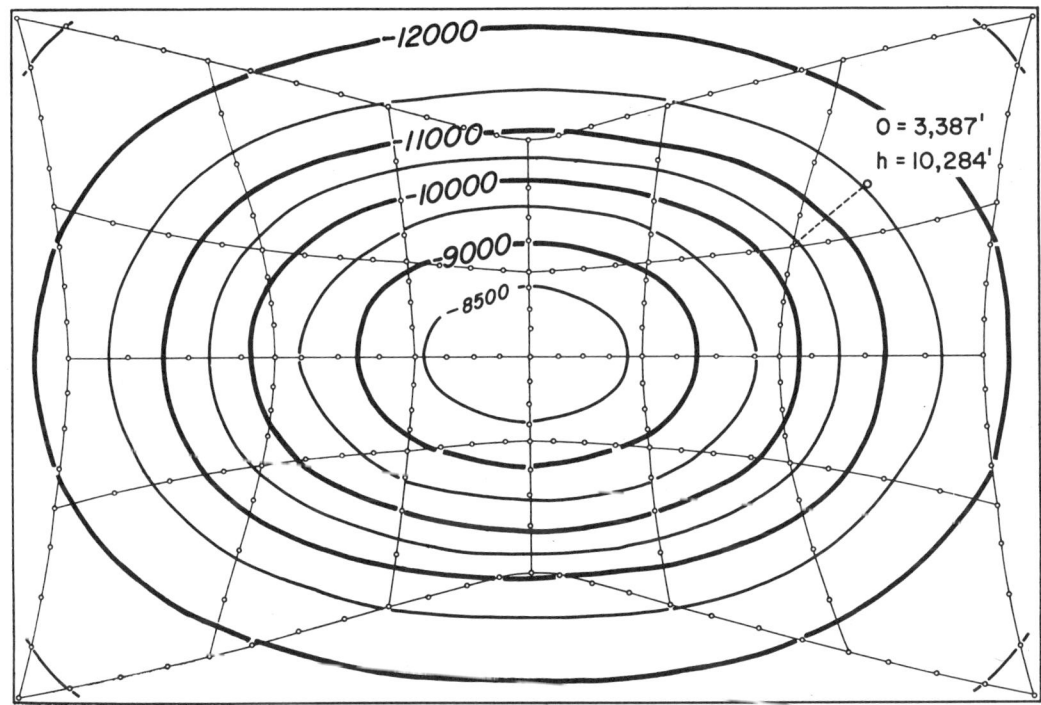

FIG. 19. A depth map made by three-dimensionally migrating the time map shown in Figure 18. The migrated location for each shotpoint is shown by the converging grid.

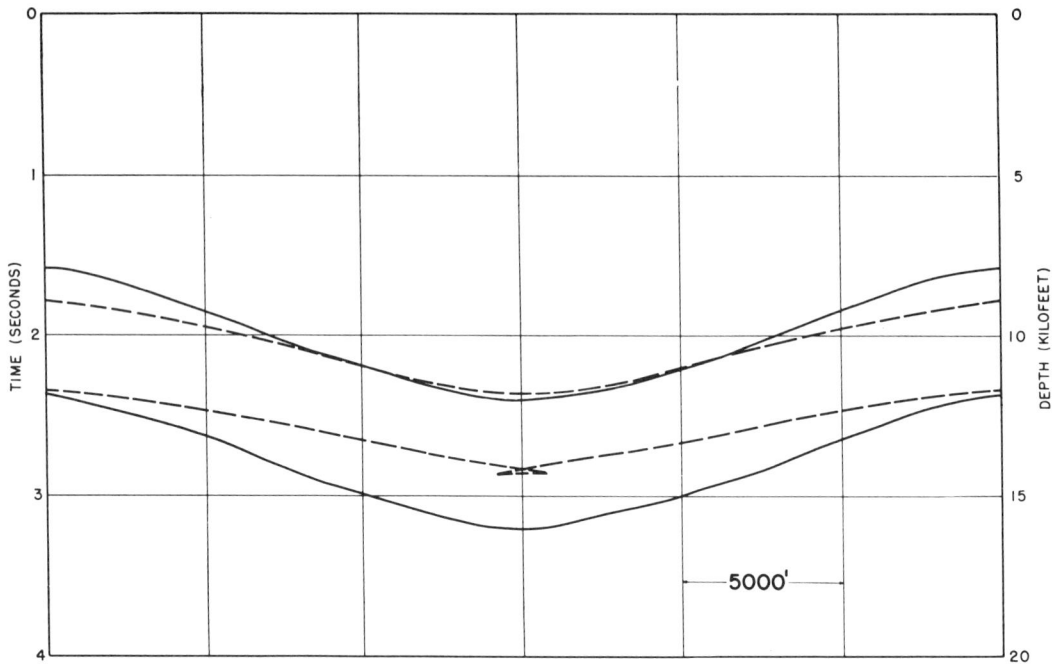

Fig. 20. An east-west cross-section across two synclines. They are referred to as a shallow and a deep syncline.

showing the migrated position of all the shotpoints. The displacement of the shotpoints is greatest in the area of steepest dip.

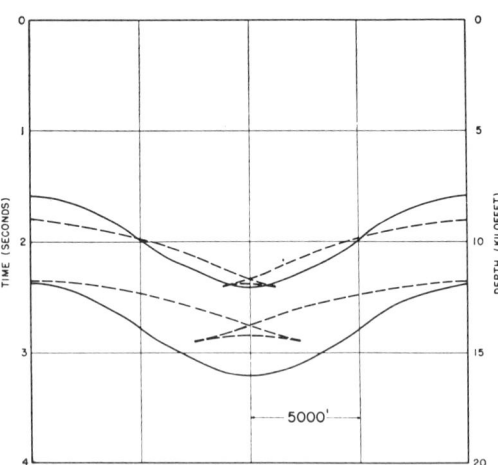

Fig. 21. A north-south cross-section of a shallow and a deep syncline. (Note the increasing effect of the buried focus crossing of the time limbs. The amount of cross-over on the time plots increases either with the narrowness of the syncline or the depth of the syncline.)

Figure 24 shows a time map of the deep syncline. The time overlap caused by the buried focus in the center gives a very complex picture. When all of these times are migrated and converted to depth, they give a very smooth syncline as shown in Figure 25.

FIELD METHOD OF THREE-DIMENSIONAL MIGRATION

The migration list made by the electronic computer can be used in any field office to migrate time maps simply and accurately. The procedure to follow is to make a time map as complete as possible. From the time map read values of travel time and ΔT. Lay off the offset on an overlay at right angles to the time contours. Post depths at the migrated points and contour the depth map. This process is simplified by making the time contours have the same values as calculated wave fronts on the list, then migrate only points on contours controlled by data. This elimination of interpolation is the secret of speed.

SUMMARY

The electronic computation of wave-front charts is an improvement over other methods

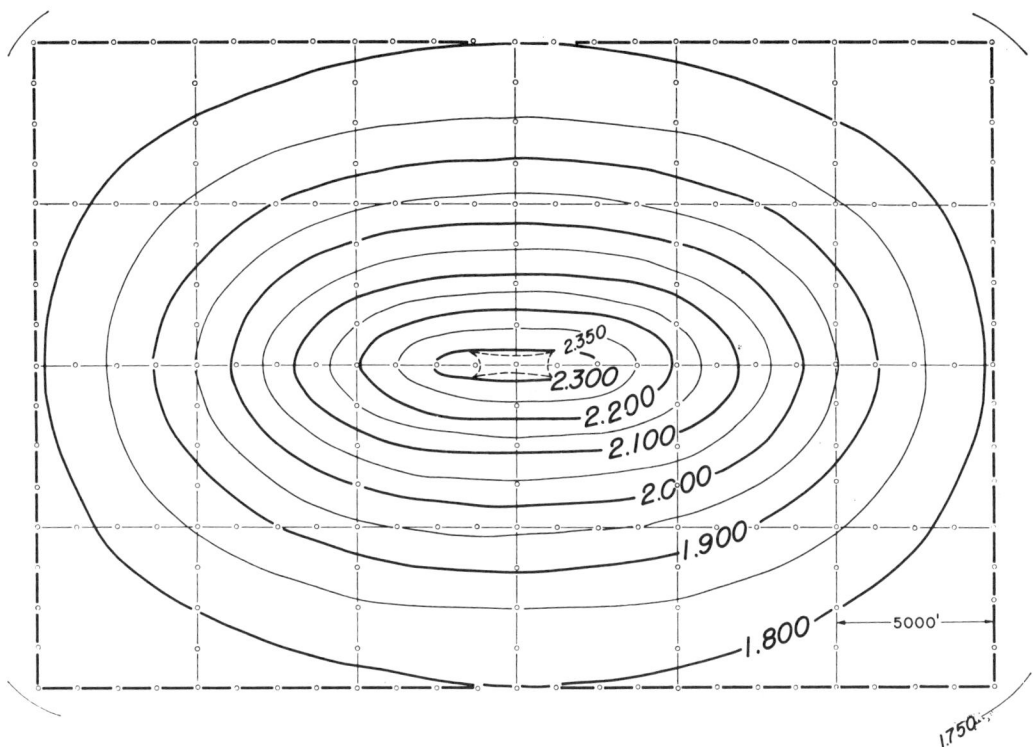

Fig. 22. A time map of the shallow syncline. Note the crossed over or hidden contour in the center (dashed).

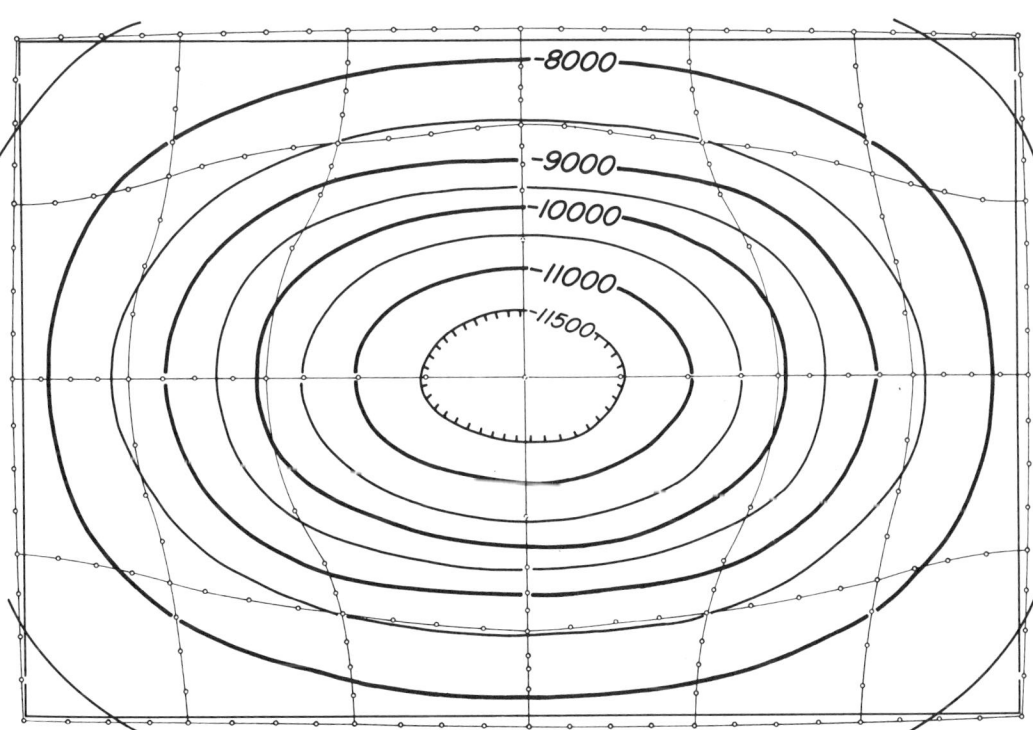

Fig. 23. A depth map of the shallow syncline showing the divergence of the shotpoint grid.

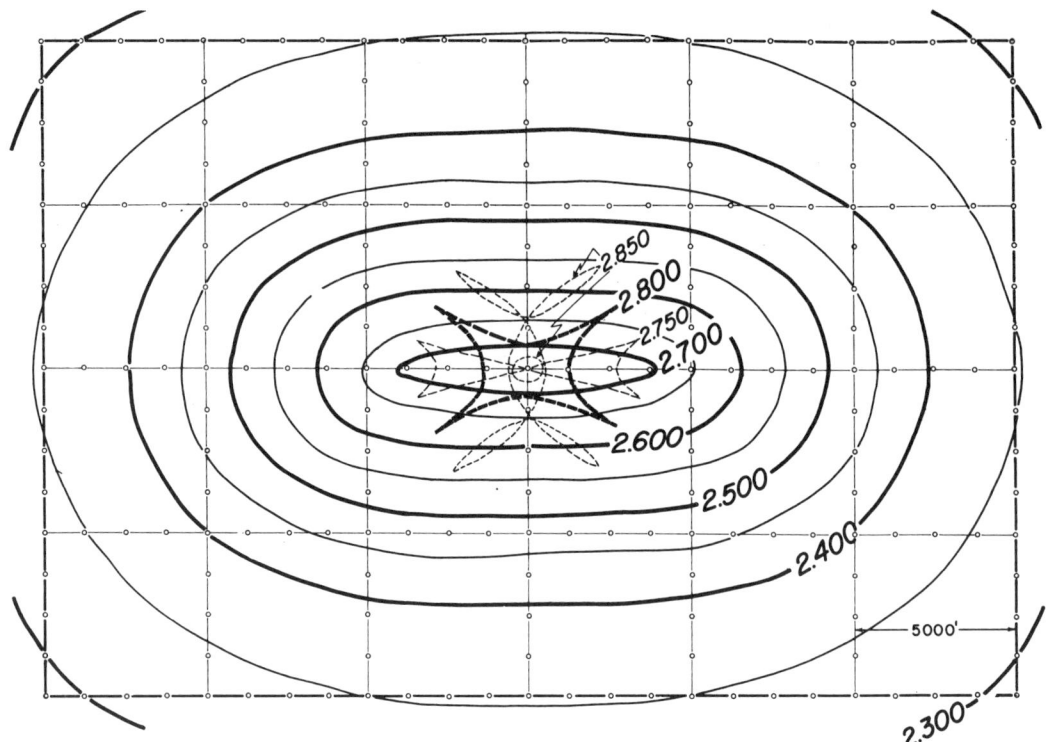

Fig. 24. A time map of a deep syncline. The crossed over or hidden contours are dashed.

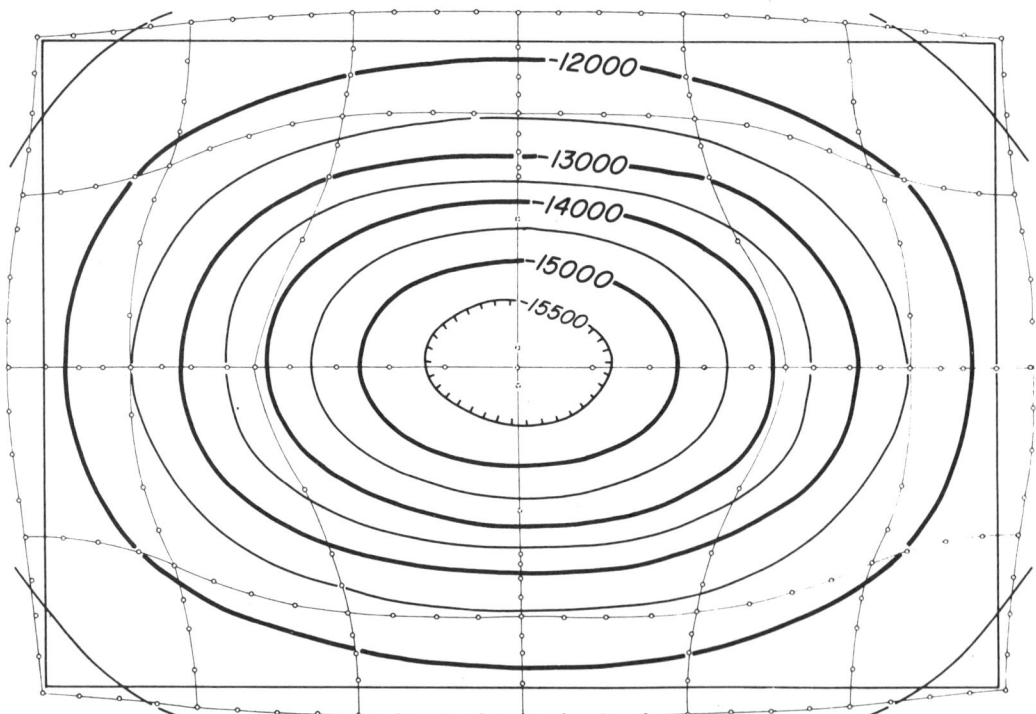

Fig. 25. A depth map of the deep syncline showing the divergence of the migrated shotpoint positions.

because: (1) Mathematical assumptions may be used that reduce the time required to determine the constants needed for calculation. (2) The speed of the computer reduces the time required to obtain a chart with increased accuracy. (3) The detailed calculations that can be handled by the machine program allow the use of complex, discontinuous velocity functions. Many layers may be included having almost exact correspondence to observed velocity data. (4) Trained personnel are not required to make the calculations after the program is completed and a check list of the input information is made. Personnel may need considerable review for manual calculations if charts are infrequently required, but the computer program retains its speed. (5) The migration lists allow easy and efficient migration in three dimensions. The lists may be made so detailed that no interpolation is necessary.

ACKNOWLEDGMENTS

The author wishes to express his appreciation to the Mobil Oil Company for permission to present this paper and to his associates for their help in preparing it. Particular thanks go to G. W. Ehlert and W. C. Woolley for their aid and cirticisms, also to Manus Foster, Charles Hickman, and Helen Gray for their part in writing the computer program.

REFERENCE

Musgrave, Albert W., 1952, Wave-front Charts and Raypath Plotters: Quarterly of the Colorado School of Mines, v. 47, n. 4.

MIGRATION BY FOURIER TRANSFORM

R. H. STOLT*

Wave equation migration is known to be simpler in principle when the horizontal coordinate or coordinates are replaced by their Fourier conjugates. Two practical migration schemes utilizing this concept are developed in this paper. One scheme extends the Claerbout finite difference method, greatly reducing dispersion problems usually associated with this method at higher dips and frequencies. The second scheme effects a Fourier transform in both space and time; by using the full scalar wave equation in the conjugate space, the method eliminates (up to the aliasing frequency) dispersion altogether. The second method in particular appears adaptable to three-dimensional migration and migration before stack.

INTRODUCTION

The migration of seismic data has been improved in recent years by application of the theory of scalar waves. Both the difference equation techniques pioneered by Jon Claerbout (1971, 1972, 1976) and integral equation techniques such as those developed by William French (1974, 1975) have been successful as applied.

Described below are two new schemes for the migration of seismic data. Both operate in momentum (i.e., wavenumber or spatial frequency) space in the horizontal (basement) direction. The first scheme is a high-accuracy, high-frequency, steep dip extension of the Claerbout finite difference algorithm. By formulating this algorithm in momentum space, we are able to (a) eliminate a matrix inversion without loss of accuracy, (b) migrate separately each momentum component using an algorithm tailor-made for each, so as to (c) reduce dispersion (within sampling limitations) to negligible proportions.

The second scheme is also based on the scalar wave equation but does not employ finite differences; rather, the exact wave equation is used to predict a transformation in frequency-momentum space. Subject to the sampling limitations of the data, dips of any angle can be migrated correctly and without dispersion.

Emphasis will be placed on digital migration of stacked seismic cross-sections. In addition, the second scheme will be shown to be adaptable to migration before stack and three-dimensional migration.

THEORETICAL FRAMEWORK

General

In what follows, we consider the earth to be a two-dimensional half-space. We assume sound to travel as a scalar field with velocity at point (x, z) of $c(x, z)$. Every point in the earth has the ability to transform downgoing sound waves into upgoing sound waves. This property is characterized by a reflection strength $R(x, z)$ whose angular dependence we ignore. Multiple reflections are also ignored.

Measurements are taken at the earth's surface by placing a source at point (x_s, z_s) and a receiver at (x_0, z_0). The reflected sound wave[1] observed at (x_0, z_0) is represented by $\psi(x_s, z_s, x_0, z_0, t)$, where

---

[1] ψ may represent a pressure, a displacement or velocity potential, or some other suitably defined parameter. Spatial derivatives of compressibility and density will be largely ignored in what follows, so the distinctions between the various fields will not be of concern here. ψ may be thought of as an impulse response or Green's function.

Manuscript received by the Editor September 8, 1976; revised manuscript received September 30, 1977.
*Continental Oil Co., Ponca City, OK 74601.
0016-8033/78/0201-0023 $0X.00. © 1978 Society of Exploration Geophysicists. All rights reserved.

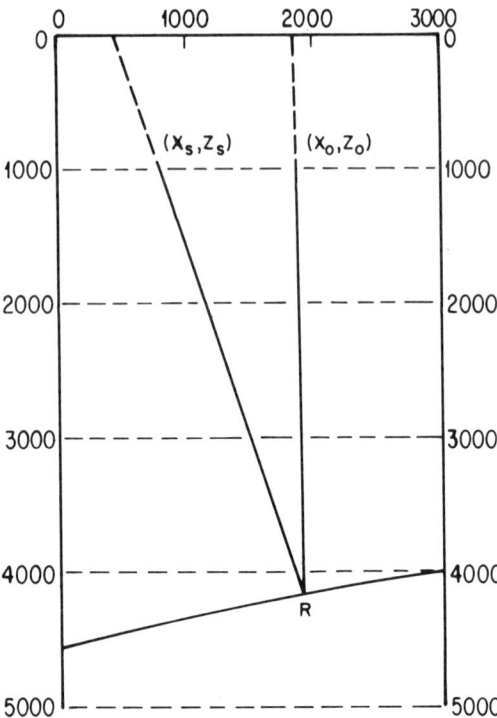

FIG. 1. Migration may be viewed as a prediction of changes in the seismic field as sources and receivers are moved into the earth.

t is the two-way traveltime from source to receiver. For a flat earth, z_s and z_0 are zero during the measurement.

By migration, one attempts to determine the reflection strength $R(x, z)$ from $\psi(x_s, 0, x_0, 0, t)$ at the earth's surface. This is done by predicting what ψ would be for sources and receivers inside the earth. Then (Claerbout, 1971),

$$R(x, z) \sim \psi(x, z, x, z, 0). \quad (1)$$

That is, (x_s, z_s) and (x_0, z_0) are extrapolated to the common point (x, z) as shown in Figure 1. As they approach each other, the traveltime between source and receiver approaches zero, and ψ, subject to limitations in source bandwidth, becomes proportional to the reflection strength at that point.

The changes in ψ as source and receiver migrate into the earth can be predicted by the scalar wave equation. We require (using subscripts to denote partial derivatives),

$$\psi_{x_s x_s} + \psi_{z_s z_s} - \psi_{tt}/c(x_r, z_s)^2 = 0, \quad (2)$$

and

$$\psi_{x_0 x_0} + \psi_{z_0 z_0} - \psi_{tt}/c(x_0, z_0)^2 = 0, \quad (3)$$

at all points in space-time. That is, the scalar wave equation governs ψ with respect to small changes in receiver or source location.

Migration of stacked sections

These two equations can be simplified considerably for the migration of stacked sections, if we pretend that "stacked" sections are equivalent to normal incidence sections, where $(x_s, z_s) = (x_0, z_0)$. We define midpoint coordinates,

$$X = (x_s + x_0)/2 \text{ and } Z = (z_s + z_0)/2, \quad (4)$$

and relative, or offset, coordinates,

$$x = (x_0 - x_s)/2 \text{ and } z = (z_0 - z_s)/2. \quad (5)$$

Setting $\psi(X, x, Z, z, t) = \psi(x_s, z_s, x_0, z_0, t)$, the stacked section in the new coordinate system corresponds to $\psi(X, 0, 0, 0, t)$ and the migrated section to $\psi(X, 0, Z, 0, 0)$. Equations (2) and (3) become

$$\psi_{XX} + \psi_{ZZ} + \psi_{xx} + \psi_{zz} - 2\psi_{xX}$$
$$- 2\psi_{zZ} - 4/c(X - x, Z - z)^2 \psi_{tt} = 0, \quad (6)$$

$$\psi_{XX} + \psi_{ZZ} + \psi_{xx} + \psi_{zz} + 2\psi_{xX}$$
$$+ 2\psi_{zZ} - 4/c(X + x, Z + z)^2 \psi_{tt} = 0.$$

If we ignore derivatives with respect to x and z,[2] we are left with the single equation

$$\psi_{XX} + \psi_{ZZ} - 4/c(X, Z)^2 \psi_{tt} = 0. \quad (8)$$

Equation (8) differs from (2) and (3) by the factor of 4 in the ψ_{tt} term. This difference is due to the fact that in (8), both source and receiver coordinates are required to migrate synchronously, whereas in (2) and (3), one set of coordinates is kept fixed. The form of (8) can be made identical to (2) and (3) by redefining t in (8) to be one-way traveltime.

The Claerbout coordinate transformation

In the Claerbout approach, the wave equation (8) is converted to a difference equation which can be

---

[2] Strictly speaking, this is hard to justify, though we can argue that as $x, z \to 0$, first derivatives with respect to x and z should vanish, and second derivatives are moveout generators which produce mainly gradual changes of amplitude with time when x and z are fixed at zero.

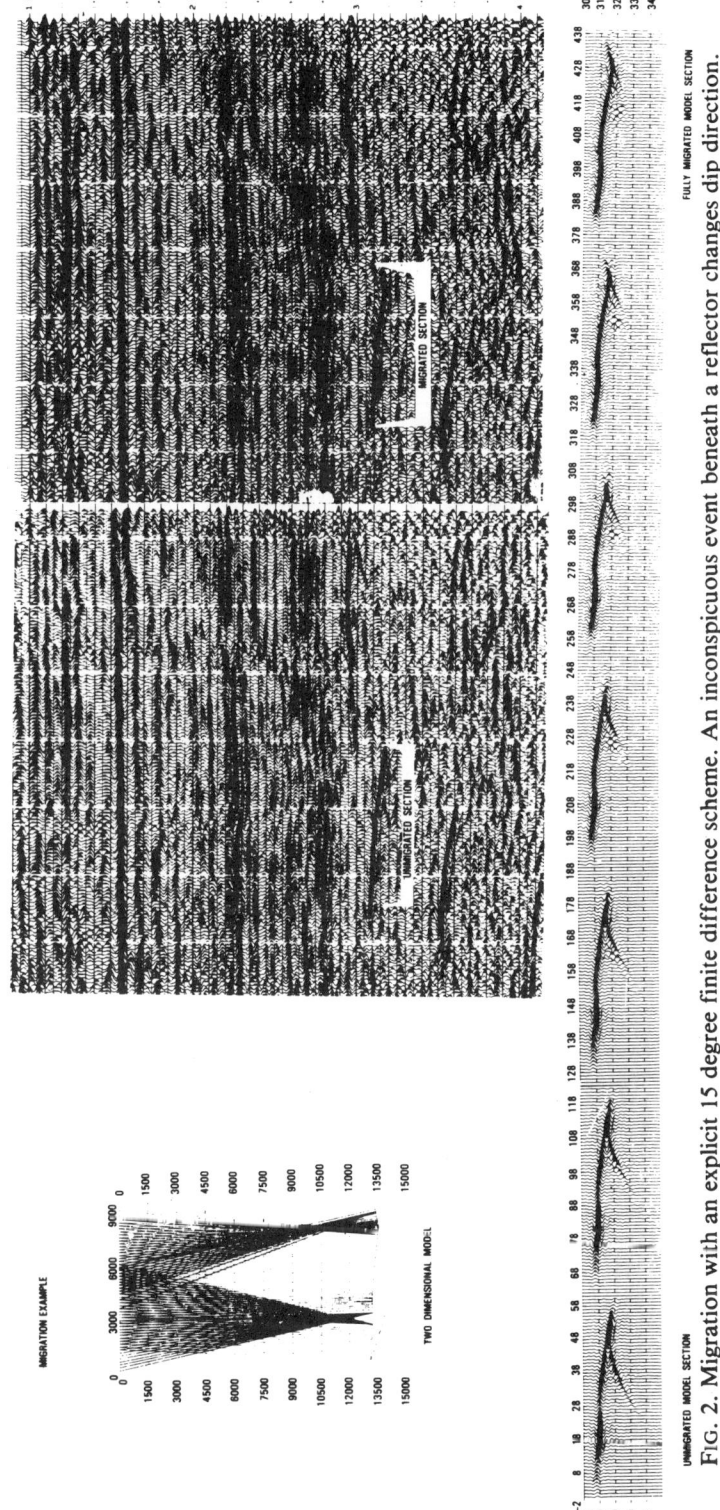

FIG. 2. Migration with an explicit 15 degree finite difference scheme. An inconspicuous event beneath a reflector changes dip direction.

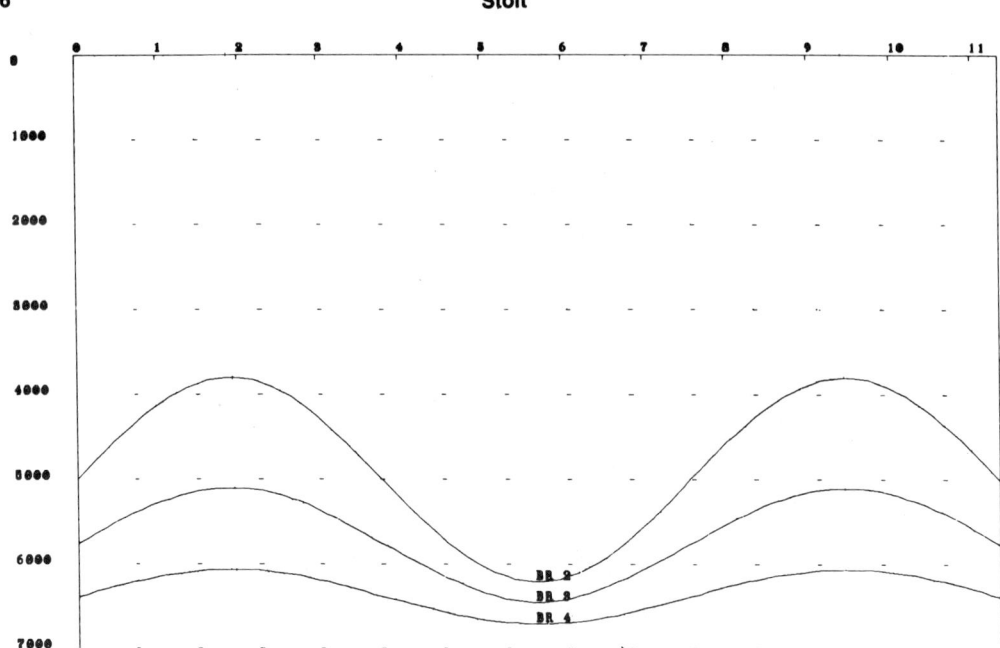

FIG. 3. A two-dimensional earth model. Velocity is 9600 fps in all layers.

used to gradually sink the source-receiver midpoint coordinates (X, Z) into the earth. To make such a scheme practical, Claerbout defines a new coordinate system in which ψ varies less rapidly with depth. If c were constant, one such transformation would be

$$D = ct/2 + Z; \, d = Z. \quad (9)$$

In this coordinate system, D is the depth of some reflection point, while d is the depth of the source-receiver midpoint.

Setting

$$\phi(x, d, D) = \psi(X, 0, Z, 0, t), \quad (10)$$

the wave equation (8) becomes (Claerbout, 1976, p. 211)

$$\phi_{XX} + \phi_{dd} + 2\phi_{dD} = 0. \quad (11)$$

The stacked section corresponds to

$$\phi(X, 0, D) = \phi(X, 0, ct/2)$$
$$= \psi(X, 0, 0, 0, t), \quad (12)$$

and the migrated section to

$$\phi(X, D, D) = \phi(X, Z, Z)$$
$$= \psi(X, 0, Z, 0, 0). \quad (13)$$

Of course, c will normally be a function of X and Z.

A coordinate system which casts the migration problem into a velocity independent form similar to that in equations (11)–(13) will be discussed later.

To help understand equation (11), consider a plane wave of angular frequency $\omega = 2\pi f$ traveling upward at angle θ to the vertical. According to equation (8), such a wave will take the form

$$\psi = e^{i2\omega(X \sin \theta - Z \cos \theta - ct/2)/c}, \quad (14)$$

or, in the new coordinate system,

$$\phi = e^{i2\omega\{X \sin \theta + d(1 - \cos \theta) - D\}/c}. \quad (15)$$

From (15) it follows that for upward traveling waves ϕ_{XX} is always greater than ϕ_{dd}. For waves traveling near the vertical, ϕ_{dd} will be negligible compared to ϕ_{XX}.

THE CLAERBOUT FINITE DIFFERENCE METHOD

Following Claerbout's approach, we now convert the wave equation (11) into a difference equation. Since $\phi_d(X, 0, D)$ is not known a priori, the equation should not involve second differences in d. Since ϕ_{dd} is very small for waves traveling near the vertical, the simplest thing to do is ignore it. This results in the so-called 15 degree approximation to the wave equation (Claerbout, 1976, p. 211)

$$\phi_{XX} + 2\phi_{dD} = 0. \quad (16)$$

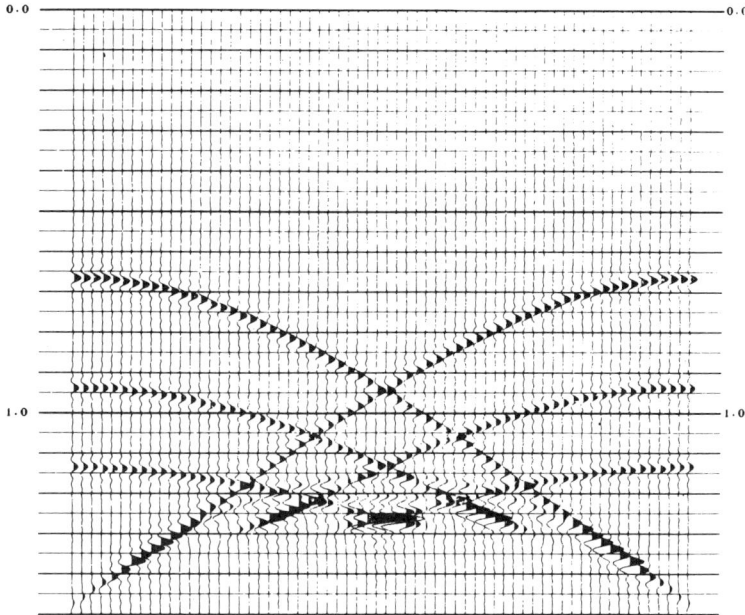

FIG. 4. A synthetic seismic section. Trace spacing is 120 ft.

We now define the discrete variables j and k by the relations

$$D_j = j\Delta D \text{ and } d_k = k\Delta d, \quad (17)$$

where ΔD and Δd are the increments in reflector depth (transformed traveltime) and source-receiver depth, respectively. We also adapt the shorthand notation

$$\phi(X)_j^k \equiv \phi(x, d_k, D_j). \quad (18)$$

The conversion of Equation (16) to a difference equation is not unique. Two possible lowest-order forms are

$$(1 - T)\phi_j^{k+1} = -(1 - T)\phi_{j+1}^k$$
$$+ (1 + T)(\phi_{j+1}^{k+1} + \phi_j^k), \quad (19a)$$

and

$$\phi_j^{k+1} = -\phi_{j+1}^k + (1 + 2T)(\phi_{j+1}^{k+1} + \phi_j^k), \quad (19b)$$

where

$$T = \frac{\Delta D \Delta d}{8} \frac{d^2}{dX^2}, \quad (20)$$

is an operator in X. In practice, T may be a second (or higher) difference operator in X. The form (19a) is referred to as an implicit solution for ϕ_j^{k+1}, since in order to solve for that quantity, it is necessary to invert the operator $1 - T$. Since no inversion is re-required in (19b), we call it an explicit solution for ϕ_j^{k+1}.

The implicit form (19a) is a more expensive algorithm than (19b) but has the capacity for greater accuracy at steep dips.

Under many circumstances, the cheap explicit form (19b) will adequately migrate a stacked section. Figure 2 is an example of such a migration in which an apparent downturn of a surface at a fault is converted into an upturn. Migration of a simple model shows the upturn to have developed from an inconspicuous event beneath the reflector. The event appears less prominent on the actual section. This is partly attributable to interference from another event beneath it and partly to losses of diffractive energy during stack. The migration depth increment Δd used to migrate the model was the equivalent of about 500 msec (that is, six steps were required to migrate an event at 3 sec). The actual section was migrated using a smaller increment.

Figures 3 to 5 provide an example of the limitations of the explicit 15 degree algorithm. Figure 3 shows a two-dimensional model consisting of three reflecting surfaces whose depths vary sinusoidally. The maximum dip of the bottom reflector is 15 degrees; that of the middle, 30 degrees; and that of the top, 45

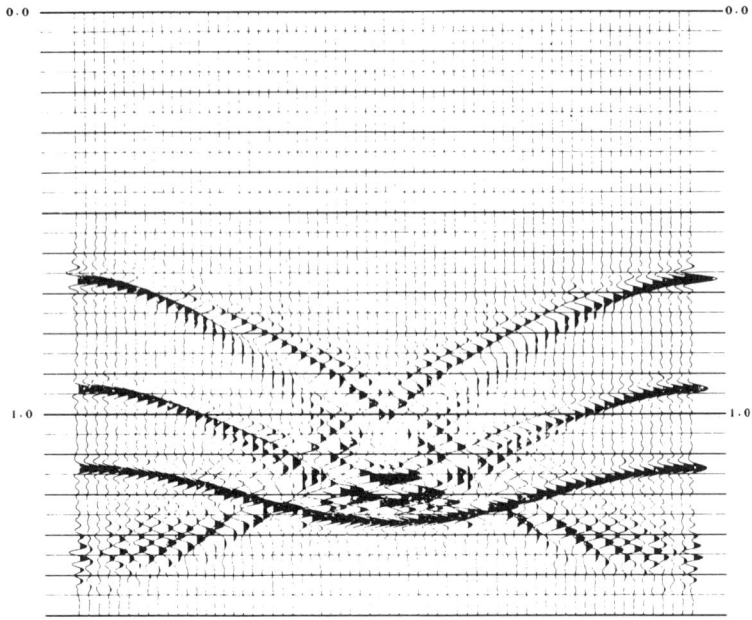

FIG. 5. A 15 degree Claerbout migration ($\Delta d = 860$ ft). Note dispersion in regions of steeper dip.

degrees. Velocity is constant at 9600 fps in all layers. Figure 4 is a synthetic normal incidence section constructed from this model, filtered 4–40 Hz. Trace spacing is 120 ft. The 15 degree Claerbout migration of this section is shown in Figure 5. As one might expect, the bottom reflection is migrated properly. The middle reflection, however, shows strong evidence of dispersion (that is, different frequencies are migrated to different places), and the top reflection is less migrated than mangled. Migration step size used was about 860 ft or 180 msec. Use of a smaller Δd step would not, in this case, improve the migration.

Higher order approximations to equation (16) are possible. Little is gained, however, unless the ϕ_{dd} term neglected in (16) is included.

MOMENTUM OR K-SPACE MIGRATION

For waves traveling at large angles to the vertical, ϕ_{dd} (though still smaller than ϕ_{XX}) is not negligible. Since it is not desirable to include ϕ_{dd} explicitly, an approximation must be found.

It is convenient to take a Fourier transform of ϕ with respect to X at this point, defining $p = 2\pi K$ to be the Fourier conjugate of X. Though not necessary for the development of a higher order approximation, there are several advantages to this step. First, the computer time disadvantage of an implicit solution to the wave equation disappears, since the operator inversion becomes a simple division in the wavenumber domain. Second, a simpler algorithm is allowed because the wave equation does not mix traces with different p values. Third, each p value can be migrated separately, using an algorithm individually tailored to it. Finally, the second derivative ϕ_{xx} is well approximated clear up to the spatial Nyquist frequency.

In the wavenumber domain, equation (11) takes the form

$$p^2 \phi = \phi_{dd} + 2\phi_{dD}. \qquad (21)$$

We can approximate the effect of ϕ_{dd} by differentiating equation (21) with respect to D,

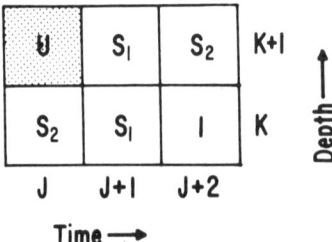

FIG. 6. The two-coeffiecint K-space migration operator.

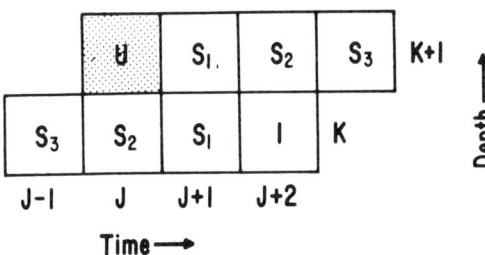

FIG. 7. The three-coefficient K-space migration operator.

45 degree approximation to the scalar wave equation (22).

A simple difference approximation to equation (24) is easily made. Using the discretization formula (18), define the following lowest-order difference operators centered at $(j + 1, k + 1/2)$,

$$\frac{1}{4\Delta D}\left(\phi_{j+2}^{k+1} + \phi_{j+2}^{k} - \phi_{j}^{k+1} - \phi_{j}^{k}\right) \sim \phi_D, \quad (25)$$

$$\frac{1}{\Delta d}(\phi_{j+1}^{k+1} - \phi_{j+1}^{k}) \sim \phi_d, \quad (26)$$

$$\frac{2}{\Delta d(\Delta D)^2}(\phi_{j+2}^{k+1} - \phi_{j+1}^{k+1} + \phi_{j+1}^{k}$$
$$- \phi_{j}^{k}) \sim \phi_{dDD} + \frac{4}{\Delta d \Delta D}\phi_D. \quad (27)$$

$$p^2\phi_D = 2\phi_{dDD} + \phi_{ddD}, \quad (22)$$

and with respect to d:

$$p^2\phi_d = 2\phi_{ddD} + \phi_{ddd}. \quad (23)$$

Neglecting ϕ_{ddd} in (23) allows us to write the single equation,

$$2p^2\phi_D - p^2\phi_d = 4\phi_{dDD}. \quad (24)$$

Equation (24) represents what is commonly called a

Substitution of (25), (26), and (27) into (24) yields the difference equation

$$\phi_j^{k+1} - \phi_{j+2}^{k} + S_1(\phi_{j+1}^{k+1} - \phi_{j+1}^{k})$$
$$+ S_2(\phi_{j+2}^{k+1} - \phi_j^k) = 0, \quad (28)$$

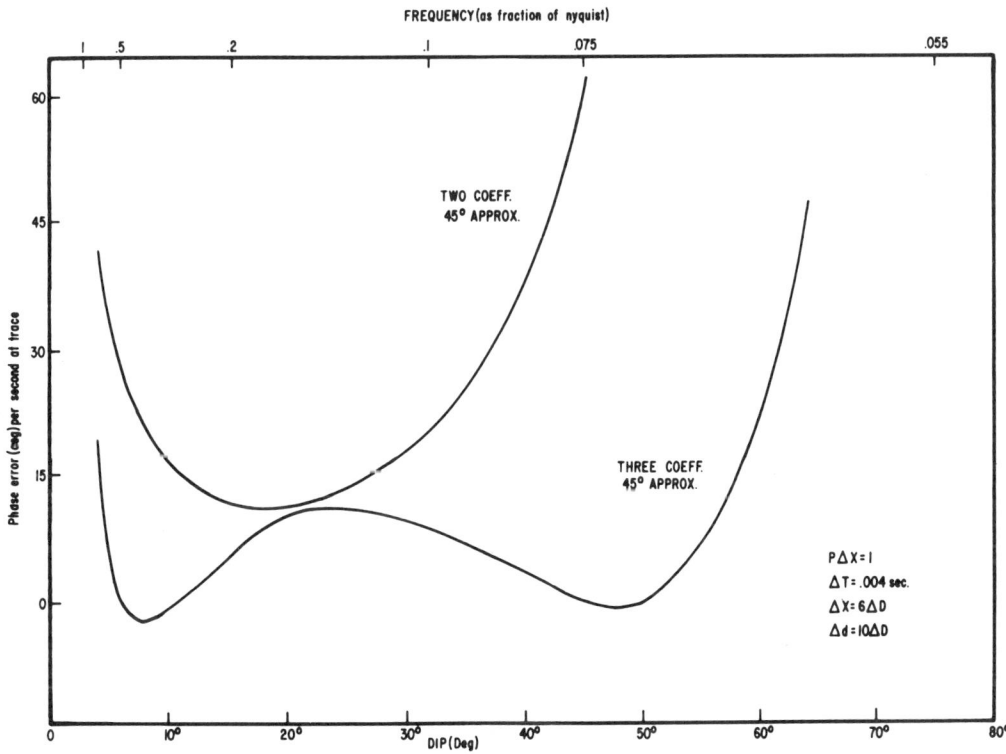

FIG. 8. An illustration of the phase error to be expected from the K-space algorithms.

$$S_1 = -2(8 - p^2 \Delta D^2)/(8 + p^2 \Delta D \Delta d), \quad (29)$$

$$S_2 = (8 - p^2 \Delta D \Delta d)/(8 + p^2 \Delta D \Delta d). \quad (30)$$

Equation (28) allows solution for an unknown ϕ_j^{k+1} in terms of the five known ϕ values illustrated in Figure 6. Stability is assured provided the polynomial,

$$1 + S_1 Z + S_2 Z^2,$$

has no roots inside the unit circle; i.e., provided its Levinson reflection coefficients (Claerbout, 1976, p. 55–57) are less than one in magnitude. This imposes the constraints

$$|S_2| < 1, \quad (31)$$

and

$$|S_1| < |1 + S_2|, \quad (32)$$

which are automatically met for any migration step size Δd larger than ΔD.

Since X is generally poorly sampled compared to D, and since equation (28) incorporates an extremely accurate approximation to ϕ_{xx}, it might be thought that equation (28) as it stands is an accurate approximation to the wave equation. Unfortunately, that is not the case, the reason being that the low-order approximations to ϕ_D and ϕ_{dDD} have retained errors of the same magnitude as those which were eliminated.

Conversion of (28) to a high accuracy equation is accomplished in two steps. First, we bring in more points along the D-axis. We write,

$$\phi_j^{k+1} - \phi_{j+2}^k + S_1(\phi_{j+1}^{k+1} - \phi_{j+1}^k)$$
$$+ S_2(\phi_{j+2}^{k+1} - \phi_j^k) + S_3(\phi_{j+3}^{k+1} - \phi_{j-1}^k) = 0. \quad (33)$$

That is, the unknown point ϕ_j^{k+1} is determined from the seven points illustrated in Figure 7. Stability is assured provided the polynomial,

$$1 + S_1 Z + S_2 Z^2 + S_3 Z^3,$$

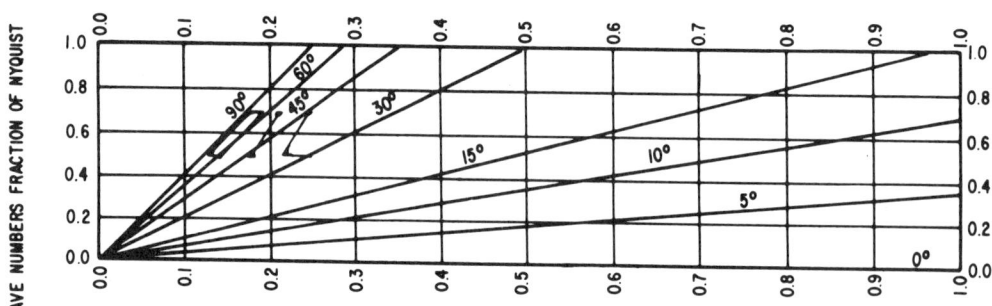

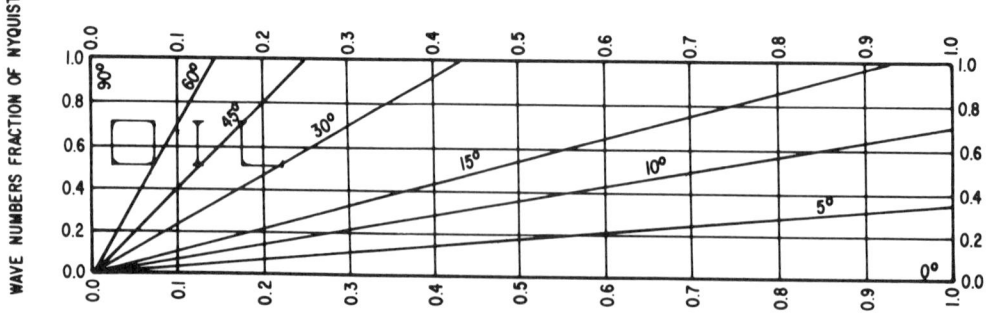

FIG. 9. Migration in F-K space.

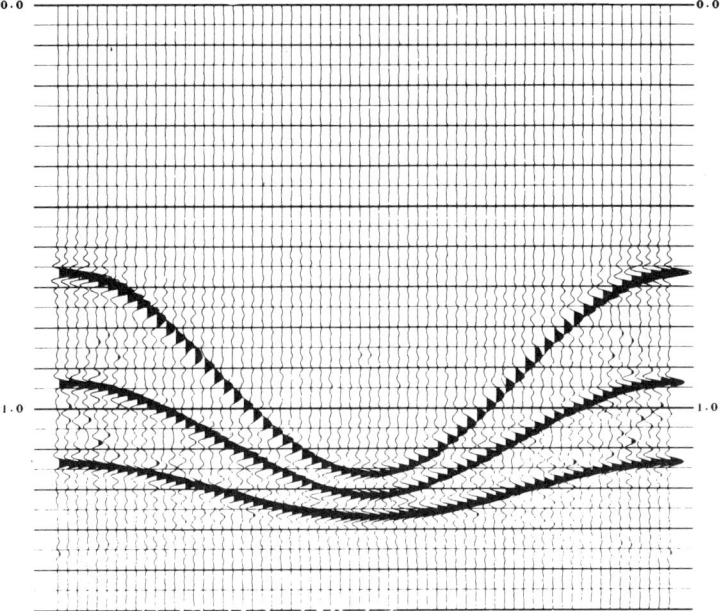

FIG. 10. Migration of a synthetic section using the 45 degree Claerbout algorithm. Maximum phase error is less than 8 degrees for frequencies below the aliasing frequency.

has no roots inside the unit circle. Constraining the Levinson reflection coefficients to be less than one in magnitude, we find

$$|S_3| < 1, \tag{34}$$

$$|S_2 - S_1 S_3| < |1 - S_3^2|, \tag{35}$$

and

$$|S_1 - S_2 S_3| < |1 - S_3^2 + S_2 - S_1 S_3|. \tag{36}$$

The appearance of the anti-causal term ϕ_{j-1}^k in equation (33) may be somewhat disquieting, but is really no cause for alarm. It merely reflects the fact that when dealing with bandlimited data, higher order approximations to D-derivatives at the point $j + 1$ will use more points on both sides of $j + 1$.

The coefficients S_1, S_2, and S_3 may be determined by choosing higher order analogs to the difference operators (25)–(27). However, greater accuracy may be achieved by choosing S_1, S_2, and S_3 so that the difference equation (33) best approximates the exact wave equation (21) rather than the 45 degree approximation (24).

To do this, we look at individual plane wave solutions to equation (21) setting

$$\phi = e^{i(qd - 2\omega D/c)} \tag{37}$$

Equation (21) then gives the dispersion relation (for upcoming waves)

$$q = \frac{2\omega}{c} - \sqrt{\frac{4\omega^2}{c^2} - p^2}. \tag{38}$$

The difference equation (33), on the other hand, gives the relation

$$\sin\left(\frac{\tilde{q}\Delta d}{2} + \frac{2\omega \Delta D}{c}\right) + S_1 \sin\frac{\tilde{q}\Delta d}{2} + S_2 \sin\left(\frac{\tilde{q}\Delta d}{2} - \frac{2\omega \Delta D}{c}\right) + S_3 \sin\left(\frac{\tilde{q}\Delta D}{2} - \frac{4\omega \Delta D}{c}\right) = 0; \tag{39}$$

or, solving for $\tilde{q}$,

$$\tilde{q} = \frac{2}{\Delta d} \arctan \frac{(S_2 - 1) \sin 2\omega \Delta D/c + S_3 \sin 4\omega \Delta D/c}{S_1 + (S_2 + 1) \cos 2\omega \Delta D/c + S_3 \cos 4\omega \Delta D/c}. \tag{40}$$

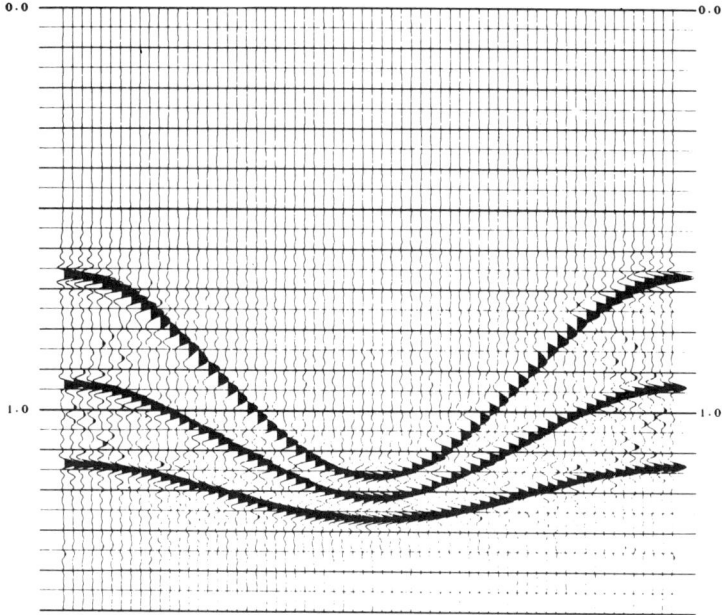

FIG. 11. Migration of a synthetic section using the *F-K* algorithm.

Since $\bar{q}$ will in general be different from the correct value q, a plane wave of frequency ω and wavenumber p [dip $\theta = \arcsin(pc/2\omega)$] will develop an error in phase ϵ proportional to the difference $\bar{q} - q$ and to the distance traveled:

$$\epsilon = (\bar{q} - q)d. \qquad (41)$$

Values for S_1, S_2, and S_3 may be chosen so as to (in some sense) minimize ϵ. ϵ may be forced to zero at any three frequencies by substituting equation (38) into equation (39) and solving the resultant linear set of equations for S_1, S_2, and S_3. Substitution of q at more than three frequencies results in an overdetermined system which can be solved by least squares. If θ_m is the maximum dip present in the data and ω_m the maximum frequency, the frequencies chosen should lie in the range

$$pc/2 \sin \theta_m < \omega < \omega_m. \qquad (42)$$

Figure 8 illustrates the phase errors to be expected from the two-coefficient difference equation (28) and the three-coefficient equation (33). In this example, $p\Delta X = 1$ radian, $\Delta X = 6\Delta D$, and $\Delta d = 10 \Delta D$. For the two-coefficient equation, $S_1 = -1.926174$, $S_2 = .932886$. For the three-coefficient equation, $S_1 = -1.913213$, $S_2 = .914654$, and $S_3 = .010563$. These values were chosen to give zero-phase error at $\omega_1 = \text{Nyquist}/2$, $\omega_3 = pc/2\sin 45$ degrees, and $\omega_2 = (\omega_1 + \omega_3)/2$. Phase error is plotted in degrees per second of trace, assuming a .004 sec sample interval. Two things are apparent in this illustration: (1) equation (33) is more accurate than equation (28) over the entire range of dips and frequencies; and (2) by forcing three zero crossings for equation (33), we have actually gotten four (the fourth zero crossing occurs at about 50 degree dip), significantly extending its region of accuracy. This suggests that (a) modification of the coefficients of equation (28) could not produce accuracy comparable to that of equation (33); and (b) adding a fourth coefficient to equation (33) is not likely to significantly improve accuracy.

In practice, equation (33) is found to be accurate and stable for dips up to 45–55 degrees. Beyond this, accuracy may require an extremely small Δd, and stability problems may be encountered. Note that in general the coefficients S_1, S_2, and S_3 and also the step size Δd will be different for every spatial frequency p.

The phase error defined in equation (41) does not include error introduced by a finite sample interval ΔX in X. In principle, these errors are zero, provided the maximum dip angle $\theta_{\max}$ obeys

$$\theta_{\max} < \arcsin(\Delta D/\Delta X). \qquad (43)$$

If the maximum frequency in the data is $f_{\max} = \nu$

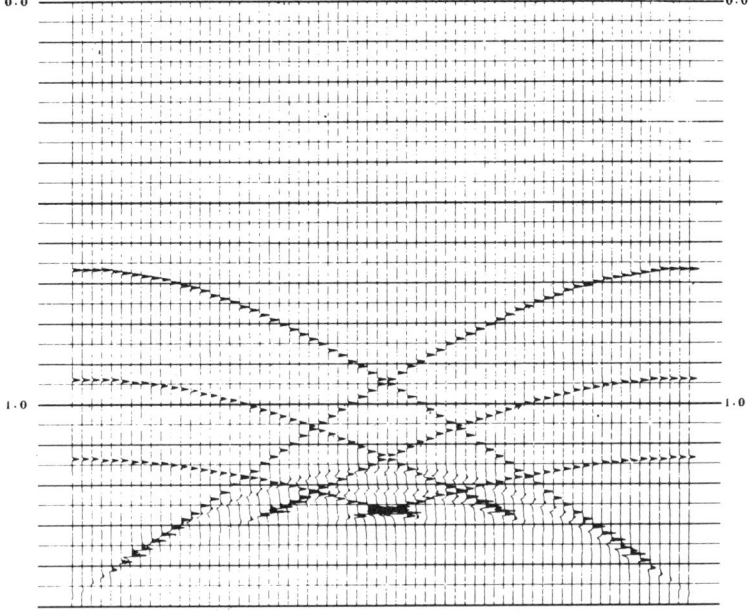

FIG. 12. A synthetic seismic section.

$fnyq$, where $fnyq$ is the Nyquist frequency (125 Hz for $\Delta t = 4$ msec), then the above relation becomes less restrictive:

$$\theta_{max} < \arc\sin(\Delta D/\nu\Delta X). \tag{44}$$

Dips θ greater than θ_{max} are migrated correctly as long as frequency obeys

$$f < fnyq \cdot \Delta D/DX \sin\theta \equiv f_a. \tag{45}$$

f_a will be referred to as the aliasing frequency.

At larger frequencies, the dip will be interpreted as smaller than θ_{max} and will be migrated incorrectly.

Examples of migration using the equation (33) algorithm will be deferred until after a discussion of the $F - K$ migration scheme.

MIGRATION IN F–K SPACE

Suppose we take a two-dimensional Fourier transform of the surface field $\phi(X, 0, D)$:

$$A(p, \omega) = \frac{1}{2\pi} \int dX \int dD \, e^{i(pX - 2D/c)} \phi(X, 0, D), \tag{46}$$

so that

$$\phi(X, 0, D) = \frac{1}{2\pi} \int dp \int d\omega \, e^{-i(pX - 2\omega D/c)} A(p, \omega). \tag{47}$$

For upcoming waves, equation (47) generalizes at positive depth to

$$\phi(X, d, D)$$
$$= \frac{1}{2\pi} \int dp \int d\omega \, e^{-i(pX + qd - 2\omega D/c)} A(p, \omega), \tag{48}$$

where, to satisfy the wave equation (11),

$$q = \frac{2\omega}{c} - \sqrt{\frac{4\omega^2}{c^2} - p^2}. \tag{49}$$

The migrated section $\phi(X, D, D)$ then has the form

$$\phi(X, D, D)$$
$$= \frac{1}{2\pi} \int dp \int d\omega A(p, \omega) e^{-i\left\{pX - \sqrt{\left(\frac{4\omega^2}{c^2} - p^2\right)}D\right\}}. \tag{50}$$

The substance of equations (46) and (50) is that migration may be accomplished by a double Fourier transform of the original data from (X, D) space into (p, ω) space (46), followed by the more complicated transformation (50). If equation (50) could be converted into a double Fourier transform, a practical migration scheme could result.

Fortunately, a simple change of variable from ω to

$$k = \sqrt{\frac{4\omega^2}{c^2} - p^2}$$

does the trick:

$$\phi(X, D, D) = \frac{1}{2\pi} \int dp \int dk\, B(p, k)\, e^{-i(pX-kD)}, \quad (51)$$

where

$$B(p, k) = \frac{1}{\sqrt{1 + p^2/k^2}} A\left(p, \frac{kc}{2}\sqrt{1 + p^2/k^2}\right). \quad (52)$$

The transformation (52) represents, for a fixed p, a shift of data from frequency ω to a lower frequency $\omega' = \sqrt{\omega^2 - p^2 c^2/4}$ (in fact, a "moveout correction" where ω takes the place of time, and p of offset), plus a change of scale $k/\sqrt{k^2 + p^2} = \omega'/\omega$. The frequency shift, depicted in Figure 9, effects only what migrators have always known, namely that an apparent dip of θ_A before migration translates into a dip,

$$\theta_M = \arcsin \tan \theta_A, \quad (53)$$

after migration.

The operations (46), (51), and (52) could easily be done in an analog system. On a digital computer, the Fourier transforms (46) and (51) will be carried out as FFTs. The transformation (52) then involves a dangerous interpolation of the data in the frequency domain. To avoid ghost events appearing on the section, it is usually necessary at least to double the trace length by adding zeros before performing the initial time FFT.

No phase error or dispersion should be seen in double Fourier transform migration, since the exact wave equation is used. The aliasing problem discussed in the last section will still be present, though it is now possible to predict exactly where aliasing may exist and conceivably even unravel it.

EXAMPLES OF FOURIER TRANSFORM MIGRATION

Migrations of the synthetic section of Figure 4 are shown in Figures 10 and 11. Figure 10 results from k-space finite difference algorithm (33). Maximum-phase error at the bottom of the section was held to less than 8 degrees. All three reflectors have assumed their proper shapes. Little dispersion is evident, except for a loss of high frequencies in the region of 45 degree dip. This would be expected, since the aliasing frequency at that dip is 28.3 Hz. Figure 11 shows a migration using the double Fourier transform (F-K) algorithm. It appears very similar to the K-space finite difference migration (Figure 10).

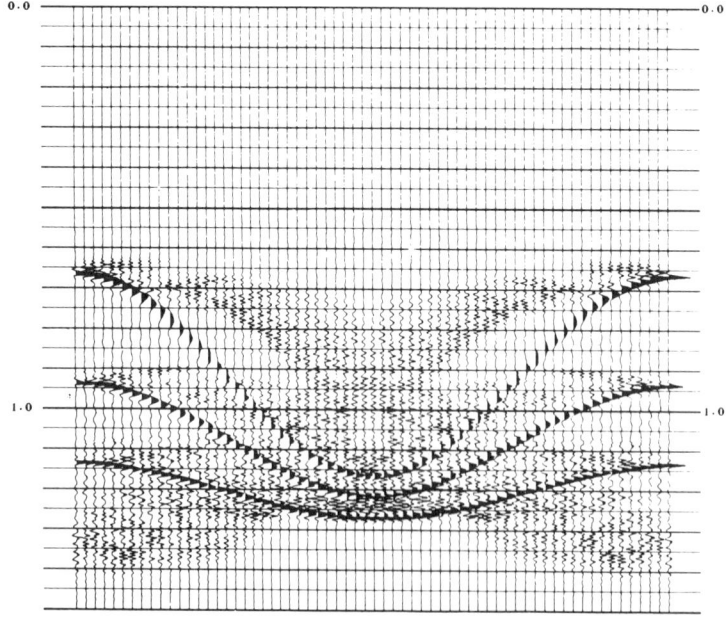

FIG. 13. A K-space migration.

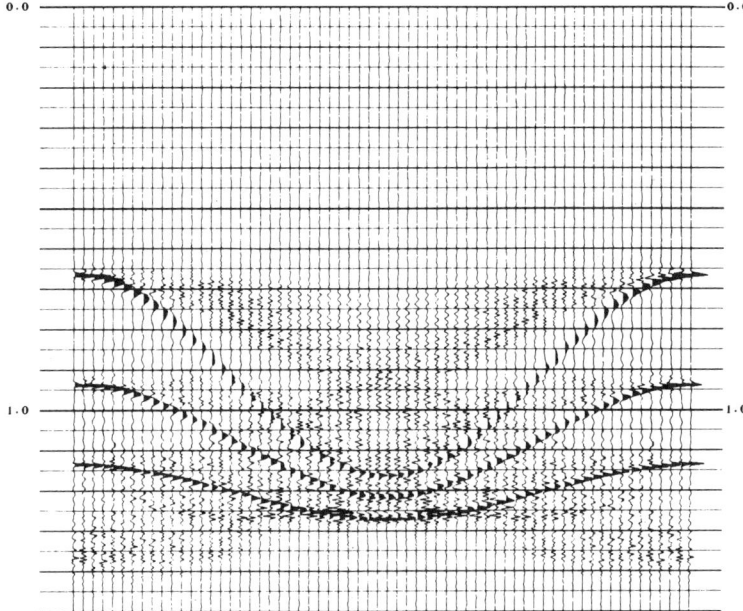

FIG. 14. An *F-K* migration.

Figures 12 to 14 show more clearly what happens to frequencies above the aliasing frequency. Figure 13 shows a *K*-space finite difference migration of the same model as before with a frequency range expanded to 0–90 Hz. Figure 14 is an *F-K* migration in the range 0–125 Hz. The similarity between the two migrations is still apparent. Frequencies below the aliasing frequency are in both cases migrated properly, while frequencies above it are undermigrated, mostly remaining close to their original position.

The remaining examples involve seismic field data. Maximum dip on Figure 15 is of the order of 35 degrees. Trace spacing is 220 ft, with rms velocities ranging from 7000 to 13,000 fps. Figure 16 is the *K*-space finite difference migration (maximum phase error <3 degrees); Figure 17 the *F-K* migration. Though some differences are apparent in the two migrations (due in part to slightly different parameterizations), they give substantively the same reasonable subsurface picture.

Figure 18 is a somewhat more complex section. The 15 degree Claerbout migration, shown in Figure 19, leaves several crossing events and some dispersion. The *K*-space and *F-K* migrations in Figures 20 and 21 give a more satisfying picture with a substantively different interpretation. Residual crossing events indicate a three-dimensional structure.

In Figure 22, the structure is actually three-dimensional, involving dips in excess of 45 degrees. To complicate matters, crucial data from the steep north flank of the structure were not recorded. Trace spacing is 220 ft, with rms velocities in the 7000–14,000 fps range. The *K*-space and *F-K* migrations in Figures 23 and 24 are again very similar. In this case, the results are predictably less than perfect but do provide some basis for interpretation.

The concluding section, Figure 25 is a regional seismic line. Its *F-K* migration, Figure 26, is a dramatic illustration of steep dip migration.

MIGRATION BEFORE STACK

The *F-K* migration scheme in principle can be used to effect moveout correction, stack, and migration of data in one process. In this case, we use a three-dimensional Fourier transform in the coordinate system specified in equations (4) and (5):

$$A(P, p, \omega) = (2\pi)^{-3/2} \int dt \int dX \cdot$$
$$\cdot \int dx\, e^{i(PX + px - \omega t)} \psi(X, x, 0, 0, t). \quad (54)$$

We assume the field to be governed by the two wave equations (2) and (3), where, for simplicity, c is assumed constant. Then,

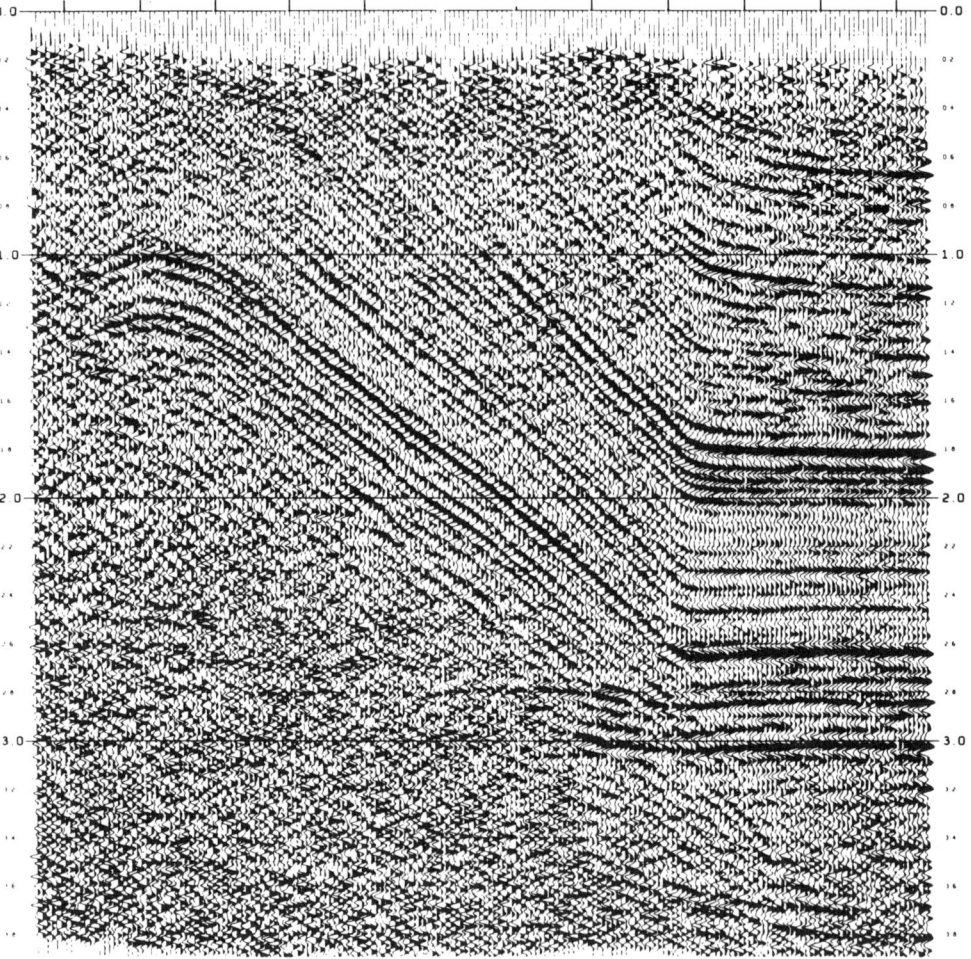

FIG. 15. Five-fold CDP section maximum dip 35 degrees.

$$\psi(X, x, (z_0 + z_s)/2, (z_0 - z_s)/2, t)$$
$$= (2\pi)^{-3/2} \int d\omega \int dP \int dp \cdot$$
$$\cdot e^{-i(PX + px - q_s z_s - q_0 z_0 - \omega t)} A(P, p, \omega), \quad (55)$$

where, from equations (2) and (3)

$$q_s = \omega/c \sqrt{1 - (P - p)^2 c^2/4\omega^2};$$
$$q_0 = \omega/c \sqrt{1 - (P + p)^2 c^2/4\omega^2}. \quad (56)$$

The migrated section is

$$\psi(X, 0, ct/2, 0, 0)$$
$$= (2\pi)^{-3/2} \int d\omega \int dP \cdot$$
$$\cdot \int dp \, e^{-i(PX - (q_s + q_0)ct/2)} A(P, p, \omega). \quad (57)$$

To put (57) in the form of a Fourier transform, a coordinate transformation is required. We define two new variables which will replace p and ω,

$$u = q_s + q_0; \; v = q_s - q_0. \quad (58)$$

From equations (56) and (58) follow the relations,

$$p = uv/P, \quad (59)$$

$$\omega = sgn(\omega)(c/2P)\sqrt{P^4 + P^2(u^2 + v^2) + u^2 v^2}, \quad (60)$$

and

$$\psi(X, 0, ct/2, 0, 0)$$
$$= (2\pi)^{-3/2} \int dP \int du \int dv\, e^{-i(PX - uct/2)} \cdot$$
$$\cdot A(P, p, \omega) c^2 (u^2 - v^2)/4\omega P. \quad (61)$$

Again, a simple transformation accomplishes the migration process. Note that the Fourier p-integral becomes a simple integration over v, since only zero offset is relevant after migration.

MIGRATION IN THREE DIMENSIONS

The three-dimensional analog of equation (8) is

$$\psi_{XX} + \psi_{YY} + \psi_{ZZ} - 4/c^2 \psi_{tt} = 0. \quad (62)$$

When c is constant, we can write

$$\psi(X, Y, Z, t) = (2\pi)^{-3/2} \int dP\, e^{iPX} \int dQ\, e^{iQY} \cdot$$
$$\cdot \int d\omega\, e^{-i\omega t} A(P, Q, \omega, Z), \quad (63)$$

where

$$A(P, Q, \omega, 0) = (2\pi)^{-3/2} \int dX \int dY \cdot$$
$$\cdot \int dt\, \psi(X, Y, 0, t) e^{-i(PX + QY - \omega t)}, \quad (64)$$

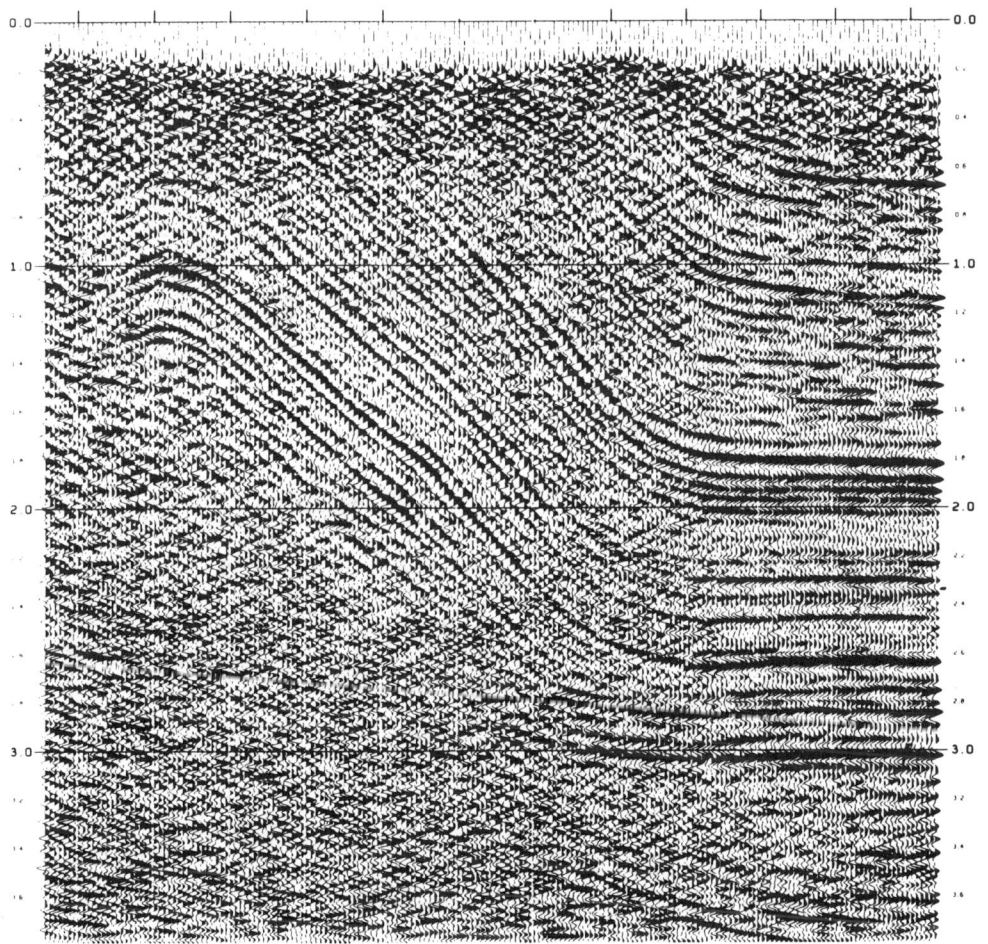

FIG. 16. A K-space migration of Figure 11.

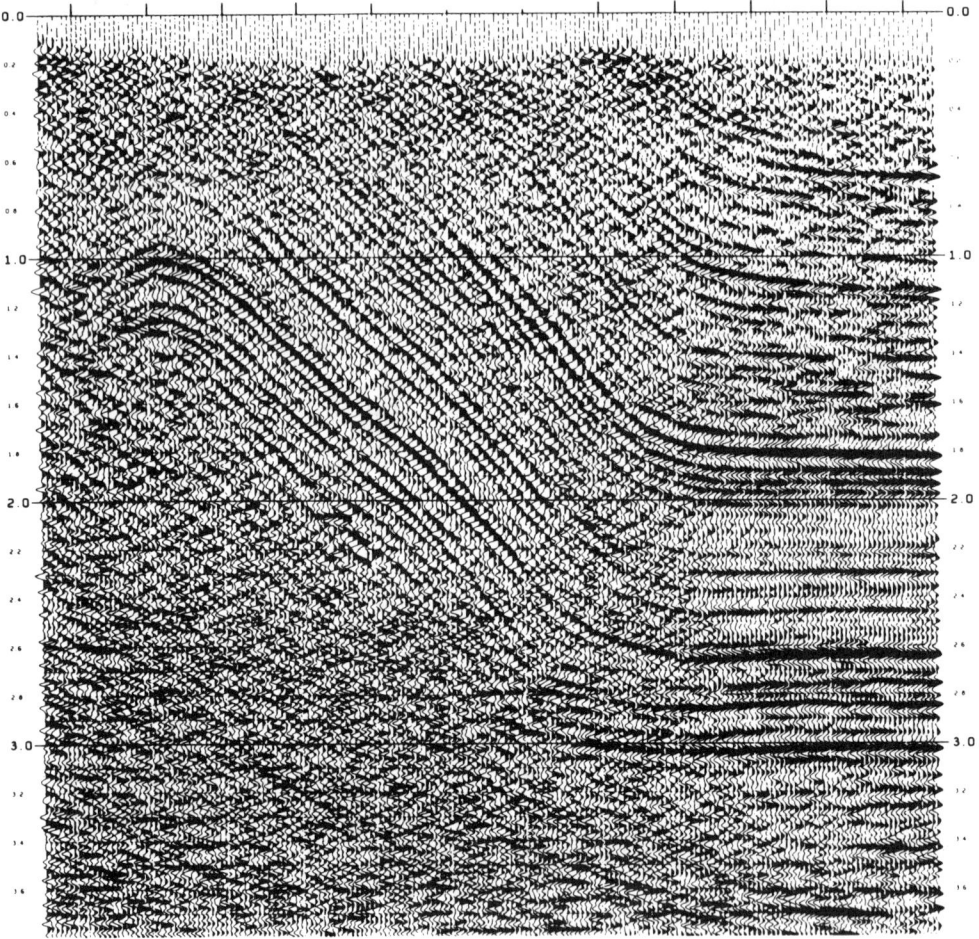

FIG. 17. An *F-K* migration of Figure 11.

is the triple Fourier transform of the unmigrated three-dimensional data. Now, $A(P, Q, \omega, Z)$ satisfies the transformed wave equation

$$A_{ZZ} = (P^2 + Q^2 - 4\omega^2/c^2)A, \qquad (65)$$

which has upcoming solutions,

$$A(P, Q, \omega, Z) = A(P, Q, \omega, 0) e^{-iZ\sqrt{4\omega^2/c^2 - P^2 - Q^2}}.$$

$$(66)$$

Hence, we can write the migrated field as,

$$\psi(X, Y, Z, 0) = (2\pi)^{-3/2} \int dP \int dQ \int d\omega\, B(P, Q, \omega) e^{i(PX + QY - 2\omega Z/c)}, \qquad (67a)$$

where

$$B(P, Q, \omega) = A\{P, Q, \omega\sqrt{1 + (P^2 + Q^2)c^2/4\omega^2}\}/\sqrt{1 + (P^2 + Q^2)c^2/4\omega^2}. \qquad (67b)$$

EXTENSION TO A VARIABLE VELOCITY

The K-space migration scheme described above relies on a velocity which is X-independent. The derivation of the F-K algorithm was even more restrictive, requiring velocity to be constant. In order to use these schemes in the presence of a variable velocity, it is necessary to transform to a coordinate

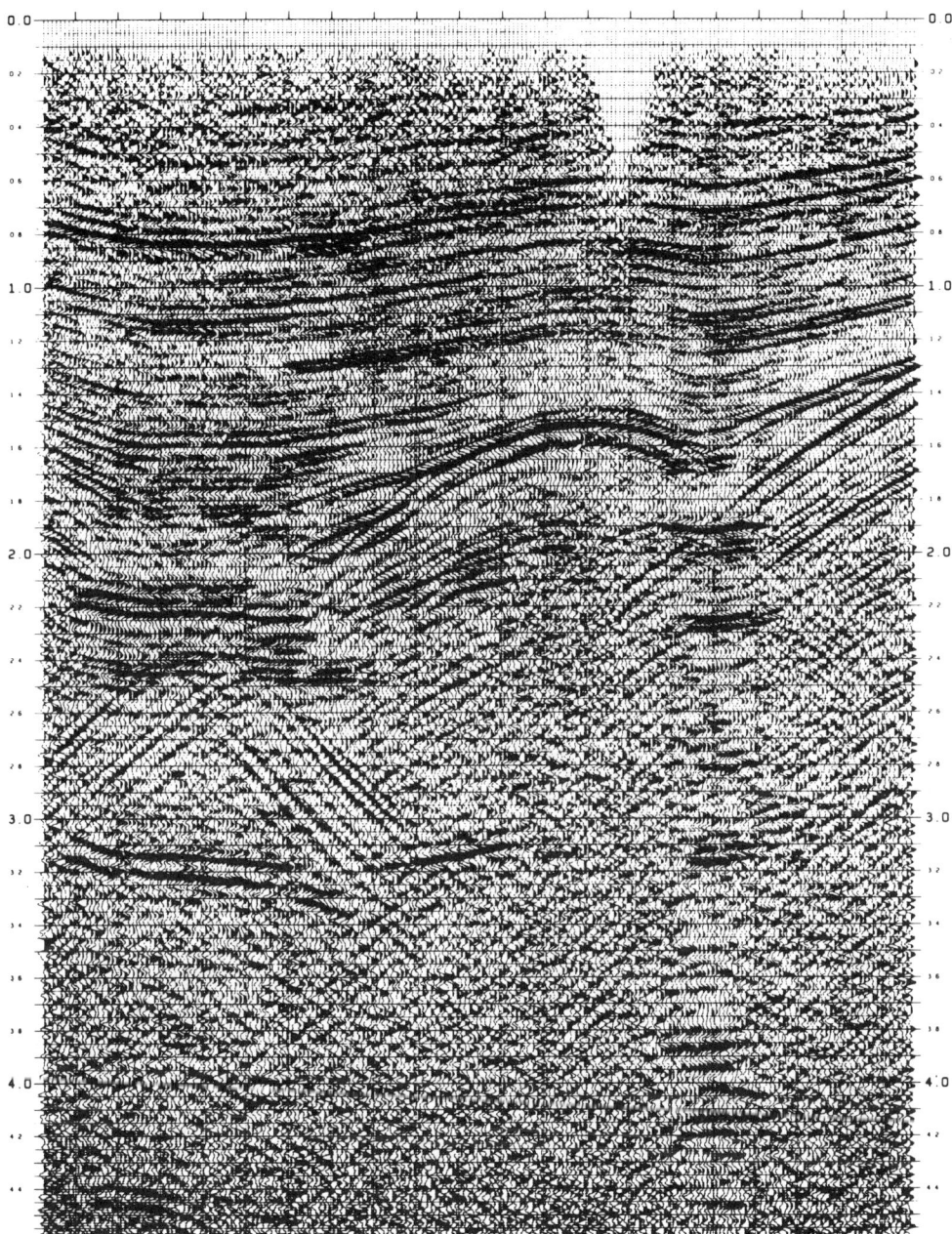

FIG. 18. A complex 10-fold CDP section.

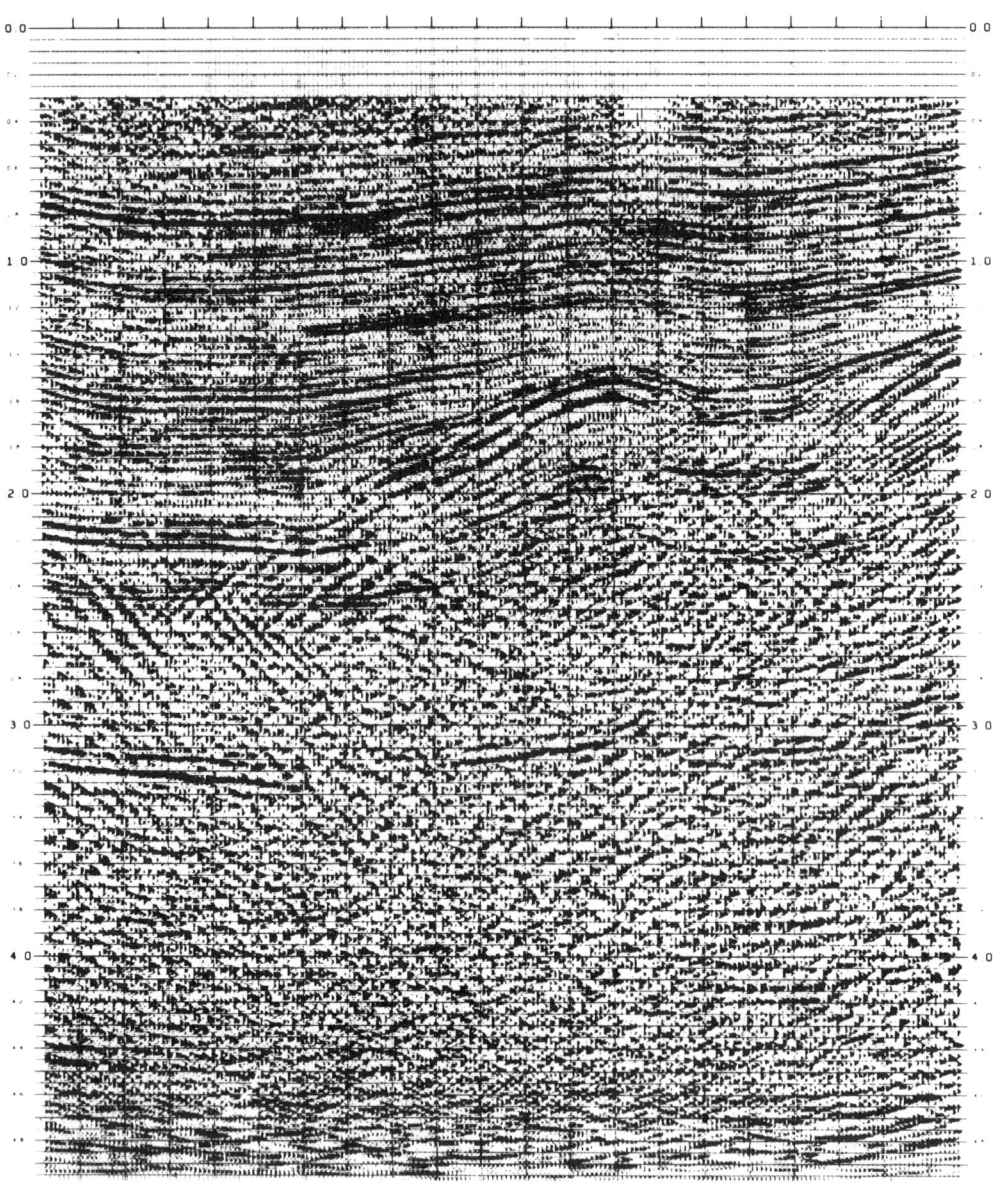

FIG. 19. A 15 degree Claerbout migration of Figure 14.

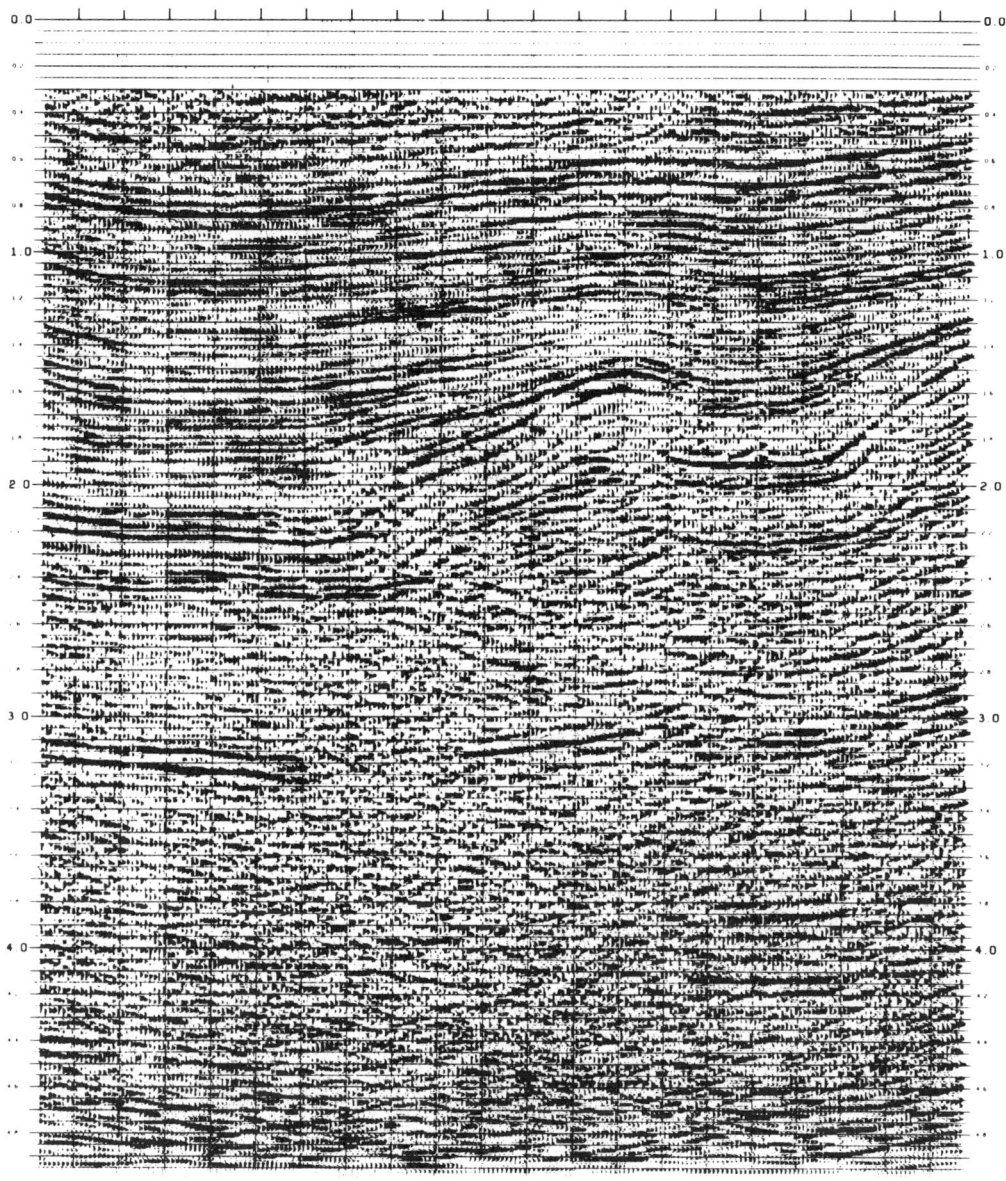

FIG. 20. A K-space migration of Figure 14.

system in which both the wave equation and the boundary conditions are velocity independent.

Beginning with the coordinate system X, Z, t of equation (8), we first define new time and depth coordinates

$$t' = \frac{t}{2} + \int_0^Z \frac{dZ}{c}, \qquad (68)$$

$$Z' = \frac{1}{c_0} \int_0^Z dZ c. \qquad (69)$$

Since Z' represents the apparent depth to a reflector at Z in a layered medium, we may expect this to be a useful coordinate system for migration. Neglecting velocity derivatives, the wave equation becomes

$$\psi'_{XX} + c^2/c_0^2 \psi_{Z'Z'} + 2/c_0 \, \psi'_{Z't'} = 0. \qquad (70)$$

FIG. 21. An *F-K* migration of Figure 14.

Note that the coefficient of the dominant ψ_{Zt} term is now constant. To define the migration limits in this coordinate system, we define two new variables $\zeta(t')$ and $\eta(t')$

$$t' = \int_0^\zeta \frac{dZ}{c}, \tag{71}$$

and

$$\eta = \int_0^\zeta c \, dZ. \tag{72}$$

Then before migration the limits are

$$Z' = 0, \ t' > 0, \tag{73}$$

and the limits upon migration are

$$Z' = \eta(t')/c_0, \ t' > 0. \tag{74}$$

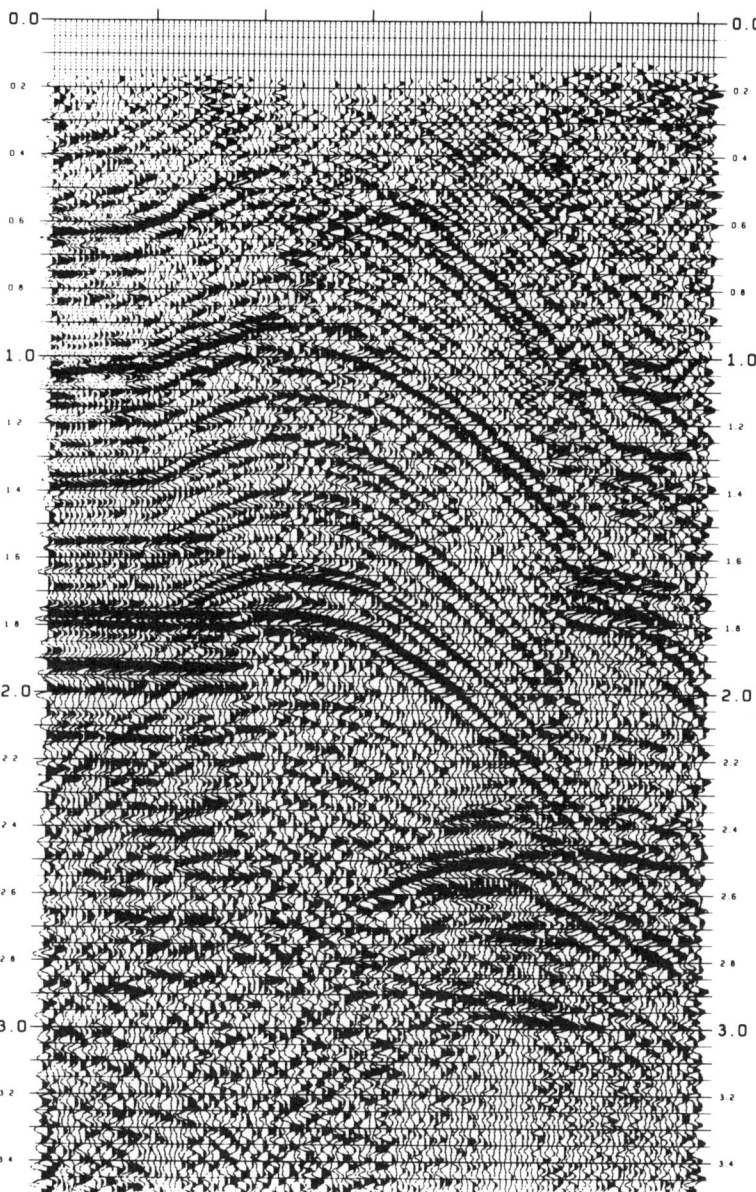

FIG. 22. A 10-fold CDP section with dips greater than 45 degress.

Because Equation (74) is time dependent, this coordinate system is not in general useable for K-space or F-K migration. However, one more change of variables will do the trick. Define

$$D = \sqrt{2 \int_0^{t'} \eta(t') dt'} \qquad (75)$$

and

$$d = Z' c_0 D / \eta. \qquad (76)$$

Setting $\phi(X, d, D) = \psi(X, Z, t)$, the wave equation takes the form

$$\phi_{XX} + W(X, d, D) \phi_{dd} + 2 \phi_{dD} = 0, \qquad (77)$$

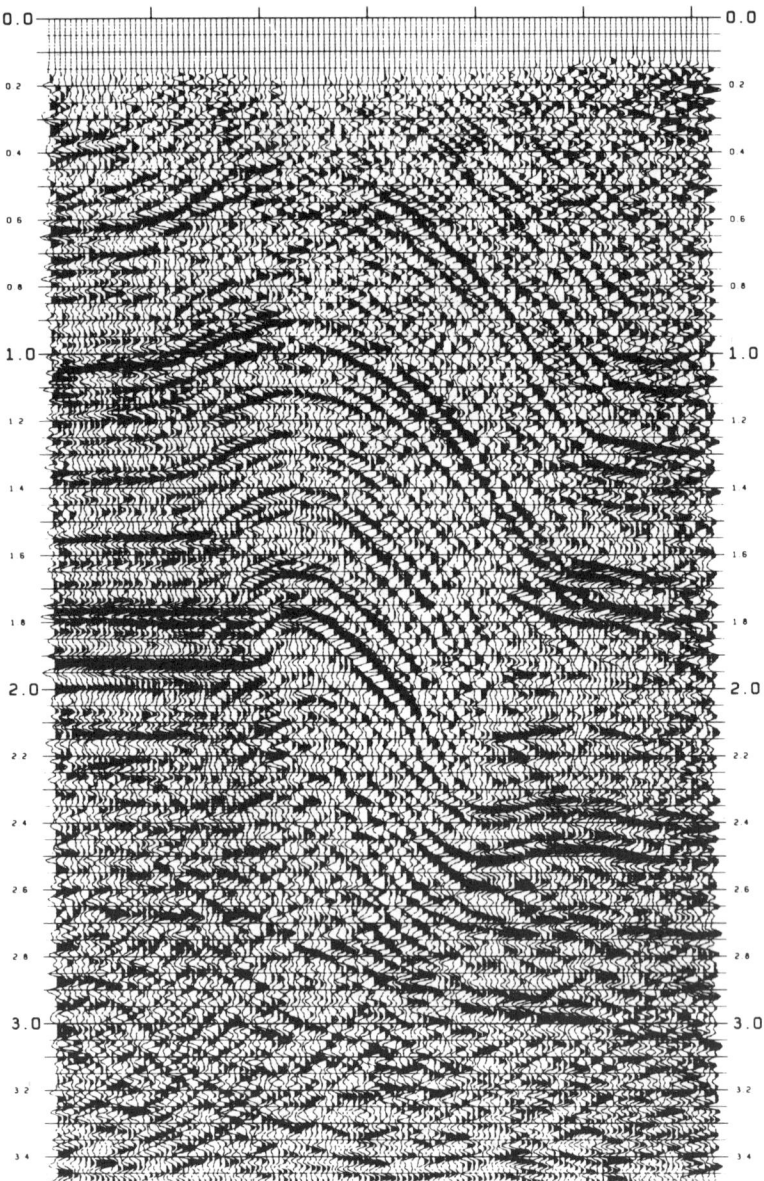

FIG. 23. A K-space migration of Figure 18.

where

$$W = \frac{c^2(X,Z)D^2}{\eta^2} + \frac{2d}{D}\left(1 - \frac{c^2(X'z)D^2}{\eta^2}\right). \quad (78)$$

Migration proceeds from the half-plane,

$$d = 0, D > 0, \quad (79)$$

to the half-plane,

$$d = D, D > 0. \quad (80)$$

All explicit dependence on X and Z now resides in the coefficient W of ϕ_{dd}. $W \neq 1$ reflects the fact that diffractions are not pure hyperbolas in a layered medium. Since ϕ_{dd} is ordinarily small, it is usually justifiable to replace W with an average constant value (usually a number between .5 and 1) for a given section. Then (77) has almost the form of equation

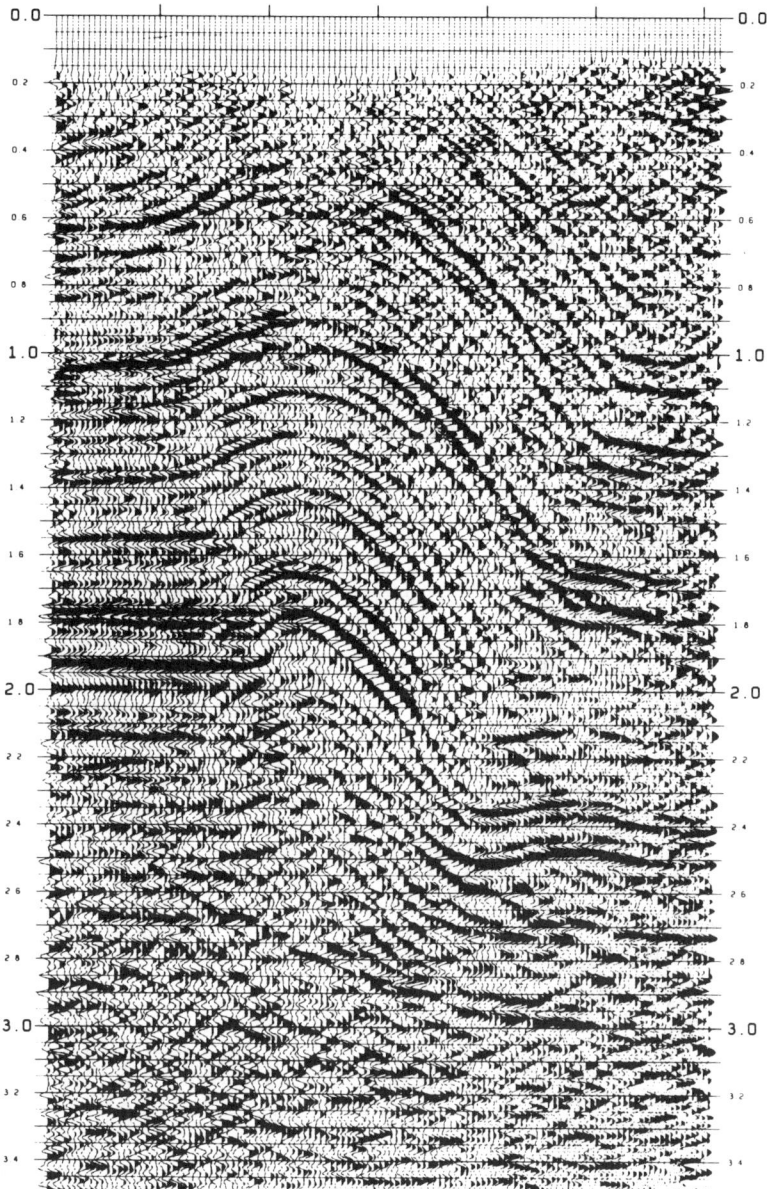

FIG. 24. An *F-K* migration of Figure 18.

(11), and the derivation of the *K*-space and *F-K* migration algorithms proceeds as outlined above, except for a slight modification of the dispersion relation (38) or (49).

Use of *K*-space or *F-K* migration in practice, then, involves a time to depth conversion (75). Even though simple to effect, this gives rise to practical difficulties in that (a) the frequency content of the data is altered; and (b) incorrect lateral velocity variations will distort the reflecting surfaces and cause improper migration. These problems can be lived with, however, and at present the *K*-space and *F-K* migration schemes appear to be practical and useful.

ACKNOWLEDGMENTS

I wish to thank the management of Continental Oil Co., for permission to publish this paper. Jerry Ware, Pierre Goupillaud, and Bill Heath have through

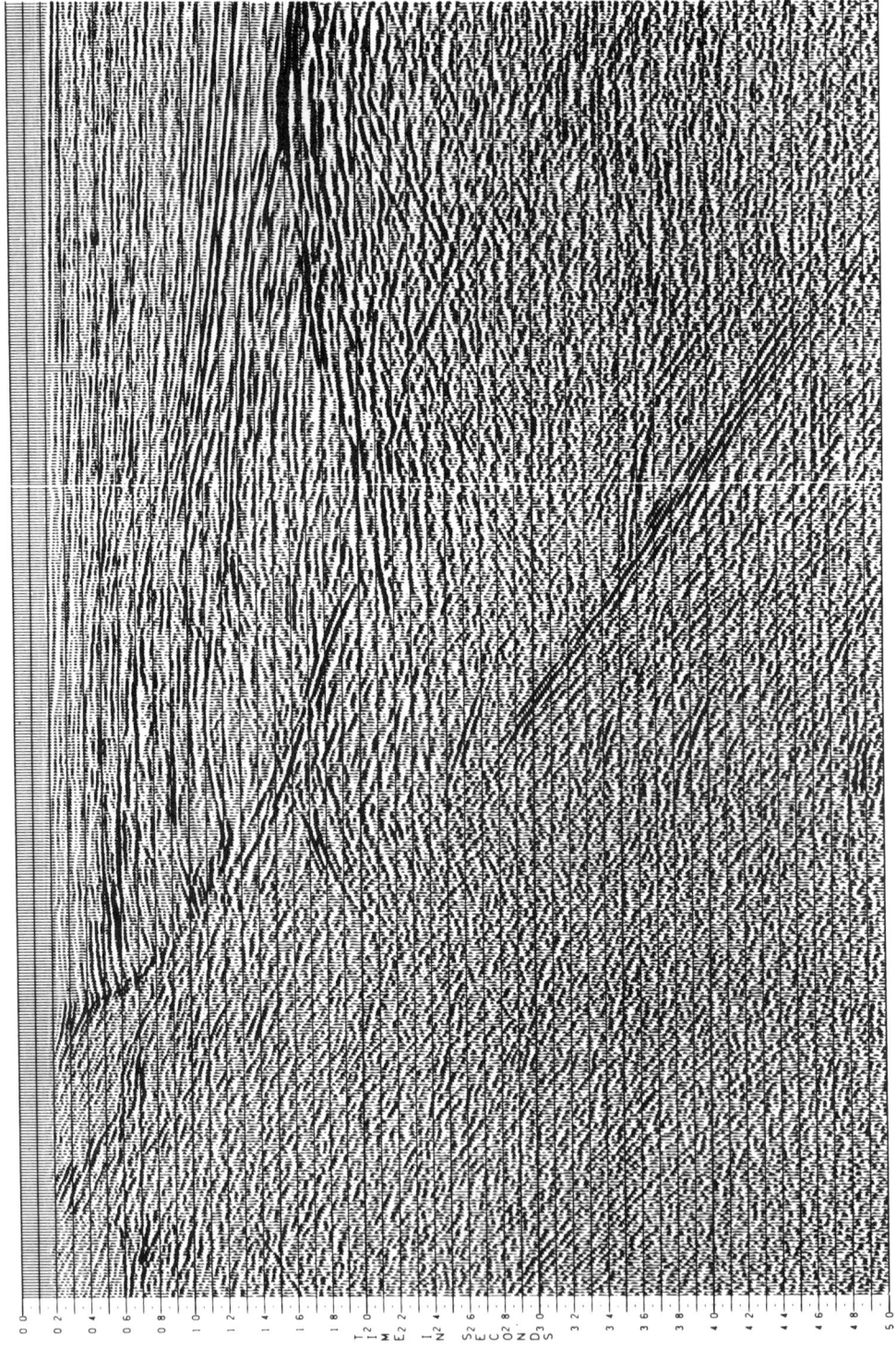

FIG. 25. A 24-fold stacked section.

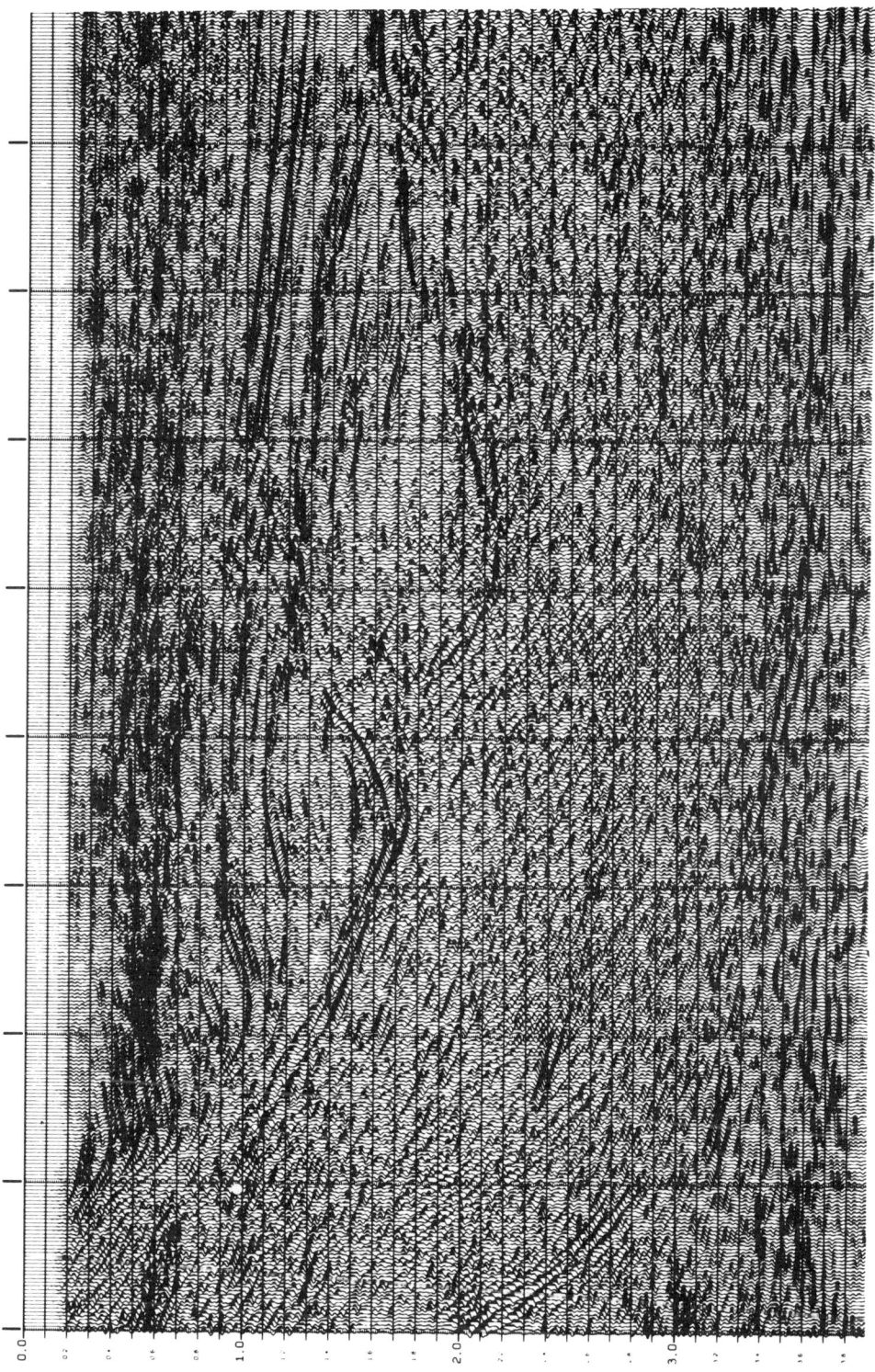

Fig. 26. An *F-K* migration.

several discussions contributed to the theoretical development of this work. Several staff members of the data processing division of Conoco's Exploration-Geophysics Department have, through discussion, criticism, and programming assistance, helped convert the theory to practice.

The development of the high-accuracy K-space migration scheme owes a great deal to the work of Bjorn Enquist at the Stanford Exploration Project. Francis Muir of Chevron has provided several helpful comments and suggestions in addition to the stability criteria and a more efficient form for the K-space algorithm. Of course, this paper would not have been possible but for the original inspiration of Jon Claerbout.

REFERENCES

Claerbout, J. F., 1971, Toward a unified theory of reflector mapping: Geophysics, v. 36, p. 467–481.
——— 1976, Fundamentals of geophysical data processing: New York, McGraw-Hill Book Co., Inc.
Claerbout, J. F., and Doherty, S. M., 1972, Downward continuation of moveout corrected seismograms: Geophysics, v. 37, p. 741–768.
French, W. S., 1974, Two-dimensional and three-dimensional migration of model-experiment reflection profiles: Geophysics, v. 39, p. 265–287.
——— 1975, Computer migration of oblique seismic reflection profiles: Geophysics, v. 40, p. 961–980.